NANOTECHNOLOGY IN THE AUTOMOTIVE INDUSTRY

Micro and Nano Technologies

NANOTECHNOLOGY IN THE AUTOMOTIVE INDUSTRY

Edited by

HUAIHE SONG
Professor, State key Laboratory of Chemical Resource Engineering, College of Materials and Engineering, Beijing University of Chemical Technology, Beijing, China

TUAN ANH NGUYEN
Principal Research Scientist, Vietnam Academy of Science and Technology, Hanoi, Vietnam

GHULAM YASIN
Academic Researcher, Institute for Advanced Study, College of Physics and Optoelectronic Engineering, Shenzhen University, Shenzhen, Guangdong, China

NAKSHATRA BAHADUR SINGH
Professor Emeritus, Chemistry Department, SBSR & Research and Technology Development Centre, Greater Noida, Uttar Pradesh, India

RAM K. GUPTA
Associate Professor, Department of Chemistry, Pittsburg State University, Pittsburg, KS, United States

ELSEVIER

Elsevier
Radarweg 29, PO Box 211, 1000 AE Amsterdam, Netherlands
The Boulevard, Langford Lane, Kidlington, Oxford OX5 1GB, United Kingdom
50 Hampshire Street, 5th Floor, Cambridge, MA 02139, United States

Copyright © 2022 Elsevier Inc. All rights reserved.

No part of this publication may be reproduced or transmitted in any form or by any means, electronic or mechanical, including photocopying, recording, or any information storage and retrieval system, without permission in writing from the publisher. Details on how to seek permission, further information about the Publisher's permissions policies and our arrangements with organizations such as the Copyright Clearance Center and the Copyright Licensing Agency, can be found at our website: www.elsevier.com/permissions.

This book and the individual contributions contained in it are protected under copyright by the Publisher (other than as may be noted herein).

Notices

Knowledge and best practice in this field are constantly changing. As new research and experience broaden our understanding, changes in research methods, professional practices, or medical treatment may become necessary.

Practitioners and researchers must always rely on their own experience and knowledge in evaluating and using any information, methods, compounds, or experiments described herein. In using such information or methods they should be mindful of their own safety and the safety of others, including parties for whom they have a professional responsibility.

To the fullest extent of the law, neither the Publisher nor the authors, contributors, or editors, assume any liability for any injury and/or damage to persons or property as a matter of products liability, negligence or otherwise, or from any use or operation of any methods, products, instructions, or ideas contained in the material herein.

ISBN: 978-0-323-90524-4

For information on all Elsevier publications
visit our website at https://www.elsevier.com/books-and-journals

Publisher: Matthew Deans
Acquisitions Editor: Carrie Bolger
Editorial Project Manager: Mariana L. Kuhl
Production Project Manager: Debasish Ghosh
Cover Designer: Mark Rogers

Typeset by STRAIVE, India

Contents

Contributors — xvii
Preface — xxix

Section A Nanocomposites for automotive application

1. Nanocomposites: An introduction — 3
Deepshikha Rathore

1. Introduction — 3
2. Nanocomposites — 4
3. Synthesis techniques of nanocomposites — 7
4. Challenges — 9
5. Nanocomposites for lightweight vehicles — 10
6. Nanocomposites in tyres — 11
7. Nanocomposites in tribology — 12
8. Nanocomposites for fuel — 12
9. Nanocomposites: Costs and benefits — 13
10. Summary — 13
References — 14

2. Using XRD technique for model composite and related materials — 15
Takashiro Akitsu

1. Motivation of this study — 15
2. Battery model — 22
3. Catalyst model — 28
4. Perspective — 33
References — 34

3. Polymeric nanocomposites — 37
V. Lakshmi, Akhil S. Karun, and T.P.D. Rajan

1. Introduction — 37
2. Perspective of polymer nanocomposites for automobile engineering — 38
3. Polymer nanocomposites classification — 43
4. Commercialization of polymer nanocomposite in automobile industry — 58
5. Conclusion — 61
References — 61

4. Enhanced synergistic effect by pairing novel inherent flame-retardant polyurethane foams with nanolayers of expandable graphite for their applications in automobile industry   65

Felipe M. de Souza, Mark Arnce, and Ram K. Gupta

1. Introduction   65
2. Relevance of flame retardants for automobiles   66
3. Concerns and importance of polyurethanes for the automobile industry   68
4. Synthesis and characterizations of novel flame-retardant polyurethane foams   69
5. Important characteristics and properties of the polyurethane foams   71
6. Conclusion   81
References   82

5. Natural fiber-reinforced nanocomposites in automotive industry   85

Abu Bin Imran and Md. Abu Bin Hasan Susan

1. Introduction   85
2. Selection of natural fiber and preparation   88
3. Natural fiber-reinforced nanocomposites   92
4. Applications of natural fiber-reinforced nanocomposites in the automotive industry   95
5. Conclusions and prospect   98
Acknowledgments   98
References   98

6. High-performance polyurethanes foams for automobile industry   105

Felipe M. de Souza, Jonghyun Choi, Tenzin Ingsel, and Ram K. Gupta

1. Introduction   105
2. Importance of renewable materials for automobiles   106
3. Significance of polyurethane foams and current issues   107
4. Synthesis and characterizations of bio-derived polyurethane foams   107
5. Important characteristics and properties of bio-derived polyurethane foams   109
6. Conclusion   126
Acknowledgment   126
References   126

7. Carbon–carbon nanocomposites for brake systems and exhaust nozzles   131

Mehmet İskender Özsoy, Serbülent Türk, Fehim Fındık, and Mahmut Özacar

1. Introduction   131
2. Carbon–carbon nanocomposites (CCNCs)   134
3. Properties of CCNCs   136
4. Application area of CCNCs   137

5. Technologies needed to advance CCNCs	138
6. CCNCs for brake systems	142
7. CCNCs for exhaust nozzles	147
8. Conclusion and future perspectives	151
References	151

8. Metallic nanocomposites: An Introduction — 155
Sandra Gajević, Slavica Miladinović, and Blaža Stojanović

1. Introduction	155
2. Metallic nanocomposites	156
3. Conclusion	160
References	161

9. Metallic nanocomposites for automotive applications — 163
A.G. Arsha, Visakh Manoj, L. Ajay Raag, M.G. Akhil, and T.P.D. Rajan

1. Introduction	163
2. Nanocomposites vs conventional composites in automotive applications	164
3. Potential nanoreinforcements	166
4. Processing of nanocomposites	167
5. Effect of nanoparticles and mechanisms on the properties of metallic nanocomposites	176
6. Characteristics of metallic nanocomposite systems	182
7. Metallic nanocomposite coatings	190
8. Automotive application of metallic nanocomposites	192
9. Conclusion	195
References	195

10. Metal matrix nanocomposites — 199
P.S. Samuel Ratna Kumar, P.M. Mashinini, and S. John Alexis

1. Introduction	199
2. Nanocomposite materials	200
3. Fabrication of nanocomposites	201
4. Characterization study	206
5. Nanocomposites in auto components	209
6. Conclusion	211
References	211

11. Fiber-reinforced nanocomposites — 215
Younes Ahmadi, Nasrin Raji Popalzai, Mubasher Furmuly, and Nangialai Azizi

1. Introduction	215
2. Characterization methods	217

3. Design and manufacturing of FNCs 222
   4. Applications of FNCs 223
   5. Concluding remarks 225
   References 225

12. Electrically conductive polymer nanocomposites for thermal comfort in electric vehicles 229

Heinrich Leicht, Eduard Kraus, Benjamin Baudrit, Thomas Hochrein, Martin Bastian, and Maurice Langer

   1. Carbon nanotubes and thermal comfort in electric vehicles 229
   2. Physical aspects of CNT/polymer nanocomposites for heating applications 231
   3. Conclusion 244
   Acknowledgment 245
   References 245

Section B   Nano-alloys for automotive application

13. Ti-based nanoalloy in automobile industry 255

Asit Behera and Ajit Behera

   1. Introduction 255
   2. What is Ti-based nanoalloy? 257
   3. History 257
   4. Basic metallurgy of Ti-based nanoalloy 258
   5. Basic manufacturing process of Ti-based nanoalloy 261
   6. Mechanical properties of Ti-based nanoalloy 262
   7. Applications of Ti nanoalloys in automobile industry 263
   8. Summary 266
   References 266

14. Applications of copper alloy nanoparticles in automotive industry 269

J. AngelinThangakani, C. Dorothy Sheela, R. Dorothy, N. Renugadevi, J. Jeyasundari, Susai Rajendran, and Ajit Behera

   1. Introduction 269
   2. Properties of Cu-NP 270
   3. Synthesis of Cu-NP 272
   4. Applications 276
   5. Limitations of Cu-NP 281
   6. Conclusion 281
   References 281

15. Nano-steels in the automotive industry — 287
Mina Madadi, Mahdi Yeganeh, and Mostafa Eskandari

1. Introduction to nanosteels and their strengthening mechanisms — 287
2. Grain refinement is a unique mechanism for improving strength and toughness — 289
3. Advanced high-strength steels (AHSS) — 293
4. Steels in automotive industry — 296
5. Conclusion — 311
References — 311

Section C Nanocoatings for automotive application

16. Magnetic nanoparticles-based coatings — 317
P. Poornima Vijayan, Archana Somadas Radhamany, Ansar Ereath Beeran, Maryam Jouyandeh, and Mohammad Reza Saeb

1. Introduction — 317
2. MNPs used to prevent corrosion in metal — 318
3. MNPs as antifouling component — 333
4. Smart coatings based on MNPs — 334
5. MNPs for electromagnetic absorbing coatings — 337
6. MNP coating for textiles — 338
7. Conclusions and future perspectives — 339
References — 340

17. Nano coatings for scratch resistance — 345
Sahar Amiri

1. Background of polymeric coating — 345
2. Introduction to scratch process — 346
3. Typical organic coatings on coil coated steel — 347
4. Introduction to sol–gel method — 348
5. Applications of sol–gel derived coating — 350
6. Concluding remarks — 366
References — 366

18. Self-healing nanocoatings — 371
Andressa Trentin, Mayara Carla Uvida, Adriana de Araújo Almeida, Thiago Augusto Carneiro de Souza, and Peter Hammer

1. Introduction — 371
2. Inorganic corrosion inhibitors — 376
3. Organic corrosion inhibitors — 384

4. Conclusions	396
Acknowledgments	396
References	396

19. Self-healing nanocoatings for automotive application — 403
Abhinay Thakur and Ashish Kumar

1. Introduction	404
2. Nanocoatings	405
3. Types of nanocontainers-based self-healing coatings	409
4. The release of nanomaterials	411
5. Self-healing process investigation	414
6. Impact on self-healing nanocoatings on various aspects	416
7. Commercially available self-healing nanocoatings	418
8. Applications in other field of the automobile industry	421
9. Advantages and disadvantages	422
10. Conclusion	422
References	424

20. Conductive nanopaints: A remarkable coating — 429
Maria Nayane de Queiroz, Antônia Millena de Oliveira Lima, Manuel Edgardo Gomez Winkler, Vanessa Hafemann Fragal, Adley Forti Rubira, Thiago Sequinel, Lucas da Silva Ribeiro, Francisco Nunes de Souza Neto, Emerson Rodrigues Camargo, Mauricio Zimmer Ferreira Arlindo, Christiane Saraiva Ogrodowski, and Luiz Fernando Gorup

1. Introduction	429
2. Conductive coating—Market value	431
3. Types, characteristics, and use of conductive nanopaints	432
4. Recent development of conductive nanopaints	437
5. Commercial conductive nanopaints	439
6. Application of conductive nanopaints in the automotive industry	441
7. General conclusions and future perspectives	443
References	444

Section D Nanodevices for energy conversion and storage in the automotive application

21. Battery-supercapacitor hybrid systems: An introduction — 453
Anuj Kumar, Shumaila Ibraheem, Ram K. Gupta, Tuan Anh Nguyen, and Ghulam Yasin

1. Introduction	453
2. Combination of battery and supercapacitor	454

3. Hybridization of battery and supercapacitor	455
4. Conclusion	457
References	457

22. Supercapacitors: An introduction — 459

Narendra Lakal, Sumit Dubal, and P.E. Lokhande

1. Introduction	459
2. Supercapacitor component development	460
3. Structural supercapacitor performance parameters	462
4. Conclusion	463
References	463

23. Nanomaterials based solar cells — 467

Ritesh Jaiswal, Anil Kumar, and Anshul Yadav

1. Introduction	467
2. Working of solar cell	468
3. Classification of solar cell	471
4. Nanotechnology in solar cells	479
5. Summary	481
Acknowledgments	482
References	482

24. Two dimensional MXenes for highly stable and efficient perovskite solar cells — 485

Sahil Gasso, Manreet Kaur Sohal, Navdeep Kaur, and Aman Mahajan

1. Introduction	485
2. Perovskite solar cells	488
3. Issues with PSCs and their solutions	491
4. Introduction to 2D materials	493
5. MXene additive in PSCs	495
6. Future scope	502
7. Conclusion	503
Acknowledgement	503
References	504

Section E   Nanocatalysts for automotive application

25. Nanocatalysts for exhaust emissions reduction — 511

Ramesh Ch. Deka, Sudakhina Saikia, Nishant Biswakarma, Nand Kishor Gour, and Ajanta Deka

1. Introduction	511

2. Methods	513
3. Theoretical studies	519
4. Conclusion and future scope	522
References	522

26. Automobile exhaust nanocatalysts 529

Kevin V. Alex, K. Kamakshi, J.P.B. Silva, S. Sathish, and K.C. Sekhar

1. Introduction	529
2. Catalytic convertor	534
3. Nanocatalysts	537
4. Wash-coat compositions and oxygen storage components (OSC)	548
5. Catalytic convertors in diesel and learn-burn gasoline engines	551
6. Conclusion	555
7. Future prospects	556
Acknowledgments	556
References	556

27. Nanofuel additives 561

Luis A. Gallego-Villada, Edwin A. Alarcón, and Gustavo P. Romanelli

1. Introduction	561
2. Synthesis of nanoparticles as nanofuel additives	562
3. Properties of nanofuel additives in blends with fuels	568
4. Application of nanofuel additives: Combustion performance	569
5. Future aspects	573
References	573

28. Nanocatalysts for fuel cells 579

Elisangela Pacheco da Silva, Vanessa Hafemann Fragal, Rafael Silva, Alexandre Henrique Pinto, Thiago Sequinel, Matheus Ferrer, Mario Lucio Moreira, Emerson Rodrigues Camargo, Ana Paula Michels Barbosa, Carlos Alberto Severo Felipe, Ramesh Katla, and Luiz Fernando Gorup

1. Introduction	579
2. Fuel cell technology	581
3. Nanocatalyst for fuel cell—Market in value	583
4. Types, characteristics, and synthesis of nanocatalysts	585
5. Recent development of nanocatalysts for fuel cells	591
6. Advantages and challenges of nanocatalysts for fuel cells	596
7. General conclusions and future perspectives	599
References	599

Section F Nanomaterials for automotive application

29. Magnetic nanomaterials for electromagnetic interference shielding application   607
Seyyed Mojtaba Mousavi, Sonia Bahrani, and Gity Behbudi

1. Introduction   607
2. Microwave absorption of magnetic carbon-based nanocomposites   609
3. Microwave absorption performance   611
4. Conclusions and outlook   615
References   616

30. Graphene in automotive parts   623
Kuray Dericiler, Nargiz Aliyeva, Hadi Mohammadjafari Sadeghi, Hatice S. Sas, Yusuf Ziya Menceloglu, and Burcu Saner Okan

1. Introduction   623
2. Transformation into lightweighting innovations   629
3. Graphene in body and structural parts   630
4. Coating applications of graphene   632
5. Graphene in tire manufacturing   634
6. Graphene in electronic parts of vehicles   635
7. Graphene as a lubricating agent in fluids   636
8. Graphene potential in electric vehicles   642
9. Conclusions and outlook   643
References   645

31. Toxicity/risk assessment of nanomaterials when used in the automotive industry   653
S. Sathish, S. Rathish Kumar, K.C. Sekhar, and B. Chandar Shekar

1. Introduction   653
2. Impact of nanomaterials in the automotive industry   654
3. Nanotoxicity   658
4. Role of nanomaterials with their toxicity   659
5. Conclusions   669
References   670

32. Nanolubricant additives   675
Mohamed Kamal Ahmed Ali, Mohamed A.A. Abdelkareem, Ahmed Elagouz, and Hou Xianjun

1. Introduction   675
2. Preparation of nanolubricants   679

3. Tribological and thermophysical performance of nanolubricant additives	686
4. Mechanisms of nanolubricant additives	692
5. Role of nanolubricants in improving vehicle engines performance	697
6. Conclusions and recommendations	704
Acknowledgments	706
References	706

33. Nanofluids as coolants — 713

Zafar Said, Maham Sohail, and Arun Kumar Tiwari

1. Introduction	713
2. Numerical and experimental studies	727
3. Challenges and future outlook	728
4. Conclusion	732
References	732

34. Nanomaterials in automotive fuels — 737

Arun Kumar Tiwari, Amit Kumar, and Zafar Said

1. Introduction	737
2. Nanomaterials impact on fuel properties	738
3. Metal oxide nanomaterials application in automotive fuels	740
4. Conclusion	746
References	746

35. Nanomaterials for electromagnetic interference shielding application — 749

Arun Kumar Tiwari, Amit Kumar, and Zafar Said

1. Introduction	749
2. EMI and their potential receptor in automotives	750
3. Electromagnetic interference shielding	750
4. Nanomaterials for EMI shielding in automotive applications	751
5. Conclusion	760
References	765

36. Automotive coolants — 773

Zafar Said, Maham Sohail, and Arun Kumar Tiwari

1. Introduction	773
2. Features of advanced cooling system	776
3. Numerical studies and correlations	779
4. Experimental Studies	781

5. Advancements in automotive cooling using nanotechnology — 786
6. Challenges and outlook — 788
7. Conclusion — 790
References — 790

Index — *793*

Contributors

Mohamed A.A. Abdelkareem
Automotive and Tractors Engineering Department, Faculty of Engineering, Minia University, El-Minia, Egypt

Younes Ahmadi
Department of Analytical Chemistry, Faculty of Chemistry, Kabul University, Kabul, Afghanistan

M.G. Akhil
CSIR-National Institute for Interdisciplinary Science and Technology, Trivandrum; Academy of Scientific and Innovative Research (AcSIR), New Delhi, India

Takashiro Akitsu
Department of Chemistry, Faculty of Science, Tokyo University of Science, Tokyo, Japan

Edwin A. Alarcón
Chemical Engineering Department, Environmental Catalysis Research Group, Universidad de Antioquia, Medellín, Colombia

Kevin V. Alex
Department of Physics, School of Basic and Applied Sciences, Central University of Tamil Nadu, Thiruvarur, Tamil Nadu, India

S. John Alexis
Department of Automobile Engineering, Kumaraguru College of Technology, Coimbatore, Tamil Nadu, India

Mohamed Kamal Ahmed Ali
Automotive and Tractors Engineering Department, Faculty of Engineering, Minia University, El-Minia, Egypt

Nargiz Aliyeva
Sabanci University Integrated Manufacturing Technologies Research and Application Center & Composite Technologies Center of Excellence; Faculty of Engineering and Natural Sciences, Materials Science and Nano Engineering, Sabanci University, Tuzla, Istanbul, Turkey

Sahar Amiri
Petroleum and Chemical Engineering Faculty, Islamic Azad University, Tehran, Iran

J. AngelinThangakani
Department of Chemistry, The American College, Madurai, Tamil Nadu, India

Adriana de Araújo Almeida
São Paulo State University (UNESP), Institute of Chemistry, Araraquara, SP, Brazil

Mauricio Zimmer Ferreira Arlindo
School of Chemistry and Food Science, Federal University of Rio Grande, Rio Grande, Rio Grande do Sul, Brazil

Mark Arnce
Department of Chemistry, Pittsburg State University, Pittsburg, KS, United States

A.G. Arsha
CSIR-National Institute for Interdisciplinary Science and Technology, Trivandrum; Academy of Scientific and Innovative Research (AcSIR), New Delhi, India

Nangialai Azizi
Department of Analytical Chemistry, Faculty of Chemistry, Kabul University, Kabul, Afghanistan

Sonia Bahrani
Department of Medical Nanotechnology, School of Advanced Medical Sciences and Technology, Shiraz University of Medical Science, Shiraz, Iran

Ana Paula Michels Barbosa
School of Chemistry and Food Science, Federal University of Rio Grande, Rio Grande, Rio Grande do Sul, Brazil

Martin Bastian
German Plastics Center—SKZ, Wuerzburg, Germany

Benjamin Baudrit
German Plastics Center—SKZ, Wuerzburg, Germany

Gity Behbudi
Department of Chemical Engineering, University of Mohaghegh ardabili, Ardabil, Iran

Ajit Behera
Department of Metallurgical & Materials Engineering, National Institute of Technology, Rourkela, India

Asit Behera
School of Mechanical Engineering, Kalinga Institute of Industrial Technology, Bhubaneswar, India

Nishant Biswakarma
Department of Chemical Sciences, Tezpur University, Tezpur, Assam, India

Emerson Rodrigues Camargo
LIEC—Interdisciplinary Laboratory of Electrochemistry and Ceramics, Department of Chemistry, UFSCar-Federal University of São Carlos, São Carlos, São Paulo, Brazil

B. Chandar Shekar
Department of Physics, Kongunadu Arts and Science College, Coimbatore, Tamil Nadu, India

Jonghyun Choi
Department of Chemistry, Pittsburg State University, Pittsburg, KS, United States

Elisangela Pacheco da Silva
Department of Chemistry, UEM—State University of Maringa Avenida Colombo, Maringá, Paraná, Brazil

Lucas da Silva Ribeiro
LIEC—Interdisciplinary Laboratory of Electrochemistry and Ceramics, Department of Chemistry, UFSCar-Federal University of São Carlos, São Carlos, São Paulo, Brazil

Ajanta Deka
Department of Physics, Girijananda Chowdhury Institute of Management and Technology, Guwahati, Assam, India

Ramesh Ch. Deka
Department of Chemical Sciences, Tezpur University, Tezpur, Assam, India

Kuray Dericiler
Sabanci University Integrated Manufacturing Technologies Research and Application Center & Composite Technologies Center of Excellence; Faculty of Engineering and Natural Sciences, Materials Science and Nano Engineering, Sabanci University, Tuzla, Istanbul, Turkey

R. Dorothy
Department of EEE, AMET University, Chennai, Tamil Nadu, India

Sumit Dubal
Symboisis Skills and Professional University, Pune, India

Ahmed Elagouz
Automotive and Tractors Engineering Department, Faculty of Engineering, Minia University, El-Minia, Egypt

Ansar Ereath Beeran
Department of Chemistry, MES Asmabi College, Thrissur, Kerala, India

Mostafa Eskandari
Faculty of Engineering, Department of Materials Science and Engineering, Shahid Chamran University of Ahvaz, Ahvaz, Iran

Carlos Alberto Severo Felipe
School of Chemistry and Food Science, Federal University of Rio Grande, Rio Grande, Rio Grande do Sul, Brazil

Matheus Ferrer
CCAF—Advanced Crystal Growth and Photonics, CDTEC—Technological Development Center, Federal University of Pelotas, UFPEL, Pelotas, RS, Brazil

Fehim Fındık
Biomaterials, Energy, Photocatalysis, Enzyme Technology, Nano & Advanced Materials, Additive Manufacturing, Environmental Applications and Sustainability Research & Development Group (BIOENAMS R & D Group), Sakarya University; Faculty of Technology, Department of Metallurgical and Materials Engineering, Sakarya University of Applied Sciences, Sakarya, Turkey

Vanessa Hafemann Fragal
Department of Chemistry, UEM—State University of Maringa Avenida Colombo, Maringá, Paraná, Brazil

Mubasher Furmuly
Department of Analytical Chemistry, Faculty of Chemistry, Kabul University, Kabul, Afghanistan

Sandra Gajević
Faculty of Engineering, University of Kragujevac, Kragujevac, Serbia

Luis A. Gallego-Villada
Chemical Engineering Department, Environmental Catalysis Research Group, Universidad de Antioquia, Medellín, Colombia

Sahil Gasso
Department of Physics, Guru Nanak Dev University, Amritsar, Punjab, India

Luiz Fernando Gorup
LIEC—Interdisciplinary Laboratory of Electrochemistry and Ceramics, Department of Chemistry, UFSCar-Federal University of São Carlos, São Paulo, São Carlos, São Paulo; School of Chemistry and Food Science, Federal University of Rio Grande, Rio Grande, Rio Grande do Sul; Materials Engineering, Federal University of Pelotas, Campus Porto, Pelotas; Institute of Chemistry, Federal University of Alfenas, Alfenas, Minas Gerais, Brazil

Nand Kishor Gour
Department of Chemical Sciences, Tezpur University, Tezpur, Assam, India

Ram K. Gupta
Department of Chemistry; Kansas Polymer Research Center, Pittsburg State University, Pittsburg, KS, United States

Peter Hammer
São Paulo State University (UNESP), Institute of Chemistry, Araraquara, SP, Brazil

Thomas Hochrein
German Plastics Center—SKZ, Wuerzburg, Germany

Shumaila Ibraheem
Institute for Advanced Study, Shenzhen University, Shenzhen, Guangdong, China

Abu Bin Imran
Department of Chemistry, Faculty of Engineering, Bangladesh University of Engineering and Technology, Dhaka, Bangladesh

Tenzin Ingsel
Department of Chemistry, Pittsburg State University, Pittsburg, KS, United States

Ritesh Jaiswal
Kamla Nehru Institute of Technology, Sultanpur, Uttar Pradesh, India

J. Jeyasundari
Department of Chemistry, SVN College, Madurai, Tamil Nadu, India

Maryam Jouyandeh
Center of Excellence in Electrochemistry, School of Chemistry, College of Science, University of Tehran, Tehran, Iran

K. Kamakshi
Department of Science and Humanities, Indian Institute of Information Technology Tiruchirappalli, Tiruchirappalli, Tamil Nadu, India

Akhil S. Karun
Materials Science and Technology Division, CSIR-National Institute for Interdisciplinary Science and Technology (NIIST), Trivandrum, Kerala, India

Ramesh Katla
School of Chemistry and Food Science, Federal University of Rio Grande, Rio Grande, Rio Grande do Sul, Brazil

Navdeep Kaur
Department of Physics, Guru Nanak Dev University, Amritsar, Punjab, India

Eduard Kraus
German Plastics Center—SKZ, Wuerzburg, Germany

Amit Kumar
Mechanical Engineering Department, Institute of Engineering & Technology, Dr. A.P.J. Abdul Kalam Technical University, Lucknow, Uttar Pradesh, India

Anil Kumar
Kamla Nehru Institute of Technology, Sultanpur, Uttar Pradesh, India

Anuj Kumar
Department of Chemistry, GLA University, Mathura, Uttar Pradesh, India

Ashish Kumar
Department of Chemistry, Faculty of Technology and Science, Lovely Professional University, Phagwara, Punjab, India

P.S. Samuel Ratna Kumar
Department of Mechanical and Industrial Engineering, University of Johannesburg, Johannesburg, South Africa

Narendra Lakal
Sinhgad Institute of Technology, Lonavala, India

V. Lakshmi
Materials Science and Technology Division, CSIR-National Institute for Interdisciplinary Science and Technology (NIIST), Trivandrum, Kerala, India

Maurice Langer
Fraunhofer IWS, Dresden, Germany

Heinrich Leicht
German Plastics Center—SKZ, Wuerzburg, Germany

P.E. Lokhande
Sinhgad Institute of Technology, Lonavala, India

Mina Madadi
Faculty of Engineering, Department of Materials Science and Engineering, Shahid Chamran University of Ahvaz, Ahvaz, Iran

Aman Mahajan
Department of Physics, Guru Nanak Dev University, Amritsar, Punjab, India

Visakh Manoj
CSIR-National Institute for Interdisciplinary Science and Technology, Trivandrum, India

P.M. Mashinini
Department of Mechanical and Industrial Engineering, University of Johannesburg, Johannesburg, South Africa

Yusuf Ziya Menceloglu
Sabanci University Integrated Manufacturing Technologies Research and Application Center & Composite Technologies Center of Excellence; Faculty of Engineering and Natural Sciences, Materials Science and Nano Engineering, Sabanci University, Tuzla, Istanbul, Turkey

Slavica Miladinović
Faculty of Engineering, University of Kragujevac, Kragujevac, Serbia

Mario Lucio Moreira
Department of Physics, Federal University of Pelotas, Pelotas, RS, Brazil

Seyyed Mojtaba Mousavi
Department of Chemical Engineering, National Taiwan University of Science and Technology, Taiwan

Tuan Anh Nguyen
Institute for Tropical Technology, Vietnam Academy of Science and Technology, Hanoi, Vietnam

Christiane Saraiva Ogrodowski
School of Chemistry and Food Science, Federal University of Rio Grande, Rio Grande, Rio Grande do Sul, Brazil

Burcu Saner Okan
Sabanci University Integrated Manufacturing Technologies Research and Application Center & Composite Technologies Center of Excellence; Faculty of Engineering and Natural Sciences, Materials Science and Nano Engineering, Sabanci University, Tuzla, Istanbul, Turkey

Antônia Millena de Oliveira Lima
Departament of Chemistry, UEM—State University of Maringa Avenida Colombo, Maringá, Paraná, Brazil

Mahmut Özacar
Biomaterials, Energy, Photocatalysis, Enzyme Technology, Nano & Advanced Materials, Additive Manufacturing, Environmental Applications and Sustainability Research & Development Group (BIOENAMS R & D Group); Faculty of Science & Arts, Department of Chemistry, Sakarya University, Sakarya, Turkey

Mehmet İskender Özsoy
Faculty of Engineering, Mechanical Engineering Department; Biomaterials, Energy, Photocatalysis, Enzyme Technology, Nano & Advanced Materials, Additive Manufacturing, Environmental Applications and Sustainability Research & Development Group (BIOENAMS R & D Group), Sakarya University, Sakarya, Turkey

Alexandre Henrique Pinto
Department of Chemistry & Biochemistry, Manhattan College, Riverdale, NY, United States

P. Poornima Vijayan
Department of Chemistry, Sree Narayana College for Women (affiliated to University of Kerala), Kollam, Kerala, India

Nasrin Raji Popalzai
Department of Analytical Chemistry, Faculty of Chemistry, Kabul University, Kabul, Afghanistan

Maria Nayane de Queiroz
Departament of Chemistry, UEM—State University of Maringa Avenida Colombo, Maringá, Paraná, Brazil

L. Ajay Raag
CSIR-National Institute for Interdisciplinary Science and Technology, Trivandrum, India

T.P.D. Rajan
Materials Science and Technology Division, CSIR-National Institute for Interdisciplinary Science and Technology (NIIST), Trivandrum, Kerala; CSIR-National Institute for Interdisciplinary Science and Technology, Trivandrum; Academy of Scientific and Innovative Research (AcSIR), New Delhi, India

Susai Rajendran
Department of Chemistry, St. Antony's College of Arts and Sciences for Women, Thamaraipady, Dindigul, Tamil Nadu, India

S. Rathish Kumar
Department of Biotechnology, Sri Ramakrishna College of Arts and Science, Coimbatore, Tamil Nadu, India

Deepshikha Rathore
Amity University Rajasthan, Jaipur, India

N. Renugadevi
Department of Zoology, GTN Arts College, Dindigul, Tamil Nadu, India

Mohammad Reza Saeb
Department of Polymer Technology, Faculty of Chemistry, Gdańsk University of Technology, Gdańsk, Poland

Gustavo P. Romanelli
Center for Research and Development in Applied Sciences "Dr. Jorge J. Ronco" (CINDECA-CCT La Plata-CONICET); Course of Organic Chemistry, CISAV, Faculty of Agricultural and Forest Sciences, National University of La Plata, La Plata, Argentina

Adley Forti Rubira
Departament of Chemistry, UEM—State University of Maringa Avenida Colombo, Maringá, Paraná, Brazil

Hadi Mohammadjafari Sadeghi
Sabanci University Integrated Manufacturing Technologies Research and Application Center & Composite Technologies Center of Excellence; Faculty of Engineering and Natural Sciences, Materials Science and Nano Engineering, Sabanci University, Tuzla, Istanbul, Turkey

Zafar Said
Sustainable and Renewable Energy Engineering Department; Research Institute for Sciences and Engineering, University of Sharjah, Sharjah, United Arab Emirates; U.S.-Pakistan Center for Advanced Studies in Energy (USPCAS-E), National University of Sciences and Technology (NUST), Islamabad, Pakistan

Sudakhina Saikia
Department of Chemical Sciences, Tezpur University, Tezpur, Assam, India

Hatice S. Sas
Sabanci University Integrated Manufacturing Technologies Research and Application Center & Composite Technologies Center of Excellence; Faculty of Engineering and Natural Sciences, Materials Science and Nano Engineering, Sabanci University, Tuzla, Istanbul, Turkey

S. Sathish
Department of Physics, School of Basic and Applied Sciences, Central University of Tamil Nadu, Thiruvarur, Tamil Nadu, India

K.C. Sekhar
Department of Physics, School of Basic and Applied Sciences, Central University of Tamil Nadu, Thiruvarur, Tamil Nadu, India

Thiago Sequinel
Faculty of Exact Sciences and Technology (FACET), Federal University of Grande Dourados, Dourados, Mato Grosso do Sul, Brazil

C. Dorothy Sheela
Department of Chemistry, The American College, Madurai, Tamil Nadu, India

J.P.B. Silva
Centre of Physics of Minho and Porto Universities (CF-UM-UP), Campus de Gualtar, Braga, Portugal

Rafael Silva
Department of Chemistry, UEM—State University of Maringa Avenida Colombo, Maringá, Paraná, Brazil

Maham Sohail
Sustainable and Renewable Energy Engineering Department, University of Sharjah, Sharjah, United Arab Emirates

Manreet Kaur Sohal
Department of Physics, Guru Nanak Dev University, Amritsar, Punjab, India

Archana Somadas Radhamany
Department of Chemistry, Sree Narayana College for Women (affiliated to University of Kerala), Kollam, Kerala, India

Felipe M. de Souza
Department of Chemistry, Pittsburg State University, Pittsburg, KS, United States

Thiago Augusto Carneiro de Souza
São Paulo State University (UNESP), Institute of Chemistry, Araraquara, SP, Brazil

Francisco Nunes de Souza Neto
LIEC—Interdisciplinary Laboratory of Electrochemistry and Ceramics, Department of Chemistry, UFSCar-Federal University of São Carlos, São Carlos, São Paulo, Brazil

Blaža Stojanović
Faculty of Engineering, University of Kragujevac, Kragujevac, Serbia

Md. Abu Bin Hasan Susan
Department of Chemistry, Faculty of Science, University of Dhaka, Dhaka, Bangladesh

Abhinay Thakur
Department of Chemistry, Faculty of Technology and Science, Lovely Professional University, Phagwara, Punjab, India

Arun Kumar Tiwari
Mechanical Engineering Department, Institute of Engineering & Technology, Dr. A.P.J. Abdul Kalam Technical University, Lucknow, Uttar Pradesh, India

Andressa Trentin
São Paulo State University (UNESP), Institute of Chemistry, Araraquara, SP, Brazil

Serbülent Türk
Biomaterials, Energy, Photocatalysis, Enzyme Technology, Nano & Advanced Materials, Additive Manufacturing, Environmental Applications and Sustainability Research & Development Group (BIOENAMS R & D Group); Biomedical, Magnetic and Semiconductor Materials Application & Research Center (BIMAS-RC), Sakarya University, Sakarya, Turkey

Mayara Carla Uvida
São Paulo State University (UNESP), Institute of Chemistry, Araraquara, SP, Brazil

Manuel Edgardo Gomez Winkler
Departament of Chemistry, UEM—State University of Maringa Avenida Colombo, Maringá, Paraná, Brazil

Hou Xianjun
Hubei Key Laboratory of Advanced Technology for Automotive Components, Wuhan University of Technology, Wuhan, China

Anshul Yadav
CSIR—Central Salt and Marine Chemicals Research Institute, Bhavnagar, Gujarat, India

Ghulam Yasin
Institute for Advanced Study, Shenzhen University, Shenzhen, Guangdong, China

Mahdi Yeganeh
Faculty of Engineering, Department of Materials Science and Engineering, Shahid Chamran University of Ahvaz, Ahvaz, Iran

Preface

The latest technologies can be not only found in health care and space applications but also used in supercars/hypercars. These cars require high-performance materials that have high strength and high stiffness and are lightweight. For higher performance, car engines should become stronger but smaller, use lower amounts of fuel, and have a cleaner exhaust. In Koenigsegg Gemera, a tiny friendly giant (TFG) engine (2 L/three cylinders) could provide 600 horsepower (hp) with 20% lesser fuel consumption than the typical engine (2 L/four cylinders) and negative CO_2 emission. In hybrid supercars such as the Lamborghini Sian Roadster and Koenigsegg Gemera, both small (1×34 hp/48 V E-motor) and giant (3×500 hp/800 V battery pack) electric engines have been used for different purposes to balance the performance and cost. Regarding the battery back used in hybrid supercars, Lamborghini Sian Roadster uses supercapacitors that can store more power (10–100 times) with fast charging time. In the near future, the hybridization of battery and supercapacitor may exhibit superior features, as compared to the pure battery or supercapacitor. This new hybrid approach can overcome the limits of a single component and even achieve the superfast charging time, more efficiency, and high mileage range on a single charge with improved safety.

This book is inspired by a true story about the *Lamborghini Centenario Roadster (blue cepheus/hera)*—one of only 20 Roadster models produced. On September 1, 2020, a young Vietnamese businessman, owner of SV Group Limited, went to the Lamborghini Newport Beach in the United States to buy the *Centenario Roadster*. This supercar costs about US$2.5 million. However, on the time its wheel runs in Ho Chi Minh City (his beloved hometown in Vietnam), he should invest more than US$10 million in total transportation costs. The young man said that he planned to buy this car because of his passion and also to help him forget his ex-lover. The prospect of watching the *Centenario Roadster* on the streets of Hanoi City was an inexhaustible source of inspiration when writing this book proposal.

This book explores how nanomaterials and nanotechnology can be used to enhance the performance of devices used in automotive industries. It discusses various approaches and materials such as nanoalloys, nanocomposites, nanocoatings, nanodevices, nanocatalysts, and nanosensors used in modern vehicles.

Tuan Anh Nguyen
Vietnam Academy of Science and Technology, Hanoi, Vietnam

ically to perform as intended.

SECTION A

Nanocomposites for automotive application

CHAPTER 1

Nanocomposites: An introduction

Deepshikha Rathore
Amity University Rajasthan, Jaipur, India

Chapter outline

1. Introduction — 3
2. Nanocomposites — 4
   2.1 Nanocomposites using ceramic matrix — 5
   2.2 Nanocomposites using metal matrix — 5
   2.3 Nanocomposites using polymer matrix — 6
   2.4 Nanofillers — 6
3. Synthesis techniques of nanocomposites — 7
   3.1 Solution casting — 7
   3.2 Sol–gel — 8
   3.3 In situ impregnation process — 8
   3.4 Covalent binding — 8
   3.5 Polymerization process — 9
4. Challenges — 9
5. Nanocomposites for lightweight vehicles — 10
6. Nanocomposites in tyres — 11
7. Nanocomposites in tribology — 12
8. Nanocomposites for fuel — 12
9. Nanocomposites: Costs and benefits — 13
10. Summary — 13
References — 14

1. Introduction

In the field of automotive industry, nanocomposites are attracting more attention and anticipation due to their variety of enhanced properties and applications. One of the most important aspects is the environmental hazard generated by automotive industry, which can be minimized by implementing nanocomposites. Some other aspects can be enhanced such as efficiency, quality, light weight, and long life, while some aspects such as scratch, wear resistance, and corrosion can be reduced with the help of nanocomposites. In order to improve the above-mentioned aspects of automotive industry, a vital role has been played by nanocomposites in introducing various nanoparticles for manufacturing several parts of an automobile as depicted in Fig. 1.

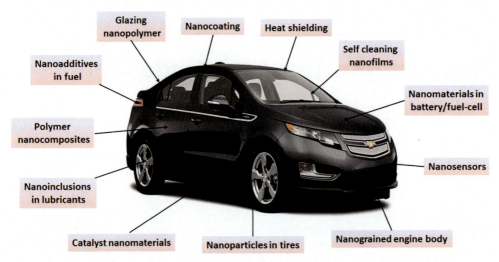

Fig. 1 Nanotechnology employed in manufacturing several parts of an automobile (not exhibiting the definite automobile) [1].

The basic concept of nanomaterials and fabrication of their nanocomposites by different techniques along with the challenges in their fabrication have been discussed in the next section. The latest applications of nanocomposites in automotive industry which include design of lightweight automobile parts, manufacturing tyres using reinforcing agents, improvement in tribology, and enhancement in fuel efficiency have also been reviewed. The estimation of cost and benefits of nanocomposites is also a great challenge as discussed in the last part of this chapter including impact and risk of nanomaterials on the environment and living things.

2. Nanocomposites

Nanotechnology is the manipulation and characterization of materials including fabrication of devices at the nanoscale, which is acting as the driving force to bring great revolution not only for humankind but also for sustainable environment and all creatures as well. Nanotechnology can be defined exactly in many ways, but generally, in scientific and engineering aspects, nanotechnology is the modification of matter at nano range from 1 to 100 nm, toward at least one dimension. Although, all dimensions have been manipulated by several researchers in terms of nanolayers (2D), nanotubes (1D), and nanoparticles (0D). After manipulation from bulk to nanomaterials, these 0D, 1D, and 2D nanomaterials can be used as nanofillers with different variety of matrices to design nanocomposites [2]. An appropriate nanofiller with matrix in proper ratio can improve the strength of nanocomposites [3]. Additionally, nanocomposite materials can be

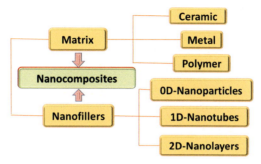

Fig. 2 Ingredients of nanocomposites.

categorized into three sets using different matrix materials including nanofillers as illustrated in Fig. 2 with its ingredients.
- Nanocomposites using ceramic matrix
- Nanocomposites using metal matrix
- Nanocomposites using polymer matrix
- Nanofillers

2.1 Nanocomposites using ceramic matrix

In these nanocomposites, one or more types of ceramic matrices are purposely incorporated to improve chemical and thermal stability with wear resistance. Some energy dissipation components like nanoparticles, nanoplatelets, and nanofibers are introduced into ceramic matrix to enhance fracture durability and minimize brittleness [4, 5]. The reinforcement at the nano scale is generally used to design nanocomposites. For crystalline reinforcement, different metal powders like silica, TiO_2, and clay have been introduced with SiC, Al_2O_3, and SiN as raw materials. Due to their better interfacial chemical interaction and small particle size, layered silicates and clays are very common nanofillers. They can enhance matrix properties by adding in a very small amount [6]. Various techniques are established to synthesize nanocomposites using ceramic matrix. The latest one in which hybrid precursor is melted by spinning and curing with pyrolysis of fillers is known as single-source precursor method. Some common techniques are also available like spray pyrolysis, conventional powder technique, chemical and physical vapor deposition (CVD & PVD), sol–gel, co-precipitation, and colloidal technique [3].

2.2 Nanocomposites using metal matrix

Various metals and their alloys possess ductility that can be used as matrix with different reinforcement materials such as nanofillers for designing nanocomposites. The metals such as Mg, Al, Pb, W, Fe, and Sn are very common metal matrices. Nanocomposites using these metal matrices exhibit high value of modulus, toughness, ductility, and strength. These properties are very useful in automotive industry and aerospace. Some very common techniques have been utilized to prepare nanocomposites using metal matrix such as electrodeposition, rapid solidification, spray pyrolysis, vapor techniques, liquid metal infiltration, sol–gel, and colloidal methods. Some other novel techniques have also been developed as one-pot synthesis of the nanocomposite using gold nanoparticles and carbon dots including falling-drop quenching process [3].

2.3 Nanocomposites using polymer matrix

Polymer-based matrix nanocomponents contain nanofillers categorized as nanoparticles (0D), nanotubes (1D), and nanolayers (2D) [7]. For fabrication of nanocomposites, the most common polymers are polypropylene, polyamides, polyethylene, vinyls, styrenics, polycarbonates, acrylics, epoxies, polyurethanes, and polybutylene terephthalate including different types of resins. The interaction at molecular level between polymer matrix and nanofillers generates nanocomposites with numerous properties. Whereas introduction of small quantity and size of nanofillers less than 100 nm can alter the properties of nanocomposites. The combination of carbon fillers and polymer resin produces carbon fiber-reinforced composites. Since nanocomposites deliver immense impact on the realm of vehicles and driving including cars, motors, and bikes, hence they have brought huge opportunities in automotive industry, which can be attributed as per the requirement to fabricate numerous auto parts. The polymer nanocomposites have been employed by intercalation (in situ and melt) and sol–gel techniques. Layered silicate polymer nanocomposites have been perfectly synthesized using in situ intercalation polarization method, in which silicate layers are distributed in an appropriate monomer trailed by polarization. The formation of polymer from monomer during polarization happens in the interlayered space of layered silicates. Due to which silicate layers start to swell in the monomer solution on exposing heat or radiation. This process can also be initiated using catalyst by cation exchange inside the interboundary part [8].

2.4 Nanofillers

Generally, inorganic fillers which contain minimum one dimension at nano range comprise the organic polymer matrix with proper dispersion result hybrid materials, which are known as nanocomposites. Due to nanometer size these nanofillers possess high aspect ratio. Hence, inclusion of very less amount of nanofillers can alter the macroscopic properties of polymer. Subsequently, as compared to individual polymer matrix and nanofillers, nanocomposites exhibit enhanced properties like better thermal stability and advanced mechanical properties. For designing nanocomposites, very frequently utilized nanofillers are layered silicates and montmorillonite clay available in almost 1 nm thick sheets with 100–1000 nm length and 1000:1 aspect ratio. These sheets are arranged in stacks and linked through van der Walls forces with consistent gap among them known as interlayer spacing. These forces are comparatively weak, due to which small molecules begin to intercalate easily between sheets or layers. Remarkably small filler's loading levels are essential for property enhancement of polymer matrix. Expected outcomes of nanocomposites comprise enhancement in Young's modulus, heat distortion temperature, flexural strength, barrier properties, and many more; these have been employed in a variety of applications including automotive industry [9].

3. Synthesis techniques of nanocomposites

Several researchers have employed numerous types of techniques and achieved great success in the synthesis of nanocomposites with their appropriate and enhanced properties. The nanocomposites have been synthesized by the following techniques as shown in Fig. 3.

3.1 Solution casting

To synthesize polymeric nanocomposites, solution casting is one of the vibrant processes. In this process, a very small amount of polymer-dissolved dilute solution including nanofillers is cast in the form of film or sheet directed by the solvent evaporation. It has the capability to introduce nanofillers directly into the polymer matrix. In this technique, a suitable amount of nanofillers is dispersed into polymer matrix followed by vital stirring. Further, the film is casted and then dried for evaporation of solvent. In the incorporation of huge number of inorganic components or nanofillers, powder of nanofillers should be utilized for preventing agglomeration, due to which thickness of casted film can be

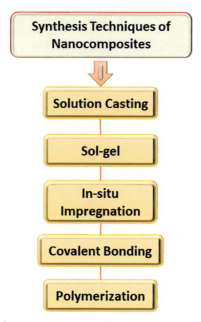

Fig. 3 Synthesis techniques of nanocomposites [10].

maintained easily. But some limitations of this technique also arise in the form of inhomogeneity and pores formation [10].

3.2 Sol–gel

A variety of nanocomposites can be prepared with the help of sol–gel technique. In this technique, nanofillers and different kinds of matrices (ceramic/metal/polymer) are directly mixed with suitable catalyst and solvent to fabricate hybrid structures at the molecular scale. Sol–gel technique is not only suitable for the preparation of polymer nanocomposites but also perfect for ceramic or metal nanocomposites. Various precursors such as tetraethyl orthosilicate (TEOS), tetramethoxysilane (TMOS), Ti(IV)-butoxide, and Zr(IV)-propoxide have been utilized during hydrolysis and condensation steps including appropriate acid, base, or catalysts [10].

3.3 In situ impregnation process

To synthesize proton-conducting, cost-effective nanocomposites, in situ impregnation process is an appropriate technique. This technique can fabricate homogeneous film or sheet at low temperatures. The incorporation of membranes into the inorganic precursors starts the impregnation process. In the step of condensation, the membranes act as essential reactants. While the presence of water in the membrane attracts atoms of inorganic precursors, it is known as nucleophilic attack, which leads to hydrolysis process. If the hydrolysis process is catalyzed by acid and acidified polymer containing $SO_3^-H^+$ group act as catalyst, then any external reactants are not required. There is a basic limitation of impregnation of polymer membranes, which creates difficulty to control concentration gradient of precursor solutions during inclusion of inorganic components into the polymer matrix at a certain level. For example, tetra ethyl orthosilicate (TEOS) is used to modify Nafion membrane [10].

3.4 Covalent bonding

The formation of nanocomposites through creation of covalent bonds between inorganic nanofillers and aromatic polymers as organic matrix is an efficient process in the production of proton-conducting membranes, which is very useful in fuel cells. The molecular precursors are needed in this method, which comprise chemical bond between element and organic moieties. The present chemical bonds are hydrolytically stable and the available element usually creates inorganic network through sol–gel technique. For attachment with inorganic nanofillers covalently, the organic matrix can be altered chemically. This process is known as silylation. After silylation, the organic matrix is available for common polycondensation and hydrolysis reactions. In the covalent bonding process, the basic goal is to prepare polymer comprising SO_3H functions and silicon moieties. The fabrication process contains three steps in sequence: (a) sulfonation,

(b) silylation, and (c) sol–gel. These steps are experimentally so simple, effective, and new processes to introduce silicon functional groups into polymeric carbon matrix by combining the features of a cross-linked polymer due to the presence of organic–inorganic (C—Si) covalent bonds in single macromolecule. This cross-linking of polymer is attained in the sulfonation step during the creation of SO_2 bridges with the successive unit of poly (ether ether ketone) (PEEK). Hence, the first step follows direct sulfonation of Poly (ether ether ketone) (SOPEEK) with chlorosulfonic acid ($ClSO_3H$) resulting in SOPEEK—sulfochlorinated PEEK. Further, the second step silylation introduces covalently linked silicon moieties in SOPEEK and the third step is hydrolysis of sulfochlorinated PEEK for finding the SO_3H functionalized polymer according to requirement [10].

3.5 Polymerization process

The development of polymer nanocomposites for automotive industry application can be accomplished by polymerization process. One of the important tasks for fuel cells is to design ideal ionic membrane, in which hierarchically well-organized ionic particles/channels should be arranged perpendicular to the membrane surface. This can be attained only via surface-initiated atom transfer radical polymerization (ATRP). With the help of ATRP through surface initiation, the uniform thick polyelectrolyte is grown in the form of brush, due to which 2D-arranged microporous silicon provides proton-conducting membrane in whole polymer matrix. For example, initially the self-assembled monolayers of 2-bromo-2-methyl-N-(3-triethoxysilylpropyl)propionamide modify the macroporous silicon scaffold. Then the modified membranes are copolymerized using monomethoxy oligo-(ethylene glycol) methacrylate (MeOEGMA) and sulfopropyl methacrylate (SPM). Hence, after this process membrane networks are entirely occupied with polySPM-co-MeOEGMA brushes, which indicates that the copolymerization process via surface initiation has been configured properly in the constricted network of membrane. This modification of membrane in terms of pores-filling via surface polarization renders hybrid membrane, which is an innovative method to generate proton-conducting networks including tailored physicochemical properties [10].

4. Challenges

Over the previous decade, the researchers have faced the biggest challenge in the fabrication of nanocomposites in terms of compatibility of nanofillers and polymer matrix. Because mostly nanofillers are hydrophilic (water absorbent) and polymer matrix is hydrophobic (water repellent), subsequently those nanofillers disperse nonuniformly into the polymer matrix, resulting in reduced mechanical strength of nanocomposites [11]. To enhance the compatibility between nanofillers and polymer matrix, some coupling reagents, additives, and compatibilizers are employed. Such reagents are as follows:

carboxylated polyethylene (CAPE), maleated polyethylene (MAPE), titanium-derived mixture (TDM), corona discharge, and Maleic anhydride polypropylene (MAPP). A great innovation in the nanocomposites field can be achieved with proper understanding of interfacial interaction and molecular structure between nanofillers and matrix including association between their structure and property [12].

Another challenge is dispersion of nanofillers in the matrix in an appropriate manner, because during processing of nanocomposites the formation of clusters of nanofillers occurs, which decreases the properties of nanocomposites. To improve the properties of nanocomposites, the separation of nanofillers plays a significant role, which can be achieved by some equipment like inline dispersers and high-pressure homogenizers. After many processes, lumps formation and aggregation of nanofillers are still possible. Thus, their inhibition must be taken into account during the synthesis of nanocomposites [12].

5. Nanocomposites for lightweight vehicles

The lighter weight of vehicle's body is attracting several characteristics such as less fuel consumption, low friction, high strength, etc. Nanocomposites have the capability to provide lightweight structure to vehicles with high strength. Because, in nanocomposites, the nanoparticles, nanotubes, and nanoplatelets are embedded as nanofillers into the polymer network. An appropriate nanofiller and polymer network with proper ratio can improve the strength according to requirement [13]. Whereas carbon bucky fibers possess 150 GPa tensile strength, which is almost 50 times greater than that of steel with 1/5th less weight, the polymer fibers can also provide light, stiff, and thinner feature for designing various vehicle's bodies [14]. The polymer network of nanocomposites offers about 25% less weight over highly filled plastic [15]. Chevrolet Volt is the electric car manufactured by General Motors, in which unreinforced polymeric materials are used to design fenders, rear deck lid, and roof, composites are used to make the doors and hood including almost 45.4 kg thermoplastics, while glass fiber reinforced sandwich composites are used to prepare lightweight horizontal body panels [16]. Evora sports car manufactured by Lotus British automaker uses aluminum for chassis and composites for high stiff roof and body panels. Moreover, carbon fiber/epoxy composite is a lightweight composite, which is used to design sink and tank of Tesla sports car to maintain power-to-weight ratio high. It is a very basic fact that the fuel consumption rate can be decreased by reducing the mass of the vehicle. The values of 1.3%–1.8% and 2.7%–3.6% reduction in CO_2 emissions have been obtained by decreasing 5% and 10% of the weight of the vehicle, respectively [17]. This feature can improve the power of engine and reduce fuel consumption and rolling resistance. Nanocomposites as structural low-density composites can reduce almost 50% weight, if trunk compartments of vehicles are designed by them [16]. In the coal-burning process a low-density nanomaterial is found, which is known as fly ash containing micro hollow cenosphere structures.

Because of its low density, fly ash can be coated with copper, nickel, or other metals and materials to generate extremely strong lightweight compositions, which are extensively used in the automotive industry.

6. Nanocomposites in tyres

Recently, implementation of nanocomposites in the manufacture of automobile tyres has brought many possibilities to reduce accidental risk and enhance safety of drivers, passengers as well as vehicles including fuel efficiency and handling performance with high level of control. Nanocomposites for the production of automobile tyres contain silica fillers with carbon black, carbon nanotubes (CNTs), graphene, and nanoclay as reinforcing agents and pigments including other additives and oils, which are imbedded before vulcanization into the network of styrene-butadiene and polybutadiene rubber, which has been represented in Fig. 4 [18]. Generally, diffusion of carbon black at nanoscale into the rubber intensifies the wear resistance and tensile strength of tyres due to the presence of double bonds between filler and polymer network. But silica nanoparticles can improve tractive force at wet surfaces with rolling resistance more than carbon additives. Different kinds of nanoadditives also have the capability to enhance tyre's lifetime significantly. Moreover, abrasion resistance of tyres can also be enhanced by adding nanographene platelets (NGPs) into tyre treads. Overall, execution of nanocomposites into tyres can

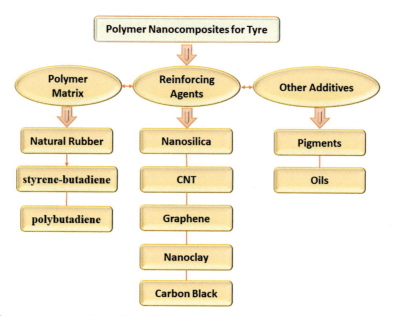

Fig. 4 Polymer nanocomposites utilized for tyres.

not only improve the wear and abrasion resistance but can also reduce air release rates and erosion of tyres with enhanced steering response, surrounding stability, and lifetime.

7. Nanocomposites in tribology

Tribology is one of the main features of automotive industry, which characterizes the relative motion of interacting surfaces in terms of science and engineering. It incorporates the research and analysis of the principles of wear, friction, and lubrication with their application. Integration of nanocomposites has brought some more fantastic features in tribology for vehicles' safety. As brake screeching is a very basic issue of vehicles instigated by variation in friction forces, small disturbance, and dynamic instability of structure of pad material. This issue can be sorted out by introducing nanomaterials as nanocomposites into the pad materials, which enhances the performance and quality of the brake materials in terms of small screech and long lifetime. The molecular dynamic approach by using nanocomposites can improve vehicle safety and comfort level. Moreover, graphene and graphene oxide (GO) are widely used nanomaterials in lubrication. The functionalization of these graphene-based nanomaterials with definite functional groups can enhance their properties and stability. Recently, various graphene-based products and their nanocomposites have been developed for lubricant additives, friction reducing, and antiwear agents (see Fig. 1). The distance between graphene layers can be enlarged by surface modification, which permits them to stay in the form of single layer in the matrix. The functionalized graphene-based nanocomposites can disperse uniformly into the base liquid medium, which is one of the most significant advantages of surface modification. The second advantage suggests that these nanocomposites can be employed in different places. Furthermore, the study on interaction between graphene layers and matrix can provide several opportunities in the automotive industrial applications [19].

8. Nanocomposites for fuel

The evolution of automotive technologies over centuries from steam power to today's wide range has been remarkable. Yet, one aspect that calls for careful investigation is the impact of different fuels and technologies on the environment. As a result, electric and hybrid technologies have gained significant momentum in the past two decades or so. This investigation has also explored the potential of hydrogen fuel cells in replacing other polluting fuels. Although polymer nanocomposites have already proved its potential in fuel cell applications, research is still going on to come up with specific nanofillers and specific spatial arrangement of the fillers for various applications. A clear understanding of all the underlying mechanisms and compatibility between nanofillers and polymers will massively contribute to the durability of fuel cells. Proton conductivity of the polymers can be increased by the introduction of nanometric metal (IV) phosphates such as zirconium phosphate (ZrP) and tin phosphate (SnP) in polymer matrices. This can be achieved by the precipitation from a solution containing M(IV) ions, within an appropriate polymer matrix. To ensure continuous fuel cell operation especially at high temperature and lower humidity, right amount of metal particles should be introduced leading to improved proton conductivity [10, 20].

9. Nanocomposites: Costs and benefits

Usually, nanomaterials are toxic for environment even though nanotechnology exhibits excessive prospects to develop novel and efficient technical products for humankind. Although, the use of nanocomposites in the automotive industry has brought lots of benefits to both environment and society, because of reduction in vehicle's weight, improvement in wear resistance, fuel efficiency, and hardness with better performance as discussed in the previous sections. Even then, the impact and risk of nanomaterials on the environment and living things must be estimated including their cost, source, and exposure. In spite of huge benefits and applications in different fields, the balance in the cost and amount of nanomaterials is also one of the significant tasks. In the automotive industry, certain financial and environmental impacts related to nanotechnology have been evaluated by many researchers [13]. In the assessment, they included the nanocomposites prepared by layered silicate and propylene in the position of aluminum or steel for lightweight vehicle's panels by considering simple material processing technique with fuel-related economic benefits. They analyzed that utilization of nanocomposites presently enhances the manufacturing and processing cost but reduces the fuel consumption and discharges in environment. Not only in automotive industry but also in aerospace, medical, and several research fields, nanotechnology has exhibited fruitful outcomes, e.g., in magneto-resonance imaging (MRI), as drug or gene delivery carriers, in phototherapy, single-domain magnetic nanoparticles for transportation of drugs in the cancer therapy. In order to understand the technical feasibility, economic implications, and environmental impact of nanotechnology at high level, a broad perspective is required. Although not much work has been done in this regard, a brisk beginning can bring all the efforts on track within no time. Although nanocomposites are costlier than aluminum, they are still very competitive as they offer a better fuel economy due to their lighter weights. Better fuel economy automatically reduces the environmental impacts. Another area where nanocomposites score over their metal counterparts is that they can easily be fabricated into any pleasing shape [13].

10. Summary

The flavor of nanocomposites in automotive industry has achieved great attention with their attractive and efficient properties. These properties can be tuned by altering the amount of ingredients required in nanocomposites. The various application of nanocomposites has been incorporated to achieve lightweight of vehicles, enhancement in fuel efficiency, improvement in tribology, manufacturing of tyres, etc. The risk and impact of nanocomposites on environment and social life with their cost and benefits have also been brought into consideration.

References

[1] J. Mathew, J. Joy, S.C. George, Potential applications of nanotechnology in transportation: a review, J. King Saud Univ. Sci. 31 (2019) 586–594.
[2] P. Nguyen-Tri, T.A. Nguyen, P. Carriere, C.N. Xuan, Nanocomposite coatings: preparation, characterization, properties, and applications, Int. J. Corros. (2018) 1–19.
[3] E. Omanović-Mikličanin, A. Badnjević, A. Kazlagić, et al., Nanocomposites: a brief review, Health Technol. 10 (2020) 51–59.
[4] F.F. Lange, Effect of microstructure on strength of Si_3N_4-SiC composite system, J. Am. Ceram. Soc. 56 (9) (1973) 445–450.
[5] M. Harmer, H.M. Chan, G.A. Miller, Unique opportunities for microstructural engineering with duplex and laminar ceramic composites, J. Am. Ceram. Soc. 75 (2) (1992) 1715–1728.
[6] X. Long, C. Shao, H. Wang, J. Wang, Single-source-precursor synthesis of SiBNC-Zr ceramic nanocomposites fibers, Ceram. Int. 42 (16) (2016) 19206–19211.
[7] T. Ogasawara, Y. Ishida, T. Ishikawa, R. Yokota, Characterization of multi-walled carbon nanotube/phenylethynyl terminated polyimide composites, Compos.—A: Appl. Sci. Manuf. 35 (1) (2004) 67–74.
[8] A. Malas, Progress in Rubber Nanocomposites Woodhead Publishing Series in Composites Science and Engineering, 2017, pp. 179–229. 6-rubber nanocomposites with graphene as the nanofiller.
[9] C.C. Okpala, The benefits and applications of nanocomposites, Int. J. Adv. Engg. Tech. 5 (4) (2014) 12–18.
[10] G.G. Kumar, K.S. Nahm, Polymer Nanocomposites - Fuel Cell Applications, 2011.
[11] A. Ashori, Wood-plastic composites as promising green-composites for automotive industries, Bioresource Biotechnol. 99 (2008) 4661–4667.
[12] S.O. Adeosun, G.I. Lawal, E.I. Akpan, Review of green polymer nanocomposites, J. Miner. Mater. Charact. Eng. 11 (2012) 385–416.
[13] M.C. Coelho, G. Torrão, N. Emami, J. Grácio, Nanotechnology in automotive industry: research strategy and trends for the future—small objects, big impacts, J. Nanosci. Nanotechnol. 12 (2012) 6621–6630.
[14] H. Presting, U. Koning, Future nanotechnology developments for automotive applications, J. Mater. Sci. Eng. 23 (2003) 737–741.
[15] J. Garcés, D. Moll, J. Bicerano, R. Fibiger, D. McLeod, Polymeric nanocomposites for automotive applications, Adv. Mater. 12 (2000) 1835.
[16] R. Stewart, Lightweighting the automotive market, Reinf. Plast. 53 (2009) 21. 14–16, 18–19.
[17] G. Fontaras, Z. Samaras, On the way to 130Â gÂ CO_2/km—estimating the future characteristics of the average European passenger car, Energy Policy 38 (4) (2010) 1826–1833.
[18] P. Nguyen, The Nanotechnology and Nanosafety in Automotive Industry, M.S. Project, Wichita State University, December, 2011.
[19] J. Sun, S. Du, Application of graphene derivatives and their nanocomposites in tribology and lubrication: a review, RSC Adv. 9 (2019) 40642–40661.
[20] A.G. Olabi, T. Wilberforce, M.A. Abdelkareem, Fuel cell application in the automotive industry and future perspective, Energy 214 (2021) 118955.

CHAPTER 2

Using XRD technique for model composite and related materials

Takashiro Akitsu
Department of Chemistry, Faculty of Science, Tokyo University of Science, Tokyo, Japan

Chapter outline

1. Motivation of this study — 15
   1.1 Introduction — 15
   1.2 XRD and automotive materials — 16
   1.3 XRD of composite materials for separated Rietveld analysis — 17
   1.4 Conventional XRD for thermal expansion — 18
2. Battery model — 22
   2.1 Li-ion battery and model composites — 22
   2.2 Co-precipitation synthesis of metal oxides — 24
   2.3 Problem of grain boundary — 25
   2.4 Proposal for application in future — 26
3. Catalyst model — 28
   3.1 TiO$_2$ photocatalysts and model composites — 28
   3.2 Proposal for other catalytic composite materials — 31
4. Perspective — 33
5. References — 34

1. Motivation of this study

1.1 Introduction

Our lives today are based on consumption, and there is no end to the problems faced by industrialized societies, such as the environmental pollution that results from the continuous production of garbage and the increase in carbon dioxide due to the large consumption of fossil fuels. However, what we can do in science is, as Buckminster Fuller states in the Spaceship Earth, "We already have the scientific and technological capabilities. With proper use, it is possible to make all humankind happy. The evolution of the world-swallowing industry has saved billions of years of energy, which have been regularly accumulated and stored on board the Spaceship Earth." It stipulates that we are not stupid enough to continue consuming in the blink of an eye in astronomical history.

These energy savings have been stored in the "Life Reproduction Security Bank" account of our spacecraft, but we should consume this account only in the self-starter sector (starting sector). We will reproduce this idea in the society of the 21st century, create a mechanism to continue to generate new energy from small-scale energy, support the foundation of a sustainable society with scientific technology, and circulate in a closed system. How to efficiently use the energy of sunlight that falls from the outside to support human life while regaining the natural circulation of the environment even in the closed system of the Earth like in the spaceship that continues research is very important.

In this context, we have studied on composite materials composed of metal complexes with other functional materials such as metal oxides. Among them, for example, organic–inorganic composite materials are for the purpose of further adding functions and improving performance. Various functions such as gas occlusion in nanopores and associated catalytic reaction and physical property control are expected, and metal complexes become metal–organic frameworks (MOFs) and coordination polymers. In addition, we have been conducting research on the control of changes in the crystal structure of metal oxides due to surface adsorption modification of metal complexes by compounding metal oxides that are widely used as battery materials and catalysts.

How to investigate such composite materials? Experimentally, it is easy to measure temperature changes of structure for model systems. In order to impart anisotropy of crystal structure change, adsorption of MOF to the surface (mixing powder) is confirmed by shifting the IR spectra. It was reported that the composite system of MOF and titanium oxide (TiO_2) has an effect on the temperature change of the crystal structure of TiO_2. At present, we have found that this effect is large in lithium-ion battery materials such as $LiMnO_2$, which have a layered crystal structure typically.

1.2 XRD and automotive materials

X-ray diffraction (XRD) may be one of the most popular measurements for obtaining not only crystal structures but also various aspects of solid-state materials. Even within "conventional" use of (powder) XRD measurements, interesting information could be obtained for composite materials as well as pure compounds such as single crystals. Indeed, nanocomposite materials for automotive applications may be composed of various categories of samples. In this context, herein, our studies using XRD for model composite materials (mainly functional materials and metal complexes) and related other materials will be introduced together with expected demands or potential applications for industrial purposes such as manufacturing of cars (Fig. 1). In this chapter, simple XRD measurements used for the evaluation of related composite materials with metal complexes will be mentioned taken from the author's publications.

Several types of batteries (e.g., Lead–acid battery) may be essential components for cars. Li-ion batteries should be composed of suitable metal oxides (e.g., $LiMnO_2$) or other porous materials containing Li^+ ions. We have prepared model composite materials of $LiMnO_2$ and cyanide-bridged, bimetallic, metal–organic frameworks and measured

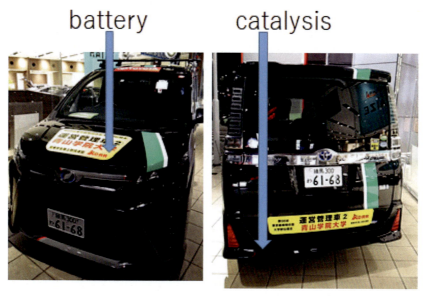

Fig. 1 A car and functional materials (battery and catalysis).

XRD patterns and IR spectra for various ratios of the mixture. Variable temperature XRD patterns and the resulting lattice distortion suggested interaction between two components as powder grains. Of course, powder engineering and the related nanomaterials may be one of the important fields of chemical engineering.

Catalysis (e.g., three-way catalyst) may be one of the important materials for automobiles. TiO_2 is a typical photocatalyst material exhibiting water splitting, decomposition of organic/biochemical (waste or toxic) compounds, or hydrophobic for a mirror. We have prepared model composite materials containing TiO_2 and transition metal complexes and measured powder XRD patterns. By curve separation of the XRD pattern, we could investigate crystal structures of both TiO_2 and transition metal complex separately using conventional programs for Rietveld analysis.

1.3 XRD of composite materials for separated Rietveld analysis

Rietveld method is one of the analysis methods for powder diffraction data. Rietveld method can directly refine structural parameters and lattice constants for the entire powder X-ray or neutron diffraction patterns. Therefore, Rietveld method is a tool that allows one to understand the physical phenomena and chemical properties of polycrystalline materials from a structural perspective. Information obtained from Rietveld method are precise lattice constant, lattice distortion, crystal size, content of each component in the mixture, and so on. Therefore, we could use Rietveld method to investigate structural changes of composite materials.

We have developed a simple method for the evaluation of peaks by composite phase using conventional XRD patterns. When peaks due to multiple phases exist in a short range, it is difficult to regard the peak intensity between the peaks as the peak intensity in the pure phase. And the quantitative value is abnormally high especially when the integrated intensity of neighboring peaks will increase. Fixing the crystal structure and orientation of the complex, in this time, we refined and analyzed the crystal structure and orientation only for $LiMnO_2$, for which one must pay attention to changes.

Specifically, normally, if a researcher analyzes XRD patterns of composite materials as it is, one can immediately see that the quantitative value is abnormal. Fluctuating the significant difference from the background may contribute to reduce the error. In this way, we have successfully analyzed each component of XRD of a composite material.

1.4 Conventional XRD for thermal expansion

Thermal expansion is important information for materials that can be detected by means of XRD. Strain of crystalline materials (expressed as 2nd rank tensor) is proportional to temperature change (scalar), whose expansion coefficient converts scalar to tensor. For example, ZnO is known to exhibit anisotropic thermal expansion, which is different from the direction against the c-axis. Moreover, in the specific temperature region, positive (namely normal) or negative thermal expansion (NTE) can be observed. As mentioned later, negative thermal expansion is expansion of volume or axes on cooling (not heating). On the other hand, in the case of $CaCO_3$, anisotropic thermal expansion surface was employed to describe such anisotropy from the viewpoint of crystal (or molecular) structures. Its thermal expansion surface exhibits circular symmetry about [001] direction (trigonal axis). The maximum expansion direction is perpendicular to the flat carbonate groups.

For about a decade, a phenomenon called NTE has been attracting attention in the fields of solid-state chemistry and structural properties. In contrast to the fact that normal volume expands as most substances heat up, the volume shrinks as most substances heat up (NTE). It is expected to be applied to composite materials that show positive and negative thermal expansion to the desired degree, and materials with a constant volume (zero thermal expansion) regardless of temperature changes. The origins of NTE are as follows: (1) Generally, the shorter the bond distance, the stronger the bonding ability, but the bond length and lattice volume may shrink when a structural phase transition occurs. (2) In the vibration mode that crosses or is parallel to the M-X-M' axis of lattice vibration (phonon), the length of the M … M' interatomic distance is different. (3) Magnetic transition in which a magnetic attraction is induced between atoms. (4) In the case of an arranged lattice frame in which rigid coordination polyhedra share vertices to form a space, the M—O and O—O bonds in each coordination polyhedron are not easily distorted even if the temperature changes, but the arrangement with shared vertices. Mechanisms such as the ability of rotational displacement between polyhedrons have been proposed.

We have given some examples of thermal expansion, but this time we will focus on NTE. Materials exhibiting NTE have received attention in recent years, both functionally and commercially. NTE properties are due to the presence of low-energy transverse vibration modes of oxide cross-linked (M-O-M') and cyanide cross-linked (M-CN-M') backbones, cross-linked species. This vibration has an effect over a wide temperature range. A hard polyhedron centered on metal is considered as a point, and the center point is called rigid unit mode, in which all polyhedra can rotate without bending. For example, in the oxide cross-linked phase ZrW_2O_8, NTE can be explained by the concerted rotation of the ZrO_6 octahedron and WO_4 tetrahedron as the temperature increases. A phase based on cyanide bridges with greater flexibility can be explained similarly [1].

For example, $Nd(DMF)_4(H_2O)_3Fe(CN)_6·H_2O$ was the first compound exhibiting photo-induced change of magnetism among 3d-4f molecule-based magnets. However, the mechanism of phase transition of crystals or valance change by photo-induced electron transfer was not clear [2–4]. Therefore, temperature dependence of crystal structures was investigated for D-substituted complexes whose water and DMF are NMR solvents (easy to obtain D-substituted ones) [5]. It should be noted that this complex has 36H atoms of three coordinated water, one crystalline solvent, and four DMF ligands. It should be noted that all isotope complexes exhibited NTE only along the a-axis around 120–160 K, almost zero thermal expansion around 160–220 K, and (normal) positive thermal expansion above 220 K. The magnitude of NTE depended on the number of D-atoms, which varied the strength of hydrogen bonds in the crystal. There are complicated hydrogen bonds in the crystal, which may be one of the important reasons for this anisotropic and NTE (Fig. 2).

It should be noted that it may be possible to use XRD for material discrimination (composite properties) due to the difference in NTE behavior. In a normal substance that expands as the temperature rises, the XRD shift becomes lower as the temperature rises. Considering a function obtained by differentiating the surface spacing d with respect to the temperature T from Bragg's equation, the coefficient of thermal expansion α is proportional to the function obtained by differentiating the diffraction angle θ with respect to the temperature T [6–9]. The amount of shift when this diffraction angle is temperature can be obtained by measuring XRD at two temperatures. It should be noted that we have proposed an evaluation function for this purpose newly. Ratios of lattice spacing (at T K and 0 K) obeys van't Hoff like relationship. In this equation, K is not equilibrium constant but was defined from the ratio of change of d namely lattice constants. The rate of change of the surface spacing $d(T)$, $d(0)$ extrapolated to the temperature T K and 0 K

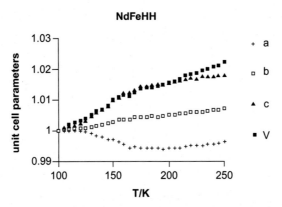

Fig. 2 NTE of the *a*-axis of Nd(DMF)$_4$(H$_2$O)$_3$Fe(CN)$_6$·H$_2$O (H isotope).

is defined as $K = d(T) - d(0)/d(0)$, and $\ln K$ vs $1/T$. The degree of deviation from the approximate anisotropy of the van't Hoff equation ($\ln K = -\Delta H/RT + \Delta S/R$) is plotted using a method that discusses the degree of temperature change of reflection in a specific plane exponential direction [10].

Thermal expansion in crystalline materials is a relatively well-known physical phenomenon. Due to the inherent an harmonic effect of bond oscillations, the average distance between atomic bonds generally increases in temperature and affects expansion on a macroscopic scale. The relative ratio of the material that expands with increasing temperature usually falls within the range of $0 \times 10^{-6} < \alpha < 20 \times 10^{-6}$ K^{-1} for the expansion coefficient α.

Although there are many reports of transition metal oxides in NTE, it has been reported that certain cyanide cross-linked complexes, which are transition metal complexes, exhibit NTE. For example, the NTE of M^{II}Pt^{IV}(CN)$_6$ (M = Mn, Fe, Co, Ni, Cu, Zn, Cd) is compositionally dependent (effect of metal substitution) and is particularly tetragonally distorted Cu^{II}Pt^{IV}(CN)$_6$. Since the Jahn–Teller extension axis of Cu^{II} ion is in the c-axis direction of the crystal lattice, the symmetry is low, and the NTE in the ab plane is positive in the *c*-axis direction, which is anisotropic. Unlike other isotropic M^{II} ions, in the case of Cu^{II} ions, the Cu^{II}-N bond is weakened in the Jahn–Teller extension axis direction, and positive thermal expansion is actually observed. It is known that the increase in the Cu … Pt interatomic distance in the c-axis direction due to heating is largely due to the thermal expansion due to the low-energy axial Cu^{II}-N expansion and contraction mode.

Furthermore, one of the major themes discussed here is the temperature change of XRD of the composite material of the oxide of the lithium-ion battery material and the metal complex MOF. Oxides containing lithium are largely distorted in the b-axis direction, whereas MOFs are characterized by being distorted in different directions.

Focusing on the shift of the specific XRD peak (010) of the entire composite material due to temperature change, in addition to observing a certain correlation, the MOF composite tends to enhance the anisotropy of the structural change [11].

In general, relationship of structural change and chemical factors is known for other inorganic materials too. From the viewpoint of Pauling's bond strength, hard and soft character of bonds affected the thermal expansion coefficients. On the other hand, organic materials (one-dimensional-shaped fiber polyethylene) exhibited clear anisotropy of the thermal expansion coefficients. In this case, structural or bond dimensionality is an important factor for structural changes. In this way, bond strength or direction (structure) results in anisotropic thermal expansion.

Relation to crystal structure was stated in International Tables for Crystallography. Covalent bonds are associated with very small thermal expansion, whereas van der Waals bonds give rise to large thermal expansion. In accordance with their relatively high elastic stiffness, (short) hydrogen bonds lead to comparably small thermal expansion.

Therefore, also for one-dimensional coordination polymers, the main molecular design guidelines for the chiral $[CuL_2][Cr_2O_7]$ complex are as follows [12]: (i) Incorporate a copper(II) complex ($S = 1/2$, Jahn–Teller, flexibility of structural change). (ii) Copper(II) It was a finding of structure-physical property correlation in crystals of a complex (vibronic coupling, Jahn–Teller). Points to be examined in detail in the future are: (iii) Function of organic ligand (chirality, radical, etc.), (iv) Function of metal ion, (v) Function of metal arrangement (bridge/ferromagnetism, spin frustration, etc.), and (vi) Molecular assembly (crystal). The function of distortion and phase transition can be mentioned (Fig. 3).

In layer-like structures, the maximum thermal expansion occurs normal to the layers (mica, graphite, and pentaerythritol). Thermal expansion decreases when the density of weak bonds decreases: therefore, expansion is greater for crystals with small molecules (many van der Waal contacts per volume) than for their larger homologues. It is observed rather frequently in anisotropic materials that an enhanced expansion occurs along one direction and a contraction (negative expansion) in directions perpendicular to that direction. Only a few crystals show negative volume expansion and usually only over a narrow temperature range.

The use of XRD diffractometer, which can change temperature and has high resolution, is indispensable, and the features of this research are the relationship with magnetism by temperature change X-ray diffraction, and the Jahn–Teller coordination structure/bridge structure of Cu(II) units. Extracting structural information of complex systems was based on the results, such as crystal lattice hierarchy, chiral symmetry effect, H/D isotope effect (hydrogen bond), polymorphism, phase transition, etc. The point is to challenge the evaluation method. It is a sample that contains metal elements and is advantageous for diffraction intensity because it is used to test the measurement method [13].

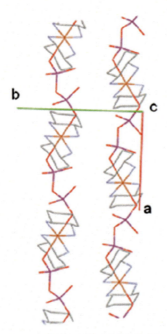

Fig. 3 One-dimensional crystal structure of chiral [CuL$_2$] [Cr$_2$O$_7$] complex.

The expected result is that the behavior of the temperature factor parameters of Cu^{II} and coordination atoms involved in the Jahn–Teller effect will be clarified in a system in which the change in crystal structure (lattice volume/coordination bond) due to temperature change is small. This corresponds to the intermediate case between a system in which both the lattice and the coordination bond contract due to a normal temperature change and a system in which the lattice contracts due to an NTE temperature change and the coordination bond contracts depending on the case. We are also considering that we will be able to obtain insights to deepen our understanding of the above-mentioned lattice vibration of phase transition or NTE mechanism and the Jahn–Teller effect [14].

2. Battery model

2.1 Li-ion battery and model composites

Aiming at preparations of composite materials of LiMnO$_2$ and MOFs, using similar measurements shown in the previous introductory section, close investigation of variable temperature XRD patterns measured in conventional methods can also provide detailed information in the case of composite materials. Prior to show the main results of the final composite, the author would like to remark the crystallographic faces of LiMnO$_2$, typical metal oxide for Li-ion battery material. The (010) face is parallel to the b-axis. (002) face is perpendicular to the c-axis. The (011) face locates bisection of axes. In this way, crystal structure of $LiMnO_2$ exhibits anisotropy clearly. Structural changes of the composite materials can be carried out by variable temperature XRD to discuss anisotropic distortion [15].

On the other hand, these are XRD patterns of variable ratios of mononuclear copper complex and $LiMnO_2$ composite materials at room temperature. The composition of mixture was changed gradually as shown here. Increasing copper(II) complex (or MOF) resulted in anisotropic elongation of the b-axis and (010) for $LiMnO_2$ in this case. Next, we examined variable temperature XRD pattern shift for composite material of Cu: $LiMnO_2 = 5:5$ at several points of temperature. Consequently, the (011) plane indicated irregular structural change. The reason can be considered as follows. Composite was elongated distance of the (011) plane and the b-axis of $LiMnO_2$. Anisotropic structural change was induced by the formation of composites in different direction. Thus, it might suggest specific intercalation of Cu(II) complexes into $LiMnO_2$ layers [16].

This is also a material design concept of the present hybrid composite materials. Metal oxides (e.g., $LiMnO_2$) for rechargeable batteries must include Li-ion (or larger Na-ion in near future). Using large pores in MOF materials such as cyanide-bridged MOF, large space for metal ions can be added. However, MOF potentially exhibits Jahn–Teller distortion of lattice, which is not good for the purpose, anisotropic structural change should be decreased. For this purpose, randomly oriented crystallites were mixed to improve surface interactions by adsorption or other effects [17].

As a secondary battery material that supports new energy use, research on sodium-ion batteries with an eye on the next generation of lithium-ion batteries that use rare metal elements is becoming more important. Secondary battery materials using metal oxides have been developed following lithium, but sodium has a larger ionic radius than lithium ions and is required to be handled safely. Therefore, it has been reported that the "Prussian blue analogues" containing alkali metal ions, which are expected as a new candidate material, have rapid charge/discharge characteristics. However, the cyanide complex site, which is the MOF skeleton having nanopores in which ions enter and exit, is deformed by the structural phase transition, and there is a problem that the characteristics and efficiency of the battery are reduced. Therefore, we use "chiral cyanide cross-linked metal complex" as a secondary battery material, which has been studied in our laboratory on the external field responsiveness of crystal structure and physical properties. The introduction of an organic ligand has nanopores suitable for large sodium ions, and the introduction of a chiral crystal structure limits symmetry and prevents unwanted crystal structure changes (distortion) due to an asymmetric structure. The purpose is to search for materials that overcome the problem of "similar" (Fig. 4).

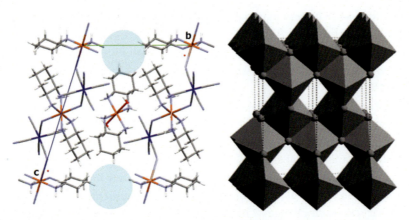

Fig. 4 Concept of composite materials of cyanide-bridged chiral Cu-MOF and LiMnO$_2$.

2.2 Co-precipitation synthesis of metal oxides

As for the preparation of metal oxides containing model systems, we have also studied on co-precipitation synthesis of metal oxides (for Li-ion battery) starting from a two-component chiral metal complex and chiral template structures. As a method for synthesizing ferrite (composite metal oxide containing iron), which is a typical ceramic material, there are several methods for introducing several kinds of metal ions of interest. One is the "solid-phase method," in which a metal salt is crushed and mixed in a solid phase and then fired. There are also difficult drawbacks. On the other hand, the "co-precipitation method" is a method in which a sparingly soluble salt is precipitated from a solution containing several types of metal ions and then fired. The pH and solubility product of the solution are complicatedly controlled, but the distribution of metal ions is uniform. It is possible to obtain a complex oxide, and nickel ferrite and the like exhibit characteristics that cannot be realized in a solid phase, such as exhibiting high-quality magnetism. In this way, in the synthesis by the co-precipitation method, the spatial distribution of several types of metal ions in the poorly soluble salt before firing as a template can be reflected in the spatial distribution of metal ions in the ceramics after firing [18].

Therefore, in this research period, when a two-component chiral metal complex is fired and the ceramics are synthesized by the co-precipitation method, it is possible to induce the spatial distribution of the metal ions of the ceramics by reflecting the chiral structure as a template. The purpose of this step is to observe the anisotropy of the temperature change rate (thermal expansion rate) of the crystal lattice and the anisotropy of the temperature change rate of the axial coordination bond of copper(II) site.

1. Synthesize a two-component chiral metal complex that serves as a precursor as follows. The usual synthesis method of precipitating powder (or single crystal) from a solution of anion and cation complex has the same overall composition and spatial distribution of metal ions as the synthesis of sparingly soluble salts by the

co-precipitation method of ceramics. This is a method for obtaining a certain precursor.
2. Measure the IR spectra, XAS spectra, XRD pattern, etc. of the precursor two-component chiral metal complex can not only confirm the presence or absence of an organic ligand, but also let us discuss about the oxidation number of metal ions, and the change in crystal structure.
3. The precursor two-component chiral metal complex is calcined to obtain a composite metal oxide.
4. Measure the IR spectrum, XRD pattern, etc. of the obtained composite metal oxide to confirm the formation of oxide and the change in crystal structure. Also, confirm the anisotropy and chiral structure of the temperature change rate from the temperature change of the XRD pattern and, if possible, the structural analysis result.

In the future, synchrotron radiation XRD measurement will clarify the chiral space distribution of metal ions in composite metal oxides and the deviation of the temperature change dependence of the crystal lattice and coordination structure, and it is expected to be a useful inorganic material.

2.3 Problem of grain boundary

In addition, grain boundary may be an important factor for composite materials. $LiMnO_2$ and several inorganic compounds in this table and composition dependence were investigated by means of IR and XRD at room temperature. In this way, we confirmed that XRD was detectable as a proof of grain boundary stress. Here are components of composite materials used. $LiMnO_2$ used was in space group *Pmnm*, and chiral metal complexes (namely, two-dimensional, cyanide-bridged Ni—Fe MOF (including both H and D isotopes) and discrete mononuclear Cu complex) were employed to prepare composite materials in this study [17].

Based on these preparations, during this research, (1) among the related complexes, the number of atoms is relatively small and the symmetry is high, $[CuL_2] [Ag(CN)_2]_2$ and $[CuL_2] [Au(CN)_2]_2$ are used as samples, and (2) a guest mononuclear Cu(II) complex is contained in a nanopore host having a chiral structure. Crystals of H/D isotropic substitutions of $[CuL_2]_2[M(CN)_6]_2 \cdot nH_2O$ (M = Cr, Co, Fe), which are molecular magnetic materials, are used as samples. We will try to observe the anisotropy of the structural temperature change of the crystal lattice and Jahn–Teller strain using both single-crystal structure analysis and ab initio crystal structure analysis by powder X-ray diffraction [15].

As an expected result, by dealing with the anisotropy of the structural temperature change, a system in which both the lattice and the coordination bond contract due to the normal temperature change, and a system in which the lattice contracts due to the temperature change of the NTE and the coordination bond (in some cases). Findings that deepen the understanding of the above-mentioned lattice vibration of the NTE mechanism and the Jahn–Teller effect can be obtained. We also want to discuss the relationship between functions such as quenching as a light-emitting material and chiral structure as a magnetic material, and temperature changes and external field responsiveness of an anisotropic crystal structure.

2.4 Proposal for application in future

Metals and intermetallic compounds are used as battery materials and magnetic materials, but the materials themselves are generally the same except that they generate anisotropy due to morphology or constituent atoms such as magnetic anisotropy as a one-dimensional nanowire or two-dimensional nanothin film according to the prior art. It has isotropic physical properties and a crystal structure. In the research on metal complexes and MOFs that have external field responsiveness (magnetism changes) such as temperature and light, by incorporating a site with chirality, an asymmetric molecular structure and crystal structure was created, and it was found that the structural change due to the temperature shows a peculiar anisotropy. Intramolecular hydrogen is mainly used for one-dimensional chain coordination polymers of chiral metal complexes and MOFs that have space in three dimensions, with the aim of distinguishing between the coordination structure and the degree of distortion of the crystal lattice with respect to temperature changes. Experiments were conducted with relatively flexible metal compound crystals such as bonds and coordination bonds. Results such as observation of anisotropic negative expansion due to deuterium substituents were obtained.

When this is measured and interpreted in detail by synchrotron radiation powder X-ray diffraction (XRD), there is no deviation from "a certain mathematical formula" with respect to the temperature change of the diffraction angle, and the distortion deformation is large or small (plane index), which can be distinguished clearly. It was found that by adsorbing chiral molecules on isotropic metal–semiconductor nanoparticles, it is possible to impart optical properties induced by chiral molecules to achiral metal–semiconductors. Applying this to other isotropic metal compounds, we are still conducting research to impart asymmetries such as optical properties for metal nanoparticles, magnetic properties for metal oxides, and changes in crystal structure, and to measure and evaluate them.

In this way, by incorporating a chiral element into an isotropic metal compound and applying a method (schematic diagram) to impart asymmetry of electronic physical characteristics and anisotropy of crystal structure and its change, redox of secondary battery material is applied. We would like to develop research with the rational control of structural phase transitions associated with the reaction in mind. In secondary battery materials based on MOFs and coordination polymers, structural phase transitions are associated with redox (repeated charging/discharging disrupts the regular crystal structure and hinders the diffusion of ions in solids). Rather than using an isotropic and strong structural

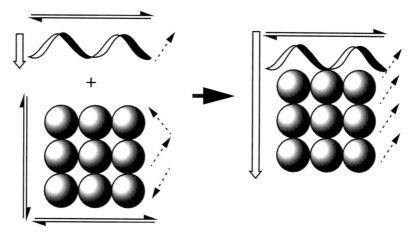

Fig. 5 Concept of composite materials of one-dimensional nanowire metal complex and functional oxide to add anisotropic structural changes.

skeleton, stopping in a specific anisotropic direction that has less adverse effect even if the structure changes should be a breakthrough. Alternatively, we will extend it to the above-mentioned optical property control and magnetic property control that induces attractive force and affects structural changes, with a view to supporting various metal materials (Fig. 5).

The expected results are as follows. In the IR spectra, the higher the composition ratio of the composite metal oxide, the higher the frequency shift of the peak around 800 cm^{-1} was observed. It was found that the peak of the chiral coordination polymer is affected by the adsorbed composite metal oxide surface, not just a mixture of the coordination polymer and the composite metal oxide.

As a result of evaluating the structural temperature change of the composite material, it should be noted that the peak near $2\theta = 20°$ shows a remarkable deviation from the isotropic property of the composite metal oxide, resulting in a chiral one-dimensional chain coordination polymer. Furthermore, it was found that as the proportion of this anisotropic coordination polymer increases, the anisotropy of the structural temperature changes of the originally isotropic composite metal oxide increases.

By the way, a two-dimensional chiral coordination polymer $[CuL_2][Co(CN)_6] \cdot 4H_2O$ containing a mononuclear complex in nanopores and its precursor chiral mononuclear $[CuL_2(H_2O)_2](NO_3)_2$, exhibiting anisotropic structural change potentially, were synthesized according to the previous report. The $LiMnO_2$ crystal was added to them at 10: 0, 9: 1, 8: 2, 7: 3, 6: 4, 5: 5, 4: 6, 3: 7, 2: 8, 1: 9, 0:10, mass ratio was mixed to form a composite material, and adsorption on the surface of oxide ($LiMnO_2$) was confirmed by means of IR spectra (Fig. 6) [15].

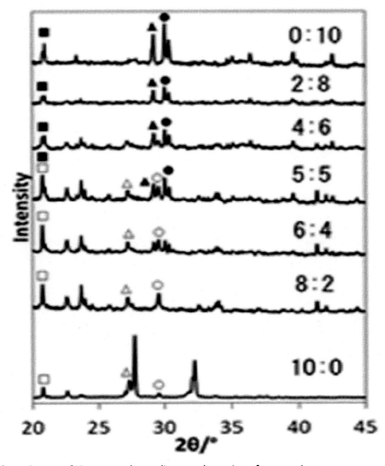

Fig. 6 Shift or change of IR spectra depending on the ratios of composites.

3. Catalyst model

3.1 TiO$_2$ photocatalysts and model composites

In recent years, for example, research has been actively conducted to approach various problems facing modern society from the field of science, and among them, research on titanium oxide (TiO$_2$) photocatalysts is particularly active as a part of tackling energy problems and environmental problems. This photocatalytic reaction is a phenomenon discovered in the Honda–Fujishima effect, which is a reaction in which current flows by irradiating TiO$_2$ with light and at the same time decomposes water to generate hydrogen and oxygen, which is a reaction of photosynthesis in plants. Similarly, the TiO$_2$ photocatalyst in the reaction of producing oxygen from water attracted attention because it reproduced the simplest system for photosynthesis. Furthermore, hydrogen can be used in fuel cells, and the theoretical value of conversion efficiency when extracting electrical energy from hydrogen is 80% or more, which is expected to be nearly twice the maximum efficiency of gasoline, and only water is discharged. Expected as an alternative energy to petroleum in the future, TiO_2 can extract hydrogen from water only by sunlight, so research on the decomposition reaction of water by a photocatalyst is also developing in order to use hydrogen as energy for fuel cells using Pt materials.

In addition, TiO_2 has high photooxidation power and superhydrophilicity, can oxidatively decompose a small amount of dirt, bacteria, viruses, etc., and has a small contact angle when water is adsorbed on the surface, so dirt can be easily removed. In addition, TiO_2 is present in large quantities in nature, is extremely stable chemically, is harmless to the human body and the environment, and is available at low cost, so it can be an important substance for maintaining a sustainable society.

In addition to these, photovoltaic power generation has been attracting attention as a new energy source in recent years, and although the conversion efficiency is currently inferior to that of silicon solar cells, the production method is simple, the manufacturing cost is low, and it can be produced with low energy. Attention is also focused on research on dye-sensitized solar cells, which are expected to have a light energy conversion efficiency of 80% or more. It is used for electrodes by utilizing its properties as a semiconductor that allows energy to flow.

In this way, the photocatalysts have the potential to solve various environmental problems in modern society at the same time, and it is an epoch-making ability to approach the two problems of energy and environmental pollution in particular. Because of its unique nature, it may be important as a technology that will support the future of humankind in the 21st century.

Of course, certain types of MOF can act as catalysts for automobiles for exhaust gas purification and so on. We have almost investigated the composite materials of $[CuL_2][Co(CN)_6] \cdot 4H_2O$ and TiO_2 by means of conventional powder XRD and IR spectroscopy similar to the composite materials of $LiMnO_2$ (Fig. 7). Furthermore, separated Rietveld analysis was also carried out for some composite samples (Fig. 8). Finally, two TiO_2 samples could be distinguished from powder XRD (Fig. 9) [19].

The procedure of preparation and measurement of the composite materials are as follows: (1) Mixing 0.1 g of TiO_2 particles (P-25) and 10 mL of the complex DMSO solution (0.5 mm), (2) Stirring at room temperature for 24 h, and (3) Washing with methanol after suction filtration. And it was measured in what form the molecule was adsorbed on the TiO_2 particle surface. For these samples, the particle size was 10–20 nm. Almost no change was seen in the appearance before and after adsorption. Powder XRD patterns were measured in this way; the measuring equipment was laboratory diffractometer using Cu K-alpha radiation at room temperature. In this way, these were quite conventional ways actually [20].

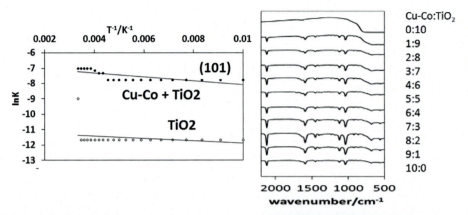

Fig. 7 (Left) XRD and (Right) IR data for composite materials of [CuL$_2$] [Co(CN)$_6$] · 4H$_2$O and TiO$_2$.

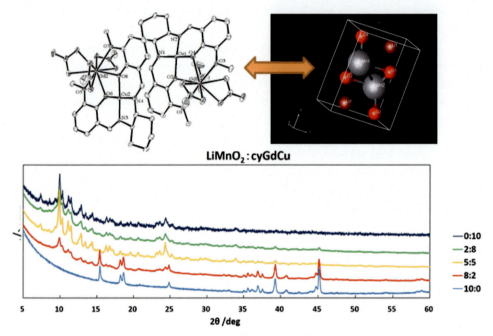

Fig. 8 (Above) components of composite materials and (below) XRD data for composite materials of different ratios.

The XRD of the mixture of the complex and TiO$_2$ was measured and structural change was investigated by Rietveld analysis (Fig. 9). Black and red lines denote TiO$_2$ only and composite, respectively. After deconvolution of XRD patterns of the composite materials, Rietveld method was applied to determine the crystal structure of TiO$_2$ separately. In this case, it should be noted that anatase and rutile types of TiO$_2$ were distinguishable from the XRD patterns. By changing the composition of

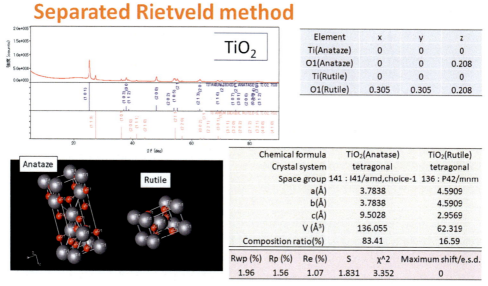

Fig. 9 Analysis of mixed XRD data for different TiO_2.

the composite materials, we determined cell volume of anatase and rutile types of TiO_2, respectively. The bonding distance and angle of TiO_2 are changed, which could be detectable by the separated analysis of XRD patterns. Especially, the volume of the unit lattice of TiO_2 was changed largely in the cases of two complexes. With some other results, we can obtain conclusion here from this study. Novel salen type Cu(II) complexes were synthesized. Adsorption of complex toward TiO_2 was identified via XPS measurement. Change (distortion) of structure of TiO_2 was seen by powder XRD measurement. In this way, separated analysis of XRD patterns of composite materials could provide deeper information. These papers were associated with separated analysis of XRD patterns for similar composite materials containing TiO_2 [15].

3.2 Proposal for other catalytic composite materials

The metal surface and the metal nanocluster surface are being studied as typical heterogeneous catalysts, while the coordination of metal complexes such as organometallic compounds and metal enzymes is being studied as a typical homogeneous molecular catalyst. Recently, research on hybrid catalysts in which a metal complex is adsorbed on the surface of a metal or metal oxide has begun, and not only the improvement of catalyst reaction efficiency, which was the main purpose of conventional supported catalysts. The trend of research is to make complex functions such as imparting molecularity and clarity peculiar to the system.

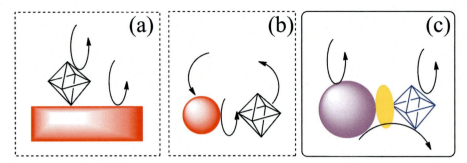

Fig. 10 Types of composite catalytic materials.

By the way, when a composite system with a homogeneous catalyst is created in a situation where it affects the electronic state of the metal atoms that are the active points of each other without completely covering the surface of the homogeneous catalyst (each reaction engineering may be well understood). (a) In addition to the advantages that the chemical reactions of both catalysts simply occur, both at the same time, or in a homogeneous catalyst system (b) in addition to that, the product is passed and occurs continuously, and (c) a composite. A synergistic effect of catalytic activity in the novel electronic state of the system and transfer of products or reactants is expected, but the current situation is that catalytic chemical studies that substantiate the hypothesis (c) have not been conducted much (Fig. 10).

Here is an outline of research on zinc(II) complex-colloidal gold. Chiral Schiff base zinc is newly synthesized and identified (element analysis, IR spectrum, powder diffuse reflection spectrum, X-ray crystal structure analysis) with a commercially available colloidal gold (particle size: 10, 40, 80 nm) aqueous solution for biotechnology. The acetone solution of the complex and the azobenzene-containing complex were mixed at a predetermined concentration, and the visible ultraviolet absorption spectrum and the CD spectrum of the constituent components and the mixed solution were measured. Colloidal gold with a particle size of 10 nm showed an absorption band near 520 nm, and no CD spectrum was observed. On the other hand, the chiral Schiff base zinc(II) complex shows an absorption band around 380 nm, a CD band around 370 nm and 410 nm, and the azobenzene-containing complex shows an absorption band around 380 nm, and a CD band around 350, 370, and 400 nm. In the mixed solution of colloidal gold and zinc(II) complex, a weak and broad CD band was observed around 450–520 nm. Furthermore, in the fluorescence spectrum, quenching due to intermolecular interaction was observed, suggesting that the electronic state of gold atoms on the surface is different from that of gold colloid alone.

Recently, we have been actively using photofunctional metalloid-containing semi-metal nanoparticles used as quantum dots in fluorescent reagents in fields such as physics and chemistry, both in Japan and overseas. Based on the size dependence of the spectral shift of conductor nanoparticles, we are currently conducting basic research on intermolecular interactions and spectroscopic detection in nanoparticle composite systems using a new concept of chiral chemistry (chiral transfer). Furthermore, it would be significant if we could propose a new spectroscopic evaluation method that uses original chiroptical characteristics rather than quantum dot fluorescence. Based on these, the UV and CD spectra of gold nanoparticles (gold colloidal solution) containing a chiral metal complex were measured, and achiral gold nanoparticles having no direct chemical bond, which had never been observed before. The purpose is to observe the chiral transfer from the chiral metal complex to the surface as an induced CD band in the plasmon band wavelength region.

When a composite system with a heterogeneous catalyst, the catalytic activity can be obtained from the novel electronic state. The originality of the hypothesis lies in the expected synergistic effect of product/reactant delivery (Fig. 10C). We are also interested in the "separation" of structural analysis of adsorbed molecules from XRD data.

4. Perspective

In this way, with the inorganic materials required for automobiles in mind, we introduced an example of researching a model of an organic–inorganic composite material incorporating further functions using a devised XRD. Although not mentioned here, inorganic materials for other uses may also be important.

It is said that fossil fuel energy sources such as oil and coal will be depleted in the latter half of the 21st century. For example, secondary damage to nuclear power plants and power shortages caused by the earthquake have become social problems. Dye-sensitized solar cells are expected to be the most promising candidates for new energy sources. In order to maximize the energy efficiency of photovoltaic conversion, it is necessary to optimize elemental technologies such as semiconductors, photocatalysts, sensitizing dyes, and thin-film lamination, and integrate them as a system. In this research, we focused on sensitizing dyes, and in order to widen the wavelength of light that can be used, we created a sensitizing dye material that combines photoresponsive organic dyes and conducted crystallographic and photochemical studies [21]. The concept of anisotropic light using composite catalytic materials is shown in Fig. 11. There seems to be a need to study both crystal structures and electronic states.

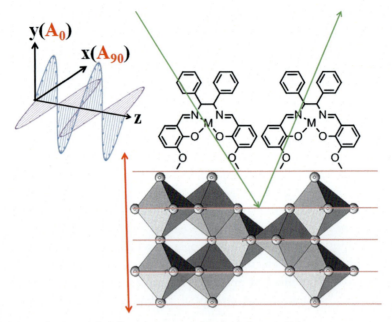

Fig. 11 Concept of anisotropic light using composite catalytic materials.

References

[1] T. Akitsu, M. Ohwa, Y. Endo, S. Sonoki, Y. Aritake, Y. Kimoto, Some factors and effects on thermally structural changes of lattice for cyanide-bridged bimetallic assemblies of cu(II), The Open Crystallogr. J. 4 (2011) 25–29.
[2] T. Akitsu, Y. Einaga, Structures and electronic properties of one-dimensional cyano-bridged Ln(DMF)4(H2O)2Cr(CN)6·H2O complexes (Ln = Tb, Dy, Ho, and Er; DMF = N,N'-dimethylformamide), Asian Chem. Lett. 10 (2006) 95–102.
[3] T. Akitsu, Y. Einaga, Crystal structures, magnetic properties, and XPS of one-dimensional cyanide-bridged 3d-4f complexes Ln(DMF)4(H2O)2Mn(CN)6·H2O (Ln = Ce, Nd, Sm, and Gd; DMF = N,N'-dimethylformamide), Asian Chem. Lett. 10 (2006) 113–120.
[4] T. Akitsu, Y. Einaga, Structure of Cyano-bridged Eu(III)-co(III) bimetallic assembly its application to Photophysical Photophysical verification of photo-magnetic phenomenon, Chem. Pap. 61 (2007) 194–198.
[5] Y. Kimoto, T. Matsui, T. Akitsu, Isotope effect and lanthanide contraction for 3d-4f cyanide-bridged complexes exhibiting negative thermal expansion, The Open Crystallogr. J. 4 (2011) 16–20.
[6] T. Akitsu, M. Okawara, K. Sano, Variable temperature powder X-ray synchrotron diffraction studies on chiral cu(II) and Dicyano ag(I), au(I) bimetallic assemblies, Asian Chem. Lett. 14 (2010) 1–20.
[7] T. Akitsu, K. Sano, Y. Kimoto, Structure of Bis(N-ethylethylenediamine)cu(II) and Dicyano ag(I) bimetallic assembly, Int. J. Curr. Chem. 1 (2010) 89–93.
[8] T. Akitsu, Y. Endo, Y. Kimoto, M. Ohwa, Novel thermally-accessible structural distortion and lattice strain of a chiral cyanide-bridged cu (II)-Ni (II) complex, The Open Crystallogr. J. 4 (2011) 21–24.
[9] T. Akitsu, S. Sonoki, K. Sano, Synchrotron XRD, soft XAS, and IR spectra of Bis(N-ethylethylenediamine)cu(II) and Dicyano ag(I) or au(I) bimetallic assemblies, Int. J. Curr. Chem. 2 (2011) 95–99.
[10] T. Akitsu, K. Sano, Analogy of van't Hoff relationship for thermally-accessible lattice strain of copper(II) complex, Netsu Sokutei 36 (2009) 244–246.

[11] S. Sonoki, A. Yamazaki, T. Akitsu, H/D isotope effect on thermally-accessible structural changes of cyanide-bridged cu(II)-co(III) bimetallic assemblies, in: Crystallography: Research, Technology and Applications, Nova Science Publishers, Inc.(NY, USA), 2012, pp. 65–90.

[12] N. Hayashi, T. Akitsu, Observation of long-range vicinal effect in chiral cu(II)-Cr(VI) or cu(II)-W(VI) bimetallic coordination polymers, Polymers 3 (2011) 1029–1035.

[13] T. Akitsu, Y. Endo, M. Okawara, Y. Kimoto, M. Ohwa, Influence of water molecules on properties of binuclear or bridged structures for chiral CuII-NiII, CuII-PdII, and CuII-PtII Tetracyano-bimetallic assemblies, The Open Crystallogr. J. 4 (2011) 2–7.

[14] S. Ehara, T. Akitsu, Novel thermally-accessible structural changes of chiral and some related cu(II) complexes, J. Chem. Chem. Eng. 8 (2014) 929–941.

[15] Y. Orii, K. Atsumi, M. Matsuno, Y. Machida, T. Akitsu, Separation of XRD patterns to determine crystal structures of $LiMnO_2$ showing anisotropic lattice strain as composites with chiral metal complexes, in: Advances in Chemistry Research, Nova Science Publishers, Inc.(NY, USA), 2015, pp. 89–110. 26, chapter 6.

[16] K. Atsumi, T. Akitsu, P.S.P. Silva, V.H.N. Rodrigues, Separated XRD analysis and non-linear optics for interaction in composite materials of chiral cyanide bimetallic complexes and $LiMnO_2$ metal oxide, in: Advances in Chemistry Research, Nova Science Publishers, Inc.(NY, USA), 2016, pp. 123–140. 30, chapter 7.

[17] M. Kobayashi, T. Akitsu, Anisotorpic lattice strain of composite materials of $LiMnO_2$ and H/D-substituted chiral cyanide-bridged Ni(II)-Fe(III) coordination polymers, J. Chem. Chem. Eng. 8 (2014) 557–563.

[18] M. Kobayashi, S. Ehara, N. Hayashi, S. Sonoki, T. Akitsu, Preparations of complex bimetallic oxides from bimetallic assemblies containing different copper(II) precursors and comparison of some related systems, J. Chem. Chem. Eng. 8 (2014) 647–653.

[19] D. Tazaki, Y. Orii, T. Akitsu, Anisotropic lattice distortion of composite materials of chiral cu(II)-co(III) or cu(II) complexes and TiO_2, in: Cobalt: Characteristics, Compounds and Applications, Nova Science Publishers, Inc.(NY, USA), 2013, pp. 315–326. chapter 13.

[20] M. Matsuno, S. Noor, T. Numata, T. Haraguchi, T. Akitsu, M. Hara, Synthesis and structural characterization of new [cu(II)-TiO_2] composites from cu(II)-salen as precursors, J. Indian Chem. Soc. 94 (2017) 1089–1098.

[21] M. Yamaguchi, Y. Tsunoda, S. Tanaka, T. Haraguchi, M. Sugiyama, S. Noor, T. Akitsu, Molecular design through orbital and molecular design of new naphthyl-salen type transition metal complexes toward DSSC dyes, J. Indian Chem. Soc. 94 (2017) 761–772.

CHAPTER 3

Polymeric nanocomposites

V. Lakshmi, Akhil S. Karun, and T.P.D. Rajan
Materials Science and Technology Division, CSIR-National Institute for Interdisciplinary Science and Technology (NIIST), Trivandrum, Kerala, India

Chapter outline

1. Introduction — 37
2. Perspective of polymer nanocomposites for automobile engineering — 38
   2.1. Nanomaterials in automobile industry — 38
   2.2. Polymer nanocomposites in automobile industry — 38
   2.3. Processing of polymer nanocomposites — 40
   2.4. Properties of polymer nanocomposites used for automobile application — 42
3. Polymer nanocomposites classification — 43
   3.1. Nanofiber-reinforced polymer composites — 43
   3.2. Nanoparticle-reinforced polymer composites — 50
   3.3. Layered material-reinforced polymer nanocomposites — 52
4. Commercialization of polymer nanocomposite in automobile industry — 58
5. Conclusion — 61
References — 61

1. Introduction

The importance of polymeric nanocomposites for the construction of advanced vehicle parts for automobile engineering is reflected in the augmented replacement of metal by the lightweight, high-performance materials, thereby reducing fuel consumption and emission of toxic gases at the same time maintaining adequate safety and comfort [1, 2]. Polymer nanocomposites are multiphase materials that incorporated nano-sized (size <100 nm) fillers, which are called as reinforcement materials such as nanoparticles, nanofibers, or nanorods, into the polymeric matrix [3]. The reinforcement of approximately 2–10 wt% of the nano-scale fillers could enhance the thermomechanical, flame-retardant, heat resistance, and electrical properties of the nanocomposites, compared to their precursor materials [4]. The dispersion of the nanofillers with in the polymeric materials can be achieved either by melt compounding or in situ polymerization. In most of the cases, the nanofillers have to be modified to enhance the adhesion between the filler and the matrix [5]. The presence of the interacting interface between the nanofiller and the polymer plays a major role in imparting the multifunctional properties of the nanocomposites. The use of polymer nanocomposite-based vehicle parts not only reduces the vehicle's weight but also improves energy efficiency and reduction in CO_2 emission [6].

2. Perspective of polymer nanocomposites for automobile engineering

The development of nanomaterials opens a new era in the automobile industry by offering advantages such as lightweight, reduction in friction and emission by the engine, enhancement in corrosion resistance, etc. [7]. Moreover, the excellent electrical properties of the nanomaterials improve the performance of the sensors used in the automobile and led a way toward advanced electronics [8]. The nanotechnology has huge opportunity in automobile industry, since 50 million cars are being produced per year according to the estimation of the United Nations. Therefore, incorporation of these nanomaterials into the polymer matrix led to significant improvement in the functional and structural properties of the polymers. For example, the addition of nanoparticles to the polymer enhances the mechanical property by reducing the crack propagation [9]. Moreover, it substantially reduces the thermal expansion of the polymer. There will be a synergic effect from both the nanomaterials as well as the polymer which could improve the performance of the composite for particular end-use applications.

2.1 Nanomaterials in automobile industry

There are a great range of nanomaterials being used in automotive parts such as carbon-based nanomaterials, nanoceramic particles, metal nanoparticles, boron nitride nanotubes, graphite and graphene, etc. For example, nano oxides such as silica, alumina, nano clays, carbon nanofibers, carbon black, graphene, polyhedral oligomeric silsesquioxanes, and nanostructured poly(alkylbenzene)-poly(diene) (PAB-PDM) are used in automobile tires [10]. These materials remarkably increase the lifetime of the tires by increasing the rolling resistance, abrasive resistance, and wet traction. The addition of nanomaterials such as nanostructured boric acid, tungsten nanospheres, copper nanoparticles, and graphene to the car fluid, which is termed as nanofluid, not only enhances the mechanical properties but also produces economic benefits. So, nanomaterials have now become an inevitable constituent of automobile industry.

2.2 Polymer nanocomposites in automobile industry

The nanomaterials having specific functional properties are suitable reinforcement materials for the polymers, as the nanoscale size as well as the special properties as a result of their high surface to volume ratio provides multifunctional properties compared to the individual constituent. They display excellent combination of optical, electrical, thermal, magnetic, and other physicochemical properties. However, it should be noted that the synergic effect of using nanomaterials in the polymer matrix is more or less influenced by

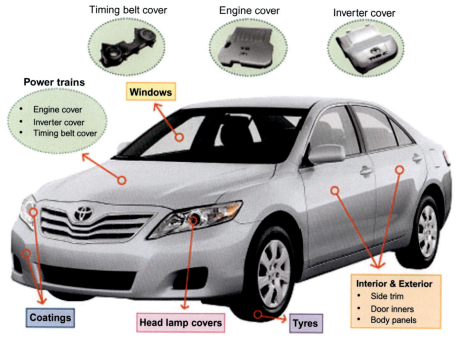

Fig. 1 Illustration on different parts of the car made of polymer nanocomposites [11].

the molecular-level interaction between them. For example, the silica nanoparticles incorporated polymer nanocomposites showed relatively high elastic modulus, low density, good abrasive resistance, and low coefficient of thermal expansion, high thermal stability, and low running costs. The key factor that drives the use of polymer nanocomposites in automobile parts is their lightweight, which reduces the overall vehicle weight, thereby increasing the engine efficiency, reducing CO_2 emission, and providing superior performance. The polymer nanocomposites are used in different parts of the vehicles which are illustrated in Fig. 1.

Polyolefins are the most popular group of plastic materials used in daily life due to their remarkable properties such as low weight, chemical and corrosion resistance, good electrical insulation, etc. The most important polymers in this group are low-density polyethylene (LDPE), high-density polyethylene (HDPE), and isotactic polypropylene (iPP). However, the polymer-based nanocomposites containing relatively low loading (<6 wt%) of nano-sized fillers are gaining more acceptance in the mainstream of automobile engineering. The nanofillers can improve the stiffness of the polymer, and also increase the dimensional stability, gas barrier property, electrical conductivity, flame retardancy, etc. The commercial debut of thermoplastic polyolefin (TPO) based nanocomposite for auto industry appeared as a step assist in 2002 Safari and Austro van which was presented by General Motors. Recently, all the industries and automotive original

Table 1 Polymer nanocomposites commercialized for different parts of vehicles.

Product	Application	Function
Polymer-CNT nanocomposite	Timing belt cover, scratch-resistant coatings in automotive and aircraft frame and body components	Structural and electrical properties
Polymer-clay nanocomposites	Use in motor compartment of vehicle for casting and connectors, auto side board	Nano clay provides strength
Polymer-carbon nanofiber nanocomposites	Aerospace structural components	Structural property

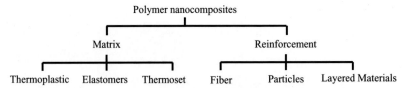

Chart 1 Composition and classification of polymer composites.

equipment manufacturers are more interested in polypropylene-based nanocomposites, as nano-PP is stiffer and easily processable than standard PP.

The addition of the nanofillers synergically enhances the properties of the nanocomposites and prepares the final composite most suitable for automobile applications. Some of the important nanofillers used so far in polymer nanocomposite for automotive fields are nano clays, nanofiber, polyhedral oligomeric silsesquioxanes, etc. Table 1 provides broad category of polymer nanocomposites used for different parts of automobiles.

The design of polymer composites for particular end-use application involves the selection of suitable reinforcement materials based on their properties and dispersing the fillers in polymer matrix. The different polymer matrices and the reinforcement materials used for automobile applications are categorized as Chart 1. The stiffness and strength of the reinforcement material combined with the lightweight polymer matrix enhances the performance of the automobile component.

2.3 Processing of polymer nanocomposites

Different synthetic routes are adopted for the fabrication of polymer nanocomposites. The modification of the nanofiller prior to the incorporation into the polymer matrix is very important to enhance the wettability and dispersion of the nanofillers in the polymer matrix. Therefore, the biggest challenge in the fabrication of polymer nanocomposite is the uniform dispersion of the nanofillers as well as the adhesion between the two components at the interface. Some of the well-known and versatile synthetic strategies are listed here.

2.3.1 Melt intercalation

In this method, the high molecular weight polymer is melted at high temperature, followed by adding the filler and mixed well to get a uniform distribution. The main advantage of this method is the lack of solvent which makes it environmentally friendly. In addition, melt intercalation is compatible with injection molding and extrusion, therefore more convenient and economical. However, the fillers have to be chemically or physically modified prior to melt blending to ensure the uniform mixing in the highly viscous polymer melt. The melt mixing is carried out in a shear mixer or a melt extruder. In an extruder, the polymer granules are mixed thoroughly under high temperature by means of a single or twin screw. It may also contain two counter-rotating screws inside the barrel house. The picture representation of the typical melt extruder is shown in Fig. 2. For the preparation of the polymer nanocomposite, the polymer is initially introduced into the extruder barrel where it got melted and mixed thoroughly [13]. Subsequently, the inorganic filler is added via separate hopper so that melt mixing takes place by the combination of shearing and kneading action, by this time the molten mixture turns into a homogenization zone and achieves significant degree of mixing. The homogenized mixture may pass into compression molding die or an injection-molding machine according to the structure required for end-use application.

However, the high temperature may damage the surface modification of the filler as well as promote the thermal degradation of the polymer, which are the factors to be concerned [14].

2.3.2 Polymer intercalation from solution

In this method, the nanofillers are dispersed in a suitable solvent into which the polymer is also miscible. The fillers are dispersed well in the solvent by means of stirring or

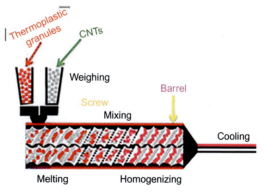

Fig. 2 Schematic representation of melt extrusion process [12]. *(Reprinted from P. Verma, P. Saini, V. Choudhary, Designing of carbon nanotube/polymer composites using melt recirculation approach: effect of aspect ratio on mechanical, electrical and EMI shielding response, Mater. Des. 88 (2015) 269–277, https://doi.org/10.1016/j.matdes.2015.08.156.)*

sonication and then the homogenous solution is added to the polymer solution, followed by solvent evaporation. The fillers got trapped in the crystallized chains of the polymer, thereby preventing the further agglomeration of the particles. Therefore, the method is suitable even with unmodified filler, provided it should be dispersed well in the selected solvent. However, the usage of large quantity of solvent makes the process environmentally unfriendly.

2.3.3 In situ synthesis

Here, the fillers were dispersed in the monomer and then the polymerization is initiated to prepare the highly dispersed polymer nanocomposite. The polymerization can be initiated by either heat, radiation, or inorganic or organic initiator or catalysts. This method also permits the grafting of the polymer onto the filler surface which can improve the final properties of the composite.

2.4 Properties of polymer nanocomposites used for automobile application

In this section, we describe some of the significant properties that need to be improved to make the polymer nanocomposite a suitable alternative for vehicle parts and systems.

2.4.1 Modulus and dimension stability

The modulus of the nanocomposites in terms of the volume fraction, aspect ratio, and particle size of the filler was described by Kerner equation, according to which using small-sized filler particles can increase the stiffness of the materials. The increase in surface-to-volume ratio of the nano-sized particle promotes the adhesion of the particle to the matrix. Therefore, the conventional micro fillers such as glass fiber limit the modulus whereas the nanoparticles could enhance the modulus tremendously. In addition, the high aspect ratio of the nanoparticles allowed the high dispersion of the particles in the polymer matrix, thereby increasing the modulus and tensile strength and decreasing the coefficient of linear expansion. This again improves the dimensional stability of the material, which is essential to manufacture large vehicles.

2.4.2 High heat distortion temperature

Automotive components may have to withstand high temperature depending on the part for which the nanocomposite is being used, for example, interior or engine compartment or elevated temperature involved in paint applications, etc. So the nanofillers that can withstand high temperature are more suitable for such kind of applications. Some examples of such nanofillers are talc, calcium carbonate, etc. which can increase the heat-distortion temperature of the composite material. The nanoparticles will be situated at the interface of the crystalline and amorphous region of the polymer and cause the softening of the crystallized chain as well at its glass transition temperature.

2.4.3 Scratch and mar resistance

Surface quality is a very important factor in many automotive applications which can be qualified by the scratch and mar resistance. The modulus of the material is a deciding factor for the scratch and mar resistance, hence if modulus in increased surface quality also increases. As the incorporation of the nanofiller enhances the modulus, there will be a substantial enhancement in the scratch resistance and mar resistance. Also, smaller particles provide less stress concentration; thereby reducing the potential for damage. The voids created by the nanofiller are very small and hence scatter less light.

2.4.4 Toughness and rheology

The toughness and rheological property of the nanocomposite material has to be good as it affects the processing of the material. The molecular structure of the polymer usually contributes to the rheological property. A unique molecular structure enhances the shear thinning, melts elasticity, and improves polymer melt processibility. Sometimes other polymers of low-molecular-weight additives are added to enhance the affinity between filler and the polymer. This strengthens the interface and manages the stress transfer during deformation. Therefore, the rheological behavior of the nanoparticle-reinforced polymer is crucial which decides the success of the material as an automobile component. So, the compatibility issue is addressed properly, the impact of the nanoparticle on the rheology can be minimized, provided the molecular architecture of the polymer should be designed properly.

3. Polymer nanocomposites classification

In terms of the type of the reinforcement material used polymer nanocomposites used for automobile applications can be broadly classified into nanofiber-reinforced polymer composites (NFRPC), nanoparticle-reinforced polymer composites (NPRPC), and layered material-reinforced polymer nanocomposites (LMRPC).

3.1 Nanofiber-reinforced polymer composites

Nanofibrous materials include both nanofibers and nanowires, which have a one-dimensional architecture with unique properties. The high specific strength and stiffness of the fiber due to their low density reduce the weight of the automobile component substantially and this is the driving force for the use of NFRPC in automobile industry. NFRPC are classified based on the length of the fiber as continuous fiber reinforcement composites and discontinuous fiber-reinforcement composites. Continuous fiber-reinforced composite consists of long fibers whereas discontinuous fiber-reinforced composites are made up of short fibers. The orientation and dispersion of the fiber in the polymer matrix define structure and properties and final composite materials. The advantages of FRPC are that they have high strength and stiffness than the matrix materials therefore acting as the load-bearing element in the composites.

Different varieties of nanofibers are available as filler for polymers that are suitable for the fabrication of structural components of vehicles. The most widely used synthetic nanofibers are cellulose nanofiber, carbon nanofibers, polymer nanofibers, etc. due to their thermal stability, strength, durability, high impact resistance, and good wear properties. Carbon nanofiber is an important synthetic fiber used as a reinforcement in polymer matrix for various applications in automotive industry. They are being used for the front and end door panels of the transportation vehicles as they provide good impact strength and high resistivity to weathering elements.

3.1.1 Manufacturing techniques

Manufacturing of NFRPC involves two steps. Initially, a nanofibrous mat was fabricated by various processes such as weaving, knitting, braiding, and stitching in mat structure. Electrospinning is the most versatile method available recently to fabricate nanofibrous mat having fiber diameter less than 100 nm. Then the as-fabricated fiber preforms are reinforced with the polymer matrix.

Conventional manufacturing technique uses prepregs, which are a combination of fiber and uncured resin and are preimpregnated with thermoset or thermoplastic materials. Dipping in an open mold can impregnate these ready-made prepregs. The impregnated polymer can be activated or cured by the application of temperature. Dow Automotive Systems has developed a technique called VORAFUSE in collaboration with different automotive companies for the fabrication of epoxy resin and carbon fiber-based prepreg. They could improve the material handling and cycle time in the compression molding process for the preparation of the polymer carbon fiber composites, which in turn reduce the weight significantly and thereby increase the efficiency of the component.

Open mold process

There are three different kinds of open mold processes for the impregnation of the polymer resin into the fiber preform such as hand lay-up, spray-up, and vacuum bag molding, which are demonstrated in Fig. 3. The most widely used method is the hand lay-up method where the fiber preform is placed in the mold, which is coated with an antiadhesive material for the east extraction of the composites. The resin is poured through a brush onto the reinforcement materials and a roller is used to force the infiltration of the resin into the fiber preform, whereas in the spray-up technique, the resin as well as the chopped fibers are sprayed on a mold simultaneously. A roller is used to fuse the fiber into the matrix. This method is widely accepted if chopped fiber is used as the reinforcement material.

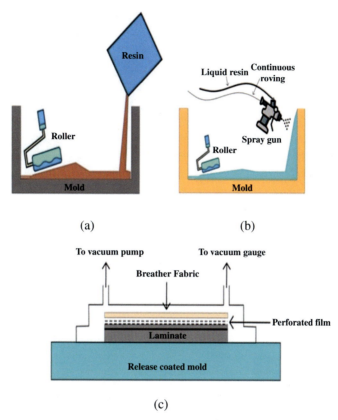

Fig. 3 (A) Hand lay-up, (B) spray-up, (C) vacuum bag molding.

In vacuum bag molding method, a laminate of the polymer such as nylon, polyethylene, and polyvinyl alcohol is placed between the vacuum bag and the mold and this will ensure the infusion of the fiber mat with the matrix. A vacuum pump connected to the vacuum bag will evacuate all the air within the bag so that the atmospheric pressure compresses the part. The advantages of this method are the fabrication of hierarchical composite materials using multiscale reinforcement fibers with minimum porosity and proper impregnation.

Resin transfer molding

Another well-known method for the fabrication of FRPC is the resin transfer molding (RTM) which is demonstrated in Fig. 4A. This method enables the fabrication of composite materials with a wide variety of fibers and even 3D reinforcements with high-quality, high-strength, composite structural material parts. A recent development in RTM uses vacuum aided infusion which is called as vacuum-assisted resin transfer molding (VARTM). In this method, the fiber preform is placed in the mold and the

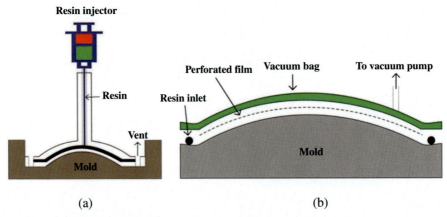

Fig. 4 (A) Resin transfer molding, (B) vacuum-assisted RTM.

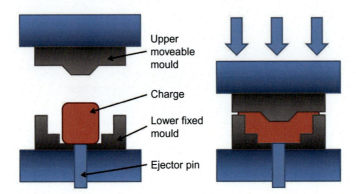

Fig. 5 Compression molding.

perforated tube is kept in between the vacuum bag and the resin container (Fig. 4B). The application of vacuum causes the resin to be infiltrating through the perforated tubes over the fiber preform to form the laminate structure.

Compression molding

Compression molding is a high-volume, high-pressure method for the fabrication of FRPC in which advanced thermoplastics can be compression molded with woven fibers, randomly oriented fiber mat, or even chopped fiber strands. The method uses a preheated mold placed in a hydraulic or mechanical press. The fibrous prepreg is placed between the two halves of the mold, which are then pressed against each other to produce the composite materials in desired shape of the mold. Different steps of the compression molding are demonstrated in Fig. 5. This method has been enormously used in automobile

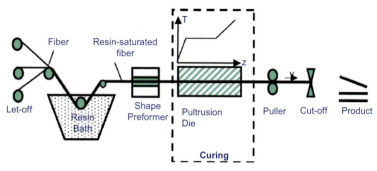

Fig. 6 Pultrusion method.

engineering for the fabrication of FRPC as they offer automation with good dimensional stability, high degree of productivity, short cycle time, etc.

Pultrusion
Pultrusion is a highly automated continuous fiber laminating process for the fabrication of FRPC with constant cross-section. It is an ideal process for the fabrication of structural composite components having high volume fraction of the fiber with a high strength-to-weight ratio. The process involves (Fig. 6) the automatic pulling of the fiber into an in-feed area where it forms the required shape and then impregnated with the resin matrix, which is further consolidated in a heated pultrusion die. The resin materials can be epoxy, polyester, or vinyl acetate which get solidified and cured within the die.

Injection molding
It is a technique used to produce molded plastic products by injecting molten plastic materials into a mold by means of heated barrel. In a typical injection molding process, the blend of fiber and the polymer are fed into the heated barrel through hopper where it is mixed thoroughly by a helical screw. When the content inside the barrel is melted and mixed, it can be injected into the mold of desired shape through a nozzle (Fig. 7).

3.1.2 Different nanofibers reinforcement in polymer nanocomposites
Carbon nanofiber
Carbon fibers are prepared by carbonizing cotton or bamboo and the largest group of carbon fibers is carbon nanofiber (CNF), which are having diameter of 50–200 nm. CNFs are mainly prepared by two approaches, catalytically vapor deposition [15] and electrospinning [16, 17], and gave different structure and properties. The structure and the scanning electron microscopic images of the carbon nanofiber are shown in Fig. 8 [18].

The CNF polymer nanocomposites can be prepared by the melt blending technique. The mechanical property of the polymer will increase with the increase in the CNF

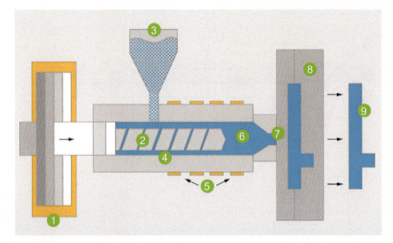

Fig. 7 Injection molding.

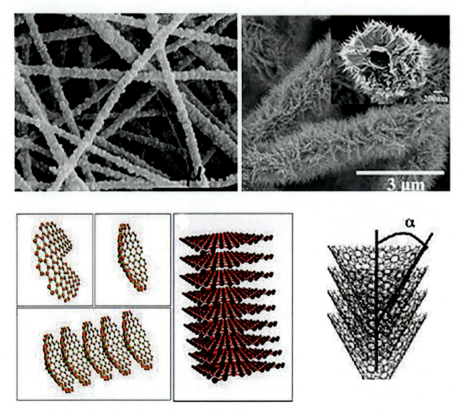

Fig. 8 Structure and SEM images of carbon nanofiber.

content. For example, the resistance to fracture increased from 68% to 78% when the CNF content in epoxy resin is increased from 0.5 to 1 wt%. However, there is dramatic reduction in the tensile strength with the concentration of the CNF due to voids and defect densities in the CNF [19]. At the same time, microhardness of the epoxy-CNF composites increased to 100% with 1 wt% of CNF content [20]. These high mechanical properties of the polymer/CNF composites made them a suitable material in automotive industry. The chassis from Lamborghini Aventador LP700-4 is made up of a carbon fiber-reinforced composite. The steel body of the car is being replaced by the carbon nanofiber-based composites, which are extremely thin and strong and are used in car roof, door, bonnet, etc. The automotive parts made up of CNF-based composite materials aid in weight reduction and double the battery life.

Nano fibrillated cellulose

Nano fibrillated cellulose (NFC) is the most widely used biomass material, which is lightweight, highly elastic, and completely eco-friendly as it is of plant origin. Cellulose nanofiber is synthesized by chemically treating wood pulp and treating it with resin in a special machine. This will break the fiber into the nanometer scale dimension. Therefore, the diameter of NFC ranges from 5 to 60 nm and has a length of several nanometers [21]. It is stronger than steel and easy to recycle. Therefore, the automakers believe that the cellulose nanofiber could reduce the vehicle weight and improve efficiency. Recently, Denso wove cellulose nanofiber into resin of large container for vehicle air conditioner. The Ford biomaterial research team has also been working with wood pulp, and the Weyerhaeuser figured out how to fabricate thermoplastic composite called THRIVE from cellulose nanofiber. These composites are currently available as a blend with polypropylene that require less energy for the fabrication and reduce the wear and tear on the processing equipment compared to the abrasive glass fiber. In addition, these composites of cellulose nanofiber meet the automaker's requirement for stiffness, durability, and temperature resistance. Many automobile companies have used cellulose nanofiber in different parts and components such as spare-wheel compartment covers, seat backrests, headliners, door trim panels, luggage compartment, speakers, carpeting, package shelves, etc. [22].

However, the dispersion of the NFC into the polymer matrix can be achieved by the effective surface modification of the NFC, which can be done through silylation [23], acetylation [24], modification by grafting polymer [25], etc. This modified NFC interacts well with the hydrophobic polymers and highly dispersed polymer nanocomposites can be synthesized. The different processing techniques adopted for the fabrication of NFC/polymer composites are compression molding, vacuum bagging, resin transfer molding, injection molding, etc. The web-like structure of the NCF plays the major role in increasing the mechanical barrier and optical properties of the polymer nanocomposites.

Fig. 9 Extraction of nanofibrillar cellulose.

The extraction of NFC is demonstrated in Fig. 9. Thermoset resins achieved most of the structural applications of NCF.

3.2 Nanoparticle-reinforced polymer composites

Nanoparticles having size of 5–100 nm has gained more attention to be used as reinforcement for polymer matrix both in academia and industry. Different types of metals and metal oxides have been adopted for the fabrication of nanoparticles, core-shell particles, etc. to be used as effective reinforcement for polymers and they can find to be an excellent material for the fabrication of automotive parts. One of the advantages of nanoparticles are their size and surface area depended properties, so they can be used in different areas such as optical, electrical, sensors, etc. There are different physical and chemical methods available for the synthesis of nanoparticles [26]. The physical methods include thermal decomposition, ball milling, spray pyrolysis, etc., whereas chemical methods are sol-gel methods, hydrothermal, precipitation, solvothermal, etc. The dispersion of the nanoparticles into the molecular level enhances the interfacial area in the composites, thereby creating a strong interracial interaction between the organic and inorganic phases in the composites. The impact of this strong interfacial interaction reflects well in the properties of the polymer nanocomposites.

The size and dispersion of the nanoparticles is very important as it influences the intrinsic properties of the polymer such as crystallization, radius of gyration, etc. It should be noted that the polymer chain dimension is greatly influenced by the nanoparticle-polymer interaction, nanoparticle-polymer size ratio, and volume fraction of the nanoparticles where all these factors are decided by the quality of the nanoparticle dispersion in the polymer matrix [27]. The entanglement length of the polymer chain reduced significantly with the volume fraction of the nanoparticles. As the nanoparticles are assumed to be having almost spherical shape, they can be considered as an isotropic material and hence can bind with the polymer in its melt state very easily. Therefore, the polymer nanoparticle composite system is easy to process either by extrusion or by injection molding. At the same time, most of the nanoparticles are polar and hydrophilic, hence to be surface modified to get effective dispersion in the organic polymer matrix. In addition, most of the nanoparticles can be incorporated into the polymer matrix only up to a threshold level, above which the nanoparticles agglomerate in the matrix and inversely affect the properties of the composite. Some important nanoparticles used as reinforcement in automobile industry are described here.

3.2.1 Carbon black

Carbon black (CB) is an amorphous material, which has disordered graphite-like structure. They were produced by the incomplete combustion of aromatic hydrocarbons [28]. They usually have a diameter between 25 and 100 nm and the van der Waal's interaction between the particles tends them to be agglomerated. The fundamental difference between the CB and the graphite or carbon nanotubes is that covalent bond exists between the primary particles of CB, whereas there is only weak interaction between the graphite or CNT particles. The incorporation of CB into the polymer matrix has led to the enhancement in many interesting properties of the composite materials such as thermal and electrical conduction as well as the dimensional stability of the composites improves. The most important composite material made from CB is the rubber/CB nanocomposites, which impart several interesting and useful properties suitable for automobile applications. Kim introduced the CB/rubber nanocomposites for automobile applications [29] especially as antivibration parts in vehicles. The interaction between the different CB particles and the CB-rubber interaction led to an increased elastic modulus and tensile strength and the properties depend also on the particles' size, concentration, and dispersion of the filler in the rubber matrix. It has been observed that nano cavitation is formed in CB filler styrene-butadiene rubber under uniaxial loading conditions and these nanocavities modify the local stress state and promote the local shear motion of filler particles [30]. Therefore, the rubber/CB composites are especially designed to be used in automotive and tire industries.

3.2.2 Silica-based nanoparticles

Silica and organosilanes are included in this category of nanoparticle which significantly improves the properties of natural rubber. The nano-level dimension of these particles as well as the cross-linking ability with the rubber play a crucial role in the tire properties. Replacement of carbon black by silica nanoparticles in tire thread lowers the rolling resistance and enhances the wet grip, therefore decrease the fuel consumption and increase the safety [31]. Other properties such as tensile strength, tear strength, chunking resistance, etc. can also be improved by the reinforcement with silica in rubber. However, poor wettability, dispersion, and distribution of the silica nanoparticles in the rubber matrix are a major shortcoming, which has been reduced by introducing sulfur-containing silane coupling agents along with silica. This could aid the silanization reaction, the extent of which determines the property of the rubber composite [32].

Polyhedral oligomeric silsesquioxane (POSS) is another silane-based organic-inorganic hybrid filler that has a cage-like, three-dimensional structure connected by Si–O–Si bond. They are prepared by the self-condensation reaction between silanes.

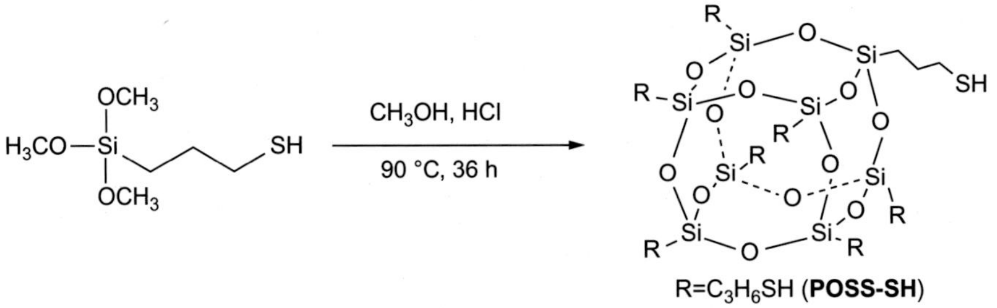

Fig. 10 Synthesis and structure of POSS.

The structure and the synthetic method are demonstrated in Fig. 10 [33]. They are approximately 1.5 nm in size along each axis and are very stable materials. There is practically no interparticle interaction and hence are highly dispersible in polymer matrix. Even weight fraction up to 50% can be incorporated into the polymer without appreciable agglomeration. Another advantage of POSS is that it is readily soluble in the polymer melt and recrystallizes during cooling the melt into a 3D network [34]. One of the most attractive characteristics of the POSS-reinforced composites is their excellent flame retardancy, which is one of the most demanding properties for the automobile components. The mechanical property of the POSS/polymer nanocomposite mainly depends on the surface functional group on the POSS structure. The surface functionality can be changed by using different silanes.

3.3 Layered material-reinforced polymer nanocomposites

Layered materials are solid materials having a strong in-plane bonding and a weak interaction between the different layers. They have lateral sizes ranging from few nanometers to micrometers while have a thickness less than 5 nm. They have large surface area with high percentage of atoms exposed toward the surface. Different kinds of layered materials are available such as graphite, clay, silicates, layered double hydroxides, etc., which are suitable fillers for different end-use applications. Even small loading of these layered materials (1–5 wt%) is enough to improve the mechanical properties up to 200% compared to the neat polymer [35]. However, the challenge in the synthesis of layered material polymer nanocomposites is the dispersion of these materials in the polymer matrix. The dispersion of the layered material depends on the mode of interaction between the different layers such as electrostatic interaction as in clay and van der Waal's interaction as in graphene. The strong electrostatic interaction in clay-like layered materials makes them difficult to be dispersed in polymer matrix. However, the most commonly used layered filler in the automotive industry is the clay or nano-clay [36], i.e., in 1980, Toyota published the great landmark in the polymer clay nanocomposites regarding the polyamide 6/organophilic clay nanocomposite which can be used as timing belt in cars. These

intercalated   intercalated and flocculated   exfoliated

Fig. 11 Morphologies of polymer layered material nanocomposites.

composites containing only 4.2 wt% clay filler can increase rapture tension by 40%, Young's modulus by 68%, and flexural modulus by 126% [37]. Also, the heat distortion temperature has been increased from 65°C to 152°C in comparison with the pure polymer. Later on, several companies introduced polyamide and polypropylene clay nanocomposites in automobile sector.

Polymer layered material (LM) nanocomposites can be prepared by (i) melt blending technique, (ii) solution blending process, or (iii) in situ polymerization process. All these processes lead to three types of morphologies in the LMPNC based on the interfacial force between the polymer and the layered material (Fig. 11).

(i) Intercalated polymer nanocomposites: This morphology is produced by the insertion of the polymer chains into the space between the layers of the LM.

(ii) Flocculated polymer nanocomposites: It is similar to intercalated except for the formation of floccus due to the interaction between the hydrophilic hydroxyl groups of the LM.

(iii) Exfoliated polymer nanocomposites: In this structure individual layers are randomly distributed in the polymer matrix to form a highly dispersed polymer nanocomposite.

The most important layer materials used for automobile applications are clay, graphene, modified graphene, carbon nanotubes, etc.

3.3.1 Clay

We have already discussed about the importance of clay in automobile sector. Clays are a broad class of layered aluminosilicates and can occur naturally or be synthetic. Different types of clay structures are available and the majority of the research has been carried out using montmorillonite clay which is a 2:1 alumina silicate, i.e., it is composed of octahedral aluminum oxide layers sandwiched between two tetrahedral silica layers. They have cations in the structure to balance the negative charge in the octahedral structure. These cations can be replaced by other organic anion of equal charge to prepare

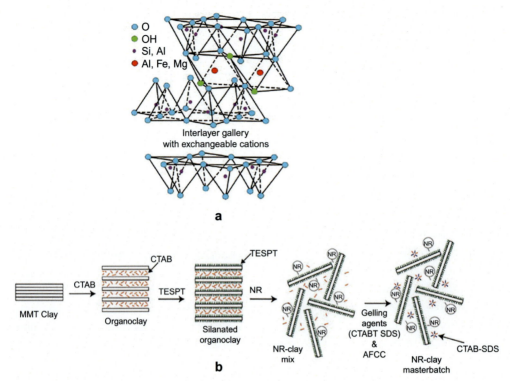

Fig. 12 (A) Structure of montmorillonite clay, (B) process of silylation of organoclay and the incorporation into the natural rubber.

organoclay, which is an effective reinforcement material for polymers. The organoclays are dispersed well in the polymer matrix. The structure and modification of clay are demonstrated in Fig. 12. The modifiers get into the interlayer spacing between the clay layers and can also be called as intercalation agents. Different intercalation agents have been used such as amino acids, primary amines, quaternary ammonium salt, and cationic surfactants and they tend to penetrate into the interlayer structure. The modification of clay with any of these molecules followed by mechanical agitation (sonication or ball milling) led to the complete exfoliation of the different clay layers. The prior exfoliated clay can be used for the preparation of highly dispersed clay polymer nanocomposite. The mechanical property of the natural rubber has been enhanced by the incorporation of silanated organoclay into the natural rubber latex [38].

As clay in the composite has improved the heat distortion temperature of the materials, GM/Blackhawk has also announced the use of polypropylene clay nanocomposites for automotive applications. In addition, the excellent flame retardancy of the clay-reinforced polymer composite could replace the traditional expensive flame-retardant

Table 2 Vehicle parts, manufacturer, and properties of some polymer clay nanocomposites which are commercially used for automotive applications.

Clay/polymer composite	Vehicle parts	Manufacturer	Properties
PP-clay	Body work	Dow Plastics/Magma	Antiscratch properties
Acetal-clay	Ceiling lights	Showa Denko	
PP-clay	Panes of doors, consoles, and interiors decoration	Ford, Volvo	Esthetics, recyclability, and weight-saving properties
Nylon6-clay	Bumpers	Toyota motors	Enhanced mechanical and weight-saving properties
Nylon6-clay	Fuel reservoir	Ube America	Air tight properties
Nylon6-clay	Engine belt covers	Toyota	
Nylon12-clay	Automotive fuel lines	Ube	

materials used for different fire safety applications in automotive industry. According to the business communication company (BCC), nano clay is the most commercially used nanomaterial in automotive sector due to its low cost and availability. They also offer reduction in heat release, reduced weight, flame retardancy, excellent dispersion and exfoliation, and polyolefin is the most common host polymer. Thermoplastic polymers such as nylon, polyphenylene sulfide, polyetheretherketone (PEEK), and polyethylene terephthalate (PET), polycarbonate, thermosets such as epoxy, and thermoset elastomers such as butadiene-styrene diblock copolymers are also used as host matrix in nano clay-reinforced polymer nanocomposites for more demanding automotive applications. An innovative invention from Toyota Motors and Ube is the modification of montmorillonite clay with amino acid to form NCH and then the NCH was intercalated with caprolactam. This NCH polymer composite exhibits excellent mechanical properties compared with the host polymer nylon6 [39].

Different commercial applications of polymer/clay composites are listed in Table 2 [40, 41].

It has been found that the mechanical properties achieved with the polymer/clay nanocomposites are very much superior compared to conventional fiber-reinforced composites with high fiber volume fraction. The best ever improved mechanical properties until present was achieved by the nylon6-clay nanocomposite having ~4 wt% of clay content. The comparison of Young's modulus of nylon6-clay composite and corresponding fiber-reinforced composite is displayed in Fig. 13 [40, 42].

The most concern regarding the polymer/clay nanocomposites is the commercialization of the polymer/clay nanotechnology and the unavailability of the commercially

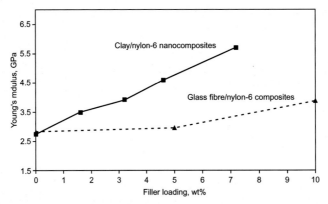

Fig. 13 Comparison of Young's modulus of clay/nylon6 nanocomposites and glass fiber-reinforced nylon6 composites with low filler loading. *(Reprinted from F. Gao, Clay/polymer composites: the story, Mater. Today 7(11) (2004) 50–55, https://doi.org/10.1016/S1369-7021(04)00509-7.)*

available organoclay. It has been seen that the organoclays are produced by the cation exchange with organic ammonium salt. However, these ammonium ions are thermally unstable and degrade below the temperature of 170°C. Therefore, new innovative techniques are being developed for the direct modification of the hydrophilic clay using multifunctional polymer and oligomers to synthesis thermally stable organoclay to be used as a reinforcement in polymer matrix.

3.3.2 Graphene

The most important application of graphene is the incorporation of these novel materials in polymers to form most efficient composite materials. These materials show substantial enhancement in the properties even at a lower loading when compared to the polymer composites with conventional microscale fillers such as glass or carbon fibers. This leads to an appreciable reduction in the component weight and simplifies the processing. It should be noted that even a small amount of graphene dispersed in the polymer matrix could enhance the tensile strength, elastic modulus, electrical and thermal conductivity, thermal stability, gas barrier properties, and flame retardancy. All these mentioned properties make them an ideal candidate for structural materials in automotive sector [43].

The innovative Graphene-based Polymer Composite materials for Automotive (*iGC Auto*) is a comprehensive task of *Concept-oriented lightweight design* to combine the novel material concepts of ultralight graphene-based polymer materials with the latest safety design, efficient fabrication, and manufacturing process and life-cycle analysis to reduce environmental impact for future vehicles. Based on this different graphene-based polymer composites are being designed, modeled, and investigated for the structural applications in automotive industry. Moreover, graphene-based polymer composites provide good dimensional stability, better flame retardancy, and superior durability.

Apart from the superior properties of graphene, many technological challenges have to be overcome to make a good dispersion of the graphene in the polymer matrix. The dispersion and the interfacial interaction between the polymer and the graphene, the network structure of graphene in the polymer matrix are some of the deciding factors for the final properties of the graphene polymer composites. Because of all these challenges no applications of polymer–graphene nanocomposites are currently marketed in automobile industry.

3.3.3 Carbon nanotubes

Carbon nanotubes (CNTs) are hollow cylinders of graphene and demonstrate remarkable electrical, thermal, and mechanical properties. The high strength and stiffness of the CNT make them an ideal reinforcement material for polymer composites. Dave Arthur, chief executive of South West Nano Technologies says, "Many traditional materials used in automobiles, inside and out, can be enhanced using carbon nanotubes." They are manufacturing three kinds of CNTs, single-walled carbon nanotubes (SWCNT), multi-walled carbon nanotubes (MWCNTs), and intermediate type (Fig. 14). One of the biggest applications of CNT is in polymer nanocomposites as the extraordinary properties of the CNT combine with the tailorable properties of the polymer to form a really versatile composite. The most common technique for the fabrication of polymer/CNT composites is the solution blending process. The CNTs are dispersed in a suitable solution and mechanically agitated to get a homogeneous solution, followed by the addition of this solution into the polymer solution and thorough agitation, which leads to the debundling and dispersion of the CNT in the polymer matrix. The composite solution can be then casted to obtain the polymer/CNT composite film. However, the choice of the solvent is

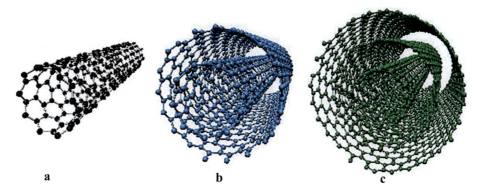

Fig. 14 Structure demonstration of (A) single-walled carbon nanotubes, (B) few-walled carbon nanotubes, (C) multiwalled carbon nanotubes [44].

very important and it should be compatible with the polymer as well. If the polymer and the CNT are dispersed in different solvents, they should be miscible with each other to prepare the dispersed polymer/CNT composite [45].

However, prolonged mechanical agitation of the CNT may lead to tube breakage and shortening. This has been addressed by the modification of the CNT physically or chemically to form organic modified CNT, which enables better dispersion in the polymer. The modification improves the interfacial interaction between the CNT and the polymer. When we consider the large-scale production of thermoplastic/CNT composite for industry, melt blending is the viable process due to its low cost and amenability [46]. In this process, an elevated temperature is subjected to the melting of the polymer and the flow of the polymer melt at high shear rate causing the debundling and dispersion of the CNT in the polymer matrix. The melt mixing can be carried out in a shear mixer or melt extruder. For the preparation of highly concentrated sample shear mixer is preferred.

CNT can introduce electrical, dielectric, thermal, rheological, and mechanical properties to the polymer composites. The intrinsic electrical conducting properties of the CNT lead to onset of electrical conductivity to the insulating polymer matrix. This is because of the formation of 3D network of the CNT within the host thermoset matrix, so as to allow the tunneling of the electron between the dispersed filler particles. The electrical property of the polymer/CNT nanocomposite is utilized to synthesis electrostatic painting for ESD protection when embedded in nylon and EMI shielding in automobile industry. Moreover, the electrical conductivity of the CNT-based polymer composites is useful for automotive fuel system line components which require electrical conductivity. However, the commercial development of CNT-based polymer composite is hindered by their high price ($20/g).

4. Commercialization of polymer nanocomposite in automobile industry

As discussed earlier, the nanofillers such as nano clays, carbon nanofibers, carbon nanotubes, and polyhedral oligomeric silsesquioxanes (POSS) are being used in polymer nanocomposites for various automobile applications. The polymers which are gaining more attention for excellent performance and processability characteristics are polyolefines, polyamides (nylon), polyetheretherketone (PEEK), polyphenylene sulfide (PPS), polycarbonate, polyethylene terephthalate (PET), etc. The thermosetting plastic epoxy resin and thermoplastic butadiene-styrene diblock copolymer are also being used for various automotive applications.

In 1991, Toyota Motor Co. in corporation with Ube first introduced nylon6/clay nanocomposite in the market to produce timing belt covers as a part of the engine for their Toyota Camry Car. They have constituted research team named as Toyota Central

Research Development Laboratories (TCRDL) to work on nylon6/clay composites and improved the methods for the production of the nanocomposite using in situ polymerization. Soon after this, Unitika Co. in Japan introduced nylon6 nanocomposite which was fabricated by injection molding for engine covers on Mitsubishi GDI engine. It is found that the replacement with nylon6/clay nanocomposite offers 20% reduction in weight with excellent surface finish. However, the nylon-based nanocomposites are found to be expensive and hence no longer used for these applications nowadays. Instead, the focus on polyamides has now shifted to other polymer nanocomposites that had high gas barrier property as well. Therefore, Ube demonstrated the use of nylon12 nanocomposites for the fuel system lines and fuel system components that require high gas barrier property, which can meet the future emission standard. Later, in 2002, General Motors in collaboration with Basell launched a polypropylene/clay composite containing 3 wt% nano clay for a step-assist automotive component for GM's Safari and Chevrolet Astro Vans, and in the doors of Chevrolet Impalas.

In 2009, General Motors used nano-enhanced sheet molding compound for one-piece compression molded rear floor assembly for Pontiac Solace, which was developed by Molded Fiber Glass Companies. Over the last 10 years, there is an upsurge in the commercialization of the polymer nanocomposites for automobile industry. Polymer nanocomposites have been used in several applications such as engine and power train, suspension and braking systems, exhaust systems and catalytic converters, frame and body parts, paints and coating, lubrication, tires, and electric and electronic equipment. It has been found that there is a significant improvement in the physicochemical properties of the polymer nanocomposites when reinforced with nano clay. The Toyota research team is now focusing on other polymers such as polystyrene, acrylic, polyimides, epoxy, and elastomers to host nano clay to understand the property enhancement to be used for different automobile components. Time line of the commercialization of polymer nanocomposite is shown in Fig. 15.

It has been reported that the automotive field will become the third largest market for polymer nanocomposite and the development of which will be driven by the emerging application possibilities of the automotive market. The future nanomaterial for the automotive sector will be the carbon nanotubes, however, the increase price limits in current applications in automobile industry.

Another important application of nanocomposite goes to the tire industry, where the elastomeric nanocomposites are gaining much attention. The elastomer nanocomposite lowers the rolling resistance and lowers weight, thereby enhances fuel saving. As already discussed, carbon black is currently used as a nanofiller in natural rubber to be used as an efficient material for vehicle's tire. Moreover, the nanocomposites used as inner-liners reduce the permeability by 50% and also improve the fuel efficiency. The demand for polyolefin-based nanocomposites is raised beyond nylon6/clay nanocomposites because of the low cost and enhanced properties of polyolefins. Although the automobile industry

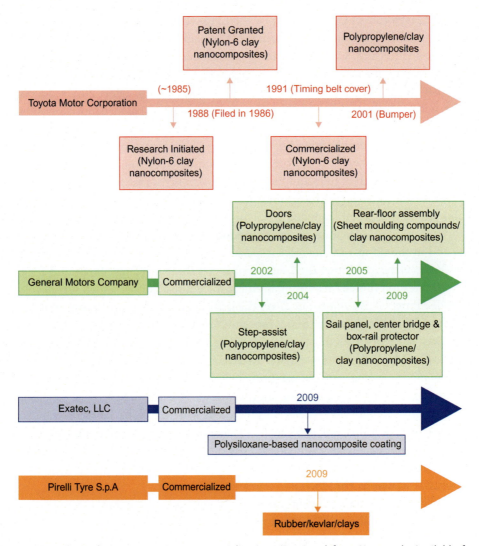

Fig. 15 Time line of nanocomposite commercialization. *(Reprinted from Nanowerk; Available from: Https://Www.Nanowerk.Com/Spotlight/Spotid=23934.Php (Accessed 20 May 2018).)*

was more depended on the nylon6-based nanocomposites in the past mainly for parts exposed to high temperature, the high cost of nylon limits its usage to under the hood applications, fuel lines, etc. It is to be noted that the future of the automobile industry will be aroused with the new generation nanomaterials like graphene, carbon nanotubes, nanofoams, and hybrid material reinforcements. The graphene is an attractive nanomaterial that could drive the market when incorporated in natural rubber.

5. Conclusion

In this chapter, the basic idea of the polymer nanocomposites for automobile application has been addressed. The processing strategies and demanding properties of the most commonly used nano filler-reinforced polymer composites for various automotive components are described. The chemistry of different fillers and the interfacial interaction of them with the polymer have been described and showed how they can influence the final properties of the material. So, we can observe a remarkable advance in polymer nanocomposites for the automotive applications. Among them, the thermoplastic polyolefins and polyamides are of particular importance, whose nanocomposites conquered a high percentage of polymer nanocomposites for the automobile industry. Nanocomposites have gained tremendous acceptance for external body parts, interior and underbonnet parts, coating, fuel lines, and fuel system component applications.

References

[1] A. Muhammad, M. Rahman, R. Baini, M.K. Bakri, Applications of sustainable polymer composites in automobile and aerospace industry, Advances in Sustainable Polymer Composites, Woodhead Publishing Series, 2020, pp. 185–207.

[2] V. Gutowski, W. Yang, S. Li, K. Dean, Lightweight Nanocomposite Materials, Lightweight and Sustainable Materials for Automotive Applications, CRS Press, Taylor & Francis Group, 2017, pp. 277–302.

[3] K.I. Winey, R.A. Vaia, Polymer nanocomposites, MRS Bull. 32 (4) (2007) 314–322, https://doi.org/10.1557/mrs2007.229.

[4] S.H. Zaferani, 1—Introduction of polymer-based nanocomposites, in: M. Jawaid, M.M. Khan (Eds.), Polymer-Based Nanocomposites for Energy and Environmental Applications, Woodhead Publishing, 2018, pp. 1–25, https://doi.org/10.1016/B978-0-08-102262-7.00001-5.

[5] R.A. Vaia, E.P. Giannelis, Polymer nanocomposites: status and opportunities, MRS Bull. 26 (5) (2001) 394–401, https://doi.org/10.1557/mrs2001.93.

[6] Council, N. R, Application of Lightweighting Technology to Military Aircraft, Vessels, and Vehicles, The National Academies Press, Washington, DC, 2012, https://doi.org/10.17226/13277.

[7] S. Komarneni, Nanocomposites, J. Mater. Chem. 2 (12) (1992) 1219–1230, https://doi.org/10.1039/JM9920201219. Feature Article.

[8] Q. Wang, L. Zhu, Polymer nanocomposites for electrical energy storage, J. Polym. Sci. B Polym. Phys. 49 (20) (2011) 1421–1429, https://doi.org/10.1002/polb.22337.

[9] S. Mallampati, M.S.R. Kumar, Effect of nanoclay on the mechanical properties of polyester and S-glass fiber (Al), Int. J. Adv. Sci. Technol. 74 (2015) 35–42, https://doi.org/10.14257/ijast.2015.74.04.

[10] A.K. Chandra, N.R. Kumar, Polymer nanocomposites for automobile engineering applications, in: D.K. Tripathy, B.P. Sahoo (Eds.), Properties and Applications of Polymer Nanocomposites: Clay and Carbon Based Polymer Nanocomposites, Springer Berlin Heidelberg, Berlin, Heidelberg, 2017, pp. 139–172, https://doi.org/10.1007/978-3-662-53517-2_7.

[11] Nanowerk, Available from Https://Www.Nanowerk.Com/Spotlight/Spotid=23934.Php. Accessed 20 May 2018.

[12] P. Saini, Conjugated polymer-based blends, copolymers, and composites: synthesis, properties, and applications, in: Fundamentals of Conjugated Polymer Blends, Copolymers and Composites, John Wiley & Sons, Ltd, 2015, pp. 1–118, https://doi.org/10.1002/9781119137160.ch1.

[13] P. Verma, P. Saini, V. Choudhary, Designing of carbon nanotube/polymer composites using melt recirculation approach: effect of aspect ratio on mechanical, electrical and EMI shielding response, Mater. Des. 88 (2015) 269–277, https://doi.org/10.1016/j.matdes.2015.08.156.

[14] S. Abedi, M. Abdouss, A review of clay-supported Ziegler–Natta catalysts for production of polyolefin/clay nanocomposites through in situ polymerization, Appl. Catal. A Gen. 475 (2014) 386–409, https://doi.org/10.1016/j.apcata.2014.01.028.

[15] K.P. De Jong, J.W. Geus, Carbon nanofibers: catalytic synthesis and applications, Catal. Rev. Sci. Eng. 42 (4) (2000) 481–510, https://doi.org/10.1081/CR-100101954.

[16] L. Feng, N. Xie, J. Zhong, Carbon nanofibers and their composites: a review of synthesizing, properties and applications, Materials (Basel) 7 (5) (2014) 3919–3945, https://doi.org/10.3390/ma7053919.

[17] M. Inagaki, Y. Yang, F. Kang, Carbon nanofibers prepared via electrospinning, Adv. Mater. 24 (19) (2012) 2547–2566, https://doi.org/10.1002/adma.201104940.

[18] J. Brunton, T. Yamada, T. Weatherford, TCAD analysis of breakdown during voltage sweep in a carbon nanofiber (CNF) interconnect, in: 2011 IEEE International Integrated Reliability Workshop Final Report, 2011, pp. 125–128.

[19] D.R. Bortz, C. Merino, I. Martin-Gullon, Carbon nanofibers enhance the fracture toughness and fatigue performance of a structural epoxy system, Compos. Sci. Technol. 71 (1) (2011) 31–38, https://doi.org/10.1016/j.compscitech.2010.09.015.

[20] S. Bal, Experimental study of mechanical and electrical properties of carbon nanofiber/epoxy composites, Mater. Des. 31 (5) (2010) 2406–2413, https://doi.org/10.1016/j.matdes.2009.11.058.

[21] A. Sharma, M. Thakur, M. Bhattacharya, T. Mandal, S. Goswami, Commercial application of cellulose nano-composites—a review, Biotechnol. Rep. 21 (2019), https://doi.org/10.1016/j.btre.2019.e00316, e00316.

[22] A. Kiziltas, E. Erbas Kiziltas, S. Boran Torun, D. Gardner, Micro-and nanocellulose composites for automotive applications, Proceedings of the SPE Automotive Composites Conference and Exhibition (ACCE), Novi, MI, USA, University of Maine, 2013, pp. 1–13.

[23] M. Abdelmouleh, S. Boufi, M.N. Belgacem, A. Dufresne, A. Gandini, Modification of cellulose fibers with functionalized silanes: effect of the fiber treatment on the mechanical performances of cellulose–thermoset composites, J. Appl. Polym. Sci. 98 (3) (2005) 974–984, https://doi.org/10.1002/app.22133.

[24] M. Bulota, K. Kreitsmann, M. Hughes, J. Paltakari, Acetylated microfibrillated cellulose as a toughening agent in poly(lactic acid), J. Appl. Polym. Sci. 126 (S1) (2012) E449–E458, https://doi.org/10.1002/app.36787.

[25] F. Zhang, W. Wu, X. Zhang, X. Meng, G. Tong, Y. Deng, Temperature-sensitive poly-NIPAm modified cellulose nanofibril cryogel microspheres for controlled drug release, Cellulose 23 (2016), https://doi.org/10.1007/s10570-015-0799-4.

[26] M.R. Vengatesan, V. Mittal, Nanoparticle- and nanofiber-based polymer nanocomposites: an overview, in: Spherical and Fibrous Filler Composites, John Wiley & Sons, Ltd, 2016, pp. 1–38, https://doi.org/10.1002/9783527670222.ch1.

[27] S. Sen, Y. Xie, S.K. Kumar, H. Yang, A. Bansal, D.L. Ho, L. Hall, J.B. Hooper, K.S. Schweizer, Chain conformations and bound-layer correlations in polymer nanocomposites, Phys. Rev. Lett. 98 (12) (2007) 128302, https://doi.org/10.1103/PhysRevLett.98.128302.

[28] A. Kausar, Contemporary applications of carbon black-filled polymer composites: an overview of essential aspects, J. Plast. Film Sheeting 34 (3) (2018) 256–299, https://doi.org/10.1177/8756087917725773.

[29] B. Kim, Prediction of mechanical behavior for carbon black added natural rubber using hyperelastic constitutive model, Elastomers Compos. 51 (2016) 308–316, https://doi.org/10.7473/EC.2016.51.4.308.

[30] H. Zhang, A.K. Scholz, J. de Crevoisier, F. Vion-Loisel, G. Besnard, A. Hexemer, H.R. Brown, E.J. Kramer, C. Creton, Nanocavitation in carbon black filled styrene–butadiene rubber under tension detected by real time small angle X-ray scattering, Macromolecules 45 (3) (2012) 1529–1543, https://doi.org/10.1021/ma2023606.

[31] H.-D. Luginsland, W. Niedermeien, New reinforcing materials for rising tire performance demands, Rubber World 228 (2003) 34–45.

[32] L. Reuvekamp, J.W. Brinke, P.J. Swaaij, J. Noordermeer, Effects of mixing conditions—reaction of TESPT silane coupling agent during mixing with silica filler and tire rubber, KGK, Kautsch. Gummi Kunstst. 55 (2002) 41–47.

[33] Y. Xia, S. Ding, Y. Liu, Z. Qi, Facile synthesis and self-assembly of amphiphilic polyether-octafunctionalized polyhedral oligomeric silsesquioxane via thiol-ene click reaction, Polymers 9 (7) (2017), https://doi.org/10.3390/polym9070251.

[34] E. Ayandele, B. Sarkar, P. Alexandridis, Polyhedral oligomeric silsesquioxane (POSS)-containing polymer nanocomposites, Nanomaterials (Basel) 2 (4) (2012) 445–475, https://doi.org/10.3390/nano2040445.

[35] W. Liu, B. Ullah, C.-C. Kuo, X. Cai, Two-dimensional nanomaterials-based polymer composites: fabrication and energy storage applications, Adv. Polym. Technol. 2019 (2019) 4294306, https://doi.org/10.1155/2019/4294306.

[36] Q.T. Nguyen, D.G. Baird, Preparation of polymer–clay nanocomposites and their properties, Adv. Polym. Technol. 25 (4) (2006) 270–285, https://doi.org/10.1002/adv.20079.

[37] S. Pavlidou, C.D. Papaspyrides, A review on polymer–layered silicate nanocomposites, Prog. Polym. Sci. 33 (12) (2008) 1119–1198, https://doi.org/10.1016/j.progpolymsci.2008.07.008.

[38] S.J. Perera, S.M. Egodage, S. Walpalage, Enhancement of mechanical properties of natural rubber–clay nanocomposites through incorporation of silanated organoclay into natural rubber latex, e-Polymers 20 (1) (2020) 144–153, https://doi.org/10.1515/epoly-2020-0017.

[39] J.M. Garcés, D.J. Moll, J. Bicerano, R. Fibiger, D.G. McLeod, Polymeric nanocomposites for automotive applications, Adv. Mater. 12 (23) (2000) 1835–1839, https://doi.org/10.1002/1521-4095(200012)12:23<1835::AID-ADMA1835>3.0.CO;2-T.

[40] F. Gao, Clay/polymer composites: the story, Mater. Today 7 (11) (2004) 50–55, https://doi.org/10.1016/S1369-7021(04)00509-7.

[41] Auto applications drive commercialization of nanocomposites, Plast. Addit. Compound. 4 (1) (2002) 30–33, https://doi.org/10.1016/S1464-391X(02)80027-7.

[42] T.D. Fornes, D.R. Paul, Modeling properties of nylon 6/clay nanocomposites using composite theories, Polymer 44 (17) (2003) 4993–5013, https://doi.org/10.1016/S0032-3861(03)00471-3.

[43] A. Elmarakbi, W. Azoti, Novel composite materials for automotive applications: concepts and challenges for energy-efficient and safe vehicles, Conference Paper, 10th International Conference on Composite Science and Technology, 2015.

[44] T.C. Hirschmann, P.T. Araujo, H. Muramatsu, X. Zhang, K. Nielsch, Y.A. Kim, M.S. Dresselhaus, Characterization of bundled and individual triple-walled carbon nanotubes by resonant Raman spectroscopy, ACS Nano 7 (3) (2013) 2381–2387, https://doi.org/10.1021/nn3055708.

[45] P. Saini, V. Choudhary, B.P. Singh, R.B. Mathur, S.K. Dhawan, Polyaniline–MWCNT nanocomposites for microwave absorption and EMI shielding, Mater. Chem. Phys. 113 (2) (2009) 919–926, https://doi.org/10.1016/j.matchemphys.2008.08.065.

[46] A.R. Bhattacharyya, T.V. Sreekumar, T. Liu, S. Kumar, L.M. Ericson, R.H. Hauge, R.E. Smalley, Crystallization and orientation studies in polypropylene/single wall carbon nanotube composite, Polymer 44 (8) (2003) 2373–2377, https://doi.org/10.1016/S0032-3861(03)00073-9.

CHAPTER 4

Enhanced synergistic effect by pairing novel inherent flame-retardant polyurethane foams with nanolayers of expandable graphite for their applications in automobile industry

Felipe M. de Souza[a], Mark Arnce[a], and Ram K. Gupta[a,b]
[a]Department of Chemistry, Pittsburg State University, Pittsburg, KS, United States
[b]Kansas Polymer Research Center, Pittsburg State University, Pittsburg, KS, United States

Chapter outline

1. Introduction — 65
2. Relevance of flame retardants for automobiles — 66
3. Concerns and importance of polyurethanes for the automobile industry — 68
4. Synthesis and characterizations of novel flame-retardant polyurethane foams — 69
    4.1 Synthesis of flame-retardant polyols and polyurethane foams — 69
    4.2 Methods for characterizations of polyols and polyurethane foams — 70
5. Important characteristics and properties of the polyurethane foams — 71
    5.1 Structural characterizations of the polyols — 71
    5.2 Physical properties of the polyurethane foams — 73
    5.3 Morphology and cellular structure of polyurethane foams — 74
    5.4 Compressive strength — 74
    5.5 Thermal behavior of the foams — 76
    5.6 Flammability behavior and flame retardancy mechanism — 79
6. Conclusion — 81
References — 82

1. Introduction

Polyurethanes (PUs) are one of the very important classes of polymers due to their several unique characteristics which make them suitable for applications in a range of industries. In general, polyurethanes are synthesized by a polyaddition reaction of polyols (a compound containing two or more hydroxyl groups) and di (or poly) isocyanates. The diversity of starting materials, fast reaction rate, and a broad range of properties are some of the unique features that make polyurethanes very attractive for industrial applications. Such a wide interest in polyurethanes for industrial applications is reflected by their huge market growth over the years. Polyurethane market was estimated at USD 60 billion in 2017 and is expected to reach over USD 80 billion by 2021 [1]. High thermal stability, decent sound insulation, tunable mechanical properties, and low density of polyurethanes make them very applicable for constructions, automobiles, footwear, packing, etc. Among many industries, the automobile industry heavily relies on polyurethanes to manufacture vehicle parts that include seats, bumpers, front panels, coating, and internal components.

Despite the versatility and broad range of applications of polyurethanes, the inherent flammability of polyurethanes is a concern for many applications. The porous microcellular structure of polyurethane foams facilitates the diffusion of oxygen which causes the propagation of fire. During the combustion of polyurethanes, heat release rate and toxic gas emission cause severe casualty especially in a closed environment [2]. The use of flame-retardants (FRs) in polyurethanes can counter this drawback. The FRs fall into two categories: additive (physical blend) or reactive FRs. The additive FRs are physically mixed with a prepolymer mixture. This method is preferred in large-scale applications due to its economical and overall effectiveness. However, blended FRs may not disperse well in the polymer matrix which may lead to deterioration of the properties over time [3]. Melamine and its derivatives [4], organophosphorus [5], metal oxides [6], and carbon-based compounds [7] are some examples of additive FRs. The additive FRs, also known as intumescent, create a char layer at the flame exposed surface to prevent further burning. This layer acts as a physical barrier between the flame and polyurethane surface to starve the flame [8]. The reactive FRs are the compounds that are bonded with the polymer's backbone. Even though it requires an extra synthesis step, the reactive FRs are usually effective and do not suffer the influence of chronological effects, maintaining their flame retardancy for a longer period [9]. The reactive FRs can be phosphorus- [10] or nitrogen-based compounds [11]. Before the implementation of eco-friendly FRs, halogen-based FRs were heavily used in industries due to their high effectiveness. However, due to their detrimental effect on nature and living beings, the use of halogen-based FRs is restricted [12, 13]. The halogenated FRs used in automobiles were also found to release damaging fumes during burning [13]. Hence, despite their effectiveness, the halogen-based FRs were judged unsuitable [14, 15]. Scientists are exploring various options such as the use of phosphorous, nitrogen, and carbon-based FRs which are non-toxic and eco-friendly. However, effective FRs are expensive, therefore combining two cost-effective FRs to create a synergy effect is an interesting approach to reduce the flammability of polyurethanes. This approach enhances the overall performance of the FRs yielding better flame retardancy and improved physicomechanical properties.

2. Relevance of flame retardants for automobiles

As a result of the competitive automobile market, there is an emphasis on improving comfort and safety without increasing the manufacturing cost of the vehicle. As mentioned, polyurethanes possess many unique properties that match well with the requirement in the automobile industry. However, the lack of fire safety in polyurethanes is a concern as it can quickly lead to life-threatening situations. Some of the FRs used in car seats and furniture are shown in Fig. 1 [16, 17]. To avoid the toxic smoke from these compounds, several studies have been made to balance cost, efficiency, and safety. A practical strategy is to use different types of FRs to create a synergy effect to enhance the flame retardancy of polyurethanes without increasing the fabrication cost. For this, Yuan et al. developed two reactive FR polyols, one containing phosphorus and the other nitrogen [18]. The synthesis of these FRs is shown in Fig. 2. It was found that the 1:1 ratio of these polyols was optimum to provide the best flame retardancy. Expandable graphite

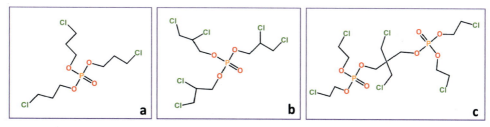

Fig. 1 Chemical structure for halogen-based FR found in automobiles (A) tris(2-chloropropyl) phosphate (TCPP), (B) tris(1,3-dichloro-2-propyl)phosphate (TCPP), and (C) 2-di(chloromethyl)-1,3 propylene diphosphate [16]. The article was printed under a CC-BY license.

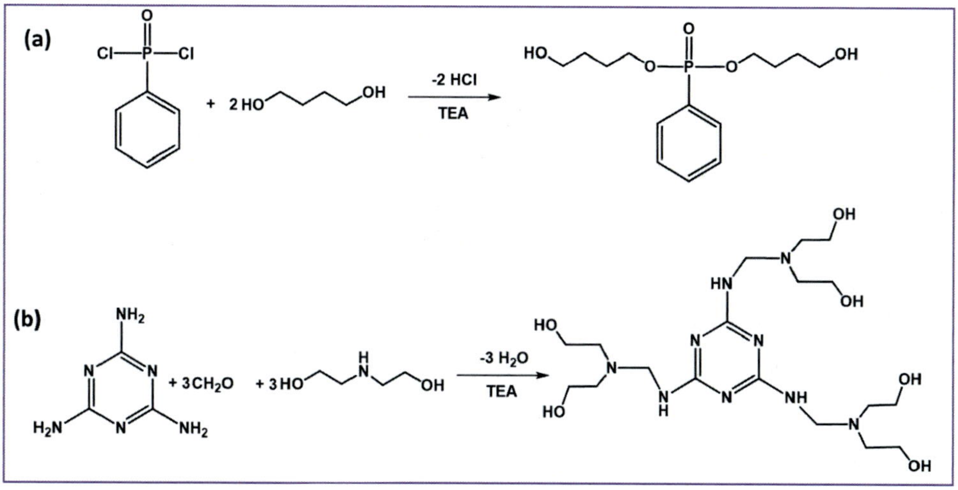

Fig. 2 Synthesis for reactive FR polyols based on (A) phosphorus and (B) nitrogen. *Adapted with permission from reference Y. Yuan, H. Yang, B. Yu, Y. Shi, W. Wang, L. Song, Y. Hu, Y. Zhang, Phosphorus and nitrogen-containing polyols: synergistic effect on the thermal property and flame Retardancy of rigid polyurethane foam composites, Ind. Eng. Chem. Res. 55 (2016) 10813–10822. Copyright 2016 American Chemical Society.*

(EG) was introduced to further enhance their flame retardancy. The combination of the reactive FRs based on nitrogen and phosphorus along with blended EG yielded satisfactory results such as improvement in limited oxygen index (LOI) from 20% to 33.5% along with 52.4% reduction in heat release rate.

In another work, Akdogan et al. conducted a detailed study on the binary combination of EG and pentaborate octahydrate (APB) to understand the influence of these materials on flame retardancy, mechanical, and thermal insulation behavior of polyurethane foams [19]. The rigid polyurethane foams prepared using 20 wt% of FRs (15 and 5 wt% of EG and APB, 15E/5A) showed a good improvement in flame retardancy. A decrease in the peak heat release rate (PHRR) as well as total heat release (THR), yielding lower values compared with 20 wt% of EG (20E) or APB (20A), respectively, were observed. In addition, the smoke production was decreased about 84% for 15E/5A and 77% for 20E compared with the neat sample. Such results are vital for the inclusion of rigid polyurethane foams in automobiles as they show a great enhancement in safety by preventing the release of heat that could trigger fuel combustion and mitigation of smoke. On top of that, an increase in the compressive strength was observed after blending with FRs. The compressive strength increased from 119 to 126 kPa after blending with FRs. Other properties related to the automobile applications such as thermal conductivity displayed a negligible increase which was mostly due to EG's inherently high thermal conductivity and the presence of trapped CO_2 in the microcellular structure of the foam through the reaction of water and isocyanate during the foaming process. The FR mechanism between EG and APB was attributed to their actions in the condensed and gas phases, respectively. During the thermal decomposition of EG, a worm-like layer was formed with almost 10 times volume expansion. This tortuous path prevented O_2 and reactive species to get into contact with the underneath layer of polyurethane foams, hence creating a dense physical barrier. Simultaneously, APB on the polyurethane surface decomposes into H_2O and NH_3 to cool down the media as well as dilutes O_2 and other gases. The concurrent action of both FRs to improve the flame resistance is known as the synergistic effect. This study demonstrated an example to obtain high-quality polyurethane foams with improved physicomechanical properties and reduced flammability.

3. Concerns and importance of polyurethanes for the automobile industry

For an ambitious market like automobiles, there is a constant technological push to develop materials that can surpass the materials/technology currently used. However, such materials must be cost-effective to create leverage among the competitors and attract customers. The inclusion of polyurethanes into the car manufacturing process yields great improvements. This chapter contextualizes the synthesis and characterizations of novel flame-retardant, rigid polyurethane foams using a cost-effective approach. The discussion begins with the importance of polyurethane foam's density for automobiles regarding the weight balance and the effect on fuel efficiency. Other properties such as cellular morphology, cellular structure, and porosity of polyurethane foams are discussed as these properties can influence the thermal and sound insulation. These properties are highly relevant in modern cars as they can provide comfort by absorbing noise from the surroundings. The influence of FRs on the compressive strength of the foams is discussed as lightweight and stronger materials are required to manufacture many parts for a vehicle. Finally, a detailed study on the flame retardancy is covered which includes thermal behavior, burning time, and weight loss.

4. Synthesis and characterizations of novel flame-retardant polyurethane foams

Two inherent reactive flame-retardants based on 1,3,5-triallyl-1,3,5-triazine-2,4,6-trione (1T) and its isomer 2,4,6-triallyloxy-1,3,5-triazine (2T) were synthesized using a facile approach that can be adapted for commercial production. The presence of triazine ring in the polyols is responsible for introducing inherent flame-retardant properties. The polyols based on these two chemicals were synthesized by reacting their terminal double bonds with 2-mercaptoethanol using a thiol-ene reaction under UV light. The EG was blended into the polyols before the foaming process to provide a synergy effect. EG has been tested and found very capable of improving flame retardancy for polyurethanes in a nontoxic way [7]. The high-performance rigid polyurethane foams were obtained that combine the flame retardancy effect of the triazine rings that act mostly in the gaseous phase forming radical scavenger species while EG acts in the solid phase forming a tortuous carbonaceous char layer to shield the underlayer against oxygen and radicals [20].

4.1 Synthesis of flame-retardant polyols and polyurethane foams

For the synthesis of 1T-based polyol, 1 mol of 1T was reacted with 3 mol of 2-mercaptoethanol (2-ME) using a thiol-ene reaction. 2-hydroxy-2-methyl propiophenone was used as a photoinitiator. The reaction was carried out for 8 h under stirring and UV light (365 nm). The same procedure was adopted for the synthesis of 2T-based polyol. The schematics of the reactions are given in Fig. 3. The polyurethane foams were prepared using the synthesized polyols, commercial polyol, EG, catalysts, blowing agent, surfactant, and diisocyanate. In a typical foaming process, 10 g of synthesized polyol, 10 g of commercial polyol (Jeffol 520), varying amounts of EG (0, 1.5, 3, 5, 8, and 10 g) were mixed using an overhead stirrer followed by the addition of catalysts, blowing agent, and surfactant. After homogenizing, the required amount of diisocyanate was added to prepare the foams. The foams containing 0, 1.5, 3, 5, 8, and 10 g of EG were named as 0EG, 1.5EG, 3EG, 5EG, 8EG, and 10EG, respectively. Foams 0EG, 1.5EG, 3EG, 5EG, 8EG,

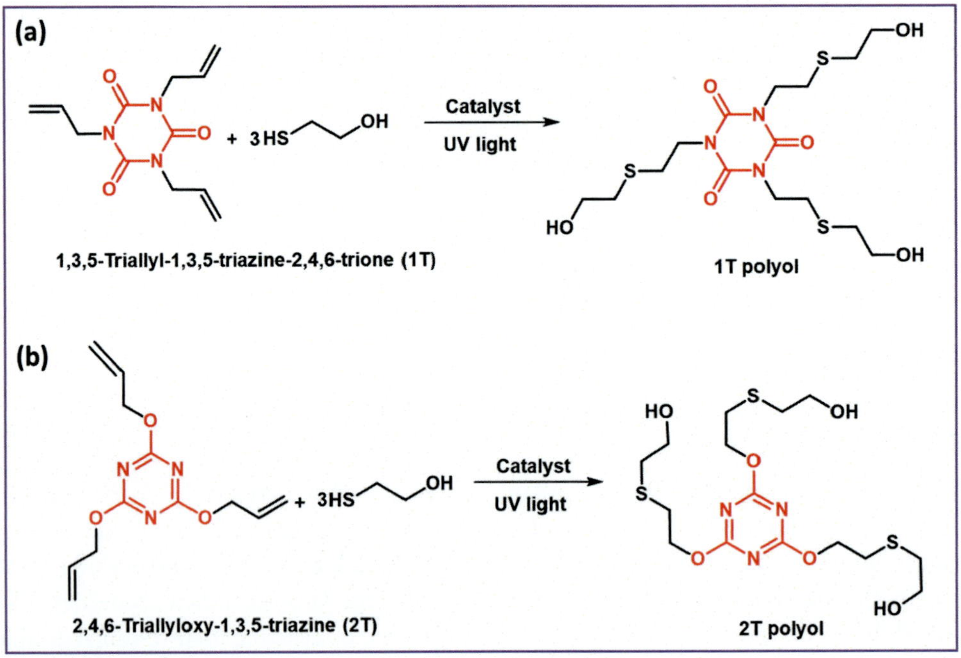

Fig. 3 Thiol-ene reaction for the synthesis of 1T- and 2T-based polyols.

and 10EG contain 0, 2.68, 5.23, 8.42, 12.82, and 15.53 wt% of EG in the final foam, respectively. The digital photos of the prepared foams are shown in Fig. 4.

4.2 Methods for characterizations of polyols and polyurethane foams

The success of the synthesis was confirmed via iodine number, hydroxy number, viscosity, gel permeation chromatography (GPC), and Fourier transform infrared spectroscopy (FT-IR). The hydroxy number was estimated using the ASTM E1899–16 method which utilizes *p*-toluenesulfonyl isocyanate and a potentiometric titration (888 Titrando, Tiamo Software system by Metrohm). The viscosity of polyols was measured using a rheometer (TA Instruments) at room temperature using a cone plate with a 2 degree angle and a 12.5 mm radius. A Water System was used to record GPC. The FT-IR spectra were recorded using PerkinElmer Spectrum-II.

The foams were cut into rectangular prisms and cylinders for various characterizations. The rectangular pieces with dimensions of $150 \times 50 \times 12.5$ (mm^3) (length × width × height) were used for the flammability test. The cylindrical pieces with dimensions of 45×30 (mm^2) (diameter × height) were used for the closed-cell content, apparent density, and compressive strength measurements. The density was determined via ASTM1622–14 standard. The closed-cell content was measured using HumiPyc.

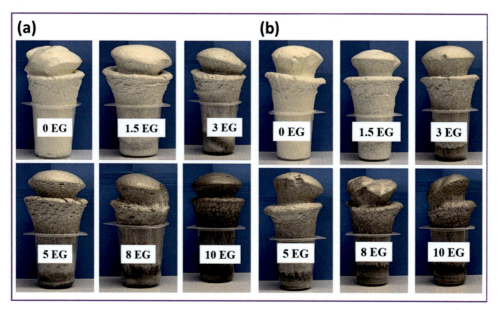

Fig. 4 Photographs of polyurethane foams having various amounts of EG prepared using (A) 1T-based and (B) 2T-based polyols.

Compressive strength was determined using an Instron tensile instrument using the Blue Hill system, with the ISO 844:2016 standard. The thermogravimetric analysis (TGA) was performed using a Q500 Discovery. The test was performed under nitrogen and air in the temperature range of 20–700°C with a ramp rate of 10°C per minute. To obtain clear SEM images, cubes of the foam with the dimensions of about 0.5 mm^3 were cut out and the top surface was gold. The flammability of the foams was studied using the ASTM D 4986–18 standard test. The foams were exposed to a flame for 10 s and then self-extinguish time was recorded. The percent weight loss for each foam was also recorded after the burn test.

5. Important characteristics and properties of the polyurethane foams
5.1 Structural characterizations of the polyols

The hydroxyl number of both polyols was calculated to be 364 mg KOH/g which is very close to the theoretical hydroxyl number (366 mg KOH/g) of these polyols. This suggested that the thiol-ene reaction was successful under the above-mentioned conditions. The gel permeation chromatograms of the starting materials and polyols are shown in Fig. 5. The chromatograms show a retention time of about 34 min for both polyols which is higher than the retention time of the starting materials. This suggests successful

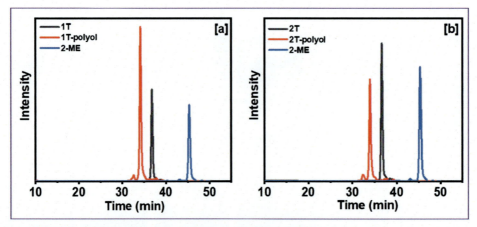

Fig. 5 GPC of the starting chemicals, 1T (A), and 2T-based polyols.

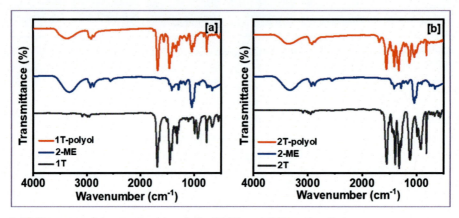

Fig. 6 FT-IR spectra of the starting chemicals, 1T (A), and 2T-based polyols.

attachment of 2-mercaptoethanol to the double bonds of 1T and 2T. FT-IR spectra for the polyols and the starting materials were taken to confirm the success of the thiol-ene reaction (Fig. 6). The peak around 3250–3500 cm^{-1} in both polyols confirms the presence of -OH group [21]. The peak around 2550 cm^{-1} in the 2-mercaptoethanol was due to the S—H bond which disappeared in the polyols, confirming conversion of S—H bond to S—C bond. Besides, the double bond stretch peak of 1T and 2T around 900 cm^{-1} disappeared which further confirms the success of the thiol-ene reaction, where double bonds reacted with the thiol group of 2-mercaptoethanol [22, 23]. The double bonds were broken to perform the addition reaction with the S—H to form C—S bonds. The weak peak around 500 cm^{-1} confirms the presence of the C—S bond in the polyols [22, 23].

5.2 Physical properties of the polyurethane foams

5.2.1 Density

The physical properties of the foams provide important information regarding their possible application in car parts such as seats, front panels, bumpers, door components, etc. The density of the foam is one of the important characteristics as it is related to mechanical strength. Fig. 7 shows the change in the density of the foams as a function of the amount of EG in the foams. It is observed that the density of the foams increases with an increase in the amount of EG in the foams. The density of foams is in the range of 30–50 kg/m^3 which is within the range of density required in the automobile industry [24–26]. A proper density is required for the applications in automobiles as heavy foams will increase the weight of the vehicle and fuel consumption. On average a reduction in 10% weight of a vehicle may lead to a decrease in about 5% fuel consumption. Therefore, obtaining low-density materials is important for the automobile industry as it provides economic benefits to the customers along with reduced emission of CO_2 due to reduced use of gasoline [27]. The density of the foams can be controlled via formulation, e.g., high amount of catalysts and blowing agents can reduce the density at the expense of mechanical strength [7].

5.2.2 Closed-cell content

The closed-cell content is one of the important characteristics to determine the suitability of polyurethane foams for thermal insulation. For better thermal insulation, a high closed-cell content is required to limit the flow of air and heat through the foam. Higher closed-cell content also improves the flame retardancy of the foams as it prevents the passage of oxygen through the closed-cellular structure. On the other hand, foams with a higher content of open-cell are used for sound insulation. Such property is crucial for an advanced vehicle as it conceives a pleasant driving environment by absorbing the noise from the engine, the friction of wheels with the road, and other surrounding sounds. For enhancement of these properties, several additives such as carbon nanotubes, magnesium

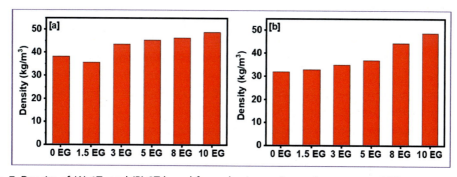

Fig. 7 Density of (A) 1T- and (B) 2T-based foams having an increasing amount of EG.

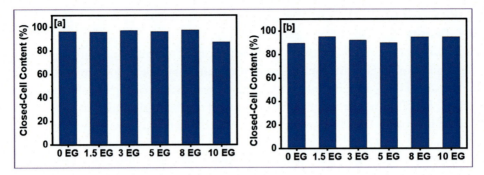

Fig. 8 Closed-cell content of (A) 1T- and (B) 2T-based foams having an increasing amount of EG.

hydroxide, nano/microsized silica, and clay are being used [28, 29]. Polyurethane foams obtained via our method using 1T- and 2T-based polyols along with EG show a high closed-cell content. The closed-cell content of these foams was in the range of 87%–95% with a negligible effect of the addition of EG on the value of closed-cell content (Fig. 8). The constancy in the closed-cell content confirms that the cellular structure of the foams is not significantly affected by the addition of EG, suggesting a homogeneous blending of EG in the polyurethane matrix.

5.3 Morphology and cellular structure of polyurethane foams

The SEM images of the foams prepared using 1T- and 2T-based polyols are shown in Figs. 9 and 10. Notably, the increasing amount of EG in the foams led to a more heterogeneous cell size distribution. However, this effect was more pronounced in the foams prepared using 1T-based polyol. As seen in the SEM images, the microcellular structure was mostly maintained despite the relatively large flakes of EG (300 μm). Chen et al. have observed a decrease in the cell size of the foams after the addition of EG [30].

5.4 Compressive strength

Mechanical properties are influenced by many factors such as density, dispersion, and interaction between the fillers and polyurethane matrix, cell size, shape, and uniformity. The compressive strength at the break for the 1T-based foams is shown in Fig. 11A. Notably, the compressive strength was observed to deteriorate after the addition of EG. An oscillatory behavior was observed that started at 170 kPa for the neat foam, decreased to the lowest of 65 kPa when 8.42 wt% EG was added, and then reached about 140 kPa for the foams containing 5.22 and 12.82 wt% of EG. On the other hand, the 2T-based foams displayed an increase in compressive strength after the addition of EG (Fig. 11B). The neat 2T-based foam nearly doubled its compressive strength after the addition of EG. The compressive strength of neat 2T-based foam increased from 90 kPa to about 180 kPa after the addition of 15.53 wt% EG. The negative influence

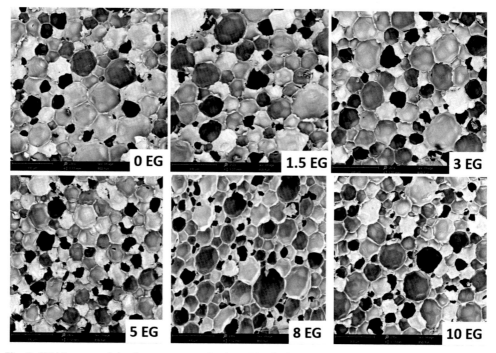

Fig. 9 SEM images of the foams prepared using 1T-polyol.

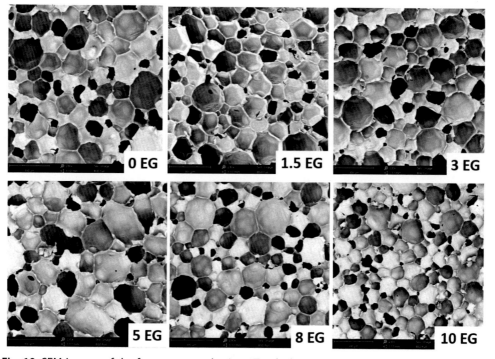

Fig. 10 SEM images of the foams prepared using 2T-polyol.

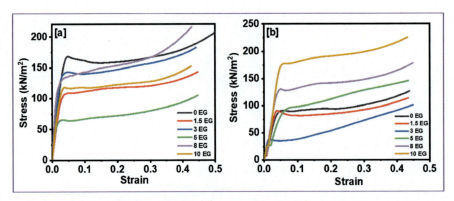

Fig. 11 Compressive strength of the foams prepared using (A) 1T- and (B) 2T-based polyols.

on the compressive strength for the 1T-based foams may be attributed to the heterogeneity of the cell size that may lead to uneven force distribution as can be seen in the SEM images. Interestingly, in the case of 2T-based foams, an opposite behavior was observed. This effect on the morphology promoted an even distribution of force during compressive strength measurements. Additionally, a decrease in cell size implies an increase of cell number per unit area, which is usually related to an increase in the mechanical property as the load applied to the foam encounters more cell walls per unit area [31]. Begum et al. conducted a study to compare the mechanical properties of rigid and flexible polyurethane foams [32]. It was concluded that flexible polyurethanes have a high impact strength compared to rigid polyurethanes. However, it was also observed that a combination of rigid and flexible polyurethane may improve the overall toughness of the material. Our results suggest that 1T- and 2T-based polyurethane foams can be applied as insulation in automobiles due to their high compressive strength, lightweight, and high closed-cell morphology.

5.5 Thermal behavior of the foams

The thermal nature of polyurethane foams depends on many factors such as NCO:OH ratio, cross-link density, chemical structure of staring chemicals, molecular weight, etc. Some approaches such as chemical structure modification to improve thermal stability are widely used. The addition of various segments such as isocyanurate, oxazolidone, imides, and triazine rings into the polyurethane backbone seems to improve the thermal properties of polyurethanes. Fillers particularly flame-retardants were found to significantly improve the thermal stability of polyurethanes as they degrade at the same or lower temperature as the polyurethane matrix and tend to form a protective layer and/or captures the reactive radical species generated during the decomposition of polyurethanes [33]. Thermogravimetric analysis (TGA) and derivative TGA (DTGA) were performed under

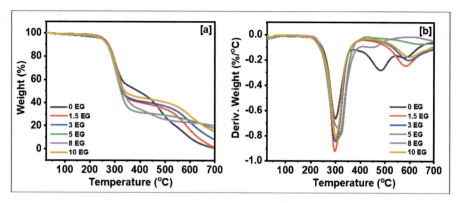

Fig. 12 (A) TGA and (B) DTGA curves of 1T-based foams under nitrogen.

nitrogen and air to understand the thermal behavior of 1T- and 2T-based foams and the effect of EG on the physicomechanical, thermal, and flammability of the foams. The TGA and DTGA graphs for the 1T-based foams under nitrogen demonstrated the first degradation step around 300°C (Fig. 12). This decomposition was related to the disruption of polyurethane foams' hard segments [34, 35]. Simultaneously, there was an irreversible expansion of EG in a temperature range of 200–350°C. It occurs through the reaction between carbon and intercalated H_2SO_4 into the graphite flakes releasing CO_2, SO_2, and H_2O [36]. This behavior is evident as the foams containing increasing amounts of EG presented higher weight loss during the first degradation step. This led to the formation of a carbonaceous char layer on the foams to improve the thermal stability as evident in the higher decomposition temperature in the second thermal degradation step. The foams without EG showed a second degradation around 500°C which shifted to around 600°C for the foams containing EG. Such improvement led to an increase (up to 20%) in residual weight at 700°C in comparison to nearly zero for the foams without EG. It suggests that the irreversible EG expansion generated a physical barrier that prevented the degradation of soft segments [36]. A similar trend in thermal degradation was observed for the 2T-based foams under nitrogen. The increasing amount of EG led to an overall enhancement in thermal stability of the second decomposition step (Fig. 13). The thermal behavior of the foams in the air was very similar to the behavior observed in the nitrogen (inert) atmosphere (Fig. 14). However, the foams without EG showed the second decomposition temperature around 550°C while the foams containing EG showed a shift in the decomposition temperature (500°C). Such effect likely occurred due to the reaction between oxygen with the carbon layer resulting in more release of CO_2 and CO gases. Despite that, higher values of char residues were obtained for the foams having a higher amount of EG compared with the neat foam. Similar behavior was observed for the 2T-based foams. There was a clear enhancement in the thermal stability after the addition of EG into the foams as seen in Fig. 15.

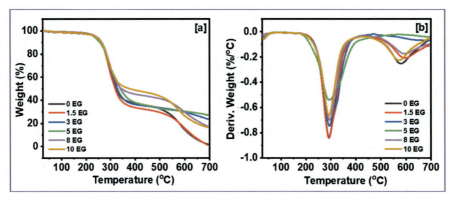

Fig. 13 (A) TGA and (B) DTGA curves of 2T-based foams under nitrogen.

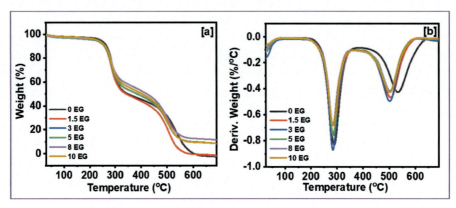

Fig. 14 (A) TGA and (B) DTGA curves of 1T-based foams under air.

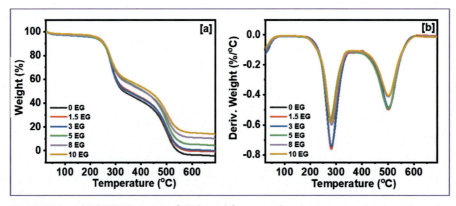

Fig. 15 (A) TGA and (B) DTGA curves of 2T-based foams under air.

5.6 Flammability behavior and flame retardancy mechanism

The capability of a material to self-quench a flame is crucial for safety as it can prevent excessive releases of toxic fumes and heat. This is particularly important for automobiles as the combustion of vehicle components can release excessive heat which can trigger an explosion once it reaches the fuel tank. Also, the combustion of internal components within automobiles releases toxic fumes such as CO, CO_2, and CN^- along with soot that can impose a life-threatening situation especially in a close environment that has limited air circulation. The flammability of 1T- and 2T-based foams was studied using the horizontal burning test (ASTM D 4986-98) method to evaluate the self-quenching of fire. This test was conducted by holding foams in a horizontal position by one end and the other end of the foam was exposed to flame for 10 s. After 10 s, the external source of the fire was removed and foam was left to burn by itself. The self-extinguishing time and weight loss after the burn test was recorded.

For the 1T-based foams, the neat foam presented a burning time of 36 s and a weight loss of 14.3%. As the amount of EG increases a drastic decrease in the burning time and consequently the weight loss was observed. The lowest burning time of 2.5 s and a weight loss of 1.94% were observed for the foam containing 12.82 wt% of EG (Fig. 16). On the other hand, the neat 2T-based foam showed the longest burning time of 108 s and weight loss of 41.2%. However, after the addition of 12.82 wt% of EG, the burning time and weight loss dropped to 2.8 s and 2.01% (Fig. 17). Such variations in the flame retardancy of neat foams could be understood by analyzing the structure of the polyurethanes. The chemical structure of 1T-based polyurethane foam is composed of an isocyanurate ring that was obtained through the cyclotrimerization of isocyanate groups (Fig. 18) [37]. During the thermal decomposition of 1T-based foams, there was a release of low-reactive radical species. Yet, its reactivity was high enough to capture reactive radicals such as H· and OH· originated from the polyurethane foam matrix, which is known to catalyze the flame due to the exothermic reaction with the

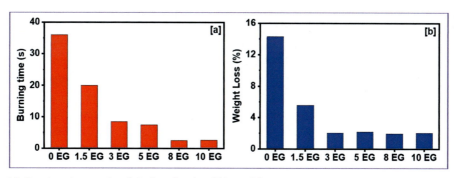

Fig. 16 Burning time and weight loss for the 1T-based foams.

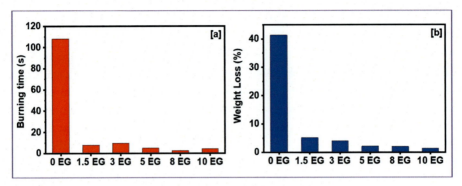

Fig. 17 Burning time and weight loss for the 2T-based foams.

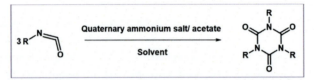

Fig. 18 Cyclotrimerization reaction for the synthesis of isocyanate to isocyanurate. *Adapted with permission from reference M. Siebert, R. Sure, P. Deglmann, A.C. Closs, F. Lucas, O. Trapp, Mechanistic investigation into the acetate-initiated catalytic Trimerization of aliphatic isocyanates: a bicyclic ride, J. Org. Chem. 85 (2020) 8553–8562. Copyright 2020 American Chemical Society.*

polyurethane matrix. Since the action of these fragments is related to capture reactive species in the gas phase, this phenomenon is named as the gas-phase mechanism. The decomposition of isocyanurate rings was studied by Qian et al. [38] as the pyrolytic route of 1T molecule was explicated in Fig. 19 [38]. As seen, many fragments were generated during the pyrolysis which can capture H· and OH· radicals. Also, it promotes a cross-linked structure with the polyurethane matrix forming a compact carbonaceous shield.

Despite 1T and 2T being isomers, their flame-retardant mechanism was different as noted by the wide variation in burning time and weight loss for the neat foams. Such difference was studied by Guo et al. [39]. They proposed a pyrolysis route for the 2T molecule. It was found that the decomposition of the triazine ring in the 2T structure leads to radical fragments that were too stable due to the presence of conjugated double bonds. The conjugation in these species resulted in stabilization and hence presenting lower reactivity. These fragments were not effective in capturing reactive radical species such as H· and OH·. The action of 2T in the gas phase was mostly as a dilutant, i.e., it formed a diffused gas layer that partially prevented the exothermic radicals from reacting with the polyurethane matrix, therefore, presenting a long burning time and more weight loss.

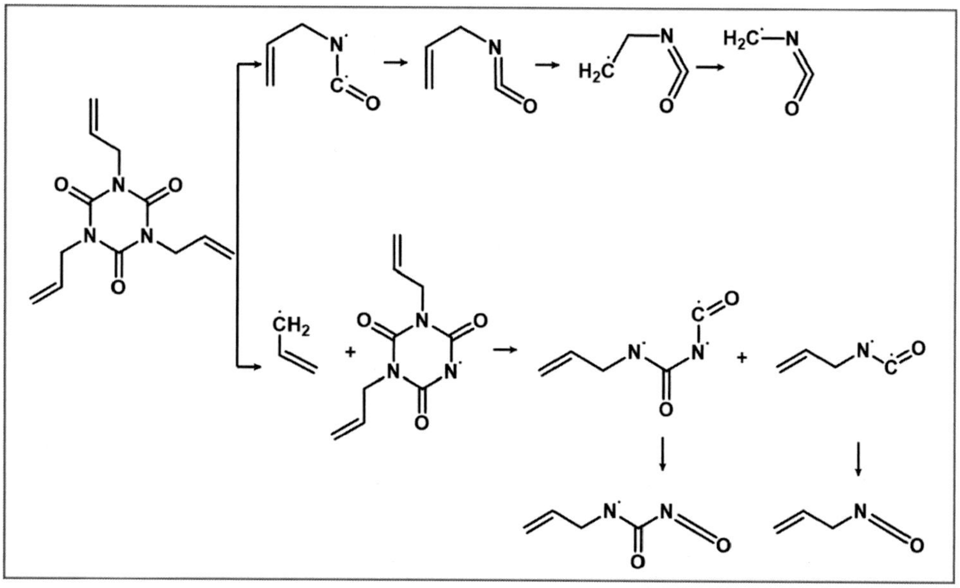

Fig. 19 An expected pyrolytic route for the 1T molecule. *Adapted with permission from reference L. Qian, Y. Qiu, N. Sun, M. Xu, G. Xu, F. Xin, Y. Chen, Pyrolysis route of a novel flame retardant constructed by phosphaphenanthrene and triazine-trione groups and its flame-retardant effect on epoxy resin, Polym. Degrad. Stab. 107 (2014) 98–105. Copyright 2014 Elsevier.*

Based on these results, is important to understand the synergy mechanism that arose after the addition of EG. As previously mentioned, most of the action from both 1T and 2T is related to the gas phase. EG, on the other hand, acted mostly in the solid phase. This flame-retardant mechanism occurs by irreversible thermal expansion of EG that led to the release of CO_2, H_2O, and SO_2 (through the edges of the graphitic layers). This effect led to a worm-like structure that prevents the diffusion of oxygen and reactive species into the inner layer of the polyurethane matrix [36]. Hence, a compact and tortuous carbonaceous char layer was formed after the thermal expansion of EG. Therefore, the solid-phase mechanism of EG in combination with the release of radical scavengers from the 1T-based foams and the release of diluting agents from the 2T-based foams were responsible for creating a satisfactory synergy effect that drastically reduced the burning time and weight loss of all the foams.

6. Conclusion

The automobile industry requires many materials with unique properties such as sound and thermal insulation, flexibility, rigidity, high mechanical strength, and flame retardancy. Fortunately, polyurethanes are capable of providing most of the desired properties required in automobiles. However, the inherently high surface area of polyurethane foams makes them susceptible to fire. Thus, chemical modification and/or addition of flame-retardants are necessary to provide safety. Expendable graphite was used as a flame-retardant in novel polyurethane foams prepared using 1T- and 2T-based polyols. The polyols were synthesized using a facile thiol-ene reaction which can be used for large-scale production as this method does not require high temperature, solvents, or costly purification steps. The foams obtained from these polyols and expendable graphite were viable materials for the automobile industry due to their many satisfactory physical properties such as high closed-cell content (>93%), decent density (average 45 kg/m^3), high compressive strength (average 120 kPa for 1T foams, and 110 kPa for 2T foams) and the remarkable flame retardancy due to the synergetic effect between the reactive polyols and expendable graphite.

References

[1] N.V. Gama, A. Ferreira, A. Barros-Timmons, Polyurethane foams: Past, present, and future, Materials (Basel) 11 (2018) 1841.
[2] M. Ionescu, Chemistry and Technology of Polyols for Polyurethanes, Shrewsbury, UK, Rapra Technology, 2007.
[3] J. Deng, W. Yuan, P. Ren, Y. Wang, D. Deng, Z. Zhang, X. Bao, High-performance hydrogen evolution electrocatalysis by layer-controlled MoS2 nanosheets, RSC Adv. 4 (2014) 34733–34738.
[4] E.D. Weil, S.V. Levchik, Flame retardants in commercial use or development for polyolefins, J. Fire Sci. 26 (2008) 5–43.
[5] E.D. Weil, S.V. Levchik, Phosphorus flame retardants, in: Kirk-Othmer Encyclopedia of Chemical Technology, 2017, pp. 1–34.
[6] Y. Wang, L. Zhang, Y. Yang, X. Cai, Synergistic flame retardant effects and mechanisms of aluminum diethylphosphinate (AlPi) in combination with aluminum trihydrate (ATH) in UPR, J. Therm. Anal. Calorim. 125 (2016) 839–848.
[7] F.M. de Souza, J. Choi, S. Bhoyate, P.K. Kahol, R.K. Gupta, Expendable graphite as an efficient flame-retardant for novel partial bio-based rigid polyurethane foams, C—J. Carbon Res. 6 (2020) 27.
[8] X.-Y. Meng, L. Ye, X.-G. Zhang, P.-M. Tang, J.-H. Tang, X. Ji, Z.-M. Li, Effects of expandable graphite and ammonium polyphosphate on the flame-retardant and mechanical properties of rigid polyurethane foams, J. Appl. Polym. Sci. 114 (2009) 853–863.
[9] L. Zhang, M. Zhang, L. Hu, Y. Zhou, Synthesis of rigid polyurethane foams with castor oil-based flame retardant polyols, Ind. Crop. Prod. 52 (2014) 380–388.
[10] S. Bhoyate, M. Ionescu, P.K. Kahol, J. Chen, S.R. Mishra, R.K. Gupta, Highly flame-retardant polyurethane foam based on reactive phosphorus polyol and limonene-based polyol, J. Appl. Polym. Sci. 135 (2018) 16–19.
[11] A. Konig, U. Fehrenbacher, T. Hirth, E. Kroke, Flexible polyurethane foam with the flame-retardant melamine, J. Cell. Plast. 44 (2008) 469–480.
[12] C. Wang, Y. Wu, Y. Li, Q. Shao, X. Yan, C. Han, Z. Wang, Z. Liu, Z. Guo, Flame-retardant rigid polyurethane foam with a phosphorus-nitrogen single intumescent flame retardant, Polym. Adv. Technol. 29 (2018) 668–676.
[13] H. Ding, K. Huang, S. Li, L. Xu, J. Xia, M. Li, Synthesis of a novel phosphorus and nitrogen-containing bio-based polyol and its application in flame retardant polyurethane foam, J. Anal. Appl. Pyrolysis 128 (2017) 102–113.
[14] M. Modesti, A. Lorenzetti, Halogen-free flame retardants for polymeric foams, Polym. Degrad. Stab. 78 (2002) 167–173.

[15] M.-J. Chen, Z.-B. Shao, X.-L. Wang, L. Chen, Y.-Z. Wang, Halogen-free flame-retardant flexible polyurethane foam with a novel nitrogen–phosphorus flame retardant, Ind. Eng. Chem. Res. 51 (2012) 9769–9776.

[16] T.T. Li, M. Xing, H. Wang, S.Y. Huang, C. Fu, C.W. Lou, J.H. Lin, Nitrogen/phosphorus synergistic flame retardant-filled flexible polyurethane foams: microstructure, compressive stress, sound absorption, and combustion resistance, RSC Adv. 9 (2019) 21192–21201.

[17] M. Fang, T.F. Webster, D. Gooden, E.M. Cooper, M.D. McClean, C. Carignan, C. Makey, H.M. Stapleton, Investigating a novel flame retardant known as V6: measurements in baby products, house dust, and Car dust, Environ. Sci. Technol. 47 (2013) 4449–4454.

[18] Y. Yuan, H. Yang, B. Yu, Y. Shi, W. Wang, L. Song, Y. Hu, Y. Zhang, Phosphorus and nitrogen-containing polyols: synergistic effect on the thermal property and flame Retardancy of rigid polyurethane foam composites, Ind. Eng. Chem. Res. 55 (2016) 10813–10822.

[19] E. Akdogan, M. Erdem, M.E. Ureyen, M. Kaya, Synergistic effects of expandable graphite and ammonium pentaborate octahydrate on the flame-retardant, thermal insulation, and mechanical properties of rigid polyurethane foam, Polym. Compos. 41 (2020) 1749–1762.

[20] L. Shi, Z.-M. Li, B.-H. Xie, J.-H. Wang, C.-R. Tian, M.-B. Yang, Flame retardancy of different-sized expandable graphite particles for high-density rigid polyurethane foams, Polym. Int. 55 (2006) 862–871.

[21] H. Liu, H. Chung, Visible-light induced thiol-Ene reaction on natural lignin, ACS Sustain. Chem. Eng. 5 (2017) 9160–9168.

[22] C.E. Hoyle, T.Y. Lee, T. Roper, Thiol-enes: Chemistry of the past with promise for the future, J. Polym. Sci., Part A Polym. Chem. 42 (2004) 5301–5338.

[23] C.R. Morgan, F. Magnotta, A.D. Ketley, Thiol/ene photocurable polymers, J Polym Sci Polym Chem Ed. 15 (1977) 627–645.

[24] Y. Li, A.J. Ragauskas, Kraft lignin-based rigid polyurethane foam, J. Wood Chem. Technol. 32 (2012) 210–224.

[25] J. Peyrton, C. Chambaretaud, A. Sarbu, L. Avérous, Biobased polyurethane foams based on new polyol architectures from microalgae oil, ACS Sustain. Chem. Eng. 8 (2020) 12187–12196.

[26] A. Wolska, M. Gozdzikiewicz, J. Ryszkowska, Thermal and mechanical behaviour of flexible polyurethane foams modified with graphite and phosphorous fillers, J. Mater. Sci. 47 (2012) 5627–5634.

[27] A. Hloiu, D. Iosif, Bio-source composite materials used in automotive industry, Sci. Bull. Automot. Ser. 24 (2013) 57–61.

[28] G. Sung, J.H. Kim, Effect of high molecular weight isocyanate contents on manufacturing polyurethane foams for improved sound absorption coefficient, Korean J. Chem. Eng. 34 (2017) 1222–1228.

[29] H. Zhou, B. Li, G. Huang, Sound absorption characteristics of polymer microparticles, J. Appl. Polym. Sci. 101 (2006) 2675–2679.

[30] Y. Chen, Y. Luo, X. Guo, L. Chen, T. Xu, D. Jia, Structure and flame-retardant actions of rigid polyurethane foams with expandable graphite, Polymers (Basel) 11 (2019) 686.

[31] A. Agrawal, R. Kaur, R.S. Walia, Investigation on flammability of rigid polyurethane foam-mineral fillers composite, Fire Mater. 43 (2019) 917–927.

[32] S. Begum, G.M.S. Ahmed, I.A. Badruddin, V. Tirth, A. Algahtani, Analysis of digital light synthesis based flexible and rigid polyurethane for applications in automobile bumpers, Mater. Express 9 (2019) 839–850.

[33] D.K. Chattopadhyay, D.C. Webster, Thermal stability and flame retardancy of polyurethanes, Prog. Polym. Sci. 34 (2009) 1068–1133.

[34] Z.S. Petrović, Z. Zavargo, J.H. Flyn, W.J. Macknight, Thermal degradation of segmented polyurethanes, J. Appl. Polym. Sci. 51 (1994) 1087–1095.

[35] J. Ferguson, Z. Petrovic, Thermal stability of segmented polyurethanes, Eur. Polym. J. 12 (1976) 177–181.

[36] G. Camino, S. Duquesne, R. Delobel, B. Eling, C. Lindsay, T. Roels, Mechanism of expandable graphite fire retardant action in polyurethanes, ACS Symp. Ser. 797 (2001) 90–109.

[37] M. Siebert, R. Sure, P. Deglmann, A.C. Closs, F. Lucas, O. Trapp, Mechanistic investigation into the acetate-initiated catalytic Trimerization of aliphatic isocyanates: a bicyclic ride, J. Org. Chem. 85 (2020) 8553–8562.

[38] L. Qian, Y. Qiu, N. Sun, M. Xu, G. Xu, F. Xin, Y. Chen, Pyrolysis route of a novel flame retardant constructed by phosphaphenanthrene and triazine-trione groups and its flame-retardant effect on epoxy resin, Polym. Degrad. Stab. 107 (2014) 98–105.

[39] S. Guo, M. Bao, X. Ni, The synthesis of meltable and highly thermostable triazine-DOPO flame retardant and its application in PA66, Polym. Adv. Technol. 32 (2021) 815–828.

CHAPTER 5

Natural fiber-reinforced nanocomposites in automotive industry

Abu Bin Imran[a] and Md. Abu Bin Hasan Susan[b]
[a]Department of Chemistry, Faculty of Engineering, Bangladesh University of Engineering and Technology, Dhaka, Bangladesh
[b]Department of Chemistry, Faculty of Science, University of Dhaka, Dhaka, Bangladesh

Chapter outline

1. Introduction   85
2. Selection of natural fiber and preparation   88
3. Natural fiber-reinforced nanocomposites   92
4. Applications of nfrns in the automotive industry   95
5. Conclusions and prospect   98
Acknowledgments   98
References   98

1. Introduction

Automobiles, the most significant and conspicuous means of transportation, have a profound influence on human civilization. Recent developments of automobile manufacture and its subsidiary industries reflect that the face of our civilization continues to undergo a rapid change. The key materials for automobile production have so far been recognized as steel, iron, magnesium, zinc, plastics, carbon, and glass fibers. Material selection, especially for different components of a vehicle like a car, has been very crucial, and this relies on several key factors, which inter alia include: thermal, chemical, or mechanical resistance, ease of manufacturing process, and durability [1]. Affordability is another major concern in production and is dictated by the expenses of manufacture, operation, and disposal relevant to the whole life cycle of an automobile.

For production, composite materials have significant advantages over other counterparts since they have the potential to render a car lighter, safer, and more efficient in energy. Composites are strong, stiff, lightweight, and do not even rust or decay to ensure their superiority over metallic objects. Although the use of plastics and polymeric composites has already a long history in the automotive industry, their widespread use in automobiles is yet to be fully materialized. The manufacturers continue to use plastics and polymer composites substantially as an excellent option to obtain lightweight while retaining safety and widening the applicability through improvement in the esthetics, aerodynamic design, and quality in many applications in the interior and exterior of vehicles.

The fiber-reinforced composite material of which at least one dimension is of nanometer scale or less than 100 nm is called fiber-reinforced nanocomposites (FRNs) [2]. The use of fiber in FRNs boosts their mechanical and chemical properties via reinforcement of the polymer matrix. The FRNs attribute greater versatility as a material in construction due to their characteristic high strength, highly specific rigidity, noncorrosive nature, durability, and lightweight. In recent years, glass, carbon, and aramid fibers along with nylon and polyester, and natural fibers have been the most popular ones to prepare composites/nanocomposites. The outstanding strength/stiffness-to-weight ratio of FRNs is due to lower fiber densities. FRNs are used in the car industry due to their capacity to sustain at high temperatures and pressures. They were developed and used for the first time in Toyota Camry car as timing belt covers of an engine in 1991 [2]. Body, chassis and tire parts, vehicle interiors, structural, mechanical, and electrical devices, IC motors, and moving mechanisms are also constructed with FRNs [3]. Because of their low prices and availability, glass fiber is the most popular fiber-reinforced composite material. Such composites are employed in the aviation, chemical and electrical resistivity industries, wind power, sports products, ships, etc. A lack of homogeneity in the dispersion of particles and a weak bond between the matrix and the particles are the major drawbacks associated with their use. With a view to resolving the issues, research is focused on the method of acquiring and adjusting related parameters and evaluation of nanocomposites for automotive industries.

Carbon nanotubes (CNTs) have superior mechanical properties as compared to carbon nanofibers (CNFs) [4]. CNFs, however, have a considerably lower cost than CNTs. The stacked cup or herringbone morphology of CNFs contributes to open edge planes on the fiber. The morphology along with the length of the CNF determines the chemical reactivity of the surface. The dynamic functionality of CNFs could then be distributed easily throughout the composite matrix. The functionalized CNFs can covalently bind between the polymer matrix and the reinforcement material. The CNFs are less impacted compared to metals for use in aerospace, oil, transportation industries, and so on. In terms of the products from FRNs, almost 99% are made of glass- or other fiber-reinforced composites, while natural fiber-reinforced nanocomposites (NFRNs) contribute only a small fraction [5]. Taking the market share into consideration, about 25% of the cost belongs to carbon fiber, and 75% belongs to glass fiber-reinforced composites. Among all composites, 37% are thermoplastics and 63% are thermosets.

Therefore, the selection of material is crucial and determines overall price, security, agency risk, weight, and vehicle emissivity. The automobile industry requires to ensure technological aspects such as weight, power, and economic and consumer strategies without any sacrifice in quality [5, 6]. The industry, therefore, greatly depends on a structured approach to select suitable materials. It is the first and the most significant concern in the construction of a vehicle. There are a number of components that can be used in the body and chassis of a car [6]. When choosing a material for a body of a car, one important consideration is its ability to endure the energy obtained from collision, known as impact toughness [7]. The cost of the materials used should be lower than commonly used materials in vehicle parts, but not at the expense of efficiency and quality. Reducing car weight ensures less dead weight on cars, which helps to lower fuel consumption. There are several ways to reduce the dead weight of a car, for example, replacement with materials of lower density, optimizing the design for load transport elements, external attachments, and optimizing the design of production process. The materials used in a car should be able to withstand different kinds of weather conditions. They must be resistant to corrosion and should be able to be changed into many shapes. The manufacturing process involves the selection of the best formation or shaping to achieve the required design. Considering the worldwide environmental demand, strength, durability, and lower price of the car components, it is important to find alternative materials originated from natural resources that are biodegradable and last longer. In the fast-evolving, lightweight vehicle market, natural fibers are expected to have a significant impact on the car industry by 2025 [8, 9]. This evolution would be reflected mainly by advancements in the vehicle design process and developments in the manufacturing processes to make them more responsive in terms of accuracy, time, and total production cost [10–12].

Natural fibers are environmentally friendly, but [13, 14] due to their high-water absorptivity, low fire resistance, and low mechanical properties, their widespread applications have been limited. In recent decades, significant attention has been focused on the use of natural fibers from both renewable and nonrenewable resources such as oil palm (oil), sisal flax, and jute in the manufacture of compost materials [15–17] (Figs. 1 and 2). The plants that generate fibers can be categorized into bast fibers (jute, flax, ramie, hemp, and kenaf), seed fibers (cotton, coir, and kapok), leaf fibers (sisal, pineapple, banana, and abaca), grass and reed fibers (rice, corn, and wheat), core fibers, as well as fibers from wood sources or crop residues. Natural fibers possess several favorable properties to facilitate reinforcement for composites [18]. They have low densities to yield composites that are incredibly lightweight. The reinforcement of biodegradable natural fibers gives green composites that are highly susceptible to rapid degradation by bacteria or enzymes. In contrast to synthetic fibers, they are often cost-effective, highly renewable, and reduce reliance on both domestic and international petroleum products. In fact, new bio-based composites have been produced due to the growing need for environmentally sustainable products and the need to reduce the expense of conventional petroleum-based reinforced fiber (i.e., carbon, glass, and aramid) composites [19–37].

Here, we review various NFRNs derived from natural fibers and their characteristic behaviors, including fiber–matrix adhesion, moisture, impact and fatigue, thermal resilience, fiber reinforcement, and fiber processing techniques. Literature has been extensively surveyed to highlight current applications of NFRNs and finally future trends of NFRNs in the automotive industries have been addressed in detail.

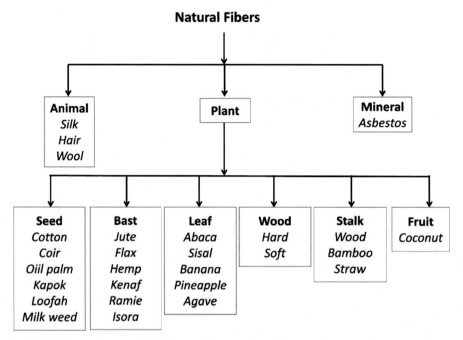

Fig. 1 Classification of natural fiber, according to the origin, with examples. (Reproduced with permission from K. Hasan, P. Horváth, T. Alpár, Potential natural fiber polymeric Nanobiocomposites: a review, Polymers 12 (2020) 1072. (Copyright 1996–2021 MDPI, Basel, Switzerland).)

2. Selection of natural fiber and preparation

The most abundant raw material in the world is wood derived from trees with an annual world production of 1.75×10^9 tons per year from more than 10,000 species. In contrast, the production of cotton is 18.5×10^6 tons per year, while those for kenaf, flax, and hemp are 9.7×10^5, 8.3×10^5, and 2.1×10^5 tons per year, respectively. Although there are various sources of plant fibers in nature, only a few are suitable for automotive applications. The key component of natural fibers is cellulose, but the quantity of pure cellulose, hemicellulose, pectin, lignin, and others can vary from fiber to fiber (Table 1) [38, 39]. The most prevalent natural fiber composites used for structural applications are flax, kenaf, and hemp due to the strength of the fibers. In Table 2, the density, elongation, tensile strength, and elastic modulus for these fibers are summarized [40]. According to the specific strength, maximum tensile strength, and modulus of elasticity, the flax fiber is competitive compared to glass fiber. However, the specific modules of hemp and kenaf are comparable with E-glass. These data suggest that the properties of the bast fiber are equal to or surpass those of glass fibers in certain instances.

Natural fibers can be harvested annually, two to three times per year in some instances, such as for kenaf, jute, and hemp. Kenaf, a plant of African origin grown in

Fig. 2 Classification of fibers. *(Reproduced with permission from D.K. Rajak, D.D. Pagar, P.L. Menezes, E. Linul, Fiber-reinforced polymer composites: manufacturing, properties, and applications, Polymers 2019;11 (10):1667 (Copyright 1996–2021 MDPI, Basel, Switzerland).)*

Table 1 Chemical composition of some common natural fibers.

Fiber	Cellulose (wt%)	Hemicellulose (wt%)	Lignin (wt%)	Waxes (wt%)
Bagasse	55.2	16.8	25.3	–
Bamboo	26–43	30	21–31	–
Flax	71	18.6–20.6	2.2	1.5
Kenaf	72	20.3	9	–
Jute	61–71	14–20	12–13	0.5
Hemp	68	15	10	0.8
Ramie	68.6–76.2	13–16	0.6–0.7	0.3
Abaca	56–63	20–25	7–9	3
Sisal	65	12	9.9	2
Coir	32–43	0.15–0.25	40–45	–
Oil palm	65	–	29	–
Pineapple	81	–	12.7	–
Curaua	73.6	9.9	7.5	–
Wheat straw	38–45	15–31	12–20	–
Rice husk	35–45	19–25	20	–
Rice straw	41–57	33	8–19	8–38

(Reproduced with permission from O. Faruk, A.K. Bledzki, H-P Fink, M. Sain, Biocomposites reinforced with natural fibers: 2000–2010, Prog. Polym. Sci. 37(11) (2012) 1552–96. (Copyright Elsevier Ltd., 2021).)

the United States for various uses, including as an oil spill absorber, can grow to a height of four meters in 4–5 months and can yield two or three harvests a year in tropical climates [7]. Jute, grown in China, India, and Bangladesh, can be grown in four to 6 months, although increased synthetic fiber production has lowered its yield. Jute and kenaf have benefits in terms of their resistance to severe climate, pests, and diseases. Hemp, which grows in most climates as an annual crop, can be traced back historically as a source of rope, cloth, and textiles for more than 10,000 years. These plants have a high assimilation rate of carbon dioxide (CO_2) and potentially clean the air by consuming large amounts of CO_2, the leading greenhouse gas. This is also true for sisal, which can produce fiber from the same plant for up to 20 years, after which the plant starts to flower and then dies. Once the fiber is harvested, it must be separated from the rest of the plant by a process called retting.

The fibers are softened and separated by a partial rotting. This can be accomplished by several techniques employing humidity, microorganisms, or chemical degradation of the bark tissue that binds the fiber and nonfiber portions to make it easier to separate the fibers. In the retting process, hemicellulose and lignin components are removed. When the stalks are left in the field, dew retting takes place so that rain, dew, or irrigation is used to keep the stems moist. This can take up to 5 weeks and creates a coarse fiber with a light brown hue. When stems are wrapped and then immersed in water, water retting happens such that the pectin is broken down by bacteria. It takes 7–10 days to create a high-quality

Table 2 Properties of selected natural and manmade fibers.

Fiber	Density (g/cm³)	Elastic modulus (GPa)	Elongation (%)	Tensile strength (MPa)
Cotton	1.5–1.6	5.5–12.6	7.0–8.0	400
Jute	1.3	26.5	1.5–1.8	393–773
Flax	1.5	27.6	2.7–3.2	500–1500
Hemp	1.47	70	2–4	690
Kenaf	1.45	53	1.6	930
Ramie	–	61.4–128	3.6–3.8	400–938
Sisal	1.5	9.4–22	2.0–2.5	511–635
Abaca	1.5	–	–	980
Coir	1.2	4.0–6.0	30.0	593
Bagasse	1.2	19.7–27.1	1.1	20–290
Pineapple	1.5	82	1–3	170–1672
Henequen	1.4	–	3–4.7	430–580
Banana	1.35	33.8	53	355
Kraft Pulp	1.5	40.0	4.4	1000
E-glass	2.5	70.0	0.5	2000–3500
S-glass	2.5	86.0	2.8	4570
Aramid (Std.)	1.4	63.0–67.0	3.3–3.7	3000–3150
Carbon (Std. PAN-based)	1.4	230–240	1.4–1.8	4000

(Reproduced with permission from S. Mahzan, M. Fitri, M. Zaleha (Eds.), UV radiation effect towards mechanical properties of natural fibre reinforced composite material: a review, IOP Conf. Ser.: Mater. Sci. Eng. 165 (2017) 012021. (Copyright 2021 IOP Publishing).)

fiber. Warm-water retting occurs when bundles are soaked for 24 h, after which water is replaced. For the next 2 or 3 days, heat is then applied to warm the batch to produce uniform and clean fiber. Green retting is a mechanical process that separates the components and is used for textiles, paper, or fiberboard products. Chemical retting happens as the pectin is dissolved using acids, causing the components to be removed. This shortens the time to as little as 48 h when it is possible to instigate the following process and produce a high-quality product. Although the natural retting process is lengthy, there are many desirable features of the resulting fibers. The chemical retting process is fast but affects several properties, including loss of tenacity, color, and luster. In order to compensate fiber deficiencies, natural fibers can be modified by a variety of physical and chemical methods; fibers can be treated to promote bonding and adhesion, dimensional stability, and thermoplasticity. The chemical structure of the fiber is not altered by physical methods, such as calendaring, stretching, thermo-treatment, and spinning or integration into yarns. Instead, they modify the dimensional and surface features of the fiber and thereby impair the mechanical bonding of polymers. To optimize the properties of the fiber–matrix interface, surface alteration of natural fibers may be used.

In general, conventional methods like compounding, extrusion, injection molding, compression molding, and resin transfer molding (RTM) are used to process natural fiber-reinforced composites. However, the nonuniformity of natural fibers produces a comparatively weaker matrix interface and different thermal behavior to obtain good-quality bio-composites compared to synthetic fibers. Injection molding and extrusion are widely being employed in the automotive industry for a thermoplastic bio-composites matrix. On the other hand, compression molding, RTM, and vacuum-assisted resin transfer molding are employed for thermoset bio-composites materials.

3. Natural fiber-reinforced nanocomposites

Nanoparticles exhibit a range of unusual and tailorable physicochemical properties [41, 42]. These features have made engineered nanoparticles as core components with widespread future applications in material sciences and engineering. Many nanomaterials are now available for commercial-scale applications due to the establishment of well-developed nanomaterial processing techniques, including chemical vapor deposition and electrospinning. Carbon nanotubes, carbon nanofibers, and polyhedral oligomeric silsesquioxane (POSS) are commercially used in nanocomposites [43, 44]. However, the most dominant industrial nanomaterials are nanoclays [45–47]. The first clay nanocomposite, PA 6/montmorillonite clay nanocomposite, was developed in 1993 by Toyota [48, 49]. Since then, as opposed to other nanomaterials such as POSS and carbon nanotubes, the idea of nanostructured material design has acquired widespread popularity in the automotive industry, primarily due to low cost and availability. It is currently estimated that the automobile and packaging industry sectors would account for about 80% of overall nanocomposite use.

Nanomaterials in current and future cars would enable environmentally sustainable materials that have better efficiency and low manufacturing costs. Nanotechnology, though, can still have adverse environmental effects, and it is essential to consider both the dangers involved with nanomaterials and the level of exposure that are likely to arise. Thus, the integration of nanoclays and natural fibers into resin systems produces two-scale reinforcements. The nanoclay increases the stiffness and hygrothermal properties of the bio-based polymer framework, whereas the principal stiffness and strength come from the natural fibers. Besides, the improved barrier properties of the nanoreinforced resin preclude moisture from the natural fibers, thus creating a crucial bio-based composite with a synergetic effect. In a nanoreinforced bio-based polymer, hybrid bio-based composites that harness the synergy between natural fibers will contribute to enhanced properties while maintaining environmental appeal. Bio-based resins obtained with epoxidized soybean oil (ESO) by partial substitution of unsaturated polyester improve hardness but sacrifice stiffness and hygrothermal properties. But bio-based resin reinforcement with nanoclays enables stiffness to be maintained without losing hardness while also enhancing barrier and thermal properties. Recent advances in the tensile strength of epoxy resin and hemp/epoxy resin composites utilizing CNTs have been reported by Longkullabutra et al. [50]. For hemp/epoxy resin composites, various volume percentages of milled CNTs were distributed, enhancing the tensile properties of composites. Elsewhere, Liu and Erhan [51] made composites based on ESO reinforced with flax fibers and organoclay-reinforced nanocomposites. The flexural and tensile modules of the composites increased proportionally with the quantity of 1,1,1-tris(p-hydroxyphenyl)ethane triglycidyl ether (THPE-GE). With fiber content lower than 10 wt%, the flexural modulus increases, but it decreases above 10 wt%. Furthermore, for fiber content of 13.5 wt%, the tensile modulus increases and then decreases. The organophilic clay in the matrix is well distributed, and an intercalated composite structure is formed. With 5–10 wt% clay content, the ESO/clay nanocomposites have a storage modulus varying from 2.0 to 2.70 MPa at 30°C. These materials are appealing as an alternative to petrochemical polymers.

Faruk and Matuana [52] integrated nanoclay into wood–plastic composites (WPCs) to improve their mechanical properties. Two separate approaches were employed to incorporate nanoclays into high-density polyethylene (HDPE)-based WPCs. The first technique included strengthening the nanoclay HDPE matrix, which was then used as a matrix in the processing of WPCs (melt blending process). The second technique was the direct insertion of nanoclay during traditional dry compounding into HDPE/wood-flour composites (direct dry blending process). The melt blending method in which nanoclay/HDPE nanocomposite was used as the matrix is considered the best way to integrate nanoclay into WPCs. With the appropriate combination of coupling agent and nanoclay, the mechanical propensities of HDPE/wood-flour composites can be enhanced significantly. $CaCO_3$/wood cellulose nanocomposite materials were formed in the presence of wood cellulose fibers and $CaCl_2$ [53] by forming an aqueous carbonate solution from organic precursor dimethyl carbonate. The hydrolysis conditions had a significant effect on the quantity and morphology of the $CaCO_3$ particles accumulated on the surface of cellulose fibers. With increasing reaction time, the volume and the dimension of $CaCO_3$ accumulated on cellulose fibers increase. In addition, nano-sized $CaCO_3$ particles with spheroid morphology were produced by reactions at room temperature, while micrometric $Ca(OH)_2$ aggregates were obtained at 70°C. Lower mass percentages of fibers favored the production of spheroid particles of $CaCO_3$ in the system. Finally, the presence of carboxylic groups at the surface of the substrates improves the precipitation selectivity of $CaCO_3$ particles on the fiber surface. The $CaCO_3$/cellulose nanocomposites could be regarded as a possible strengthener filled in polyethylene (PE) based composites.

The reinforcement factor of silane coupler is well aligned with the polymer matrix, or even the covalent bonds between them [54]. Only prehydrolyzed silanes undergo reactions on a cellulosic surface; the response of the silane couplers with lignocellulose fibers (primarily: cellulose and lignin) was quite different compared with the glass surface. The work of Abdelmouleh et al. [55] demonstrated very explicitly that the usage of cellulose fibers can be successfully improved in low-density PE and natural rubber matrices. The g-methacryloxypropyltrimethoxy (MPS), g-mercaptopropyltrimethoxy silane (MRPS), and hexadecyltrimethoxy silanes (HDS) silane couplers were used to modify the surface of cellulose fibers. The mechanical properties of these composites have been improved by increasing average fiber length, and good mechanical strength has been demonstrated by the composite materials developed using both MPS and MRPS-treated cellulose.

The cellulose whisker was chemically changed by grafting organic acid chlorides and various lengths of the aliphatic chain by an esterification reaction [56]. The crystallinity of the particulates was not compromised by chain grafting, although the cellulose surface may be crystallized with covalent grafting chains when C18 was used. The low-density polyethylene (LDPE) was extruded to prepare nanocomposite products with both unmodified and functionalized nanoparticles. With the duration of the grafted chains, the homogeneity of the resulting nanocomposites was improved. The elongation at break increased significantly when long chains were grafted on the nanoparticle surface.

Cellulose nanofibers are strongly reinforced than microfibers because of the interactions between the nanosized elements that form a percolated net with hydrogen connections or junctions, where good dispersion was achieved in the matrix. As a reinforcement medium in composites, microfibrillated cellulose (MFC) offers improved composite toughness as compared to pulp fibers. In recent years, a new MFC/polylactic acid (PLA) composite was developed using a process similar to papermaking [57] and said to be fairly straightforward to be readily available on an industrial scale. The tensile modulus, strength, and fracture strain of the composite improved linearly, depending on the MFC material. The modulus was doubled, and the mechanical strength was tripled, with the increase in the MFC content from 10 to 70 wt%. Improved mechanical properties over clean PLA resulted from the addition to MFC, particularly in terms of toughness. In related work, the characterization of various hybrid composites confirmed the synergetic behavior, maintaining the original stiffness, strain failure, and hygrothermal properties using 10% ESO and 1.5 wt% nanoclay [58].

The fatigue and impact tolerance remains a significant drawback in many future uses of natural fiber-reinforced composites. Although many new nanocomposites in different fields have been developed, optimization of their energy absorption capacity is still in the rudimentary stage [59]. The technological challenges to be concerned with are [60] (i) absence of suitable energy absorbance criteria and methods for nanocomposites such as assessing indicators, methods, and conditions, (ii) absence of theoretical models to measure the energy absorption capacity, (iii) absence of a systemic comparison of weaknesses and benefits between current methods of study, (iv) absence of basic knowledge of nanocomposite energy absorption mechanisms, and (v) need to identify possible applications of nanocomposites that absorb energy.

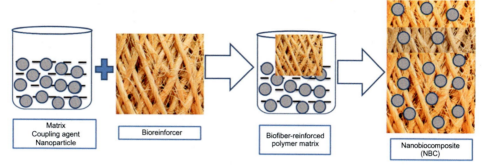

Fig. 3 Formation mechanism of nanobiocomposites. *(Reproduced with permission from K. Hasan, P. Horváth, T. Alpár, Potential natural fiber polymeric Nanobiocomposites: a review, Polymers 2020;12:1072. (Copyright 1996–2021 MDPI, Basel, Switzerland).)*

Guigo et al. [61] fabricated nanocomposites consisting of isolated lignin, natural fibers, and natural additives and obtained wood-like mechanical behaviors with several features resembling polyamides. But the thermal stability and fire tolerance of nanocomposites need to be further improved in many functional applications like the automotive industry.

The nitrogen-rich polysaccharide, chitin, is the second abundant organic material in the environment next to cellulose. Chitosan, *N*-deacetylated derivative of chitin, is an environmentally safe, nontoxic, biodegradable, and antibacterial derivative. When coupled with nanofibrous mats, the inherently beneficial effects of chitosan can be used in a wide variety of areas, including medicinal, packaging, agriculture, and automotive applications [62–64]. In contrast to cellulose, it is more quickly dissolved in acidic media. Wang et al. developed chitosan nanocomposites utilizing a multiwall carbon nanotube (MWNT) and clay. Likewise, using various other inorganic nanoparticles, nanobiocomposites have been prepared [65].

Various types of nanobiocomposite have so far been reported based on graphene oxide, silver, TiO_2, ZnO, SiO_2, and POSS. Natural fibers are hydrophilic, while epoxy resins are hydrophobic. This results in weaker interactions in the NFRNs matrix, providing poor mechanical properties. To overcome this challenge, there have been several attempts, including plasma and alkali treatment [66–80]. The incorporation of nanofillers in the NFRNs could eliminate the drawbacks in terms of mechanical performances, thermal stability, and barrier resistance (Fig. 3) [15].

4. Applications of natural fiber-reinforced nanocomposites in the automotive industry

Many automobile companies in Germany (BMW, Audi, Ford, Opel, Volkswagen, Daimler Chrysler, and Mercedes), Malaysia (Proton), and USA (Cambridge Industry)

Fig. 4 Flax, hemp, sisal, wool, and other natural fibers are used to make 50 Mercedes-Benz E-Class components. *(Reproduced with permission from S.N. Monteiro, K. Satyanarayana, A. Ferreira, D. Nascimento, F. Lopes, I. Silva, A.B. Bevitori, W.P. Inacio, J.B. Neto, T.G. Portela, Selection of high strength natural fibers, Rev. Mater. 2011;15(4):488–505. (Copyright SciELO, 2021).)*

are using different forms of natural fibers in various automotive applications (Table 3) [81]. Apart from the automobile sectors, natural fibers have also been used in building construction, sports, aerospace, and others, for example, panels, window frame, decking, and bicycle frames [82].

Fiber-reinforced composite applications may be divided into three categories in the automobile sector: body components, chassis components, and engine components. Exterior body components need a high degree of rigidity, dent resistance, and esthetic surface finish. The Corvette rear leaf spring was introduced first in 1981 [89] as chassis components. Other structures of chassis parts, such as drive shafts and road wheels, have also been introduced successfully. Unfortunately, fiber-reinforced composites have not proven effective in applying engine components as body and chassis components due to poor fatigue loads at extremely high temperatures.

Most of the cars use bumpers at the rear or front to shield passengers from the effects during a crash. The bumper absorbs energy during a collision and mitigates the injuries affected by automobiles. Carbon-reinforced fibers and glass-reinforced fibers are mostly used in automotive bumpers [90, 91]. Two major considerations to select bumper materials are greater efficiency than previous materials and affordability and improvement of strength at the expense of reduced weight [92, 93].

The engine hood, dashboards, and storage tanks are made by reinforcing natural fibers such as flax strengthened, cotton, jute, sisal, and ramie. The weight of the composite is

Table 3 The application of natural fiber composites in automotive industry [16, 81–88].

Manufacturer	Model	Application
Audi	A2, A3, A4, A4 Avant, A6, A8, Roadstar, Coupe	Boot-liner, spare tyre-lining, side and back door panel, seat back, and hat rack
BMW	3, 5, and 7 series and other Pilot	Seat back, headliner panel, boot-lining, door panels, noise insulation panels, and molded foot well linings
Citroen	C5	Interior door paneling
Daimler Chrysle	A, C, E, and S classes, EvoBus (exterior)	Pillar cover panel, door panels, car windshield/car dashboard, and business table
Ford	Mondeo CD 162, Focus	Floor trays, door inserts, door panels, B-pillar, and boot-liner
Fiat	Punto, Brava, Marea, Alfa Romeo 146, 156, 159	Door panel
General Motors	Cadillac De Ville, Chevrolet Trail Blazer	Seat backs, cargo area floor mat
Lotus	Eco Elise (July 2008)	Body panels, spoiler, seats, and interior carpets
Mercedes Benz (Fig. 4) [88]	C, S, E, and A classes	Door panels (flax/sisal/wood fibers with epoxy resin/UP matrix), glove box (cotton fibers/wood molded, flax/sisal), instrument panel support, insulation (cotton fiber), molding rod/apertures, seat backrest panel (cotton fiber), trunk panel (cotton with PP/PET fibers), and seat surface/backrest (coconut fiber/natural rubber)
	Trucks	Internal engine cover, engine insulation, sun visor, interior insulation, bumper, wheel box, and roof cover
Mitsubishi		Cargo area floor, door panels, and instrumental panel
Opel	Vectra, Astra, Zafira	Door panels, pillar cover panel, head-liner panel, and instrumental panel
Peugeot	406	Front and rear door panels, seat backs, and parcel shelf
Renault	Clio, Twingo	Rear parcel shelf
Rover	2000 and others	Rear storage shelf/panel, and insulations
Saab	9S	Door panels
Saturn	L300	Package trays and door panel
Toyota	ES3	Pillar garnish and other interior parts
Toyota	Raum, Brevis, Harrier, Celsior,	Floor mats, spare tyre cover, door panels, and seat backs
VAUXHALL	Corsa, Astra, Vectra, Zafira	Headliner panel, interior door panels, pillar cover panel, and instrument
Volkswagen	Passat Variant, Golf, A4, Bora	Seat back, door panel, boot-lid finish panel, and boot-liner
Volvo	V70, C70	Seat padding, natural foams, and cargo floor tray

thus decreased and performance is strengthened. The head impact criteria of natural fiber-reinforced composites for vehicle body parts are reasonable [94–98]. The incorporation of bamboo fibers into polyurethane-composite structure increases the cell wall thickness, enhancing the absorption coefficient of sound and vibration in automobile door panels [99–103]. Bending stiffness of 2.5 GPa was demonstrated by the composite laminate with bamboo, cotton, and flax fibers with PLA fibers [104–107].

5. Conclusions and prospect

Natural fiber-reinforced composites from organic and ecologically benign substances such as biofibers and bio-based soybeans, pure cellulose acetate, citrate-based plasticizers, and bio-modified montmorillonite nanofillers have already been employed to produce a high number of car parts which were otherwise manufactured with other metals or composites in the past. The careful use of these automotive materials helps manufacturers to increase reliability and minimize vehicle weight, fulfill load-bearing specifications. The research on composite materials, reinforced plastics, and polymers have significantly enhanced to tune properties of materials to render them suitable for internal, external, and under bonnet components. On the other hand, though natural fiber-based nanocomposite is advantageous in the automotive industry, their research and employment in the industry are still in the rudimentary stage.

The NFRNs are usually used in different vehicle elements, including bumpers, seats, dashboards, internal and external trims. NFRNs have enormous potentials in terms of financial, commercial, and environmental aspects, and the supply chains need to take the advantage in the coming days. The car manufacturers are constantly upgrading the interior and exterior parts with a view to enhancing the esthetics and acoustics, reducing weight, improving performance, and reducing time and cost. The current status, as reviewed by existing literature, also suggests that NFRNs have the necessary prospect to replace conventional materials for automotive ingredients, parts, and devices and the whole industry must work together with academicians and researchers to produce future NFRN-based automobiles.

Acknowledgments

A.B.I. gratefully acknowledges the support from the Ministry of Education while M.A.B.H.S. acknowledges financial support from the Ministry of Science and Technology, People's Republic of Bangladesh.

References

[1] S. Suresh, N. Shenbaga Vinayaga Moorthi, S. Vettivel, N. Selvakumar, Mechanical behavior and wear prediction of stir cast Al–TiB2 composites using response surface methodology, Mater. Des. 59 (2014) 383–396.

[2] J. Holbery, D. Houston, Natural-fiber-reinforced polymer composites in automotive applications, JOM 58 (11) (2006) 80–86.
[3] J. Mathew, J. Joy, S. George, Potential applications of nanotechnology in transportation: a review, J. King Saud Univ. Sci. 31 (2019) 586–594.
[4] M.S. Konsta-Gdoutos, G. Batis, P.A. Danoglidis, A.K. Zacharopoulou, E.K. Zacharopoulou, M.G. Falara, P.Shah S., Effect of CNT and CNF loading and count on the corrosion resistance, conductivity and mechanical properties of nanomodified OPC mortars, Constr. Build. Mater. 147 (2017) 48–57.
[5] J. Osborne, Automotive composites – in touch with lighter and more flexible solutions, Met. Finish. 111 (2013) 26–30.
[6] T. Mihai-Paul, B. Ciprian, K. Imre, Systematic approach on materials selection in the automotive industry for making vehicles lighter, safer and more fuel–efficient, Appl. Eng. Lett. 1 (4) (2016) 91–97.
[7] D. Ayre, Technology advancing polymers and polymer composites towards sustainability: a review, Curr. Opin. Green Sustain. Chem. 13 (2018) 108–112.
[8] A. Benyahia, A. Merrouche, Effect of chemical surface modifications on the properties of alfa fiber-polyester composites, Polym.-Plast. Technol. Eng. 53 (2014) 403–410.
[9] S. Su, C. Wu, The processing and characterization of polyester/natural fiber composites, Polym.-Plast. Technol. Eng. 49 (2010) 1022–1029.
[10] L. Miller, K. Soulliere, S. Sawyer-Beaulieu, S. Tseng, E. Tam, Challenges and alternatives to plastics recycling in the automotive sector, Materials 7 (2014) 5883–5902.
[11] A. Faiz, Automotive emissions in developing countries-relative implications for global warming, acidification and urban air quality, Transp. Res. A Policy Pract. 27 (1993) 167–186.
[12] G. Keoleian, D. Menerey, Sustainable development by design: review of life cycle design and related approaches, Air Waste 44 (1994) 645–668.
[13] A. May-Pat, A. Valadez-González, P. Herrera-Franco, Effect of fiber surface treatments on the essential work of fracture of HDPE-continuous henequen fiber-reinforced composites, Polym. Test. 32 (2013) 1114–1122.
[14] M. Norhidayah, A. Hambali, M. Bin Yaakob, M. Zolkarnain, S.H. Taufik, A review of current development in natural fiber composites in automotive applications, Appl. Mech. Mater. 564 (2014) 3–7.
[15] K. Hasan, P. Horváth, T. Alpár, Potential natural fiber polymeric Nanobiocomposites: a review, Polymers 12 (2020) 1072.
[16] A.K. Mohanty, M. Misra, L.T. Drzal (Eds.), Natural Fibers, Biopolymers, and Biocomposites, CRC Press, 2005 Apr 8.
[17] D.K. Rajak, D.D. Pagar, P.L. Menezes, E. Linul, Fiber-reinforced polymer composites: manufacturing, properties, and applications, Polymers 11 (10) (2019) 1667.
[18] A. Shalwan, B.F. Yousif, Design. In state of art: mechanical and tribological behaviour of polymeric composites based on natural fibres, Mater. Des. 48 (2013) 14–24.
[19] D. Ray, B.K. Sarkar, A. Rana, N.R. Bose, The mechanical properties of vinylester resin matrix composites reinforced with alkali-treated jute fibres, Compos. Part A 32 (1) (2001) 119–127.
[20] A.K. Behera, S. Avancha, R.K. Basak, R. Sen, B. Adhikari, Fabrication and characterizations of biodegradable jute reinforced soy based green composites, Carbohydr. Polym. 88 (1) (2012) 329–335.
[21] K. Goda, M. Sreekala, A. Gomes, T. Kaji, J. Ohgi, Improvement of plant based natural fibers for toughening green composites—effect of load application during mercerization of ramie fibers, Compos. A: Appl. Sci. Manuf. 37 (12) (2006) 2213–2220.
[22] J. Müssig, M. Schmehl, H.-B. Von Buttlar, U. Schönfeld, K. Arndt, Exterior components based on renewable resources produced with SMC technology—considering a bus component as example, Ind. Crop. Prod. 24 (2) (2006) 132–145.
[23] H.L. Bos, The Potential of Flax Fibres as Reinforcement for Composite Materials, S.l., Technische Universiteit Eindhoven, 2004, p. 192.
[24] H. Akil, M. Omar, A.M. Mazuki, S. Safiee, Z.M. Ishak, A. Bakar, Kenaf fiber reinforced composites: a review, Mater. Des. 32 (8–9) (2011) 4107–4121.
[25] S. Ochi, Mechanical properties of kenaf fibers and kenaf/PLA composites, Mech. Mater. 40 (4–5) (2008) 446–452.

[26] M. Zampaloni, F. Pourboghrat, S. Yankovich, B. Rodgers, J. Moore, L. Drzal, A.K. Mohanty, M. Misra, Kenaf natural fiber reinforced polypropylene composites: a discussion on manufacturing problems and solutions, Compos. A: Appl. Sci. Manuf. 38 (6) (2007) 1569–1580.
[27] Y. Li, Y.-W. Mai, Ye L sisal fibre and its composites: a review of recent developments, Compos. Sci. Technol. 60 (11) (2000) 2037–2055.
[28] K. Oksman, L. Wallström, L.A. Berglund, R.D.T. Filho, Morphology and mechanical properties of unidirectional sisal–epoxy composites, J. Appl. Polym. Sci. 84 (13) (2002) 2358–2365.
[29] P. Antich, A. Vázquez, I. Mondragon, C. Bernal, Mechanical behavior of high impact polystirene reinforced with short sisal fibers, Compos. A: Appl. Sci. Manuf. 37 (1) (2006) 139–150.
[30] R. Arib, S. Sapuan, M. Ahmad, M. Paridah, H.M.D.K. Zaman, Mechanical properties of pineapple leaf fibre reinforced polypropylene composites, Mater. Des. 27 (5) (2006) 391–396.
[31] H. Liu, Q. Wu, Q. Zhang, Preparation and properties of banana fiber-reinforced composites based on high density polyethylene (HDPE)/Nylon-6 blends, Bioresour. Technol. 100 (23) (2009) 6088–6097.
[32] N.M. Stark, R.E. Rowlands, Effects of wood fiber characteristics on mechanical properties of wood/polypropylene composites, Wood Fiber Sci. 35 (2) (2003) 167–174.
[33] Y. Yu, Y. Yang, M. Murakami, M. Nomura, H. Hamada, Physical and mechanical properties of injection-molded wood powder thermoplastic composites, Adv. Compos. Mater. 22 (6) (2013) 425–435.
[34] D. Dai, M. Fan, Wood Fibres as Reinforcements in Natural Fibre Composites: Structure, Properties, Processing and Applications. Natural Fibre Composites, Elsevier, 2014, pp. 3–65.
[35] S. Sapuan, A. Leenie, M. Harimi, Y.K. Beng, Mechanical properties of woven banana fibre reinforced epoxy composites, Mater. Des. 27 (8) (2006) 689–693.
[36] C. Alves, A. Silva, L. Reis, M. Freitas, L. Rodrigues, D.E. Alves, Ecodesign of automotive components making use of natural jute fiber composites, J. Clean. Prod. 18 (4) (2010) 313–327.
[37] A. Rana, A. Mandal, S. Bandyopadhyay, Short jute fiber reinforced polypropylene composites: effect of compatibiliser, impact modifier and fiber loading, Compos. Sci. Technol. 63 (6) (2003) 801–806.
[38] A. Bismarck, I. Aranberri-Askargorta, J. Springer, T. Lampke, B. Wielage, A. Stamboulis, I. Shenderovich, H.-H. Limbach, Surface characterization of flax, hemp and cellulose fibers; surface properties and the water uptake behavior, Polym. Compos. 23 (5) (2002) 872–894.
[39] O. Faruk, A.K. Bledzki, H.-P. Fink, M. Sain, Biocomposites reinforced with natural fibers: 2000–2010, Prog. Polym. Sci. 37 (11) (2012) 1552–1596.
[40] S. Mahzan, M. Fitri, M. Zaleha, UV radiation effect towards mechanical properties of natural fibre reinforced composite material: a review, IOP Conf. Ser.: Mater. Sci. Eng. 165 (2017) 012021.
[41] A. Leszczyńska, J. Njuguna, K. Pielichowski, J.R. Banerjee, Polymer/montmorillonite nanocomposites with improved thermal properties: part I. Factors influencing thermal stability and mechanisms of thermal stability improvement, Thermochim. Acta 453 (2) (2007) 75–96.
[42] A. Leszczyńska, J. Njuguna, K. Pielichowski, J.R. Banerjee, Polymer/montmorillonite nanocomposites with improved thermal properties: part II. Thermal stability of montmorillonite nanocomposites based on different polymeric matrixes, Thermochim. Acta 454 (1) (2007) 1–22.
[43] J. Markarian, Automotive and packaging offer growth opportunities for nanocomposites, Plast. Addit. Compd. 7 (6) (2005) 18–21.
[44] C. Edser, Auto applications drive commercialization of nanocomposites, Plast. Addit. Compd. 4 (1) (2002) 30–33.
[45] A. McWilliams, Nanocomposites, Nanoparticles, Nanoclays, and Nanotubes, BCC Research, Norwalk, CT, 2006.
[46] J. Njuguna, P. Wambua, K. Pielichowski, K. Kayvantash, Natural Fibre-Reinforced Polymer Composites and Nanocomposites for Automotive Applications. Cellulose Fibers: Bio-and Nano-Polymer Composites, Springer, 2011, pp. 661–700.
[47] A. Downing-Perrault, Polymer Nanocomposites Are the Future, Alyssa Downing-Perrault University of Wisconsin-Stout, 2005. March 1.
[48] Y. Kojima, A. Usuki, M. Kawasumi, A. Okada, Y. Fukushima, T. Kurauchi, O. Kamigaito, Mechanical properties of nylon 6-clay hybrid, J. Mater. Res. 8 (5) (1993) 1185–1189.

[49] Y. Kojima, A. Usuki, M. Kawasumi, A. Okada, T. Kurauchi, O. Kamigaito, Sorption of water in nylon 6-clay hybrid, J. Appl. Polym. Sci. 49 (7) (1993) 1259–1264.

[50] H. Longkullabutra, W. Thamjaree, W. Nhuapeng, Improvement in the tensile strength of epoxy resin and hemp/epoxy resin composites using carbon nanotubes, Adv. Mater. Res. 93 (2010) 497–500.

[51] Z. Liu, S.Z. Erhan, "Green" composites and nanocomposites from soybean oil, Mater. Sci. Eng. A 483 (2008) 708–711.

[52] O. Faruk, L.M. Matuana, Nanoclay reinforced HDPE as a matrix for wood-plastic composites, Compos. Sci. Technol. 68 (9) (2008) 2073–2077.

[53] C. Vilela, C.S. Freire, P.A. Marques, T. Trindade, C.P. Neto, P. Fardim, Synthesis and characterization of new $CaCO_3$/cellulose nanocomposites prepared by controlled hydrolysis of dimethylcarbonate, Carbohydr. Polym. 79 (4) (2010) 1150–1156.

[54] Y. Xie, C.A. Hill, Z. Xiao, H. Militz, C. Mai, Silane coupling agents used for natural fiber/polymer composites: a review, Compos. A: Appl. Sci. Manuf. 41 (7) (2010) 806–819.

[55] M. Abdelmouleh, S. Boufi, M.N. Belgacem, A. Dufresne, Short natural-fibre reinforced polyethylene and natural rubber composites: effect of silane coupling agents and fibres loading, Compos. Sci. Technol. 67 (7–8) (2007) 1627–1639.

[56] A.J. De Menezes, G. Siqueira, A.A. Curvelo, A. Dufresne, Extrusion and characterization of functionalized cellulose whiskers reinforced polyethylene nanocomposites, Polymers 50 (19) (2009) 4552–4563.

[57] A.N. Nakagaito, A. Fujimura, T. Sakai, Y. Hama, H. Yano, Production of microfibrillated cellulose (MFC)-reinforced polylactic acid (PLA) nanocomposites from sheets obtained by a papermaking-like process, Compos. Sci. Technol. 69 (7–8) (2009) 1293–1297.

[58] M. Haq, R. Burgueño, A.K. Mohanty, M. Misra, Hybrid bio-based composites from blends of unsaturated polyester and soybean oil reinforced with nanoclay and natural fibers, Compos. Sci. Technol. 68 (15–16) (2008) 3344–3351.

[59] J. Njuguna, S. Michałowski, K. Pielichowski, K. Kayvantash, A.C. Walton, Fabrication, characterization and low-velocity impact testing of hybrid sandwich composites with polyurethane/layered silicate foam cores, Polym. Compos. 32 (2011) 6–13.

[60] L. Sun, R.F. Gibson, F. Gordaninejad, J. Suhr, Energy absorption capability of nanocomposites: a review, Compos. Sci. Technol. 69 (14) (2009) 2392–2409.

[61] N. Guigo, L. Vincent, A. Mija, H. Naegele, N. Sbirrazzuoli, Innovative green nanocomposites based on silicate clays/lignin/natural fibres, Compos. Sci. Technol. 69 (11–12) (2009) 1979–1984.

[62] S.-F. Wang, L. Shen, W.-D. Zhang, Y.-J. Tong, Preparation and mechanical properties of chitosan/carbon nanotubes composites, Biomacromolecules 6 (6) (2005) 3067–3072.

[63] S. Wang, L. Shen, Y. Tong, L. Chen, I. Phang, P.Q. Lim, T.X. Liu, Biopolymer chitosan/montmorillonite nanocomposites: preparation and characterization, Polym. Degrad. Stab. 90 (1) (2005) 123–131.

[64] H. Huang, Q. Yuan, X. Yang, Morphology study of gold–chitosan nanocomposites, J. Colloid Interface Sci. 282 (1) (2005) 26–31.

[65] J. Huang, I. Ichinose, T. Kunitake, Nanocoating of natural cellulose fibers with conjugated polymer: hierarchical polypyrrole composite materials, Chem. Commun. 13 (2005) 1717–1719.

[66] B. Han, S. Sharma, T.A. Nguyen, L. Li, K.S. Bhat, Fiber-Reinforced Nanocomposites: Fundamentals and Applications, Elsevier, 2020.

[67] N. Rajini, J. Winowlin Jappes, I. Siva, A. Varada Rajulu, S. Rajakarunakaran, Fire and thermal resistance properties of chemically treated ligno-cellulosic coconut fabric–reinforced polymer eco-nanocomposites, J. Ind. Text. 47 (1) (2017) 104–124.

[68] P. Ramu, C.J. Kumar, K. Palanikumar, Mechanical characteristics and terminological behavior study on natural fiber nano reinforced polymer composite–a review, Mater. Today Proc. 16 (2019) 1287–1296.

[69] V. Prasad, K. Sekar, S. Varghese, M.A. Joseph, Enhancing mode I and mode II interlaminar fracture toughness of flax fibre reinforced epoxy composites with nano TiO_2, Compos. A: Appl. Sci. Manuf. 124 (2019) 105505.

[70] D. Pinto, L. Bernardo, A. Amaro, S. Lopes, Mechanical properties of epoxy nanocomposites using titanium dioxide as reinforcement–a review, Constr. Build. Mater. 95 (2015) 506–524.

[71] V. Prasad, M. Joseph, K. Sekar, Investigation of mechanical, thermal and water absorption properties of flax fibre reinforced epoxy composite with nano TiO_2 addition, Compos. A: Appl. Sci. Manuf. 115 (2018) 360–370.
[72] S.M. Mousavi, S.A. Hashemi, S. Jahandideh, S. Baseri, M. Zarei, S. Azadi, Modification of phenol novolac epoxy resin and unsaturated polyester using sasobit and silica nanoparticles, Polym. Renew. Resour. 8 (3) (2017) 117–132.
[73] M.R. Rahman, M.M. Rahman, S. Hamdan, J. Lai, Impact of maleic anhydride, nanoclay, and silica on jute fiber-reinforced polyethylene biocomposites, BioResources 11 (3) (2016) 5905–5917.
[74] S. Siengchin, R. Dangtungee, Polyethylene and polypropylene hybrid composites based on nano silicon dioxide and different flax structures, J. Thermoplast. Compos. Mater. 27 (10) (2014) 1428–1447.
[75] C.I. Idumah, A. Hassan, Hibiscus cannabinus fiber/PP based nano-biocomposites reinforced with graphene nanoplatelets, J. Nat. Fibers 14 (5) (2017) 691–706.
[76] S. Gong, H. Ni, L. Jiang, Q. Cheng, Learning from nature: constructing high performance graphene-based nanocomposites, Mater. Today 20 (4) (2017) 210–219.
[77] Z. Razak, A.B. Sulong, N. Muhamad, C.H.C. Haron, M.K.F.M. Radzi, N.F. Ismail, D. Tholibona, I. Tharazia, Effects of thermal cycling on physical and tensile properties of injection moulded kenaf/carbon nanotubes/polypropylene hybrid composites, Compos. Part B Eng. 168 (2019) 159–165.
[78] S. Fu, P. Song, H. Yang, Y. Jin, F. Lu, J. Ye, Q. Wu, Effects of carbon nanotubes and its functionalization on the thermal and flammability properties of polypropylene/wood flour composites, J. Mater. Sci. 45 (13) (2010) 3520–3528.
[79] L. Tzounis, S. Debnath, S. Rooj, D. Fischer, E. Mäder, A. Das, M. Stamm, G. Heinrich, High performance natural rubber composites with a hierarchical reinforcement structure of carbon nanotube modified natural fibers, Mater. Des. 58 (2014) 1–11.
[80] A. Turlybekuly, A. Pogrebnjak, L. Sukhodub, L.B. Sukhodub, A. Kistaubayeva, I. Savitskaya, D.H. Shokatayevac, O.V. Bondar, Z.K. Shaimardanov, S.V. Plotnikov, B.H. Shaimardanov, I. Digel, Synthesis, characterization, in vitro biocompatibility and antibacterial properties study of nanocomposite materials based on hydroxyapatite-biphasic ZnO micro-and nanoparticles embedded in alginate matrix, Mater. Sci. Eng. C 104 (2019) 109965.
[81] H. Chandekar, V. Chaudhari, S. Waigaonkar, A review of jute fiber reinforced polymer composites, Mater. Today 26 (2020) 2079–2082.
[82] S. Shinoj, R. Visvanathan, S. Panigrahi, M. Kochubabu, Oil palm fiber (OPF) and its composites: A review, Ind. Crop. Prod. 33 (1) (2011) 7–22.
[83] B. Alcock, N. Cabrera, N. Barkoula, T. Peijs, Direct forming of all-polypropylene composites products from fabrics made of co-extruded tapes, Appl. Compos. Mater. 16 (2) (2009) 117–134.
[84] B.C. Suddell, Industrial fibres: recent and current developments, in: Proceedings of the Symposium on Natural Fibres, 2008.
[85] K. Pickering, Properties and Performance of Natural-Fibre Composites, Elsevier, 2008.
[86] R.A. Ilyas, S.M. Sapuan, Biopolymers and biocomposites: chemistry and technology, Curr. Anal. Chem. 16 (5) (2020) 500–503.
[87] M. Zimniewska, M. Wladyka-Przybylak, J. Mankowski, Cellulosic Bast Fibers, Their Structure and properties Suitable for Composite Applications. Cellulose fibers: Bio-and Nano-Polymer Composites, Springer, 2011, pp. 97–119.
[88] S.N. Monteiro, K. Satyanarayana, A. Ferreira, D. Nascimento, F. Lopes, I. Silva, A.B. Bevitori, W.P. Inacio, J.B. Neto, T.G. Portela, Selection of high strength natural fibers, Rev. Mater. 15 (4) (2011) 488–505.
[89] P. Beardmore, Composite structures for automobiles, Compos. Struct. 5 (3) (1986) 163–176.
[90] Y. Zhu, J. Valdez, I. Beyerlein, S. Zhou, C. Liu, M. Stout, D.P. Butt, T.C. Lowe, Mechanical properties of bone-shaped-short-fiber reinforced composites, Acta Mater. 47 (6) (1999) 1767–1781.
[91] A.S. Khan, O.U. Colak, P. Centala, Compressive failure strengths and modes of woven S2-glass reinforced polyester due to quasi-static and dynamic loading, Int. J. Plast. 18 (10) (2002) 1337–1357.
[92] S. Mazumdar, Composites Manufacturing: Materials, Product, and Process Engineering, CRC Press, 2001.

[93] G. Dieter, Engineering Design: A Materials and Processing Approach, McGraw. Hill Publishers, New York, 2000.
[94] C. Kong, H. Lee, H. Park, Design and manufacturing of automobile hood using natural composite structure, Compos. Part B Eng. 91 (2016) 18–26.
[95] M.H. Shojaeefard, A. Najibi, M.R. Ahmadabadi, Pedestrian safety investigation of the new inner structure of the hood to mitigate the impact injury of the head, Thin-Walled Struct. 77 (2014) 77–85.
[96] G. Koronis, A. Silva, M. Fontul, Green composites: a review of adequate materials for automotive applications, Compos. Part B Eng. 44 (1) (2013) 120–127.
[97] C. Kong, H. Park, J. Lee, Study on structural design and analysis of flax natural fiber composite tank manufactured by vacuum assisted resin transfer molding, Mater. Lett. 130 (2014) 21–25.
[98] R.A. Ilyas, S.M. Sapuan, N.M. Nurazzi, M.N.F. Norrrahim, R. Ibrahim, M.S.N. Atikah, M.R.M. Huzaifa, A.M. Radzi, S. Izwan, A.M.N. Azammi, R. Jumaidini, Z.M.A. Ainun, A. Atiqah, M.R.M. Asyraf, L.K. Kian, C.S. Hassan, Macro to nanoscale natural fiber composites for automotive components: research, development, and application, in: S.M. Sapuan, R.A. Ilyas (Eds.), Biocomposite and Synthetic Composites for Automotive Applications, Woodhead Publishing, 2021, pp. 51–105.
[99] S. Ashworth, J. Rongong, P. Wilson, J. Meredith, Mechanical and damping properties of resin transfer moulded jute-carbon hybrid composites, Compos. Part B Eng. 105 (2016) 60–66.
[100] J. Flynn, A. Amiri, C. Ulven, Hybridized carbon and flax fiber composites for tailored performance, Mater. Des. 102 (2016) 21–29.
[101] J. Zhang, A.A. Khatibi, E. Castanet, T. Baum, Z. Komeily-Nia, P. Vroman, X. Wang, Effect of natural fibre reinforcement on the sound and vibration damping properties of bio-composites compression moulded by nonwoven mats, Compos. Commun. 13 (2019) 12–17.
[102] M. Farid, A. Purniawan, A. Rasyida, M. Ramadhani, S. Komariyah (Eds.), Improvement of acoustical characteristics: wideband bamboo based polymer composite, IOP Conference Series: Materials Science and Engineering, 2017.
[103] A. Jambor, M. Beyer, New cars—new materials, Mater. Des. 18 (4–6) (1997) 203–209.
[104] G. Belingardi, E.G. Koricho, Design of a composite engine support sub-frame to achieve light-weight vehicles, Int. J. Autom. Comput. 1 (1) (2014) 90–111.
[105] W. Hou, X. Xu, X. Han, H. Wang, L. Tong, Multi-objective and multi-constraint design optimization for hat-shaped composite T-joints in automobiles, Thin-Walled Struct. 143 (2019) 106232.
[106] R.S. Rahman, Z.F.S. Putra, 5-Tensile properties of natural and synthetic fiber-reinforced polymer composites, in: Woodhead Publishing Series in Composites Science and Engineering, Woodhead Publishing, 2019, pp. 81–102.
[107] S. Kalia, B. Kaith, I. Kaur, Cellulose Fibers: Bio-and Nano-Polymer Composites: Green Chemistry and Technology, Springer Science & Business Media, 2011.

CHAPTER 6

High-performance polyurethanes foams for automobile industry

Felipe M. de Souza[a], Jonghyun Choi[a], Tenzin Ingsel[a], and Ram K. Gupta[a,b]
[a]Department of Chemistry, Pittsburg State University, Pittsburg, KS, United States
[b]Kansas Polymer Research Center, Pittsburg State University, Pittsburg, KS, United States

Chapter outline

1. Introduction — 105
2. Importance of renewable materials for automobiles — 106
3. Significance of polyurethane foams and current issues — 107
4. Synthesis and characterizations of bio-derived polyurethane foams — 107
5. Important characteristics and properties of bio-derived polyurethane foams — 109
   5.1 Structural characterizations — 109
   5.2 Density of polyurethane foams — 111
   5.3 Morphology and cellular characteristics of polyurethane foams — 112
   5.4 Sound absorption behavior — 114
   5.5 Mechanical properties — 115
   5.6 Flame-retardant behavior of polyurethane foams — 119
   5.7 Coatings for automobiles — 124
6. Conclusion — 126
Acknowledgment — 126
References — 126

1. Introduction

The automobile industry is a competitive market that has been evolving rapidly in recent decades. The automobile industry requires advanced technological instruments to test the vehicles manufactured to provide customers with higher performance efficiencies, ergonomic designs for comfort, and lightweight robust cars for use in different environments. Automobile companies are diligent in pursuing affordable vehicles and are investing tremendously in research and development for low-cost, high-performance car parts. For the construction of a car, the basic requirement includes metals for the engine and body; oil for lubrication; and polymeric materials for tires, bumper, and wipers; and car paint for physical protection and esthetic purposes. On top of that, environmental rules and regulations require car engines to stay below the upper limit of greenhouse gas generation per mile. Two types of vehicles derived from environmental pact are the ultralow emission vehicles (ULEV) and the partnership for a new generation vehicle (PNGV). The former limits organic gas emissions of 0.004, 1.7, and 0.2 g/mile for CO and NO_x. The latter is based on an agreement between Ford, GM, Chrysler, and the Department of Energy, which demands an efficiency of 82.5 miles/gal. Therefore to meet such demands, continuous exploration of greener materials for automobiles is required [1, 2].

A widely used tactic in meeting environmental and safety requirements is utilizing materials that can decrease the vehicle's overall weight. Polymeric materials like polyurethane (PU) gained traction, owing to their desirable properties such as low density, a wide range of mechanical properties, facile synthesis, and excellent thermal and sound insulation properties. The automotive industry quickly ventured into exploiting PU for its broad set of advantageous properties [3]. PU, when first discovered by Otto Bayer in 1937, their practical applications were not foreseen. It was not until a year later when Rinke and co-workers filed another patent for PU synthesis using a diol and diisocyanate, did it revamp PU in its practical and commercial use [4–6]. Since then, PU with different polyols and isocyanates have been used to obtain versatile materials for applications in civil constructions, footwear, electrical housings, thermal insulations, and automobiles.

2. Importance of renewable materials for automobiles

The current tendency of most industries, including automobiles, is to use bio-based renewable materials because these materials participate in a sustainable supply–demand chain and are of relatively low price. With such incentive associated with the utilization of bio-derived materials, bio-renewable composites have been used for reinforcement or as a filler in durable goods. Low-cost sources such as corn, soybean, castor oil, and sugarcane can be used as hydrocarbon sources to obtain PU. Bio-derived PUs, like any other conventionally derived ones, can be processed via injection molding, extrusion, expansion, and thermoforming. Some examples include door and floor panels, armrests, and seats made up of industrial hemp and flax. Another case is the bio-foam made from coconut fibers that have been used for headrests and chair support. Also, a production line was developed in 2009 based on lignocellulose named *Lignolight*, with a composition of 70% wood fibers and 30% of the resin. It decreased the weight of door panels by around 40%. These bio-derived materials provide an edge over conventional filler-like glass fibers by lowering the cost and the weight.

On top of that, wood fiber provides a cushioning effect and better shock absorption properties than mineral fiber. The mineral fiber generally exhibits rigidity and brittleness. Few investigators have noted that natural fibers consume less energy during their production than glass fibers, i.e., 11.4 for natural fibers and 48.3 MJ/kg for glass fibers of the same quantities [7]. Several automobile companies adopted natural fibers from hemp, jute, soy, and corn based on those factors. Automobile models like BMW 7 Series, Chrysler Sebring, Ford Fiesta, and Ford Focus contain interior door panels partly made up of natural fibers. Ford Fusion and Lincoln MKZ models utilize around 13%–26% of soy-based fibers to reinforce PU material for a car part like the headrest. On the other hand, Toyota has developed a 40%–60% bio-derived nylon used as a cooling vessel [8].

3. Significance of polyurethane foams and current issues

Obtaining unique, renewable, value-added materials of low cost is the general objective of automobile companies. Since this chapter focuses on the use of PUs in the automobile industry, a discussion of PUs' structure–property relation is drawn to describe the significance of PU applications in car parts. Here is the general overview of how this chapter will progress. First, the effect of density on the vehicle's weight distribution is described; and how that optimizes fuel consumption and performances, which in turn help mitigate environmental concerns. Second, the correlation between morphology and cellular structure is described. These microstructural properties affect the thermal and sound insulation properties of PU. Therefore, optimizing microstructural properties can cater to improving the comfort and ergonomics of cars. Third, the ease of mechanical property manipulation in PU enables their use in different applications (e.g., rigid PU types for bumper, body parts, panels, and flexible types for cushioning and seats). Fourth, PU's application in coatings is described for the protection against the environment and exterior beauty. Finally, enhancement of flame resistance of PU with the incorporation of flame-retardants is discussed. PU is otherwise flammable and is susceptible to fire.

Flame-retardant rigid polyurethane foams (FR-RPUF) can provide numerous merits in the automotive and transportation sector. Here in this work, carvone, a natural source, was utilized as a precursor for producing FR-RPUF with the assistance of flame-retardant (FR) chemicals, aluminum trihydroxide (ATH), and aluminum hypophosphite (AHP). The resulting foams were characterized using industrial standards. The results indicate that the synthesized RPUF can be a viable candidate for vehicle parts.

4. Synthesis and characterizations of bio-derived polyurethane foams

Unsaturated oils such as corn oil, soybean oil, castor oil, and essential oils can be converted into a polyol via a facile thiol-ene reaction with the help of hydroxyl-containing mercaptan compounds. The thiol-ene reaction method has been utilized in previously reported papers for prospective large-scale bio-polyols production [9–14]. In harmony with previous bio-based polyol studies, this chapter describes the utilization of carvone oil (an extract from bay leaf) as a bio-derived alternate for FR-RPUF and its merits in its application in-vehicle components [15]. It was observed in this work that a low concentration of AHP (2.85 wt%) in RPUFs facilitated enhanced flame retardancy, resulting in a low extinguishing time of 18 s with a 10% weight loss of the original material after the burning test.

Here is the detailed synthesis part of the carvone-derived FR-RPUF. The natural source carvone oil was subjected to room temperature thiol-ene reaction with 2-mercaptoethanol (2ME). The resulting functionalized carvone oil showcased a hydroxyl number of 365 mg KOH/g, enough to promote reaction with methylene diisocyanate (MDI) to synthesize RPUF. As previously mentioned, the flame-retardant properties were enhanced by adding varying concentrations of eco-friendly FRs (ATH and AHP). As for the synthesis of AHP, the facile wet chemistry method was employed. The AHP synthesis was carried out in a 250 mL, three-necked flask. Sodium hypophosphite monohydrate ($NaH_2PO_2 \cdot H_2O$) of 50.88 g (0.48 mol) was dissolved in 30 mL of distilled water. The solution was left to stir for 15 min at a temperature of 50°C. After its complete dissolution, the temperature was raised to 85 °C and maintained for 25 min. In another flask, a solution containing 38.54 g (0.16 mol) of $AlCl_3 \cdot 6H_2O$ was prepared, which was then added dropwise to the flask containing $NaH_2PO_2 \cdot H_2O$. The reaction mixture eventually became blurry, producing a white precipitate. The temperature of the reaction mixture was maintained at 85°C for 1 h. The white product was filtered at room temperature and washed with distilled water. After that, it was dried in the oven at 100°C [16]. The carvone-based polyol was synthesized by the reaction of 1 mol of carvone with 2 mol of 2-mercaptoethanol in the presence of 2-hydroxy-2-methyl propiophenone as photoinitiator and UV light (wavelength = 365 nm) at room temperature.

Various industrial methods were used to characterize the polyol and PU foams. At first, the closed-cell content of the synthesized foams was measured using a Humi Pycnometer instrument (ASTM D 285). For this, the foam samples were cut in a cylindrical shape of 45 × 30 mm (diameter × height). By measuring the volume and weight of these samples, the apparent density was calculated following ASTM D 1622. The compression strength of the foams was measured using a universal testing machine provided by the Instron Tensile instrument. Thermalgravimetric analysis (TGA) was performed using a TGA Q500 Discovery machine by a Trios system. The test was carried out in a temperature range from 25 to 700°C at the ramp rate of 10 °C/min under the N_2 atmosphere. The horizontal burning test was performed according to ASTM D 4986–98 at which the specimens were cut in a paving stone shape of 150 × 50 × 12.5 mm^3 (height × length × width) and exposed directly into a flame for 10 s. The self-extinguishing time and weight loss were measured. Scanning electron microscopic (SEM) images were taken using the Hitachi Tabletop Microscope TM3030 series to analyze the cellular structure of the foams. For this, the samples were sputtered with gold to create a thin conductive layer to prevent the charging effect during the SEM analysis.

5. Important characteristics and properties of bio-derived polyurethane foams

5.1 Structural characterizations

The structural analysis of synthesized AHP was performed using FT-IR spectroscopy. Fig. 1 shows that the peak near 1100–1200 cm^{-1} is related to P=O [17]. The stretch around 1085–1075 cm^{-1} is pertaining to PH$_2$ stretch in both NaHP and AHP spectra [18]. The peak around 500 cm^{-1} is Al—O stretch in both AlCl$_3$.6H$_2$O and AHP [19, 20]. Other peaks were observed for the AlCl$_3$.6H$_2$O structure, such as Al-OH stretch at 1200 cm^{-1} and two peaks related to OH stretch at 1600 and 2500 cm^{-1} [18]. The characteristics peaks of AHP observed in the FT-IR spectrum confirm the successful synthesis of AHP from its starting reagents.

The synthesized polyol and the starting materials were analyzed using various methods. The carvone-based polyol exhibited an increased viscosity and molecular weight. The carvone oil's viscosity and resulting carvone polyol were determined to be 0.006 and 2.7 Pa.s, respectively. The change in molecular weight of carvone oil during the polyol synthesis was analyzed using a gel permeation chromatography (GPC) technique. As shown in Fig. 2A, the carvone polyol shows reduced elution time (36 min) as compared to the pristine carvone oil (42 min) and 2ME (47 min). The reduced elution time represents an increased molecular weight in the carvone polyol and confirms all the reactants' consumption during the synthesis. The suggested mechanism and side products were described in our previous work [21]. The resulting carvone polyol obtained a hydroxyl number close to its theoretical value (367 mg KOH/g) [22, 23].

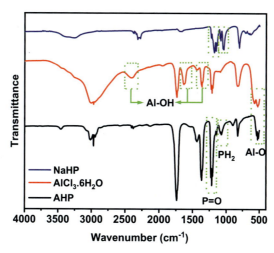

Fig. 1 FT-IR spectra for AHP and starting reagents.

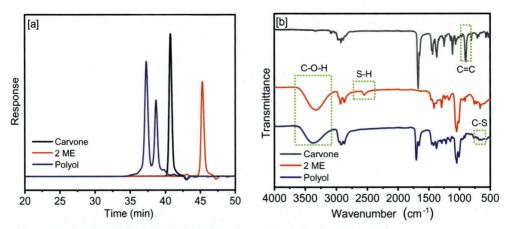

Fig. 2 (A) GPC chromatograms and (B) FT-IR spectra for carvone, 2ME, and polyol.

The structural fingerprints of carvone oil containing double bonds and 2ME containing thiol and hydroxyl groups were analyzed using the FT-IR spectra (Fig. 2B). The peak near 960 cm^{-1} represents an alkene stretch (—C═C—) for carvone oil structure [24]. In comparison, the peaks near 2560 cm^{-1} and 3500 cm^{-1} represent S—H (thiol group) and C—O—H (hydroxyl group) stretch in 2ME structure, respectively [10, 24, 25]. Upon reaction, the S—H groups in 2ME reacted with unsaturated bond —C═C— in carvone oil. This resulted in a —C—S— bond near 600–700 cm^{-1} in the carvone polyol. The disappearance of S—H and —C═C— groups and the appearance of —C—S- and C—O—H peaks confirm the polyol's formation as desired [26–29].

The FT-IR spectra for all the polyurethane foams containing different concentrations of ATH and AHP are shown in Fig. 3. During the synthesis of polyurethane foams, the

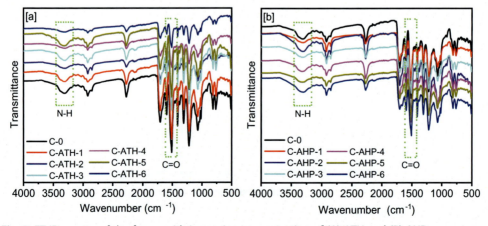

Fig. 3 FT-IR spectra of the foams with increasing concentration of (A) ATH and (B) AHP.

isocyanate group from MDI reacts with a hydroxyl group of polyols, resulting in a polyurethane group of —OC(O)NH— bonds. The bands near 1510 cm^{-1} represent hydrogen-bonded urethane groups (—OC(O)NH—), confirming the desired polyurethane reaction completion [30–32]. Small peaks were observed for N—H bond and C=O stretch around 3300 and 1700 cm^{-1}, respectively. These peaks correspond to the polyurea structure obtained from the reaction between distilled water and the MDI during the synthesis of PUFs [33]. The following subsections discuss some carvone-derived bio-based FR-RPUF and how those properties assist this polymer in enabling their versatile use in the automotive industry.

5.2 Density of polyurethane foams

Density is an important parameter attended to in most industries, including automobiles. PUs can aid in lightweighting cars and hence help improve vehicle's fuel efficiency. PUs can provide such lightweighting because of their low density enabled by their porous morphology. Generally, the improvement of performance and fuel consumption is provided by an average reduction of 10% of weight, which leads to a decrease of roughly 3%–7% of gas usage. A reduction of 100 kg in vehicle weight is comparable to reducing around 10 g of CO_2 emission per km. Polymeric composites represent 10% of the car's weight and 50% of its volume approximately. Composites are not as affordable, which challenges investigators in both academia and industry in finding simple, scalable synthetic routes and cheaper materials that can decrease the production cost. Most of these materials are polymeric composites based on polypropylene, polyesters, PU, and glass fibers and are used indoors, panels, roof, rear hatch, and others. An example of weight reduction by utilizing PU was demonstrated in Los Angeles Auto Show in 2010 that featured the Mazda Design Team. They designed a concept for the MX-0 model that had its weight reduced from 900 kg to less than 500 kg. This improvement was made possible by employing PU-based composites [8]. Practical solutions in lightweighting in part were achieved by understanding the morphology and cellular structure of RPUFs that influence the polyurethane foams' physical and mechanical properties. And how easy it is to manipulate PU's mechanical properties reflects in the vast area of applications inside the automobile industry.

As shown in Fig. 4, the addition of ATH and AHP showed a systematic increase in the foams' density. Although the increase in density was apparent, all the foams still maintained the average density range between 30 and 50 kg/m^3, a suitable range for the automobile industry's applications as bumpers or body parts. A significant increase in the ATH samples' density is observed because of the strong intermolecular interactions that the filler and the polyurethane foams experience. The hydroxyl groups in ATH can easily interact with urethane groups in PUFs via hydrogen bonding. This allows proper packing

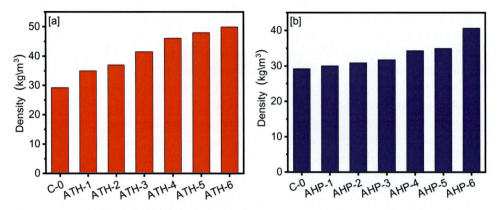

Fig. 4 Density of PU foams with (A) ATH and (B) AHP.

of the structure and better dispersion of ATH in the PUF, as demonstrated by SEM images (Figs. 5–6).

In the case of polyurethane foams containing AHP, the presence of flexible (O=P—O—) groups and the absence of hydroxyl groups seem to enable AHP to leave a plasticizing effect on the polyurethane foams. The plasticizing effect of AHP on the polyurethane foams can be confirmed by a noticeable reduction in the cells' wall thickness [34]. A contradictory relationship between the density and compressive strength of the foams is observed here.

5.3 Morphology and cellular characteristics of polyurethane foams

The SEM micrographs for the carvone-based RPUF containing different concentrations of ATH and AHP are provided in Figs. 5 and 6, respectively. For ATH foams, the pores' size decreased from 150 μm to around 80 μm with increased loading of ATH within the foams. A decrease in pore size while maintaining the cell wall thickness increases the amount of solid region within the polyurethane foam and therefore increases the apparent density of PUFs. The cell wall thickness for all the ATH containing foams was maintained around 5–8 μm. While in the case of AHP, the size of pores decreased from 150 to 50 μm as the amount of AHP increased. Here, with increasing concentration of AHP from 0.96 to 12.92 wt%, the cell wall thickness was slightly reduced from 5 to 3 μm, respectively. This behavior could result in a slight difference in the physical and mechanical properties of AHP foams when compared to ATH foams. In both cases, the cellular structure was preserved without much disruption. With the addition of ATH and AHP, a decrease in the cellular size and an increase in the foams' apparent density were observed.

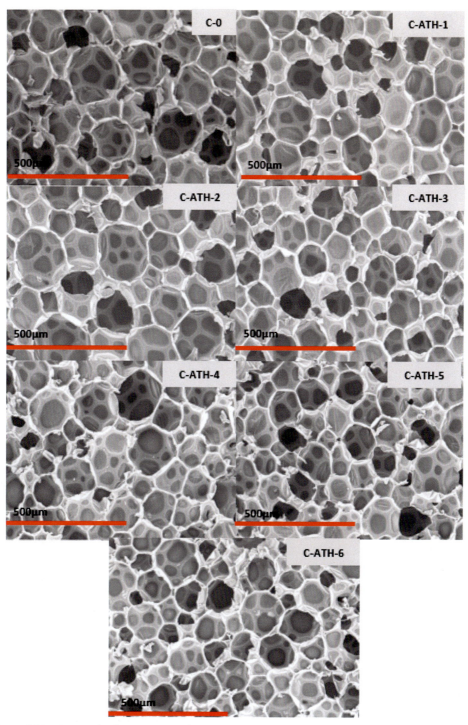

Fig. 5 SEM micrographs for neat PU and PU with an increasing amount of ATH.

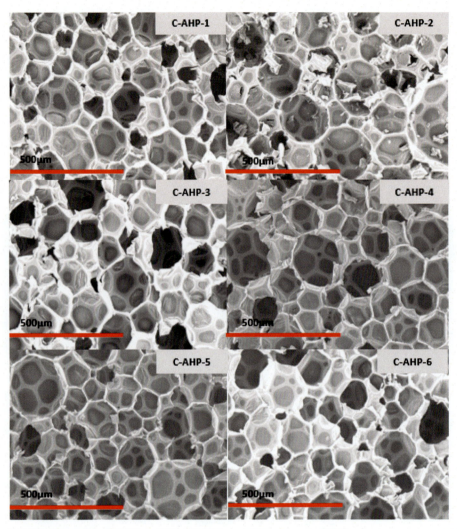

Fig. 6 SEM micrographs for PU and PU with an increasing amount of AHP.

5.4 Sound absorption behavior

To increase the customer base and to attract new customers, the automobile industry pays an incredible amount of attention to the noise, vibration, and harshness (NVH) reduction in the vehicle. Polyurethane foams are employed in car parts such as the engine, the wheels with the roads, and exterior airflow, which generally cause automotive-associated noise pollution. Some characteristics that allow sound absorption in polyurethane foams are their open cell content, tortuosity, and resistance to static airflow [35]. Previous reports have mentioned the use of carbon nanotubes (CNTs), polymer microparticles, nanosilica, nanoclay, and rice hull in polyurethane composites [36]. For instance, nanosilica and nanoclay are known to improve sound absorption at lower frequencies of 30–500 Hz, while rice hull improves it at higher frequencies of 500–8000 Hz. Utilizing composites of rice hull and nanosilica and clay can aid polyurethane foams in covering a broader range of frequencies in sound absorption.

Another report demonstrates a facile route in obtaining a polyurethane foam blended with magnesium hydroxide [37]. The magnesium hydroxide and the PU matrix exhibit superior compatibility as the hydroxyl groups in the magnesium hydroxide are free to interact with the urethane linkages. Such structural compatibility, in turn, improves the mechanical and thermal stability of the polyurethane foam without disrupting the integrity of the PU foam's pores. Sound absorption efficiencies in polyurethane foams and materials, in general, are affected by their porosity. The open porosity is determined by the relative number of pores of pore types like open, partially open, and closed. The types of pores impact sound wave penetration and material's resistance to airflow [38].

The acoustic properties of the PU foams measured demonstrated an improved sound absorption by increasing the concentration of magnesium hydroxide added. An optimum result was obtained with the addition of 1 wt% of the inorganic filler. The combined effect of the increase in the number of partially opened pores and the magnesium hydroxide damping effect in PU foams showcased an improved sound absorption coefficient. The results are demonstrated in Fig. 7. The concentration of just 1 wt% of magnesium hydroxide provided the most efficient results as the noise reduction coefficient was 0.53 [37]. Fig. 8 explains the schematics of the damping mechanism in the PU foams after the filler was added. Such results demonstrate the filler's role in damping the energy of the sound waves. In general, PU foams' sound pressure gets converted to vibrational energy in the cell walls and the air inside the cells. Before vibrational energy could accumulate and radiate more sound, fillers can aid in damping that energy and dissipating that into heat [39]. Incorporating magnesium hydroxide in PU foams proves an effective strategy for improving the material's sound absorption efficiencies. Such a facile green approach with potentially low production cost and high scalability makes this tactic an attractive route to producing efficient, sound-absorbing PU foams.

5.5 Mechanical properties

The automobile bumper is an essential component of a car placed in the front and rear sides of a vehicle to provide safety primordially. Thus, bumpers must be designed to properly absorb mechanical shocks to prevent the driver and passengers from injuries. When creating and testing vehicles, the use of suitable material, crashworthiness, aerodynamics, and functionality are analyzed. Polymers such as PU are appropriate materials for this application by presenting sufficient toughness to withstand mechanical shocks and enough flexibility to absorb the energy and prevent it from transferring to the driver or

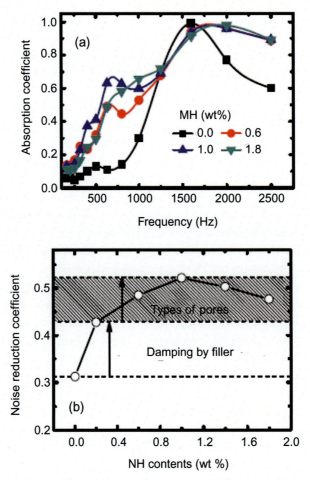

Fig. 7 (A) Increase of sound absorption in a broad range of frequencies for the PU foams with increasing concentration of magnesium hydroxide (B) Noise reduction coefficient in the addition of magnesium hydroxide expresses the range of different mechanisms for sound absorption through damping by filler and type of pores. *(Adapted with permission G. Sung, J.W. Kim, J.H. Kim, Fabrication of polyurethane composite foams with magnesium hydroxide filler for improved sound absorption, J. Ind. Eng. Chem. 44 (2016) 99–104. Copyright (2016) Elsevier).*

passengers. Polyurethanes can be processed through reaction injection molding (RIM), which enables its large-scale production. These polymers can attain different designs and provide a minimum of 0.5 GPa Young's modulus. However, since most of the PU used for shock absorption applications are thermosetting, remolding and recycling parts are not feasible. Therefore, studies are being carried out to develop more recyclable shock absorber PUs [40].

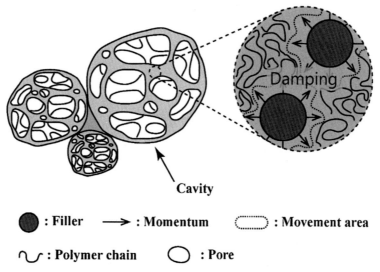

Fig. 8 Scheme for the mechanism of sound damping in the PU by adding inorganic filler. *(Adapted with permission G. Sung, J.W. Kim, J.H. Kim, Fabrication of polyurethane composite foams with magnesium hydroxide filler for improved sound absorption, J. Ind. Eng. Chem. 44 (2016) 99–104. Copyright (2016) Elsevier).*

A group of researchers selected high-performance flexible PU and rigid PU for the design of car bumpers. Small specimens of these two PU car bumpers were 3D printed and compared with two other commercial car bumpers for their performance [41]. Both the flexible and rigid PUs showcased superior performance when compared to the retail counterparts in terms of properties like Von Mises stresses and deflection and impact strength, among other properties. Compared to flexible and rigid PU alone, the latter surpassed the former in terms of deflection, strength, and modal frequencies. On the other way, flexible PU proved to possess higher impact strength. Such phenomena were expected because flexible materials can adequately absorb and dissipate mechanical impact. After all, the polymeric chains' increased mobility in the flexible polymer provides free volume for either linear or low cross-linking.

Additionally, flexible PUF is used for automobile parts such as seats providing users immense comfort. On the other hand, rigid PUs are highly cross-linked, providing them with high rigidity. These types of materials have excellent strength but can fracture if an impact is more robust than the impact materials can absorb. Rigid PUFs are suitable for their application in the vehicle's internal parts to improve safety and resistance to mechanical shock [41].

Our group analyzed the compressive strength for the bio-based rigid PUF blended with ATH and AHP individually by performing a similar study. The stress–strain curves for the compressive strength of the ATH and AHP foams are shown in Fig. 9. It was

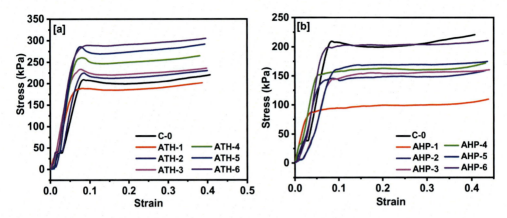

Fig. 9 Compression strength for (A) ATH and (B) AHP based rigid PU foams.

noticed that with increasing amounts of ATH from 0.96 to 16.92 wt%, the tensile strength required to deform the rigid PUF and induce 10% strain, increased from 210 to 290 kPa, respectively (Fig. 9A). This could be correlated to an increased cellular density of the foams while maintaining the cells' thickness. However, due to the plasticizing effect presented by flexible (O=P—O—) structure in AHP, the decrease in cell wall thickness adversely affected the compression strength of AHP containing foams (Fig. 9B). Even though there was a slight decrease in the AHP foams' compressive properties, the observed compression strength was still in an acceptable range for commercially applicable rigid PUF [42]. Furthermore, the effect of increasing FR concentration on the closed-cell content of PUFs was measured and shown in Fig. 10. All the FR foams containing AHP and ATH maintained the closed-cell structure >95%, which was higher than the neat sample, suggesting an advantageous property for better thermal insulations

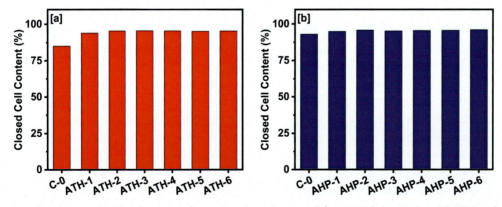

Fig. 10 Average closed-cell content for neat polyurethane and foams with increasing (A) ATH and (B) AHP.

[43, 44]. Although the addition of AHP and ATH presented notable effects on the foams' density and compression strength, no impact on the closed-cell structure of the foams was observed. This suggests that AHP and ATH can be used as useful FR additives for rigid PUF.

5.6 Flame-retardant behavior of polyurethane foams

The high porosity of PU is one reason why these materials present this variety of properties such as low density, high mechanical properties, sound, and thermal insulation. However, the porous structure increases its surface area, allowing the permeation of oxygen within it. This factor makes the PU's structure susceptible to fire, combined with the high carbon and oxygen availability in the polymeric chains that can facilitate combustion reactions. Hence, chemical or physical modification can be performed to revert this scenario. The cellular structure properties directly affect PU's fire susceptibility. Therefore, a high closed-cell structure is desirable to enhance PU's resistance to fire. Based on that, rigid PUFs generally present a high closed-cell structure, preventing the diffusion of air through its structure and making these materials one of the best thermal insulators. Although rigid PUF provides high mechanical properties, their porous morphology and chemical nature engender poor flame-retardant properties [45].

A small fire spark poses a fire hazard around materials like PUs, reported as the primary cause of fire incidents in the United States by the National Fire Protection Association [11, 12, 14]. Hence, fire propagation in cars can quickly escalate into life-threatening situations as the flammable plastics, including PU, can release toxic smoke and trigger explosions. To prevent such issues, additive or reactive type of FR compounds are added during the foams' synthesis process [13, 14]. The addition of FR can suppress fire in two ways: physical and chemical actions [46]. The physical activity involves FR to undergo an endothermic reaction producing stable thermal by-products such as CO_2, H_2O, or NH_3, which dilutes the oxygen in the environment by forming gases that surround the flame. A protective solid layer is formed due to the thermal degradation, which isolates the material's flammable surface. On the other hand, the chemical reaction involves releasing reactive radicals such as $Cl·$ and $Br·$ to react with $H·$ and $OH·$ species in flame to produce inert molecules that can prevent a highly exothermic reaction.

Out of two general types of FR, physical FR and reactive FR, physical FRs are preferred because of better flame retardancy and smoke safety. For example, aluminum trihydroxide and aluminum hypophosphite can be used as physically active FR. Upon combustion, they promote a char layer's formation on top of the PU and produce radicals that can absorb flammable gases from the surrounding like oxygen [47, 48]. On top of that, both the aluminum-based FRs are environmentally friendly. And unlike halogen-based FRs, they do not release toxic smoke during their thermal decomposition [49, 50].

Fire incidents in closed spaces are hazardous, like in the interior of an automobile. Therefore, a quick response to quenching the fire as soon as it appears is vital to improving a vehicle's safety [51]. Another issue that requires attention in the case of a fire incident is the release of toxic smoke from materials' combustion. Reports of the significant cause of death in fire incidents point to people inhaling toxic fumes [11, 12, 14]. A previous report showed that high quantities of two FR materials known as 2,2-bis(chloromethyl)-propane-1,3-diyltetrakis(2-chloroethyl) biphosphate, named as V6 and tris (2-chloroethyl) phosphate (TCEP), were found mainly in automobile foams, making about 4.6 wt% of PU foam's mass [52]. Despite being effective FRs, these materials are carcinogenic and are a major health hazard. Hence, using nontoxic and eco-friendly FRs is a necessity for the vehicle components. The chemical structure of both V6 and TCEP are given in Fig. 11. Industries and academia alike are in search of low-cost, safe, and effective FRs. New PUF that utilizes natural resources such as essential extracts from seeds, leaves, and roots is under development and highly anticipated for the automotive and transportation industry [12, 53].

To understand the thermal behavior and the mechanism of action of eco-friendly FR, our group performed a detailed study on the thermal decomposition of the FR-RPUF containing ATH and AHP through TGA and burning tests. Hence, the thermal characteristics of all the foams containing ATH and AHP were analyzed using TGA and DTGA

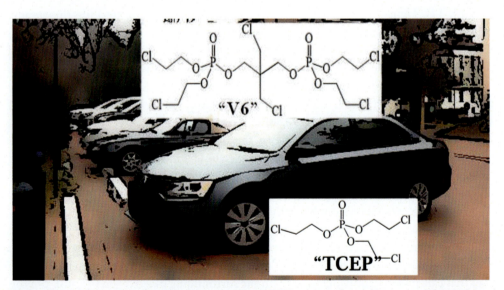

Fig. 11 Chemical structure of two FRs known as V6 and TCEP commonly used in automobiles. *(Adapted with permission M. Fang, T.F. Webster TF, G. Gooden, E.M. Cooper, M.D. McClean, C. Carignan, C. Makey, H.M. Stapleton, Investigating a novel flame retardant known as V6: measurements in baby products, house dust, and Car dust, Environ. Sci. Technol. 47 (2013) 4449–4454. Copyright (2013) American Chemical Society).*

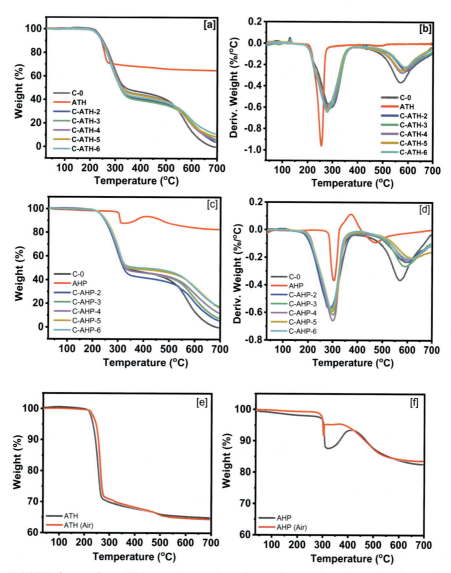

Fig. 12 (A) TGA for ATH foam, (B) D-TGA for ATH foam, (C) TGA for AHP foam, (D) D-TGA for AHP foams, (E) TGA for ATH in N_2 and air, and (F) TGA for AHP in N_2 and air.

analysis, shown in Fig. 12. As observed from TGA and DTGA peaks in Fig. 12A, B, the decomposition of ATH (Al(OH)$_3$) occurs at 260°C resulting in the release of water and formation of alumina (Al$_2$O$_3$) char. All the foams containing ATH showed higher weight loss near 260–300°C, corresponding to the decomposition of ATH by forming a char, which prevents further weight loss during the second degradation stage near 550°C. To understand this behavior in detail, we performed FT-IR analysis for ATH and ATH-6 foam, before and after TGA analysis described in Fig. 13A. Based on that, a peak near 3500 cm^{-1} for ATH before TGA analysis indicates O—H groups' presence along with

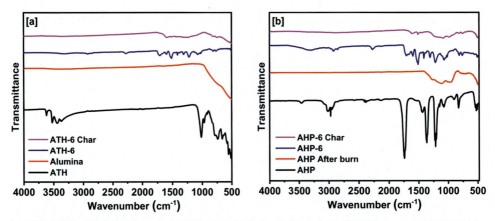

Fig. 13 FT-IR for (A) ATH foams and ATH before and after TGA analysis, (B) AHP foams and AHP before and after TGA analysis.

two peaks around 1080 and 510 cm^{-1} relating to the Al—O stretch [54]. After the TGA analysis, the ATH converted into alumina (Al_2O_3), resulting in the disappearance of the O—H stretch bond. This behavior was in continuance with the ATH-6 sample and could be correlated to the increase of char formation.

Unlike ATH, AHP containing foams showed improved weight loss during the first phase of decomposition near 280–300°C. The increase in weight loss corresponds to a higher decomposition temperature of AHP near 310°C observed from TGA and DTGA curves in Fig. 12C, D. According to the previous reports, AHP presents two thermal transitions around 310°C and 450°C [55, 56]. To complement the thermal decomposition behavior for AHP, we performed the FT-IR analysis of AHP and AHP-6 foams before and after the TGA analysis. As shown in Fig. 13B, the decomposition of AHP resulted in the formation of $Al_4(P_2O_7)_3$. This is supported by the presence of P-O-P stretch near 1120 cm^{-1}, which is also present in the char of the AHP-6 foam after TGA analysis [55, 57]. This char formation increased the thermal stability of the foams.

To understand the difference in oxygen environment during the TGA analysis, we performed TGA in N_2 and air (Fig. 12E). No noticeable changes were observed in weight loss curves for ATH during TGA analysis in the air compared to the N_2 atmosphere. However, the reaction mechanism differs in the case of AHP samples. The decomposition of AHP occurs near 310°C resulting in slight fluctuations in TGA curves performed under the N_2 atmosphere. The O_2 in the air reacts with AHP to form phosphoric acid, which causes noticeable absorption of the surrounding O_2 during the decomposition process of AHP and results in negative weight loss after 310°C (Fig. 12F) [58, 59]. The resulting phosphoric acid allows accelerated char formation in foams resulting in increased char weight during the TGA analysis [55, 60, 61].

The flame retardancy of the RPUFs was determined using a horizontal burning test. All the foams were subjected to an open flame for 10 s. The resulting self-extinguishing time, weight loss, and optical images of the burnt samples were recorded (Figs. 14, 15). As shown in Fig. 14, the foams containing ATH with increasing concentration showed a systematic reduction in weight loss during the flammability test. However, there was slight unevenness in self-extinguishing time with an increasing concentration of ATH. This could be due to the solid phase reaction of ATH, as proved by our TGA analysis. For ATH to act as an FR, it needs to decompose and form a protective char layer on the surface to prevent further combustion beneath the surface.

To further improve the flame retardancy, both solid- and gas-phase combustion reactions need to be inhibited. As presented by our TGA and FT-IR analysis, the flame retardancy in AHP foams in the air can be achieved due to the decomposition of AHP by absorbing the surrounding oxygen to form phosphoric acid that promoted the formation

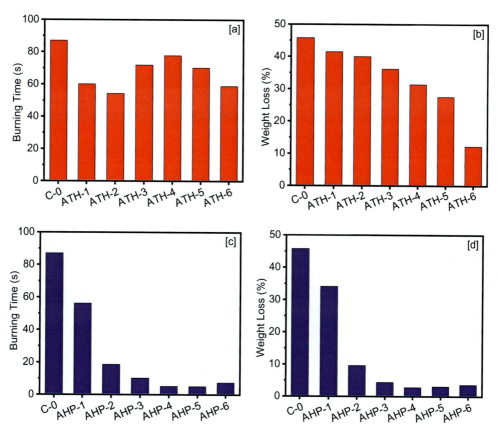

Fig. 14 Histogram graphs for (A) burning time, (B) weight loss percentage for ATH foams, (C) burning time, and (D) weight loss percentage for AHP foams.

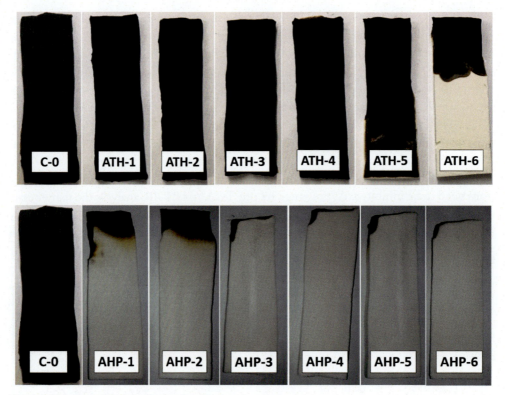

Fig. 15 Photographs of the foam after the burning test.

of an insulating char [9, 10, 12–14, 62]. Such a mechanism results in preventing the gaseous phase and solid-phase reactions resulting in significantly lower self-extinguishing time and weight loss of the AHP containing foams (Fig. 14C, D). At a very low concentration of AHP (5.54 wt%), the resulting self-extinguishing time and weight losses were 10 s and 4.39%, respectively. Similar behavior can be observed from Fig. 15, where increasing AHP concentration in foams enhances the flame retardancy and reduces the burnt area of the foam. Our results confirm that the synergistic flame retardancy mechanism controlled by solid- and gas-phase reactions can prove efficiency to improve overall flame retardancy in PUFs.

5.7 Coatings for automobiles

Composite coatings derived from PU hold great importance in the automobile industry. These coatings are responsible for providing materials with durability by improving their corrosion and scratch resistance. Coatings also make vehicles more appealing to customers with the gloss effect of paint and color retention [63, 64]. The current challenge of developing materials with a low cost and superior properties has propelled scientists to utilize raw materials abundant in nature and readily available. With that motive, a study looked at anticorrosion PU coating with CO_2 as a sustainable precursor [65]. The carbonate group present in the CO_2-based polyol introduced rigidity to the overall structure. The CO_2-based, polyol-based hard coating showcased excellent properties such as chemical stability in corrosive media, high glass transition temperature (Tg), and self-healing properties. These desirable characteristics make such material fit for automobiles because vehicles are regularly exposed to corrosive environments, extreme temperature, and mechanical friction.

The synthetic procedures involved in obtaining thermoplastic coatings are explained in Fig. 16 [65]. The mechanical properties measured through tensile strength (Fig. 16 left) showed that when CO_2 was used in a high amount in the formulation, the coating

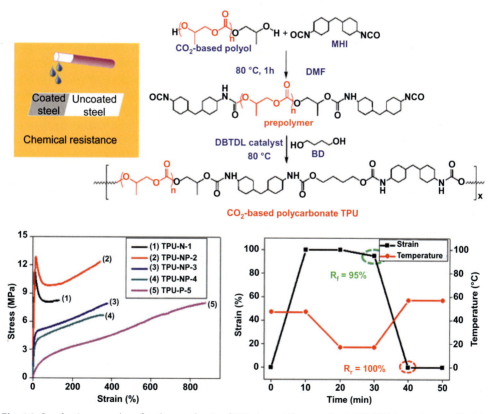

Fig. 16 Synthetic procedure for the synthesis of CO_2-based thermoplastic PU (TPU) coating applicable for automobile body protection. Tensile strength (left) and strain–time–temperature diagram. (Adapted with permission P. Alagi, R. Ghorpade, Y.J. Choi, U. Patil, I. Kim, J.H. Baik, S.C. Hong, Carbon dioxide-based polyols as sustainable feedstock of thermoplastic polyurethane for corrosion-resistant metal coating, ACS Sustain. Chem. Eng. 5 (2017) 3871–3881. Copyright (2017) American Chemical Society).

became brittle. The C=O bond contributes such brittleness because of its high rigidity property. However, by introducing poly(tetrahydrofuran) diol (PTMEG) and higher amounts of the chain extender, 1,4-butanediol, the coating became more rigid and less brittle. On the other hand, the shape memory properties for the CO_2-based PU coating (Fig. 16 right) showed 100% effective healing after heating and applied strain cycles. This exhibits satisfactory reversibility. The effect likely occurred because of the decrease of soft segments in the polymeric backbone.

6. Conclusion

Opportunities in exploring and testing cost-effective materials of superior properties propel numerous researches in different sectors of research. The automotive and transportation sector is slowly adopting bio-derived PU in the vehicle's internal components, bumpers, cushioning, seats, and thermal insulators as a sustainable alternative approach. These polymers' versatility encompasses properties such as low density accompanied by mechanical rigidity or flexibility and thermal and sound insulation with the introduction of flame-retardant chemicals into the polymer. This set of properties improves the overall efficiency, reduces fuel consumption, and improves comfort and safety. For successful large-scale production of high-value materials, cost reduction and simplified processibility are required. The addition of fillers can help reduce the cost and enhance several other properties such as mechanical strength, flexibility, sound dissipation, and flame retardancy. All in all, smart utilization of bio-based materials and fillers in PU can provide the automobile industry with growth opportunities.

Acknowledgment

Dr. Ram K. Gupta manifests his sincere gratefulness for the Polymer Chemistry Program and Kansas Polymer Research Center, and Pittsburg State University for providing financial and research support.

References

[1] T.C. Moore, A.B. Lovins, Vehicle design strategies to meet and exceed PNGV goals, SAE Trans. 104 (1995) 2676–2718.
[2] W. Taylor, W. Kellermeyer, R. Napier, B. Ralston, M. Snoberger, C.M. Atkinson, Development and Testing of a Second Generation ULEV Series HEV at West Virginia University, 1998.
[3] M. Szycher, Szycher's Handbook Ofpolyurethanes, CRC Press, 2012.
[4] C.K. Ranaweera, M. Ionescu, N. Bilic, X. Wan, P.K. Kahol, R.K. Gupta, Biobased polyols using thiol-Ene chemistry for rigid polyurethane foams with enhanced flame-retardant properties, J. Renew. Mater. 5 (2017) 1.
[5] J. Feng, B. Huang, M. Zhong, Fabrication of superhydrophobic and heat-insulating antimony doped tin oxide/polyurethane films by cast replica micromolding, J. Colloid Interface Sci. 336 (2009) 268–272.

[6] H. Yeganeh, M. Razavi-Nouri, M. Ghaffari, Investigation of thermal, mechanical, and electrical properties of novel polyurethanes/high molecular weight polybenzoxazine blends, Polym. Adv. Technol. 19 (2008) 1024–1032.

[7] A. Le Duigou, J.-M. Deux, P. Davies, C. Baley, PLLA/flax Mat/Balsa bio-Sandwich—environmental impact and simplified life cycle analysis, Appl. Compos. Mater. 19 (2012) 363–378.

[8] A. Hăloiu, D. Iosif, Bio-source composite materials used in automotive industry, Sci. Bull. Automot. Ser. 24 (2013) 57–61.

[9] S. Bhoyate, M. Ionescu, P.K. Kahol, J. Chen, S.R. Mishra, R.K. Gupta, Highly flame-retardant polyurethane foam based on reactive phosphorus polyol and limonene-based polyol, J. Appl. Polym. Sci. 135 (2018) 16–19.

[10] C. Zhang, S. Bhoyate, M. Ionescu, P.K. Kahol, R.K. Gupta, Highly flame retardant and bio-based rigid polyurethane foams derived from orange peel oil, Polym. Eng. Sci. 58 (2018) 2078–2087.

[11] S. Bhoyate, M. Ionescu, P.K. Kahol, R.K. Gupta, Castor-oil derived nonhalogenated reactive flame-retardant-based polyurethane foams with significant reduced heat release rate, J. Appl. Polym. Sci. (2018) 47276.

[12] S. Bhoyate, M. Ionescu, P.K. Kahol, R.K. Gupta, Sustainable flame-retardant polyurethanes using renewable resources, Ind. Crop. Prod. 123 (2018) 480–488.

[13] S. Ramanujam, C. Zequine, S. Bhoyate, B. Neria, P. Kahol, R. Gupta, Novel biobased polyol using corn oil for highly flame-retardant polyurethane foams, C—J. Carbon. Res. 5 (2019) 13.

[14] S. Bhoyate, M. Ionescu, D. Radojcic, P.K. Kahol, J. Chen, S.R. Mishra, R.K. Gupta, Highly flame-retardant bio-based polyurethanes using novel reactive polyols, J. Appl. Polym. Sci. 135 (2018) 46027.

[15] H.C. Kolb, M.G. Finn, K.B. Sharpless, Click chemistry: diverse chemical function from a few good reactions, Angew. Chemie.—Int. Ed. 40 (2001) 2004–2021.

[16] B. Zhao, Z. Hu, L. Chen, Y. Liu, Y. Liu, Y.-Z. Wang, A phosphorus-containing inorganic compound as an effective flame retardant for glass-fiber-reinforced polyamide 6, J. Appl. Polym. Sci. 119 (2011) 2379–2385.

[17] P. Noisong, C. Danvirutai, A new synthetic route, characterization and vibrational studies of manganese hypophosphite monohydrate at ambient temperature, Spectrochim. Acta A Mol. Biomol. Spectrosc. 77 (2010) 890–894.

[18] K. Wilpiszewska, T. Spychaj, W. Paździoch, Carboxymethyl starch/montmorillonite composite microparticles: properties and controlled release of isoproturon, Carbohydr. Polym. 136 (2016) 101–106.

[19] G. Tang, X. Wang, R. Zhang, W. Yang, Y. Hu, L. Song, X. Gong, Facile synthesis of lanthanum hypophosphite and its application in glass-fiber reinforced polyamide 6 as a novel flame retardant, Compos. Part A Appl. Sci. Manuf. 54 (2013) 1–9.

[20] M. González Gómez, S. Belderbos, S. Yáñez, Y. Piñeiro, F. Cleeren, G. Bormans, C. Deroose, W. Gsell, U. Himmelreich, J. Rivas, Development of superparamagnetic nanoparticles coated with Polyacrylic acid and aluminum hydroxide as an efficient contrast agent for multimodal imaging, Nano 9 (2019) 1626.

[21] F.M. de Souza, J. Choi, S. Bhoyate, P.K. Kahol, R.K. Gupta, Expendable graphite as an efficient flame-retardant for novel partial bio-based rigid polyurethane foams, C—J. Carbon. Res. 6 (2020) 27.

[22] M.A. Carey, S.L. Wellons, D.K. Elder, Rapid method for measuring the hydroxyl content of polyurethane polyols, J. Cell. Plast. 20 (1984) 42–48.

[23] H. Liu, H. Chung, Visible-light induced thiol-Ene reaction on natural lignin, ACS Sustain. Chem. Eng. 5 (2017) 9160–9168.

[24] M. Desroches, S. Caillol, V. Lapinte, R.M. Auvergne, B. Boutevin, Synthesis of biobased polyols by thiol–Ene coupling from vegetable oils, Macromolecules 44 (2011) 2489.

[25] Z. Yang, Y. Feng, H. Liang, Z. Yang, T. Yuan, Y. Luo, P. Li, C. Zhang, A solvent-free and scalable method to prepare soybean-oil-based polyols by thiol-Ene photo-click reaction and biobased polyurethanes therefrom, ACS Sustain. Chem. Eng. 5 (2017) 7365–7373.

[26] L. Zhang, M. Zhang, L. Hu, Y. Zhou, Synthesis of rigid polyurethane foams with castor oil-based flame retardant polyols, Ind. Crop. Prod. 52 (2014) 380–388.

[27] M.J.B. Kade, D.J. Burke, C.J. Hawker, The power of thiol-ene chemistry, J. Polym. Sci., Part A Polym. Chem. 48 (2010) 743.

[28] C.E. Hoyle, C.N. Bowman, Thiol-ene click chemistry, Angew. Chem. Int. Ed. 49 (2010) 1540.
[29] N. Elbers, C.K. Ranaweera, M. Ionescu, X. Wan, P.K. Kahol, R.K. Gupta, Synthesis of novel bio-based polyol via thiol-Ene chemistry for rigid polyurethane foams, J. Renew. Mater. 5 (2017) 74–83.
[30] S. Adnan, T.N.M. Tuan Ismail, N. Mohd Noor, N.S.M. Nek Mat Din, N. Hanzah, Y. Shoot Kian, H. Abu Hassan, Development of flexible polyurethane nanostructured biocomposite foams derived from palm olein-based polyol, Adv. Mater. Sci. Eng. (2016) 2016.
[31] T.-K. Chen, Y.-I. Tien, K.-H. Wei, Synthesis and characterization of novel segmented polyurethane/clay nanocomposites, Polymer (Guildf) 41 (2000) 1345–1353.
[32] S. Radice, S. Turri, M. Scicchitano, Fourier transform infrared studies on deblocking and crosslinking mechanisms of some fluorine containing monocomponent polyurethanes, Appl. Spectrosc. 58 (2004) 535–542.
[33] B. Czupryński, J. Paciorek-Sadowska, J. Liszkowska, Modifications of the rigid polyurethane—polyisocyanurate foams, J. Appl. Polym. Sci. 100 (2006) 2020–2029.
[34] C. Formicola, A. de Fenzo, M. Zarrelli, A. Frach, M. Giordano, G. Camino, Synergistic effects of zinc borate and aluminium trihydroxide on flammability behaviour of aerospace epoxy system, Express Polym. Lett. 3 (2009) 376–384.
[35] D.L. Johnson, J. Koplik, R. Dashen, Theory of dynamic permeability and tortuosity in fluid-saturated porous media, J. Fluid Mech. 176 (1987) 379–402.
[36] Y. Wang, C. Zhang, L. Ren, M. Ichchou, M.-A. Galland, O. Bareille, Influences of rice hull in polyurethane foam on its sound absorption characteristics, Polym. Compos. 34 (2013) 1847–1855.
[37] G. Sung, J.W. Kim, J.H. Kim, Fabrication of polyurethane composite foams with magnesium hydroxide filler for improved sound absorption, J. Ind. Eng. Chem. 44 (2016) 99–104.
[38] N. ATALLA, R. PANNETON, F.C. SGARD, X. OLNY, Acoustic absorption of macro-perforated porous materials, J. Sound Vib. 243 (2001) 659–678.
[39] S. Chen, Y. Jiang, The acoustic property study of polyurethane foam with addition of bamboo leaves particles, Polym. Compos. 39 (2018) 1370–1381.
[40] J.O. Akindoyo, M.D.H. Beg, S. Ghazali, M.R. Islam, N. Jeyaratnam, A.R. Yuvaraj, Polyurethane types, synthesis and applications-a review, RSC Adv. 6 (2016) 114453–114482.
[41] S. Begum, G.M.S. Ahmed, I.A. Badruddin, V. Tirth, A. Algahtani, Analysis of digital light synthesis based flexible and rigid polyurethane for applications in automobile bumpers, Mater. Express 9 (2019) 839–850.
[42] S. Sérgio Augusto Mello Da, C. André Luis, G. Raquel, L. Francisco Antonio Rocco, Strength properties of medium density fiberboards (MDF) manufactured with Pinus Elliottii wood and polyurethane resin derived from Castor oil, Int. J Compos. Mater. 3 (2013) 7–14.
[43] F. Feng, L. Qian, The flame retardant behaviors and synergistic effect of expandable graphite and dimethyl methylphosphonate in rigid polyurethane foams, Polym. Compos. 35 (2014) 301.
[44] R.K. Gupta, M. Ionescu, D. Radojcic, X. Wan, Z.S. Petrovic, Novel renewable polyols based on limonene for rigid polyurethane foams, J. Polym. Environ. 22 (2014) 304–309.
[45] Z. Petrovic, Polyurethanes from vegetable oils, Polym. Rev. 48 (2008) 109.
[46] F. Laoutid, L. Bonnaud, M. Alexandre, J.M. Lopez-Cuesta, P. Dubois, New prospects in flame retardant polymer materials: from fundamentals to nanocomposites, Mater. Sci. Eng. R. Rep. 63 (2009) 100–125.
[47] F. Luo, K. Wu, M. Lu, L. Yang, J. Shi, Surface modification of aluminum hypophosphite and its application for polyurethane foam composites, J. Therm. Anal. Calorim. 129 (2017) 767–775.
[48] Z.S. Xu, L. Yan, L. Chen, Synergistic flame retardant effects between aluminum hydroxide and halogen-free flame retardants in high density polyethylene composites, Procedia Eng. 135 (2016) 631–636.
[49] Y. Liu, J. He, R. Yang, Effects of dimethyl Methylphosphonate, aluminum hydroxide, ammonium polyphosphate, and expandable graphite on the flame Retardancy and thermal properties of Polyisocyanurate-polyurethane foams, Ind. Eng. Chem. Res. 54 (2015) 5876–5884.
[50] M. Modesti, A. Lorenzetti, Improvement on fire behaviour of water blown PIR–PUR foams: use of an halogen-free flame retardant, Eur. Polym. J. 39 (2003) 263–268.
[51] J. Troitzsch, Flame Retardants, Kunststoffe—Ger. Plast. 77 (1987) 90–91.

[52] M. Fang, T.F. Webster, D. Gooden, E.M. Cooper, M.D. McClean, C. Carignan, C. Makey, H.M. Stapleton, Investigating a novel flame retardant known as V6: measurements in baby products, house dust, and Car dust, Environ. Sci. Technol. 47 (2013) 4449–4454.

[53] A. Gandini, Polymers from renewable resources: a challenge for the future of macromolecular materials, Macromolecules 41 (2008) 9491–9504.

[54] M.A. González-Gómez, S. Belderbos, S. Yañez-Vilar, Y. Piñeiro, F. Cleeren, G. Bormans, C.M. Deroose, W. Gsell, U. Himmelreich, J. Rivas, Development of superparamagnetic nanoparticles coated with polyacrylic acid and aluminum hydroxide as an efficient contrast agent for multimodal imaging, Nano 9 (2019) 1–20.

[55] B. Yuan, C. Bao, Y. Guo, L. Song, K.M. Liew, Y. Hu, Preparation and characterization of flame-retardant aluminum hypophosphite/poly(vinyl alcohol) composite, Ind. Eng. Chem. Res. 51 (2012) 14065–14075.

[56] Y.W. Yan, J.Q. Huang, Y.H. Guan, K. Shang, R.K. Jian, Y.Z. Wang, Flame retardance and thermal degradation mechanism of polystyrene modified with aluminum hypophosphite, Polym. Degrad. Stab. 99 (2014) 35–42.

[57] X. Qian, L. Song, Y. Hu, R.K.K. Yuen, L. Chen, Y. Guo, N. Hong, S. Jiang, Combustion and thermal degradation mechanism of a novel intumescent flame retardant for epoxy acrylate containing phosphorus and nitrogen, Ind. Eng. Chem. Res. 50 (2011) 1881–1892.

[58] H. Zhang, J. Lu, H. Yang, J. Lang, H. Yang, Comparative study on the flame-retardant properties and mechanical properties of PA66 with different dicyclohexyl hypophosphite acid metal salts, Polymers (Basel) 11 (2019).

[59] X. Cheng, J. Wu, C. Yao, G. Yang, Aluminum hypophosphite and aluminum phenylphosphinate: a comprehensive comparison of chemical interaction during pyrolysis in flame-retarded glass-fiber-reinforced polyamide 6, J. Fire Sci. 37 (2019) 193–212.

[60] S. Wu, D. Deng, L. Zhou, P. Zhang, G. Tang, Flame retardancy and thermal degradation of rigid polyurethane foams composites based on aluminum hypophosphite, Mater. Res. Express 6 (2019).

[61] Y. Yuan, C. Ma, Y. Shi, L. Song, Y. Hu, W. Hu, Highly-efficient reinforcement and flame retardancy of rigid polyurethane foam with phosphorus-containing additive and nitrogen-containing compound, Mater. Chem. Phys. 211 (2018) 42–53.

[62] S. Bhoyate, M. Ionescu, P.K. Kahol, R.K. Gupta, Castor-oil derived nonhalogenated reactive flame-retardant-based polyurethane foams with significant reduced heat release rate, J. Appl. Polym. Sci. 136 (2019) 1–7.

[63] D.K. Chattopadhyay, K. Raju, Structural engineering of polyurethane coatings for high performance applications, Prog. Polym. Sci. 32 (2007) 352–418.

[64] A. König, A. Malek, U. Fehrenbacher, G. Brunklaus, M. Wilhelm, T. Hirth, Silane-functionalized flame-retardant aluminum Trihydroxide in flexible polyurethane foam, J. Cell. Plast. 46 (2010) 395–413.

[65] P. Alagi, R. Ghorpade, Y.J. Choi, U. Patil, I. Kim, J.H. Baik, S.C. Hong, Carbon dioxide-based polyols as sustainable feedstock of thermoplastic polyurethane for corrosion-resistant metal coating, ACS Sustain. Chem. Eng. 5 (2017) 3871–3881.

CHAPTER 7

Carbon–carbon nanocomposites for brake systems and exhaust nozzles

Mehmet İskender Özsoy[a,b], Serbülent Türk[b,c], Fehim Fındık[b,d], and Mahmut Özacar[b,e]

[a]Faculty of Engineering, Mechanical Engineering Department, Sakarya University, Sakarya, Turkey
[b]Biomaterials, Energy, Photocatalysis, Enzyme Technology, Nano & Advanced Materials, Additive Manufacturing, Environmental Applications and Sustainability Research & Development Group (BIOENAMS R & D Group), Sakarya University, Sakarya, Turkey
[c]Biomedical, Magnetic and Semiconductor Materials Application & Research Center (BIMAS-RC), Sakarya University, Sakarya, Turkey
[d]Faculty of Technology, Department of Metallurgical and Materials Engineering, Sakarya University of Applied Sciences, Sakarya, Turkey
[e]Faculty of Science & Arts, Department of Chemistry, Sakarya University, Sakarya, Turkey

Chapter outline

1. Introduction — 131
2. Carbon–carbon nanocomposites (CCNCs) — 134
   2.1 CCNCs production methods — 135
3. Properties of CCNCs — 136
4. Application area of CCNCs — 137
5. Technologies needed to advance CCNCs — 138
6. CCNCs for brake systems — 142
7. CCNCs for exhaust nozzles — 147
8. Conclusion and future perspectives — 151
References — 151

1. Introduction

Carbon–carbon nanocomposites (CCNCs) are a material family consisting of carbon-based matrices reinforced with carbon (or graphite) fibers. CCNCs show variety from simple, one-way, fiber-reinforced structures to complex 3D woven structures. Composite contains two components, matrix and reinforcement. The reinforcing element can be either fiber, particle, or layered. However, in CCNCs, fibers are most commonly used as a reinforcing element and are called CFRP, they also provide the strength of the composites. Matrix is an epoxy or similar polymer resin often used to bind reinforcements together.

Carbon fiber (CF) was discovered when Edison came up with the idea of carbonizing an ordinary cotton thread loop that glowed in a vacuum in 1879, after experimenting with more than 1600 types of materials. Carbon filaments were then produced by dissolving cellulosic materials in zinc chloride solvent, passing from a mold to a bath containing liquid, and regenerating the cellulose as a thread or filament. However, in the production of composite materials, CF began to be widely used again about 30 years ago as a miracle material [1].

The application of CF was limited to the production of very small quantities until the 1970s and to utilization only in large-budget aviation and sports equipment. Since the market volume and production capacity are two important factors that directly affect the product cost, prices have decreased after the manufacturers increased their CF production capacity and their widespread use in other areas such as machinery and automobiles [2]. Therefore, CF prices have been dropped remarkably over the past two decades [3].

After the exploration of carbon–carbon composites with high strength and thermal resistance by Brennan Chance Vought Aircraft in 1958, the use of these basic materials in heat preservation devices generated great opportunities [4]. Rayon carbon fabric-reinforced phenolic composites, which are developed by taking advantage of the rayon fabric's fallen thermal conductivity and the phenolic resin's lofty char efficiency, are widely utilized in thermal preservation systems. Carbon phenolic composites, which form a thinner ablative composite structure, generally show better ablation resistance and better load and fuel efficiency due to their constantly increasing ablative properties [5]. The Space Shuttle Columbia catastrophe occurred on February 1, 2003, owing to insufficient impact resistance to air of the thermal insulation foam in the outer tank during the spacecraft reentered the Earth's planetary atmospheric space. After the unfortunate accident that occurred as a result of the ruptured reinforcement foam damaging the left reinforced carbon–carbon panels of Columbia, the way of detailed researches has been opened to increase the fracture toughness, impact tolerance, and thermal resistance of the reinforced carbon–carbon panels [6]. Nanocomposites can withstand the together actions of mechanical impact loads and thermal stresses. The use of different nanoparticles including montmorillonite, nanosilica, polyhedral oligomeric silsesquioxane, and nanoclay, and surface modification act as thermal insulating elements to improve char layer toughness and integrity. The three-phase composite system consisting of a nonhomogeneous composition of fiber reinforcement, matrix, and nanofillers displays complex ablation behavior [7].

CCNCs stand out as materials that can be used with the best performance under extreme temperature conditions. CCNCs applications range from biomedical devices to brake pads and rocket exhaust nozzles. Due to such wide range of usage areas, the increasing worldwide interest in CCNCs has led to the development of basic carbon science. Also, it has created important literature in the articles about CCNCs published in technical and scientific periodicals. This information needs to be examined comprehensively and systematically to inspire new applications of CCNCs in brake pads and rocket exhaust nozzles and enable further research. Therefore, more extensive information is needed on the current state of CFs and CCNCs researches and forecasts for the other "high-tech" materials and their applications.

In CCNCs, CFs are frequently utilized as the primary carbon component. The gap amid CFs is occupied by impregnating carbon raw material, mainly in gaseous or liquid form. Of these, the liquid overfilling technique is more widely used. While liquid-phase filling technique is used in the fabrication of thick composites, gas-phase filling method is preferred in the production of thin composites. Properties such as viscosity, carbon content as a result of carbonization, matrix microstructure, and matrix crystal structure should be considered in the choosing of raw materials.

The factors to be considered in the selection of the thermoset resin required for densification of CCNCs can be summarized as follows:
- Carbon formation is 50%–70% of the resin weight. Experimental data show that this amount does not increase with the pressure application through carbonization.
- The carbon matrix structure is glassy and graphitization does not occur up to 3000°C.
- The stresses applied during the heat treatment can cause the matrix to graphitize [8].

CCNCs are much stronger than graphite, and their three-dimensional structures further enhances CCNCs' properties. Therefore, CCNCs are very gorgeous for high-temperature using as well as brake systems and exhaust nozzles. This chapter is written to give the novel knowledge and the general idea about the use of CCNCs in brake systems and exhaust nozzles using books and recent journal papers. This chapter contains eight subheadings including introduction, carbon–carbon nanocomposites (CCNCs), properties of CCNCs, application area of CCNCs, technologies needed to advance CCNCs, CCNCs for brake systems, CCNCs for exhaust nozzles, and conclusion and future perspectives.

In this chapter, the historical development of CCNCs as well as the requirements and the selection criteria for those materials are introduced. The classifications are done for CCNCs, and production methods are highlighted. The properties such as thermal properties, density and composition, microstructures, elemental composition, specific heat, optical properties, porosity, and permeability needed to advance CCNCs are briefly explained. The expected properties from the brake systems as well as test techniques are explained. Also, frictional properties of the CCNCs and oxidation protection of CCNCs are distinguished. Working principles and types of brake systems and used materials are described. The anticipated properties of CCNCs for exhaust nozzles are highlighted. Also, the types of exhaust nozzle systems used materials, and the test techniques are explained. Oxidation protection and thermo-oxidative studies are clarified for better performance. Then, working principles of exhaust systems are illuminated. Finally, the chapter has been completed by briefly discussing the results drawn and future perspectives.

2. Carbon–carbon nanocomposites (CCNCs)

There are three categories of CCNCs presently used as mats in braking systems: carbon fabric laminates, semirandomly chopped CFs, and cross-layer reinforced laminated CFs (Fig. 1). The matrix contains either a combination of pyrolytic and glassy carbons or pyrolytic carbon. Pyrolytic carbon matrix is attained through the chemical vapor deposition (CVD) technique. Glassy carbon is generally produced by the carbonization of phenolic resins that possessed high char efficiency. The resins are utilized to densify the porous disc and consolidate CFs into disc shape by impregnation [9].

Four types of reinforcements are used in CCNCs: oxidized PAN fiber, PAN-essenced CFs, cellulose-essenced CFs, and pitch-essenced CFs. A special vantage of utilizing oxidized PAN fiber is that a preform can be more easily handled by traditional textile procedures rather than by utilizing the alternate CF. In CCNCs two types of matrices are used such as thermosetting resin (including furan, phenolic, and polyimide resins) and thermoplastic matrix (including pitch and other matrices such as PEEK and PEI). Choice of an appropriate precursor matrix is essential for CCNCs and should have the following four properties: (i) Carbon yield strength should be high, (ii) Low viscosity molten precursor should be used, (iii) Carbon fiber reinforcement should easily wet the molten precursor, and (iv) A suitable microstructured matrix carbonization should be established [1].

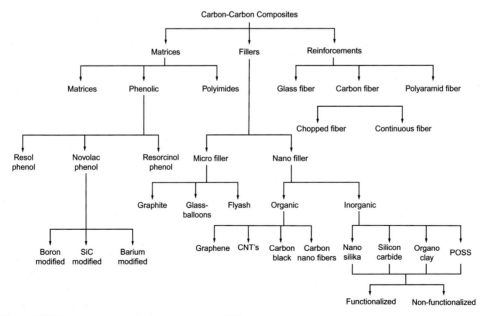

Fig. 1 CCNCs systems and their components [10].

2.1 CCNCs production methods

The part and orientation of the fiber, used in the manufacture of CCNCs, inside of composite control the stages of the production process. The shape of fiber explicates the improvement of preforms in easy cylinders, blocks, contours, cones, revolution surfaces, and complexed shapes and geometries. During the fabrication of CCNCs, two ways are used to fill the spaces between the fibers: the gas-phase path (CVD) and the liquid-phase path (thermosetting pitch or resin) [11].

In filling the spaces by CVD process, various low molecular weight hydrocarbons such as benzene, propane, and methane are utilized as precursors. Hydrocarbons decompose in a hot environment and are coated in the form of carbon accumulation in the heated floor regions of the CFs. In the liquid-phase path, impregnation is carried out using liquid impregnates such as oil pitches or coal tar, and thermosetting gums. The manufacturing of CCNCs can include various impregnations straggled by hot isostatic pressing at 750°C temperatures and 100 MPa pressure. The process is completed by carbonization at 1000°C and graphitization at 2750°C [12].

To manufacturing of CCNCs with various carbon species (CSs) including carbon nanotubes (CNTs), graphene, fullerene, CF, and others or for some special implementations, many novel methods have been evolved in recent years: mixing of solution, in situ polymerization, latex technology, spinning of coagulant, pulverization, densification, and deposition of layer by layer. The solution mixing method, which is suitable for working with small sample sizes, is the extensively utilized method for fabrication of CCNCs [13,14]. Solution blending is carried out in three steps: dispersing the CSs in a suitable solvent, dissolving the polymer resins in the solvent, and mixing the dispersed CSs with the polymer matrix at room or high temperatures. CCNCs are produced by finally settling or pouring the mixture. CCNCs are produced by eventually casting or precipitating the mixture. In the melt blending method, another widely used method for manufacturing CCNCs, thermoplastic polymers like poly (ethylene 2,6-naphthalate), polystyrene, and polypropylene can be processed as matrix materials. Melt blending uses high temperatures and high shear forces to disperse CSs in polymer matrices and is highly friendly with existing industrial applications [13]. In the in situ polymerization method, CSs are admixed with monomers in the absence or existence of solvents. Then these monomers are polymerized by condensation or addition reactions with hardeners or curing agents at high temperatures. In this method, covalent bonds form between functionalized CSs and polymer matrix such as epoxy, thus are produced CCNCs with much enhanced mechanical features via strong interfacial bonds [15]. CCNCs can also be produced using latex technology to incorporate CSs into polymer matrices [13,14]. Latex is a colloidal dispersion of individual polymer particles, generally in an aqueous environment. In this method, CSs are added after the synthesis of polymer to produce the CCNCSs. For this, the CSs are first exfoliated or dispersed/stabilized in an aqueous surfactant solution.

A polymer latex is then mixed into stable dispersions of surfactant-treated CSs. Finally, the prepared mixture is freeze-dried and then melt-processed to obtain CCNCs of CSs dispersed in the polymer matrix. CCNCs are produced by wet spinning of CSs into a polymer solution and coagulating the web into a solid fiber through a slow drawing process. CCNCs are manufactured by mixing and pulverizing polymers and CSs with a pan mill or twin screw. In the densification method, firstly, CSs dispersions are prepared and transferred to the uncured epoxy pool. The matrix is then grafted into CSs dispersions and cured to produce CCNCs. In the layer-by-layer deposition method, CCNCs are prepared by dipping a solid substrate (glass slides, silicon wafers) into CSs/polymer solutions and curing [16].

3. Properties of CCNCs

CCNCs are a group of composites whose microstructures and properties can be adapted for different applications. There are many combinations of CCNCs available, the properties of which vary greatly depending on the type of fibers and matrices, the fiber volume ratio, the architecture of the fiber preform, production processes, and protective coatings. CCNCs preserve their strengths even at extremely elevated temperatures, as their low thermal expansion and lofty thermal conductivity properties provide superb resistance to thermal shocks. CCNCs have various features including low coefficient of thermal expansion, high sublimation temperature, chemical resistance, biocompatibility, shape stability, and high-temperature wear [17].

The mechanical properties are highly dependent upon the architecture and fiber content. Usually, the highest assessment of the content of fiber is 60% by volume. The modulus of elasticity and strength against standard tensions will be maximum along the fiber direction and minimum in the transverse direction following the mixing rule. CSs are ideal reinforcement materials for CCNCs with elevated aspect ratios, large surface areas, low densities, and outstanding mechanical features. Contrary to microparticulate composites, the mechanical features of nanocomposites are largely dependent on the dispersion of the nanofiller, different from the features of matrix and fillers. As well as the dispersion of nanofillers, the elevated aspect ratios, alignment, and interface interactions between CSs and the polymer matrices also significantly affect the mechanical properties of CCNCs. Mechanical properties such as hardness, compressive, and shear for brakes, and compression and tensile strength for nozzles are essential [16].

The utmost significant features of CCNCs are thermal expansion and conductivity. CCNCs produce different microstructures for the matrix by offering the occasion to "adapt" thermophysical features to carbon materials owing to several fiber preforms, dissimilar matrices, and processings [17].

The role of the matrix depends on its type and varies with the type of polymer. Since the thermal behavior of CCNCs is related to crystallinity, heat conduction increases as the crystallinity increases. The thermoplastic and CVD matrices have higher thermal conductivity than thermoset carbon matrices.

The thermal expansion property of materials to be used at elevated temperatures is very significant. The thermal expansion coefficients of CCNCs are critically exaggerated by fiber processing. The measured thermal expansion coefficient values are betwixt 0 and 1×10^{-6}/K in the fiber direction and $6-8 \times 10^{-6}$/K in the vertical direction to the fibers. Thermal properties including thermal expansion, temperature resistance, and thermal shock resistance are important for nozzle systems [18].

They can use in abrasive implementations owing to the principal tribological features of these materials, as well as their high strength and thermal conductivity properties. The friction coefficients of CCNCs are low enough to be 0.5–0.8 in the fiber direction (0.3–0.5) and the vertical direction. Frictional properties of CCNCs and wear resistance are important for brake and nozzle systems.

When a thermoset-essenced CCNC is subjected to heat treatment over 2800°C, the electrical conductivity increases as there is a crystal growth in all directions. The development of a three-dimensional structure is hardly seen in carbon fibers. The density, porosity, and matrix microstructure of a CCNC control the composite's electrical conductivity [9].

Numerous previous studies have shown that elemental carbon is the best known biocompatible material. Elemental carbon is biocompatible with soft tissues, blood, and bone. Bioactive CCNCs attract attention in many implant surgery areas. CCNCs are used as materials that can promote bone formation at the interface instead of soft tissue [19].

Above 400°C temperature, carbon fibers easily react with oxygen and burn quickly, causing the composite to deteriorate partially or completely. The oxidative activities of CCNCs begin with the gasification of CFs, after which the expansion of naturally occurring pores in the composite worsens the oxidation. The speed of oxidation of CCNCs in air rises quickly with increasing temperature. Oxidation resistance of CCNCs is important for brake systems [17].

4. Application area of CCNCs

CCNCs are used in turbine discs, heat exchangers, and pipe fabrication. At temperatures over 1500°C, the use of CCNCs is an indispensable factor in achieving Mach 6 flight speed, and this is because CCNCs are the only composites with unsurpassed mechanical features. In addition, CCNCs have also been implemented in high-temperature regions, as heat resistance is partially required in heat exchangers [20]. Also, CCNCs are utilized in aircraft and race car brakes and clutches, heat shields, rocket motors, industrial applications, and biomedical devices. The braking action in aircraft is carried out using a hydraulic actuator system by pressing the discs against each other. CCNCs are preferred in rocket engines due to their superior ablation resistance. High temperatures of up to 1400°C on its front are reached when a rocket launches into space at speed exceeding 27,000 km/h. Reentry temperatures can approach 1700°C, and this temperature is well above the working temperatures of metals. The thermal shock resistance of CCNCs provides fast passing from −160°C to 1700°C in the cool environment without breaking during reentry. CCNCs are utilized as heat shields in space shuttles and ballistic missiles due to their thermal expansion and conduction features as well as ablation and thermal shock resistance.

As industrial applications, it is possible to mention glass-making, high-temperature mechanical fasteners, furnace heating elements, hot gas ducts and press dies, and molds for forming superplastic metals. On the other hand, elemental carbon is used in biomedical fields as it is compatible with soft tissue, blood, and bone and has the best biocompatibility among all materials [9].

5. Technologies needed to advance CCNCs

In a previous study, the dense and homogeneous graphite/phenolic resin, graphite-CF/phenolic resin, and graphite-CF-CNTs/phenolic resin composites were effectively produced by hot compression molding. TGA data (Fig. 2) showed that the existence of CNTs in the graphite/phenolic resin composite increased its thermal stability. These results showed that CNTs strengthen the bond between the phenolic matrix and CF and graphite, ensuing in dense CCNCs [21].

The inclusion of CNTs in different CNTs/polyamide 6 (PA6) significantly increases the thermal conductivity of nanocomposites and laminates. For example, it has been shown that the thermal conductivities of 4% CNTs/PA6 nanocomposites and laminate with 2% CNTs by weight increased by 180% and 42%, respectively, compared to pure PA6 resin [22].

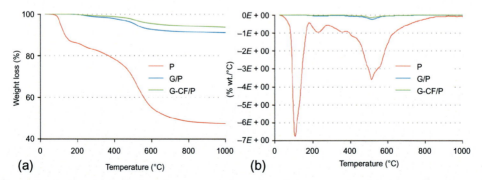

Fig. 2 Thermal analysis of phenolic resin, graphite/phenolic resin, and graphite-CF/phenolic resin composites (A) TGA and (B) DTG diagrams [21].

The thermal conductivity of parallel silicon carbide nanofibers (SiCNFs)-CCNCs is 2.3 times that of CCNCs. The vertical thermal conductivity of SiCNFs-CCNCs is about 1.7 times that of CCNCs. Modification of SiCNFs increased the vertical thermal conductivity of CCNCs [23].

In a previous study, the porosity and density of all CCNCs samples were measured before and after pyrolysis. The existence of CNTs in the CCNCs significantly decreased its open porosity after pyrolysis. Also, the apparent density of pyrolyzed and unpyrolyzed CCNCs is the lowest for graphite-CF-CNT/phenolic resin, followed by graphite-CF/phenolic resin and graphite/phenolic resin CCNCs. Its significant reduction in open porosity (from 15% to 10%) was due to the inclusion of CNTs in the graphite/phenolic resin CCNCs [21]. The influence of pressures on the density and porosity of CCNCs before and after pyrocarbon deposition was shown in Table 1 [24].

The CVD experiments were carried out at total pressure of 16 kPa for 10 h at various deposition temperatures of 1000; 1050; 1100; and 1150°C. In the experimental condition, the efficiency of pyrocarbon deposition into the pore of CCNCs depends on the two main factors: diffusion process of gasses into the porous structure and the decomposition of methane. The densities of composites insignificantly increased with 0.89% and 1.05% at temperatures of 1000 and 1050°C [25].

A 3D-linked CF-CNT hybrid structure was produced to develop the thermal properties of polymer-derived silicon carbonitride matrix-based CCNCs across the thickness. Compared to plain-woven CF, the cross-positioned pores in the CCNCs structure were filled with CNT. CNT has formed CCNCs with a 3D-linked CF-CNT hybrid network structure integrated between the flat woven CF. Fig. 3 displays SEM micrographs of the CF bundle and the CCNCs structure. Fig. 3A shows that there are macrosized pores betwixt mutually perpendicular CF bundles. In CCNCs structure, these pores in intact CF layers are filled with CNT (Fig. 3B and C). On the magnified images (Fig. 3D and E), it is seen that the macrosized pore is alienated into multiple nanosized pores by CNTs, and the microstructure of this region appears to contain CNTs of different lengths. The long CNT zone appears less rough than the short CNT zone. It can also be observed from Fig. 3B that the structure of CCNCs still has some gaps between opposing vertical CF bundles and between CFs and the surface is rough [24].

The elemental composition has an important place in both CCNCs and the coatings of these structures. The isothermal oxidation behavior of the coated CCNCs at 1500°C was investigated in a study. The coating could prevent CCNCs from oxidation at 1500°C for 550 h. By determining the elemental composition formed as a result of oxidation, the phases formed on the surface were analyzed and their effects on the result were evaluated. Similarly, to change the ratio of oxygen-containing groups between CCNCs, elemental distribution is used, and their thermal stability can change according to the oxygen amount they contain [26].

Table 1 Variation of density and porosity properties of CCNCs before and after pyrocarbon deposition according to total pressures [24].

Sample	Pressure (kPa)	Before CVD			After CVD		
		Density (g/cm^3)	Open porosity (vol%)	Close porosity (vol%)	Density (g/cm^3)	Open porosity (vol%)	Close porosity (vol%)
C1	6.6	1.604	15.10	14.09	1.642	13.37	14.14
C2	15.0	1.606	15.39	13.71	1.674	11.88	14.22
C3	30.0	1.605	15.08	14.05	1.707	8.55	16.08
C4	60.0	1.612	14.71	14.12	1.715	8.21	16.07
C5	93.3	1.612	14.83	14.00	1.715	8.32	16.97

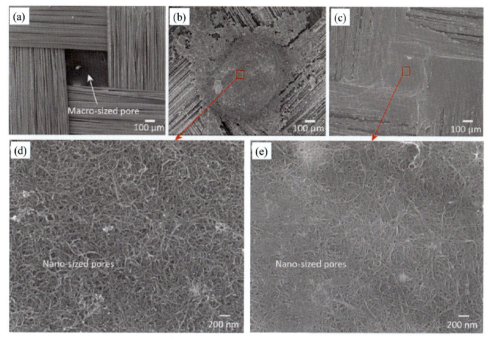

Fig. 3 SEM micrographs of CF and CCNCs structure. (A) CFs bundle, (B) the surface of CCNCs structure, (C) the surface of CF-long CNT CCNCs structure, (D) short CNT zone, and (E) long CNT zone [24].

Since specific heat enhancement is leisurely at lofty temperatures, thermal conductivities tend to decrease with boosting temperature under the influences of specific heat and thermal dissipation. The structure of composites possesses a very important influence on thermal transmission. The superficial carbon matrix and long fibers in the composite plane ease the diffusion of phonons and the formation of a continual transportation way for electrons. However, thermal conductivity depends on the fibers and matrix and is discontinuous because of intertwined cracks, interfaces, and pores that occur during composite production [27].

As for optical properties, some of the components of CCNCs showed unique properties, including unprecedented exciton binding energies, large nonlinear sensitivities, and emission/absorption spectra, allowing each optical property to be assigned a specific chirality [28]. For example, SWCNTs have also been well integrated into electro-optical devices like field effect transistors for light emission or sensing [29]. In another study, the optical properties of CCNCs with which CNT was added and containing gelatin were examined, and the great sensitivity of the nanotubes to the environment was found to be remarkable in terms of optical properties [30].

The shielding CCNCs have displayed superior performance in absorption and protection of microwave. Various CCNCs have been evolved with carbon types, including graphene, CNT, CFs, carbon nanoparticles, carbon black, carbon, and aerogels for electromagnetic interference shielding due to their superior mechanical and electrical features. It has been ascertained that the structure, electrical conductivity, and permeability of the shielding CCNCs play a significant role in obtaining lofty electromagnetic interference shielding [31]. The complex permeability of manufactured CCNCs increases the permeability compared to composites of nanoparticles with the addition of expanded carbon microspheres. Also, interfacial interplays occurring at the grain boundaries in the structure of CCNCs amplify polarization, electron jumps, and microcurrents while inducing magnetic moments [32].

6. CCNCs for brake systems

A brake system is a system that applies artificial frictional resistance to stop or slow the motion of a moving element. Friction is the resistance to the movement of two objects on each other. During this process, the brake absorbs the kinetic energy of the moving element, converts it into heat energy by friction, and dissipates it to the surrounding atmosphere. Therefore, the brake system must meet the following requirements:
- The driver should be able to have good control of the vehicle during braking.
- The brakes should be potent adequate to stop the vehicle at minimum distances in emergencies.
- Brakes must have decent anti-wear properties.
- Brake effectiveness should not decrease in long-term applications [33].

Brakes and clutches are devices that transfer rotating energy. Energy loss during operation causes an increase in temperature. The following four factors are important in the performance of this type of device: (i) transmitted torque, (ii) actuation force, (iii) increase in temperature, and (iv) loss of energy. Transmitted torque and coefficient of friction should be examined separately for the geometry and operation of the device. On the other hand, the increase in temperature is due to the loss of energy from the heat radiating surfaces.

Disc-disc brake systems utilized in aircraft and clutches of Formula 1 racing cars are examples of mating friction elements being moved in a parallel direction to the shaft. The benefits of the disc system comprise a large friction area placed in a small area, a more efficient heat dissipation from the friction surfaces, and an even distribution of applied pressure [9].

Lately, heavy payload and high velocity are important reasons to prefer commercial and military aircraft. Therefore, the brake materials used in aircraft must meet various requirements like excellent thermomechanical features, high specific heat, high coefficients of friction, and superb wear characteristics. Moreover, brake materials should also serve well in diverse ambient conditions like wetness and salt fog [34].

CCNCs are mostly used as brake materials in aircraft and Formula 1 cars because of their superior thermal properties, excellent frictional performance, lightweight, and dimensional stability in harsh environments. The frictional features and general wear mechanisms of CCNCs have been examined for a long time [35,36].

CCNCs-based brake pads or discs are fabricated from the best carbon fiber or long-chain carbon compounds. The first step in disc making is the heating of the white polyacrylonitrile fibers until they turn black, during which the fibers are oxidized first and are arranged in mat-like layers. After a cut to shaping, very pure carbon fibers are obtained by the carbonization process.

They then pass through two condensation heat cycles lasting hundreds of hours, during which a hydrocarbon-rich gas is injected into the furnace at about 1000°C. The carbonization of the gas fed into the furnace fills the gaps between the layers of felt-like material, allowing it to fuse and form a solid material. The produced disc is processed and made ready to be mounted on the vehicle [37].

Formula 1 cars sometimes have to slow down within a few seconds from 350 to 70 km/h or come to a complete stop. In heavy braking situations, the temperature of the pads and brake rotor can rise from 400°C to over 1000°C. The temperature of 1000°C, the highest temperature a carbon brake disc can take, is reached at the end of braking. When the brake pedal is first depressed, braking does not occur because the caliper/disc tandem has not yet reached the operating temperature. However, in the first half-second of braking, the temperature can reach 1200°C, after which the deceleration occurs suddenly. When the optimum braking temperature is achieved, the friction coefficient between the discs and pads can reach up to 0.6.

C-C brake materials have more than twice the heat capacity and one-fourth the density of steel cermet brakes. These good physical properties, together with the high thermal strength and friction resistance of CCNCs provide CCNCs with almost four times the stopping power of steel or copper brakes. For example, the Airbus A320 can make 1500 descents with the metal brake, while 2500 landings can be made with C-C brakes [3].

CCNCs brakes have very special features. The performance of a CCNC brake is relatively poor at temperatures below 400°C, while it reaches an optimum level over 650°C. While conventional brakes wear with normal wear mechanism like any friction material, a CCNC brake not only wears by this mechanism but also undergoes oxidation. Oxidation takes place in the form of burning of the disc surface, and at temperatures over 600°C it becomes the main mechanism of wear by accelerating. During a braking event, the temperature of the brake discs reaches about 1200°C, thus oxidation plays an extremely important role in the brake wear process. Air ducts in the brake discs feed

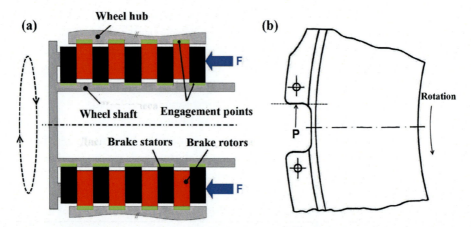

Fig. 4 Airplane brake system: (A) the brake system operational scheme, (B) CCNC brake disc engagement zone [42].

air to the brakes and thus cool down lower to the oxidation temperature, but paradoxically, as they uphold their high temperature for a fairly long time, the high amount of oxygen in the air used to cool them accelerates oxidation. If a carbon brake disc overheats or wears out, it can fail extraordinarily by a sudden and momentary explosion [38].

CCNCs have been used as brake materials in the Formula 1 cars and aerospace industry for a long time owing to their low thermal expansion coefficients, good thermal shock resistances, high strengths and hardness at high temperatures, thermal stabilities, excellent friction performances, low densities, lightness, superb chemical resistance, and dimensional stabilities in harsh environments. These composites have gradually substituted steel in the brake production of aerospace and Formula 1 and are the most innovative materials for these types of brakes [3,35,39]. During braking, most of the kinetic energy is converted into heat energy, resulting in high temperatures, stresses, and thermal deformations formed by thermal changes on the discs, causing severe oxidation of CCNCs, which is the most important factor in damage to disc materials. Transfer of friction heat is a complex process that can straightly influence the tribological features of brake discs [40,41].

Figs. 4 and 5 show the rotor/stator discs in the aircraft braking systems. Rotors are attached to the rim and rotate with wheels. Rotors rub against the stators for braking. Fig. 4B shows the engagement zone of CCNCs brake disc.

CCNCs are very important because of their high toughness and strength properties at lofty temperatures over 2000°C. They are considered unique in lofty temperatures in many implementations like engines and aviation. However, these composites undergo oxidation in the exposure to an oxidizing atmosphere over 450°C. The various

Fig. 5 CCNCs used to make aircraft brake discs (stator and rotor) [43].

approaches described in the literature for the protection of CCNCs can be considered into two categories: (i) use of inhibitors to slow the oxidation rate and block active sites, and (ii) make of coatings to prevent carbon outflow and oxygen ingress. The first method involves adding some oxidation inhibitors (such as B, Si) to the C-C matrix. These additives can significantly reduce the oxidation rate of CCNCs, but they will lose their efficiency in the long-term implementations at temperatures over 1000°C.

The second method is to make oxidation-resistant coatings on their surface to protect CCNCs at high temperatures. SiC ceramic is extensively utilized as a coating material due to its superior oxidation resistance and good suitability with CCNCs. However, the coating of CCNCs may be insufficient for missiles and aircraft brakes [3,44].

Despite having many superior features, CCNCs brakes have low oxidation resistance, and the friction coefficients decrease in humid environments. CCNCs brakes perform poorly during landing in the morning after aircraft fly in wet conditions overnight. Further, since CCNCs brakes exhibit considerably lower performance in the marine environment due to salt fog, these disadvantages should cope over for better implementations [34].

C/SiC aircraft C-C brake materials consist of short fiber mesh, nonwoven fiber fabric, 90° nonwoven fiber fabric, and fiber layers. Si and SiC are generally dispersed on short fiber mesh layers. The SiC matrix contains two kinds of SiC particles, nanosized and micron-sized, at the C—SiC interfaces. Nano-SiC particles can reduce the wear rate and stabilize the friction features with the help of the friction film they form. Micro-SiC particles can increase friction resistance with increased rubble plowing action [45].

High friction brake shoes were produced by various carbon materials for the emergency braking system of the crane used in the ultra-deep coal mine. The addition of graphene improved the brake shoe material friction coefficient by reducing the friction plate surface temperature rise rate and the brake shoe wear rate. The reduction in wear rate may be due to the lubricating influence of graphene. Due to the good heat conduction property of graphene, the higher graphene content decreases the heat conduction coefficient, so the heat buildup rate during friction and the temperature of the friction disk surface increase [46].

Developed to overcome the disadvantages of CCNCs brakes, CC-SiC nanocomposites have a higher coefficient of friction, better resistance to oxidation, and lower sensitivity to humid environments compared to CCNCs. CC-SiC nanocomposites are potential next-generation brake materials for aircraft but have higher densities than CCNCs. Since CC-SiC nanocomposite brake assemblies are heavier than CCNCs brakes, their usage in some weight-sensitive aircraft are limited [34]. Also, multilayer ceramic coating systems like C-SiC, C-SiC-$MoSi_2$-Al_2O_3, and C-SiC-B_4C have been developed to protect the exposed frictionless CCNCs surface from high-temperature service deterioration. With these systems, oxidation studies were carried out in 60% humid climates up to 1200°C in both dynamic and static conditions. It has been determined that the C-SiC-B4C coating performs better in the oxidizing media [47].

The advantages of CCNCs in brake systems are as follows:
- Lower density and better specific features.
- Better thermal stability in an oxygen-free environment.
- Low thermal expansion coefficient.
- Available in a wide variety of features and forms.
- High modulus, particularly pitch-essenced fiber.
- Good strength, particularly PAN-essenced fiber.
- Good fatigue properties and high thermal conductivity.
- Excellent frictional resistance.
- No inhalation problem with filament diameters.
- Good chemical resistance.
- Biocompatibility.
- Low electrical resistance.

The drawbacks of CCNCs in brake systems are as follows:
- Low strain against failure.
- Relatively high cost.
- Weak impact resistance of composites.
- Compressive strength is lower than the tensile strength.
- Damage in electrical systems since they are electrically conductive.
- It exhibits anisotropy in transverse and axial directions.
- Oxidizes in the air at temperatures above 450°C.

7. CCNCs for exhaust nozzles

To bring the exhaust gases to ambient pressure, the nozzle operates according to the Venturi effect, thus making them a harmonious jet. If the pressure is high, the flow can be choked, and the jet can be supersonic [3]. The role of the nozzle in applying counter-pressure to the rocket motor is very important to convert the static high pressure and temperature gas into a rapidly moving gas at a pressure close to ambient pressure.

A nozzle is an appliance designed to regulate fluid flow properties when exiting or entering a closed compartment or tube. A nozzle is made from a tube or pipe whose cross-sectional area alters to be utilized to alter the flow of a fluid. Therefore, as the exhaust nozzle is subjected to high temperature, pressure, friction and wear, it is estimated to have the following properties:
- High temperature resistance,
- Low friction and wear,
- High compression and tensile strength,
- Low thermal expansion,
- Thermal shock resistance,
- Oxidation resistance,
- Reasonable cost,
- No toxicity (environmentally friendly).

Nozzles can be used in the exhaust, rockets, jet motors, as well as other areas. Also, the nozzle forms can be similar to the conical, bell, or spike-like shapes. The system that performs the following four basic functions is called the nozzle:
- Measures fluid at a specified flow rate.
- Atomizes the liquid in droplets.
- Provides hydraulic momentum.
- It distributes droplets in a certain pattern.

Many factors need to be considered in order to choose the appropriate material to be used in spray nozzles. The maximum temperature limits are determined by the melting or softening of the material. However, if there is a chemical attack, corrosion, and oxidation, these temperature limits must be reduced [48].

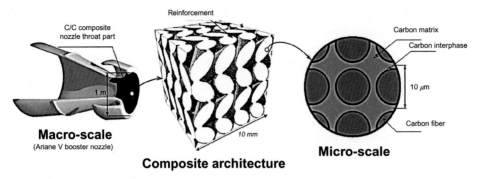

Fig. 6 Microstructure of CCNCs used in the nozzle throat pieces [49].

Since nozzles are exposed to high temperature, pressure, friction and wear, thermal expansion, and shock, the selected material for the nozzles should have sufficient properties to resist these effects. Therefore, CCNs can be chosen since they have excellent properties for severe nozzle conditions and show splendid performance within nozzle operations such as exhaust, jet motors, and missile systems.

Similar to brake systems, the oxidation protection of CCNCs may be the same for thermo-oxidative operating conditions of exhaust nozzles. So to prevent the oxidation or slow down the oxidation rate, either inhibitors and sealants should be used, or the exhaust nozzle should be coated.

Nozzles are frequently utilized to check up the flow direction, rate, shape, pressure, and mass of the emerging flow. The higher the speed of the fluid in a nozzle, the higher the pressure energy. The CCNCs used in the construction of the nozzle throat consist of a matrix and reinforcements as shown in Fig. 6. The reinforcement architecture can show a complex multiaxis and multidimensional geometry. As shown in the last image of Fig. 6, CCNCs can be formed when the reinforcing fiber bundles are densified in a matrix by the method of liquid- or vapor-phase deposition [49].

The aviation field proceeds to be one of the prime implementation areas for CCNCs. Classic examples of CCNCs used as thermal protection systems are heating elements, heat shields, furnace fixturing, load plates, X-ray targets, exit cones, and reentry devices of rocket nozzles, the nose tips of space shuttles, etc. In these applications, all of the structural, thermal, thermomechanical, thermophysical, refractory, and mechanical features of CCNCs are used. Rocket nozzles have to stand an immensely swift temperature increment in a pretty much abrasive atmosphere while protecting a high degree of their entirety. Since nozzle materials are subjected to vigorous thermal stresses, they must exhibit predictable and uniform performance to ensure the necessary correctness and range [50].

CCNCs are made by weaving sets of CFs by carefully orienting them in two, three, or four directions in a carbon matrix. By injecting an organic liquid resin, the spaces betwixt the CFs are filled and converted to carbon charcoal by heating. Pressure is applied to the system to compact the CCNCs by densification, and the process is repeated with more resin injection. The graphitization is carried out at temperatures over 2000°C. CCNCs exhibit excellent performance in high heat transfer zones like the nozzle throat, as their versatile CF reinforcements enable them to better stand the high thermal stresses caused by steep temperature gradients within the material. Carbonized ablative materials can be fabricated by thermosetting resin, like epoxies, phenolics, or silicon reinforced by fibers, like carbon, silica, glass, or quartz [51].

The inner walls of the thermal protection materials that make up the rocket nozzles are exposed to very high temperatures up to 3000°C. The number of materials that can remain solid under such severe conditions is very few. Among these, CCNCs have many important advantages such as low thermal expansion, the ideal ratio of mechanical features against the density at lofty temperatures, cost-influential machining of large and thick parts. Despite their superior performance, thermochemical restrictions cause recessive degradations on the surfaces of CCNCs by loss of mass or ablation. The main cause of this mass loss is oxidation, but mechanical erosion also contributes to a small extent [49].

The inner surface is made of C-SiC layer, and the outer surface is made of CF-reinforced plastic to strengthen the structure of the nozzle by reducing the oxidation and mechanical erosion of CCNCs. Since CF-reinforced plastic probably loses its strength at high temperatures, a ceramic insulator layer was used to keep the surface temperature of this layer below 180°C by preventing heat conduction. C-SiC has erosion resistance, high specific strength, and high heat capacity. The specific strength reduces abruptly at high temperatures over 800°C in the case of metal, while in the case of C-SiC it can be protected at a temperature over 1400°C, so it is utilized for the combustor liner, turbine disk, high-temperature nozzle throat. In addition, C-SiC is commonly used in nozzles because of its respectable erosion resistance [52].

The image of the redesigned pusher with the C-C composite nozzle extension attached is presented in Fig. 7A. The flange fasteners used vary depending on the material being tested. Orbital ATK C-C nozzles are machined to vary in final thickness near the flange to permit for aft stiffener ring and split ring. A Grafoil gasket interfaces with the chamber, while the connection is provided by a carbon phenolic split ring. Zirconia-essenced insulator was implemented by air plasma sprayed method between the outer surface of the C-C cones and the carbon phenolic split ring fitting. The orbital ATK C-C nozzle installation in the test station is shown in Fig. 7B. The hot flame test of orbital ATK nozzles is presented in Fig. 8 [53].

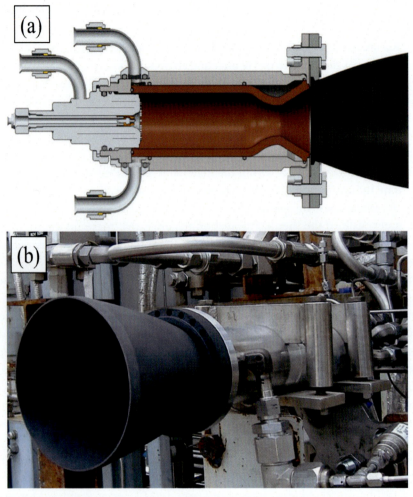

Fig. 7 (A) 5340 N propellant with additives-fabricated GRCop-84 lined and C-C nozzle extension (B) OATK C-C extension located in the test stand-115 at Marshall Space Flight Center of NASA [53].

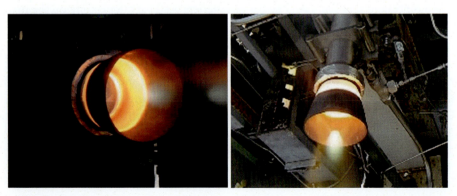

Fig. 8 Hot fire test of OATK C-C nozzle extensions at test stand-115 in the Marshall Space Flight Center of NASA [53].

8. Conclusion and future perspectives

CCNCs are composite structures consisting of hydrocarbons or polymers as matrix precursors reinforced with CFs in diverse directions. Their densities and features vary according to the species and volume ratios of the matrix precursors used with applied heat treatment temperatures. CCNCs are very important for many applications due to their toughness and high strength properties at high temperatures above 2000°C. CCNCs can be fabricated by conventional production methods as well as also advanced manufacturing techniques. Due to their chemical resistance, good wear resistance, shape stability, low thermal expansion, and high thermal conductivity characteristics, CCNCs show outstanding resistance to thermal shocks and keep their strength even at very lofty temperatures. Therefore, CCNCs stand out as materials that can be used with the best performance in extreme temperature conditions. Brake pads and discs of racing cars and aircraft, space shuttles, heat shields, ballistic missiles, rocket engines, and nozzle systems are areas where CCNCs are widely used. CCNCs made of carbon fabric laminates, semirandomly chopped CFs, and cross-ply reinforced laminated CF mats are widely used in brake systems. Due to their superior properties, CCNCs have almost four times better-stopping performance than steel or copper brakes. CCNs are widely used in nozzle systems as they perform exceptionally under harsh nozzle conditions such as exposure to high temperature, pressure, friction and wear, thermal expansion, and shock. CCNCs undergo oxidation in oxidizing atmospheres above 450°C. CCNCs can be protected from oxidation in two ways: (i) using inhibitors to slow the rate of oxidation and block active sites and (ii) making coatings to prevent carbon outflow and oxygen ingress. The high production costs of CCNCs have limited their use, and therefore cheaper production techniques should be developed to allow them to be used more widely without sacrificing their superior properties.

References

[1] P. Morgan, History and early development of carbon fibers, in: P. Morgan (Ed.), Carbon Fibers Their Compos, Taylor & Francis Group, Danvers, MA, 2005, pp. 65–115.
[2] E. Fitzer, M. Heine, Carbon fibre manufacture and surface treatment, in: R.B. Pipes (Ed.), Fiber Reinf. Compos. Mater., Elsevier, New York, USA, 1988, pp. 17–146.
[3] E. Fitzer, L.M.M. Manocha, Carbon Reinforcements and Carbon/Carbon Composites, Springer-Verlan Berlin Heidelberg GmbH, New York, USA, 1998.
[4] T.C. Chen, C.C. Liu, Inverse estimation of heat flux and temperature on nozzle throat-insert inner contour, Int. J. Heat Mass Transf. 51 (2008) 3571–3581, https://doi.org/10.1016/j.ijheatmasstransfer.2007.10.029.
[5] Y. Tong, S. Bai, H. Zhang, Y. Ye, Laser ablation behavior and mechanism of C/SiC composite, Ceram. Int. 39 (2013) 6813–6820, https://doi.org/10.1016/j.ceramint.2013.02.012.
[6] A.P. Mouritz, S. Feih, E. Kandare, Z. Mathys, A.G. Gibson, P.E. DesJardin, S.W. Case, B.Y. Lattimer, Review of fire structural modelling of polymer composites, Compos. Part A Appl. Sci. Manuf. 40 (2009) 1800–1814, https://doi.org/10.1016/j.compositesa.2009.09.001.

[7] Y. Chen, P. Chen, C. Hong, B. Zhang, D. Hui, Improved ablation resistance of carbon-phenolic composites by introducing zirconium diboride particles, Compos. Part B Eng. 47 (2013) 320–325, https://doi.org/10.1016/j.compositesb.2012.11.007.

[8] Y. Mansuroglu, C-C Composites, 2001, p. 1. Date Accessed: 2021-04-29 Http://Www.Yusufmansuroglu.Com.Tr/.

[9] G. Savage, Carbon–Carbon Composites, First ed., Springer-Science+Business Media, B.V., London, 1993, https://doi.org/10.1201/b15551-8.

[10] P. Sanoj, B. Kandasubramanian, Hybrid carbon-carbon ablative composites for thermal protection in aerospace, J. Compos. 2014 (2014) 1–15, https://doi.org/10.1155/2014/825607.

[11] G.R. Devi, K.R. Rao, Carbon-carbon composites—an overview, Def. Sci. J. 43 (1993) 369–383, https://doi.org/10.14429/dsj.43.4291.

[12] S.B.V. Kumar, N.V. Londe, A.O. Surendranathan, G.J. Rao, Fabrication methods, recent developments and applications of carbon-carbon composites (CCC): a review, Int. Res. J. Eng. Technol 5 (2018) 1252, https://www.researchgate.net/publication/329267753.

[13] M. Moniruzzaman, K.I. Winey, Polymer nanocomposites containing carbon nanotubes, Macromolecules 39 (2006) 5194–5205, https://doi.org/10.1021/ma060733p.

[14] J.H. Du, J. Bai, H.M. Cheng, The present status and key problems of carbon nanotube based polymer composites, Express Polym. Lett. 1 (2007) 253–273, https://doi.org/10.3144/expresspolymlett.2007.39.

[15] M. Ma, W. Ye, X.X. Wang, Effect of supersaturation on the morphology of hydroxyapatite crystals deposited by electrochemical deposition on titanium, Mater. Lett. 62 (2008) 3875–3877, https://doi.org/10.1016/j.matlet.2008.05.009.

[16] P.C. Ma, N.A. Siddiqui, G. Marom, J.K. Kim, Dispersion and functionalization of carbon nanotubes for polymer-based nanocomposites: a review, Compos. Part A Appl. Sci. Manuf. 41 (2010) 1345–1367, https://doi.org/10.1016/j.compositesa.2010.07.003.

[17] C. Scarponi, Carbon-carbon composites in aerospace engineering, in: Adv. Compos. Mater. Aerosp. Eng., Elsevier, 2016, pp. 385–412.

[18] L.M. Manocha, High performance carbon–carbon composites, Sadhana 28 (2003) 349–358.

[19] L. Zhang, H. Li, K. Li, S. Zhang, J. Lu, W. Li, S. Cao, B. Wang, Carbon foam/hydroxyapatite coating for carbon/carbon composites: Microstructure and biocompatibility, Appl. Surf. Sci. 286 (2013) 421–427, https://doi.org/10.1016/j.apsusc.2013.09.110.

[20] H. Hatta, K. Goto, T. Sato, N. Tanatsugu, Applications of carbon-carbon composites to an engine for a future space vehicle, Adv. Compos. Mater. 12 (2003) 237–259, https://doi.org/10.1163/156855103772658588.

[21] T.H. Nguyen, M.T. Vu, V.T. Le, T.A. Nguyen, Effect of carbon nanotubes on the microstructure and thermal property of phenolic / graphite composite, Int. J. Chem. Eng. 2018 (2018) 1–8.

[22] Z. Shen, S. Bateman, D.Y. Wu, P. McMahon, M. Dell'Olio, J. Gotama, The effects of carbon nanotubes on mechanical and thermal properties of woven glass fibre reinforced polyamide-6 nanocomposites, Compos. Sci. Technol. 69 (2009) 239–244, https://doi.org/10.1016/j.compscitech.2008.10.017.

[23] J. Chen, P. Xiao, X. Xiong, The mechanical properties and thermal conductivity of carbon/carbon composites with the fiber/matrix interface modified by silicon carbide nanofibers, Mater. Des. 84 (2015) 285–290, https://doi.org/10.1016/j.matdes.2015.06.085.

[24] J. Yang, J. Sprengard, L. Ju, A. Hao, M. Saei, R. Liang, G.J. Cheng, C. Xu, Three-dimensional-linked carbon fiber-carbon nanotube hybrid structure for enhancing thermal conductivity of silicon carbonitride matrix composites, Carbon N. Y. 108 (2016) 38–46, https://doi.org/10.1016/j.carbon.2016.07.002.

[25] D. The, L. Van Thu, N. Tuan, N. Trung, Effects of temperature and pressure on the density of carbon-carbon nanocomposite fabricated by chemical vapor deposition, Chem. Eng. Trans. 78 (2020) 211–216, https://doi.org/10.3303/CET2078036.

[26] X. Ren, H. Li, Y. Chu, Q. Fu, K. Li, Preparation of oxidation protective ZrB2-SiC coating by in situ reaction method on SiC-coated carbon/carbon composites, Surf. Coat. Technol. 247 (2014) 61–67, https://doi.org/10.1016/j.surfcoat.2014.03.017.

[27] X. Liu, H.L. Deng, J.H. Zheng, M. Sun, H. Cui, X.H. Zhang, G.S. Song, Mechanical and thermal conduction properties of carbon/carbon composites with different carbon matrix microstructures, New Carbon Mater. 35 (2020) 576–584, https://doi.org/10.1016/S1872-5805(20)60511-X.

[28] J. Maultzsch, R. Pomraenke, S. Reich, E. Chang, D. Prezzi, A. Ruini, E. Molinari, M.S. Strano, C. Thomsen, C. Lienau, Exciton binding energies in carbon nanotubes from two-photon photoluminescence, Phys. Rev. B: Condens. Matter Mater. Phys. 72 (2005) 1–4, https://doi.org/10.1103/PhysRevB.72.241402.

[29] J.A. Misewich, R. Martel, P. Avouris, J.C. Tsang, S. Heinze, J. Tersoff, Electrically induced optical emission from a carbon nanotube FET, Science 300 (2003) 783–786, https://doi.org/10.1126/science.1081294.

[30] S. Berger, F. Iglesias, P. Bonnet, C. Voisin, G. Cassabois, J.S. Lauret, C. Delalande, P. Roussignol, Optical properties of carbon nanotubes in a composite material: the role of dielectric screening and thermal expansion, J. Appl. Phys. 105 (2009) 7–12, https://doi.org/10.1063/1.3116723.

[31] S. Gupta, N.H. Tai, Carbon materials and their composites for electromagnetic interference shielding effectiveness in X-band, Carbon N. Y. 152 (2019) 159–187, https://doi.org/10.1016/j.carbon.2019.06.002.

[32] R. Peymanfar, F. Fazlalizadeh, Fabrication of expanded carbon microspheres/ZnAl2O4 nanocomposite and investigation of its microwave, magnetic, and optical performance, J. Alloys Compd. 854 (2021) 157273, https://doi.org/10.1016/j.jallcom.2020.157273.

[33] P. Shiva Shanker, A review on properties of conventional and metal matrix composite materials in manufacturing of disc brake, Mater. Today Proc. 5 (2018) 5864–5869, https://doi.org/10.1016/j.matpr.2017.12.184.

[34] X. Xu, S. Fan, L. Zhang, Y. Du, L. Cheng, Tribological behavior of three-dimensional needled carbon/silicon carbide and carbon/carbon brake pair, Tribol. Int. 77 (2014) 7–14, https://doi.org/10.1016/j.triboint.2014.04.008.

[35] S. Ozcan, P. Filip, Microstructure and wear mechanisms in C/C composites, Wear 259 (2005) 642–650, https://doi.org/10.1016/j.wear.2005.02.112.

[36] S. Wu, Y. Liu, Y. Ge, L. Ran, K. Peng, M. Yi, Structural transformation of carbon/carbon composites for aircraft brake pairs in the braking process, Tribol. Int. 102 (2016) 497–506, https://doi.org/10.1016/j.triboint.2016.06.018.

[37] S. De Groote, Brake system, 2008, p. 1. Date Accessed: 2021-04-29 Www.F1technical.Net.

[38] https://www.racecar-engineering.com/tech-explained/f1-2014-explained-brake-systems. (Accessed 29 April 2021).

[39] X. Xiong, B.Y. Huang, J.H. Li, H.J. Xu, Friction behaviors of carbon/carbon composites with different pyrolytic carbon textures, Carbon N. Y. 44 (2006) 463–467, https://doi.org/10.1016/j.carbon.2005.08.022.

[40] S. Zhao, G.E. Hilmas, L.R. Dharani, Numerical simulation of wear in a C/C composite multidisk clutch, Carbon N. Y. 47 (2009) 2219–2225, https://doi.org/10.1016/j.carbon.2009.04.012.

[41] H.J. Xu, B.Y. Huang, M.Z. Yi, X. Xiong, B.L. Lei, Influence of matrix carbon texture on the temperature field of carbon/carbon composites during braking, Tribol. Int. 44 (2011) 18–24, https://doi.org/10.1016/j.triboint.2010.09.004.

[42] A.A. Stepashkin, D.Y. Ozherelkov, Y.B. Sazonov, A.A. Komissarov, Fracture toughness evolution of a carbon/carbon composite after low-cycle fatigue, Eng. Fract. Mech. 206 (2019) 442–451, https://doi.org/10.1016/j.engfracmech.2018.12.018.

[43] P. Chowdhury, H. Sehitoglu, R. Rateick, Damage tolerance of carbon-carbon composites in aerospace application, Carbon N. Y. 126 (2018) 382–393, https://doi.org/10.1016/j.carbon.2017.10.019.

[44] Q.G. Fu, H.J. Li, X.H. Shi, K.Z. Li, J. Wei, M. Huang, Oxidation protective glass coating for SiC coated carbon/carbon composites for application at 1773 K, Mater. Lett. 60 (2006) 431–434, https://doi.org/10.1016/j.matlet.2005.09.006.

[45] S. Fan, L. Zhang, Y. Xu, L. Cheng, G. Tian, S. Ke, F. Xu, H. Liu, Microstructure and tribological properties of advanced carbon/silicon carbide aircraft brake materials, Compos. Sci. Technol. 68 (2008) 3002–3009, https://doi.org/10.1016/j.compscitech.2008.06.013.

[46] D. Wang, J. Yin, Z. Zhu, D. Zhang, D. Liu, H. Liu, Preparation of high friction brake shoe material and its tribological behaviors during emergency braking in ultra-deep coal mine hoist, Wear 458–459 (2020) 203391, https://doi.org/10.1016/j.wear.2020.203391.

[47] R.M. Mohanty, Climate based performance of carbon-carbon disc brake for high speed aircraft braking system, Def. Sci. J. 63 (2013) 531–538, https://doi.org/10.14429/dsj.63.3932.

[48] Spray Nozzle Catalogue, 2021, p. 1. Date Accessed: 2021-04-29 https://www.spray-nozzle.co.uk/.

[49] G.L. Vignoles, Y. Aspa, M. Quintard, Modelling of carbon-carbon composite ablation in rocket nozzles, Compos. Sci. Technol. 70 (2010) 1303–1311, https://doi.org/10.1016/j.compscitech.2010.04.002.

[50] S.-J. Park, M.-K. Seo, Interface Science and Composites, first ed., UK, Oxford, 2011.

[51] J.H. Koo, J. Langston, Polymer nanocomposite ablative technologies for solid rocket motors, in: Nanomater. Rocket Propuls. Syst., Elsevier Inc., 2018, pp. 423–493, https://doi.org/10.1016/B978-0-12-813908-0.00012-5.

[52] S.J. Kim, Y.R. Kim, Y. Kim, M.H. Kim, M.S. Lee, 2D exhaust nozzle with multiple composite layers for IR signature suppression, Results Phys. 19 (2020) 103395, https://doi.org/10.1016/j.rinp.2020.103395.

[53] P.R. Gradl, P.G. Valentine, Carbon-carbon nozzle extension development in support of in-space and upper-stage liquid rocket engines, in: 53rd AIAA/SAE/ASEE Jt. Propuls. Conf., 2017, 2017, pp. 1–27, https://doi.org/10.2514/6.2017-5064.

CHAPTER 8

Metallic nanocomposites: An Introduction

Sandra Gajević, Slavica Miladinović, and Blaža Stojanović
Faculty of Engineering, University of Kragujevac, Kragujevac, Serbia

Chapter outline

1. Introduction — 155
2. Metallic nanocomposites — 156
3. Conclusion — 160
References — 161

1. Introduction

Materials have had a significant role in the improvement and development of civilization, from the beginning until today. Importance of materials is best reflected throughout history, the prehistoric ages were named after the most commonly used materials such as stone, bronze, and iron. Today's construction requirements from the aspect of increasing the service life and reducing the weight and price of the product are achieved by constant development and application of new materials. Innovative engineering products often require the use of multifunctional materials with a wide range of different properties. Such technological demand is difficult to satisfy with the use of single-phase materials. In order to achieve the required technological goals, a synergistic combination of materials with different technical and technological properties is needed. Such needs are addressed by the development of new materials. Composite materials are usually made of a certain base and one or more reinforcements which can be in the form of particles, short and long fibers. The use of composite materials is primarily conditioned by their improved physical-mechanical and tribological properties compared to the properties of the base material. Improvement of properties can be achieved in density, ductility, modulus of elasticity, tensile strength, hardness, wear resistance, friction coefficient, deformability, and thermal and electrical conductivity [1–3]. In this way, engineers are provided with greater freedom in construction, the ability to combine the characteristics of materials, and their adaptation to requirements that were not feasible using conventional monolithic materials.

The development of composite materials and technology based on their application is among the most important advances in the history of materials. From the beginning of the application of composites around 1950 until today, the successful use of these materials in the automotive, aerospace, and many other industries is evident [4]. During the development of metallic composite materials, some composites had a problem in homogeneity and ductility and then in toughness, hardness, and stiffness. Overcoming the mentioned problems has been achieved by developing hybrid composites which contain two or more reinforcements in addition to the base material. In this way, the properties of classic composites are improved, such as low toughness, insufficient rigidity, and high production cost. These properties were possessed by composites that have microlevel reinforcements in their composition, and due to the problems that arose, research was continued and focused on composites with reinforcements at the nanolevel. Development of composites with reinforcements at the nanolevel has led to the production of nanocomposites which became the base for the development of nanotechnology. Nanotechnology has a very important role in the development of industry. Research on nanotechnology is focused on the analysis of materials and systems whose structures and components exhibit new and improved physical, chemical, and biological properties. In the early 1980s the events that marked the beginning of the development of nanotechnology took place. These events refer to the development of scanning tunneling microscopes (*STM*), atomic force microscopes (*AFM*), and scanning probe microscopes (*SPM*) [5] enabling experimental techniques and methods necessary for measuring and manipulating nanostructures. *SPM* has opened a new world of nanotechnology that enables the observation and manipulation of individual atoms and molecules on solid surfaces. Nanotechnology is constantly evolving around the world to change and affect the quality of life in the future. The fascinating world of nanomaterials and their various applications is becoming a part of our lives. It is estimated that today nanotechnology is at a similar level of development as information technology was in the fifties of the last century [5, 6].

The increased demand for nanocomposite materials in industry indicates their great application in the future, and this is a challenge for scientists and engineers. The use of nanomaterials strongly influences the development of the automotive industry, aerospace industry, food industry, medicine, and pharmacy. From an engineering point of view, the nanoscale is extremely small, so many tools that are routinely used in microproduction cannot be used for nanoproduction. The motivation in nanoscience is to try, among other things, to understand the behavior of materials when the sample size is close to atomic dimensions.

2. Metallic nanocomposites

Nanocomposites are, usually, defined as the combination of base material and reinforcement with at least one dimension in nanolevel. As a base of metallic nanocomposites in automotive industry, usually, low-density metals such as aluminum (*Al*), magnesium (*Mg*), and titanium (*Ti*) are used. Reinforcements in metallic nanocomposites can be classified by the type of material and shape of reinforcement. Mostly used materials for nanoreinforcements are: oxides, carbides, nitrides, and borides. Nanoreinforcement can be in the shape of particles, whiskers, platelets, and fibers. Particles can be powder or dispersion in the order of 10 or 100 nm, spherical, cubic, tetragonal, flat, or irregular in shape, and are usually approximately equiaxial, while the fibers are cylindrical in shape, and can be continuous and short (whiskers). Platelets are rectangular, sanidic, or circular in shape. Among all the reinforcements of metallic nanocomposites, the particle reinforcements are mostly used. The most commonly used types of nanoparticles are *SiC, TiC, WC, TaC, TiB$_2$, AlN, B$_4$C,* and *Al$_2$O$_3$* [7, 8].

Aluminum and aluminum alloys use for the base of metallic nanocomposites in automotive industry has increased due to their good properties, such as low density, good thermal conductivity and corrosion resistance, relatively low production cost, and good recyclability. The parts that are made based of aluminum alloy are pistons, cylinder head (engine head), engine blocks, rims, intake manifolds, crankshaft connecting rods, camshaft, cardan shafts, helicopter propellers, and production of brake discs and drums for cars and trains [9–11]. The use of aluminum castings in the field of automotive industry is accompanied by the development of casting and heat treatment. Great efforts are made in order to improve the mechanical and tribological characteristics through the control of the microstructure, more precisely the improvement of the procedures for obtaining composites, treatment of metal melting, and heat treatment. Excessive research of metallic base nanocomposites indicates potential applications in automotive parts such as piston ring, piston, connecting rod, brake discs and brake drums, camshafts, and valves [12–17].

In automotive industry there is constant effort to develop lightweight parts so the fuel consumption would be reduced. Vehicle's body shell is 40% of vehicle's total weight. High-strength steel is used for European cars and it makes 50%–60% of body. In automotive industry, today, composites with base of high-strength steel, aluminum, and composites of carbon fiber and polymer are mostly used. Compared to the conventional steel constructions, application of high-strength steel has reduced the weight of the vehicle by 20%, while use of aluminum and the application of composites of carbon fibers have reduced weight by 40% and 50%, respectively. Another example of reduction of weight by the usage of lightweight materials, compared with conventional material, is that the weight of aluminum ties in the suspension is less for 15%, hollow composite shaft is 18% lighter, and aluminum wheels are up to 36% lighter compared to steel wheels [8, 9].

Particle-reinforced nanocomposites are widely used in the automotive industry because of their ability to withstand high temperatures and pressures. Disadvantages of nanocomposites with a metallic base are inhomogeneity, weak bonds between the base and particles, and agglomeration of nanoparticles. In the case of ceramic nanoparticles, their thermal characteristic changes as a consequence of the growth of surface energy and different interatomic distances. Also, the application of nanoparticles in the nanocomposite leads to technological difficulties in testing and high costs of production of nanocomposites with a metallic base. The reason for this is that it is necessary to consume a large amount of energy in order to achieve an even distribution of nanoparticles in the base of the nanocomposite material.

In addition to all the advantages provided by nanoparticles, there are some disadvantages. Nanoparticles can be found in air, water, and soil because they are the most common by-products of the combustion of hydrocarbon compounds and exhaust gases. In human and animal bodies nanoparticles are present, which has led to the discussion of their toxicity and impact on the environment. Diseases that accumulation of nanoparticles in the human body can cause are the inflammatory airway disease, Crohn's disease, bronchitis, asthma, cardiovascular disease, lung cancer, neurodegenerative diseases, liver cancer, Parkinson's disease, Alzheimer's disease, and others. A great attention needs to be given to the protection of people during manufacture, transport, handling, use, disposal, and recycling of parts and machines which in themselves have nanocomposites [18].

The need to overcome the observed shortcomings makes the research and testing of nanocomposites still relevant, primarily in the process of obtaining and the possibility of variation of influential parameters.

Nanocomposite fabrication methods can be classified based on the state of the nanocomposite's base during the preparation process, which can be liquid, solid, and semisolid states. Solid-state preparation processes include different powder metallurgy techniques with modifications in the processing steps such as high-energy ball mill, hot pressing, hot isostatic pressing, cold pressing followed by sintering treatment, and extrusion and friction stir processing. Liquid-state processes include different processes such as casting, stir casting, and squeeze casting, while rheocasting technique with its variants such as compo casting or in combination with squeeze casting are among semi-solid processes [7].

The most important production methods that are used for aluminum nanocomposites are stir casting, squeeze casting, compo casting, and powder metallurgy. Casting is a process of obtaining nanocomposites which is performed by heating the base material to a temperature higher than the temperature of the molten material and then inserting the reinforcing particles into the base and mixing the mixture. Stir casting is a process that is very similar to the conventional casting process. The difference is that in stir casting there is a slow mixing of the molten alloy in order to achieve an even distribution of the reinforcements in the base. Stir casting process is the most common method of fabrication because of the low cost and possibility of using a wide range of materials. The compo-casting method is similar to the stir casting, and the main difference is that the base is in a semi-solid state, and in essence, it is a variant of the reocasting or thixocasting method used for the production of nanocomposites. The characteristic of this method is that the semi-solid metal must be mixed more intensively than it is the case in stir casting.

Powder metallurgy is achieved by mixing metal powder and ceramic particles in a defined series of steps. The first step involves converting the base material into powder particles, so that the base powders and the reinforcements are mixed in an appropriate ratio. The resulting mixture is poured into a mold and pressed in order to achieve a cohesive structure with dimensions that are close to the dimensions of the final element. Final product is obtained by pressing under high temperatures. When mechanically mixing powders, it is recommended to use ball milling in order to achieve better mechanical characteristics. Second step can be extrusion, heat treatment, machining, and others, and it is applied after the final stages in this process. The process of obtaining nanocomposites by powder metallurgy is very expensive, but leads to a significant improvement in the characteristics of the material, such as strength and stiffness.

Mechanical milling is the process of milling or shredding a mixture of powders by various methods. All these methods are based on the same principle of inserting particles of material into the mill where they are subjected to high-energy collisions with other particles and with added steel balls that speed up the process itself. Depending on the type of material from which the powders are obtained, different types of mills are used. The most commonly used are ball milling, vibrating, vortex, and planetary. Ball milling is usually used to make fine powders.

Friction stir processing (*FSP*) is a process for obtaining nanocomposites in the solid state and it is used for surface treatment of aluminum nanocomposites and homogenization of aluminum alloys obtained by powder metallurgy. This technique is based on friction welding, with the aim of obtaining a surface layer without porosity, with a homogeneous distribution of reinforcement particles in the base, and strong bonding between the reinforcement and the base. In this way, a localized modification of the microstructure is produced with a specific surface improvement of the metal properties.

It can be noticed that during the development of nanocomposites, in addition to conventional methods for obtaining nanocomposites, new methods for their production were developed with the aim of achieving an even distribution of reinforcements in the base. Each new process requires certain modifications of the previous ones in order to be applied for the production of nanocomposites depending on the base material and the type of reinforcement. Researches are still on going in order to find an appropriate procedure for the production of nanocomposites in terms of justified economy, the possibility of application for serial production, and achieving an even distribution of reinforcements in the base. Scientists and researchers in recent years have been combining different procedures to achieve improvements and are doing so with more or less success [19]. With the development of materials, there is a need for their analysis and characterization. *SEM* and energy dispersive spectroscopy (*EDS*) are used as standard methods for material analysis and characterization. These methods are also used in microcomposites. Of the newer methods, Transmission electron microscopy (*TEM*) has been used because of its ability to better represent events at the nanolevel.

Aluminum-based composite parts of automobiles are manufactured by combining aluminum with different reinforcement, such as for: pistons $SiC-Al_2O_3$, Al_2O_3,

MoS_2, TiC, Gr, piston rings Al_2O_3–Gr, drum brakes TiB_2–SiC, shaft SiC, camshafts SiC–CNT, Al_2O_3–TiC, and valves Al_2O_3. In the following text, a brief overview of these researches is given in order to show the combinations of base materials and nanoreinforcers for the production of nanocomposites in demanding parts in the automotive industry. The application of hybrid Al nanocomposite for cylinder fabrication was analyzed by Tiruvenkadam et al. The hybrid composite consisted of Al6061 base and nanoreinforcements (2.25 wt%), zirconium dioxide (ZrO_2), silicon carbide (SiC), and graphite (Gr) with sizes of 100 nm, 220 µm, and 100 µm, respectively. They observed improvement of mechanical and thermal characteristics and homogeneous structure of the material [12]. Combination of aluminum base and CNT reinforcement (2–6 wt%) for the production of piston rings was investigated by Carvalho et al. Optimal combination of mechanical and tribological characteristics was obtained for 2 wt% of CNT [13]. Next to the piston rings, nanocomposites with CNT reinforcements were used for disc and drum brakes by Sundaram and Mahamani. They combined A356 base and 2–8 vol% of CNT, and the best mechanical characteristics were obtained for 8 vol% of CNT [15]. Different combination of reinforcements SiC and Al_2O_3 in ratio 0.5, 1.0, 1.5, and 2 wt% in $LM6$ alloy for disc brakes was investigated by Muley et al. [20]. Particle reinforcement of SiC (1–4 wt% and size of 40 nm) in Al base of nanocomposite can be used for camshaft which was proved by Rana et al. [16]. Nanocomposites with ZrO_2 nanoparticles with size of 25 nm, were used for connecting rod as reported by Ramachandra et al. [14].

The main reasons for the increase in demand for nanocomposites are the increased production of cars and therefore light materials. Their application in structural and drive parts is expected to increase as composites act as a good replacement for conventional metal parts. Great importance is given to reducing the weight of vehicles precisely because of improving fuel economy, improving vehicle performance, and reducing emissions, so in the years to come it is expected to increase demand and application of nanocomposites in the automotive and other industries.

3. Conclusion

From the available literature and scientific papers, it can be seen that nanocomposites are attractive materials for researchers, both from practical and theoretical aspects, due to possible combinations of required properties. Over the last three decades, great efforts have been made to use nanotechnology knowledge and nanoscience to produce nanomaterials with the necessary, predefined functionality.

Although there is great potential for the use of metallic nanocomposites in various fields, their greater use is slowed down by the high cost of production, problems in the production of large and complex parts, as well as their increased stiffness. Another reason that slowed down the application of nanocomposites is that they can be hazardous if they are not properly handled. The toxicity of these materials needs to be more thoroughly investigated so they can be safely applied for the production of parts in automobile industry.

References

[1] P.K. Rohatgi, B. Schultz, Lightweight metal matrix nanocomposites–stretching the boundaries of metals, Mater. Matt. 2 (4) (2007) 16–21.
[2] C. Zweben, Composite Materials Mechanical Engineers' Handbook, John Wiley & Sons, Inc, 2015.
[3] Ebrahimi, F. (Ed.),, Nanocomposites: New Trends and Developments, BoD–Books on Demand, 2012, ISBN: 978-953-51-0762-0.
[4] Hashim, A. A. (Ed.),, Advances in Nanocomposite Technology, BoD–Books on Demand, 2011, ISBN: 978-953-307-347-7.
[5] I. Capek, Nanocomposite structures and dispersions: science and nanotechnology, in: D. Mobius, R. Miller (Eds.), Fundamental Principles and Colloidal Particles Studies in Interface Science, Vol. 23, Elsevier, Amsterdam, 2006. ISBN-13: 978-0-44-52716-5, ISBN-10: 0-444-52716-8.
[6] J.P. Davim, C.A. Charitidis, Nanocomposites: Materials, Manufacturing and Engineering, Springer, Heidelberg New York Dordrecht London, 2013 (ISSN 2192-8983).
[7] S. Veličković, S. Garić, B. Stojanović, A. Vencl, Tribological properties of aluminium matrix nanocomposites, Appl. Eng. Lett. 1 (3) (2016) 72–79.
[8] S. Veličković, B. Stojanović, L. Ivanović, S. Miladinović, S. Milojević, Application of nanocomposites in the automotive industry, MVM 45 (3) (2019) 51–64.
[9] B. Stojanovic, J. Glisovic, Automotive engine materials, in: S. Hashmi (Ed.), Reference Module in Materials Science and Materials Engineering, Elsevier, Oxford, 2016, pp. 1–9.
[10] B. Stojanović, L. Ivanović, Application of aluminium hybrid composites in automotive industry, Tehnicki Vjesnik—Technical Gazette 22 (1) (2015) 247–251.
[11] A. Vencl, A. Rac, New wear resistant Al based materials and their application in automotive industry, MVM 30 (Special Edition) (2004) 115–139.
[12] N. Tiruvenkadam, P.R. Thyla, M. Senthilkumar, M. Bharathiraja, A. Murugesan, Synthesis of new aluminum nano hybrid composite liner for energy saving in diesel engines, Energy Convers. Manag. 98 (2015) 440–448.
[13] O. Carvalho, M. Buciumeanu, S. Madeira, D. Soares, F.S. Silva, G. Miranda, Optimization of AlSi–CNTs functionally graded material composites for engine piston rings, Mater. Des. 80 (2015) 163–173.
[14] M. Ramachandra, A. Abhishek, P. Siddeshwar, V. Bharathi, Hardness and wear resistance of ZrO_2 nano particle reinforced Al nanocomposites produced by powder metallurgy, Procedia Mater. Sci. 10 (2015) 212–219.
[15] M.U. Sundaram, A. Mahamani (Eds.), Development of carbon nanotube reinforced aluminum matrix composite brake drum for automotive applications, in: Research and Innovation in Carbon Nanotube-Based Composites, 2015.
[16] R.S. Rana, R. Purohit, V.K. Soni, S. Das, Development and wear analysis of Al-nano SiC composite automotive cam, Mater. Today Proc. 2 (4–5) (2015) 3586–3592.
[17] S. Dhanabal, S. Vetrivel, R.M. Vimal, An overview of hybrid metal matrix composites – characterization, directed applications, and future scope, Int. J. Sci. Eng. Appl. Sci. 1 (9) (2015) 344–350.
[18] Z. Djordjević, Kompozitne konstrukcije, Fakultet inženjerskih nauka Univerziteta u Kragujevcu, Kragujevac, 2018.
[19] S. Veličković, B. Stojanović, M. Babić, A. Vencl, I. Bobić, G.V. Bognár, F. Vučetić, Parametric optimization of the aluminium nanocomposites wear rate, J. Braz. Soc. Mech. Sci. Eng. 41 (1) (2019) 1–10.
[20] A.V. Muley, S. Aravindan, I.P. Singh, Mechanical and tribological studies on nano particles reinforced hybrid aluminum based composite, Manuf. Rev. 2 (2015) 26.

CHAPTER 9

Metallic nanocomposites for automotive applications

A.G. Arsha[a,b], Visakh Manoj[a], L. Ajay Raag[a], M.G. Akhil[a,b], and T.P.D. Rajan[a,b]
[a]CSIR-National Institute for Interdisciplinary Science and Technology, Trivandrum, India
[b]Academy of Scientific and Innovative Research (AcSIR), New Delhi, India

Chapter outline

1. Introduction — 163
2. Nanocomposites vs conventional composites in automotive applications — 164
3. Potential nanoreinforcements — 166
4. Processing of nanocomposites — 167
   4.1 Solid-state processing — 168
   4.2 Deposition processes — 171
   4.3 Liquid-state processing — 172
5. Effect of nanoparticles and mechanisms on the properties of metallic nanocomposites — 176
   5.1 Strength — 176
   5.2 Ductility — 178
   5.3 Hardness — 178
   5.4 Wear behavior — 179
   5.5 Corrosion behavior — 180
   5.6 Thermal properties — 181
6. Characteristics of metallic nanocomposite systems — 182
   6.1 Aluminum — 182
   6.2 Magnesium — 183
   6.3 Copper — 185
   6.4 Titanium — 187
   6.5 High entropy materials — 187
7. Metallic nanocomposite coatings — 190
8. Automotive application of metallic nanocomposites — 192
9. Conclusion — 195
References — 195

1. Introduction

Metallic nanocomposite (MNC) has been paid significant attention by many researchers to their wide range of potential applications in the automotive and aerospace industries. Nanoreinforcements can remarkably improve mechanical strength, creep resistance at elevated temperature, better machinability, and higher fatigue life. Improvement in the properties of MMCs is accredited to fine particle size, uniform particle distribution, interparticle spacing hardening mechanism, and thermal and mechanical stability at high temperature. The fabrication of such type of composites is good for desired needs of advancement and improvements in automotive industry. Metal matrix composites such as aluminum and magnesium reinforced with continuous carbon, SiC, or boron fibers are used in aerospace, automotive other applications due to their low weight and excellent physical properties. There is much interest in producing similar composites that incorporate nanoparticles and CNT for structural applications, as these materials are capable to yield greater improvements than micron-sized reinforcements [1]. Lightweighting becomes an issue for energy efficiency in automobiles. It arises the need for developing a novel generation of materials that will combine both weight reduction and safety issues. Throughout this work, the applicability of nanoparticle-based composite materials is discussed with regard to the fulfilment of these requirements. The application of nanocomposites for development of automotive components is reflected in the improvement of the production rate, environmental and thermal stability, and the reduction in weight in the automotive industry, less wear parts, and indirectly to reduce emissions and environmental pollution. The hardest part of the vehicle is the body shell, which makes 40% of its total weight. Therefore, the most common approach to lightweight design incorporated within the structure of the vehicle makes use of combination of different materials, depending on their engineering properties and functional characteristics. The materials used in the automotive industry are composites which are made of high-strength steel, aluminum, and composites of carbon fiber and plastic. Applying high-strength steel is achieved by decreasing the weight by 20% compared to the conventional steel structures, use of aluminum of 40% and 50% at the application of composites of carbon fiber. Therefore, using light materials reduces the weight of the vehicle since the weight of aluminum ties in the suspension is smaller by 15% than the conventional hollow composite shaft [2, 3].

2. Nanocomposites vs conventional composites in automotive applications

The advantage of nanocomposites over conventional composites is that their mechanical, electrical, thermal, barrier, and chemical properties such as increased tensile strength, improved heat deflection temperature, flame retardancy, etc. can be achieved with typically 3–5 wt% loading of the nanomaterials such as clays, nanotubes, and nanofibers while the latter requires a high content of the inorganic fillers from 10 wt% to as much as 50 wt% in general, to impart the desired properties. Owing to their nanoscale size features and very high surface-to-volume ratios, they possess unique combination of multifunctional properties not shared by their more conventional composite counterparts reinforced with microsized fillers.

Another advantage of nanocomposites is that the strength, shrinkage, warpage, viscosity, and optical properties of the polymer matrix are not significantly affected. The enhanced properties are attributed to the structure and morphology of the nanocomposite, as they (clays/polymer) contain organically treated clays such as hectorite, montmorillonite, and synthetic mica as well as nanotubes (carbon nanotubes, halloysite nanotubes). These nanomaterials have a large aspect ratio (1000:1) and each one is approximately 1 nm thick and hundreds or thousands of these layers are stacked together with weak van der Waals forces to form a clay particle, resulting in subsequent exfoliation in which the individual layers are peeled apart and then dispersed throughout the polymer matrix. The excellent degree of exfoliation, which results in smaller particle sizes and provides the greater surface area to interact with the host polymer, results in improved performance. CNT-enabled nanocomposites are also receiving attention as a mechanical reinforcement and electrically conductive additive for automotive fuel system line components requiring electrical conductivity [4].

The advantage of nanoparticles is that, because of the high specific surface area, already at low concentrations major effects on the macroscopic properties can be obtained [5].

The main advantages of nanocomposites over other composite materials are:

(a) high surface/volume ratio allows small filler size and distance between fillers;
(b) better mechanical properties—high ductility without strength loss, scratch resistance;
(c) improved optical properties (light transmission depends on particle size)

However, there are still many limitations and challenges for nanocomposites production over conventional composites. These include:

Processing: Compatibility, dispersion, and exfoliation between nanomaterials and polymer matrices. Only a limited number of plastic matrices (mostly thermoplastics) are compatible with nanoclays/nanotubes/nanofibers as intercalation of clays with the precursor of a polymer can change the functionality of the polymer and inhibit its properties.

Cost: The production of nanocomposites on a commercial scale at viable prices, as polymer matrix price depends on crude oil prices and CNT price is also high.

Consistency and reliability in volume production: It is possible to get consistency and reliability in volume production materials to a great extent. However, particle size distribution and control in volume manufacturing are not so easy.

High lead time: Commercializing the end-use products would take a longer time, mainly due to stringent approval and OEMs acceptance.

Oxidative and thermal instability of nanoclays: Commonly used organoclays are thermally unstable due to exchange of metal cations in clay galleries with organic ammonium salts and can degrade at temperatures as low as 170°C. It is clear that such organoclays are not suitable for most engineering plastics that are fabricated by melt processing technology [6].

3. Potential nanoreinforcements

Typical nanoparticles are oxides (Al_2O_3, SiO_2, B_2O_3, and $TiO.$), nitrides (BN, Si_3N_4, AlN, TiN), silicides ($TiSi_2$), borides (TiB_2), and carbides (SiC, TiC, B_4C), out of which silicon carbide (SiC) and aluminum oxide (Al_2O_3) are most often found in research works as the selected reinforcements. Fig. 1 shows the possible distribution of the matrix and nanosized reinforcement particles in the composite. SiC particles have become one of the popular reinforcing phases for many aluminum alloy-based metal matrix composites. They are hard and brittle ceramic particles with high strength, high modulus of elasticity, and high thermal and electrical resistance. The use of nanoscale Al_2O_3 has been motivated by its wide availability and low tendency to dissociate into elemental Al and O, avoiding recombination with elements contained in the alloying matrix and the formation of any undesirable phase. However, its strengthening effect is found to level off when the volume fraction is above 4 vol%, which is attributed to the clustering of nano Al_2O_3 particulates due to its low wettability with molten metals and alloys, also TiC_7 and TiB_2 ceramic nanoparticles exhibit good wettability and thermodynamic stability are often chosen as reinforcing phases in metal matrix nanocomposites. Table 1 shows the major reinforcements used in metal matrix nanocomposites.

Moreover, different allotropes of carbon, fullerenes, and carbon nanotubes have been investigated as fillers for several research works published in literature. CNT offers very high mechanical properties to the metal matrix and, meanwhile, they lead to increased electrical conductivity, which makes MMnCs very attractive materials for electrical and automobile applications. Single-wall carbon nanotubes (SWCNT) and multiwall

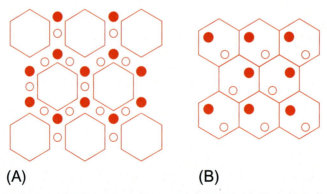

(A) (B)

Fig. 1 Possible distribution of the matrix and nanosized reinforcement particles in the composite, where (A) the reinforcements are distributed along the grain boundaries of the matrix phase and the reinforcements are inside the matrix grains. (B) the open hexagons represent the matrix grains and the open and filled circles represent the reinforcements.

Table 1 Major reinforcements used in metal matrix nanocomposites [7].

Nanoparticles	Density ρ (g/cm^3)	Role of nanoparticles in properties
Silicon carbide, SiC	3.21	Enhancement in hardness, tensile strength, wear resistance, and thermal properties
Boron carbide, B$_4$C	2.52	Higher impact strength and hardness, excellent chemical and abrasion resistance, high neutron absorption capacity
Titanium carbide, TiC	4.93	Good wettability, thermodynamic stability, thermal conductivity, and high wear resistance
Tungsten carbide, WC	15.63	High hardness, high melting point good electrical and thermal conductivity, high corrosion resistance
Alumina, Al$_2$O$_3$	3.9	Improved specific strength, high temperature strength, high wear resistance
Silicon dioxide, SiO$_2$	2.65	Improved compressive and flexural strength
Carbon nanotube	1.3	Enhanced three-point bending strength, mechanical properties, thermal conductivity, and low density
Graphene	2.26	Low density, superior thermal and electrical conductivity, high light transmittance, and corrosion resistance
Fullerene, C$_{60}$	1.65	High tensile strength and excellent conductor of heat and electricity

carbon nanotubes (MWCNT) both are used for MMnCs production. For example, copper-0.1 wt% MWCNT composites shown a 47% increase in hardness and bronze-0.1 wt% SWCNT showed a 20% improved electrical conductivity. The intermetallic compounds (NiAl, Al$_3$Ti) have also been successfully used as reinforcement phase in MMNCs.

4. Processing of nanocomposites

Processing of nanocomposite includes solid- and liquid-state processing. Solid-state includes powder metallurgy, immersion plating, diffusion bonding electroplating, spray deposition, physical vapor deposition, chemical vapor deposition, etc. Liquid-state processing includes stir casting, melt oxidation processing, melt infiltration, squeeze casting, compo casting, or rheo casting, etc. due to high surface energy nanoparticles that have the tendency of agglomeration and clustering. Attractive van der Waals bonding, and electrostatic and moisture adhesiveness affect its uniform distribution during processing. Various processing methods have been reported in literatures to fabricate nanocomposites that improve the uniform distribution of nanoparticles thereby avoiding agglomeration and clustering in the matrix material [8].

4.1 Solid-state processing

Solid-state processing techniques are taking place below the solidus temperature, where both phases remain in the solid state that minimizes diffusivity and undesirable reactions between the nanoparticles and the metal matrix.

4.1.1 Powder metallurgy

The most common solid-state processes are based on powder metallurgy techniques. This method of manufacturing has been used to produce nanoparticle reinforced metallic materials like Al, Mg, Ti, Cu, etc. PM is a low-temperature manufacturing that not only avoids strong interfacial reaction but also minimizes undesired reaction between matrix and reinforcement. Fig. 2 shows the schematic of typical powder metallurgy processing outline [9].

Powder processing includes mixing of nanoparticles and metal powders by mechanical alloying that is ball milling followed by compacting and sintering. An advantage of solid-state processing of powders is that it allows introducing a high volumetric content of reinforcement in the MMNC.

4.1.2 Severe plastic deformation (SPD)

Based on this grain refinement effect, bulk nanostructured materials processed by several methods of severe plastic deformation (SPD), such as ECAP, HPT, accumulative roll

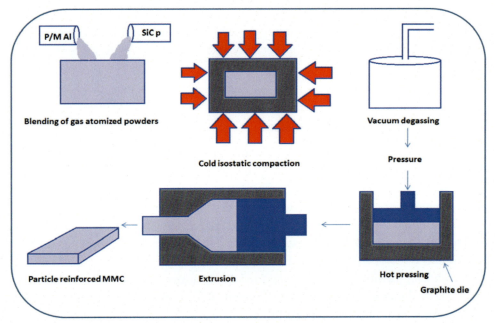

Fig. 2 Schematic of typical powder metallurgy processing outline.

bonding (ARB), friction stir processing (FSP), cyclic extrusion compression, torsion compression, and multiaxial forging have been adapted to fabricate bulk nanocomposite materials. ECAP and ARB are the most widely used methods among the various SPD techniques. The main advantage of SPD processed materials is the possibility of overcoming a number of difficulties connected with residual defects and powder contaminations in the powder compacted samples.

Equal channel angular pressing (ECAP)

Among the various SPD processes, ECAP is one of the convenient procedures for obtaining ultrafine-grained materials by extruding metallic materials through specially designed channel dies without a significant change in geometries. Fig. 3 shows schematic of ECAP setup showing the sample rotations. In ECAP the deformation takes place by simple shear limited to a narrow zone at the plane of intersection of the die channels. The sample in the form of bar or rod is machined to fit within the channel and the die is placed in the form of press so that the sample can be pressed through the die using a plunger.

For example, Al–Al$_2$O$_3$ nanocomposite was pressed at 400 °C with a pressure of 200 MPa and four ECAP passes. The interaction between severe shear deformation and in situ oxidation during ECAP attributed to the formation of nanocomposite. The ultimate strength of the nanostructured material reached 740 MPa in compression with a plastic strain to fracture of the order of 1%.

High pressure torsion (HPT)

During HPT deformation a disc-shaped specimen is pressed between two anvils which have a cylindrical cavity with a depth smaller than the thickness of the specimen. One of the anvils is rotated, whereas the other is fixed and the specimen is deformed ideally by

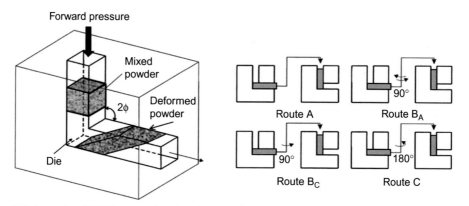

Fig. 3 Schematic of ECAP setup showing the sample rotations [10].

simple shear. Usually a pressure of several GPa is applied during HPT deformation. Due to the large applied hydrostatic pressure HPT is the suitable method to induce very high strains continuously and to deform high strength materials.

Accumulative roll bonding (ARB)
ARB is considered as the unique solid-state process, which can be utilized for particle-reinforced metal matrix composites manufacturing with highly improved mechanical properties. Fig. 4 shows the schematic illustration showing the principles of the ARB process. In this method, reinforcement particles are added between Al strips during initial ARB cycle, after that the produced sheets are cut and stacked together and rolled.

Uniform distribution of the particles will be achieved after certain number of ARB cycles by following these procedures. Owing to its ability to reduce the grain size it is considered as one of the severe plastic deformation methods for production of ultrafine-grained and nanostructured metallic nanocomposites. This technique has been extensively applied for improving the strength of different metals and their alloys through grain refinement. It has been reported that the mechanical properties of MMNCs can be improved using ARB method but considerable decrease in elongation and ductility is also observed.

Friction stir processing (FSP)
The basic concept of FSP is simple, a rotating tool with pin and shoulder is inserted in a single piece of material for microstructural modification and traversed along the desired line to cover the region of interest. Friction between the tool and workpieces results in localized heating that softens and plasticizes the workpiece. A volume of processed

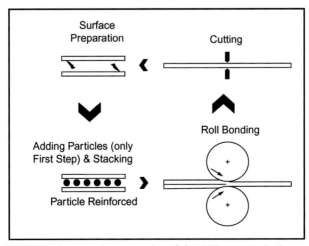

Fig. 4 Schematic illustration showing the principles of the ARB process [11].

material is produced by movement of materials from the front of the pin to the back. During this process the material undergoes extreme plastic deformation which results in significant grain refinement.

High-energy ball milling

High-energy ball milling is a mechanical deformation process that is used in the processing of nanocrystalline materials in powder form. In the high-energy ball milling process, coarse-grained structures undergo disassociation as the result of severe cyclic deformation induced by milling. This process has been successfully used to produce metals with minimum particle sizes from 4 to 26 nm. The high-energy ball milling technique is simple and has high potential to scale up to produce large quantities of materials. Bulk metallic materials produced by this approach have achieved the theoretical densities of nanocrystalline materials and greatly improved mechanical properties compared to their conventional, micron-grained counterparts [12].

4.2 Deposition processes

Generally, the coatings produced through metal-based matrixes comprising the second phase (as reinforcement) of metallic, ceramic, or polymeric material are known as metallic matrix composite coatings (MMCC). Similarly, the coatings fabricated with both (metallic matrix and strengthening phase) or at least one of them having a distinctive nanometers (1–100 nm) length scale are called metallic nanocomposite coatings (MNCC). Various deposition techniques such as thermal spray, plasma spray, laser processing, electrolytic and electroless coating methods are developed and used for the formation of MNCC. Fig. 5 shows the schematic of thermal spray coating.

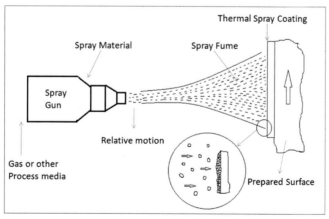

Fig. 5 Schematic of thermal spray coating [13].

Thermal spray composite coatings have long been established as high-performance functional coatings for applications such as wear resistance, corrosion resistance, and thermal barriers. Thermal spray nanocomposite coatings are much thicker than those achievable by other methods. Thermal spray processes have the unique capability for dimensional restoration of damaged or worn-out engineering and structural components without replacement. This is a sustainable approach for incorporating nanocomposite materials as thermal barrier, wear-resistant, corrosion-resistant, and abradable coatings. The micro hardness, wear resistance, and corrosion resistance of the MNCCs are remarkably improved after incorporation of the nanoparticles [14].

Electroless process is a coating technique in which the deposition of metal, alloy, or a composite on an activated surface by autocatalytic reduction of metallic ions from the salt solution containing a reducing agent. Most of the substrates can be plated by this method, irrespective of their conductivity due to the nature of deposition, without an external electrical current. Also, it is a better way to plate complex parts completely over sharp edges and deep recesses [15].

4.3 Liquid-state processing

The new developments in liquid-state processes are to overcome difficulties associated with the low wettability issues of nanoparticles and the molten metal. The formation of undesirable compounds from reactions between the melt, nanoparticles, and the environment should be also avoided. To resolve these difficulties numerous processes have been developed, such as casting the melt in a crucible while is stirring via mechanical or ultrasonic probe or infiltration by the molten metal of a porous ceramic structure aided by high-pressure casting.

4.3.1 Stir casting

Stir casting method is economical and is always preferred. Stir casting has some remarkable advantages. It is more flexible and simple method compared to other methods. Stir casting method is highly suitable for near net shape components and is applicable to large volume production. Stirring process has some important advantages, e.g., the wide selection of materials, better matrix–particle bonding, easier control of matrix structure, simple and inexpensive processing, flexibility and applicability to large quantity production, and excellent productivity for near-net-shaped components. However, there are some problems associated with stir casting of AMCs such as poor wettability and heterogeneous distribution of the reinforcement material. Fig. 6 shows the schematic of stir casting technique.

Studies suggested that reinforcement in particulate form up to 30% by weight can be added in molten alloy to achieve better distribution of the reinforcement. Homogeneity of the reinforcement particles added during solidification of composite depends on following factors: (a) Pouring temperature and solidification rate, (b) Stirring blade angle, (c) Stirring speed and time, and (d) percentage, reinforcement's size, and its relative density.

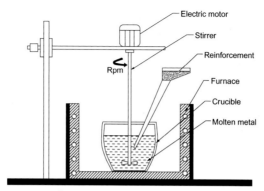

Fig. 6 Schematic of stir casting technique [16].

4.3.2 Compo casting

Compo casting is one of the economic and efficient methods for fabricating metal matrix nanocomposites over other conventional methods due to its advantages like low casting temperature, uniform distribution of reinforcement without agglomeration, good wettability, and better matrix-reinforcement bonding. Compo casting is a solid–liquid state method in which a vortex is created in the semisolid molten metal is using an impeller driven by an electric motor, and the reinforcements are added into the vortex under stirring. During compo casting process, the higher viscosity of semisolid molten metal slurry transmits shear force over the agglomerated reinforcements which leads to better dispersion of reinforcement in the matrix. Nanosize reinforcements can be successfully incorporated into the matrix by combo casting process.

4.3.3 Ultrasonic cavitation technique

Ultrasonic processing technique is more reliable for producing nanocomposites. The application of high-intensity ultrasonic waves has proven to be an effective means for the uniform dispersion of nanocomposites. Fig. 7 shows the schematic representation of cavitation and steaming effects for nanoparticle dispersion. By this method the nanoparticle dispersion and wetting in the metal melts and agglomerates have been greatly improved when compared with the severe agglomerates in composites fabricated by traditional methods [17].

4.3.4 Spray casting and disintegrated melt deposition

The synthesis of nanoparticle-reinforced composites using the DMD technique comprises superheating the metallic alloy and the reinforcement powder in a multilayered arrangement under an inert gas atmosphere in a graphite crucible by a resistance heating furnace. Fig. 8 shows schematic diagram of DMD process. Spray deposition or spray forming is a method similar to DMD. During spray deposition the molten metal with

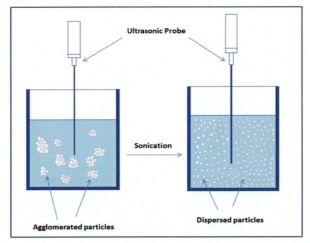

Fig. 7 Schematic representation of cavitation and steaming effects for nanoparticle dispersion.

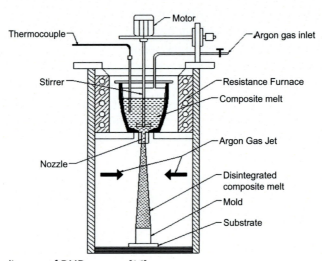

Fig. 8 Schematic diagram of DMD process [16].

the nanoparticles is conveyed through a nozzle into a gas atomizer, where it is atomized with inert gas into fine molten droplets and collected on a substrate where metal solidification is completed. The reinforcement may be placed on the substrate and the molten metal may be sprayed into it. Advantages of spray deposition are the detrimental reaction products that are usually avoided because the time of contact of the nanoparticles and the molten metal is extremely short and microstructure achieved is very fine due to the very high cooling rate.

The disintegration of composite melt assures higher solidification rate and fine-grain structure. This processing technique can be mainly employed for the fabrication of aluminum and magnesium nanocomposites.

4.3.5 Squeeze infiltration process

Infiltration techniques involve infiltrating a preform or partial matrix containing the reinforcements with a liquid metal. The preform consists of particles formed in a particular shape with some binding agent, and a pore former and can be composed of the additives and binding agent alone or with some portion of the matrix added as filler material. Fig. 9 shows the schematic representation of Squeeze infiltration process.

Infiltration methods that have been used include ultrahigh pressure, where the pressure is used to infiltrate a high-density preform of nanoparticles. Where in pressure less infiltration a block of metal is melted on top of a lower density preform of nanoparticles and allowed to seep into the preform [17].

4.3.6 In situ process

An alternative approach to embed nanoparticles inside an aluminum melt is by in situ process. In this process reinforcement is introduced in the matrix as a result of precipitation from the melt while it solidifies by some chemical reaction during processing. For example, copper matrix nanocomposite reinforced by in situ TiB_2 nanoparticle was prepared by reactions of B_2O_3, carbon, and titanium in copper–titanium melt. The in situ-formed TiB_2 particles' size of about 50 nm are uniformly distributed throughout copper matrix [18].

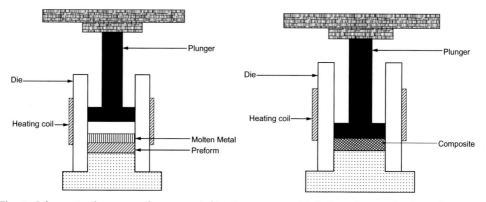

Fig. 9 Schematic diagrams of squeeze infiltration process (A) before the application of squeeze pressure and (B) after the application of squeeze pressure as a driving force to infiltrate molten metal into the preform.

5. Effect of nanoparticles and mechanisms on the properties of metallic nanocomposites

Nanoparticles of different species impart different effects on the MMNCs. For the end applications, strength is an important selection criterion because it determines the structural efficiency and is usually a targeted property for composites. Other than the strength of the ceramics, the distribution of nanoparticles is responsible for the structural efficiency. When nanoparticles are blended, their interaction with matrices determines the distribution of themselves. They may act as nucleation sites or nonnucleation sites, locating inside the matrix grains or in between the grain boundaries.

5.1 Strength

The introduction of nanoparticles causes grain refinement in the metal matrix due to grain growth constraint caused by pinning at the grain boundary or the introduction of nucleation sites [19]. The Hall–Petch effect is taken into account as the strength of materials depends on the grain size of matrix. The empirical relationship is given as follows [20]:

$$\sigma_{GR} = \sigma_0 + \frac{K_y}{\sqrt{d}}, \tag{1}$$

where σ_0 is the yield strength of a single crystal, regardless of any strengthening mechanisms other than the solid solution effect, d is the average matrix grain size, and k_y is a constant that varies with the materials. In MMNCs, this contribution is determined by the matrix grain size, which is influenced by the size and volume fraction of nanoparticles. According to the Hall–Petch effect, there is a strengthening limit and there is a refinement limit. Furthermore, a higher volume fraction causes nanoparticle aggregation and degradation of refinement. The dislocations suffer bowing, reconnecting, and leaving a dislocation loop around the nanoparticles when they move through the small hard particles. This resistance against the motion of dislocations results in considerate strength increments and accounts for a primary mechanism in particle-reinforced MMNCs, known as Orowan strengthening mechanism (Fig. 10) [21].

HRTEM images of interface bonding, and dislocation looping and bowing of Al-0.5 wt% Al_2O_3 nanocomposite by modified compocasting process are shown in Fig. 11 [27]. This mechanism is negligible in the MMCs reinforced with microscale particles and is activated only when particle size is below 1 μm as both the size of particles and the spacing between them should be comparable to the dimension of dislocation [22]. Also there emerges a strong bonding at atomic level between dispersed nanoparticles and the matrix due to the nanoscale connection and sound synthesizing methods. Strong coherent bonding facilitates the load transfer at the interface and increases the strength of the materials [23].

Metallic nanocomposites for automotive applications 177

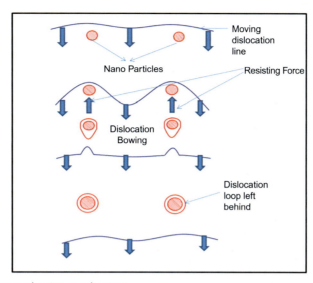

Fig. 10 Orowan strengthening mechanism.

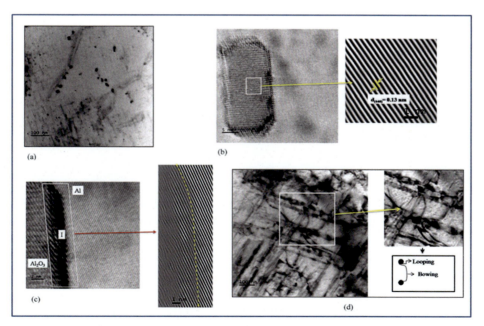

Fig. 11 HRTEM images of Al-0.5 wt% Al$_2$O$_3$ nanocomposite by modified compocasting process showing (A) Al$_2$O$_3$ nanoparticles in aluminum matrix; (B) Crystalline lattice of Al$_2$O$_3$ nanoparticle; (C) Al-Al$_2$O$_3$ interface bonding; (D) Dislocation looping and bowing in the Al-Al$_2$O$_3$ nanocomposite.

5.2 Ductility

When it comes to strengthening, ductility is normally sacrificed. Because of the large size of the particles and their aggregations, MMCs with micrometer reinforcement face a trade-off between tensile strength and ductility. The stress concentration rises as the number of particles increases [24]. This problem could be solved by nanocomposites with smaller reinforcement and stronger interfacial bonding. The effects of nanoparticles on the ductility of MMNCs appear to differ depending on the fabrication process. The grain boundaries and the distribution of nanoparticles are the most important factors [25]. In several liquid-state processes, nanoparticles help to refine the matrix. Despite this beneficial impact on microstructure, declining ductility is still a common occurrence in MMNCs, highlighting the significance of nanoparticle distribution. The mechanisms that cause nanoparticles to boost ductility are unknown. The conditions of grain boundaries and particle dispersion are two possible hypotheses. The fracture is caused by flaws in grain boundary bonding and decohesion. The grain boundaries between matrix grains are replaced by the interface between matrix and nanoparticles, making the MMNCs less ductile. The interface fraction increases as the nanoparticle content increases. The ease of dislocation motion is related to particle dispersion. During deformation, the CTE mismatch and elastic modulus mismatch result in a high dislocation density. The ductility of Mo alloys with intragranular and intergranular La_2O_3 particles was compared by Liu et al. [26]. The former nanoparticle distribution pattern reduces grain boundary cracking, which is caused by larger intergranular particles that help to nucleate and pin down dislocations. The effect of interior nanoparticles is similar to that of heat-treated alloys with nanometer-scale precipitates. However, achieving this particle distribution, especially with exogenous reinforcement particles, is difficult. Except for certain smaller particles that do not agglomerate, intergranular particles appear to fracture and decohere from the matrix. The negative effects of nanoparticles decrease as their size increases.

5.3 Hardness

In comparison to the base alloy, nanoparticles distributed in aluminum alloy greatly increase the hardness properties. The hard nanoparticles in the aluminum matrix act as barriers to the movement of dislocations produced in the matrix, and the higher particle density results in the Orowan mechanism. The effect of Al_2O_3 nanoparticles on the hardness of A356 alloy was investigated by Sree Manu et al. In heat-treated condition, composite reinforced with 0.5 wt% Al_2O_3 nanoparticles manufactured using a changed compocasting method has a higher hardness (147 BHN) than other systems. Because of the agglomeration of nanoparticles, as the weight percentage (1 wt%) of Al_2O_3 nanoparticles in the matrix increased, the hardness value decreased compared to the previous

Table 2 Hardness of MMNCs obtained through various processing techniques [28].

Sl. No.	Composite system	Processing technique	Micro hardness (Hv)
1	A206 + 5 vol% Al_2O_3 (APS ≈ 100 nm)	Compo casting	183 (For A206–174)
2	A2024 + 0.6 wt% Al_2O_3 (APS ≈ 13 nm)	Master powder feeding (MPF)	130 (For A2024–108)
3	A7XXX + 0.76 wt% CNF (APS: Dia ≈ 130 nm and Length ≈ 10 μm)	Stir + MPF + T6 + hot extrusion	120 (For A7XXX – 98)
4	A2024 + 0.6 wt% Al_2O_3 (APS ≈ 13 nm)	Stir casting	112
5	A356 + 3.5 Vol% SiC (APS ≈ 50 nm)	Compo casting	80 (For A356–53.1)
6	A356 + 4 wt% Al_2O_3 (APS ≈ 50 nm)	Compo casting	78
7	A356 + 2.5 vol% MgO (APS ≈ 500 nm)	Stir casting	70
8	Mg + 6 vol% SiC (APS ≈ 50 nm)	Semisolid + ultrasonic-assisted casting	180 (For Mg–40)
9	AZ61 + 6.5 vol% SiO_2 (APS ≈ 20 nm)	Friction stir processing	97 (For AZ61–60)
10	Mg + 5.6 Ti + 1.5 B_4C (wt%) (APS ≈ 50 nm)	Disintegrated melt deposition + hot extrusion	71
11	Mg + 1.1 vol% Al_2O_3 (APS ≈ 50 nm)	Disintegrated melt deposition + hot extrusion	65
12	Mg + 0.66 vol% ZrO_2 (APS ≈ 29–68 nm)	Disintegrated melt deposition + hot extrusion	51
13	Mg + 1.3 wt% CNT (APS: Dia ≈ 30–50 nm and length ≈ 1–2 μm)	Disintegrated melt deposition + hot extrusion	46

ones. Since nanoparticles have a high surface energy due to their large surface area to volume ratio, their tendency to bind together and form agglomerated zones in the matrix increases as the number of nanoparticles added exceeds the cap. The composite's hardness is also determined by the manufacturing method used and the accuracy of the uniform reinforcement distribution [27]. Hardness of metal matrix nanocomposites obtained through various processing techniques are shown in Table 2.

5.4 Wear behavior

The composite showed signs of abrasion, delamination, and oxide formation. The impression surface coverage for each mechanism is determined by the load applied normal to the worn surface. The wear function is determined by the worn specimen's surface morphology. Under mild wear conditions, abrasion wear dominates, while delamination takes over in the transition region. Oxidation formation becomes more evident under heavy loads [29]. At slow sliding speeds, abrasive wear is marked by multiple grooves and shallow scratches running parallel to the sliding direction. The features associated with abrasion were discovered after a thorough inspection of the worn track on the pin surface. At low sliding velocity, the mechanism of abrasive wear observed is very similar to the results of several other researchers [30]. Wear debris and wear tracks were compacted on the sliding floor. The wear rate was lower for the composite because it is stronger than pure magnesium, and the ZnO reinforcement increased the composite's hardness [31]. Debris formation due to wear was observed at low sliding speeds, and delamination theory was used to explain it. This delamination process resulted in adhesive wear and fatigue wear. When a material's microstructure includes hard particles, crack nucleation occurs around these hard particles when the material slides against another surface [32]. Ploughing is primarily caused by wear debris that forms as a result of the delamination theory's final stage.

The surfaces of the pins maintain their metallic luster as they slip under heavy loads and at high speeds. The formation of an oxide layer occurs when the pin surface slides against the disc surface under heavy loads and at high sliding speeds. Oxidation is a form of wear mechanism that occurs as a result of frictional heating of surfaces during sliding. The valleys of the pin surfaces accumulated oxide debris as the sliding continued, and the surface developed a protective layer. Since the layer is formed above the pin surface, direct contact between the pin and disc surfaces is avoided, which reduces wear [33].

5.5 Corrosion behavior

The nanoparticles in the composite act as an inert material to the corrosive liquid, shielding the matrix from the chemical reaction and improving corrosion resistance. The corrosion resistance of Al alloys and nanocomposites is due to the fine grains, which results in a high grain boundary density, which can create more passivation sites [34]. According to Sree Manu et al., the uniform distribution of nonconductive Al_2O_3 nanoparticles in the matrix obstructs electron mobility to the electrolyte during the reduction process and causes accumulation of negative charges at the particle-matrix interface of the composite, shifting the potential of Al-0.5 wt percent Al_2O_3 nanocomposite in the negative direction [27].

Fig. 12 shows a schematic representation of the corrosion prevention process involving the addition of diamond nanoparticles to a Ni-P matrix. Long and straight grain boundaries were found in the Ni-P coating cross-section, reducing the time it takes for corroding agents to hit the metallic surface. Co-deposited nanoparticles, on the other

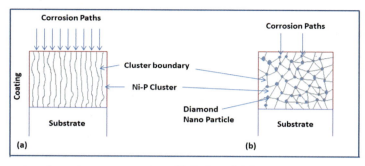

Fig. 12 Schematics of coatings cross-section: (A) Ni-P and (B) Ni-P/DNP elucidates the corrosion prevention by diamond nanoparticle.

hand, act as barriers in corrosion pathways, preventing columnar formation. The corrosion paths become less straight as the nanoparticle content in the coating increases, resulting in increased corrosion resistance of the composite coating [35].

5.6 Thermal properties

The effective thermal conductivity of a composite material with nanosized nonmetal particles embedded in a metal matrix is influenced by the e–ph coupling as well as the size and volume fraction of the particles. So many studies indicate that (i) the effect of the e-ph interaction has a simple interpretation as thermal resistance, so compared with the perfect coupling, namely that the electrons and phonons are in equilibrium, the effect of the e-ph coupling in nonmetal-metal composite materials leads to a reduction in thermal conductivity; (ii) the thermal conduction of the nanosized particle composites is dependent on particle size, whereby the thermal conductivity decreases with decreases in particle size; (iii) the particle volume fraction has an obvious effect on thermal conductivity, whereby when the average particle size is in the range of 0–150 nm the thermal conductivity decreases with the increasing particle volume fraction; (iv) when the interfacial thermal resistance is large enough it can cause the effective thermal conductivity of composites to be lower than that of both the matrix and particles; and (v) a reduction in the thermal conductivity of composites with nanosized nonmetal particles embedded in a metal matrix is the combined result of interfacial thermal resistance and e–ph coupling [36].

The coefficient of thermal expansion (CTE) of nanoparticle reinforced metal matrix composites is mainly influenced by the bonding between matrix and reinforcement. The uniformly distributed nanoparticle hinders the expansion of the matrix material and thereby providing improved CTE. The factors which negatively influence the CTE are the conflict in values due to misfit strain within the constituents of composite material and due to clusters of particulates, residual stresses, fractures, and formation of interfacial-layer between matrix and reinforcements [37].

6. Characteristics of metallic nanocomposite systems
6.1 Aluminum

Aluminum and aluminum-based alloys find a wide range of applications in aerospace, automotive, and defense sectors because of their distinctive properties such as lightweight, high electrical conductivity, and good corrosion resistance. Al matrix composites can be used for the automobile products such as engine piston, cylinder liner, and brake disc/drum. They are the most used engineering alloys, next only to steel. But these composites and alloys have some constraints too due to their low stiffness, low yield strength, and poor wear and tear resistance. The strength and wear resistance at both ambient and elevated temperatures can be improved by adding ceramic hard particles to soft aluminum alloys. But it is necessary that uniform distribution of the reinforcement is necessary to gain effective load-bearing capacity of reinforcement. Heterogeneous distribution of reinforcement particles and their agglomeration can lead to lower ductility, strength, and toughness of the composites. Improved mechanical properties can be achieved when the reinforcement particles are very small in size. Aluminum-based MMCs with nanometer-range (<0.1 lm) discontinuous hard phases as reinforcements have attracted considerable research interest in recent years due to the potential for the development of novel composites with unique mechanical and physical properties. Homogeneous dispersion of the reinforcement particles is the major challenge while fabrication of nanocomposites by adding nanoparticles into the matrix. An eventual technique to achieve homogeneous dispersion is the powder metallurgy route involving high-energy milling. The most important reinforcements added to aluminum and its alloys are Al_2O_3, SiC, carbon nanotubes, AlN, BN, CuO, graphite, TiO_2, and some intermetallic phases.

Al–Al_2O_3 composites have a high potential for very specific applications such as in aerospace and military weaponry but their use has been limited due to the high cost of their fabrication. Al–Al_2O_3 nanocomposites can be prepared either by in situ through chemical reactions or by ex situ methods in which particles are added to Al and the mixture is milled together to produce the nanocomposite. The in situ composites have superior mechanical properties (not significantly different.) than ex situ composites since the former is having stronger interfacial bond.

The exceptional mechanical properties of CNTs make them ideal candidates as reinforcements in composite materials to increase both stiffness and strength while also contributing to weight savings. The major challenge in reinforcing metallic matrices with carbon nanotubes is the agglomeration and poor dispersion and lack of control over alignment of CNTs within the matrix which results in inferior mechanical properties. The tendency of CNTs to agglomerate is due to their high aspect ratio and van der Waals bonding. The agglomeration of the CNTs can be inhibited by following mechanical alloying processing technique [38]. By increasing the milling time and weight fraction of the CNTs higher strength and hardness can be achieved [39]. CNT-reinforced Al matrix composites show excellent mechanical properties (hardness, yield strength, compressive strength, and Young's modulus) due to the formation of an amorphous layer between the Al matrix and the carbon nanotubes which led to a better wetting behavior of CNTs by the Al matrix [40]. SiC is another important reinforcement of Al because of its high strength and modulus of elasticity. Further, its presence in Al increases the wear resistance. The addition of nanometer-sized SiC further improves the mechanical properties and also increases the ductility of the composites. Large difference in the coefficients of thermal expansion between Al and SiC and the poor wettability between Al and SiC are the major difficulties in fabricating Al–SiC composites by the conventional solidification methods. Also the formation of brittle phases like Al_4C_3 and Si due to the undesirable reactions between SiC and molten Al adversely affects the mechanical properties. Al–SiC composites can be widely used in the aerospace and automobile industries for applications such as electronic heat sinks, automotive drive shafts, ground vehicle brake rotors, jet fighter aircraft fins, or explosion engine components.

Al–TiC composites combine the ductility and toughness of the aluminum matrix with the strength, stiffness, hardness, and wear resistance of TiC particles, and are attractive candidates for structural applications, especially in the automotive sector. These composites are also potential grain refiners [41]. Table 3 shows the mechanical properties of aluminium matrix nanoparticle reinforced composites.

6.2 Magnesium

Particulate reinforced magnesium matrix composites are highly potential materials for aerospace, automotive, and other industrial applications due to their low density and high specific mechanical properties. Powder metallurgy technique can be used to produce a heterogeneous Mg-based nanocomposite with significantly increased yield strength and retained malleability where we can see heterogeneously distributed nanosized ceramic particles in the magnesium matrix. The outcome will be a continuous "hard region" with reinforcement and an isolated "soft region" without reinforcement. Heterogeneous Mg-based nanocomposites have superior combined strength and malleability/toughness properties when compared to homogeneous nanocomposites. The malleability/toughness properties of heterogeneous Mg-based nanocomposite are also affected by the density of soft region and the milling time. Malleability and toughness increase with increase in milling time. However, the plastic deformation ability of the heterogeneous Mg-based nanocomposites is sacrificed as other composites. It is the grain size and ratio of hard region together determines the strengths of heterogeneous Mg-based nanocomposites. The strength, stiffness, damping behavior, wear behavior, creep, and fatigue properties of Mg alloys can be remarkably improved by the addition of reinforcing particles into the Mg matrices. It is the alloying elements and their concentration that determine the yield strength, elongation, hardness, grain refinement, intermetallic formation, and

Table 3 Mechanical properties of aluminum matrix nanoparticle reinforced composites.

Sl. No.	System	Hardness	Ultimate tensile strength	Yield strength	Compressive strength	Wear
1	Al356-AlN (by in situ chemical reaction, AlN particles 60 to 80 nm in diameter)	Hardness increases 150 HV at 10 wt% AlN (75 Hv for Al356)	With 0.5 vol% AlN particles, the ultimate strength increased by 35% (234 MPa for A356 alloy)	With 0.5 vol% AlN particles, yield strength is increased by 25% (165 MPa for A356 alloy)	Improved compressive strength with increase in AlN	Improved wear resistance due to pinning effect of AlN particle at the grain boundaries
2	Al-Al$_2$O$_3$ (ball milling technique, size of Al$_2$O$_3$ particles <25 nm)	Hardness increases 5 vol% Al2O3 had a hardness of 190. (hardness of pure Al47.2)	UTS increases with increase in % volume fraction of Al2O3 whereas the ductility decreases with increase in volume fraction	Improved yield strength 461 MPa for 10 vol% Al$_2$O$_3$ (yield strength of pure Al 11 MPa)	Improved compressive strength 600 MPa for 10 vol% Al$_2$O$_3$ (compressive strength for pure Al 58 Mpa)	Improved wear resistance
3	Al-C (ball milling technique, nanocomposites with <100 nm grain size)	Significantly higher hardness, with increasing C content	Improved tensile strength whereas the ductility decreases with carbon content	Improved yield strength with increase in volume fraction of C	Improved compressive strength 260 MPa with 5% C (compressive strength for pure Al 58 Mpa)	Improved wear resistance
4	Al2024-CNT (ball milling technique, grain particle size 30 nm)	Hardness increases (245 Hv for 2 vol% MWCNTs whereas 137 for Al2024)	21% increase in tensile strength over that of pure Al (90 MPa)	Improved yield strength (yield strength of 770 MPa for 2 vol % MWCNTs whereas 324 MPa for Al2024)	Improved compressive strength (810 MPa for 2 vol% MWCNTs)	Improved wear resistance
5	Al-SiC ball milling technique, particle size 31 nm	Hardness increases 163 HV (hardness of pure Al47.2)	Improved tensile strength of 162.61 MPa (90 MPa for pure Al)	Improved yield strength 370 MPa 15 vol% SiC. (yield strength of pure Al 11 MPa)	Improved compressive strength	Improved wear resistance, wear resistance increase with increase in % of SiC

corrosion properties. Better dispersion of nanoscale particles in Mg matrices inhibits dislocation motions which lead to the strengthening of Mg matrices. It has been generally known that better dispersion of nanoscale particles in Mg matrices leads to strengthening of the Mg matrices due to their inhibition of dislocation motions. For example, the addition of alumina (Al_2O_3) nanoparticles as reinforcements can result in grain refinement which improves the yield strength of the Mg alloy. But uniform dispersion of nanoscale reinforcements in metal matrices is challenging because of their large surface areas, nanoscale particle sizes, and strong van der Waals forces. Addition of various nanoreinforcements in Mg matrices, such as alumina (Al_2O_3), zirconia (ZrO_2), yttria (Y_2O_3), graphene nanoplatelets, and calcium phosphate ceramic can improve the mechanical properties of Mg matrices via synergistic strengthening mechanisms such as Orowan Stregthening, dispersion strengthening, grain refinement strenghgthening, work hardening strengthening, solid solution strengthening and difference in thermal expansion strengthening. Normally difference in elastic modulus between the reinforcement nanoparticles and Mg matrices, difference in thermal expansion coefficients, load transfer strengthening from matrix to reinforcement, precipitation strengthening of nanoparticles, and grain refinement strengthening conclude the strengthening of magnesium matrix nanocomposites. Even though the nanoscale reinforcements enhance the yield strength of the composites, higher concentrations of these particles which may lead to agglomeration adversely affect the mechanical properties of Mg matrices [42, 43]. Table 4 shows mechanical properties of magnesium matrix nanoparticle reinforced composites.

6.3 Copper

One of the major groups of commercial materials constitutes copper and copper-based alloys and because of their excellent properties such as outstanding resistance to corrosion, easy fabricability, and good fatigue resistance they are widely used in engineering applications. However, the main disadvantage of copper alloys is their low intrinsic strength. Addition of hard ceramic particles to soft copper alloys provides a combination of properties of both metallic matrix and ceramic reinforcement components and this may result in improvement of physical and mechanical properties of the composite.

The presence of fine Al_2O_3 particles in copper matrix leads not only to improvement in the hardness of this material but also to a decrease in the grain growth rate at temperatures even close to the melting point of the copper matrix. A wide range of properties in the composite for different applications can be achieved by controlling the amount, size, and distribution of the reinforcing particles. For many applications such as electronic packaging or manufacturing of electrodes and contact terminals, the materials must have both high strength and high conductivity. Cu–Al_2O_3 MMCs combine the high electrical and thermal conductivity of the copper phase and the high strength and high thermal and chemical stability of the Al_2O_3 phase. Thus, Cu–Al_2O_3 MMCs have the potential to offer

Table 4 Mechanical properties of Magnesium matrix nanoparticle reinforced composites.

Sl. No.	System	Hardness	Ultimate tensile strength	Yield strength	Compressive strength	Wear
1	Mg-SiC (ultrasonic-cavitation based casting method particle size 50 nm)	microhardness increase from 19% to 34% (for pure Mg, 49 Hv)	Improved tensile strength when compared to unreinforced Mg	Improved yield strength when compared to unreinforced Mg	Improved compressive strength with increasing amount of SiC	Superior wear resistance at lower and higher wear rate at high load
2	Mg-BN (powder metallurgy)	Hardness increases with increasing amount of BN (58 Hv with 1.5 wt% BN whereas 49 Hv for pure Mg)	UTS is improved (205 MPa with 1.5 wt% BN whereas 177 MPa for pure Mg)	Slight increase in yield strength with 1.5 wt% BN (146 MPa whereas 144 MPa for pure Mg)	Improvement in compressive strength with 1.5 wt%	0.5 wt% of boron nitride as reinforcement improves wear resistance
3	Mg-graphene (Powder metallurgy, Particle size 5–15 nm)	Hardness increases (70.5 Hv with 0.3 wt % graphene nanoparticles whereas 70.5 for pure Mg)	UTS is improved (244 MPa with 0.3 wt% graphene nanoparticles whereas 224 MPa for pure Mg)	Improvement in yield strength (200 MPa with 0.3 wt% graphene whereas 190Mpa for pure Mg)	Improved compressive strength with increase in graphene content	Improved wear resistance
4	Mg-Cu (powder metallurgy technique, particle size 50 nm)	Significant improvement in hardness	Best tensile properties were obtained in the Mg–0.6 vol% Cu composite	Improvement in yield strength but ductility is adversely affected	Improvement in compressive strength	

both high strength and high electrical conductivity. In order to achieve high fracture toughness and low processing cost, it is necessary that the Al_2O_3 phase in the microstructure is in particulate form, and that the particle size is small. The high hardness of the Cu–Al_2O_3 nanocomposite is due to the combination of grain size strengthening, solid solution strengthening, and dispersion hardening [44]. Hardness of the Cu–B_4C composites synthesized from blended elemental powder mixtures of electrolytic copper, amorphous boron, and graphite powders offers better increases with increasing volume percentage of B_4C and also in comparison to other particle-dispersion-strengthened coppers, the thermal stability of the B_4C-dispersion-strengthened composites is higher [45]. Table 5 shows mechanical properties of copper matrix nanoparticle reinforced composites.

6.4 Titanium

Titanium and its alloys find extensive applications in the industry due to their lightweight, high specific strength, acceptable mechanical properties at elevated temperatures, excellent oxidation resistance, and good corrosion resistance (at least up to about 1073 K). But, these alloys are not used widely because of their low ambient temperature ductility. A number of attempts have been made to increase the room-temperature ductility of these alloys. Refinement of grain size down to nanometer level, crystal structure modification to a more symmetric structure, and heat treatment to alter the nature and proportion of phases in the alloy have been tried by some researchers [46]. Addition of a reinforcement phase to produce composites has been another approach to improve the properties and performance of these alloys. Different types of reinforcement have been added including Al_2O_3, Nd_2O_3, SiC, TiB, TiB_2, TiC, TiN, Ti_5Si_3, $TiSi_2$, and hydroxyapatite. These reinforcements are incorporated into the matrix either in situ during heat treatment of the mechanically alloyed powder mixture or by adding the ceramic powder before milling has started dissolution of substantial amounts of interstitial elements such as oxygen, nitrogen, and carbon in titanium makes the titanium matrix very brittle. Therefore, it is desirable that we choose an alloying element that does not have large solubility in the titanium matrix but can form compounds, which can be uniformly dispersed in the titanium matrix. Boron is one such element. Table 6 shows mechanical properties of titanium matrix nanoparticle reinforced composites.

6.5 High entropy materials

HEAs are considered to be new "avatar" in physical metallurgy and materials engineering. The field of alloy design ventured into unknown compositional space with the advent of multiprincipal component HEAs, which comprise of five or more elements with wide difference in the melting point, in equal or near-equal proportions. These alloys are characterized by their unusually high configurational entropy of mixing and have attracted worldwide attention due to their unique characteristics, making them

Table 5 Mechanical properties of Copper matrix nanoparticle reinforced composites

Sl. No.	System	Hardness	Yield strength	Compressive strength	Wear
1	Cu-Al_2O_3 (by powder metallurgy, size 30 nm)	Hardness increases (118 Hv for 15 wt% Al_2O_3 whereas 58.9 Hv for Cu)	Improved yield strength 860 MPa for 4 vol% Al_2O_3 (33.3 MPa for pure copper)	67% increase in compressive strength as Al_2O_3 content increases to 12.5%	Improved wear resistance. Abrasive wear rate decreases with increasing amount of Al_2O_3
2	Cu-B_4C (powder metallurgy particle size 100 nm)	Hardness of the composites increased with increasing volume percentage of B_4C	Improved yield strength	Increase in compressive strength as B_4C content increases	Improved wear resistance. Abrasive wear rate decreases with increasing amount of B_4C
3	Cu-Diamond (mechanical alloying technique, nanodiamond particles have a mean diameter of 4–5 nm)	Improvement in hardness (HV of 280 for 11 vol% diamond)	Improved yield strength	Increase in compressive strength as diamond content increases	Improved wear resistance
4	Cu-Sic (mechanical alloying)	Improvement in hardness. Hardness increased by 48% by adding 20% SiC (127 Hv for copper)	Improved yield strength as SiC content increases	Increase in compressive strength as SiC content increases	Wear is reduced from 45% to 77% due to the addition of SiC from 10% to 20%

Table 6 Mechanical properties of Titanium matrix nanoparticle reinforced composites.

Sl. No.	System	Hardness	Yield strength	Compressive strength	Wear
1	Ti-BN (mechanical alloying)	Significant improvement in microhardness (216% increase, 60 HV for Ti)	Improved yield strength (1.66GPa) (240 MPa for non-reinforced Ti)	Increase in compressive strength as BN content increases	Wear resistance is improved (97.8%)
2	Ti-SiC (powder metallurgy)	Significant improvement in microhardness (369 Hv for 10% SiC, 60 HV for Ti)	Improved yield strength	Increase in compressive strength as SiC content increases	Improved wear resistance with increase in Vol. of SiC
3	Ti-B_4C (powder metallurgy)	Significant improvement in microhardness	Improved yield strength	Increase in compressive strength as B_4C content increases	20% B_4C improve friction characteristic

appealing from a technological as well as scientific point of view. The design and development of bulk nanostructured HEAs and their composites possess many processing challenges as these are multicomponent alloys with a large number of possibilities of phase formation and microstructural evolution.

Cu-based and HEA-based nanocrystalline alloys are considered as potential candidates for wear resistance applications in bearings [47]. The design and development of materials for bearings have gone through various stages in the last 50 years. While the traditional use of steel bearings remains predominant, the nonoxide ceramics, like SiC, Si_3N_4 have also been developed for specific bearing applications [48]. The next level of development was realized in the use of hybrid bearings with ceramic balls being enclosed within metallic (steel) raceways. But high cost and lack of reliable properties along with poor machinability of ceramics have been major disadvantages in the widespread use of the hybrid bearings. As the use of softer dispersion in Cu- or HEA-matrix is expected to provide better wear resistance, other metallic materials like Cu-Pb, HEA-Pb, and HEA-Bi system appear very effective in this perspective. However, the hardness of the conventionally processed Cu- or HEA-matrices is very low and hence, it does not enable them to be used in demanding applications, requiring higher wear resistance [49]. This necessitates the property enhancement of these bearing materials by incorporating novel phases. In this regard, Cu- or HEA-based nanocomposites appear to be an important material system.

The intrinsic HEAs exhibit superior mechanical properties (strength, ductility, and toughness) and hence, they are considered as potential candidates for wear resistance, irradiation resistant application. By engineering microstructure using various novel design concepts wear resistance can be improved further. One of the novel concepts involves incorporation of soft dispersoids (Sn, Pb, Bi, Sb, Ag, etc.) or lubricating phases (MoS_2, BaF_2/CaF_2) in the nanostructured HEA matrix, so that these dispersoids can act as lubricants and improve wear resistance [50].

However, the challenge involves engineering the microstructure consisting of soft dispersoids and lubricating phase, distributed uniformly in the HEA matrix. Similarly, irradiation resistance of HEAs can substantially be improved by incorporation of uniform dispersion of oxide nanoparticles in the HEA matrix. The refractory high entropy alloys are another class of HEAs, which contain alloying elements of Mo, Nb, Ta, Zr, W, Ti, V, etc. with near equiatomic concentrations of the high melting points metals (Mo, Nb, W). One of the most promising and exciting outcomes originated from this new alloy family is the use of HEAs as structural materials. Interestingly, these HEAs show BCC structure, exhibiting good combination of mechanical properties—high hardness, ultimate tensile strength, moderate toughness and ductility, and high density.

The final aim for the design and development of novel HEAs is high strength with sufficient ductility. It has been increasingly evident that this is not possible to achieve for the single-phase HEAs. Nanocrystallization of the multicomponent HEA is not sufficient enough to obtain sufficiently high strength and ductility in FCC alloys. On the other hand, nanocrystalline BCC HEAs exhibit superior strength with low ductility and toughness, not useful enough for engineering applications [51]. An effective approach to improve mechanical properties of the alloys is by precipitation strengthening, which is widely used in commercial aluminum alloys, steels, and super alloys. The annealing or aging of the supersaturated solid solution at intermediate temperature is normally carried out to engineer a microstructure consisting of fine-scale precipitates in a matrix which provides the considerable strengthening, whereas softer matrix leads to sufficient ductility.

7. Metallic nanocomposite coatings

In terms of corrosion resistance, surface hardness, antibacterial activity, and mechanical stability against wear and erosion, a material's surface properties determine its applicability and durability. As a result, chemical or physical surface modifications to achieve superior efficiency have been commonly used to set the bar for various applications. Surface hardening, heat treatment, surface coatings, and other technologies are now available to enhance the exterior characteristics. Surface coating, on the other hand, is a significant process with widespread acceptance for improving appearance, properties, and advanced functionalities. The ability to integrate nanosized particles into a metal/alloy matrix has ushered in a new age of composite coatings known as electroless nanocomposite coatings.

Due to broad interface and nanoscale effects, these coatings have improved mechanical, physical, and functional properties [52]. The hardness and wear resistance of Cu-P-Cg-SiC composite coatings obtained by incorporating hard SiC particles with graphite were increased further [53]. The addition of reduced graphene oxide to the Cu matrix can cause a compressive microstrain, which causes lattice shortening and improves mechanical efficiency [54].

Carbon nanotubes with single-walled or multiwalled walls have unique electrical, magnetic, and physical properties, making them an effective reinforcement for electroless composite coatings. Cu-MWCNT composite films with varying MWCNT dimensions were formed by Arai and Kanazawa on acrylonitrile butadiene styrene (ABS) resin substrates. As compared to metallic copper coatings, the composite coatings had lower coefficients of friction and similar electrical resistivity [55].

The addition of hard diamond particles improves the composite coating's mechanical properties. Fig. 13 shows some SEM images of Ni-P, Ni-P/nanometer diamond depositions at various concentrations of nanodiamond in electroless solution. For the development of abrasive materials for surface treatment, the effect of cobalt ions ($CoCl_2$) and

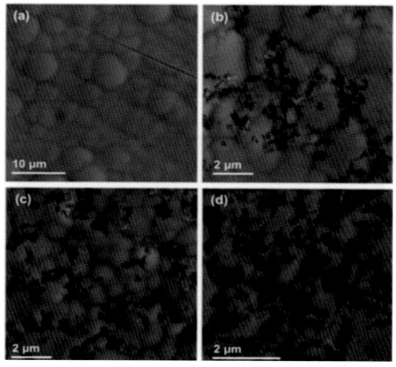

Fig. 13 Backscatter scanning electron microscopy images of (A) Ni-P, Ni-P/nanometer diamond depositions at various concentrations of nanodiamond in electroless solution: (B) 0.5, (C) 1.0, (D) 4.0 g L^{-1}.

the reducer (NaH_2PO_2) concentrations, as well as bath parameters (pH and temperature), on the kinetics of co-deposition of diamond dispersoid and cobalt layers was investigated [56].

The addition of hard ceramic carbides such as SiC, B_4C, and WC to electroless composite coatings improves the system's mechanical, tribological, and anticorrosive properties. The corrosion resistance of carbon steel was improved by SiC particles in an electroless Cu-P deposit. Electrochemical tests and weight loss measurements revealed that the Cu-P-SiC composite coating outperformed Cu-P coatings and uncoated carbon steel in terms of anticorrosion efficiency [57].

Metallic composite coatings with high wear resistance, hardness, excellent corrosion protection, and solderability were developed on various engineering components such as hydraulic pistons, ship propellers valves, pumps, splines, gears, bushings, thrust washers, automotive and truck transmissions, moulds and dies, engine shafts, and electrical equipment. Electroless nickel-based composites are used in the automotive industry to protect parts made of iron, steel, aluminum, copper, brass, and other alloys from corrosion, increase abrasion and wear resistance, and reduce friction resistance. As a result, it is used on brakes, injectors, and clutches, among other things.

8. Automotive application of metallic nanocomposites

Compared to air or water transportation, most of the researches and developments based on nanotechnology are in automobile sector as we depend on it more frequently. Nanotechnology is applied to body parts, emissions, chassis and automobile interiors, electrics and electronics, engines, and drive trains. One of the most discussed topics in the automobile research field is weight reduction of vehicles. By reducing the weight we can increase the fuel efficiency, reduce CO_2 emissions and production cost. It is estimated that a fuel economy of 7% can be attained by reducing the weight of an automobile by 10% [58]. Magnesium and its alloy (lightweight) dispersed with nanosize reinforcements have shown improved mechanical properties without much reduction in its other properties [59].

Nanoparticles of Y_2O_3 or Al_2O_3 are added to magnesium either by agitating molten melt or by mixing the magnesium powder and nanoreinforcements in the required composition by mechanical alloying or simple blending. Cu nanoparticles were also imparted to magnesium which gave an improvement of 104% in 0.2% yield strength with a slight reduction in its ductility. The main reasons for these increments are (a) increase in the dislocation density in magnesium matrix due to elastic modulus and CTE mismatch between the reinforcement and matrix; (b) presence of harder ceramic/metallic nanoparticles as reinforcement; and (c) reduction in grainsize. These magnesium nanocomposites provide improved tensile, hardness, dynamic, compressive, high temperature, fatigue and wear properties over conventional composites of magnesium and can be widely used in engine blocks and gear housings of vehicles which help to reduce the weight and improve other properties compared to conventional materials.

In modern automobiles around 10%–15% of the fuel energy is consumed by the friction of the moving mechanical parts of the engine such as piston, cylinder wall, crank drive elements like connecting rod, crank shaft and bearings, and valve drive system including the valves and the cam shaft. Among these, piston and cylinder wall aggregate are the major parts of mechanical frictional loss. Another problem that we face in the automobiles is heat generation by the engine and its cooling. At present, to overcome these problems we use engine oils in between the moving parts to reduce the friction, radiator and coolant to reduce the heat generated and cool the engines. The functions of the engine oils, radiator, and coolant can be increased by the application of the nanotechnology.

By coating the cylinder wall with nanocrystalline materials we can reduce abrasion and friction and in turn the fuel consumption. Iron carbide and boride nanocrystals with sizes 50–120 nm are used to coat the engine parts which result in extremely hard surface with very low friction [60]. Figs. 14 and 15 show some examples of metallic nanocoatings for automotive components.

Radiator is used to remove heat from the engine. Conventional heat transfer fluids like water, ethylene glycol, and mineral oil are used for this purpose. But these are not as efficient to remove all the heat generated. Researches and discussions are going on how to increase the heat transfer rate. By adding nanofluids into the coolant its heat transfer rate is improved significantly. Nanofluids of CuO, alumina, carbon nanotube, silica, and titanium oxide dispersed in carrier liquid enhance heat transfer rate of the resultant coolant compared to the carrier liquid alone [61]. Al_2O_3 nanoparticles in water

Fig. 14 Photograph of electroless Nickel-CeO_2 nanocomposite coated cylinder liner.

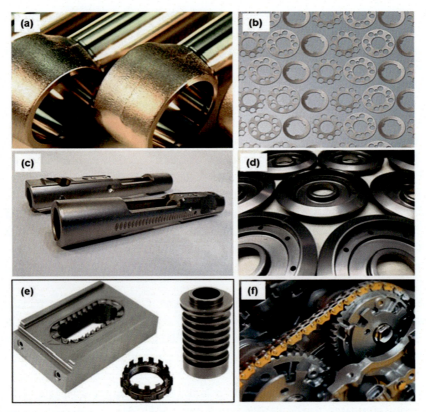

Fig. 15 Photographs of electroless nickel composite coated (A) Nickel-Boron nitride plated connecting rods, (B) Nickel Diamond coated friction shims, (C) Nickel-Boron nitride plated gun barrel, (D) Nickel-PTFE coated rotors, (E) Nickel-Teflon coated mould components, and (F) Nickel-PTFE-nanodiamond coated gear.

coolant improve the heat-carrying capacity rate and it depends on the amount of nanoparticles adding to the coolant.

We interact mostly with the interior parts of the automobile such as seats, door paddings, dashboard, air bags, seat belts, boot carpets, etc. where microbial and bacterial infections are most common. Since interior is the place where we interact mostly, it should be free from all bacterial and microbial infections. Because of our hygiene concerns and problems caused by these infections various antimicrobial agents like oxidizing agents (aldehydes, halogens), radical formers (isothiazones, peroxo compounds), and chitosan and ammonium compounds have been used. As most of these agents have some side effects and are toxic in nature they cannot be used. Nanotechnology has promised good agents that can be used without any side effects. The most important nanostructured antibacterial and antimicrobial agents are silver, gold, titanium oxide, zinc oxide, titania nanotubes, gallium, liposomes loaded nanoparticles, and copper nanoparticles [62, 63, 64].

9. Conclusion

The use of nanomaterials has already yielded families of composites with superior properties which will find growing uses in automobile applications. Metallic nanocomposites are expected to play a critical role in many industries with high-performance materials which today make use of conventional composites. In this context, nanocomposites are suitable materials to meet the emerging demands arising from scientific and technologic advances. They may potentially be manufactured at low cost and may offer advantages on density and processing with respect to metal matrix composites currently used in the fabrication of parts for automotive applications. Processing methods for different types of nanocomposites are explained, some of these pose challenges thus giving opportunities for researchers to overcome the problems being encountered with nanosize materials. They offer enhanced performance over monolithic and micro composite counterparts and are consequently suitable candidates to overcome the limitations of many currently existing materials and devices. In view of their unique properties such as very high mechanical properties even at low loading of reinforcements, these materials have been projected in various automobile sectors. Thus metallic nanocomposites provide opportunities creating new worldwide interest in these new materials.

References

[1] G. Yasin, M. Arif, T. Mehtab, M. Shakeel, M.A. Khan, W.Q. Khan, Metallic nanocomposite coatings, in: Corrosion Protection at the Nanoscale, Elsevier, 2020, pp. 245–274.

[2] S. Veličković, B. Stojanović, L. Ivanović, S. Miladinović, S. Milojević, Application of nanocomposites in the automotive industry, Int. J. Mobil. Veh. Mech. 45 (2019) 51–64.

[3] V. Gutowski, W. Yang, S. Li, K. Dean, X. Zhang, Lightweight nanocomposite materials, in: Lightweight and Sustainable Materials for Automotive Applications, CRC Press, 2017.

[4] S. Komarnenei, Nanocomposites, J. Mater. Chem. 2 (1992) 1219–1230. 1www.nanowerk.com/spotlight/spotid=23934.php2.

[5] S.M.R. Billah, Composites and nanocomposites, in: Functional Polymers, Springer, 2019, pp. 447–512.

[6] E. Omanović-Mikličanin, A. Badnjević, A. Kazlagić, M. Hajlovac, Nanocomposites: a brief review, Health Technol. 8 (2019) 1–9, https://doi.org/10.1007/s12553-019-00380-x.

[7] V. Provenzano, R. Holtz, Nanocomposites for high temperature applications, Mater. Sci. Eng. A 204 (1995) 125–134.

[8] D. Li, Y. Ye, X. Liao, Q.H. Qin, A novel method for preparing and characterizing graphene nanoplatelets/aluminum nanocomposites, Nano Res. 11 (2018) 1642–1650.

[9] C.S. Goh, et al., Development of novel carbon nanotube reinforced magnesium nanocomposites using the powder metallurgy technique, Nanotechnology 17 (1) (2005).

[10] C.M. Cepeda-Jiménez, M.T. Pérez-Prado, Processing of Nano Particulate Metal Matrix Composites, IMDEA Materials Institute, Getafe, Spain, 2018.

[11] J. Vaziri, A. Jahan, E. Borhani, M. Yousefieh, K.L. Edwards, Evaluating promising applications of a new nanomaterial produced by accumulative roll bonding process: a preliminary multiple criteria decision-making approach, Proc. Inst. Mech. Eng. L 233 (2016) 1023–1032.

[12] D.Y. Ying, D.L. Zhang, Processing of Cu–Al2O3 metal matrix nanocomposite materials by using high energy ball milling, Mater. Sci. Eng. A 286 (1) (2000) 152–156.

[13] S.A. Mahmoudi, Mohammed Ammar GPU-Low-Energy Tracking of the Left Ventricle in the Cloud-Conference: 6th International Work-Conference on Bioinformatics and Biomedical Engineering, 2018.
[14] H. Li, K.A. Khor, P. Cheang, Impact formation and microstructure characterization of thermal sprayed hydroxyapatite/titania composite coatings, Biomaterials 24 (6) (2003) 949–957.
[15] W.M. Daoush, et al., Electrical and mechanical properties of carbon nanotube reinforced copper nanocomposites fabricated by electroless deposition process, Mater. Sci. Eng. A 513 (2009) 247–253.
[16] M.G. Akhil, K.M.S. Manu, T.P.D. Rajan, B.C. Pai, Liquid phase processing of metal matrix composites, in: Reference Module in Materials Science and Materials Engineering, 2021.
[17] A.J. Cook, P.S. Werner, Pressure infiltration casting of metal matrix composites, Mater. Sci. Eng. A 144 (1–2) (1991) 189–206.
[18] M. Estruga, et al., Ultrasonic-assisted synthesis of surface-clean TiB2 nanoparticles and their improved dispersion and capture in Al-matrix nanocomposites, ACS Appl. Mater. Interfaces 5 (17) (2013) 8813–8819.
[19] J.H. Martin, B.D. Yahata, J.M. Hundley, J.A. Mayer, T.A. Schaedler, T.M. Pollock, 3D Printing of High-Strength Aluminium Alloys. ©, Macmillan Publishers Limited, part of Springer Nature, 2017. All rights reserved.
[20] C.-S. Kim, I. Sohn, J.B.F. MarjanNezafati, B.F. Schultz, Z. Bajestani-Gohari, P.K. Rohatgi, K. Cho, Prediction models for the yield strength of particle-reinforced unimodal pure magnesium (Mg) metal matrix nanocomposites (MMNCs), J. Mater. Sci. 48 (2013) 4191–4204.
[21] J. Ye, B.Q. Han, Z. Lee, B. Ahn, S.R. Nutt, J.M. Schoenung, A tri-modal aluminum based composite with super-high strength, Scr. Mater. 53 (2005) 481–486.
[22] Q. Nguyen, M. Gupta, Increasing significantly the failure strain and work of fracture of solidification processed AZ31B using nano-Al2O3 particulates, J. Alloys Compd. 459 (2008) 244–250.
[23] J. Jue, D. Gu, K. Chang, D. Dai, Microstructure evolution and mechanical properties of Al-Al2O3 composites fabricated by selective laser melting, Powder Technol. 310 (2017) 80–91.
[24] X. Wang, N. Wang, L. Wang, X. Hu, K. Wu, Y. Wang, et al., Processing, microstructure and mechanical properties of micro-SiC particles reinforced magnesium matrix 140 composites fabricated by stir casting assisted by ultrasonic treatment processing, Mater. Des. 57 (2014) 638–645.
[25] K. Nie, X. Wang, K. Wu, L. Xu, M. Zheng, X. Hu, Processing, microstructure and mechanical properties of magnesium matrix nanocomposites fabricated by semisolid stirring assisted ultrasonic vibration, J. Alloys Compd. 509 (2011) 8664–8669.
[26] G. Liu, G. Zhang, F. Jiang, X. Ding, Y. Sun, J. Sun, et al., Nanostructured high-strength molybdenum alloys with unprecedented tensile ductility, Nat. Mater. 12 (2013) 344–350.
[27] K.M. Sree Manu, S. Arun Kumar, T.P.D. Rajan, M. Riyas Mohammed, B.C. Pai, Effect of alumina nanoparticle on strengthening of Al-Si alloy through dendrite refinement, interfacial bonding and dislocation bowing, J. Alloys Compd. 712 (2017).
[28] L. Ceschini, et al., Aluminum and Magnesium Metal Matrix Nanocomposites, Springer, 2017.
[29] S. Natarajan, R. Narayanasamy, S.P. KumareshBabu, G. Dinesh, B. Anil Kumar, K. Sivaprasad, Sliding wear behaviour of Al 6063/TiB2 in situ composites at elevated temperatures, Mater. Des. 30 (2009) 2521–2531.
[30] Z. Zhang, L. Zhang, Y.W. Mai, The running-in wear of a steel/SiCp–Al composite system, Wear 194 (1996) 38–43.
[31] R.N. Rao, S. Das, Effect of sliding distance on the wear and friction behavior of as cast and heat-treated Al–SiCp composites, Mater. Des. 32 (2011) 3051–3058.
[32] R.L. Deuis, C. Subramanian, J.M. Yellup, Dry sliding wear of aluminium composites-A review, Combust. Sci. Technol. 57 (1997) 415–435.
[33] C.Y.H. Lim, S.C. Lim, M. Gupta, Wear behaviour of SiCp reinforced magnesium metal matrix composites, Wear 255 (2003) 629–637.
[34] K.D. Ralston, N. Birbilis, Effect of grain size on corrosion: a review, Corrosion 66 (2010) e075005–e075013.
[35] H. I-Sorkhabi, M. Es' haghi, Corrosion resistance enhancement of electroless Ni-P coating by incorporation of ultrasonically dispersed diamond nanoparticles, Corros. Sci. 77 (2013) 185–193.

[36] Y. Liu, et al., The thermal conductivity of a composite material with nanosized nonmetal particles embedded in a metal matrix, J. Alloys Compd. 801 (2019) 136–141.
[37] S.S. Sidhu, et al., Metal matrix composites for thermal management: a review, Crit. Rev. Solid State Mater. Sci. 41 (2016) 132–157.
[38] A. Esawi, K. Morsi, Dispersion of carbon nanotubes (CNTs) in aluminum powder, Compos. A: Appl. Sci. Manuf. A38 (2007) 646–650.
[39] R. Perez-Bustamante, I. Estrada-Guel, W. Antunez-Flores, M. Miki-Yoshida, P.J. Ferreira, R. Martı́nez-Sanchez, Novel Al-matrix nanocomposites reinforced with multi-walled carbon nanotubes, J. Alloys Compd. 450 (2008) 323–326.
[40] R. Perez-Bustamante, I. Estrada-Guel, P. Amezaga-Madrid, M. Miki-Yoshida, J.M. Herrera-Ram'ırez, R. Mart'ınez-Sanchez, Microstructural characterization of Al–MWCNT composites produced by mechanical milling and hot extrusion, J. Alloys Compd. 495 (2010) 399–402.
[41] Y. Birol, Response to thermal exposure of the mechanically alloyed Al–Ti/C powders, J. Mater. Sci. 42 (2007) 5123–5128.
[42] T. Lin, K. Zhao, J. Liu, X. Luo, X. He, L. An, Heterogeneous Mg-based nanocomposites with simultaneously improved strength and toughness, Mater. Lett. 276 (2020) 128231.
[43] M. Shahin, K. Munir, C. Wen, Y. Li, Magnesium matrix nanocomposites for orthopedic applications: a review from mechanical, corrosion, and biological perspectives, Acta Biomater. 96 (2019).
[44] S. Sabooni, T. Mousavi, F. Karimzadeh, Mechanochemical assisted synthesis of Cu(Mo)/Al2O3 nanocomposite, J. Alloys Compd. 497 (2010) 95–99.
[45] T. Takahashi, Y. Hashimoto, Preparation of boron carbide-dispersion-strengthened copper by the application of mechanical alloying, J Jpn Soc Powder Powder Metall 36 (1989) 85–89.
[46] F.H. Froes, C. Suryanarayana, D. Eliezer, Synthesis, properties, and applications of titanium aluminides, J. Mater. Sci. 27 (1992) 5113–5140.
[47] Y.E.H. Jien-Wei, Recent progress in high entropy alloys, Ann. Chim. Sci. Mater. 31 (2006) 633–648.
[48] B. Basu, M. Kalin, Tribology of Ceramics and Composites: Materials Science Perspective, John Wiley & Sons, 2011.
[49] F.P. Bowden, D. Tabor, The Friction and Lubrication of Solids, Oxford university press, 2001.
[50] S. Yadav, A. Kumar, K. Biswas, Wear behavior of high entropy alloys containing soft dispersoids (Pb, Bi), Mater. Chem. Phys. 210 (2017) 222–232.
[51] E. Pickering, N. Jones, High-entropy alloys: a critical assessment of their founding principles and future prospects, Int. Mater. Rev. 61 (2016) 183–202.
[52] M. Roy, Nanocomposite films for wear resistance applications, in: Surf. Eng. Enhanc. Perform. Against Wear, Springer, Vienna, 2013, pp. 45–78.
[53] S. Faraji, A. Abdul Rahim, N. Mohamed, et al., A study of electroless copper–phosphorus coatings with the addition of silicon carbide (SiC) and graphite (Cg) particles, Surf. Coat. Technol. 206 (2011) 1259–1268.
[54] Q. Zhang, Z. Qin, Q. Luo, et al., Microstructure and nanoindentation behavior of Cu composites reinforced with graphene nanoplatelets by electroless co-deposition technique, Sci. Rep. 7 (2017) 1338.
[55] S. Arai, T. Kanazawa, Electroless deposition and evaluation of Cu/multiwalled carbon nanotube composite films on acrylonitrile butadiene styrene resin, Surf. Coat. Technol. 254 (2014) 224–229.
[56] D. Stoychev, N. Koteva, M. Stoycheva, et al., Electroless deposition of cobalt matrix for co-deposition of highhard nano and microparticles, Mat. Bulgaration Sci. Fund. 2 (102) (2009) 7. Project ID.
[57] S. Faraji, A.A. Rahim, N. Mohamed, et al., Corrosion resistance of electroless Cu-P and Cu-P-SiC composite coatings in 3.5% NaCl, Arab. J. Chem. 6 (2013) 379–388.
[58] M.C. Coelho, G. Torrao, N. Emami, J. Gracio, Nanotechnology in automotive industry: research strategy and trends for the future small objects, big impacts, J. Nanosci. Nanotechnol. 12 (2012) 1–10.
[59] T. Luo, X. Wein, X. Huang, L. Huang, F. Yang, Tribological properties of Al2O3 nanoparticles as lubricating oil additives, Ceram. Int. 40 (2014) (2014) 7143–7149.
[60] W. Matthias, K. Wolfram, S. Mirjana, Nanotechnologies, Automobiles—Innovative Potentials in Hesse for the Automotive Industries and Its Subcontractors, J. King Saud Univ. Sci. (2008) 3.

[61] G. Satyamkumar, S. Brijrajsinh, M. Sulay, T. Ankur, R. Manoj, Analysis of radiator with different types of nano fluids, J. Eng. Res. Stud. 6 (1) (2015) 1–2.
[62] T.C. Dakal, A. Kumar, R.S. Majumdar, V. Yadav, Mechanistic basis of antimicrobial actions of silver nanoparticles, Front. Microbiol. 1 (2016) 1831.
[63] S. Prabhu, E.K. Poulose, Silver nanoparticles: mechanism of antimicrobial action, synthesis, medical applications, and toxicity effects, Int. Nano Lett. 2012 (2) (2012) 32.
[64] Mohsen Mohseni, Bahram Ramezanzadeh, Hossein Yari, Mohsen MoazzamiGudarzi, The Role of Nanotechnology in Automotive Industries. (2012). https://doi.org/10.5772/49939.

CHAPTER 10

Metal matrix nanocomposites

P.S. Samuel Ratna Kumar[a], P.M. Mashinini[a], and S. John Alexis[b]
[a]Department of Mechanical and Industrial Engineering, University of Johannesburg, Johannesburg, South Africa
[b]Department of Automobile Engineering, Kumaraguru College of Technology, Coimbatore, Tamil Nadu, India

Chapter outline

1. Introduction — 199
2. Nanocomposite materials — 200
3. Fabrication of nanocomposites — 201
4. Characterization study — 206
5. Nanocomposites in auto components — 209
6. Conclusion — 211
References — 211

1. Introduction

The automobile industry actively works to produce and apply lightweight components with outstanding mechanical, corrosion, and tribological properties. Today's automobile industry should satisfy several criteria to lower fuel consumption while maintaining sufficient vehicle comfort and protection. Advanced automobile materials are assisting manufacturers in the creation of lightweight fuel-efficient cars. Thanks to lightweight materials like aluminum and magnesium components, the new crop of cars and trucks will follow robust crash safety standards [1–3]. The commonly used materials in the automobile components are composites which are made of high-strength steel, aluminum, and carbon fiber/plastic composites. Using high-strength steel results in a 20% reduction of weight relative to traditional steel constructions and 40% reduction in aluminum utilization. As a result, by applying lightweight materials, the vehicle's weight is reduced. Aluminum links in the suspensions were 15% lightweight than traditional ties. Hollow composite shafts become 18% lighter than traditional shafts. The use of cast aluminum rims reduces up to 36% when comparing to rims [4]. Aluminum is being used as the major element in the manufacture of composite materials known as aluminum matrix composites (AMCs) since this is an extremely lightweight material. Aluminum composites have a strong balance of mechanical and tribological properties as compared to aluminum alloy. Furthermore, these properties can be customized to meet unique requirements. AMCs have good physical and mechanical properties like higher strength-to-weight ratio, higher strength, better ductility, better wear resistance, lower thermal expansion coefficient, high temperature creep resistance, corrosion resistance, and better fatigue resistance [5, 6]. In late 1940s, the engine casing and air-cooled engine of a Volkswagen Beetle car weighed almost 25 kg of magnesium castings, and the US Convair B-36 bomber planes contained 8600 kg of magnesium [7, 8]. Magnesium is becoming more common as a lightweight material in automobile, sporting, and electronic products. Magnesium also has a number of other advantages, like superior castability, high absorption capability, effective electrical insulation, which is the most readily machinable of any structural metals and requiring lower energy to manufacture than aluminum. Magnesium has certain constraints like poor modulus of elasticity and durability, as well as weak creep and abrasion resistance and a high corrosion rate. The introduction of new magnesium alloys as well as the incorporation of reinforcements to produce magnesium-based composites to overcome similar constraints [9, 10].

Nanocomposites are a new type of material with superior mechanical and thermal properties. They are ideally being used in the aviation, automotive, chemical, and transportation industries due to their reliability. Also, the need for lightweight materials in automobile parts manufacturing has helped reduce emission levels [11]. The manufacture of timing belt covers as part of the engine for Toyota Camry cars was the first commercialization of nanocomposites in the automobile sector in 1991. The automotive industry is implementing most of the nanotechnology-based product development. Body parts, frames, and tires, vehicle interiors, electrical and electronics, IC engines, and drive systems all use nanotechnology. In recent decades, substantial progress has been achieved in a variety of industries, motivating scientists/researchers to build new structural materials for improved engineering efficiency [12–14].

Recent research on aluminum/magnesium-based nanocomposites have shown that adding nanosize reinforcements enhances the strength of aluminum/magnesium while having no negative impact on ductility. Furthermore, it has been demonstrated that using a minimal weight/volume fraction of nanolevel reinforcements performs better that are equal to or even better than metal matrix composites reinforced with an equal or greater weight/volume fraction in micron-level reinforcements. The addition of nanoscale reinforcements is thus an appealing option for improving mechanical properties without compromising aluminum/magnesium's ductility [11, 15, 16]. Finally, the opportunities of nanocomposites in automobile sector are discussed in detail.

2. Nanocomposite materials

Based on the chosen matrix material and reinforcement function, composite structures are created by applying basic elements to the reinforcement to enhance those characteristics. Composites are made up of a deformable sheet of metal/solid ceramic support in macro/microsize and can be used as a common building element in the automotive components. Metal Matrix Composite (MMC) is the name given to these composites with a metal core [17, 18]. The condition of nanocomposites emerges where a composite substance contains reinforcement of at least one direction in the nanometer range or below 100 nm [19]. In addition to the interesting exposure into the atom environment and its control, different physical properties are influenced at this length range such as increase in hardness, ductility, reactivity, and stability. The surface-to-volume ratio is high at the nanoparticle, which will decrease the thermal conductivity, density, and melting point of the nanocomposite material. As a result, nanocomposites are composite materials made up of the matrix material and nanoscale reinforcement which does not dissolve in each other. Due to their ability to sustain extreme temperatures and stresses, nanocomposite-hardened particles are widely used in the automobile industry. The efficiency of nanomaterials will be such that about 0.5%–5.0% in weight or volume fraction of material is restricted. The properties of the nanocomposites are greater than the traditional macro- or microlevel-reinforced composites and can be fabricated with simple low-cost processes [20–22]. Just few nanocomposites are rolled out in the market, although few other materials are looking for a possible way and many more are still being developed in the labs of numerous scientific institutions and industries.

3. Fabrication of nanocomposites

Due to their ability to sustain extreme temperatures and stresses, nanocomposites hardened and low-density materials are widely used in the automobile industry. Nanocomposites were produced using a wide range of manufacturing processes. The major limitation of nanocomposite materials is their lack of homogeneous reinforcement dispersion and low bonding between the matrix and reinforcement particles. These limitations were overcome by controlling the process parameters of manufacturing techniques. Metal-based nanocomposites are made up of a ductile matrix and hard nano particle reinforcement. Nano sized materials replaced the macro and micro sized reinforcement materials in metal matrix composites to improve the properties of the materials. Based on the reinforcement and the matrix material, different properties of these nanocomposites were increased, such as density, stiffness, wear, fracture, and resistance to corrosion. The oxides, carbides, nitrides, and borides are the primary divisions of the reinforcement material. Some of the commonly used materials are TiC, MoS_2, MWCNT, SWCNT, Al_2SiO_5, Al_2O_3, SiC, TiB_2, TiO_2, B_4C, AlN, BN, TiN, ZrB_2, eggshell, and flyash [20, 21, 23–30]. Low-density metals such as aluminum (Al) and magnesium (Mg) were used as matrix materials to develop metal nanocomposites for their ease of use and high efficiency. In combination with carbon and oxide-based reinforcement, aluminum and magnesium alloys have seen numerous uses in the automobile industry. In general, there are three types of basic classification processes for producing metal-based nanocomposites: (i) solid-state, (ii) liquid-state, and (iii) semisolid state processes [31].

Aluminum alloy is favored as a matrix, due to its benefits, such as strength-to-weight ratio, low cost, and simple workability. Some of the commonly used aluminum

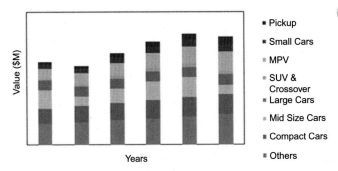

Fig. 1 Aluminum application in automobiles from 2017 to 2022 [33].

nanocomposite fabrication techniques are stir casting, compo-casting, powder metallurgy, and friction stir processing. Stir casting is the most cost-effective of all the techniques because of the low cost, the ability to use a wide variety of components, and the manufacturing of bulk components. A cylinder head, pistons, crankcase, cylinder head cover, exhaust manifold, intake system, drive shaft, gears, gear housing, etc. [32]. These are examples of standard automobile components made of aluminum-based composite materials. Fig. 1 shows the aluminum applications in automobiles from 2017 to 2022.

The production of the casting process and heat treatment led to the use of aluminum castings in the automotive industry. Extensive research has been conducted to improve mechanical, corrosion, and tribological properties by controlling the grain arrangement, matrix, and reinforcement bonding, by controlling the process parameter for casting such as stirring speed, time, and casting temperature of the stir casting technique as shown in Fig. 2. Researchers have examined the nanocomposite, to replace the automotive parts like piston rings, pistons, connecting rods, disc brakes, drum brakes, camshaft, shafts, and valves [32].

Magnesium alloys used in automobiles have shown potential, and there has been a lot of study into their effectiveness. Die-cast components, such as four-wheel-drive transfer cases, transmission cases, engine cradles, steering-wheel modules, seats, and instrument panels, are mostly made of that same alloy. Magnesium-based nanocomposites with nanoparticulates are normally treated in the same way as composites with macro/micro size reinforcements are using either solid- or liquid- or semisolid-state manufacturing processes. The magnesium matrix material is introduced into the crucible and melted into the liquid state and then nano reinforcement should be added to obtain a uniform dispersion of reinforcement material into the matrix. Various techniques are used to distribute the nanoreinforcements when they reach the liquid stage to ensure a uniform distribution before casting. The most popular technique is to stir up the liquid metal using a mechanical stirrer, although others have used ultrasonic energy to distribute the nano-sized particles. Likewise, the liquid metal is injected under pressure or squeezed into a

Metal matrix nanocomposites 203

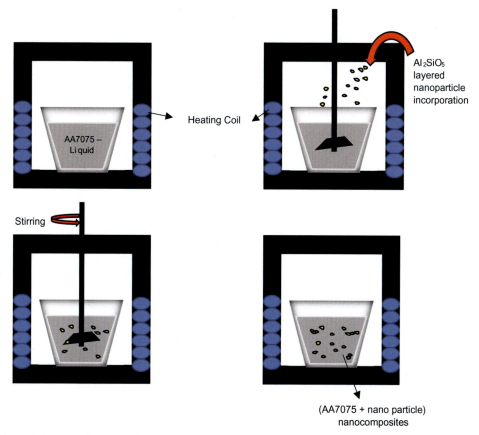

Fig. 2 Schematic diagram of Stir casting processes.

preform comprising the reinforcement material. Solid-state manufacturing, also known as powder metallurgy route, requires merely blending or mechanical alloying metal particles and nanoreinforcements into the calculated volume. This method is similar to the development of aluminum nanocomposites solid-state processes. Friction stir processing and semisolid stirring are two other techniques for developing magnesium-based nanocomposites. Disintegrated Melt Deposition (DMD) is a cost-effective method that incorporates the benefits of spray manufacturing and typical casting to manufacture mass composite materials at higher temperatures as shown in Fig. 3 [34]. This method as similar to the stir casting processes used to develop the aluminum-based nanocomposites.

Friction stir processing (FSP) is a commonly used manufacturing method for both aluminum and magnesium-based nanocomposites. The FSP method is appealing and offers many advantages. In comparison to liquid metallurgy processes, this does not consume a lot of energy (Fig. 4). The process of the method will not affect the chemical and physical properties of the reinforcement material. The overall distribution of the

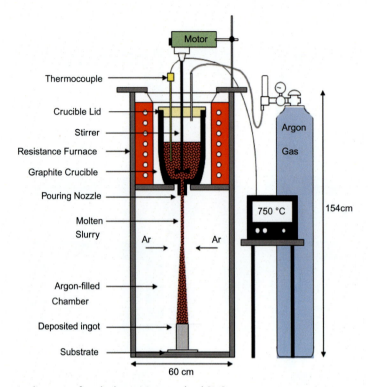

Fig. 3 Schematic diagram of melt deposition method [34].

reinforcement materials in the matrix is not affected by the density variation. As a result, the possibilities of using reinforcement through FSP are endless. In practice, any possible material can be effectively reinforced. The temperature at which the matrix and reinforcement are processed does not induce adverse interfacial changes [35].

The detailed description of the touch probe experimental design for the ultrasonic method is shown in Fig. 5. This test's main goal is to deagglomerate nanomaterials in a matrix material. The ultrasonic touch cavitation process was used by Li et al. to fabricate Al- and Mg-related nanocomposite materials. When nanomaterials are mixed with matrix material using an ultrasonic vibrating device, a localized hot spot is created with a rise in temperature of about 5000 K and a pressure rise of over 1000 atm. Those micro spots travel at speeds of up to 10^7 m/s. As a result, these can spread the nanomaterials in the matrix material evenly. During the manufacturing of nanocomposite, nanomaterials that are found in cluster form were broken down into smaller particles and distributed uniformly across the matrix material owing to the quick heating and cooling rate. Nanomaterials are sprayed onto the top of molten material in this process. The molten

Metal matrix nanocomposites 205

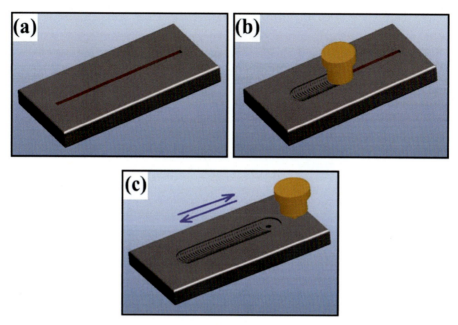

Fig. 4 Schematic diagram of friction stir processing [35].

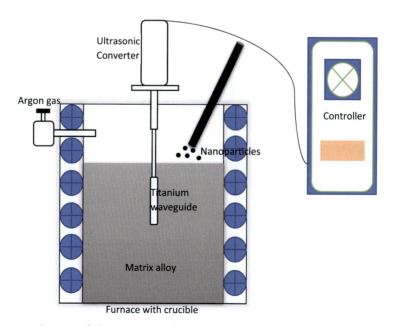

Fig. 5 Schematic diagram of ultrasonic cavitation [36].

material is combined with the nanomaterials for a fixed period of time using an ultrasonic vibrating method until all of the particles are almost dispersed in the molten melt. Researchers have created a high volume of Al-based nanocomposites with dispersed nanosized SiC particles. Nanostructured SiC particles have been found to be uniformly distributed in the matrix material by adding 2 wt% Nanostructured SiC into A356 alloy [36].

4. Characterization study

This section describes the microstructural characterization study of developed metal matrix nanocomposites. Large quantities of metal matrix nanocomposite samples were assembled, physically polished to a 50 nm surface quality, and chemically processed to eliminate physical cleaning compound compounds for surface characterization. The surface morphology or characterization, on the other hand, is critical in evaluating the nanocomposite materials feasibility for application purposes. The arrangement of reinforcement inside the matrix material characteristics is explored in microstructural analysis of developed nanocomposites. Microstructural analysis of developed Al–nano ZrO_2 nanocomposites showed a uniform reinforcement distribution with restricted clustering, strong matrix-reinforcement interfacial bonding, substantial grain refinement with increasing reinforcement percentage, and the absence of porosity. Also, a good wettability between matrix and reinforcement material was occurred due to melt deposition parameters selection [37].

Fig. 6A represents the surface morphology of an AA7075 (T6) metal matrix with balanced porosity due to stir casting process parameter. Increasing the process speed and time allows atmospheric gases to enter the molten metal, causing a volatile impact and a spike in porosity. Fig. 6B and C shows a balanced porosity and homogenous dispersion of nanoclay in the matrix material. Fig. 6D illustrates the 3 wt% nanoclay-reinforced composite material (T6), which has weak bonding, wettability, and cluster development along the material's surface. This occurs due to a greater proportion of nanoclay reinforcement in the matrix material, resulting in poor bonding [38]. The developed nanoclay-reinforced composite material was characterized using a scanning electron microscope as shown in Fig. 7. It is to comprehend the dispersed pattern and cluster formation of reinforcement into the matrix material [39]. Fig. 8 shows a transmission electron microscopy (TEM) image of a SiC Mg nanocomposite manufactured using ultrasonic vibrations. Most of the SiC nanoparticles are well scattered, while others are clustered, as shown in Fig. 8A. As shown in Fig. 8B with higher magnification, a uniform dispersed of SiC nanoparticles occurs just outside of the cluster formation.

Metal matrix nanocomposites 207

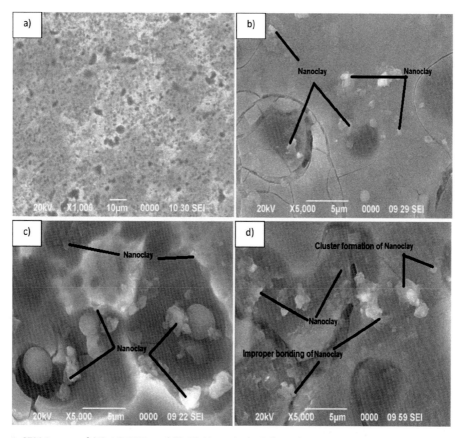

Fig. 6 SEM image of (A) AA7075 and (B–D) Nanoclay-reinforced nanocomposites [38].

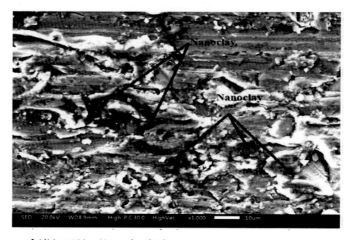

Fig. 7 SEM image of AlMg4.5Mn–Nanoclay [39].

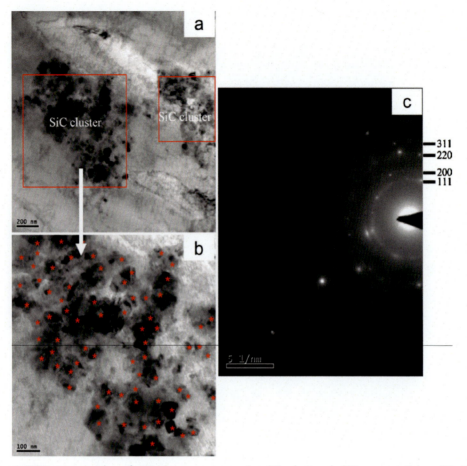

Fig. 8 TEM micrographs of SiC-Mg nanocomposite: (A) clustered SiC nanoparticles; (B) SiC nanoparticles with higher magnification; (C) pattern regime of SiC-Mg nanocomposite [40].

Furthermore, the matrix alloy spreads the clustered SiC nanoparticles. The pattern regime in Fig. 8C shows that the particle cluster is made up of SiC nanoparticles [40].

The micrographs of Mg-Y_2O_3 nanocomposite synthesized by a melt deposition technique, as shown in Fig. 9, revealed that the reinforcements formed individually and that small clusters were noticeable in the metal matrix [41]. Fig. 10 shows a descriptive scanning electron microscope image of grain morphology for Mg1.0Cu nanocomposite with a reduction in the matrix material's grain size [34].

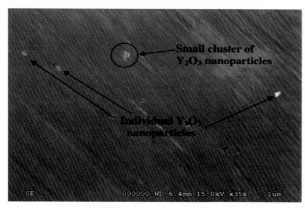

Fig. 9 SEM image of Mg-Y$_2$O$_3$ nanocomposite [41].

Fig. 10 SEM image of Mg1.0Cu nanocomposite [34].

5. Nanocomposites in auto components

Using the solid-state process, only 10% volume fraction of 50-nm alumina (Al$_2$O$_3$) particles were added to an aluminum/magnesium alloy matrix, raising yield stress to 515 MPa. The developed material is 15 times greater than the matrix alloy property, 6 times greater than the matrix alloy on addition of nano size reinforced Al$_2$O$_3$, and about 1.5 times stronger than AISI 304 stainless steel. However, certain processing issues must be addressed, as well as difficulties associated with expanding the technology. In connecting rod, Cast aluminum/magnesium-matrix nanocomposites might become suitable for producing relatively close components to substitute steel, wrought aluminum, and titanium components while minimizing reciprocating weight of components requiring good strength [42].

Due to the low wear characteristics of aluminum, cast iron cylinder liners are usually required. Porsche utilizes metal matrix composites for cylinder liners by introducing a porous silicon particle into the cast aluminum block, and Honda follows a similar process in die cast aluminum bores by adding alumina and carbon fibers. In comparison to cast iron liners, these reinforcement particles increase wear properties and heat transfer performance. Pistons and cylinder liners are made of aluminum alloy with uniformly scattered graphite particles to improve the solid lubrication property. The graphite-reinforced aluminum has a better coefficient of friction, wear rate, and it does not seize when lubricated at the boundary. In gasoline and diesel engines and sportscars, aluminum-reinforced graphite particles pistons and liners were found to have lower coefficient of friction and wear rates. The coefficient of friction of aluminum-reinforced graphite has been calculated to be as low as 0.2. The use of these materials for cylinder liners in lightweight aluminum engine blocks allows engine cylinders to hit working temperature earlier while also improving wear resistance, lowering cold start gases, and lowering weight [43–45].

Many automobile manufacturers began using aluminum and light gauge material for suspension parts to minimize unsprung weight and boost braking system, but many parts are still made of cast iron. Auto parts made of hard silicon carbide (SiC) particles reinforced with aluminum or aluminum nanocomposites, like wheel hubs or control arms, can further enhance aluminum alloy structures by improving strength properties comparable to cast iron that uses low material than relative to aluminum arms. Self-lubricating material like graphite reinforced with aluminum bushings could also be integrated into cast control arms, allowing for maintenance-free parts that will last lifespan of the automobile [46].

Aluminum/magnesium matrix composites without the addition of lead-containing carbon-based reinforced particles can be used to replace copper-lead bearings in crankshaft main bearing caps. The nontoxic carbon-based reinforcement replaces lead-based copper with aluminum reinforced with carbon-based particle composite bearings to reduce weight. The bearings will enhance wear properties due to deformation of the carbon particles leads to the formation of a stable carbon film, which allows the component to self-lubricate, resulting in increased component durability. Almost every journal bearing with in drive train might gain from such materials [47].

The density of automobile disc brakes and brake calipers, which are generally made of cast iron, could be significantly reduced. Vehicles like the Lotus Elise, Chrysler Prowler, General Motors EV1, Volkswagen Lupo 3 L, and Toyota RAV4-EV utilizing SiC particle reinforced aluminum composite brake pads [11]. The extensive use of aluminum composite brake pads generally requires lower costs and better machinability. To effectively address cost and machinability constraints, researchers have developed aluminum-reinforced SiC-nano size graphite nanocomposites and aluminum-reinforced Al2O3-nano size graphite [48].

Incorporating fly ash into components that aren't subjected to excessive loading will save money and weight in the metal matrix (aluminum and magnesium). A/C pump frames, belt/chain synchronization covers, alternator housings, drive housing, intake manifolds, and valves cover can be replaced with aluminum/magnesium reinforced fly ash composites to minimize automobile size and weight, improving emissions and saving energy. In addition to making aluminum lighter and less costly, including fly ash decreases its thermal expansion coefficient and enhances its wear resistance. The chassis' strength and durability affect vehicle performance, as well as occupant adaptability in serious crashes. When porous ceramic particles are embedded into matrix materials, a syntactic foam product is created, approximately half as heavy as the base material, and can absorb massive amounts of energy per unit of weight compared to monolithic alloys and open-cell foams. Aluminum-fly ash chemosphere syntactic foams are being produced that can be used to strengthen tubular frame parts in crash protection to enhance rotational toughness and energy storage during vehicle impact [46, 48, 49].

6. Conclusion

Automobile manufacturers must depend on advancements in metal matrix nanocomposites to achieve established fuel efficiency and pollution standards while manufacturing vehicles with both the reliability and facilities that customers demand. The automotive manufacturers can tailor enhance mechanical strength, less wear rate, and self-lubricating lightweight metal-based composites for applications using nanoreinforced materials, resulting in substantial weight savings and improved fuel economy. The increased investigations for metal matrix nanocomposites are intended to increase their use in functional and operational components, boosting future growth.

References

[1] S.B. Singh, Metal Matrix Composites: A Potential Material for Futuristic Automotive, SAE Technical Paper 2003-26- 0038, 2003.
[2] A. Sherman, P. Sklad, Collaborative Development of Lightweight Metal and Alloys for Automotive Applications, SAE Technical Paper, 2002-01-1938, 2002.
[3] M. Kim, et al., Development of Cast-Forged Knuckle Using High Strength Aluminum Alloy, SAE Technical Paper 2011-01-0537, 2011.
[4] V. Gutowski, W. Yang, S. Li, K. Dean, X. Zhang, Lightweight nanocomposite materials, in: O. Faruk, J. Tjong, M. Sain (Eds.), Lightweight and Sustainable Materials for Automotive Applications, CRC Press, 2017, pp. 277–302.
[5] P.K. Rohatgi, Use of Lightweight Metal-Matrix Composites for Transportation Applications in India, SAE Technical Paper 2004-28-0070, 2004.
[6] S.T. Mavhungte, E.T. Akinlabi, M.A. Onitiri, F.M. Varachia, Aluminum matrix composites for industrial use: advances and trends, Proc. Manuf. 7 (2017) 178–182.

[7] H.E. Friedrich, B.L. Mordike, Magnesium Technology: Metallurgy, Design Data, Applications, Springer-Verlag, Berlin Heidelberg, 2006, https://doi.org/10.1007/3-540-30812-1.
[8] G.D. Cole, Magnesium, Chem. Eng. News Arch 81 (36) (2003) 52, https://doi.org/10.1021/cen-v081n036.p052.
[9] M.P. Staiger, A.M. Pietak, J. Huadmai, G. Dias, Magnesium and its alloys as orthopedic biomaterials: a review, Biomaterials 27 (2006) 1728–1734, https://doi.org/10.1016/j.biomaterials.2005.10.003.
[10] N. Li, Y. Zheng, Novel magnesium alloys developed for biomedical application: a review, J. Mater. Sci. Technol 29 (2013) 489–502, https://doi.org/10.1016/j.jmst.2013.02.005.
[11] W.H. Hunt, D.B. Miracle, Automotive applications of metal matrix composites, in: D.B. Miracle, S.L. Donaldson (Eds.), ASM Handbook: Composites, vol 21, ASM International, Materials Park, Ohio, 2001, pp. 1029–1032.
[12] J. Mathew, J. Joy, C.S. George, Potential applications of nanotechnology in transportation: a review, J. King Saud Univ. Sci. 31 (4) (2019) 586–594.
[13] S. Veličković, B. Stojanović, L. Ivanović, S. Miladinović, S. Milojević, Mobility & vehicle mechanics, Int. J. Mobil. Veh. Mech. 45 (3) (2019) 51–64, https://doi.org/10.24874/mvm.2019.45.03.05.
[14] N. Tiruvenkadam, P.R. Thyla, M. Senthilkumar, M. Bharathiraja, A. Murugesan, Synthesis of new aluminum nano hybrid composite liner for energy saving in diesel engines, Energ. Conver. Manage. 98 (2015) 440–448.
[15] M. Wong, W.L.E. Gupta, Effect of hybrid length scales (micro + nano) of SiC reinforcement on the properties of magnesium, Solid State Phenom. 111 (2006) 91–94.
[16] S.F. Hassan, M. Gupta, Effect of particulate size of Al2O3 reinforcement on microstructure and mechanical behavior of solidification processed elemental Mg, J. Alloys Compd. 419 (2006) 84–90, https://doi.org/10.1016/j.jallcom.2005.10.005.
[17] P.S.S.R. Kumar, D.S.R. Smart, S.J. Alexis, MWCNT Reinforced Aluminium 5-Series With Improved Damping Corrosion, LAP LAMBERT Academic Publishing, 2018. (July 27, 2018), ISBN: 978-6139850549.
[18] P.S.S.R. Kumar, S.J. Alexis, Damping study on MWCNT-reinforced Al composites, in: Hysteresis of Composites, Li Longbiao, IntechOpen, 2018, https://doi.org/10.5772/intechopen.81190. December 19th.
[19] I. Khan, K. Saeed, I. Khan, Nanoparticles: properties, applications and toxicities, Arab. J. Chem. 12 (7) (2019) 908–931.
[20] P.S.S.R. Kumar, P.M. Mashinini, Dry sliding Wear behaviour of AA7075—Al2SiO5 layered nano-particle material at different temperature condition, Silicon (2020), https://doi.org/10.1007/s12633-020-00728-3.
[21] P.S. Samuel Ratna Kumar, S. Jyothi, S.J. Alexis, Corrosion behavior of aluminum alloy reinforced with MWCNTs, micro and nano technologies, in: Corrosion Protection at the Nanoscale, Elsevier, 2020, pp. 47–61. ISBN 9780128193594 https://doi.org/10.1016/B978-0-12-819359-4.00004-0.
[22] F.J. Heiligtag, M. Niederberger, The fascinating world of nanoparticle research, Mater. Today 16 (7–8) (2013) 262–271.
[23] O. Yilmaz, S. Buytoz, Abrasive wear of Al2O3-reinforced aluminium-based MMCs, Compos. Sci. Technol. 61 (2001) 2381–2392.
[24] P. Pradeep, P.S.S.R. Kumar, I.D. Lawrence, S. Jayabal, Characterization of particulate reinforced aluminium 7075/TiB$_2$ composites, Int. J. Civil Eng. Technol. 8 (2017) 178–190.
[25] A. Atrian, G.H. Majzoobi, M.H. Enayati, H. Bakhtiari, A comparative study on hot dynamic compaction and quasi-static hot pressing of Al7075/SiC$_{np}$ nanocomposite, Adv. Powder Technol. 26 (2015) 73–82.
[26] S.E. Hernández Martínez, J.J. Cruz-Rivera, C.G. Garay-Reyes, C.G. Elias-Alfaro, R. Martínez-Sánchez, J.L. Hernández-Rivera, Application of ball milling in the synthesis of AA 7075–ZrO$_2$ metal matrix nanocomposite, Powder Technol. 284 (2015) 40–46.
[27] N.G. Siddesh Kumar, R. Suresh, G.S. Shiva Shankar, High temperature wear behavior of Al2219/n-B4C/MoS2 hybrid metal matrix composites, Compos. Commun. 19 (2020) 61–73.
[28] S.K. Dwiwedi, A.K. Srivastava, M.K. Chopkar, Wear study of chicken eggshell-reinforced Al6061 matrix composites, Trans. Indian Inst. Met. 72 (2019) 73.

[29] S. Rajesh Ruban, K. Leo Dev Wins, J. David Raja Selvam, R.S. Rai, Experimental investigation and characterization of *in situ* synthesized sub micron ZrB_2-ZrC particles reinforced hybrid AA6061 aluminium composite, Mater. Res. Express 6 (2019) 1050e1, https://doi.org/10.1088/2053-1591/ab4311.

[30] J.D.R. Selvam, I. Dinaharan, P.M. Mashinini, High temperature sliding wear behavior of AA6061/fly ash aluminum matrix composites prepared using compocasting process, Tribol. Mater. Surf. Interf. 11 (1) (2017) 39–46, https://doi.org/10.1080/17515831.2017.1299324.

[31] P.S. Samuel RatnaKumar, S. JohnAlexis, Synthesized carbon nanotubes and their applications, in: Carbon Nano-Objects, Carbon-Based Nanofillers and Their Rubber Nanocomposites, 2019, pp. 109–122, https://doi.org/10.1016/B978-0-12-813248-7.00004-3.

[32] J.E. Allison, G.S. Cole, Metal-matrix composites in the automotive industry: opportunities and challenges, JOM 45 (1993) 19–24, https://doi.org/10.1007/BF03223361.

[33] Lucintel—Global Management Consulting & Market Research Firm, Available from. https://www.lucintel.com/. Accessed 10.06.2018.

[34] M. Gupta, W.L.E. Wong, Magnesium-based nanocomposites: lightweight materials of the future, Mater Charact 105 (2015) 30–46.

[35] R. Palanivel, I. Dinaharan, R.F. Laubscher, J. Paulo Davim, Influence of boron nitride nanoparticles on microstructure and wear behavior of AA6082/TiB_2 hybrid aluminum composites synthesized by friction stir processing, Mater. Des. 106 (2016) 195–204.

[36] Y. Yang, X. Li, Ultrasonic Cavitation Based Casting of Aluminum Matrix Nanocomposites for Automobile Structures, SAE Technical Paper 2006-01-0290, 2006, https://doi.org/10.4271/2006-01-0290.

[37] J. Hemanth, Development and Property Evaluation of Aluminum-Alloy Reinforced with Nano-$ZrO2$ Metal Matrix Composites (NMMCs) for Automotive Applications, SAE Technical Paper 2009-01-0218, 2009, https://doi.org/10.4271/2009-01-0218.

[38] P.S. Samuel Ratna Kumar, M. Peter Madindwa, Investigation on tribological behaviour of aluminosilicate reinforced AA7075 composites for aviation application, Trans. Indian Inst. Met. 74 (2021) 79–88, https://doi.org/10.1007/s12666-020-02112-6.

[39] P.S. Samuel Ratna Kumar, S. John Alexis, M. Saravana Mohan, P. Edwin Sudhagar, S.R. Madara, Vibration—Impact study on AlMg4.5Mn reinforced nanoclay composites, Mater. Today: Proc. 28 (2) (2020) 1140–1143, https://doi.org/10.1016/j.matpr.2020.01.096.

[40] K.B. Nie, X.J. Wang, X.S. Hu, L. Xu, K. Wu, M.Y. Zheng, Microstructure and mechanical properties of SiC nanoparticles reinforced magnesium matrix composites fabricated by ultrasonic vibration, Mater. Sci. Eng. A 528 (2011) 5278–5282, https://doi.org/10.1016/j.msea.2011.03.061.

[41] C.S. Goh, J. Wei, L.C. Lee, M. Gupta, Properties and deformation behaviour of Mg–Y_2O_3 nanocomposites, Acta Mater. 55 (2007) 5115–5121, https://doi.org/10.1016/j.actamat.2007.05.032.

[42] Q. Jun, A. Linan, P.J. Blau, Sliding friction and wear characteristics of Al2O3-Al nanocomposites, in: Proceedings of STLE/ASME IJTC, 2006, pp. 59–60.

[43] P.K. Rohatgi, S. Ray, Y. Liu, Tribological properties of metal matrix graphite particle composites, Int. Mater. Rev. 37 (3) (1992) 129.

[44] P.K. Rohatgi, et al., Solidification during Casting of Metal- Matrix Composites, Invited paper, ASM Handbook, vol. 15, ASM International, 2008, pp. 390–397.

[45] M. Gumus, Reducing cold-start emission from internal combustion engines by means of thermal energy storage system, Appl. Therm. Eng. 29 (2009) 652–660.

[46] B.F. Anthony Macke, Schultz, Pradeep Rohatgi, metal matrix composites offer the automotive industry an opportunity to reduce vehicle weight, improve performance, advanced materials and processes, ASM Int. 170 (3) (2012) 19–23.

[47] M. Kestursatya, J.K. Kim, P.K. Rohatgi, Friction and wear behavior of centrifugally cast lead free copper alloy containing graphite particles, Metall. Mater. Trans. A 32A (8) (2001) 2115–2125.

[48] G. Withers, P.D.W. Tilakaratna, Performance Evaluation of ULTALITE® Low Cost Aluminum Metal Matrix Composite Based Brake Drums, SAE Technical Paper 2005-01-3936, 2005.

[49] P.K. Rohatgi, D. Weiss, Casting of Aluminum-Fly Ash Composites for Automotive Applications, SAE Technical Paper 2003-01-0825, 2003.

CHAPTER 11

Fiber-reinforced nanocomposites

Younes Ahmadi, Nasrin Raji Popalzai, Mubasher Furmuly, and Nangialai Azizi
Department of Analytical Chemistry, Faculty of Chemistry, Kabul University, Kabul, Afghanistan

Chapter outline

1. Introduction — 215
2. Characterization methods — 217
   2.1 Analysis of physical properties — 218
   2.2 Mechanical characterization — 218
   2.3 Failure analysis — 220
   2.4 Rheological characterization — 220
   2.5 Morphological characterization — 220
3. Design and manufacturing of FNCs — 222
4. Applications of FNCs — 223
   4.1 Automobile industries — 223
   4.2 Airplane industries — 223
   4.3 Chemical industry — 224
   4.4 Military devices — 224
   4.5 Other applications — 224
5. Concluding remarks — 225
References — 225

1. Introduction

Nanocomposites are engineered materials that are assembled by two or more components having different qualities. In these systems, at least one of the incorporated parts has dimension below 100 nm (Fig. 1) [1]. Nanocomposite formulation process helps in improving one or more characteristics of polymers. Fiber-reinforced nanocomposite (FNC) is a class of nanocomposites, which is made by incorporation of fibrous materials within the polymer network [2]. FNCs can be formulated by two means: through (i) the incorporation of nanomaterials within the fiber-reinforced composites and (ii) the utilization of nanofibers to strengthen nanocomposite. The reinforcement using nanofibers can offer several advantages over other nano-reinforcements methods because of the dual nano–macro roles and the continuity of fibers [2]. In this regard, different nanofibers such as metallic nanofibers, organic, oxide, and ceramic were used as reinforcing agents within the composite networks [3–6]. Based on literature, organic, inorganic, and metallic composites can be reinforced by nanofibers [2, 6–8].

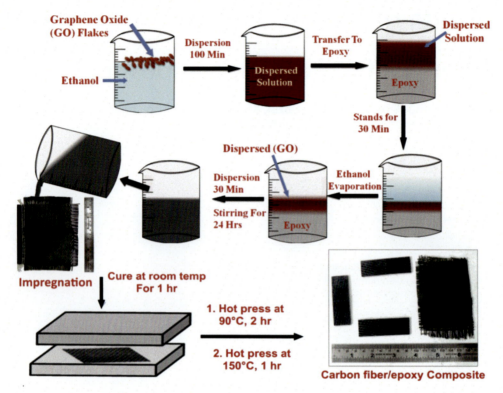

Fig. 1 Schematic representation for incorporation of graphene oxide into the epoxy resin to form epoxy-carbon-based FNCs [1].

FNCs have been generally used in biomedical, construction, automobile, transportation, aerospace, and sports materials [9–11]. They can be good alternatives to replace the metallic materials, especially in aerospace and automobile industries. Their peculiarities can be altered by the incorporation of high-strength fibers within the matrices. As a result, different fibers like glass, carbon, basalt, natural fibers, ceramic, and metallic fibers have been employed to reinforce the composite matrices [4, 12, 13] (Fig. 2). However, the difference in various properties (like structural, thermal, density, strain, etc.) of matrices and fibers result in variability in thermomechanical properties of the resultant FNCs. Consequently, the thermomechanical properties of FNCs strongly depend on the fabrication methodologies and the interaction between the matrix phase and the reinforcing phase. Recently multiscale (hierarchical) FNCs were formulated using nanofillers/nanomaterials (at the nanoscale) and fibers (at the microscale). The presence of nanofillers within the matrix resulted in enhancing the strength, stiffness, and thermal properties of FNCs. Many types of nanomaterials including graphene, carbon nanotubes, nano-TiO_2, nano-SiO_2, nano-CuO, nano-Al_2O_3, and nanoclay have been incorporated

Fig. 2 Extraction of natural fibers (sisal), where (A), (B), (C), and (D) are sisal plant, decortication process, drying process, and raw fiber, respectively [4].

into the FNCs [14–16]. Accordingly, FNCs are found to be promising materials for various applications. Therefore, present chapter has provided a brief note on FNCs and addressed various methods used for their characterization. Further, the chapter explored the design and manufacturing of FNCs along with their effective applications in various realms.

2. Characterization methods

In order to study the effect of incorporation of nanofibers, various characterization techniques have been employed [17]. This chapter has made an effort to integrate the adopted procedures to study physical, rheological, and mechanical properties of FNCs. Besides, some sophisticated characterization tools are emphasized to understand the microstructural and morphological properties of such systems. The development of novel product (to meet the market demand) constantly force the researchers to explore novel evaluation and characterization techniques to analyze the products for diversified applications [18]. These applications can range from simple construction (such as sports items, leisure's, interior furnishing, etc.) to advanced engineering constructions (e.g., aerospace, aircraft, advanced armory systems, war tanks, etc.) [2].

2.1 Analysis of physical properties

Physical characterization of FNCs is employed to analyze their dimensional stability and help understand their shelf-life and applicational areas. Such characterization techniques mostly include water absorption, fiber loading, density, hardness, void content, swelling, hydrothermal aging, scratch resistance, etc., some of which are briefed as follows:

One of the parameters that depends on physical properties of FNCs is density, which plays an important role to investigate their dimensional stability. Generally, on the basis of density, composites can be classified into low, medium, and high densities. High-density composites have lower porosity, voids, and intercellular spaces compared to low-density composites [4]. As a result, the moisture adsorption and retention capacity increase with increasing the density of composites. Pores are generally classified into three different types [19]. Type 1 pores are also known as "gas evoluted pores," which are mostly produced during the manufacturing of feed-stock filaments at extrusion time. One of the commonly used methods employed for the layer-by-layer assembly of composites is fused deposition modeling where the inappropriate dispersion of carbon fibers within the filaments results in the inconstant fusion of layers, causing the generation of such pores. Type 2 pores are also defined as "physical gap at layer/layer interfaces." They are produced by the presence of voids at the interface of different layers. Type 3 pores are commonly formed because of the pulled-out carbon fibers located at the cracked interface of samples. However, these are not considered as intrinsic porosity of the FNCs since they are generated at the time of post-processing fracture of composite samples. The 3D observation of mesostructure of porous materials as well as finding the porosity of FNCs can be done with the help of X-ray computed tomography [7].

2.2 Mechanical characterization

Recently FNCs have widely been used for various high-performance applications, which are automotive, aerospace sports goods, and construction. These composites replaced metals mainly in aerospace and automobile industries [2]. The recent progresses in aircraft suggested 50% of reduction in weight since most of the primary structural components have been developed from FNCs (e.g., epoxy glass fiber composites and epoxy carbon fiber composites) [2, 20].

Exploring the mechanical properties of FNCs is a very important requirement for their design and manufacturing and predicting their lifetime. Nanocomposites formulated through the application of very low amount (5 wt%) of nano-fillers like nano-clays, chitin nanofibrils, cellulose nanoparticles, cellulose nanofibrils, and nano-whiskers had shown great enhancement in mechanical properties. The presence of these nanofillers mostly governs the mechanical characteristics of FNCs. The high aspect ratio, excessive rigidity, and affinity of nanofillers for the polymer matrices are accountable for the increased mechanical properties (such as tensile strength, Young's modulus, and elongation at break) of the FNCs. FNCs possess superior mechanical properties compared to their pristine counterparts and those of conventional materials. However, there is still room for further improvement in their manufacturing procedures, reducing the cost of the manufacturing processes, their applications, and enhancing their properties [20].

As mentioned earlier, the tensile strength is a crucial factor and its determination helps in understanding the mechanical performance of the FNCs. The standard methods are generally employed to test the in-plane tensile properties of high modulus FNC materials [21]. The tensile properties of FNC materials depend on the direction of the test since these materials have continuous or discontinuous fiber network [22]. The tensile tests of FNCs are determined in three directions (with reference to the directions of the fiber), which are 0 and 90 (ASTM D3039) and ∓ 45 degrees (ASTM D3518) [23].

Other important parameters for the mechanical properties of FNCs are their hardness, flexural strength, fracture toughness, and wear strength. The hardness is "the resistance of materials to the plastic/localized deformation" [24]. Such deformation may occur due to indentation, cutting, or bending [25]. Measuring the hardness of materials is useful to evaluate their quality and to control their manufacturing processes as well as research point of view. Hardness can be correlated to the tensile strength and used as the indicator for the determination of wear ductility, resistance, and toughness of materials. The flexural strength (i.e., modulus of rupture/transverse rupture strength/bend strength) is also a property of materials and can be defined as the stress in materials just before they yield in a flexure test (Fig. 3) [1, 26].

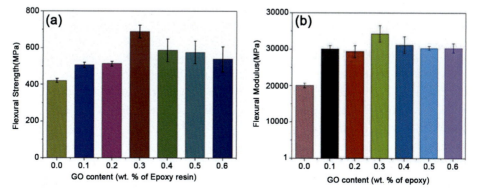

Fig. 3 Variation observed in (A) flexural strength and (B) flexural modulus after the incorporation of graphene oxide in epoxy resin [1].

2.3 Failure analysis

The failure analysis of FNCs is a topic of interest for many scientists [27]. Some of the main reasons for failure in composites are fracture of fiber, shear cracking of the matrix, and cracking of matrix in transverse direction. When FNCs are subjected to the pure shear test, most parts of the FNC network experience plane-strain and the uniaxial tensions along the direction of applied force. Some theoretical model has been proposed for the determination of fracture energy of composite materials using fiber-reinforced hydrogels as study material [27]. The hydrogel-based composites were fabricated via impregnation of 3D-printed thermoplastic fiber into the crosslinked alginate and polyacrylamide-based hydrogels. The formulated FNCs exhibited superior failure resistance compared to those of plain material.

2.4 Rheological characterization

Studying the flow behavior of fluid under isothermal condition or isostress. FNCs are generally synthesized in film, sheet, and bar forms, development of which initially necessitates the fabrication of nanofibers. Recently, composites have been formed by means of transfer molding process where flow rate should be controlled in order to synthesize an effective composite. The rheology affects our understanding to employ FNC materials and allows us to determine magnetic, thermomechanical, and electrical responses of nanocomposites.

2.5 Morphological characterization

Determination of morphology of FNCs exposes numerous structural characteristics in 2D as well as 3D images. In addition, extra information including fracture analysis, interface failure, crystalline characteristics, elemental analysis, and surface roughness can be obtained via this method. Therefore, micro-structural properties of FNCs were generally analyzed by various microscopical techniques, viz., atomic force microscopy, scanning electron microscopy (Fig. 4) [28], optical microscopy, transmission electron microscopy (Fig. 5) [1], etc. In addition to these microscopy techniques, fluorescence microscopy has been employed to investigate the morphology of fibers possessing fluorescent compounds, especially dyed fibers [29]. In this technique, confocal microscopy was used for FNCs, which offers various advantages including reduction in background information and field depth control [2].

Molecular structure and estimating the molecular weight, grafting percentage, and identifying the fiber–polymer interactions within FNCs networks are determined with the help of spectroscopic techniques such as Raman Spectroscopy, Fourier-transform infrared spectroscopy (FTIR), and CP/MAS ^{13}C NMR spectroscopy [7]. In addition, the crystallographic techniques are employed to analyze the crystal pattern of FNCs and effect of

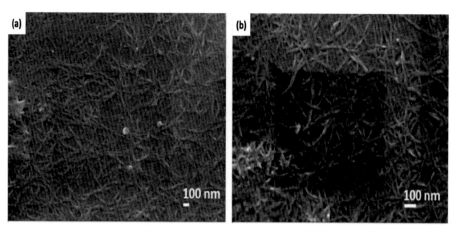

Fig. 4 Field emission-scanning electron microscopy images of cellulose nanofibers at (A) low and (B) high magnifications [28].

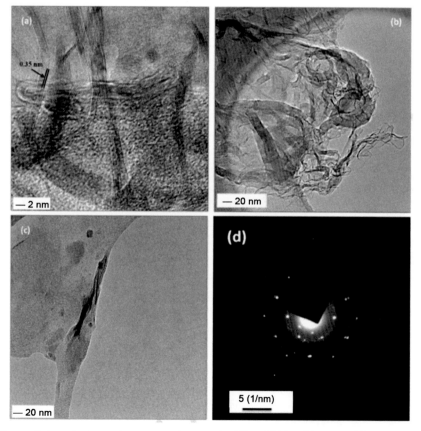

Fig. 5 Transmission electron microscopy micrographs (A) exhibiting fringes and layering, (B) folded and multilayer structure of graphene oxide, (C) dispersed state of graphene oxide, and (D) SAED pattern [1].

incorporation of nanofiber on crystallite size. Determination of this parameter is very important since it causes the decline of mechanical properties of the FNC systems.

3. Design and manufacturing of FNCs

In recent years, the formulation of FNCs has attracted excessive attention among researchers as well as composite industries [30]. The FNCs were manufactured by addition of small volume of nanofillers in the polymer matrix [21]. Present section highlights the design and manufacturing of some of FNC systems that are extensively used in various industries.

One of the important FNC systems is ceramic-reinforced FNCs, which possess high-thermal shock resistance, mechanical, electrical, and thermal properties. During the fabrication of fiber-reinforced ceramic nanocomposites, the size, type, shape, and percentage of the fiber were decided prior to the infiltration of matrix. In such nanocomposites, desired metal fibers were employed to improve toughness while the ceramic fibers were used to induce higher resistance to crack (to the formulated FNCs). Several methodologies have been applied to incorporate the ceramic matrix within the fibers. These techniques are high-temperature sintering, preceramic polymer pyrolysis, chemical vapor reaction, and electrophoretic deposition [31]. Following these approaches, oxide and non-oxide ceramics can be percolated into the ceramic fibers, which results in the formation of a porous structure, necessitating the surface treatment (of the end product). After attaining the final structure at ambient temperature, the surface of the product undergoes machining grinding/coating, lapping, milling, or drilling [32].

Due to the increasing demand for sustainability and environmental protection, number of scientists and technologists have focused on the development of fiber-reinforced bio-composites with similar or superior properties compared to their synthetic equivalents [33]. Recently, the evolution of bio-based FNCs has attracted a lot of attention because of their excellent performance in structural, mechanical, electrical, pharmaceutical, and biomedical applications. Bio-based FNCs are nanocomposite materials that are made up of either bio-polymers or fillers derived from bio-resources (cellulose and cellulose nanofibers) [28]. These materials have been extensively used in different applications such as automotive, furniture, energy storage devices, biosensors, etc., since they possess great versatility [34]. Compared with inorganic fillers, eco-friendly nanofibers exhibit many advantages like sustainability, nontoxicity, biodegradability/biocompatibility, lower density, low cost, and good thermal properties. For instance, cotton or jute fibers are types of biofibers that are used in textile industry. They are bundle of individual nanofibers, which are held together by thin layers of lignins, pectins, and polysaccharides. These materials have successfully substituted the extensive utilization of synthetic nanofibers, which were mainly derived from petroleum products [31].

4. Applications of FNCs

FNCs have extensive potential to be used in wide areas from construction of aircraft (structures and engines) and weapons to home appliances. Some of the important uses have been reviewed as follows:

4.1 Automobile industries

The automobile industries have adopted FNCs faster than other industries due to the higher strength-to-weight ratio of these materials. FNCs have been used in high-end sports cars and started to find their way into other vehicles. Carbon-based fiber materials have been used in sports cars for many years. For example, filters made up of metal fiber-sintered sheets are employed for diesel, gasoline, and ventilation filters. In addition, heat-resistant textiles are produced from metal fibers and used for bending the glass in automobiles, which protects the glass during the bending process under high temperature and pressure. The adBlue tanks, selective catalytic reduction tubes, and heating cables (used in car seats) are also made up of metal-nanofiber composites. Metal-fiber heating cables exhibit high durability and flexibility compared to copper wires [35]. The heat-resistant FNCs are desired materials to manufacture fire-resistant cloths (for fire fighters) and lightweight fabrics (e.g., covers of automobiles and skin of aircraft) that can be used under extreme temperature and stress. FNCs can also be as suitable alternates to replace corrosion-prone metals with lighter weight and higher abilities. The lightweight bodies made from polymer or metal nanocomposites are reported that showed very high strength and low density [33]. The variety of uses of FNCs in automotive industry is not limited to the aforementioned materials and can further be highlighted like their applications in tyre manufacturing, fuel system, gas separation membranes, engine covers, door handles, and timing belt covers.

4.2 Airplane industries

Composite materials have been used to construct (almost 50% of) Boeing 787 and Airbus A350 XWB. Carbon-based fibers are also employed in the manufacturing of helicopters, turbine engines, rocket motors, and satellites. In addition, boron fiber-reinforced aluminum and aluminum-reinforced graphite fibers were used in the fabrication of space shuttle orbiter struts due to their high compression strength and stiffness as well as low density [33]. Such materials were preferred owing to their low coefficient of thermal expansion. Due to the exceptional thermal properties, FNCs were also applied in the construction of high-temperature fighter engines and structures of aircraft, missiles, spacecraft, and electronic packaging materials. FNCs are capable of reducing the sound of airplanes with nanofibers [36].

4.3 Chemical industry

Various chemical industries require materials that can withstand corrosion. Therefore, most of the polymer fibers/metal fibers play a prominent role in corrosion resistance properties of FNCs. For example, Ni/TiO_2, Fe/MgO, and Fe/TiN are utilized in petrochemical industries since they exhibit anticorrosive properties [33].

4.4 Military devices

Military devices are required to possess high strength-to-weight ratio. Among various materials used to manufacture military equipment, carbon fiber composites are most desired and common materials. As it is well known, fiberglass is a common material for boats/yachts since they have lower price compared to those of carbon fibers. Glass fibers are oriented randomly that induce more uniform structure and properties to the materials. While, metal fiber filters can form highly porous structures with very low pore sizes, making them appropriate candidates for highly efficient air filters [37]. Such air filters are utilized in nuclear power plants to prevent releasing of radioactive pollutants [37].

4.5 Other applications

Carbon-based FNCs provide lightweight and high stiffness hence they have found extensive use in the manufacturing of golf balls and tennis rackets. Metal-based FNCs (iron and its derivatives like $Fe_{23}C_6$/Fe_2B and Cu/Nb) have induced long life span to structural materials. CNT-ceramic composites have robust mechanical, barrier, electrical, and thermal properties, which ensure their applications in manufacturing of gas storage devices, toxic gas sensors, flat panel displays, Li^+ batteries, and conducting coatings [38]. For instance, Al_2O_3-CNT composites exhibited high contact damage resistance without a showing a noticeable increase in hardness and toughness (compared to that of pristine materials). The excellent barrier characteristics along with suitable flexibility and rigidity of clay nanocomposites made them suitable candidates to be used in food industry. Specific examples are packaging for meats, cereals, sweets, boil foods, coatings/paperboard for dairy products and fruit juice, beer, as well as carbonated drinks bottles. In general, the utilization of FNCs in packaging industry can enhance considerably the shelf-life of food products.

The applications of FNC systems are numerous, including manufacturing of new generation materials and enhancing the performance of existing devices like sensors, fuel cells, and coatings. Dielectric FNC systems (i.e., two-phase heterogeneous nanodielectrics) are widely used in electric/electronic industries [18, 35, 39]. For example, TiO_2, Al_2O_3, Ag, and AlN nanofibers are used in microelectronic devices [40]. Lastly, CNT-based nanocomposites are reported as potential candidates for photovoltaic cells and photo diodes, drums for printers, data storage media, optical limiting devices, etc.

5. Concluding remarks

This chapter provided a brief note on introduction, different characterization techniques, and various applicational areas of FNCs. In addition, as environmental concerns are growing with increasing usage of non-biodegradable composites, sustainable manufacturing of alternative nanocomposites having high energy efficiency has become an urgent need. Therefore, the importance and effective utilization of bio-based FNCs have also been briefed in this chapter. Bio-based FNCs possess outstanding UV resistance, thermal, electrical, mechanical, and barrier properties and can serve as exceptional materials in the fabrication of high-performance devices and equipment. Furthermore, the production of FNCs can be made more cost-effective, rebut, and high-ending in various applicational areas. The future of these materials is very wide and can be applied in wide areas including preparation of paints/coatings (for automobile and metal) to prevent corrosion and manufacturing of aircraft (and its parts).

References

[1] A.K. Pathak, M. Borah, A. Gupta, T. Yokozeki, S.R. Dhakate, Improved mechanical properties of carbon fiber/graphene oxide-epoxy hybrid composites, Compos. Sci. Technol. 135 (2016) 28–38, https://doi.org/10.1016/j.compscitech.2016.09.007.

[2] A. Dey, S. Mandal, S. Bhandari, C. Pal, J. Tersur Orasugh, D. Chattopadhyay, Characterization methods, Fiber-Reinforced Nanocomposites: Fundamentals and Applications, Elsevier, UK, 2020, pp. 7–67.

[3] N. Adrus, M. Ulbricht, Rheological studies on PNIPAAm hydrogel synthesis via in situ polymerization and on resulting viscoelastic properties, React. Funct. Polym. 73 (2013) 141–148.

[4] J. Naveen, M. Jawaid, P. Amuthakkannan, M. Chandrasekar, Mechanical and Physical Properties of Sisal and Hybrid Sisal fiber-Reinforced Polymer Composites, Elsevier Ltd, 2018, https://doi.org/10.1016/B978-0-08-102292-4.00021-7.

[5] K. Dutta, B. Das, J.T. Orasugh, D. Mondal, A. Adhikari, D. Rana, R. Banerjee, R. Mishra, S. Kar, D. Chattopadhyay, Bio-derived cellulose nanofibril reinforced poly(N-isopropylacrylamide)-g-guar gum nanocomposite: an avant-Garde biomaterial as a transdermal membrane, Polymer (Guildf). 135 (2018) 85–102, https://doi.org/10.1016/j.polymer.2017.12.015.

[6] R. Yahaya, S.M. Sapuan, M. Jawaid, Z. Leman, E.S. Zainudin, Effect of fibre orientations on the mechanical properties of kenaf–aramid hybrid composites for spall-liner application, Def. Technol. 12 (2016) 52–58, https://doi.org/10.1016/j.dt.2015.08.005.

[7] J.A. Rodríguez-González, C. Rubio-González, C.A. Meneses-Nochebuena, P. González-García, L. Licea-Jiménez, Enhanced interlaminar fracture toughness of unidirectional carbon fiber/epoxy composites modified with sprayed multi-walled carbon nanotubes, Compos. Interfaces 24 (2017) 883–896, https://doi.org/10.1080/09276440.2017.1302279.

[8] A. Dey, R. Bera, S. Ahmed, D. Chakrabarty, Smart superabsorbent UV resistant etherified PVA gel: synthesis and characterization, J. Ind. Eng. Chem. 21 (2015) 1219–1230, https://doi.org/10.1016/j.jiec.2014.05.038.

[9] X. Li, W. Liu, L. Sun, K.E. Aifantis, B. Yu, Y. Fan, Q. Feng, F. Cui, F. Watari, Resin composites reinforced by nanoscaled fibers or tubes for dental regeneration, Biomed. Res. Int. 2014 (2014), https://doi.org/10.1155/2014/542958.

[10] Z.S. Metaxa, M.S. Konsta-Gdoutos, S.P. Shah, Carbon nanofiber-reinforced cement-based materials, Transp. Res. Rec. (2010) 114–118, https://doi.org/10.3141/2142-17.

[11] S. Lina, D. Zhen, The carbon fiber composite materials application in sports equipment, Adv. Mater. Res. 341–342 (2012) 173–176, https://doi.org/10.4028/www.scientific.net/AMR.341-342.173.

[12] S. Tang, C. Hu, Design, preparation and properties of carbon fiber reinforced ultra-high temperature ceramic composites for aerospace applications: a review, J. Mater. Sci. Technol. 33 (2017) 117–130, https://doi.org/10.1016/j.jmst.2016.08.004.

[13] T.A. Nguyen, B. Han, S. Sharma, L. Longbiao, K.S. Bhat, Fiber-reinforced nanocomposites: an introduction, in: B. Han, S. Sharma, T.A. Nguyen, L. Longbiao, K.S. Bhat (Eds.), Fiber-Reinforced Nanocomposites: Fundamentals and Applications, Elsevier, UK, 2020, pp. 3–6.

[14] W. Qin, F. Vautard, L.T. Drzal, J. Yu, Mechanical and electrical properties of carbon fiber composites with incorporation of graphene nanoplatelets at the fiber-matrix interphase, Compos. Part B Eng. 69 (2015) 335–341, https://doi.org/10.1016/j.compositesb.2014.10.014.

[15] Y.J. Kwon, Y. Kim, H. Jeon, S. Cho, W. Lee, J.U. Lee, Graphene/carbon nanotube hybrid as a multifunctional interfacial reinforcement for carbon fiber-reinforced composites, Compos. Part B Eng. 122 (2017) 23–30, https://doi.org/10.1016/j.compositesb.2017.04.005.

[16] R.K. Nayak, K.K. Mahato, B.C. Ray, Water absorption behavior, mechanical and thermal properties of nano TiO_2 enhanced glass fiber reinforced polymer composites, Compos. Part A Appl. Sci. Manuf. 90 (2016) 736–747, https://doi.org/10.1016/j.compositesa.2016.09.003.

[17] Y. Ahmadi, M.T. Siddiqui, Q.M.R. Haq, S. Ahmad, Synthesis and characterization of surface-active antimicrobial hyperbranched polyurethane coatings based on oleo-ethers of boric acid, Arab. J. Chem. 13 (2020) 2689–2701, https://doi.org/10.1016/j.arabjc.2018.07.001.

[18] Y. Ahmadi, S. Ahmad, Recent progress in the synthesis and property enhancement of waterborne polyurethane nanocomposites: promising and versatile macromolecules for advanced applications, Polym. Rev. (2019) 1–41, https://doi.org/10.1080/15583724.2019.1673403.

[19] F. Ning, W. Cong, J. Qiu, J. Wei, S. Wang, Additive manufacturing of carbon fiber reinforced thermoplastic composites using fused deposition modeling, Compos. Part B Eng. 80 (2015) 369–378, https://doi.org/10.1016/j.compositesb.2015.06.013.

[20] K. Schulte, S. Chandrasekaran, C. Viets, B. Fiedler, New Functions in Polymer Composites Using a Nanoparticle-Modified Matrix, Elsevier Inc., 2015, https://doi.org/10.1016/B978-0-323-26434-1.00030-1.

[21] V. Dikshit, S.C. Joshi, Manufacturing of Multiscale Interlaminar Interface Composites and Quantitative Analysis of Interlaminar Fracture Toughness, Elsevier, UK, 2020.

[22] H. Takagi, A. Nakagaito, K. Nishimura, T. Matsui, Mechanical characterisation of nanocellulose composites after structural modification, High. Perform. Optim. Des. Struct. Mater. II 166 (2016) 335–341.

[23] J. Vázquez-Moreno, R. Sánchez-Hidalgo, E. Sanz-Horcajo, J. Viña, R. Verdejo, M. López-Manchado, Preparation and mechanical properties of graphene/carbon Fiber-reinforced hierarchical polymer composites, J. Compos. Sci. 3 (2019) 30, https://doi.org/10.3390/jcs3010030.

[24] A.C. Fischer-Cripps, The Handbook of Nanoindentation, Fischer-Cripps Laboratories Pty Ltd, Forestville, Australia, 2009. www.fclabs.com.au.

[25] S. Alam, F. Habib, M. Irfan, W. Iqbal, K. Khalid, Effect of orientation of glass fiber on mechanical properties of GRP composites, J. Chem. Soc. Pakistan 32 (2010) 265–269.

[26] M. Ashby, E. Cope, D. Cebon, Materials selection for engineering design, in: Informatics for Materials Science and Engineering: Data-Driven Discovery for Accelerated Experimentation and Application, Elsevier Inc., 2013, pp. 219–244, https://doi.org/10.1016/B978-0-12-394399-6.00010-2.

[27] S. Lin, C. Cao, Q. Wang, M. Gonzalez, J.E. Dolbow, X. Zhao, Design of stiff, tough and stretchy hydrogel composites via nanoscale hybrid crosslinking and macroscale fiber reinforcement, Soft Matter 10 (2014) 7519–7527, https://doi.org/10.1039/c4sm01039f.

[28] J.T. Orasugh, N.R. Saha, D. Rana, G. Sarkar, M.M.R. Mollick, A. Chattoapadhyay, B.C. Mitra, D. Mondal, S.K. Ghosh, D. Chattopadhyay, Jute cellulose nano-fibrils/hydroxypropylmethylcellulose nanocomposite: a novel material with potential for application in packaging and transdermal drug delivery system, Ind. Crop Prod. 112 (2018) 633–643, https://doi.org/10.1016/j.indcrop.2017.12.069.

[29] T. Patwary Plateau, Evaluation of tensile strength of jute fiber reinforced polypropylene composite, Adv. Mater. 6 (2017) 149, https://doi.org/10.11648/j.am.20170606.15.

[30] V. Dikshit, S.K. Bhudolia, S.C. Joshi, Multiscale polymer composites: a review of the interlaminar fracture toughness improvement, Fibers 5 (2017), https://doi.org/10.3390/fib5040038.

[31] J.T. Orasugh, G. Sarkar, N.R. Saha, B. Das, A. Bhattacharyya, S. Das, R. Mishra, I. Roy, A. Chattoapadhyay, S.K. Ghosh, D. Chattopadhyay, Effect of cellulose nanocrystals on the performance of drug loaded in situ gelling thermo-responsive ophthalmic formulations, Int. J. Biol. Macromol. 124 (2019) 235–245, https://doi.org/10.1016/j.ijbiomac.2018.11.217.

[32] F. Khalil, S. Galland, A. Cottaz, C. Joly, P. Degraeve, Polybutylene succinate adipate/starch blends: a morphological study for the design of controlled release films, Carbohydr. Polym. 108 (2014) 272–280, https://doi.org/10.1016/j.carbpol.2014.02.062.

[33] J.T. Orasugh, S.K. Ghosh, D. Chattopadhyay, Nanofiber-reinforced biocomposites, in: Fiber-Reinforced Nanocomposites: Fundamentals and Applications, Elsevier, 2020, https://doi.org/10.1016/b978-0-12-819904-6.00010-4.

[34] Y. Ahmadi, K.-H. Kim, S. Kim, M. Tabatabaei, Recent advances in polyurethanes as efficient media for thermal energy storage, Energy Storage Mater. 30 (2020) 74–86, https://doi.org/10.1016/j.ensm.2020.05.003.

[35] Y. Zhao, J. Wei, R. Vajtai, P.M. Ajayan, E.V. Barrera, Iodine doped carbon nanotube cables exceeding specific electrical conductivity of metals, Sci. Rep. 1 (2011) 1–5, https://doi.org/10.1038/srep00083.

[36] R. Asmatulu, W. Khan, M.B. Yildirim, Acoustical properties of electrospun nanofibers for aircraft interior noise reduction, ASME Int. Mech. Eng. Congr. Expo. Proc. 15 (2010) 223–227, https://doi.org/10.1115/IMECE2009-12339.

[37] R.V. Kurahatti, A.O. Surendranathan, S.A. Kori, N. Singh, A.V.R. Kumar, S. Srivastava, Defence applications of polymer nanocomposites, Def. Sci. J. 60 (2010) 551–563, https://doi.org/10.14429/dsj.60.578.

[38] S.I. Bhat, Y. Ahmadi, S. Ahmad, Recent advances in structural modifications of hyperbranched polymers and their applications, Ind. Eng. Chem. Res. 57 (2018) 10754–10785, https://doi.org/10.1021/acs.iecr.8b01969.

[39] Y. Ahmadi, K.H. Kim, Functionalization and customization of polyurethanes for biosensing applications: a state-of-the-art review, Trends Anal. Chem. 126 (2020) 115881, https://doi.org/10.1016/j.trac.2020.115881.

[40] A. Makino, X. Li, K. Yubuta, C. Chang, T. Kubota, A. Inoue, The effect of cu on the plasticity of Fe-Si-B-P-based bulk metallic glass, Scr. Mater. 60 (2009) 277–280, https://doi.org/10.1016/j.scriptamat.2008.09.008.

CHAPTER 12

Electrically conductive polymer nanocomposites for thermal comfort in electric vehicles

Heinrich Leicht[a], Eduard Kraus[a], Benjamin Baudrit[a], Thomas Hochrein[a], Martin Bastian[a], and Maurice Langer[b]

[a]German Plastics Center—SKZ, Wuerzburg, Germany
[b]Fraunhofer IWS, Dresden, Germany

Chapter outline

1. Carbon nanotubes and thermal comfort in electric vehicles — 229
2. Physical aspects of CNT/polymer nanocomposites for heating applications — 231
   2.1 Dispersion of CNTs in polymers and general percolation theory — 231
   2.2 Temperature coefficient of resistance of CNT nanocomposites — 235
   2.3 Aspects for Joule heating of CNT/polymer nanocomposites — 239
   2.4 Long-term stability of CNT/polymer nanocomposites in the scope of Joule heating — 242
3. Conclusion — 244
Acknowledgment — 245
References — 245

1. Carbon nanotubes and thermal comfort in electric vehicles

Conventional vehicles primarily use waste heat of their internal combustion engine for heating the passenger compartment. Full electric vehicles must use other energy sources as they generate less waste heat. One possibility is heat generation by burning fuels like bioethanol, but the necessity of a fuel tank and other components significantly increases the vehicle weight [1]. Another possibility is to convert the electrical energy, e.g., of the traction battery, into heat. The easiest way to do so is convective heating, which uses electrical heating elements that heat the air being transported into the cabin. At temperatures below 0°C convective heating has a high power consumption (several kW), which is in the order of the driving power of a small vehicle at a constant speed of 50 km/h, thus significantly reducing the vehicle range [1, 2]. The power consumption can be cut in half by additionally using the so-called infrared (IR) radiation systems based on electrically conductive coatings [1]. These systems are lightweight and have low power consumption (several hundred W). Simulations as well as experiments show that positioning of the IR systems is crucial to create a uniform comfort for all body parts [1].

Fig. 1 Exemplary overview over automotive heating applications.

Electrically conductive coatings are one way to produce such IR systems and one example for the application of carbon nanotube (CNT) composites for heating in vehicles (see Fig. 1). Due to their flexibility and low thickness it is possible to apply them in several parts of the vehicle's interior, e.g., door coverings, pillars, and headliner [1, 3, 4]. By that, the wall surface temperature (IR heating) is increased and the necessary air temperature (convection heating) in the cabin may be lower to achieve a high thermal comfort for the passengers [1, 5].

There are more applications shown in Fig. 1, which could be addressed by CNT materials and their composites, e.g., heating of mirrors and windows using transparent CNT-films or the beforementioned CNT-coatings for heating of seats, steering wheel, and other surfaces.

CNTs can be divided into single-walled CNTs (SWCNTs) and multiwalled CNTs (MWCNTs) as shown in Fig. 2. MWCNTs have been discovered by Iijima in 1991 [6], followed by SWCNTs, discovered by Endo in 1993 [7]. Since then there has been a lot of research to understand this new class of carbon materials. Due to their extraordinary properties, e.g., thermal conductivity up to 6600 W $(mK)^{-1}$ electrical conductivity up to 10^6 S m^{-1} as well as tensile strength of 50–200 GPa and Young's modulus as high as 1.2 TPa [8–11], a lot of effort has been put into research to evaluate possible applications. Considered applications include sensorics [12–17], electronics [18], electrostatic dissipation and electromagnetic interference shielding [19–22] as well as heating. Heat-related applications are, for example, micro heaters, flexible de-icing units, transparent heating glasses or coatings (for vehicles) [15, 23–26].

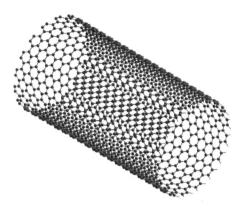

Fig. 2 Schematic cutout from a SWCNT, representing the tubular structure built from hexagons of carbon atoms (like in graphene). MWCNTs can be seen as nested SWCNTs.

CNTs can be dispersed in a (polymeric) matrix to create CNT/polymer nanocomposites, for example, to protect the CNTs from environmental influences (or vice versa) and improve the mechanical integrity of the CNT network. The high aspect ratio of CNTs allows the production of percolating networks in polymer nanocomposites at low CNT loadings, thus the composites may maintain the mechanical integrity and flexibility of the polymer matrix [27, 28]. But their behavior is different from pure CNT networks, as, for example, in CNT films without an additional matrix.

The following chapters will highlight certain aspects to understand CNT/polymer composites in the scope of heating applications. First, the effect of dispersion and volume fraction of CNTs on the electrical resistivity of CNT/polymer composites will be explained in the scope of percolation theory. Then, models will be introduced to describe the effect of composite morphology on their temperature coefficient of resistivity as well as their thermal stability, which are key aspects to allow a safe use as Joule heating materials.

2. Physical aspects of CNT/polymer nanocomposites for heating applications

2.1 Dispersion of CNTs in polymers and general percolation theory

The electrical conductivity of a composite consisting of an insulating matrix (e.g., a polymer) and a conducting filler (e.g., CNTs) strongly depends on the filler's volume fraction within the composite (see Fig. 3). A schematic representation of the CNT-network is shown in Fig. 4. If the volume fraction is too low the conducting filler will not be able to create conducting pathways throughout the material and the electric conductivity of the overall composite is on the same order as the insulating matrix. When increasing the volume fraction of the filler, the electric conductivity of the composite will significantly

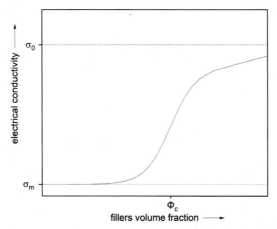

Fig. 3 Schematic representation of the relationship between electrical conductivity and fillers volume fraction for a composite (consisting of an electrically isolating matrix with conductivity σ_m and a filler with higher electrical conductivity σ_0) with percolation behavior at its percolation threshold Φ_c.

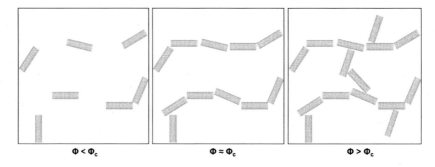

Fig. 4 Schematic representation of CNTs, creating a percolating network with rising volume fraction.

increase at a certain threshold value. This volume fraction is denoted as percolation threshold and describes the lowest filler-matrix ratio, around which the dispersed filler forms electrically conductive pathways throughout the material independent of its thickness. Adding more conducting filler to the composite increases the number of the conductive pathways and thus the overall electric conductivity of the composite, but the slope is usually smaller compared to the steep increase around the percolation threshold.

Several theories have been developed to describe the percolation behavior as well as electrical conductivity of composites and there are several reviews giving an overview of modeling approaches [29–33]. The main factors influencing percolation are the fillers conductivity, geometry and size (aspect ratio), volume fraction, and dispersion [29]. Some approaches for CNT/polymer composites even consider morphology of the CNTs (e.g., waviness) or quantum-mechanical aspects [34–37]. It has been shown in calculations that the percolation threshold, for example, increases with CNTs waviness. However as the increase is well below of a factor of two, the effect can be considered small (taking into account typical percolation thresholds of CNT/polymer composites below 1.0 wt%) [38].

In the following, a simple model, the general percolation theory, will be described instead of the more complex models, as it is commonly used in literature and shows good results for CNT/polymer nanocomposites. It does not take into account any structural information. The general percolation theory for conductive fillers in an insulating matrix above percolation threshold can be derived from Broadbent and Hammersley's work describing fluid flow through a statistical medium, where the phrase percolation theory was first introduced [39, 40]. The composites electrical conductivity σ as a function of the fillers volume fraction Φ is described by a power law

$$\sigma = \sigma_0 \cdot (\Phi - \Phi_c)^\beta \qquad (1)$$

using the fillers conductivity σ_0, the percolation threshold Φ_c, and the critical exponent β, which is an index of the dimensionality of the statistically percolating system [38]. Bauhofer and Kovacs showed in their review that the critical exponents reported in literature are between 1.3 and 4.0 with peak at 2.0 [38]. In their opinion, the critical exponent contains no reliable information about the CNT network (at least for most of the experimental data), while Foygel et al. found that the critical exponent is dependent on the fillers aspect ratio [38, 41]. For ideal two-dimensional and three-dimensional systems, theoretical values for β of 1.30 and 1.94 have been predicted [39].

The excluded volume concept yields $\Phi_c \approx \eta^{-1}$ for a statistical distribution of fillers in the limit of large aspect ratios η [42]. A typical aspect ratio of 1000 for CNTs gives a realistic magnitude of the percolation threshold around 0.1 wt%. Much lower percolation thresholds are usually attributed to kinetic percolation, which is caused by flocculation of homogeneously dispersed particles due to particle movement, where the application of general percolation theory is questionable. This can lead to the observation of two percolation thresholds, a kinetic and statistical one. Much higher percolation thresholds suggest that the filler particles might not have been dispersed homogeneously [38].

Bauhofer and Kovacs concluded that there might be an indirect proportionality between percolation threshold and maximum conductivity, especially for composites based on the same matrix. While the conductivity of composites with identical filler concentration varies by one or two magnitudes for identical matrices, it can vary by 10 or more orders of magnitude for different matrices [38]. Thus, the choice of matrix material is also a crucial factor for the maximum conductivity of the composite.

The electrical conductivity of CNT/polymer nanocomposites does usually not reach that of the associated CNT network, even at high loadings. While nanocomposites usually reach up to several thousand S m^{-1} (see Table 1) [46–49], pure CNT films can have

Table 1 Key figures of exemplary CNT/polymer composites, arranged alphabetically with respect to polymer matrices.

Matrix	Filler	Φ_c (wt %)	β (-)	$\rho_{max}(\Omega\ cm)$	TCR behavior	τ (s)	References
HDPE	MWCNT	2.0	–	10^5	PTC	–	[43]
m-Aramid	MWCNT	0.085	2.2	10^0	–	<13	[27]
PDMS	MWCNT	0.050	4.0	3×10^{-1}	NTC	–	[28]
PDMS	MWCNT	0.270	2.3	10^0	–	<7	[44]
PDMS	SWCNT	<0.250	–	10^{-2}	–	–	[20]
PEEK	MWCNT	3.5	2.4	10^5	NTC	–	[45]

A larger overview can, for example, be found in [38].

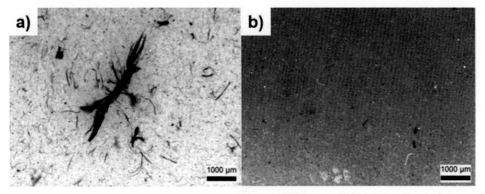

Fig. 5 Direct light microscopy of a thin film of polyurethane resin mixed with 0.05 wt% SWCNTs by (A) centrifugal mixing in a dual-asymmetric centrifuge and (B) calendaring.

conductivities up to 10^6 S m^{-1} [50, 51]. Composites tend to show an increasing degree of agglomeration with increasing CNT content [52–54], giving a physical limitation to changes in conductivity [55]. Thus, to reach the maximum potential of nanocomposites, improvement of filler dispersion in the matrix is crucial, while the process should not damage the CNTs. Some examples of dispersion processes are sonication, calendaring, ball milling, and melt blending which might also be assisted by solvents, surfactants, or functionalization of the CNTs [38, 56–58]. The example in Fig. 5 shows that the high shear during calendaring is necessary to debundle the agglomerates of CNTs in the polyurethane resin, while it is not possible to properly disperse the CNTs by low-shear centrifugal mixing.

On the other hand, good dispersion usually implies the formation of an insulating polymer layer around each CNT, thus the most thorough dispersion does not necessarily result in the highest conductivity. Furthermore, it seems that some polymer types and dispersion techniques favor the formation of polymer coatings of different thicknesses on CNTs, while other type and production method of the CNT seems to be less important [38].

2.2 Temperature coefficient of resistance of CNT nanocomposites

The electrical resistance of a material is a crucial variable for Joule heating (which will be discussed in detail in Section 2.3), however it is usually a function of temperature. This dependency is described by the temperature coefficient of resistance (TCR). It may be positive (PTC), when resistivity increases with temperature, negative (NTC), when resistivity decreases with temperature or zero (ZTC), when the resistivity is independent of temperature. The TCR may also change its behavior in different temperature regimes.

The following chapter will first describe how the TCR of single CNTs is influenced by the band structure of different types of CNTs, then take into account contact resistance between CNTs for CNT films and finally add the effect of the polymer matrix, which will account for glass transition and melt temperature, to derive a model for the TCR of CNT/polymer composites.

CNTs have been found to be either semiconducting, small-bandgap or metallic due to their band structure, which is defined by their morphology (e.g., chirality, diameter). Disorder and defects also influence the properties of CNTs [59]. Electrical transport in metallic SWCNTs is ballistic (mean free path of charge carriers is higher than the structure size) and varies between diffusive and ballistic for MWCNTs [60–63]. Metallic CNTs usually show PTC behavior, but NTC behavior can be observed for semiconducting SWCNTs and MWCNTs with large diameter at certain conditions [64–68].

CNT films show NTC behavior due to semiconducting properties [15, 69] or their granular structure in which conductive grains are connected by multiple random junctions. Impurities, disorder of the bulk material and contact type can influence the electrical resistance [63]. The conduction mechanism can be described by variable range hopping (VRH), for example, using the Fogler Tebler Shklovskii VRH model [55, 63, 70]. The film resistance exponentially decreases with temperature, although the changes are quite small above room temperature [63]. However also PTC behavior has been observed [23].

For different CNT/polymer nanocomposites, different types of temperature coefficient of resistance, i.e., PTC, NTC, and ZTC, have been reported [12, 28, 43, 45, 71–78] and several explanations have been discussed in literature, for example, changes in composite morphology due to glass transition temperature [72] or thermal expansion of the polymer [71, 74], as well as changes in tunneling resistance during (thermal fluctuation-induced) tunneling between conductive particles [76, 79–81].

Gong et al. suggested a mechanism involving the contributions of thermal expansion of the matrix, thermally assisted tunneling and thermally activated hopping to explain the observed behavior of TCR. Depending on the conditions, one contribution might

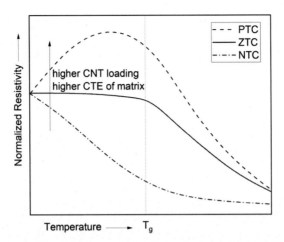

Fig. 6 Schematic representation of the temperature-dependent resistivity of CNT/polymer composites with glass transition temperature T_g according to the model of Gong et al. [81].

dominate the others, but they all have at least a minor effect in all temperature regimes. This leads to the schematic temperature-resistivity curves shown in Fig. 6. Simulation based on the derived model showed good agreement with experimental data of MWCNT/epoxy composites acquired between 300 and 480 K [81].

As for thermal expansion, Gong et al. made several assumptions for their model [81]:
1. Thermal expansion of CNTs is neglected due to the low coefficient of thermal expansion (CTE) of CNTs in comparison to polymers in the temperature range and low CNT loadings.
2. Thermal expansion leads to an isotropic volumetric strain causing positional changes of the CNTs.
3. Aggregation of CNTs is neglected and CNTs are not stretchable or compressible in axial direction, as their Young's modulus is much higher compared to polymers.

Based on these assumptions thermal expansion causes the CNT network to deform in the same way as its polymer matrix [81, 82], resulting in an average increase of intermediate distance between individual CNTs with increasing temperature.

The electrical resistance of CNT/polymer nanocomposites with randomly dispersed CNTs is a sum of intrinsic CNT resistance and contact resistance at CNT junctions [14, 81, 83]. The latter usually dominates the overall resistance [12, 28, 84, 85] and is the quantum tunneling resistance of electrons at absolute zero temperature, which can be described by thermally assisted tunneling for intermediate elevated temperatures (below the matrix polymers glass-transition temperature T_g) due to excited levels of tunneling across the barrier [80, 81].

The Landauer-Buettiker formula is usually used for the mathematical description of these two types of quantum tunneling [81, 82, 86]. From this, the following relation between contact resistance R_c, tunnel parameter $\mathfrak{A}$, and tunnel distance d can be derived using approximations [81, 82, 86, 87]:

$$R_c \propto \exp(\mathfrak{A} \cdot d) \; (T \leq T_g) \tag{2}$$

This relation contains two variables that are differently influenced by temperature. The contact resistance increases with increase of intertube space (e.g., caused by thermal expansion). The tunnel parameter $\mathfrak{A}$ depends on the height of the tunneling barrier, which decreases with increasing temperature, following a lower contact resistance [81, 86].

CNT loading of the composites significantly influences which of the two factors is dominating the TCR below T_g. At low loadings (i.e., initially large intertube distance), decrease of the tunneling barrier height is dominating and NTC behavior is observed. At high loadings (i.e., initially small intertube distance) increase of the intermediate CNT distance is dominating and PTC behavior is observed. In the intermediate loading range, one may achieve ZTC behavior with CNT/polymer nanocomposites by adjusting the competing mechanisms. For example, Gong et al. reported a transition with ZTC around 3 wt% MWCNT in an epoxy matrix [81].

To summarize the situation below T_g, a high PTC might be achieved using polymers with high coefficient of thermal expansion and higher CNT loadings, whereas high NTC can be achieved using polymers with low coefficient of thermal expansion and lower CNT loadings. To achieve ZTC, polymers with high coefficient of thermal expansion should be combined with sparse CNT networks [81].

Around T_g, the TCR shows a sudden change in its behavior. For temperatures above T_g TCR is dominated by the transport of thermally activated hopping, which occurs at the top energy levels of the barrier. It can be expressed by an Arrhenius approach using the thermal activation energy for electron hopping ε, Boltzmann constant k_b and the contact resistance at the glass transition temperature R_{T_g}. This leads to the following expression, which is inversely logarithmically proportional to temperature [81, 88]:

$$R_c = R_{T_g} \exp\left(\frac{\varepsilon}{k_b} \cdot \frac{T - T_g}{T \cdot T_g}\right) \; (T \geq T_g) \tag{3}$$

In this range, the TCR is mainly dependent on the activation energy for electron hopping, which can be influenced by the polymer matrix (i.e., work function), CNT loading (i.e., hopping length), and glass transition temperature. By that, CNT/polymer nanocomposites exhibit NTC behavior at temperatures above T_g, which is stronger for higher CNT loadings and low T_g of the polymer. To achieve a high resistance change ratio, polymers with low T_g and high thermal activation energy should be used, e.g., PA, PET, and PVA [81].

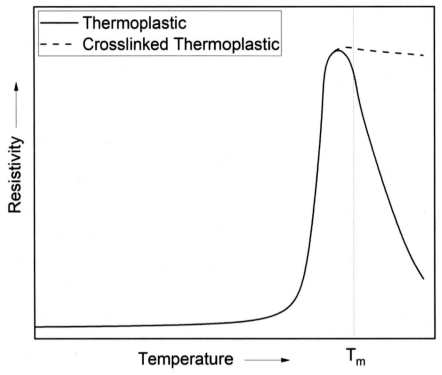

Fig. 7 Schematic of the PTC-peak for electrically conductive composites based on thermoplastic matrices with a melt temperature T_m, according to [43, 72].

Gong et al. also noted that the influences of other factors, for example, morphology and electrical properties of CNTs, on the TCR properties are limited according to their model [81].

The model discussed so far did not specifically consider melting of semicrystalline thermoplastic matrices, thus the effect of melting on the composites resistivity will be addressed in the following. Around the polymers melt temperature T_m a significant increase in resistivity (PTC peak), which can, for example, reach up to eight orders of magnitude [76] can be observed [71, 72]. After this point the resistivity shows a more or less pronounced reduction [43, 72], as shown in Fig. 7. Several examples are reported in the literature [43, 71–77, 89, 90].

At T_m, the polymer crystallites melt and lead to a significant change in the morphology of the polymer matrix, which is associated with a sharp increase in polymer volume. This leads to a distortion of the percolating CNT network and an increase of the intertube spacing (see Fig. 8), which causes peaking of the resistivity.

Meyer made an observation confirming this effect, when he compared the resistivity-temperature–dependency for semicrystalline *trans*-polybutadiene and amorphous *cis*-polybutadiene. While the semicrystalline polybutadiene showed a peak in

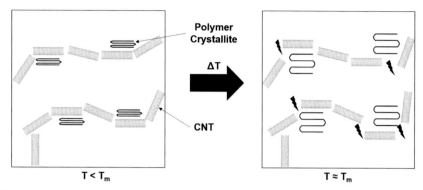

Fig. 8 Nonscale representation of the morphological changes inside CNT/polymer composites causing the PTC peak by distortion of the CNT network. Loss of contact between CNTs is visualized by small flashes.

resistivity around the melting temperature, amorphous polybutadiene had monotonous PTC behavior without a peak [72].

NTC behavior after the peak in the resistivity-temperature curve might also be explained by morphological changes, i.e., a rearrangement of the conductive fillers in the composites' microstructure due to an increased mobility of the fillers after surpassing the matrix melting temperature [43]. Several observations in literature support this explanation. The resistivity drop tends to be more profound for low aspect-ratio fillers (e.g., CB) compared to high aspect-ratio fillers (e.g., CNTs) [43, 72]. Also cross-linking of the thermoplastic matrix (e.g., by γ-radiation) eliminates the NTC effect, as the restricted movement of polymer chains hinders reagglomeration of conductive fillers [43]. On the other hand, the NTC effect could be promoted by application of an electrical field at high temperatures, which is beneficial for rearrangement, orientation or reaggregation of the conductive fillers [74].

Due to the sensitivity of the composite's percolating network on morphological changes, it should be considered that the resistivity also changes with deformations caused by mechanical loading. This is, for example, used in strain-sensing applications [13, 14, 91]. If this behavior is undesirable, one can, for example, apply CNT sheets on prestrained elastomeric sheets. By that, the film resistance remains almost constant in the prestrain range [92].

2.3 Aspects for Joule heating of CNT/polymer nanocomposites

In the previous sections, it has been discussed how the resistivity of CNT/polymer composites is influenced by filler loading, conductive network morphology, matrix material, and temperature. Now their impact will be examined from the view of electric heating. By electric Joule heating, also called resistive heating, electrical energy can be converted into heat energy.

The electric power P which is converted into heat as a result of the current flow I in the material can be represented by applied voltage U and electrical resistance R by following Joule's law together with Ohm's law ($U = I \cdot R$) [27]:

$$P = U \cdot I = \frac{U^2}{R} \tag{4}$$

Thus, heating power is quadratically proportional to the applied voltage and inversely proportional to the materials' resistance. As discussed in the previous chapter, the electrical resistance might be a function of temperature which influences the heating characteristics of the used material.

PTC materials are extensively used for self-regulated heaters, temperature sensors, overcurrent protectors and current limiters [73], as their resistance increases with temperature, limiting the current flow and heat generation at a fixed voltage. Therefore, the heating power will be reduced and strong PTC behavior significantly limits the achievable maximum temperature of the material [26, 93].

NTC materials decrease their resistance to temperature, thus heating using a constant voltage easily causes overheating and damaging of the heating element or the surroundings (thermal runaway) [26].

ZTC materials allow precise control of temperature, as the generated power is only sensitive to changes of applied voltage due to their temperature independent resistance. This behavior might be an intrinsic property of the material but can also be achieved by combination of PTC and NTC materials [26, 81, 94]. For example, Chu et al. produced ZTC polymer nanocomposite heating systems by combining a CB/PDMS composite with PTC behavior and a MWCNT/PDMS composite with NTC behavior [26].

A typical heating run for heating elements is shown in Fig. 9. Time-dependent temperature curves can be divided into three different regions: temperature growth (heating; I) region, equilibrium (maximum temperature; II) region, and temperature decay (cooling; III) region as shown in Fig. 9 [27].

Temperature dynamics during heating of a plate can be described by the differential equation

$$c \cdot m \cdot \frac{dT}{dt} = P - A \cdot \alpha \cdot (T(t) - T_a) \tag{5}$$

which leads to the following expression of the time-dependent temperature [95]

$$T(t) = T_a + (T_e - T_a)\left(1 - e^{-t/\tau}\right) \tag{6}$$

using the plate's specific heat capacity c, mass m, surface area A, heat transfer coefficient α, and time constant of the temperature change τ as well as the ambient temperature T_a and maximum equilibrium temperature T_e.

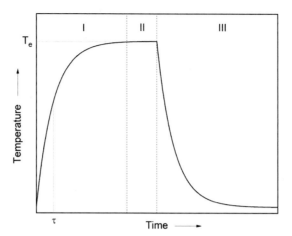

Fig. 9 Schematic time-temperature curve for a material with time constant τ and maximum equilibrium temperature T_e. The material is heated in regions I and II with a constant power P and freely cools down back to room temperature in region III.

Temperature dynamics during cooling can be expressed by Newton's law of cooling [96, 97]. The mathematical expression can be derived from the same differential equation, when the electric power for Joule heating is set to zero:

$$T(t) = T_a + (T(0) - T_a)e^{-t/\tau} \tag{7}$$

Thus, the heating element's temperature exponentially decays with time.

The equations include the maximum equilibrium temperature.

$$T_e = T_a + \frac{P}{\alpha \cdot A} \tag{8}$$

So high electrical power and a low heat transfer coefficient allow a higher equilibrium temperature of the heating element. The heat transfer coefficient is a combination of the contributions of heat dissipation to the surrounding by conduction, convection, and radiation.

The time constant of the temperature changes and can in both cases, heating as well as cooling, be calculated by

$$\tau = \frac{m \cdot c}{\alpha \cdot A} \tag{9}$$

After the time constant, the temperature difference reaches approximately 63% of the end value for heating or cooling [95]. Therefore, a small time constant, which can be achieved by small mass and specific heat capacity of the material or a high heat transfer coefficient, means fast temperature response of the heating element.

In conclusion, ideal materials for rapid heating are composites that allow high heating power (i.e., low electrical resistance below 10^{-1} Ω m [85]) and exhibit low heat capacity (i.e., mass and specific heat capacity) [28, 85, 98], which makes CNT/polymer composites promising candidates (see Table 1).

Several studies reported CNT/polymer nanocomposite thin film heaters, which achieved surface temperatures of several hundred °C within less than 30–100 s, by applying voltages of a few 10 volts [94].

One major advantage of CNT/polymer nanocomposites as heaters, for example, in comparison to conventional wire-wound film heaters, is the high temperature uniformity due to the microscopically percolating network in the polymer volume [27, 44, 92]. They also allow a higher heating efficiency (i.e., lower energy consumption) in comparison to metallic film heaters. Wu et al. reported up to 30% less energy consumption of double-walled CNT and MWCNT films compared to commercial (metallic) film heaters [99].

2.4 Long-term stability of CNT/polymer nanocomposites in the scope of Joule heating

As shown in the previous section, CNT/polymer composites have favorable characteristics for heating applications, but for industrial and consumer applications it is necessary that they also maintain these characteristics over a long period of use. Thus, they need a high long-term stability, which is mostly limited by the (thermal) stability of the polymer matrix. The following chapter will highlight the effect of composite microstructure and impurities on the thermal stability of the polymer matrix and how to improve the long-term stability, independent of whether it is thermoplastic or not.

As stated by the percolation theory, most of the electric current concentrates in the percolating network of highly conductive fillers (e.g., CNTs) embedded in an electrically insulating matrix. Thus, heat is first generated in the vicinity of the CNTs and is then transferred to the polymer matrix, limited by thermal conductivity of the polymer, which is usually four orders of magnitudes smaller than for CNTs. This causes higher local temperatures at the CNT-polymer interface in comparison to the surface temperature of the composite, which locally accelerates thermal aging of the polymer and might additionally cause local damage by overheating the polymer matrix (see Fig. 10) [85].

Chu et al. compared the thermal stability of MWCNT/polydimethylsiloxane (PDMS) nanocomposites during electrical heating (i.e., internal heating of the conductive network) and convective heating (i.e., external heating of the composite) over a time span of 40 h. Composites subjected to electrical heating showed significantly higher thermal degradation than the ones subjected to convective heating at comparable surface temperature. Electrically heated composites developed void defects and severe matrix degradation around the CNTs and fragments of the matrix could be detected using pyrolysis gas mass spectrometry. Thus, electrically heated composites grew fragile and exhibited increased resistance after aging at surface temperatures over 180°C while convective

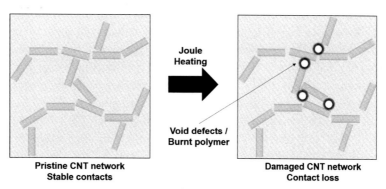

Fig. 10 Schematic representation of the overheating mechanism causing local damaging of CNT/polymer composites upon excessive Joule heating according to [85].

heating even led to enhanced mechanical properties without resistance changes up to 250°C [85]. This leads to the conclusion that the maximum operating temperature of nanocomposite heating devices should be defined with caution. Higher CNT loadings or pulsed electrical heating can help to relieve thermal degradation as spreading of thermal concentrations in the matrix is facilitated, thus allowing higher operation temperatures [28, 85].

The choice of CNTs might also alter the long-term stability of the nanocomposites. Several studies have been conducted on the effect of CNTs on the thermal stability of CNT/polymer nanocomposites. The majority reported improved thermal stabilities, which have been accounted to the intrinsically good thermal stability of CNTs [100], their network structure which can shield the composite from heat [101, 102], and their radical scavenging capability due to a high electron affinity [103–105].

But there are also contradictory results [106, 107]. Kashiwagi et al. reported that SWCNTs and MWCNTs enhance the thermal stability of polymethylmethacrylate (PMMA) composites and may act as flame retardants [101, 102], while Liu et al. observed decreased thermal stability of MWCNT/PMMA composites (at high loading levels) [108]. Similar discrepancies can, for example, be found for MWCNT/PP composites [109–111]. Several hypotheses have been put forward to explain the faster thermal degradation, like agglomeration of CNTs in the composite, participation of the CNTs in the radical initiation [112] or effects of catalyst residues [110, 113], without providing solid evidence or mechanisms [106].

Li et al. attribute the differences to various CNTs used by diverse research groups, which adds complexity as the CNTs are usually far from their perfect pristine state. They usually contain impurities, like amorphous carbon or metal catalyst residues, or disordered structures such as dangling bonds, sp^3-carbon or oxygen-containing functional groups. These might influence the degradation mechanisms [106, 107].

Li et al. reported that CNTs synthesized with cobalt catalysts severely decreased the thermal stability of CNT/PDMS composites, as the cobalt nanoparticles in the CNTs accelerate the thermal degradation of silicone through free radical generation. The radicals then lead to chain scission or molecular splitting mechanisms, which produce methane and silicon-oxycarbide or cyclic siloxane oligomers [106]. They also examined the effect of chrome, manganese, iron, nickel, copper, palladium, and zinc nanoparticles and observed that only chrome, cobalt, and palladium show an obvious decrease in thermal stability of PDMS. Furthermore, carboxylic acid groups, which might be formed during acid purification of CNTs, showed no significant decrease in thermal stability [107].

Thus, reduction of thermal stability of the polymer matrix by CNTs might be mitigated by removal of critical catalysts (e.g., by acid purification), use of impurity-free CNTs or CNTs produced with uncritical catalysts as well as addition of radical scavengers [106, 107]. It is important to state that different polymer matrices might have different degradation mechanisms, so other factors might dominate the degradation in comparison to the shown examples.

If these aspects are considered, a good long-term stability of the composites can be achieved. A good repeatability of resistance during long-period thermal cycling tests by electric heating has been reported, so the conductive network seems to be tolerant against temperature changes [27, 28]. Furthermore, CNT nanocomposites have a high tolerance toward damaging the film due to the microscopically percolating network in the polymer volume, whereas conventional wire-wound heaters would fail upon damaging the wire. In case of the nanocomposites the current has the possibility to bypass the damaged area [92].

3. Conclusion

Electromobility creates the need for heating solutions with higher efficiency and lower energy consumption. Electrically conductive CNT/polymer nanocomposites are a promising solution to this problem. They can achieve high heating rates combined with high heating efficiency in Joule heating due to their low weight and low heat capacity, as well as good long-term and cycling stability. They have a high damage tolerance and allow a high degree of design freedom. Thus, they can be used in several automotive heating applications, as infrared layer heaters. But their behavior is different from conventional heating materials and pristine CNTs or CNT networks.

The composites show percolation behavior, thus their electrical conductivity shows a significant increase at a certain volume fraction of CNTs, called percolation threshold. The percolation threshold of statistically distributed high aspect ratio fillers is inversely proportional to the aspect ratio, so lower percolation threshold might be achieved by higher aspect ratios. The maximum achievable electrical conductivity of the composites depends on the matrix material and dispersion state of the CNTs. Good dispersion is crucial to create a conducting network, but it also implies the formation of a (insulating) polymer layer around the CNTs, which might limit the electrical conductivity. Thus, not necessarily the best dispersion leads to the highest conductivity.

The electrical conductivity of the composites is usually a function of temperature, which is expressed by the temperature coefficient of resistance. CNT/polymer nanocomposites can show PTC, NTC, or ZTC behavior depending on their morphology (i.e., CNT loading) and the temperature range (i.e., above or below glass transition temperature). A suggested mechanism to explain this behavior considers the influences of thermally assisted tunneling, thermally activated hopping, and thermal expansion. These factors influence the contact resistance between adjacent CNTs in the network, which usually dominates over the contribution of the CNTs intrinsic resistance. With this knowledge the TCR of CNT/polymer nanocomposites can be tailored by altering the CNT loading, thermal expansion and thermal activation energy of the polymer. Furthermore, thermoplastic matrices usually show a strong PTC peak at the polymers melting temperature due to a volume increase upon melting of polymer crystallites.

As most of the current is concentrated in the percolating CNT network, also most of the heat is generated in the vicinity of the CNTs. This can lead to thermal degradation of the polymer matrix at composites surface temperatures lower than the thermal degradation temperature of the polymer, due to microscopical overheating. Pulsed operation of the heater, enhancement of thermal spreading (e.g., by higher CNT loading), or limitation of the heating power can minimize this problem. Impurities in CNTs can also alter the thermal stability of the polymer matrix. Thus, critical catalysts (e.g., chrome, cobalt, palladium) should be removed or impurity-free CNTs be used. Another possibility is the addition of radical scavengers to the composite.

Acknowledgment

The IGF-Project ESDBond (Grant number 20459 BG) of the research association FSKZ e.V. has been funded by the AiF in the program IGF of the Federal Ministry of Economic Affairs and Energy based on a resolution of the German Bundestag.

References

[1] T. Bauml, D. Dvorak, A. Frohner, D. Simic, Simulation and measurement of an energy efficient infrared radiation heating of a full electric vehicle, in: 2014 IEEE Vehicle Power and Propulsion Conference (VPPC), Coimbra, Portugal, IEEE, Piscataway, NJ, 2014, pp. 1–6.
[2] F.A. Wyczalek, Heating and cooling battery electric vehicles-the final barrier, IEEE Aerosp. Electron. Syst. Mag. 8 (1993) 9–14, https://doi.org/10.1109/62.242054.
[3] R. Neugebauer (Ed.), Ressourceneffizienz: Schlüsseltechnologien für Wirtschaft und Gesellschaft, first ed., Springer Vieweg, Berlin, Heidelberg, 2017.
[4] D. Nemec, S. Shakalkan, A. Gerten, S. Schesler, Dünnschichtige Heizungen auf Basis von Kohlenstoff-Nanoröhren, wt Werkstatttechnik online vol. 106 (2015) 370–371.
[5] T. Bedford, Basic Principles of Ventilation and Heating, H. K. Lewis, London, 1948.

[6] S. Iijima, Helical microtubules of graphitic carbon, Nature 354 (1991) 56–58, https://doi.org/10.1038/354056a0.

[7] M. Endo, K. Takeuchi, S. Igarashi, K. Kobori, M. Shiraishi, H.W. Kroto, The production and structure of pyrolytic carbon nanotubes (PCNTs), J. Phys. Chem. Solids 54 (1993) 1841–1848, https://doi.org/10.1016/0022-3697(93)90297-5.

[8] S. Berber, Y.K. Kwon, D. Tomanek, Unusually high thermal conductivity of carbon nanotubes, Phys. Rev. Lett. 84 (2000) 4613–4616, https://doi.org/10.1103/PhysRevLett.84.4613.

[9] M. Fujii, X. Zhang, H. Xie, H. Ago, K. Takahashi, T. Ikuta, H. Abe, T. Shimizu, Measuring the thermal conductivity of a single carbon nanotube, Phys. Rev. Lett. 95 (2005) 65502, https://doi.org/10.1103/PhysRevLett.95.065502.

[10] Y. Ryu, L. Yin, C. Yu, Dramatic electrical conductivity improvement of carbon nanotube networks by simultaneous de-bundling and hole-doping with chlorosulfonic acid, J. Mater. Chem. 22 (2012) 6959, https://doi.org/10.1039/c2jm16000e.

[11] D. Qian, A.G.J. Wagner, W.K. Liu, M.-F. Yu, R.S. Ruoff, Mechanics of carbon nanotubes, Appl. Mech. Rev. 55 (2002) 495–533, https://doi.org/10.1115/1.1490129.

[12] H.C. Neitzert, L. Vertuccio, A. Sorrentino, Epoxy/MWCNT composite as temperature sensor and electrical heating element, IEEE Trans. Nanotechnol. 10 (2011) 688–693, https://doi.org/10.1109/TNANO.2010.2068307.

[13] S. Khan, L. Lorenzelli, R.S. Dahiya, Technologies for printing sensors and electronics over large flexible substrates: a review, IEEE Sensors J. 15 (2015) 3164–3185, https://doi.org/10.1109/JSEN.2014.2375203.

[14] N. Hu, Y. Karube, C. Yan, Z. Masuda, H. Fukunaga, Tunneling effect in a polymer/carbon nanotube nanocomposite strain sensor, Acta Mater. 56 (2008) 2929–2936, https://doi.org/10.1016/j.actamat.2008.02.030.

[15] D. Jung, D. Kim, K.H. Lee, L.J. Overzet, G.S. Lee, Transparent film heaters using multi-walled carbon nanotube sheets, Sensors Actuators A Phys. 199 (2013) 176–180, https://doi.org/10.1016/j.sna.2013.05.024.

[16] M. Mohiuddin, S. van Hoa, Electrical resistance of CNT-PEEK composites under compression at different temperatures, Nanoscale Res. Lett. 6 (2011) 419, https://doi.org/10.1186/1556-276X-6-419.

[17] P. Pötschke, K. Kobashi, T. Villmow, T. Andres, M.C. Paiva, J.A. Covas, Liquid sensing properties of melt processed polypropylene/poly(ε-caprolactone) blends containing multiwalled carbon nanotubes, Compos. Sci. Technol. 71 (2011) 1451–1460, https://doi.org/10.1016/j.compscitech.2011.05.019.

[18] E. Frackowiak, V. Khomenko, K. Jurewicz, K. Lota, F. Béguin, Supercapacitors based on conducting polymers/nanotubes composites, J. Power Sources 153 (2006) 413–418, https://doi.org/10.1016/j.jpowsour.2005.05.030.

[19] M.H. Al-Saleh, U. Sundararaj, Electromagnetic interference shielding mechanisms of CNT/polymer composites, Carbon 47 (2009) 1738–1746, https://doi.org/10.1016/j.carbon.2009.02.030.

[20] H. Leicht, E. Kraus, B. Baudrit, T. Hochrein, M. Bastian, M. Langer, A. Klotzbach, A. Wanielik, Modifikation von Siliconen mit elektrisch leitfähigen Füllstoffen zur Ableitung elektrischer oder elektrostatischer Ladungen, GAK – Gummi Fasern Kunststoffe 73 (2020) 370–377.

[21] J.-M. Thomassin, X. Lou, C. Pagnoulle, A. Saib, L. Bednarz, I. Huynen, R. Jérôme, C. Detrembleur, Multiwalled carbon nanotube/poly(ε-caprolactone) nanocomposites with exceptional electromagnetic interference shielding properties, J. Phys. Chem. C 111 (2007) 11186–11192, https://doi.org/10.1021/jp0701690.

[22] J.-I. Lee, S.-B. Yang, H.-T. Jung, Carbon nanotubes–polypropylene nanocomposites for electrostatic discharge applications, Macromolecules 42 (2009) 8328–8334, https://doi.org/10.1021/ma901612w.

[23] D. Janas, K.K. Koziol, Rapid electrothermal response of high-temperature carbon nanotube film heaters, Carbon 59 (2013) 457–463, https://doi.org/10.1016/j.carbon.2013.03.039.

[24] X. Yao, S.C. Hawkins, B.G. Falzon, An advanced anti-icing/de-icing system utilizing highly aligned carbon nanotube webs, Carbon 136 (2018) 130–138, https://doi.org/10.1016/j.carbon.2018.04.039.

[25] S.-L. Jia, H.-Z. Geng, L. Wang, Y. Tian, C.-X. Xu, P.-P. Shi, Z.-Z. Gu, X.-S. Yuan, L.-C. Jing, Z.-Y. Guo, J. Kong, Carbon nanotube-based flexible electrothermal film heaters with a high heating rate, R. Soc. Open Sci. 5 (2018) 172072, https://doi.org/10.1098/rsos.172072.

[26] K. Chu, S.-C. Lee, S. Lee, D. Kim, C. Moon, S.-H. Park, Smart conducting polymer composites having zero temperature coefficient of resistance, Nanoscale 7 (2015) 471–478, https://doi.org/10.1039/C4NR04489D.

[27] Y.G. Jeong, G.W. Jeon, Microstructure and performance of multiwalled carbon nanotube/m-aramid composite films as electric heating elements, ACS Appl. Mater. Interfaces 5 (2013) 6527–6534, https://doi.org/10.1021/am400892k.

[28] K. Chu, D. Kim, Y. Sohn, S. Lee, C. Moon, S. Park, Electrical and thermal properties of carbon-nanotube composite for flexible electric heating-unit applications, IEEE Electron Dev. Lett. 34 (2013) 668–670, https://doi.org/10.1109/LED.2013.2249493.

[29] F. Lux, Models proposed to explain the electrical conductivity of mixtures made of conductive and insulating materials, J. Mater. Sci. 28 (1993) 285–301, https://doi.org/10.1007/BF00357799.

[30] M. Weber, M.R. Kamal, Estimation of the volume resistivity of electrically conductive composites, Polym. Compos. 18 (1997) 711–725, https://doi.org/10.1002/pc.10324.

[31] M.L. Clingerman, E.H. Weber, J.A. King, K.H. Schulz, Development of an additive equation for predicting the electrical conductivity of carbon-filled composites, J. Appl. Polym. Sci. 88 (2003) 2280–2299, https://doi.org/10.1002/app.11938.

[32] J.-C. Huang, Carbon black filled conducting polymers and polymer blends, Adv. Polym. Technol. 21 (2002) 299–313, https://doi.org/10.1002/adv.10025.

[33] M.L. Clingerman, J.A. King, K.H. Schulz, J.D. Meyers, Evaluation of electrical conductivity models for conductive polymer composites, J. Appl. Polym. Sci. 83 (2002) 1341–1356, https://doi.org/10.1002/app.10014.

[34] F. Deng, Q.-S. Zheng, L.-F. Wang, C.-W. Nan, Effects of anisotropy, aspect ratio, and nonstraightness of carbon nanotubes on thermal conductivity of carbon nanotube composites, Appl. Phys. Lett. 90 (2007) 21914, https://doi.org/10.1063/1.2430914.

[35] F. Deng, Q.-S. Zheng, An analytical model of effective electrical conductivity of carbon nanotube composites, Appl. Phys. Lett. 92 (2008) 71902, https://doi.org/10.1063/1.2857468.

[36] H.M. Ma, X.-L. Gao, A three-dimensional Monte Carlo model for electrically conductive polymer matrix composites filled with curved fibers, Polymer 49 (2008) 4230–4238, https://doi.org/10.1016/j.polymer.2008.07.034.

[37] L. Berhan, A.M. Sastry, Modeling percolation in high-aspect-ratio fiber systems. II. The effect of waviness on the percolation onset, Phys. Rev. E Stat. Nonlin. Soft Matter Phys. 75 (2007) 41121, https://doi.org/10.1103/PhysRevE.75.041121.

[38] W. Bauhofer, J.Z. Kovacs, A review and analysis of electrical percolation in carbon nanotube polymer composites, Compos. Sci. Technol. 69 (2009) 1486–1498, https://doi.org/10.1016/j.compscitech.2008.06.018.

[39] D. Stauffer, A. Aharony, Introduction to Percolation Theory, second ed., Taylor & Francis Group, London, 1994.

[40] S.R. Broadbent, J.M. Hammersley, Percolation processes, Math. Proc. Camb. Philos. Soc. 53 (1957) 629–641, https://doi.org/10.1017/S0305004100032680.

[41] M. Foygel, R.D. Morris, D. Anez, S. French, V.L. Sobolev, Theoretical and computational studies of carbon nanotube composites and suspensions: electrical and thermal conductivity, Phys. Rev. B 71 (2005), https://doi.org/10.1103/PhysRevB.71.104201.

[42] I. Balberg, C.H. Anderson, S. Alexander, N. Wagner, Excluded volume and its relation to the onset of percolation, Phys. Rev. B 30 (1984) 3933–3943, https://doi.org/10.1103/PhysRevB.30.3933.

[43] X.J. He, J.H. Du, Z. Ying, H.M. Cheng, Positive temperature coefficient effect in multiwalled carbon nanotube/high-density polyethylene composites, Appl. Phys. Lett. 86 (2005) 62112, https://doi.org/10.1063/1.1863452.

[44] J. Yan, Y.G. Jeong, Multiwalled carbon nanotube/polydimethylsiloxane composite films as high performance flexible electric heating elements, Appl. Phys. Lett. 105 (2014) 51907, https://doi.org/10.1063/1.4892545.

[45] M. Mohiuddin, S.V. Hoa, Temperature dependent electrical conductivity of CNT–PEEK composites, Compos. Sci. Technol. 72 (2011) 21–27, https://doi.org/10.1016/j.compscitech.2011.08.018.
[46] I.D. Rosca, S.V. Hoa, Highly conductive multiwall carbon nanotube and epoxy composites produced by three-roll milling, Carbon 47 (2009) 1958–1968, https://doi.org/10.1016/j.carbon.2009.03.039.
[47] H.S. Lee, C.H. Yun, S.K. Kim, J.H. Choi, C.J. Lee, H.-J. Jin, H. Lee, S.J. Park, M. Park, Percolation of two-dimensional multiwall carbon nanotube networks, Appl. Phys. Lett. 95 (2009) 134104, https://doi.org/10.1063/1.3238326.
[48] K.-Y. Chun, Y. Oh, J. Rho, J.-H. Ahn, Y.-J. Kim, H.R. Choi, S. Baik, Highly conductive, printable and stretchable composite films of carbon nanotubes and silver, Nat. Nanotechnol. 5 (2010) 853–857, https://doi.org/10.1038/nnano.2010.232.
[49] V. Skákalová, U. Dettlaff-Weglikowska, S. Roth, Electrical and mechanical properties of nanocomposites of single wall carbon nanotubes with PMMA, Synth. Met. 152 (2005) 349–352, https://doi.org/10.1016/j.synthmet.2005.07.291.
[50] D.S. Hecht, A.M. Heintz, R. Lee, L. Hu, B. Moore, C. Cucksey, S. Risser, High conductivity transparent carbon nanotube films deposited from superacid, Nanotechnology 22 (2011) 75201, https://doi.org/10.1088/0957-4484/22/7/075201.
[51] Z. Wu, Z. Chen, X. Du, J.M. Logan, J. Sippel, M. Nikolou, K. Kamaras, J.R. Reynolds, D.B. Tanner, A.F. Hebard, A.G. Rinzler, Transparent, conductive carbon nanotube films, Science 305 (2004) 1273–1276, https://doi.org/10.1126/science.1101243.
[52] M.B. Bryning, D.E. Milkie, M.F. Islam, J.M. Kikkawa, A.G. Yodh, Thermal conductivity and interfacial resistance in single-wall carbon nanotube epoxy composites, Appl. Phys. Lett. 87 (2005) 161909, https://doi.org/10.1063/1.2103398.
[53] W. Lu, T.-W. Chou, E.T. Thostenson, A three-dimensional model of electrical percolation thresholds in carbon nanotube-based composites, Appl. Phys. Lett. 96 (2010) 223106, https://doi.org/10.1063/1.3443731.
[54] Y. Martinez-Rubi, J.M. Gonzalez-Dominguez, A. Ansón-Casaos, C.T. Kingston, M. Daroszewska, M. Barnes, P. Hubert, C. Cattin, M.T. Martinez, B. Simard, Tailored SWCNT functionalization optimized for compatibility with epoxy matrices, Nanotechnology 23 (2012) 285701, https://doi.org/10.1088/0957-4484/23/28/285701.
[55] O.A. Bârsan, G.G. Hoffmann, L.G.J. van der Ven, G. de With, Single-walled carbon nanotube networks: the influence of individual tube-tube contacts on the large-scale conductivity of polymer composites, Adv. Funct. Mater. 26 (2016) 4377–4385, https://doi.org/10.1002/adfm.201600435.
[56] P.-C. Ma, N.A. Siddiqui, G. Marom, J.-K. Kim, Dispersion and functionalization of carbon nanotubes for polymer-based nanocomposites: a review, Compos. A: Appl. Sci. Manuf. 41 (2010) 1345–1367, https://doi.org/10.1016/j.compositesa.2010.07.003.
[57] J. Menjivar, K. Kirane, Surfactant assisted dispersion of MWCNT's in epoxy nanocomposites and adhesion with aluminum, Polym. Test. 82 (2020) 106308, https://doi.org/10.1016/j.polymertesting.2019.106308.
[58] H. Hu, L. Zhao, J. Liu, Y. Liu, J. Cheng, J. Luo, Y. Liang, Y. Tao, X. Wang, J. Zhao, Enhanced dispersion of carbon nanotube in silicone rubber assisted by graphene, Polymer 53 (2012) 3378–3385, https://doi.org/10.1016/j.polymer.2012.05.039.
[59] M.J. Biercuk, S. Ilani, C.M. Marcus, P.L. McEuen, Electrical transport in single-wall carbon nanotubes, in: Carbon Nanotubes, Springer, Berlin, Heidelberg, 2007, pp. 455–493.
[60] A. Bachtold, M. de Jonge, K. Grove-Rasmussen, P.L. McEuen, M. Buitelaar, C. Schönenberger, Suppression of tunneling into multiwall carbon nanotubes, Phys. Rev. Lett. 87 (2001) 166801, https://doi.org/10.1103/PhysRevLett.87.166801.
[61] S. Frank, P. Poncharal, Z.L. Wang, W.A. de Heer, Carbon nanotube quantum resistors, Science 280 (1998) 1744–1746, https://doi.org/10.1126/science.280.5370.1744.
[62] C. Schönenberger, A. Bachtold, C. Strunk, J.-P. Salvetat, L. Forró, Interference and interaction in multi-wall carbon nanotubes, Appl. Phys. A Mater. Sci. Process. 69 (1999) 283–295, https://doi.org/10.1007/s003390051003.

[63] K. Kędzierski, K. Rytel, B. Barszcz, A. Gronostaj, Ł. Majchrzycki, D. Wróbel, On the temperature dependent electrical resistivity of CNT layers in view of Variable Range Hopping models, Org. Electron. 43 (2017) 253–261, https://doi.org/10.1016/j.orgel.2017.01.037.

[64] F. Arai, C. Ng, P. Liu, L. Dong, Y. Imaizumi, K. Maeda, H. Maruyama, A. Ichikawa, T. Fukuda, Ultra-small site temperature sensing by carbon nanotube thermal probes, in: 2004 4th IEEE Conference on Nanotechnology, Munich, Germany, IEEE Operations Center, Piscataway, NJ, 2004, pp. 146–148.

[65] C.T. Avedisian, R.E. Cavicchi, P.M. McEuen, X. Zhou, W.S. Hurst, J.T. Hodges, High temperature electrical resistance of substrate-supported single walled carbon nanotubes, Appl. Phys. Lett. 93 (2008) 252108, https://doi.org/10.1063/1.3052867.

[66] A. Naeemi, J.D. Meindl, Physical modeling of temperature coefficient of resistance for single- and multi-wall carbon nanotube interconnects, IEEE Electron Dev. Lett. 28 (2007) 135–138, https://doi.org/10.1109/LED.2006.889240.

[67] A. Maffucci, F. Micciulla, A.E. Cataldo, G. Miano, S. Bellucci, Modeling, fabrication, and characterization of large carbon nanotube interconnects with negative temperature coefficient of the resistance, IEEE Trans. Compon. Packag. Manufact. Technol. 7 (2017) 485–493, https://doi.org/10.1109/TCPMT.2016.2643007.

[68] S. Moriyama, K. Toratani, D. Tsuya, M. Suzuki, Y. Aoyagi, K. Ishibashi, Electrical transport in semiconducting carbon nanotubes, Physica E Low Dimens. Syst. Nanostruct. 24 (2004) 46–49, https://doi.org/10.1016/j.physe.2004.04.022.

[69] N. Koratkar, A. Modi, E. Lass, P. Ajayan, Temperature effects on resistance of aligned multiwalled carbon nanotube films, J. Nanosci. Nanotechnol. 4 (2004) 744–748, https://doi.org/10.1166/jnn.2004.109.

[70] M.M. Fogler, S. Teber, B.I. Shklovskii, Variable-range hopping in quasi-one-dimensional electron crystals, Phys. Rev. B 69 (2004), https://doi.org/10.1103/PhysRevB.69.035413.

[71] F. Kohler, US3243753A, 1966.

[72] J. Meyer, Glass transition temperature as a guide to selection of polymers suitable for PTC materials, Polym. Eng. Sci. 13 (1973) 462–468, https://doi.org/10.1002/pen.760130611.

[73] S.P. Bao, G.D. Liang, S.C. Tjong, Positive temperature coefficient effect of polypropylene/carbon nanotube/montmorillonite hybrid nanocomposites, IEEE Trans. Nanotechnol. 8 (2009) 729–736, https://doi.org/10.1109/TNANO.2009.2023650.

[74] M. Ferrara, H.-C. Neitzert, M. Sarno, G. Gorrasi, D. Sannino, V. Vittoria, P. Ciambelli, Influence of the electrical field applied during thermal cycling on the conductivity of LLDPE/CNT composites, Physica E Low Dimens. Syst. Nanostruct. 37 (2007) 66–71, https://doi.org/10.1016/j.physe.2006.10.008.

[75] T.-M. Wu, J.-C. Cheng, M.-C. Yan, Crystallization and thermoelectric behavior of conductive-filler-filled poly(ε-caprolactone)/poly(vinyl butyral)/montmorillonite nanocomposites, Polymer 44 (2003) 2553–2562, https://doi.org/10.1016/S0032-3861(03)00106-X.

[76] K. Ohe, Y. Naito, A new resistor having an anomalously large positive temperature coefficient, J. Appl. Phys. 10 (1971) 99–108, https://doi.org/10.1143/JJAP.10.99.

[77] H. Nakano, K. Shimizu, S. Takahashi, A. Kono, T. Ougizawa, H. Horibe, Resistivity–temperature characteristics of filler-dispersed polymer composites, Polymer 53 (2012) 6112–6117, https://doi.org/10.1016/j.polymer.2012.10.046.

[78] Z.-D. Xiang, T. Chen, Z.-M. Li, X.-C. Bian, Negative temperature coefficient of resistivity in lightweight conductive carbon nanotube/polymer composites, Macromol. Mater. Eng. 294 (2009) 91–95, https://doi.org/10.1002/mame.200800273.

[79] R.D. Sherman, L.M. Middleman, S.M. Jacobs, Electron transport processes in conductor-filled polymers, Polym. Eng. Sci. 23 (1983) 36–46, https://doi.org/10.1002/pen.760230109.

[80] P. Sheng, Fluctuation-induced tunneling conduction in disordered materials, Phys. Rev. B 21 (1980) 2180–2195, https://doi.org/10.1103/PhysRevB.21.2180.

[81] S. Gong, Z.H. Zhu, Z. Li, Electron tunnelling and hopping effects on the temperature coefficient of resistance of carbon nanotube/polymer nanocomposites, Phys. Chem. Chem. Phys. 19 (2017) 5113–5120, https://doi.org/10.1039/C6CP08115K.

[82] S. Gong, Z.H. Zhu, Giant piezoresistivity in aligned carbon nanotube nanocomposite: account for nanotube structural distortion at crossed tunnel junctions, Nanoscale 7 (2015) 1339–1348, https://doi.org/10.1039/c4nr05656f.

[83] S. Gong, Z.H. Zhu, S.A. Meguid, Carbon nanotube agglomeration effect on piezoresistivity of polymer nanocomposites, Polymer 55 (2014) 5488–5499, https://doi.org/10.1016/j.polymer.2014.08.054.

[84] W.S. Bao, S.A. Meguid, Z.H. Zhu, G.J. Weng, Tunneling resistance and its effect on the electrical conductivity of carbon nanotube nanocomposites, J. Appl. Phys. 111 (2012) 93726, https://doi.org/10.1063/1.4716010.

[85] K. Chu, D.-J. Yun, D. Kim, H. Park, S.-H. Park, Study of electric heating effects on carbon nanotube polymer composites, Org. Electron. 15 (2014) 2734–2741, https://doi.org/10.1016/j.orgel.2014.07.043.

[86] S. Gong, Z.H. Zhu, On the mechanism of piezoresistivity of carbon nanotube polymer composites, Polymer 55 (2014) 4136–4149, https://doi.org/10.1016/j.polymer.2014.06.024.

[87] S. Maiti, S. Suin, N.K. Shrivastava, B.B. Khatua, A strategy to achieve high electromagnetic interference shielding and ultra low percolation in multiwall carbon nanotube–polycarbonate composites through selective localization of carbon nanotubes, RSC Adv. 4 (2014) 7979, https://doi.org/10.1039/c3ra46480f.

[88] H. Qiu, T. Xu, Z. Wang, W. Ren, H. Nan, Z. Ni, Q. Chen, S. Yuan, F. Miao, F. Song, G. Long, Y. Shi, L. Sun, J. Wang, X. Wang, Hopping transport through defect-induced localized states in molybdenum disulphide, Nat. Commun. 4 (2013) 2642, https://doi.org/10.1038/ncomms3642.

[89] J.-F. Gao, Z.-M. Li, S. Peng, D.-X. Yan, Temperature-resistivity behaviour of CNTs/UHMWPE composites with a two-dimensional conductive network, Polym. Plast. Technol. Eng. 48 (2009) 478–481, https://doi.org/10.1080/03602550902725480.

[90] J.-F. Gao, D.-X. Yan, H.-D. Huang, K. Dai, Z.-M. Li, Positive temperature coefficient and time-dependent resistivity of carbon nanotubes (CNTs)/ultrahigh molecular weight polyethylene (UHMWPE) composite, J. Appl. Polym. Sci. 114 (2009) 1002–1010, https://doi.org/10.1002/app.30468.

[91] H. Devaraj, T. Giffney, A. Petit, M. Assadian, K. Aw, The development of highly flexible stretch sensors for a robotic hand, Robotics 7 (2018) 54, https://doi.org/10.3390/robotics7030054.

[92] Y. Lee, V.T. Le, J.-G. Kim, H. Kang, E.S. Kim, S.-E. Ahn, D. Suh, Versatile, high-power, flexible, stretchable carbon nanotube sheet heating elements tolerant to mechanical damage and severe deformation, Adv. Funct. Mater. 28 (2018) 1706007, https://doi.org/10.1002/adfm.201706007.

[93] K. Sau, T. Chaki, D. Khastgir, Carbon fibre filled conductive composites based on nitrile rubber (NBR), ethylene propylene diene rubber (EPDM) and their blend, Polymer 39 (1998) 6461–6471, https://doi.org/10.1016/S0032-3861(97)10188-4.

[94] S. Isaji, Y. Bin, M. Matsuo, Electrical conductivity and self-temperature-control heating properties of carbon nanotubes filled polyethylene films, Polymer 50 (2009) 1046–1053, https://doi.org/10.1016/j.polymer.2008.12.033.

[95] J. Adam, Thermische Betrachtungen in Bezug auf Temperatur und Zeit, 22, FED-Konferenz Vortragsband, Bamberg, 2014, pp. 276–284.

[96] VII. Scala graduum caloris, Phil. Trans. R. Soc. 22 (1701) 824–829, https://doi.org/10.1098/rstl.1700.0082.

[97] S. Maruyama, S. Moriya, Newton's law of cooling: follow up and exploration, Int. J. Heat Mass Transf. 164 (2021) 120544, https://doi.org/10.1016/j.ijheatmasstransfer.2020.120544.

[98] L. Liu, S. Peng, X. Niu, W. Wen, Microheaters fabricated from a conducting composite, Appl. Phys. Lett. 89 (2006) 223521, https://doi.org/10.1063/1.2400065.

[99] Z.P. Wu, J.N. Wang, Preparation of large-area double-walled carbon nanotube films and application as film heater, Physica E Low Dimens. Syst. Nanostruct. 42 (2009) 77–81, https://doi.org/10.1016/j.physe.2009.09.003.

[100] E. Thostenson, C. Li, T. Chou, Nanocomposites in context, Compos. Sci. Technol. 65 (2005) 491–516, https://doi.org/10.1016/j.compscitech.2004.11.003.

[101] T. Kashiwagi, F. Du, J.F. Douglas, K.I. Winey, R.H. Harris, J.R. Shields, Nanoparticle networks reduce the flammability of polymer nanocomposites, Nat. Mater. 4 (2005) 928–933, https://doi.org/10.1038/nmat1502.

[102] T. Kashiwagi, F. Du, K.I. Winey, K.M. Groth, J.R. Shields, S.P. Bellayer, H. Kim, J.F. Douglas, Flammability properties of polymer nanocomposites with single-walled carbon nanotubes: effects of nanotube dispersion and concentration, Polymer 46 (2005) 471–481, https://doi.org/10.1016/j.polymer.2004.10.087.

[103] P.C.P. Watts, P.K. Fearon, W.K. Hsu, N.C. Billingham, H.W. Kroto, D.R.M. Walton, Carbon nanotubes as polymer antioxidants, J. Mater. Chem. 13 (2003) 491–495, https://doi.org/10.1039/b211328g.

[104] A. Galano, Carbon nanotubes as free-radical scavengers, J. Phys. Chem. C 112 (2008) 8922–8927, https://doi.org/10.1021/jp801379g.

[105] I. Fenoglio, M. Tomatis, D. Lison, J. Muller, A. Fonseca, J.B. Nagy, B. Fubini, Reactivity of carbon nanotubes: free radical generation or scavenging activity? Free Radic. Biol. Med. 40 (2006) 1227–1233, https://doi.org/10.1016/j.freeradbiomed.2005.11.010.

[106] Z. Li, W. Lin, K. Moon, S.J. Wilkins, Y. Yao, K. Watkins, L. Morato, C. Wong, Metal catalyst residues in carbon nanotubes decrease the thermal stability of carbon nanotube/silicone composites, Carbon 49 (2011) 4138–4148, https://doi.org/10.1016/j.carbon.2011.05.042.

[107] Z. Li, S.J. Wilkins, K.S. Moon, C.P. Wong, Carbon nanotube/polymer nanocomposites: improved or reduced thermal stabilities? Mater. Sci. Forum 722 (2012) 77–86, https://doi.org/10.4028/www.scientific.net/MSF.722.77.

[108] J. Liu, A. Rasheed, M.L. Minus, S. Kumar, Processing and properties of carbon nanotube/poly(methyl methacrylate) composite films, J. Appl. Polym. Sci. 112 (2009) 142–156, https://doi.org/10.1002/app.29372.

[109] J. Yang, Y. Lin, J. Wang, M. Lai, J. Li, J. Liu, X. Tong, H. Cheng, Morphology, thermal stability, and dynamic mechanical properties of atactic polypropylene/carbon nanotube composites, J. Appl. Polym. Sci. 98 (2005) 1087–1091, https://doi.org/10.1002/app.21206.

[110] T. Kashiwagi, E. Grulke, J. Hilding, R. Harris, W. Awad, J. Douglas, Thermal degradation and flammability properties of poly(propylene)/carbon nanotube composites, Macromol. Rapid Commun. 23 (2002) 761–765, https://doi.org/10.1002/1521-3927(20020901)23:13<761:AID-MARC761>3.0.CO;2-K.

[111] D. Bikiaris, A. Vassiliou, K. Chrissafis, K.M. Paraskevopoulos, A. Jannakoudakis, A. Docoslis, Effect of acid treated multi-walled carbon nanotubes on the mechanical, permeability, thermal properties and thermo-oxidative stability of isotactic polypropylene, Polym. Degrad. Stab. 93 (2008) 952–967, https://doi.org/10.1016/j.polymdegradstab.2008.01.033.

[112] S. Yang, J. Rafael Castilleja, E.V. Barrera, K. Lozano, Thermal analysis of an acrylonitrile–butadiene–styrene/SWNT composite, Polym. Degrad. Stab. 83 (2004) 383–388, https://doi.org/10.1016/j.polymdegradstab.2003.08.002.

[113] Y. Xu, G. Ray, B. Abdel-Magid, Thermal behavior of single-walled carbon nanotube polymer-matrix composites, Compos. A: Appl. Sci. Manuf. 37 (2006) 114–121, https://doi.org/10.1016/j.compositesa.2005.04.009.

SECTION B

Nano-alloys for automotive application

CHAPTER 13

Ti-based nanoalloy in automobile industry

Asit Behera[a] and Ajit Behera[b]
[a]School of Mechanical Engineering, Kalinga Institute of Industrial Technology, Bhubaneswar, India
[b]Department of Metallurgical & Materials Engineering, National Institute of Technology, Rourkela, India

Chapter outline

1. Introduction — 255
2. What is Ti-based nanoalloy? — 257
3. History — 257
4. Basic metallurgy of Ti-based nanoalloy — 258
5. Basic manufacturing process of Ti-based nanoalloy — 261
6. Mechanical properties of Ti-based nanoalloy — 262
7. Applications of Ti nanoalloys in automobile industry — 263
8. Summary — 266
References — 266

1. Introduction

The choice of titanium and its alloys in automobiles has high demand that can take crucial part to enhance the fuel efficiency. Primarily Ti and its alloys are used in various components like connecting rod, valve, and valve spring because they have low modulus of elasticity, high resistance to corrosion, high creep strength, and low density [1–4]. Generally, Ti is very expensive and it is very difficult to meet the financial benefit and life cycle benefit out from it in automobile industry in the past. Gradually, the concentration was moved to enhance the properties of alloy. In fact, to achieve financial benefit, attempt was made (1) to explore low-cost manufacturing methods and (2) to perform posttreatments to enhance mechanical strength, wear resistance as well as other predetermined properties [5, 6].

Table 1 represents the comparison between Ti alloy and steel. Here many properties like relative weight, strength, shear modulus, density, etc. are compared with Ti having same tensile strength. The table considers the optimal comparison between the two materials. In steel it is very common to apply extra material so as to increase corrosion resistance. Hence some allowance must be provided to steel in its periphery which makes it bulkier. But allowance is not required in case of Ti as it is self-resilient to corrosion.

Minerals like ilmenite ($FeTiO_3$), rutile (TiO_2), arizonite ($Fe_2Ti_3O_9$), perovskite ($CaTiO_3$), and titanite ($CaTiSiO_5$) contain Ti in it. The atomic weight of Ti is 47.90

Table 1 Properties of Ti (metal springs) vs steel equivalents [7–9].

Materials	Properties			
	Allowable stress (N/mm^2)	Shear modulus (10^3 N/mm^2)	Density (g/cm^3)	Relative weight
Steel	1000	80	7.82	100
Titanium	1000	43	4.78	33

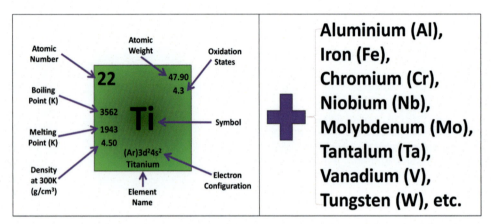

Fig. 1 Titanium information in the periodic table and their alloying elements in automobile industries.

and all other information is mentioned in Fig. 1. Titanium has many good properties which makes it more convenient to be used in various components. It has very high melting point of 1725 °C. This melting point is about 1000 °C more than aluminum and 200 °C more than steel. This high temperature operating range makes it convenient to be used in aerospace industry. Moreover, Ti is very compatible with human body. It is noncorrosive and nontoxic. Hence it has extensive use in biomedical applications. Generally, it is more bio-compatible to human tissues and bones. Ti is more resistant to attack of acids and chlorides as it can be passivated. Ti and its alloys have high degree of corrosion resistance that makes them useful in petrochemical applications and marine applications. Its corrosion resistance along with high strength enhances its usefulness in automobile industry.

Titanium is nonmagnetic and has good heat-transfer properties. It has low density and conductivity is very less. Compared to aluminum, Ti has only 3.1% electrical conductivity. The low conductivity of Ti makes it useful for being used as good resistor [10]. But somehow Ti has good thermal conductivity. It has very low thermal expansion, i.e., less than steel and less than half of aluminum. It has higher modulus of elasticity. It is lightweight as it has low density. Ti alloy can be strengthened by solute hardening, precipitation hardening, and plastic deformation. Titanium and its alloys possess tensile strengths ranging between 210 and 1380 MPa, It enables the material to be used in structural materials [11].

Ti at low temperature is generally found in hexagonal close-packed (HCP) structure (α-Ti). It is allotropic in nature and can undergo allotropic transformation at 882 °C. At 882 °C it undergoes transformation to high temperature phase (β-Ti). β-Ti has body-centered cubic (BCC) structure. Hence at any instant Ti can be visualized at any of the three phases namely α-Ti, β-Ti, and α-β-Ti. Depending upon the need, Ti can be stabilized at any of these phases. Ti alloy can be stabilized by the use of solutes having strong effect on transformation temperature. The solutes which increase transformation temperature leading to increase in α-Ti phases in alloy are called α-stabilizers. Al, N, O, and C are the examples of α-stabilizers. The solutes which increase β-Ti phases in alloy are called β-stabilizers [12, 13]. Fe, Cr, Nb, V, and W are examples of β-stabilizers. But in chemical industry and aerospace applications whatever Ti is used that consists of mixtures of α and β phases. They are mixed with different morphologies and relative volume fractions. By thermomechanical processing morphology and size of grains are controlled in the alloys. Mechanical properties of alloys are very sensitive to morphology of the two-phase Ti alloys as like dual-phase steel [14].

2. What is Ti-based nanoalloy?

It is evident that nanostructure state of material has enhanced physiomechanical properties [15]. As the technology advances preparation of nanostructured materials and their applications also increased extensively. Ti nanostructure has higher fatigue life, strength, and plasticity. Ti6Al4V alloy in its nanoform has greater demand due to enhanced mechanical properties [16]. In Ti6Al4V ultrafine grain (UFG) state is obtained using severe plastic deformation (SPD) method. Also, by SPD method, nonequilibrium grain boundaries, nanodispersed particles of secondary phases, and dislocation substructures are formed [17]. UFG VT6 alloy has higher mechanical strength and has higher efficiency of the application in manufacturing. It has enhanced service parameters as seen in compressor blades of turbine. In recent years, there was achieved a significant progress both in the theory and in optimization technique to predict the metal nanoparticle properties. The practical significance of the study of phase transformations in nanoparticles based on Ti relates to the need for the development of the theoretical bases of obtaining nanocrystals and their possible practical application in automobiles [18].

3. History

Titanium is the fourth most abundant structural metal after Al, Fe, and Mg and is the ninth most abundant element on the earth. However, world production of titanium is very low, hundreds of thousands of tons, compared to 750 million tons of steel per year. 80% of all titanium produced is used in the aerospace industry. Car suspension springs can easily be fabricated from titanium, greatly reducing weight, but titanium does not meet the needs and certainly not at the price required for a car [19]. The target price for titanium should be lowered to about 30% of its present value for serious use in mass-market cars. The main applications are cold wound springs made from cheap beta alloys and exhaust systems made from commercially pure titanium [20]. Both classes of components are currently manufactured for the automotive industry using titanium, using processes and tools for the production of steel parts. Table 2 mentions the gradual development of various patented parts in automobile industries.

4. Basic metallurgy of Ti-based nanoalloy

Titanium can exhibit different crystal structures at different temperatures as it is allotropic in nature. Titanium exists with α-phase (hcp), β-phase (bcc), and ω (hexagonal) phase. Martensitic transformation takes place between these phases. External elements are generally added to Ti so as to stabilize in a phase with a greater tolerance. These external elements can be classified into α-stabilizers and β-stabilizers. α → β transition temperature (882 °C) depends upon the amount of α stabilizers and β stabilizers.

Al is the main α-stabilizing element that is used in Ti. As Al has greater solubility in α-phase so it is more suitable to be used as α-stabilizer. β isomorphic stabilizers and β eutectoid forming elements are two groups of β stabilizing elements. These are classified depending upon the effect of resulting binary phase. B-phase has body-centered cubic crystal structure and it possesses ductile-brittle transition temperature. α-Alloys are tough at cryogenic temperatures [34] and they are easily welded. Aluminum is the main alloying element apart from Zr and Sn. The combined effect is expressed in Eq. (1) [36]:

$$\text{Aluminium equivalent (wt\%)} = \text{Al} + \frac{1}{3}\text{Sn} + \frac{1}{6}\text{Zr} + 10(\text{O} + \text{C} + 2\text{N}) \quad (1)$$

If this exceeds about 9 wt% then there may be detrimental precipitation reactions (generally Ti_3X which has an ordered HCP crystal structure). The presence of a small amount of the more ductile β-phase in nearly α-alloys is advantageous for heat treatment and the ability to forge. The alloys may therefore contain some 1 wt% of Mo (Eq. 2), where the Zr and Sn give solid solution strengthening. Ti-5Al-2.5Sn wt% is an α alloy that is available commercially in many forms [35]. Because it is stable in the α condition, it cannot be hardened by heat treatment. It is therefore not particularly strong, but can easily be welded. The toughness at cryogenic temperatures increases when the O, C, and N concentrations are reduced. The fact that the strength increases at low temperatures, without any deterioration in toughness, makes the alloy particularly suitable for the

Table 2 Various Ti alloy and Ti nanoalloy parts in the automobile parts.

Date of patent	Parts developed	References
1986	Connecting rods to connect automobile engine piston pins and cross-head pins with the crank	Sadayuki Nakamura, Atsuyoshi Kimura, Free-cutting ti alloy, EP0199198A1, 1986 [21]
1990	Engine valve of titanium alloy	Yoshiharu C/O Chuo-Kenkyusho Mae, Engine valve of titanium alloy, EP0438164B1 [22]
1991	Titanium-base alloys, namely Ti-6Al-4V and Ti-6Al-2Sn-4Zr-2Mo, have been used in automotive engines designed for racing car	Roy E. Adams, Warran M. Parris, Paul J. Bania, Alpha-beta titanium-base alloy and method for processing thereof, US5219521A, 1991 [23]
1995	Automobile engines	Masakatsu Hosomi Hisashi Maeda, Heat resistant titanium alloy, JPH09165634A, 1995 [24]
1998	Engine valve	Tadahiko Furuta, Takashi Saito Hiroyuki Takamiya, Toshiya Yamaguchi, Titanium-based composite material, method for producing the same and engine valve, WO2000005425A1, 1998 [25]
1999	Titanium aluminide alloys in automobile engines	Ian P Jones, Tai-Tsui Cheng, Titanium aluminide alloys, US5997808A, 1999 [26]
2000	Automobile or a structural component	Nozomu Ariyasu, Satoshi Matsumoto, Titanium alloy excellent in ductility, fatigue strength and rigidity and method for producing the same, EP1295955A1 [27]
2003	Engine of an automobile, and structural component of high speed rail vehicle	Nozomu Ariyasu, Satoshi Matsumoto, Titanium alloy excellent in ductility, fatigue strength and rigidity and method for producing the same, EP1295955A1, 2003 [27]
2003	Titanium alloy and automotive exhaust systems thereof	Yoji Kosaka, Stephen P. Fox, Titanium alloy and automotive exhaust systems thereof, US8349096B2, 2003 [28]
2009	Suspension springs	Shang Xue Li Fangshu Ono Kazuya Inumi Matsumoto Yoaki Chiba Akihiko, Nanocrystal titanium alloy and production method for same, WO2011037127A2, 2009 [29]

Continued

Table 2 Various Ti alloy and Ti nanoalloy parts in the automobile parts—cont'd

Date of patent	Parts developed	References
2011	Titanium-coated screw tap	Zhang Yaoliang Zhang Meibao Yang Yongjin Zhuang Jian, Process for processing titanium-coated screw tap, CN102234758B, 2011 [30]
2013	Automobile engine valve	Kuang-O Yu, Ernest M. Crist, Fusheng Sun, Titanium alloy having good oxidation resistance and high strength at elevated temperatures, EP2687615A2, 2013 [31]
2015	Automobile engine part	Tetsu Kazuhiro, Kawakami Taka, Hashi Fujii Hideki, A + β type cold-rolled and annealed titanium alloy sheet having high strength and high Young's modulus, and method for Producing same, WO2015156356A1, 2015 [32]
2016	Brake disc material	Shang Haofeng Li Guangquan Song Liwei, Method for preparing titanium-based alloy automobile brake disc material, CN105755312A, 2016 [33]

manufacture of cryogenic storage vessels, for example, to contain liquid hydrogen. A near-α alloy can be developed with good elevated temperature properties ($T < 590$ °C) [36]:

$$Ti - 6Al - 2Sn - 4Zr - 2Mo \qquad (2)$$

$$Ti - 6Al - 4Sn - 3.5Zr - 0.5Mo - 0.35Si - 0.7Nb - 0.06C \qquad (3)$$

The Nb is added for oxidation resistance and the carbon to allow a greater temperature range over which the alloy is a mixture of α + β, in order to facilitate thermomechanical processing. This particular alloy is used in the manufacture of aeroengine discs and has replaced discs made from much heavier nickel-base superalloys. The final microstructure of the alloy consists of equiaxed primary-α grains, Widmanstätten α plates separated by the β-phase. Most α + β alloys have high-strength and formability, and contain 4–6 wt% of β-stabilizers which allow substantial amounts of β-phase to be retained on quenching from the β → α + β phase fields, e.g., Ti-6Al-4V. Al reduces density, stabilizes and strengthens α-phase while vanadium provides a greater amount of the more ductile β-phase for hot-working [37]. This alloy, which accounts for about half of all the titanium that is produced, is popular because of its strength (1100 MPa), creep resistance at 300 °C, fatigue resistance, and castability [38].

5. Basic manufacturing process of Ti-based nanoalloy

Generally, by two different processes nanostructured materials are prepared. One is by severe plastic deformation method and another by compaction. By severe plastic deformation a bulk sample is refined by its microstructure and turned into nanosized fine particles in bulk samples. By compaction method initially nanopowders and nanoparticles are formed by various physiochemical processes and then the bulk is subjected to greater compression which leads to a bulk nanostructured material. Large workpieces made of VT6 alloy are manufactured by warm rolling, drawing, and sheet rolling to form a rod-like nano structure. High pressure torsion (HPT), overall multistage forging, and equal channel angular pressing (ECAP) are the some of the SPD methods used to form nanostructure in Ti-alloys [39–42]. In SPD methods, it is very much crucial to determine temperature, applied force, deformation scheme, etc. But HPT method is more convenient than SPD method in forming nanosized grains (<100 nm).

For example, by HPT method whatever Ti-based nanostructured samples were prepared at first that was like thin discs having diameter of 10 mm and thickness of 0.3 mm with the microstructure in the nanosize range that is having grain size of about 80 nm. Generally, UFG metals and alloys that are fabricated by SPD methods have grain size of about 100–500 nm. It contains nanostructural elements like nanoparticles, nanotwins, segregations, etc., inside the grain [43]. These elements affect material properties of the grain significantly. So these materials are categorized into nanostructured bulk materials. The incorporation of UFG structure in Ti alloy enhances its durability, specific strength, and fatigue resistance. These tasks can be achieved by improving ECAP in parallel nanostructured Ti alloys [44]. The ECAP treated VT 6 experiences higher stress when deformed at simple shearing strain. Morphological structure, temperature, and angle of channel greatly affect the structure and properties of the nanoalloy. Increase in the level of accumulated deformation leads to intensive grinding of the alloy structure. By increasing workpiece deformability this problem may be solved. Angle of channels crossing $\phi = 120$ and 135 degrees reduces the deformation intensity when compared to $\phi = 90$ degree.

It has been found in experiment that when deformation temperature reaches below recrystallization temperature of about 600–700 °C it provides greater plasticity [45]. In ECAP passes as the low degree of deformation increases ($e \geq 2$), α-phase globular particle refinement occurs. It develops twinning and dislocations are accumulated. Subgrains and boundaries of twins are converted into high angle ones at high deformation degrees ($e \geq 6$). More homogeneous distribution and increase in the number of reflections indicate the evolution of high angle disorientations after many passes of ECAP. This is indicated by concentric circles on electron diffraction patterns which depict UFG metals fabricated by SPD methods. It is also observed that parameters of the crystal lattice of the β-phase are changed with the increase of the deformation degree in the course of ECAP. It is pretty sure that it happens due to diffusion V and Al in redistribution in α- and β-phases [46].

6. Mechanical properties of Ti-based nanoalloy

When nanoscale range is achieved in grain, the formation of the nanostructured state in titanium and the alloy VT6 by the SPD method is found. As high pressure tension is developed in the alloy, the alloy becomes very brittle because plasticity is very low. If forging is done in alloy VT 6 then strength is increased. In a submicron range of 0.4 μm strength increases up to 1360 MPa that is 25%–30% larger than for the initial large grain state. VT6 alloy with ECAP has increased strength of 30% that is lower than ECAP strengthening (twofold). Combination of ECAP and warm rolling in two phase alloy VT6 improves mechanical properties. The properties depend upon the initial morphology of α- and β-phases. Upon warm rolling of the alloy VT6 having initial globular structure, subgrain structure with low angle boundaries is formed. This β-phase is formed in interfacial interlayers. It results in increment of strength up to 1300 MPa and elongation of 9%. So here in this case ECAP equipped with rolling enables increment of alloy deformation capacity compared to regular large grain alloy. In fact, by the combination of ECAP and drawing upon annealing, strength and plasticity can be increased in the alloy UFG Ti Grade 4 [47]. Like Hall–Petch equation for the yield stress, the effect of the fatigue limit (σ_R) on the grain size is often expressed in a formula:

$$\sigma_R = \sigma i_R + \frac{K}{\sqrt{R_d}}, \qquad (4)$$

where σi_R and KR are constants for a specific material. But the result obtained for fatigue properties is somehow contradictory. Because the end value depends upon the type of testing equipment, and versatility of fatigue tests parameters. But it is not always correct that by the decrement of grain size fatigue limit of pure grade titanium VT10 in different structural states will always increase. In UFG VT10 that is fabricated by ECAP equipped with cold rolling, the ultimate strength increases significantly but fatigue limit does not exceed 400 MPa. The UFG alloy predominantly enhances the fatigue life in comparison with the regular alloy. The fatigue strength of the UFG alloy in the low cycle range is almost 30% higher. This property is very suitable for many engineering applications [48].

7. Applications of Ti nanoalloys in automobile industry

Ti-nanoalloy is mostly used in lightweight automotive applications due to its excellent corrosion resistance, high strength, and low density. But cost of Ti restricts its use in many parts of automobile industry. Its cost is higher than steel and aluminum. Increasing general requirements, engine downsizing with turbo-charging, and new processing routes attract the use of nanoalloys. Majority of engine components of superbikes and sports cars use β-titanium alloys. The use of Ti targets for weight savings, a high strength alloy Ti-5Al-5Mo-5V-3Cr is newly introduced in Boeing 787 aircraft which substitutes high strength steels like 300M and 4340 for some of the applications where weight consideration and performance evaluation is important. Ti-6Al-2Sn-4Zr-6Mo (Ti6246) is also a high strength β-Titanium alloy. This alloy is used in aero engine gas turbines and in hot sections of racing car engines [49] (Fig. 2).

For high performance conrods (Fig. 3), Ti nanoalloys are very much suitable. For half of the weight and same strength as steel, Ti nanoalloy is very effective for manufacturing lightweight conrods. But it is very difficult to substitute steel completely as it has low modulus of elasticity. Ti has low friction and low wear resistance. Hence sometimes for some applications coating in Ti is needed. But this problem can be resolved using Ti nanoalloy. The alloys can also be utilized in production lines more easily. But these Conrods do not require any coating. 140 GPa is the modulus of elasticity for Titanium-Boron-SB2 and 132 GPa for the Titanium-nanoalloy. The modulus of elasticity is about 35% greater than for conventional Titanium alloys. High performance and reasonably priced Ti conrods manufacturing is very much necessary that will be suitable for production lines. By the application of conrods again weight reduction can be experienced for new generation "Downsizing" engines [50]. Since 1960 Titanium valves are used for

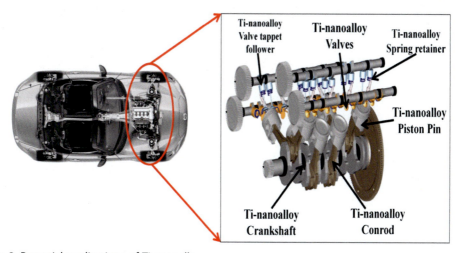

Fig. 2 Potential applications of Ti nanoalloys.

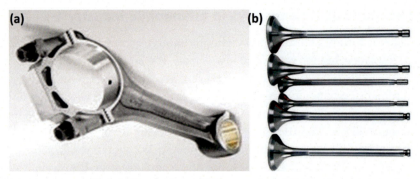

Fig. 3 (A) Titanium-nanoalloyed (Ti-6Al-4V + 12%TiC) conrod and (B) Titanium intake valve.

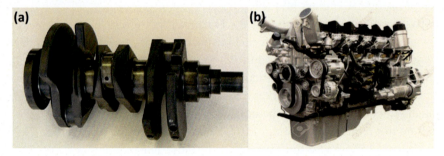

Fig. 4 Ti nanoalloy (A) (Ti-6Al-4V + 12%TiC) crankshaft and (B) commercial truck engine.

high-performance engines. Lack of high temperature strength and high price limits its wide usefulness in various industries. Conventional high-temperature Ti-alloys and new-generation Ti-nanoalloys show reduction in price and have better properties. Intermetallic orthorhombic titanium nanoalloys are also used for extreme performance for some special application. For low fuel consumption cars and new "Downsize" engines, Ti-nanoalloy valves are used. They help to lessen the mass of finger follower, rocker arm, or valve-tappet high performance engines effectively [51].

Even if a Ti crankshaft (Fig. 4) is an extreme application in terms of costs, it allows learning new manufacturing and processing processes. The feasibility of making large molded components near the net is another reason. Using Ti-nanoalloy, the crankshaft has full functionality with 40% weight reduction, excellent wear resistance, high strength, and high modulus [52]. For engine parts, there are still many variables to be worked out to perfect the process. With nitriding, higher heat and furnace pressures generate harder materials. If the material gets too hard, it can become brittle, losing some of its attractiveness for sliding engine parts. So far, automotive and commercial trucks companies have stuck to more proven materials. There is a need to adjust engine material selection to meet upcoming fuel economy requirements, and adding Ti to the toolbox of solutions

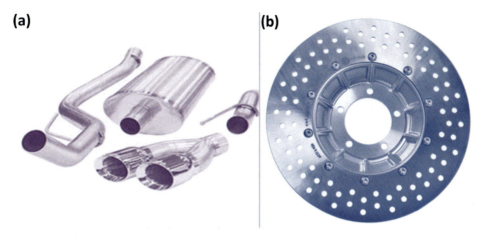

Fig. 5 Ti nanoalloy (A) Automotive exhaust systems and (B) brake disk rotor.

Table 3 Use of Ti nanoalloys in exhaust system by various manufacturers.

Composition	Manufacturer
Roll bonded CP Ti	Deutsche Titan
Ti-1.5Al	Kobe Steel
Al dipped CP Ti	Kobe Steel
Ti-0.45Si-Fe	TIMET
Ti-0.6Si-0.5Al-Nb	Kobe Steel
Ti-1Cu-Nb	Nippon Steel

is a positive. It is proved that Ti nanoalloys could be an ideal solution for many applications, but engine designers will approach use of the material very carefully [53].

Due to its availability and excellent cold formability, CP Ti has been a material of choice in automotive exhaust systems (Fig. 5A). As can be seen, CP Ti grade 1 or 2 exhibits inferior properties to ferritic stainless steel in oxidation resistance and strength particularly at elevated temperatures [54]. Several new alloys have been developed in the last several years to improve oxidation resistance of Ti to expand the application to exhaust systems with higher gas temperatures as summarized in Table 3. The iron disc rotor to a large extent has a negative effect on fuel consumption, acceleration performance, and inertia force. In addition, corrosion of the rotor discs tends to reduce the braking force and braking mechanism. In order to overcome such disadvantages, an improved disc brake rotor has been proposed (Fig. 5B), for example, the surface of the main rotor body, which is made of titanium or titanium alloys [55]. Automobile industries continue working to expand adoption of Ti-nanoalloy in various products as a part of the solutions it offers to enhance the value of motorbikes and four-wheelers.

8. Summary

The main advantages of Ti alloys for self-tapping applications are 45% lighter than steel, virtually total resistance to acid attack, corrosion resistance four times higher than stainless steel, high specific strength, good ductility, and easy to form by forging and processing operations. But nowadays, the costs of extraction, alloying with other metals, casting and metal forming processes make it difficult to enter the automotive industry. This chapter gives a brief idea of how the market is expanding the use of Ti-nanoalloy in automobile industries. The R&D efforts are essential between academia, research institutes, and industry to make effective utilization of titanium nanoalloys for motorcycles and four-wheelers worldwide in future.

References

[1] M. Javidani, D. Larouche, Application of cast Al–Si alloys in internal combustion engine components, Int. Mater. Rev. 59 (3) (2014) 132–158, https://doi.org/10.1179/1743280413Y.0000000027.

[2] L.L. Myagkov, K. Mahkamov, N.D. Chainov, I. Makhkamova, 11—Advanced and conventional internal combustion engine materials, in: R. Folkson (Ed.), Alternative Fuels and Advanced Vehicle Technologies for Improved Environmental Performance, Woodhead Publishing, 2014, ISBN 9780857095220, pp. 370–408e, https://doi.org/10.1533/9780857097422.2.370.

[3] A.K. Sachdev, K. Kulkarni, Z.Z. Fang, Titanium for automotive applications: challenges and opportunities in materials and processing, JOM 64 (2012) 553–565, https://doi.org/10.1007/s11837-012-0310-8.

[4] J.E. Allison, A.M. Sherman, M.R. Bapna, Titanium in engine valve systems, JOM 39 (1987) 15–18, https://doi.org/10.1007/BF03258873.

[5] C. Dredge, R. M'Saoubi, B. Thomas, O. Hatt, M. Thomas, M. Jackson, A low-cost machinability approach to accelerate titanium alloy development, Proc. Inst. Mech. Eng. B J. Eng. Manuf. (2020), https://doi.org/10.1177/0954405420937865.

[6] Y. Xu, Y. Lu, K.L. Sundberg, Effect of annealing treatments on the microstructure, mechanical properties and corrosion behavior of direct metal laser sintered Ti-6Al-4V, J. Mater. Eng. Perform. 26 (2017) 2572–2582, https://doi.org/10.1007/s11665-017-2710-y.

[7] K. Chen, Z. Jiang, F. Liu, J. Yu, Y. Li, W. Gong, C. Chen, Effect of quenching and tempering temperature on microstructure and tensile properties of microalloyed ultra-high strength suspension spring steel, Mater. Sci. Eng. A 766 (2019) 138272, https://doi.org/10.1016/j.msea.2019.138272.

[8] E.J. Graesser, Effect of intrinsic damping on vibration transmissibility of nickel-titanium shape memory alloy springs, Metall. Mater. Trans. A 26 (1995) 2791–2796, https://doi.org/10.1007/BF02669637.

[9] F.H. Froes, H. Friedrich, J. Kiese, Titanium in the family automobile: the cost challenge, JOM 56 (2004) 40–44, https://doi.org/10.1007/s11837-004-0144-0.

[10] E.O. Ezugwu, R.B. Da Silva, J. Bonney, Á.R. Machado, The effect of argon-enriched environment in high-speed machining of titanium alloy, Tribol. Trans. 48 (1) (2008) 18–23, https://doi.org/10.1080/05698190590890290.

[11] A. Razmi, R. Yeşildal, M.N. Khalaji, Experimental and numerical investigation of heat transfer of TiN/TiCN/TiC deposited on Cp-Ti and Ti6Al4V substrates materials, Int. J. Innov. Res. Rev. 3 (1) (2019) 1–5.https://dergipark.org.tr/en/pub/injirr/issue/51806/689064.

[12] A. Boyne, D. Wang, R.P. Shi, Y. Zheng, A. Behera, S. Nag, J.S. Tiley, H.L. Fraser, R. Banerjee, Y. Wang, Pseudospinodal mechanism for fine α/β microstructures in β-Ti alloys, Acta Mater. 64 (2014) 188–197, https://doi.org/10.1016/j.actamat.2013.10.026.

[13] C. Sauer, G. Lütjering, Influence of α layers at β grain boundaries on mechanical properties of Ti-alloys, Mater. Sci. Eng. A 319–321 (2001) 393–397, https://doi.org/10.1016/S0921-5093(01)01018-8.

[14] M.A. Murzinova, G.A. Salishchev, D.D. Afonichev, Formation of nanocrystalline structure in two-phase titanium alloy by combination of thermohydrogen processing with hot working, Int. J. Hydrogen Energy 27 (7–8) (2002) 775–782, https://doi.org/10.1016/S0360-3199(01)00155-0.

[15] A.D. Pogrebnjak, S.O. Bor'ba, Y.O. Kravchenko, Effect of the high doze of N+(1018 cm^{-2}) ions implantation into the (TiHfZrVNbTa)N nanostructured coating on its microstructure, elemental and phase compositions, and physico-mechanical properties, J. Superhard Mater. 38 (2016) 393–401, https://doi.org/10.3103/S1063457616060034.

[16] A.M. Soufiani, F. Karimzadeh, M.H. Enayati, Formation mechanism and characterization of nanostructured Ti6Al4V alloy prepared by mechanical alloying, Mater. Des. 37 (2012) 152–160, https://doi.org/10.1016/j.matdes.2011.12.044.

[17] G. Serra, L. Morais, C.N. Elias, I.P. Semenova, R. Valiev, G. Salimgareeva, M. Pithon, R. Lacerda, Nanostructured severe plastic deformation processed titanium for orthodontic mini-implants, Korean J. Couns. Psychother. 33 (7) (2013) 4197–4202, https://doi.org/10.1016/j.msec.2013.06.012.

[18] A. Amanov, I. Cho, Y. Pyun, Microstructural evolution and surface properties of nanostructured Cu-based alloy by ultrasonic nanocrystalline surface modification technique, Appl. Surf. Sci. 388 (Part A) (2016) 185–195, https://doi.org/10.1016/j.apsusc.2016.01.237.

[19] B. Koch, B. Skrotzki, Strain controlled fatigue testing of the metastable β-titanium alloy Ti-6.8Mo-4.5Fe-1.5Al (Timetal LCB), Mater. Sci. Eng. A 528 (18) (2011) 5999–6005, https://doi.org/10.1016/j.msea.2011.04.031.

[20] A.M. Sherman, C.J. Sommer, F.H. Froes, The use of titanium in production automobiles: potential and challenges, JOM 49 (1997) 38–41, https://doi.org/10.1007/BF02914682.

[21] S. Nakamura, A. Kimura, Free-cutting Ti alloy, EP0199198A1, (1986).

[22] Y. Chuo-Kenkyusho Mae, Engine valve of titanium alloy, EP0438164B1 n.d.

[23] R.E. Adams, W.M. Parris, P.J. Bania, Alpha-beta titanium-base alloy and method for processing thereof, US5219521A, (1991).

[24] M. Hosomi, H. Maeda, Heat resistant titanium alloy, JPH09165634A, (1995).

[25] T. Furuta, T.S.H. Takamiya, T. Yamaguchi, Titanium-based composite material, method for producing the same and engine valve, WO2000005425A1, (1998).

[26] I.P. Jones, T. Cheng, Titanium aluminide alloys, US5997808A, (1999).

[27] N. Ariyasu, S. Matsumoto, Titanium alloy excellent in ductility, fatigue strength and rigidity and method for producing the same, EP1295955A1, (2003).

[28] Y. Kosaka, S.P. Fox, Titanium alloy and automotive exhaust systems thereof, US8349096B2, (2003).

[29] S. Xue, L. Fangshu, O. Kazuya, I.M. Yoaki, C. Akihiko, Nanocrystal titanium alloy and production method for same, WO2011037127A2, (2009).

[30] Z. Yaoliang, Z. Meibao, Y. Yongjin, Z. Jian, Process for processing titanium-coated screw tap, CN102234758B, (2011).

[31] K. Yu, E.M. Crist, F. Sun, Titanium alloy having good oxidation resistance and high strength at elevated temperatures, EP2687615A2, (2013).

[32] T. Kazuhiro, K. Taka, H.F. Hideki, A+β type cold-rolled and annealed titanium alloy sheet having high strength and high Young's modulus, and method for producing same, WO2015156356A1, (2015).

[33] S. Haofeng, L. Guangquan, S. Liwei, Method for preparing titanium-based alloy automobile brake disc material, CN105755312A, (2016).

[34] S. Jin, S.K. Hwang, J.W. Morris, The effect of grain size and retained austenite on the ductile-brittle transition of a titanium-gettered iron alloy, Metall. Mater. Trans. A 6 (1721) (1975), https://doi.org/10.1007/BF02642299.

[35] Q.Y. Sun, H.C. Gu, Tensile and low-cycle fatigue behavior of commercially pure titanium and Ti–5Al–2.5Sn alloy at 293 and 77 K, Mater. Sci. Eng. A 316 (1–2) (2001) 80–86, https://doi.org/10.1016/S0921-5093(01)01249-7.

[36] .

[37] A. Momeni, S.M. Abbasi, Effect of hot working on flow behavior of Ti–6Al–4V alloy in single phase and two phase regions, Mater. Des. 31 (8) (2010) 3599–3604, https://doi.org/10.1016/j.matdes.2010.01.060.

[38] S. Djanarthany, J.C. Viala, J. Bouix, An overview of monolithic titanium aluminides based on Ti3Al and TiAl, Mater. Chem. Phys. 72 (3) (2001) 301–319, https://doi.org/10.1016/S0254-0584(01)00328-5.
[39] A. Panigrahi, B. Sulkowski, T. Waitz, K. Ozaltin, W. Chrominski, A. Pukenas, J. Horky, M. Lewandowska, W. Skrotzki, M. Zehetbauer, Mechanical properties, structural and texture evolution of biocompatible Ti–45Nb alloy processed by severe plastic deformation, J. Mech. Behav. Biomed. Mater. 62 (2016) 93–105, https://doi.org/10.1016/j.jmbbm.2016.04.042.
[40] M. Ashida, P. Chen, H. Doi, Y. Tsutsumi, T. Hanawa, Z. Horita, Superplasticity in the Ti–6Al–7Nb alloy processed by high-pressure torsion, Mater. Sci. Eng. A 640 (2015) 449–453, https://doi.org/10.1016/j.msea.2015.06.020.
[41] C. Haase, R. Lapovok, H. Pang Ng, Y. Estrin, Production of Ti–6Al–4V billet through compaction of blended elemental powders by equal-channel angular pressing, Mater. Sci. Eng. A 550 (2012) 263–272, https://doi.org/10.1016/j.msea.2012.04.068.
[42] M. Chandrasekaran, 10—Forging of metals and alloys for biomedical applications, in: M. Niinomi (Ed.), Woodhead Publishing Series in Biomaterials, Metals for Biomedical Devices, second ed., Woodhead Publishing, 2019, pp. 293–310, https://doi.org/10.1016/B978-0-08-102666-3.00010-9.
[43] I.P. Semenova, G.I. Raab, R.Z. Valiev, Nanostructured titanium alloys: new developments and application prospects, Nanotechnol. Russia 9 (2014) 311–324, https://doi.org/10.1134/S199507801403015X.
[44] R. Naseri, H. Hiradfar, M. Shariati, A comparison of axial fatigue strength of coarse and ultrafine grain commercially pure titanium produced by ECAP, Archiv. Civ. Mech. Eng 18 (2018) 755–767, https://doi.org/10.1016/j.acme.2017.12.005.
[45] I. Karaman, J. Robertson, J.T. Im, The effect of temperature and extrusion speed on the consolidation of zirconium-based metallic glass powder using equal-channel angular extrusion, Metall. Mater. Trans. A 35 (2004) 247–256, https://doi.org/10.1007/s11661-004-0125-5.
[46] B. Zhang, T. Yang, M. Huang, D. Wang, Q. Sun, Y. Wang, J. Sun, Design of uniform nano α precipitates in a pre-deformed β-Ti alloy with high mechanical performance, J. Mater. Res. Technol. 8 (1) (2019) 777–787, https://doi.org/10.1016/j.jmrt.2018.06.006.
[47] Y.T. Zhu, T.C. Lowe, Observations and issues on mechanisms of grain refinement during ECAP process, Mater. Sci. Eng. A 291 (1–2) (2000) 46–53, https://doi.org/10.1016/S0921-5093(00)00978-3.
[48] H. Matsumoto, S. Lee, Y. Ono, Y. Li, A. Chiba, Formation of ultrafine-grained microstructure of Ti–6Al–4V alloy by hot deformation of α′ Martensite starting microstructure, Adv. Eng. Mater. 13 (6) (2011) 470–474, https://doi.org/10.1002/adem.201000317.
[49] S. Hémery, P. Villechaise, Comparison of slip system activation in Ti-6Al-2Sn-4Zr-2Mo and Ti-6Al-2Sn-4Zr-6Mo under tensile, fatigue and dwell-fatigue loadings, Mater. Sci. Eng. A 697 (2017) 177–183, https://doi.org/10.1016/j.msea.2017.05.021.
[50] K. Weinert, S. Bergmann, C. Kempmann, Machining sequence to manufacture a γ-TiAl-conrod for application in combustion engines, Adv. Eng. Mater. 8 (1–2) (2006) 41–47, https://doi.org/10.1002/adem.200500200.
[51] K. Faller, F.H. Froes, The use of titanium in family automobiles: current trends, JOM 53 (2001) 27–28, https://doi.org/10.1007/s11837-001-0143-3.
[52] K. Bobzin, T. Brögelmann, Minimizing frictional losses in crankshaft bearings of automobile powertrain by diamond-like carbon coatings under elasto-hydrodynamic lubrication, Surf. Coat. Technol. 290 (2016) 100–109, https://doi.org/10.1016/j.surfcoat.2015.08.064.
[53] S. Abkowitz, S.M. Abkowitz, H. Fisher, CermeTi® discontinuously reinforced Ti-matrix composites: manufacturing, properties, and applications, JOM 56 (2004) 37–41, https://doi.org/10.1007/s11837-004-0126-2.
[54] Y. Kosaka, K. Faller, S.P. Fox, Newly developed titanium alloy sheets for the exhaust systems of motorcycles and automobiles, JOM 56 (2004) 32–34, https://doi.org/10.1007/s11837-004-0249-5.
[55] M. Duraiselvam, A. Valarmathi, S.M. Shariff, G. Padmanabham, Laser surface nitrided Ti-6Al-4V for light weight automobile disk brake rotor application, Wear 309 (1–2) (2014) 269–274, https://doi.org/10.1016/j.wear.2013.11.025.

CHAPTER 14

Applications of copper alloy nanoparticles in automotive industry

J. AngelinThangakani[a], C. Dorothy Sheela[a], R. Dorothy[b], N. Renugadevi[c], J. Jeyasundari[d], Susai Rajendran[e], and Ajit Behera[f]

[a]Department of Chemistry, The American College, Madurai, Tamil Nadu, India
[b]Department of EEE, AMET University, Chennai, Tamil Nadu, India
[c]Department of Zoology, GTN Arts College, Dindigul, Tamil Nadu, India
[d]Department of Chemistry, SVN College, Madurai, Tamil Nadu, India
[e]Department of Chemistry, St. Antony's College of Arts and Sciences for Women, Thamaraipady, Dindigul, Tamil Nadu, India
[f]Department of Metallurgical & Materials Engineering, National Institute of Technology, Rourkela, India

Chapter outline

1. Introduction — 269
2. Properties of Cu-NP — 270
3. Synthesis of Cu-NP — 272
   3.1 Cu and copper oxide (CuO_x) NPs — 272
   3.2 Metal oxide-supported Cu-NPs — 273
   3.3 Polymer-supported Cu-NPs — 275
   3.4 Biosynthesis of Cu-NPs using different plant extracts — 275
4. Applications — 276
   4.1 Heat transfer systems and heat exchanger — 276
   4.2 Automotive hydraulic brake tube — 277
   4.3 Gear box and bearings — 278
   4.4 Electrical driving controls — 278
   4.5 Gas separation and storage — 279
   4.6 Nano-infused fluids and engine coatings — 279
   4.7 Carbon emission control — 279
   4.8 Antimicrobial fabrics — 279
   4.9 Aircraft materials processes and hardware — 280
   4.10 Some other applications — 280
5. Limitations of Cu-NP — 281
6. Conclusion — 281
References — 281

1. Introduction

Nanotechnology is considered as one of the key technologies of the future automotive industry to support its competitiveness. The main topics of the automotive industry are the reduction of fuel consumption, environmental impact, safety information to the driver, comfort and alternatives to toxic and/or expensive materials [1, 2]. In recent decades, nanoscale particles have received a lot of attention due to their different physical, chemical, and mechanical properties. Compared to their bulk counterparts, simple or composite nanoparticles find their way into different disciplines of science and technology due to their different properties [3]. Various nanostructures provide a variety of tunable properties due to the effects of quantum confinement [4]. Copper nanoparticles (Cu-NP) in automobiles are of great interest due to their catalytic, optical, thermal, magnetic, antibacterial, electron, and electrical conduction properties [5, 6]. For example, Cu-NPs have been used in automotive fluids. One of the earliest uses of Cu-NP was to color glass and pottery in Mesopotamia in the ninth century. This was done by making a copper and silver salt glaze and applying it to clay pottery. When ceramic was fired at high temperature under reducing conditions, metal ions moved out of the glaze and were reduced to metal. The end result was a double layer of metallic nanoparticles with a small amount of enamel between them. When the finished ceramic is exposed to light, the light enters the first layer and is reflected. The light that enters the first layer is reflected by the second layer of nanoparticles, causing an interference effect with the light reflected by the first layer, producing a glowing effect that results from both constructive and destructive interference. Currently, Cu-NP has a variety of applications in the field of heat transfer systems, "anticancer" cytotoxic activity, antibacterial activity, antifungal activity, ultra-resistant materials, sensors, chemicals, catalysts, antioxidant activity, light-emitting diodes (LEDs), solar cells, single-electron transistors and lasers, dye decomposition catalysts, and anti-heat mosquitoes such as Aedesaegypti species, etc. [7, 8].

2. Properties of Cu-NP

Copper is one of the most common metals that have reddish color and is second only to silver in electrical conductivity. Its use as a structural material is limited by its weight. However, some of its superior characteristics, such as high electrical and thermal conductivity, often outweigh the weight factor [9]. Copper is highly malleable and ductile, making it ideal for wire fabrication. It is corroded by saltwater, but it is not affected by freshwater. The maximum tensile strength of copper varies greatly. For molten copper, the tensile strength is approximately 25,000 psi, and for cold rolled or cold drawn, the tensile strength increases in the range of 40,000–67,000 psi [10]. In nanoscale, Cu-NP is an important semiconductor. Semiconductors depend on the forbidden band. The forbidden band plays a fundamental role in the electrical and optical properties of the material. As the size changes from bulk to nano, the bandgap energy (E_g) increases. In metals, the conduction bands are divided into single electron levels, and the spacing and gap between these levels increase with decreasing particle size (Fig. 1). In the case of semiconductors, the phenomenon is slightly different because the band gap already exists in the ground state. If the energy of the photon is less than the forbidden band, the electron cannot reach the conduction band. If the energy of the photon is too high, the electron

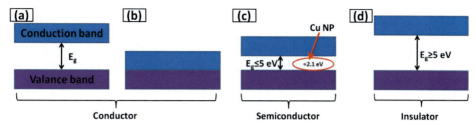

Fig. 1 Fermi level comparison of bulk copper to nanocopper: (A and B) conductor, (C) semiconductor, and (D) insulator.

enters the conduction band, but part of the energy is lost when the electron returns to the edge of the conduction band (Fig. 1C) [11]. If the energy of light is greater than or equal to the energy of the conduction band minus the energy of the valence band ($E_{light} \geq E_c - E_v$), the electrons are photoexcited from the valence band of the conduction band. The Cu-NP bandgap is reported to be 2.1 eV. This bandgap energy range of Cu-NP is due to various synthetic methods and the underlying particle size [12].

The photoluminescence (PL) of copper nanoparticles has a wide range of applications including impurity levels and defect detection, interpretation of the recombination mechanism, material quality control, molecular structure, modeling, and crystallinity. The photoluminescence of copper used can be explained by the structure of the band. This emission is due to the direct radiative recombination of electrons excited in the sp. band with gaps in the d band. The visible photoluminescence of metallic nanoparticles is usually due to the transition between bands between the sp. conduction band and the d band. The electronic structure of the Cu atom can explain the photoluminescence spectrum of Cu nanoparticles. In the case of Cu, the 3d and 4sp valence conduction electrons play an important role in the fluorescence phenomenon. The outermost electrons of all the atoms that make up the nanoparticles form six and five of these are below the Fermi level called the d-band, the sixth band is above the Fermi level, and the last band is known as the driving-band d or sp. band. The photoluminescence spectrum of Cu-NPs showed an excitation peak at 562 nm with excitation at 435 nm. The fluorescence peak was obtained at 530 nm when excited from the source at 350 nm [13]. Copper has a large interband loss in most of the visible spectrum. Band transitions, which form an important loss mechanism for material at optical frequencies, occur when electrons jump to higher energy levels from the sky caused by the absorption of incident photons. In metals, when bonded electrons absorb incident photons, the electrons can move from one energy level below or near the Fermi surface to the next higher energy level in the sky. Both processes result in high losses in optical frequencies. In semiconductors, the valence electrons absorb energy from the photons and pass through the conduction band, causing losses. This loss manifested itself as an increase in the imaginary part of the permittivity [14].

Several studies have been conducted that Cu-NPs is promising and has high antimicrobial activities either as antibacterial agent for Gram-positive bacteria (*Staphylococcus aureus, B. subtilis,* and *B. cereus*), Gram-negative bacteria (*E. coli, Pseudomonas aeruginosa, Salmonella typhi,* Klebsiella pneumonia, Enterobactor, Micrococcus, rice pathogen *X. oryzae* pv. oryzae) or as antifungal agent (Trichoderma viride, Aspergillus Niger, *C. albicans,* and Curvularia). Cu-NPs activity showed promising results of clear inhibition zone when compared with some of antibiotics (Chloramphenicol, Plantomycin, Streptomycin, Ampicillin, Cefepime hydrochloride monohydrate L-arginine, and Ofloxin) and antifungal drugs (Fluconazole and Ketoconazole) [15, 16].

3. Synthesis of Cu-NP

The synthesis of high-performance copper nanostructures is highly dependent on the method used, and proper control of the size, shape, and spatial distribution of the particles is very important. Therefore, to increase Cu-NP production at an industrial level, it is necessary to develop a new simple and inexpensive method to improve Cu-NP characteristics. Metallic copper is very unstable because it is easily oxidized under atmospheric conditions and can form Cu_2O and CuO on the surface during and after preparation. Therefore, nanoparticles must be protected by adding surface protection stabilizers such as organic ligands, surfactants, or polymers that can form a complex with copper ions.

3.1 Cu and copper oxide (CuO_x) NPs

The procedure for synthesizing Cu-based NPs is generally based on the same procedure used to prepare other metallic NPs. Most of the methods are "bottom-up" or "top-down." Both approaches have their strengths and weaknesses, but the bottom-up approach is becoming more common because it provides more room to control the shape and size of the resulting NP. Furthermore, Cu-NPs are susceptible to oxidation, which can often result in the formation of more stable copper oxide NPs. The difficulty of classifying synthetic techniques to form such nanomaterials is mainly due to the similarities in the associated chemical processes [17]. The synthesis of NP from Cu and copper oxide basically revolves around four types of chemical reactions: reduction, hydrolysis, condensation, and oxidation. Depending on the choice of final material, one or a combination of the above chemicals can be applied. The synthesis of Cu-NP often results in a reduction of the sources of Cu(I) or Cu(II) [18]. The synthesis of NP from copper oxide, on the other hand, essentially requires hydrolysis of the precursor followed by a dehydration process, which is the final material. Furthermore, an oxidation process (which may be unavoidable with Cu-based NPs) can be developed to prepare Cu-based NPs with higher oxidation numbers of each lower oxidation state precursor. In the synthetic process, the applied techniques provide the appropriate environment and energy to facilitate the selection process and are subject to additional restrictions to adjust the stability, properties, and morphology of the final NP [19]. Copper-based nanomaterials can be divided into five main categories: chemical treatment, electrochemical synthesis, photochemical technology, sonochemical methods, and heat treatment. The most common procedure is based on chemical treatment and some of the more modern methods are considered extensions of this protocol that offer advantages in terms of size and shape selectivity. Depending on the material, the reaction environment and the associated synthetic method, it can be divided into several subcategories of wet chemistry, reverse micelles, WM-assisted, biosynthetic, and IL-assisted methods [20]. The wet chemistry technology is mainly for the preparation of metallic Cu-NPs containing reducing agents that provide electrons for the reduction of Cu salts ($CuSO_4$, copper (II) acetylacetonate, $CuCl_2$, or $Cu(NO_3)_2$, etc.). Reducing agents used for this purpose often include sodium borohydride, hydrazine, 1,2-hexadecanediol, glucose, ascorbic acid, CO, or recently introduced borane compounds. Various styling products have also been used to stabilize the resulting Cu-NP and control particle growth [21]. Another approach is the so-called reverse micelle method, in which a surfactant is added to a container containing polar and other nonpolar solvents to form an oil-in-water microemulsion. Inverted micelles act as nanoreactors of uniform size for NP production, allowing precise control over the shape and size of the resulting nanomaterials. The size of the resulting NP can be adjusted by adjusting the reaction conditions to produce micelles of different sizes [22].

3.2 Metal oxide-supported Cu-NPs

Various metal oxides have been studied as support materials for Cu-based NPs and related nanomaterials. The choice of support largely depends on its ease of synthesis, compatibility with different substrates/reagents, and, above all, on the properties it confers on the resulting supported nanomaterial. Ceria is an important carrier material for various catalysts due to its particularly high oxygen content on its surface. A hybrid Cu/CeO_2 nanocatalyst prepared by depositing Cu-NP on ceramic nanocubes covered with hexamethylenediamine was used to promote CO oxidation [23]. Furthermore, CuO NP was prepared on a CeO_2 carrier and used as a highly efficient and environmentally friendly recyclable catalyst. Rothenberg et al. [24] synthesized a supported Cu-ceria catalyst by co-melting, drying, and sequentially pyrolyzing a mixture of $Ce(NO_3) \cdot H_2O$ and $Cu(NO_3) \cdot 3H_2O$. Another important support widely used as a support for NP is alumina [25]. Choudary and colleagues reported the synthesis of Cu-NP on an alumina carrier from precursors of copper (II) acetylacetonate and aluminum isopropoxide using an airgel-based method developed a Cu catalyst by immobilizing Cu-NP on aluminum hydroxide (oxy) [AlO (OH)] fibers [26]. Due to the nature of Lewis acid, zinc oxide NP has been used primarily for catalysis as a catalyst itself, but it can also be used as a vehicle for other nanocatalytic species. For example, a Cu/ZnO catalyst was prepared from the hydroxycarbonate precursor. In modelless synthesis technology, ZnO nanoarrays are grown in a microchannel reactor, an aqueous Cu (II) salt solution is added, and the resulting mixture is reduced to synthesize ZnO nanorar-supported Cu-NP (NR) [27]. Recently, Cu/ZnO microbead catalysts have been prepared by a two-step process in which the ZnO nanorods are hydrothermally cultured and then combined with preformed Cu-NP. In another study, Carrelovich et al. [28] has been used Cu/ZnO catalyst that synthesized by citrate decomposition. Another metal oxide that has been extensively studied as a support for Cu-NP is MgO. Burri et al. prepared a MgO carrier by calcining $Mg(NO_3)_2·6H_2O$ with K_2CO_3 at 723 K for 12 h and impregnated the resulting material with Cu-NP to obtain a Cu per MgO catalyst. Titanium oxide has been widely used to support a variety of nanomaterials, mainly due to its unique photochemical properties. For example, titanium oxide-supported NP synthesized from Cu using a wet alkoxide chemical process. Pasini et al. [29] reported the preparation of titanium-supported Cu-NP by immobilizing a prefabricated monometallic Cu species on TiO_2, a technique that is also applicable to copper oxide NPs. Cu oxide NPs (composed mainly of Cu_2O) are prepared on titanium oxide and other materials by adding carriers (TiO_2, MgO, etc.) to the freshly prepared NP suspension. Liu et al. [30] Cu-NP decorated synthetic TiO_2 nanotubes using Pd activation to promote the deposition of Cu or copper oxide. The nanotubes were activated by first immersing them in a 10% acetic acid solution of 8.9×10^{-4} MPd $(OAc)_2$, followed by a 0.2 NaH_2PO_2 solution. After the activation step, Cu-NPs were deposited by mixing the nanotubes with a solution of $CuSO_4·5H_2O$, $KNaC_4H_4O_6·4H_2O$ and formaldehyde under sonication. The resulting material was washed with water and dried at 393 K. Zirconia-supported Cu-NP has also received a lot of attention lately because it can be prepared using a similar protocol as titanium oxide-supported materials. Rao et al. [31] reported that the ZrO_2 carrier was impregnated with an adequate amount of $Cu(NO_3)_2·3H_2O$ to prepare Cu-filled zirconium oxide. Although expensive, lanthanum oxide has unique Lewis acid properties that are beneficial in a variety of catalytic applications. For example, Cu-NP is prepared with the support of La_2O_3 using the sedimentation method. Likewise, Azambre et al. [32] reported on the synthesis of multiwalled Cu-VO_x nanotubes by exchanging the protonation matrix of the original dodecylamine-VOx nanotubes with Cu(II) salts. Recent developments have identified a set of new support materials that can host Cu-NP. The organic metal framework (MOF) is a porous crystalline material that contains a three-dimensional framework of metal ions or metal aggregates that are held together by a rigid bipod or multilegged organic linker. They have many interesting properties, such as porosity, crystallinity, and relatively strong metal-ligand coordination bonds [33]. Jiang et al. prepared [34] Cu-NP loaded into a MIL-101(Cr) MOF and used the resulting material as a heterogeneous catalyst for the reduction of nitroaromatic compounds. Another interesting supporting document has been identified by Mitsudome et al. [35] Cu-NP prepared with hydrotalcite (HT) support. This material was synthesized by mixing HT with an aqueous solution of Cu(II) trifluoromethanesulfonate, adding an aqueous solution of ammonia to adjust the pH to 8.0, and stirring the resulting mixture at room temperature for 1 h. The obtained HT-supported Cu(II) NP was used to catalyze dehydrogenation without alcohol.

3.3 Polymer-supported Cu-NPs

Polymer-backed NPs can in principle be considered a subclass of carbon-based material-backed NPs. Kang et al. [36] reduced Cu_2O NP and Cu supported by a particle size of 30–120 nm on Cu-NP on a cellulose acetate polymer reducing hydrogen gas and a Cu(II) complex in the presence of cellulose acetate. Cu-NP encapsulated in the dendrimer was prepared by chemically reducing the Cu complex entrapped in the dendrimer using sodium borohydride (NaBH4). Yang et al. [37] reported a simple procedure to prepare Pd-CuNP with the support of one-dimensional hybrid nanofibers using hydrated bacterial cellulose nanofibers (BCF) as a template. This material was prepared by immersing BCF in an aqueous solution of $PdCl_2$ and $CuCl_2$, after which the metal ions adsorbed on BCF were reduced with potassium hydride to give the BCF-supported metal Pd-CuNP. Synthesis of Cu-NP in nitrogen-rich copolymer micro-layers can be prepared by incomplete condensation of melamine and cyanuric chloride. The polymer microplate was then suspended in a Cu(II) acetate solution and sonicated for 1 h. The composite was then treated with hydrazine hydrate to produce the desired material [38]. The nitrogen dopant in the material enhanced the interaction with Cu and allowed the microplate to function as an excellent support material. By mixing melan 50 nm Cu-NP with poly(3,4-(ethylenedioxy)-thiophene)/poly(styrene sulfonic acid) (PEDOT/PSS), Cu-NP supported by a composite polymer can be prepared. Cu-NP is synthesized by reducing copper salts with hydrazine in the presence of cetyltrimethylammonium bromide (CTAB) surfactant and stabilizer of poly(vinylpyrrolidone) (PVP) [39].

3.4 Biosynthesis of Cu-NPs using different plant extracts

Valodkar et al. [40] peptide-coated Cu-NPs were prepared using common milk blanket ivulia euphorbic stem latex as a reducing and styling agent. Lee et al. [41] obtain Cu-NP using Magnoliakobus leaf extract. Cu-NPs were synthesized using an aqueous latex extract of *C. procera* and its cytotoxicity to tumor cells was studied. Ginger *Zingiber officinale* and *Syzygium aromaticum* Clove extracts have been studied the antimicrobial activity of formed NPs. In Capparis zeylanica, *Vitis vinifera*, *Nerium oleander* leaf, leaf extracts were also explored to synthesize Cu-NPs and their antibacterial action has been investigated. Various plant extracts have been explored by Pineapple, *C. grandis* peel, Hibicus Rosa-sinensis leaf, Guava, Phyllanthus embilica Gooseberry, *G. biloba* L. leaf, Lemon fruit, and *A. vera* flower [42, 43]. *Tridax procumbens* leaves, Arevalanata leaves, *D. innoxia* aqueous leaves, *Punica granatum* peel, *Cassia Auriculata* leaves, and *Allium Sativum* "chopped garlic leaves" were used as reducing and capping agents to bio-synthesize Cu-NPs in 2016. Nasrollahzadeh et al. [44] utilized *Plantago asiatica* leaf to get Cu-NPs and their application for the cyanation of aldehydes using $K_4Fe(CN)_6$ have been performed. In another study, Triumfettarotundifolia extract mediated synthesis of Cu-NPs and its pharmacological activities have been carried out. Cu-NPs were synthesized using Ripened *D. erecta* fruit extract to reduce toxic azo dyes congo red and methyl orange from water. The flower, leaf, and stem of Gnidia glauca and *Plumbago zeylanica* extract is used to get Cu-NPs which are applied as an antidiabetic agent. *Tinosporia cardifolia* extract was also reported to produce nature-friendly Cu nano-coated fabric and its antimicrobial activities [45].

4. Applications

Today, nanocopper alloys play an important role in automobiles for their functionality, efficiency, comfort, and safety. Copper has been a versatile material whose properties have been an important component of the vehicle since the dawn of the automotive industry with the Ford Model T in 1916. Even the most basic models include approximately 1 km of wiring, which is used primarily for vehicle transport. Gross vehicle weights range from 15 kg for small cars to 28 kg for luxury cars. Some of the potential applications for different automotive purposes are outlined here.

4.1 Heat transfer systems and heat exchanger

Scientists and engineers have been working for decades developing fluids with improved heat transfer properties. Such fluids have applications to improve the efficiency of automobiles, heating/air conditioning systems, and industrial equipment. For metal nanoparticles based on copper nanoparticles, adding just 0.3% by volume of copper nanoparticles to ethylene glycol increased thermal conductivity by up to 40%. They state that these fluids are one of the most advanced heat transfer media in the world, with hyperpyrexia and long-term stability [46]. They dramatically improve heat transfer and their ability to do so is far superior (over 10 times better) than the best competitive fluids. It also outperforms competing fluids for long-term stability. Nanoparticles have been shown to dramatically improve the thermal conductivity of the base fluid. The improvement is more than 10 times greater than expected. Nanofluids can provide significant savings, as even a slight increase in heat transfer can save pump energy. Nanofluids, which have approximately three times the thermal conductivity of the base fluid, can double the heat transfer coefficient of the fluid. To double it without adding nanoparticles, you must multiply the pump output by 10. Metallic nanoparticles improve heat transfer over oxide nanoparticles [47].

The new heat exchanger was developed using cuproblaze technology, a low investment cost, economical, and environmentally friendly process. It manufactures strong and reliable brazed copper/brass radiators, offering performance and cost advantages compared to Al radiators. Due to its ability to withstand high pressures and low temperatures, this technology is increasingly selected in more market segments. It demands absolute reliability in extreme conditions, such as long-distance trucks, agricultural and construction equipment, generators, and off-road vehicles. At the end of the radiator's useful life, the copper in the radiator can be recycled to produce new radiators or other copper products [48].

4.2 Automotive hydraulic brake tube

The braking system of a vehicle, as well as the engine and transmission, are very important to the performance of the vehicle. The tube that carries pressurized air or fluid through the system is an important link between the wheel's master and slave cylinders. The brake system tubing is vulnerable to pressure from the passage of air and fluids, corrosion from mud and road salt, and damage to the protective surface layer from exposed sharp stones under the chassis. Many other auto parts operate in the same harsh environment, but few are less tolerant in the event of failure. Therefore, one of the main considerations in the design of automobile hydraulic braking systems is the integrity of the brake tubes that distribute the system pressure [49]. For many years, tubes for automobile brake systems have been made from low-carbon steel. Since steel does not have the inherent corrosion resistance of the road environment, one or more surface coatings are applied after brazing to protect the steel substrate from corrosion. Although the composition of the coating has changed since the use of the original molten tin lead coating, coating imperfections are still a problem. Copper-Nickel tubing (UNS C70600) was manufactured with a common braking system. Copper-nickel alloy C70600, which is an alloy of 90% copper and 10% nickel, is inherently resistant to road salt corrosion, increasing its use as a brake tube: (1) Evolution of life expected from automobiles. (2) Global service experience data on brake tube wear. (3) Increased cost of anti-corrosion coating on steel brake pipes. This product complies with ASTM B466 (American Society for Testing and Materials), which specifies dimensions, tensile strength, and yield strength. Castability and internal cleanliness meet SAE J527, ASTM A254, and SMMT C5B (Society of Engine Manufacturers and Traders) specifications. Additionally, this alloy meets ISO 4038 (International Organization for Standardization) and SAE J1047 requirements for pressure containment, fabrication, and corrosion resistance. Today's production steel tubes meet the 60 cycle requirement, but failed long before 120 cycles [50]. The Cu90Ni10 tube completed 200 cycles without substantially reducing the original burst strength. Volvo started using Cu90Ni10 tubes in vehicle models in 1976 and has been using them ever since. Audi, Porsche, and Aston Martin are also known to use it.

4.3 Gear box and bearings

When it comes in contact with the moving parts of the engine, the Cu-nano alloy provides a surface that is strong enough to provide support, yet resistant to sticking and wear. In this application, it acts as a bearing. Copper alloy bearing material is used for the selector fork and heavy duty bearing. Graphite is nontoxic and weight is reduced using aluminum and copper-graphite composite bearings instead of leaded copper. The bearings also improve wear properties as the deformation of the graphite particles results in the formation of a continuous graphite film, which provides self-lubrication of the component and allows better component life. These materials can be useful for almost any sleeve bearing in a propulsion system [51].

4.4 Electrical driving controls

AICC (Autonomous Intelligent Cruise Control) is not a simple cruise control, it allows the driver to maintain the previously set speed without having to step on the accelerator. With available technology in addition to the automatic transmission, AICC monitors the speed of the vehicle in front and changes its speed accordingly, maintaining a safe distance. Today, cars are more than just a means of transportation. It is designed to be comfortable and has work and leisure space. In applications that take advantage of nano-copper alloys, the seat can automatically memorize and adjust passengers with a small motor. Automatic temperature control keeps the interior at a comfortable level and allows communication access from the most discreet cabins (hands-free mobile phones, navigation aids, security trackers, DVD players, and even Internet connections). Copper has opened a new window in the world. In the future, accelerometer-driven autopilot systems could be combined with actuators that control handling, speed, and braking. Radar and 360° cameras constantly monitor traffic and vehicle position [52].

Electric motors, alternators, actuators, inductors, and wire harnesses depend on highly reliable conductivity. As the electrical development of the automobile increases awareness, safety, and automation, more copper will be needed. High-quality brass is ideal for electrical connections due to its long-lasting flexibility and resistance to corrosion. Automotive designers and manufacturers are constantly developing electronic applications that depend on the electrical conductivity of copper. (1) Sensors: Sensors (pressure, temperature, speed) for the automotive industry account for about a third of the global sensor market. With the current sensory system of the vehicle, the sensors allow, for example, the detection of danger on the road, the adjustment of the brakes, the control of the interior temperature of the vehicle and the execution of self-diagnostic tests of the vehicle. In particular, copper is used for coils and cables. (2) Antilock Braking System: This was one of the first "smart" technologies designed to make driving safer by adjusting the brakes based on grip. Since then, a number of innovations have expanded the solutions available in terms of braking and grip. In particular, EBD (Electronic Brake Force Distribution) or

ESC (Electronic Stability Control) is used for the better brake force distribution in the vehicles [53].

4.5 Gas separation and storage

Interest in using hydrogen as a fuel for automobiles, perhaps by combustion in fuel cells, has led to the study of ways to store hydrogen at ambient pressure. Copper-containing compounds with an organometallic structure (MOF) can store up to 2.5% by weight of hydrogen at 77 K (liquid nitrogen temperature). Its compound, MOF-505, is composed of $Cu_2(CO_2)_4$ driver-type units linked by biphenyltetracarboxylic dianbodies [54].

4.6 Nano-infused fluids and engine coatings

Nanotechnology is at the forefront of next-generation engine fluids, lubricants, and coatings. Nanofluids are designed to provide the perfect particle shape, size, and concentration—three key determinants that affect vehicle wear and friction—and provide better lubricity than traditional fluids. This provides additional protection against heat, friction, and wear. Gold and copper nanoparticles have been shown to be effective in developing new types of liquid protective films to replace traditional fossil fuel-based oils, making car engines more durable [55].

4.7 Carbon emission control

Nano copper alloys play an important role in systems designed to reduce gasoline consumption and CO_2 emissions. The direct injection system enables more precise control of the air–fuel ratio, reducing fuel consumption and emissions. Traditional camshafts have also been gradually replaced by electronic valves that further improve engine efficiency [56]. Nanocopper alloys play a role in hybrid vehicles and fuel cell vehicles, and researchers are actively investigating new powertrains. First, a hybrid that combines a traditional combustion engine with an electric motor can provide an interim solution. Traditional fuels are used for long-distance travel and electric motors are used in urban environments. Secondly, work is underway on the development of fuel cell engines. This is a virtually contamination-free solution. Both different systems are equipped with powerful electric motors and can hold up to 12 kg of nano-copper alloy [57].

4.8 Antimicrobial fabrics

There is a demand for antibacterial materials that contain antibacterial fibers amid growing public anxiety about the degree of microbial infection. Nanotechnology has provided new solutions for the development of antibacterial fabrics. In this study, copper (Cu) and silver (Ag) nanoparticles and Cu and Ag alloy nanoparticles were microwaved using chitosan as a stabilizer to reduce their respective nitrates with ascorbic acid. In this study, copper (Cu) and silver (Ag) nanoparticles and CuAg alloy nanoparticles were microwaved using chitosan as a stabilizer to reduce their respective nitrates with ascorbic acid. Cu nanoparticles show higher inactivation of bacteria such as *B. subtilis*, *E. coli* and methicillin-resistant *Staphylococcus aureus* (MRSA) than did Ag nanoparticles at the same concentration [58]. The nanoparticles showed a more potent antibacterial effect than did ions of the same metal. Nanoparticle-impregnated fabrics reduced microbial viability by 80%–90%, but this decreased in relation to the number of washes the fabric was subjected to and indicated a leached out of the nanoparticles. Pretreatment of cotton fabrics with tannic acid and citric acid enhanced the durability of the antimicrobial effect when washed and this increased with concentration of the acid. Citric acid treated fabrics showed higher durability than tannic acid treated fabrics [59].

4.9 Aircraft materials processes and hardware

In aircraft, copper is used primarily in electrical systems for interlocking cables, connections, and bus bars. Beryllium copper is one of the best performing copper-based alloys. It is a newly developed alloy containing approximately 97% copper, 2% beryllium, and enough nickel to increase elongation. The most valuable characteristic of this metal is that the heat treatment dramatically improves its physical properties, increasing the tensile strength from 70,000 psi in the annealed state to 200,000 psi in the heat treated state [60]. The fatigue and wear resistance of beryllium copper makes it suitable for diaphragms, precision bearings and rings, ball cages, and spring washers. Brass is an alloy of zinc and copper that contains small amounts of aluminum, iron, lead, manganese, magnesium, nickel, phosphorus, and tin. Brass with 30%–35% zinc is very ductile, while 45% zinc has relatively high strength. Muntz metal is brass made up of 60% copper and 40% zinc. It has excellent resistance to corrosion in saltwater. Its resistance can be increased by heat treatment. The maximum tensile strength of this metal during casting is 50,000 psi, which can be stretched 18%. It is used in the manufacture of bolts, nuts, and parts that come into contact with saltwater [61]. Bronze is a copper alloy that contains tin. True bronze contains up to 25% tin, but less than 11% bronze is more useful, especially for items like airplane pipe fittings. These copper-based alloys contain up to 16% (typically 5%–11%) aluminum, to which other metals such as iron, nickel, and manganese can be added. Aluminum bronze has excellent tear quality, high strength, toughness, and resistance to impact and fatigue. Due to these properties, they are used in membranes, gears, and pumps. Aluminum bronze is available in rods, bars, plates, sheets, strips, and forgings [62].

4.10 Some other applications

Nanofluids can solve many problems affecting the heating, ventilation, and air-conditioning (HVAC) industries. However, its use can go beyond heat transfer in automobiles and manufacturing equipment. Nanofluids improve the efficiency of high heat flux devices such as supercomputers, EMI shields, and the provision of new technologies for cancer treatment. Conductive inks and pastes containing Cu nanoparticles are very expensive precious metals used in printed electronics, displays, permeable conductive thin film applications, spark plugs, electrical wiring, busbars and cables, high conductivity cables, and electrodes [63, 64].

5. Limitations of Cu-NP

Copper nanoparticles are classified as highly flammable solids and must be stored away from the ignition source. They are also known to be very toxic to aquatic life. Cu-NP is dose-dependent for human lung cancer cell line (A549), human liver tumor (HepG2), Chinese hamster ovary (CHO), human osteosarcoma (Saos), and mouse embryonic fibroblast (3T3L1). It is toxic. This study revealed that Cu-NP coated with a nontoxic aqueous latex extract can be used directly for in vivo delivery/delivery of nanoparticles for cancer treatment. Cancer studies have demonstrated in vitro cytotoxicity of Cu-NP against human colon cancer Caco-2 cells, human liver cancer HepG2 cells, and human breast cancer Mcf-7 cells tested. Cu albumin NP has been shown to be more toxic and induce cell death in prepared NPs than in normal cells [65, 66].

6. Conclusion

Cu-NPs are of particular interest for their historical use as fluid dispersants in heat transfer systems for dyes and their latest biomedical agents. Nanomaterials are being applied in more areas of engineering and technology. This chapter briefly described the synthesis procedure and the various types of applications in the automotive industry.

References

[1] I. Shancita, H.H. Masjuki, M.A. Kalam, I.M. Rizwanul Fattah, M.M. Rashed, H.K. Rashedul, A review on idling reduction strategies to improve fuel economy and reduce exhaust emissions of transport vehicles, Energ. Conver. Manage. 88 (2014) 794–807, https://doi.org/10.1016/j.enconman.2014.09.036.

[2] A.M. Omer, Energy, environment and sustainable development, Renew. Sustain. Energy Rev. 12 (9) (2008) 2265–2300, https://doi.org/10.1016/j.rser.2007.05.001.

[3] M. Auffan, J. Rose, J.Y. Bottero, et al., Towards a definition of inorganic nanoparticles from an environmental, health and safety perspective, Nat. Nanotechnol. 4 (2009) 634–641, https://doi.org/10.1038/nnano.2009.242.

[4] M. Kole, T.K. Dey, Enhanced thermophysical properties of copper nanoparticles dispersed in gear oil, Appl. Therm. Eng. 56 (1–2) (2013) 45–53, https://doi.org/10.1016/j.applthermaleng.2013.03.022.

[5] M.K.A. Ali, X. Hou, M.A.A. Abdelkareem, Anti-wear properties evaluation of frictional sliding interfaces in automobile engines lubricated by copper/graphene nanolubricants, Friction 8 (2020) 905–916, https://doi.org/10.1007/s40544-019-0308-0.

[6] Y. Kamikoriyama, H. Imamura, A. Muramatsu, et al., Ambient aqueous-phase synthesis of copper nanoparticles and nanopastes with low-temperature sintering and ultra-high bonding abilities, Sci. Rep. 9 (2019) 899, https://doi.org/10.1038/s41598-018-38422-5.

[7] S. Dey, G. Chandra Dhal, Controlling carbon monoxide emissions from automobile vehicle exhaust using copper oxide catalysts in a catalytic converter, Mater. Today Chem. 17 (2020) 100282, https://doi.org/10.1016/j.mtchem.2020.100282.

[8] V.F. Ribeiro, D.N. Simões, M. Pittol, D. Tomacheski, R.M.C. Santana, Effect of copper nanoparticles on the properties of SEBS/PP compounds, Polym. Test. 63 (2017) 204–209, https://doi.org/10.1016/j.polymertesting.2017.07.033.

[9] M.S. Usman, M.E. El Zowalaty, K. Shameli, N. Zainuddin, M. Salama, N.A. Ibrahim, Synthesis, characterization, and antimicrobial properties of copper nanoparticles, Int. J. Nanomedicine 8 (2013) 4467–4479, https://doi.org/10.2147/IJN.S50837.

[10] J.L. Viesca, A. Hernández Battez, R. González, R. Chou, J.J. Cabello, Antiwear properties of carbon-coated copper nanoparticles used as an additive to a polyalphaolefin, Tribol. Int. 44 (7–8) (2011) 829–833, https://doi.org/10.1016/j.triboint.2011.02.006.

[11] F. Parveen, B. Sannakki, M.V. Mandke, H.M. Pathan, Copper nanoparticles: synthesis methods and its light harvesting performance, Solar Energy Mater. Solar Cells 144 (2016) 371–382, https://doi.org/10.1016/j.solmat.2015.08.033.

[12] R.K. Swarnkar, S.C. Singh, R. Gopal, Effect of aging on copper nanoparticles synthesized by pulsed laser ablation in water: structural and optical characterizations, Bull. Mater. Sci. 34 (2011) 1363–1369, https://doi.org/10.1007/s12034-011-0329-4.

[13] B. Sharma, M.K. Rabinal, Ambient synthesis and optoelectronic properties of copper iodide semiconductor nanoparticles, J. Alloys Compd. 556 (2013) 198–202, https://doi.org/10.1016/j.jallcom.2012.12.120.

[14] D. Gupta, S.R. Meher, N. Illyaskutty, Z.C. Alex, Facile synthesis of Cu2O and CuO nanoparticles and study of their structural, optical and electronic properties, J. Alloys Compd. 743 (2018) 737–745, https://doi.org/10.1016/j.jallcom.2018.01.181.

[15] C. Khurana, P. Sharma, O.P. Pandey, B. Chudasama, Synergistic effect of metal nanoparticles on the antimicrobial activities of antibiotics against biorecycling microbes, J. Mater. Sci. Technol. 32 (6) (2016) 524–532, https://doi.org/10.1016/j.jmst.2016.02.004.

[16] J.P. Ruparelia, A.K. Chatterjee, S.P. Duttagupta, S. Mukherji, Strain specificity in antimicrobial activity of silver and copper nanoparticles, Acta Biomater. 4 (3) (2008) 707–716, https://doi.org/10.1016/j.actbio.2007.11.006.

[17] A. Waris, M. Din, A. Ali, M. Ali, S. Afridi, A. Baset, A.U. Khan, A comprehensive review of green synthesis of copper oxide nanoparticles and their diverse biomedical applications, Inorganic Chem. Commun. 123 (2021) 108369, https://doi.org/10.1016/j.inoche.2020.108369.

[18] E. Tomaszewska, S. Muszyński, K. Ognik, P. Dobrowolski, M. Kwiecień, J. Juśkiewicz, D. Chocyk, M. Świetlicki, T. Blicharski, B. Gładyszewska, Comparison of the effect of dietary copper nanoparticles with copper (II) salt on bone geometric and structural parameters as well as material characteristics in a rat model, J. Trace Elem. Med. Biol. 42 (2017) 103–110, https://doi.org/10.1016/j.jtemb.2017.05.002.

[19] V.K. Sharma, R.A. Yngard, Y. Lin, Silver nanoparticles: green synthesis and their antimicrobial activities, Adv. Colloid Interface Sci. 145 (1–2) (2009) 83–96, https://doi.org/10.1016/j.cis.2008.09.002.

[20] P.K. Khanna, S. Gaikwad, P.V. Adhyapak, N. Singh, R. Marimuthu, Synthesis and characterization of copper nanoparticles, Mater. Lett. 61 (25) (2007) 4711–4714, https://doi.org/10.1016/j.matlet.2007.03.014.

[21] J. Ramyadevi, K. Jeyasubramanian, A. Marikani, G. Rajakumar, A.A. Rahuman, Synthesis and antimicrobial activity of copper nanoparticles, Mater. Lett. 71 (2012) 114–116, https://doi.org/10.1016/j.matlet.2011.12.055.

[22] A.K. Gupta, M. Gupta, Synthesis and surface engineering of iron oxide nanoparticles for biomedical applications, Biomaterials 26 (18) (2005) 3995–4021, https://doi.org/10.1016/j.biomaterials.2004.10.012.

[23] P. Sudarsanam, B. Hillary, M.H. Amin, N. Rockstroh, U. Bentrup, A. Brückner, S.K. Bhargava, Heterostructured Copper–Ceria and Iron–Ceria nanorods: role of morphology, redox, and acid properties in catalytic diesel soot combustion, Langmuir 34 (8) (2018) 2663–2673, https://doi.org/10.1021/acs.langmuir.7b03998.

[24] J. Albadi, M. Keshavarz, F. Shirini, M. Vafaie-nezhad, Copper iodide nanoparticles on poly(4-vinyl pyridine): a new and efficient catalyst for multicomponent click synthesis of 1,4-disubstituted-1,2,3-triazoles in water, Cat. Commun. 27 (2012) 17–20, https://doi.org/10.1016/j.catcom.2012.05.023.

[25] C. Saldías, D.D. Díaz, S. Bonardd, C. Soto-Marfull, A. Cordoba, S. Saldías, C. Quezada, D. Radic, Á. Leiva, In situ preparation of film and hydrogel bio-nanocomposites of chitosan/fluorescein-copper with catalytic activity, Carbohydr. Polym. 180 (2018) 200–208, https://doi.org/10.1016/j.carbpol.2017.10.018.

[26] J. Chaudhary, G. Tailor, B.L. Yadav, O. Michael, Synthesis and biological function of Nickel and Copper nanoparticles, Heliyon 5 (6) (2019), https://doi.org/10.1016/j.heliyon.2019.e01878, e01878.

[27] L.-C. Wang, Y.-M. Liu, M. Chen, Y. Cao, H.-Y. He, G.-S. Wu, W.-L. Dai, K.-N. Fan, Production of hydrogen by steam reforming of methanol over Cu/ZnO catalysts prepared via a practical soft reactive grinding route based on dry oxalate-precursor synthesis, J. Catal. 246 (1) (2007) 193–204, https://doi.org/10.1016/j.jcat.2006.12.006.

[28] M. Lakshmi Kantam, V. Swarna Jaya, B. Sreedhar, M. Mohan Rao, B.M. Choudary, Preparation of alumina supported copper nanoparticles and their application in the synthesis of 1,2,3-triazoles, J. Mol. Catal. A Chem. 256 (1–2) (2006) 273–277, https://doi.org/10.1016/j.molcata.2006.04.054.

[29] W. Lipińska, K. Grochowska, J. Karczewski, J. Ryl, A. Cenian, K. Siuzdak, Thermally tuneable optical and electrochemical properties of Au-Cu nanomosaic formed over the host titanium dimples, Chem. Eng. J. 399 (2020) 125673, https://doi.org/10.1016/j.cej.2020.125673.

[30] X. Li, J. Yao, F. Liu, H. He, M. Zhou, N. Mao, P. Xiao, Y. Zhang, Nickel/Copper nanoparticles modified TiO2 nanotubes for non-enzymatic glucose biosensors, Sens. Actuators B 181 (2013) 501–508, https://doi.org/10.1016/j.snb.2013.02.035.

[31] D. Das, J. Llorca, M. Dominguez, S. Colussi, A. Trovarelli, A. Gayen, Methanol steam reforming behavior of copper impregnated over CeO2–ZrO2 derived from a surfactant assisted coprecipitation route, Int. J. Hydrogen Energy 40 (33) (2015) 10463–10479, https://doi.org/10.1016/j.ijhydene.2015.06.130.

[32] B. Azambre, M.J. Hudson, Growth of copper nanoparticles within VOx nanotubes, Mater. Lett. 57 (20) (2003) 3005–3009, https://doi.org/10.1016/S0167-577X(02)01421-0.

[33] S. Cheng, N. Shang, F. Cheng, S. Gao, C. Wang, Z. Wang, Efficient multicomponent synthesis of propargylamines catalyzed by copper nanoparticles supported on metal-organic framework derived nanoporous carbon, Cat. Commun. 89 (2017) 91–95, https://doi.org/10.1016/j.catcom.2016.10.030.

[34] L. Huang, H. Jiang, J. Zhang, Z. Zhang, P. Zhang, Synthesis of copper nanoparticles containing diamond-like carbon films by electrochemical method, Electrochem. Commun. 8 (2) (2006) 262–266, https://doi.org/10.1016/j.elecom.2005.11.011.

[35] T. Mitsudome, M. Matoba, T. Mizugaki, K. Jitsukawa, K. Kaneda, Core–shell AgNP@CeO$_2$ nanocomposite catalyst for highly chemoselective reductions of unsaturated aldehydes, Chemistry 19 (17) (2013) 5255–5258, https://doi.org/10.1002/chem.201204160.

[36] S. Zhang, P. Kang, M. Bakir, A.M. Lapides, C.J. Dares, T.J. Meyer, Polymer-supported CuPd nanoalloy as a synergistic catalyst for electrocatalytic reduction of carbon dioxide to methane, Proc. Natl. Acad. Sci. 112 (52) (2015) 15809–15814, https://doi.org/10.1073/pnas.1522496112.

[37] K. Wang, L. Yang, W. Zhao, L. Cao, Z. Sun, F. Zhang, A facile synthesis of copper nanoparticles supported on an ordered mesoporous polymer as an efficient and stable catalyst for solvent-free Sonogashira coupling reactions, Green Chem. 19 (2017) 1949–1957, https://doi.org/10.1039/C7GC00219J.

[38] N. Zohreh, S.H. Hosseini, A. Pourjavadi, C. Bennett, Immobilized copper(II) on nitrogen-rich polymer-entrapped Fe$_3$O$_4$ nanoparticles: a highly loaded and magnetically recoverable catalyst for aqueous click chemistry, Appl. Organomet. Chem. 30 (2) (2016) 73–80, https://doi.org/10.1002/aoc.3398.

[39] A.A. Athawale, P.P. Katre, M. Kumar, M.B. Majumdar, Synthesis of CTAB–IPA reduced copper nanoparticles, Mater. Chem. Phys. 91 (2–3) (2005) 507–512, https://doi.org/10.1016/j.matchemphys.2004.12.017.

[40] M. Valodkar, R.N. Jadeja, M.C. Thounaojam, R.V. Devkar, S. Thakore, Biocompatible synthesis of peptide capped copper nanoparticles and their biological effect on tumor cells, Mater. Chem. Phys. 128 (1–2) (2011) 83–89, https://doi.org/10.1016/j.matchemphys.2011.02.039.

[41] H.-J. Lee, J.Y. Song, B.S. Kim, Biological synthesis of copper nanoparticles using Magnolia kobus leaf extract and their antibacterial activity, Chem. Technol. Biotechnol. 88 (November 2013) 1971–1977, https://doi.org/10.1002/jctb.4052.

[42] M. Nilavukkarasi, S. Vijayakumar, S. Prathip Kumar, Biological synthesis and characterization of silver nanoparticles with Capparis zeylanica L. leaf extract for potent antimicrobial and anti proliferation efficiency, Mater. Sci. Energy Technol. 3 (2020) 371–376, https://doi.org/10.1016/j.mset.2020.02.008.

[43] R.C. Kasana, N.R. Panwar, R.K. Kaul, et al., Biosynthesis and effects of copper nanoparticles on plants, Environ. Chem. Lett. 15 (2017) 233–240, https://doi.org/10.1007/s10311-017-0615-5.

[44] Y.T. Prabhu, K. Venkateswara Rao, V. Sesha Sai, T. Pavani, A facile biosynthesis of copper nanoparticles: A micro-structural and antibacterial activity investigation, J. Saudi Chem. Soc. 21 (2) (2017) 180–185, https://doi.org/10.1016/j.jscs.2015.04.002.

[45] D.A. Jamdade, D. Rajpali, K.A. Joshi, R. Kitture, A.S. Kulkarni, V.S. Shinde, J. Bellare, K.R. Babiya, S. Ghosh, Gnidia glauca- and *Plumbago zeylanica*-mediated synthesis of novel copper nanoparticles as promising antidiabetic agents, Adv. Pharm. Sci. (2019), https://doi.org/10.1155/2019/9080279, 9080279.

[46] M.M. Elias, M. Miqdad, I.M. Mahbubul, R. Saidur, M. Kamalisarvestani, M.R. Sohel, A. Hepbasli, N.A. Rahim, M.A. Amalina, Effect of nanoparticle shape on the heat transfer and thermodynamic performance of a shell and tube heat exchanger, Int. Commun. Heat Mass Transfer 44 (2013) 93–99, https://doi.org/10.1016/j.icheatmasstransfer.2013.03.014.

[47] A.H. Ramin Ranjbarzadeh, M. Isfahani, M. Afrand, A. Karimipour, M. Hojaji, An experimental study on heat transfer and pressure drop of water/graphene oxide nanofluid in a copper tube under air cross-flow: applicable as a heat exchanger, Appl. Therm. Eng. 125 (2017) 69–79, https://doi.org/10.1016/j.applthermaleng.2017.06.110.

[48] R. Du, W. Li, T. Xiong, et al., Numerical investigation on the melting of nanoparticle-enhanced PCM in latent heat energy storage unit with spiral coil heat exchanger, Build. Simul. 12 (2019) 869–879, https://doi.org/10.1007/s12273-019-0527-3.

[49] G. Perricone, V. Matějka, M. Alemani, J. Wahlström, U. Olofsson, A test stand study on the volatile emissions of a passenger car brake assembly, Atmos. 10 (2019) 263, https://doi.org/10.3390/atmos10050263.

[50] H.W. Xian, N.A.C. Sidik, G. Najafi, Recent state of nanofluid in automobile cooling systems, J. Therm. Anal. Calorim. 135 (2019) 981–1008, https://doi.org/10.1007/s10973-018-7477-3.

[51] S. Bhaumik, M. Kamaraj, V. Paleu, Tribological analyses of a new optimized gearbox biodegradable lubricant blended with reduced graphene oxide nanoparticles, Proc. Inst. Mech. Eng. Pt J: J. Eng. Tribol. 235 (5) (2021) 901–915, https://doi.org/10.1177/1350650120925590.

[52] S. Jang, Y. Seo, J. Choi, T. Kim, J. Cho, S. Kim, D. Kim, Sintering of inkjet printed copper nanoparticles for flexible electronics, Scr. Mater. 62 (5) (2010) 258–261, https://doi.org/10.1016/j.scriptamat.2009.11.011.

[53] M.J. Kao, C.H. Lo, T.T. Tsung, Y.Y. Wu, C.S. Jwo, H.M. Lin, Copper-oxide brake nanofluid manufactured using arc-submerged nanoparticle synthesis system, J. Alloys Compd. 434–435 (2007) 672–674, https://doi.org/10.1016/j.jallcom.2006.08.305.

[54] G.H. Hong, J.H. Oh, D. Ji, S.W. Kang, Activated copper nanoparticles by 1-butyl-3-methyl imidazolium nitrate for CO_2 separation, Chem. Eng. J. 252 (2014) 263–266, https://doi.org/10.1016/j.cej.2014.05.012.

[55] Vijaya K Rangari, Ghouse M Mohammad, Shaik Jeelani, Angel Hundley, Komal Vig, Shree Ram Singh and Shreekumar Pillai, Synthesis of Ag/CNT hybrid nanoparticles and fabrication of their Nylon-6 polymer nanocomposite fibers for antimicrobial applications, Nanotechnology, 21, 9, https://doi.org/10.1088/0957-4484/21/9/095102.

[56] G. Wang, G. Ran, G. Wan, P. Yang, Z. Gao, S. Lin, F. Chuan, Y. Qin, Size-selective catalytic growth of nearly 100% pure carbon nanocoils with copper nanoparticles produced by atomic layer deposition, ACS Nano 8 (5) (2014) 5330–5338, https://doi.org/10.1021/nn501709h.

[57] R.V. Gonçalves, R. Wojcieszak, H. Wender, C.S.B. Dias, L.L.R. Vono, D. Eberhardt, S.R. Teixeira, L.M. Rossi, Easy access to metallic copper nanoparticles with high activity and stability for CO oxidation, ACS Appl. Mater. Interfaces 7 (15) (2015) 7987–7994, https://doi.org/10.1021/acsami.5b00129.

[58] P. Sharma, S. Pant, V. Dave, K. Tak, V. Sadhu, K.R. Reddy, Green synthesis and characterization of copper nanoparticles by Tinospora cardifolia to produce nature-friendly copper nano-coated fabric and their antimicrobial evaluation, J. Microbiol. Methods 160 (2019) 107–116, https://doi.org/10.1016/j.mimet.2019.03.007.

[59] Q. Xu, X. Ke, N. Ge, et al., Preparation of copper nanoparticles coated cotton fabrics with durable antibacterial properties, Fibers Polym. 19 (2018) 1004–1013, https://doi.org/10.1007/s12221-018-8067-5.

[60] B. Bashir, A. Rahman, H. Sabeeh, M.A. Khan, M.F. Aly Aboud, M.F. Warsi, I. Shakir, P.O. Agboola, M. Shahid, Copper substituted nickel ferrite nanoparticles anchored onto the graphene sheets as electrode materials for supercapacitors fabrication, Ceram. Int. 45 (6) (2019) 6759–6766, https://doi.org/10.1016/j.ceramint.2018.12.167.

[61] J. Lin, H. Zhang, H. Hong, et al., A thermally conductive composite with a silica gel matrix and carbon-encapsulated copper nanoparticles as filler, J. Elec. Mater. 43 (2014) 2759–2769, https://doi.org/10.1007/s11664-014-3159-5.

[62] D.E. Betsy Pugel, J.R. Rummel, C. Conley, Tiny houses: Planetary protection-focused materials selection for spaceflight hardware surfaces, in: 2016 IEEE Aerospace Conference, 2016, pp. 1–14, https://doi.org/10.1109/AERO.2016.7500727.

[63] See https://www.copper.org/publications/newsletters/innovations/2006/01/copper_nanotechnology.html, 05.05.2021..

[64] V. Beedasy, P.J. Smith, Printed electronics as prepared by inkjet printing, Materials 13 (2020) 704, https://doi.org/10.3390/ma13030704.

[65] M. Hejazy, M.K. Koohi, A.B.M. Pour, D. Najafi, Toxicity of manufactured copper nanoparticles - A review, Nanomedicine Res. J. 3 (1) (2018) 1–9, https://doi.org/10.22034/NMRJ.2018.01.001.

[66] See https://www.azonano.com/article.aspx?ArticleID=3271#:~:text=The%20properties%20of%20nanoparticles%2C%20for,%2C%20shape%2C%20and%20chemical%20environment.&text=Copper%20nanoparticles%20are%20graded%20as,very%20toxic%20to%20aquatic%20life. 05.05.2021.

CHAPTER 15

Nano-steels in the automotive industry

Mina Madadi, Mahdi Yeganeh, and Mostafa Eskandari
Faculty of Engineering, Department of Materials Science and Engineering, Shahid Chamran University of Ahvaz, Ahvaz, Iran

Chapter outline

1. Introduction to nanosteels and their strengthening mechanisms — 287
2. Grain refinement is a unique mechanism for improving strength and toughness — 289
   2.1 Mechanism of grain refinement — 289
3. Advanced high-strength steels (AHSS) — 293
   3.1 Sustainability — 293
   3.2 The effect of using advanced high-strength steels in the automotive industry — 295
4. Steels in automotive industry — 296
   4.1 First-generation advanced high-strength steels — 297
   4.2 Second-generation advanced high-strength steels — 299
   4.3 Third-generation steels in automotive industry — 300
5. Conclusion — 311
References — 311

1. Introduction to nanosteels and their strengthening mechanisms

Nanostructured materials include a group of materials that have at least one dimension less than 100 nm in size. These materials can contain thin layers that are less than 100 nm thick, while their lateral dimensions are not in this dimension range. Among these, we should mention nanocrystalline particles or bulk materials that have grains with a size of less than 100 nm. Of course, these materials are called nanocrystalline materials when their dimensions in all dimensions (crystalline grains or particles) are less than 100 nm. It goes without saying that the numbers reported will vary depending on the type of measurement tool used [1].

In 1990s, nanomaterials have become a focal point of materials science due to their unique physical, chemical, and mechanical properties that destine these materials to novel and promising applications. The small dimension of the grains or particles in nanomaterials and their specific processing methods affect their defect structure (vacancies, dislocations, disclinations, stacking, and twin faults as well as grain boundaries) that has a significant influence on the properties of these materials [1].

There is always competition between achieving high strength and a high and uniform deformation rate. While achieving both parameters in the manufacture of parts is of great importance. Nanocrystalline and ultrafine-grained metals exhibit high strength compared with their coarse-grained counterparts [2]. Conventional strengthening methods, such as grain refinement, solid solution, and dispersion strengthening, have been proved to sacrifice the plastic deformation capacity [3].

In principle, the strength of steel is contributed by a number of strengthening mechanisms as follows:
- intrinsic strength of ferritic iron,
- solid solution strengthening,
- precipitation strengthening,
- grain size strengthening, and
- dislocation strengthening.

The yield strength ($\Delta\sigma_y$) can then be expressed as

$$\sigma_y = \Delta\sigma_{Fe} + \Delta\sigma_{ss} + \Delta\sigma_{gs} + \Delta\sigma_{dis} + \Delta\sigma_{ppt} \qquad (1)$$

In Eq. (1), the strength has been expressed as a simple linear sum of individual strengthening mechanisms and has been applied in many cases extracting the different strengthening mechanisms in steels with bainitic microstructures [4].

In Eq. (1), the share of five mechanisms is considered. Fig. 1 shows the contribution and limitation increase of each mechanism to the strength of a steel part to achieve a yield stress of 700 MPa. Of course, it goes without saying that achieving a strength higher than 1030 MPa is possible in high-strength, all-binary steels [5].

Steel has always been one of the most widely used metals in the industry. The high-strength and high-toughness steel is the most desirable product. However, in fact, strength and toughness are normally contrary. If the strength of steel increases its toughness normally decreases and vice versa. The composition and microstructure of the steel are primary factors controlling the strength and the toughness. Thus the grain refinement is a highly effective method to improve both the strength and the toughness [6–8].

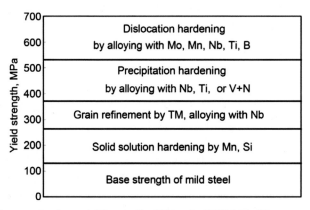

Fig. 1 Strengthening mechanisms yielding to the YS level of 700 MPa [5].

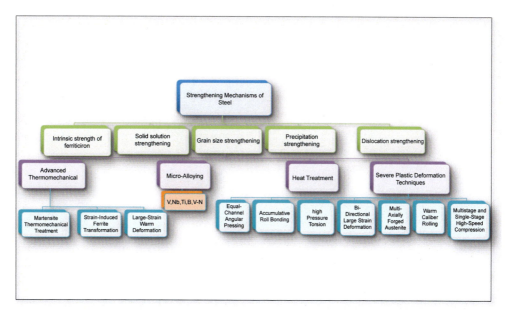

Fig. 2 Classification of strengthening mechanisms.

Therefore, it can be concluded that to have a good combination of strength and toughness, fine-graining is the best choice. Because there is the ability to simultaneously increase the strength and toughness in this method. Fig. 2 shows a schematic of fine-graining methods.

2. Grain refinement is a unique mechanism for improving strength and toughness

2.1 Mechanism of grain refinement

In general, the production processes of super granular metal materials (or nanostructures) are divided into two general categories: top-down and bottom-up. The first category includes methods for the production of nanomaterials by bonding atoms and molecules, such as electrical deposition, condensation of nanostructured powders and vapor deposition, and mechanical alloying. One of the limitations of this category is the production of small parts that are suitable for the electronics industry. Also, their production process causes the presence of contamination and porosity during the production process.

The second method will be through severe plastic strain by controlling the metallurgical process. To compare these two methods (Severe plastic deformation processes and Different thermomechanical processes), their second advantage includes (i) objects with larger dimensions, (ii) no contamination and porosity, and (iii) a wide range of alloys and

Table 1 Summary of major SPD processes [9].

Process name	Schematic representation	Equivalent plastic strain
Equal channel angular extrusion (ECAE) (by Segal in 1977)		$\varepsilon = n\frac{2}{\sqrt{3}}\cot(\varphi)$
High-pressure torsion (HPT) (by Valiev et al. in 1989)		$\varepsilon = \frac{\gamma(r)}{\sqrt{3}}, \gamma(r) = n\frac{2\pi r}{t}$
Accumulative roll-bonding (ARB) (by Saito, Tsuji, Utsunomiya, Sakai, in 1998)		$\varepsilon = n\frac{2}{\sqrt{3}}\ln\left(\frac{t_0}{t}\right)$

metals. The second category is divided into two main groups: Micro-Alloying and Heat Treatment.

In the following sections, two processes with different methods are examined (Table 1).

2.1.1 Severe plastic deformation (SPD) techniques

Severe plastic deformation (SPD) techniques impose large accumulated plastic strains at room or elevated temperatures, e.g., mainly in the temperature regime of warm deformation. Many different severe plastic deformation (SPD) processing techniques have been proposed, developed, and evaluated. These techniques include:
- Equal Channel Angular Pressing (ECAP),
- Accumulative Roll Bonding (ARB),
- High Pressure Torsion (HPT),
- Multi-Directional Forging (MDF),
- Twist Extrusion (TE),
- Constrained Groove Pressing and Rolling (CGP and CGR),

- Cyclic Channel Die Compression (CCDC),
- Torsion Extrusion and Torsion after Extrusion (TE and TAE),
- Hydrostatic Extrusion (HE),
- Repetitive Corrugation and Straightening (RCS),
- Cylinder Covered Compression (CCC),
- Friction Stir Processing (FSP), and
- Submerged Friction Stir Processing (SFSP)

All of these procedures are capable of introducing large plastic straining and significant microstructural refinement in bulk crystalline solids [4]. Some of these methods are ECAP, HPT, MDF, and ARB fully developed methods for the production of UFG bulk materials, which can be used to achieve grain sizes of 70–500 nm [10–12].

2.1.2 Thermomechanical

The production of nano-grain steel with sizes of 100 and 1000 nm, respectively, and with excellent mechanical properties has been one of the most important goals of researchers in the last decade [1–3]. In order to produce nano/ultrafine steels, the method of severe plastic deformation and thermomechanical process of martensite is used.

Thermomechanical processes in comparison with the methods of severe plastic deformation, advantages such as low cumulative strain, cohesion, control of thermal regime during the process, simple design of the strain path, and the ability to industrialize more are considered [4].

There is usually competition between the size of the grain that can be produced, the amount and size of materials produced, and the cost of the process. For construction applications that require metals with significant dimensions and quantities, the thermomechanical method, which involves deformation and annealing of bulk alloys, can be an optimal method of production.

Conventional methods of thermomechanical process of alloys result in a grain size between 250 and 300 μm, which requires strict control of the process parameters to achieve a finer grain size. The advanced thermomechanical process involves the modification of the conventional large-scale process by optimizing the deformation conditions that improve and profitability for solid-state processes such as recrystallization, deposition, and phase transformation [13–15].

In order to deform austenite, two methods, conventional and advanced thermomechanical process are used. The first involves a program of several deformations to change the piece, while the second also works to control the evolution of the microstructure.

The controlled thermomechanical process has been successfully integrated into modern industrial rolling mills. The process mentioned can also have a regular deformation schedule under cooling and controlled conditions in order to achieve fine-grained microstructures. This type of process can be divided into four stages in terms of

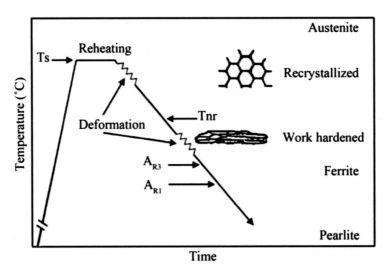

Fig. 3 Schematic of thermodynamic process [16].

temperature reduction. Fig. 3 is a schematic of thermomechanical operations for microalloy steel with display of critical temperatures (T_S, T_{nr}, A_{R3}, and A_{R1}).

In the first stage, the coarse grains were shredded due to immersion at high TS temperatures due to deformation and frequent recrystallization.

The second stage occurs when, due to the strain induced by the sediments, the recrystallization cannot take place completely in the time between passes, below the temperature without recrystallization (T_{nr}).

The grains at this stage are pancreatic in shape and a misplaced infrastructure is formed within the grains.

In the third stage, the deformation occurs in the phase of fuzzy transformation and in the temperature range between the beginning of ferrite formation (A_{R3}) and the end of ferrite transformation (A_{R1}). Ferrite nucleates in and on the grain boundaries. At this point, the austenite is constantly deformed and the transformed ferrites begin to strain.

In stage four, microstructures occur during cooling, depending on the cooling rate and the stage at which the deformation is repeated.

During the hot rolling process, several microstructural changes occur that ultimately have a significant effect on the austenite grain size and final properties of the steel. Fig. 2 shows the grain size changes of austenite during hot rolling, between passes, work hardening, dynamic (meta dynamic) and static recovery, and dynamic (meta dynamic) and static recrystallization are the most important phenomena that occur in the thermomechanical process of hot rolling. The changes that occur during the hot rolling process are called dynamic changes and the changes that occur after hot rolling or between passes are called static changes [16–20].

2.1.3 Production of ultrafine-grained steels by heat treatment

Grain refinement is always an effective way to simultaneously increase strength and toughness. Since reducing the grain size of primary austenite can reduce the grain size of austenite decomposition products, any method that can make these changes is useful. By performing controlled thermomechanical operations, one of these methods is to perform a series of heat treatments. In the Thermo Mechanical Controlled Processes (TMCP), recrystallization and non-recrystallization rolling are used to refine austenite grain, of which the minimum size or thickness is 10–20 µm in C-Mn steels and 5–10 µm in steels microalloyed with Ti and Nb [21]. Controlled austenitization is another effective way to obtain ultrafine-grained austenite in heat treatment steels, such as rapid cyclic transformation, reversion from tempered and cold rolled martensite. The former, which includes 2–4 cycles of rapid austenite–martensite transformation, was employed to obtain austenite grain of 1–10 µm. The latter including tempering and further heavily cold rolling was attempted to refine austenite grain size to 1–5 µm [22, 23].

2.1.4 Production of ultrafine-grained steels by micro-alloying

The presence and development of the use of alloying elements such as Nb, Ti, and V in microstructures in order to achieve a fine-grained structure is of great importance (UFG-steels During the solidification process, it will be a good opportunity for the alloying elements to form carbide and nitride deposits by combining with nitrite and carbon. These fine precipitates play an effective role by retarding recrystallization (and therefore, increasing the recrystallization stop temperature) that usually follows deformation and thus, helps to retain the accumulated strain and deformed structures of austenite grains. Table 2 lists some of the important effects of these alloying elements. There have been several studies on the recrystallization of carbon steels and Nb microalloyed steel [24–27], and various models have been proposed for predicting the recrystallization of austenite. The resulting decrease in the grain boundary energy reduces their potential for the nucleation of non-martensitic products as was observed in boron steels. It is well known that Nb, V, and Ti microalloying elements play important roles in it [24–27].

3. Advanced high-strength steels (AHSS)

3.1 Sustainability

The global automotive industry must comply with government regulations, customer needs, and corporate development to stay ahead of international competition. Environmental, government, and customer demands to reduce fuel consumption, reduce human losses, and reduce the cost of the product can be met by choosing a lightweight material. Hence, the first two spectra are low-density engineering (aluminum, magnesium, titanium, and composite alloys) and the second spectrum is high-strength advanced steels. Among these two ranges of materials, advanced high-strength steels due to their higher

Table 2 Role of micro-alloy elements [24–27].

Alloy elements	Role of micro-alloy elements
V	1. Increases hardenability and mechanical properties after tempering of HSLA steels 2. Increase of hardenability at low austenite grain size
Nb	1. Precipitates NbC with pin dislocations can be intensively retarded that recovery and recrystallization of deformed austenite 2. Niobium has a threefold influence on the mechanical properties of steel which are as grain size refinement during thermomechanical hot forming, precipitation hardening, and lowering the γ to α transition temperature 3. Solution of niobium in austenite are retarding of austenite (γ) to ferrite (α) transformation
Ti	1. Low price of titanium compared to V and Nb 2. The main forms of titanium precipitate are TiN, Ti4C2S2, and TiC or Ti(C, N) composite particles 3. Control of austenite grain size through TiN particles
Nb-Ti	1. The use of niobium and titanium as microalloy elements in steel for grain refinement is widespread 2. TiN precipitates, forming during solidification, can powerfully inhibit motion of austenite grain boundaries 3. NbC, which usually precipitates after deformation of austenite, can pin dislocations so that recovery and recrystallization of deformed austenite can be intensively retarded

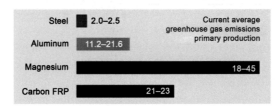

Fig. 4 Average greenhouse gas emissions during material production (in kg CO_2e9/kg) [29].

mechanical properties, lower production costs and lower raw materials, and easier production and processing, have attracted the attention of the automotive industry. They seek to replace as much as possible ordinary steels with high-strength steels [28].

A good way to compare these materials and their impact on the atmosphere is to look at the relative GHG emissions produced during the material production phase. Fig. 4 shows the amount of CO_2 gases emitted during production of 1 kg of each material. Steel is the cleanest material compared to each of the others by at least a factor of 4–5. It is argued however that steel does not achieve the same amount of mass reduction as the other materials and therefore these figures are unbalanced. Fig. 5 takes a more balanced look at material

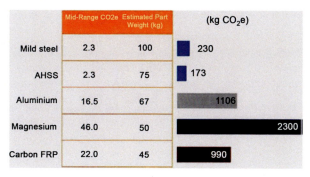

Fig. 5 Material production greenhouse gas emissions normalized by average mass reduction potential [30].

production emissions by accounting for the average amount of mass reduction for each. The numbers are reduced for all materials relative to a mild steel baseline with AHSS still showing a significant advantage over all other body materials [29, 30].

3.2 The effect of using advanced high-strength steels in the automotive industry

Average of 4.9% fuel economy improvement could be achieved for every 10% reduction in total vehicle weight but a 3%–4.5% increase in the safety risk could be caused with a decrease of 100 kg in the weight of a car. It is possible to save 25% of the weight and 14% of the cost when conventional steels in a four-door car body are replaced by AHSSs. In this section, new-generation steels are examined in terms of their history, variety, microstructure, characteristics, and applications.

According to the 2018 survey by Steel Market Development Institute (SMDI), 75% of consumers regard safety as the most crucial factor in their decision to purchase or lease a vehicle. However, just because consumers prioritize safety, that does not mean they want extra bulkiness added to their car. As automakers strive to build sturdy yet lightweight vehicles, steel companies are trying to match the automakers' such demands. AHSS is increasingly getting recognized as the material to satisfy such consumer demands. AHSS is stronger than the conventional automotive steel sheet. With AHSS, thinner steel sheet can be used compared to the conventional steel sheet, which enables automakers to build vehicles that are both strong and lightweight. AHSS is usually applied as structural steel like side sills, which are typically installed at the bottom of the vehicle's door to minimize external impact.

The use of AHSS steels has always improved the quality of cars. The following are some of the positive changes in the use of AHSS steels:

2019 Ram 1500, the winner of the 2019 North American Truck of the Year, includes 54% AHSS in the truck bed and cab, and 98% in the frame. AHSS eliminated 45 kg (approx.100 pounds) while gaining stiffness and durability of the frame, making the new 2019 Ram the strongest model thus far.

As for the 2020 Toyota Corolla, the increased use of AHSS enhanced body rigidity and occupant safety. The torsional rigidity is improved by 60% than its predecessor. 2019 Subaru Forester applied 31% AHSS for long-lasting quality, enhanced crash safety, and a quieter interior.

To achieve lightweight in the body, 2019 Jeep Cherokee used 65% hot stamped, high-strength steels and AHSS. In addition, thanks in part to the AHSS used in the body, the 2019 Mercedes-Benz G-class lost 170 kg (approx. 375 pounds). Steel aided in rigidity and load-bearing functions of the vehicle.

More than 65 AHSS-intensive vehicles debuted at major auto shows across North America in 2018 and in the NAIAS 2019, achieving lightweight in the body, higher tensile strength, and durability [31].

4. Steels in automotive industry

These steels combine an increased formability with a high-strength level at a wide temperature and strain rate spectrum. Advanced high-strength steels, among them especially dual phase and TRIP steels, feature promising results in this field, while their extraordinary mechanical properties can be tailored and adjusted by alloying and processing. Different phases with a wide variety of strength levels together with an extraordinary fine microstructure contribute to the excellent cold formability and crash behavior of AHSS [32].

In the following and in Fig. 6, we can see the course of changes of steels made in terms of comparison in the range of strength and percentage of deformation. As of today, automotive steels are classified into five main groups:
- Mild Steels (abbreviated as MS, not to be confused with Martensitic Steels).
- (Conventional) High-Strength Steels (abbreviated as HSS).

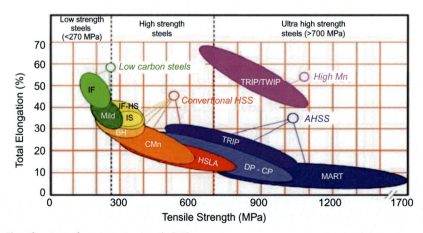

Fig. 6 Classification of automotive steels [33].

Table 3 Conventional high-strength steels, advanced high-strength steels, and high manganese steels [32].

HSS	AHSS	HMS
BH = Bake hardening IF-HS = High-strength IF P = Rephosphorized IS = Isotropic CMn = Carbon-manganese HSLA = High-strength low alloy	DP = Dual phase TRIP = Transformation-induced plasticity CP = Complex phase PM = Partly martensitic	HMS-TRIP = High Mn-transformation-induced plasticity HMS-TWIP = High Mn-twinning-induced plasticity

- First-generation Advanced High-Strength Steels (abbreviated as AHSS).
- Second-generation Advanced High-Strength Steels.
- Third-generation Advanced High-Strength Steels [33].

AHSS are multiphase steels that contain phases like martensite, bainite, and retained austenite in quantities sufficient to produce unique mechanical properties. Compared to conventional high-strength steels, AHSS exhibit higher strength values or a superior combination of high strength with good formability. Recently, a new group of austenitic steels with high manganese contents have been developed for automotive use. These high manganese steels (HMS) combine excellent mechanical properties with an alloying concept which is less costly than conventional or new high-strength austenitic stainless steels. This group is divided into transformation-induced plasticity steels (HMS-TRIP) and twinning-induced plasticity steels (HMS-TWIP) due to the characteristic phenomena occurring during plastic deformation. The different steel grades for car body use can be characterized by their microstructure or their alloying concept as shown in Tables 3 and 4. Typical mechanical property ranges of these different steels are indicated in Fig. 1. It is obvious that AHSS offer higher strength values than HSS [2, 3].

4.1 First-generation advanced high-strength steels

Advanced high-strength steels of the first generation do not have a higher percentage of alloying elements than high-strength low-alloy steels, but the structure of these steels is engineered by special thermomechanical processes and generally has a combination of ferrite, bainite, martensite, and austenite structures. From this generation of steels, we can mention two-phase steels, ferritic-bainite, transformable ductile steels, complex phase steels, and martensitic steels. The final strength and ultimate elongation of this generation of steels are 1500–1600 MPa and 5%–30%, respectively [24]. Here are some of the most important and widely used first-generation steels explained.

Table 4 Steels, microstructure, and characteristic features of steel grades for car body use [32, 33].

Steels			Microstructure	Characteristic features
Mild steels			α	– LC: unalloyed Al-killed low carbon steels; extra deep-drawing grades – IF: interstitial free steels; microalloyed extra deep-drawing grades
Conventional High-Strength Steels (HSS)			α	– BH: bake hardening steel grades, which show additional strengthening during paint bake treatment by controlled C aging – IF-HS: high-strength interstitial free steels, strengthened by Mn and P addition – P: P alloyed high-strength steels – IS: steels with medium yield strength and isotropic flow behavior, microalloyed with Ti or Nb – CMn: high-strength steels with increased C, Mn, and Si contents for solid solution strengthening – HSLA: high-strength low alloy steels, strengthened by microalloying with Nb or Ti
Advanced High-Strength Steels	First generation		$\alpha + \alpha'$ $\alpha + \alpha B + \gamma R$ $\alpha' + \alpha$ $\alpha + \alpha B + \acute{\alpha}$	– DP: dual phase steels with a microstructure of ferrite and 5–30 vol.% martensite islands – TRIP: transformation-induced plasticity steels with a microstructure of ferrite, bainite, and retained austenite – PM: partly or fully martensitic steels – CP: complex phase steels with a mixture of strengthened ferrite, bainite, and martensite
	Second generation		γ or high fractions of γ	– HMS-TRIP: steels with an alloying concept that strain-induced $\gamma \in \alpha'$ transformation occurs – HMS-TWIP: steels with an alloying concept that mechanical twinning occurs during straining
	Third generation		$\gamma + M$	– Q&P steels: Quenching and Partitioning (QP or Q&P) steels – TBF steels: TRIP-aided Bainitic Ferrite (TBF) steels – NanoSteel (NS) – MMn Steels: medium-Mn steels

4.1.1 Dual phase steels

Dual phase (DP) steels are widely used in automotive industry due to a good combination of continuous yielding behavior, high strength, high strain hardening rate, low yield stress-to-tensile strength ratio and good formability. DP steels have high tensile strength in the range of 500–1200 MPa and total elongation in the range of 12%–34%, which depend on fractions of ferrite, martensite, and bainite. Traditional microstructures of DP steels consist of polygonal ferrite and martensite. To satisfy custom requirements, ferrite-bainite-martensite and ferrite-bainite steels were produced in order to modify mechanical properties: bainite instead of martensite were shown to improve formability with a little decrease of strength.

The effect of martensite fraction, distribution, and martensite region size, and the effect of ferrite fraction and grain size on mechanical behavior of DP steels have been intensively studied. With increasing the martensite fraction, the yield strength and ultimate tensile strength increase while uniform and total elongations decrease. The distribution of martensite also affects the mechanical behavior. Martensite regions existing as isolated areas within ferrite matrix result in a better combination of strength and ductility than martensite regions forming a chain-like network structure surrounding ferrite. Refinement of ferrite or/and martensite regions simultaneously enhances strength and ductility. Ultrafine-grained DP steels with the average ferrite grain size of ~1.2 μm exhibit a high ultimate tensile strength up to 1000 MPa [34]. The 2015 Ford Edge also is a great example of efficient design with various steel grades for optimizing both performance and mass. The material distribution shown in Fig. 7 is made up of 50% AHSS with about 20% high-strength steel. The AHSS grades used for the Edge are also mainly dual phase, martensitic, or hot stamped steels as shown [35].

4.2 Second-generation advanced high-strength steels

Advanced second-generation steels, unlike first-generation steels, have a high percentage of alloying elements and an austenitic structure at ambient temperature. In these steels, the addition of 15%–30% manganese has caused the defective energy in the arrangement of these steels in the range of 20–40 (mJ/m^2) and has led to a change in the formation

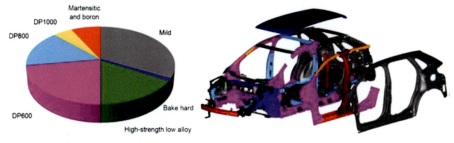

Fig. 7 2015 Ford edge body-in-white [35].

Table 5 Summarizing the use of steels in the automotive industry [36].

Steel grade	YS (MPa)	UTS (MPa)	Total EL (%)	n-value (5%–15%)	r-bar	Application code
Mild 140/270	140	270	38–44	0.23	1.8	A, C, F
BH 210/340	210	340	34–39	0.18	1.8	B
BH 260/370	260	370	29–34	0.13	1.6	B
IF 260/410	260	410	34–38	0.20	1.7	C
DP 280/600	280	600	30–34	0.21	1.0	B
IF 300/420	300	420	29–36	0.20	1.6	B
DP300/500	300	500	30–34	0.16	1.0	B
HSLA 350/450	350	450	23–27	0.22	1.0	A, B, S
DP 350/600	350	600	24–30	0.14	1.1	A, B, C, W, S
DP 400/700	400	700	19–25	0.14	1.0	A, B
TRIP 450/800	450	800	26–32	0.24	0.9	A, B
HSLA 490/600	490	600	21–26	0.13	1.0	W
DP 500/800	500	800	14–20	0.14	1.0	A, B, C, W
SF 570/640	570	640	20–24	0.08	1.0	S
CP 700/800	700	800	10–15	0.13	1.0	B
DP 700/1000	700	1000	12–17	0.09	0.9	B
Mart 950/1200	950	1200	5–7	0.07	0.9	A, B
MnB	1200	1600	4–5	n/a	n/a	S
Mart 1250/1520	1250	1520	4–6	0.07	0.9	A

Application Code: A—Ancillary parts, B—Body structure, C—Closures, F—Fuel Tank, S—Suspension/Chasis, W—Wheels.

mechanism of these steels from slip to twin and this choice of combination has created better properties than first-generation steels. The final tensile strength and ultimate elongation in advanced high-strength steels of the second generation are 900–160 MPa and 45%–70%, respectively. Among these groups, we can mention Twinning-Induced Plasticity (TWIP), Micro band Induced Plastisity (MIP), and Shear band Induced Plastisity (SIP) steels. It should be noted that due to the high percentage of manganese in these steels, the cost of these steels is higher than advanced steels of high strength of the first generation [24]. Some applications of steels in the automotive industry are summarized in Table 5 [36].

4.3 Third-generation steels in automotive industry

First-generation steels have been developed in limited combinations and have become a viable alternative to high-strength and high-strength low-alloy steels. As the strength increases, the thickness of the sheet used in the car body decreases and this reduces the weight of the car. The main problem of first-generation steels is the reduction of elongation with increasing strength, which reduces the ductility and limits the use of these steels in the automotive industry. Second-generation steels have a significant combination of strength and elongation due to the high amount of manganese to create stable austenite at room temperature, but still second-generation steels due to the high percentage of alloying elements with high cost and welding low ductility is the main reason for the lack of expansion of these steels in the automotive industry. Also, in the forming processes of second-generation steels, especially tube steels, there are delayed cracks, which cause problems in the production and assembly process. Due to the mentioned issues, advanced high-strength steels are a good choice for the automotive industry due to their high strength and ductility and reasonable cost. The third-generation advanced high-strength steels (AHSS 3. Gen.) have been developed in order to meet the requirements of the automotive industry for weight reduction, improved fuel efficiency, and CO_2 mitigation. The alloy design of AHSS 3. Gen., their microstructure characteristics, mechanical properties, potential applications as well as the need for thorough process parameter control are discussed in this paper. The cold-rolled AHSS 3. Gen. can be differentiated as quenching and partitioning (Q&P) steels, TRIP assisted bainitic-ferritic (TBF) steels, Medium-Mn steels (MMnS) and Nano Steel (NS). Quenching and partitioning (Q&P) steel and medium-Mn steel (MMnS) are the most promising candidates of the AHSS 3. Gen. Specific to these materials is the occurrence of different phases with typical sizes on the nm scale, the elemental partitioning between these phases and the impact of the metastable austenite phase on the mechanical properties by means of transformation-induced-plasticity (TRIP) as well as of twinning-induced-plasticity (TWIP) effects, (TBF) steels and Nano Steel (NS) [37].

4.3.1 Q&P steels

The Q&P processing was originally proposed by Speer et al. [38] as a new approach to produce steel microstructures consisting of a martensitic matrix containing considerable amounts of retained austenite [38]. Since first proposed in 2003, Q&P steel has gained interest for its potential to enhance properties of strength and ductility with compositions similar to transformation-induced plasticity (TRIP) steel and has been proposed as a third-generation automotive steel [39]. The considerable interest in steels processed by quenching and partitioning (Q&P) is due to their ability to achieve a good combination of strength and ductility with no or very limited alloying additions. In earlier studies, the Q&P processing of advanced high-strength steels (AHSS) has been shown to improve their mechanical properties considerably [40–45]. These steels contain carbon, manganese, silicon, nickel, and molybdenum alloying elements. Depending on the strength level, alloying elements can be as high as 4%, which is much lower than that of second-generation AHSS [33]. The purpose of Q&P steel in the context of automotive structures is to obtain a new type of ultrahigh-strength steel with good ductility to improve fuel economy while promoting passenger safety. With a final microstructure of ferrite (in the case of partial austenitization), martensite, and retained austenite, Q&P steel exhibits an excellent combination of strength and ductility, which permits its use in a new generation of advanced high-strength steels (AHSS) for automobiles [39].

4.3.2 Ultrafine-grained microstructure
The microstructure of commercial Q&P steels is composed primarily of martensite (50%–80%) formed during quenching, and ferrite (20%–40%) that under partial austenitization condition formed from the austenite phase during slow cooling, as well as dispersed retained austenite (5%–10%) stabilized by carbon enrichment during partitioning. Reduced fractions of ferrite can be used in higher strength products. Q&P process allows for creating a fine acicular microstructure of lath martensite interwoven with carbon enriched retained austenite, as displayed in Fig. 8 [39, 46].

4.3.3 Elemental partitioning phenomenon
The quenching and partitioning process consists of four steps which are shown in Fig. 9:
- Complete or partial austenitization.
- Quick quenching to temperature lower than the M_s (starting temperature of martensite) but higher than M_f (end temperature of martensite) to produce a controlled volume fraction of supersaturated martensite and unmodified austenite.
- Subsequent partitioning at quench temperature (single-stage operation) or higher than Ac3 (two-stage operation) for complete penetration of carbon from martensite to the remaining austenite. And allows the austenite to stabilize at room temperature.

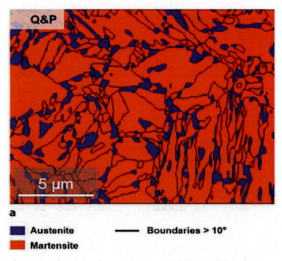

Fig. 8 Microstructure of a quenching and partitioning (Q&P) steel [39, 46].

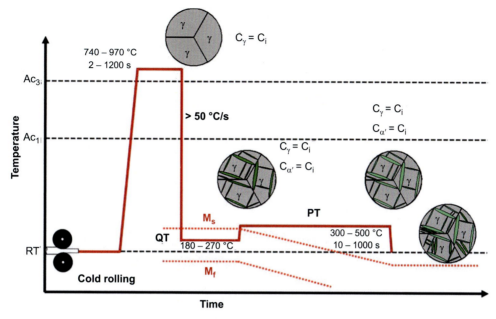

Fig. 9 Schematic illustration of heat treatment profile of quenching and partitioning process [46, 47].

- Quench at room temperature where austenite converts to martensite with less stability, while austenite with sufficient carbon content remains at room temperature [48–50].

Partitioning temperature and quenching temperature are the three most important factors in the amount of austenite remaining. As the partition temperature increases, the volume fraction of the remaining austenite also increases. At higher partitioning temperatures and close to the starting temperature of bainite, there is a possibility of bainite metamorphism and reduction of the residual austenite fraction. In other words, during partitioning operations at high temperatures, the remaining austenite can be decomposed into carbides, bainite and even perlite. The movement of the austenite/martensite interface reduces the volume fraction of austenite. At low partition temperatures, the effect of transfer carbide deposition also greatly reduces the maximum available austenite residue. In fact, in Q-P operations, the maximum residual austenite is obtained in short times and at high partition temperature [46]. This process of change can be seen in Fig. 10.

Regarding the effect of quenching time, it should be said that in this case there is a temperature for which we will have the maximum residual austenite. At temperatures above or below this temperature the volumetric residual austenite will decrease. The reason for this phenomenon can be found in the percentage of conversion of austenite to martensite [38, 46].

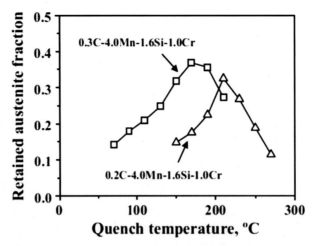

Fig. 10 Measured TQ-dependence of the volume fraction of retained austenite for two Q&P processed steels with different C contents [38].

At temperatures above the optimum quenching temperature after the initial quenching, significant unmodified austenite remains, and in the final quench to room temperature it is converted to martensite. At temperatures below the optimum quenching temperature by the initial quenching, too much austenite is converted to martensite. In fact, at optimal quench temperatures, martensite formation is prevented during the final quenching.

Fig. 11 displays the microstructure of the Q&P processed steel (Fe-0.2C-4.0Mn-1.6Si-1.0Cr). For TQ ¼ 170 °C (Fig. 11A and B), the microstructure consisted of a primary lath martensite matrix and retained austenite. The average grain size of the retained austenite and the block size of the primary martensite were measured by EBSD analysis to be approximately 200 and 490 nm, respectively (Fig. 11B). Carbides with a diameter in the range of 100–200 nm were occasionally observed in the primary martensite (Fig. 11A). For TQ ¼ 250 °C (Fig. 11C and D), the microstructure consisted of a primary lath martensite matrix, retained austenite, and secondary martensite.

4.3.4 Medium-Mn steels (MMnS)

Ultrafine-grained MMnS were first reported by Miller [52] in 1972. Since the 2010s, the MMnS have attracted intensive interest by materials scientists due to the excellent combination of high strength and good ductility of MMnS. A large amount of austenite can be stabilized by carbon and mainly manganese partitioning in MMnS [53].

The high percentages of residual austenite are obtained by enriching the elements carbon and manganese in the austenite phase. The austenite return process from the martensite phase is usually applied to medium (3%–8%) manganese steels. However, it can

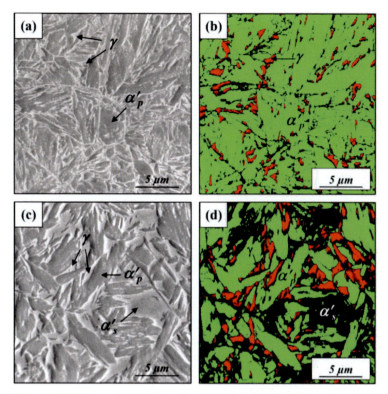

Fig. 11 (A) SEM and (B) EBSD analysis of the microstructure of Q&P processed steel (Fe-0.2C-4.0Mn-1.6Si-1.0Cr) obtained by quenching to 170 °C and partitioning treatment at 450 °C for 300 s (C) SEM and (D) EBSD analysis of the microstructure of Q&P processed steel (Fe-0.2C-4.0Mn-1.6Si-1°Cr) obtained by quenching to 250 °C and partitioning treatment at 450 °C for 300 s [51].

also be applied to low-manganese steels based on the specified targets for mechanical properties.

Medium manganese steels are considered as third-generation steels due to their excellent mechanical properties and low cost (compared to high manganese steels).

Hu et al. recently proposed a new strategy for designing medium-manganese steels with similar properties to TWIP steels. The mechanical properties of medium manganese steels are strongly dependent on the austenite phase properties such as quantity, size, and stability, so controlling the austenite phase properties of medium manganese steels is of particular importance. Initially, manganese and nickel elements were used for the austenite phase stability, but due to the high price of the nickel element, most attention has been paid to the austenite phase stability in steels with manganese (without high amounts of nickel).

In the process of returning austenite from the martensite phase, the return process infiltrates carbon and manganese elements during annealing to the grain boundaries and leads to the formation of austenite phase blades at the martensite phase boundary. The significant amount of manganese and carbon infiltration returned to the austenite stabilizes this phase due to the stabilizing properties of the carbon and manganese elements that stabilize the austenite phase [53].

Medium manganese steels (3%–8% manganese) with ductility due to transformation due to having the right combination of strength, ductility, and reasonable price are good candidates for advanced high-strength steels of the third generation, so much research in this field done. The presence of manganese in these steels causes lower cooling rates to create a martensitic structure, and in higher amounts (5% manganese) in thin sections with air cooling, a martensitic structure is created.

The heat treatment cycle of MMnS consists of austenitization and ART annealing [25]. Before ART annealing, the steel strip is heated above A3 temperature to achieve a fully austenitic microstructure. Subsequently, the austenite transforms to a thermal α'-martensite by quenching to room temperature. Alternatively, conventional rolling and ART-annealing processes might be employed to manufacture this steel grade. Hot rolling is conducted above A3 temperature in combination with an austenitization treatment. During the cold rolling process, some retained austenite from hot-rolled MMnS might transform into martensite. Afterward, the cold-rolled steel strip is reheated to the intercritical annealing region between Ac1 and Ac3, and is maintained at this elevated temperature for some time, followed by cooling to room temperature, as shown in Fig. 12 [46].

For MMnS, austenite/ferrite duplex microstructures with either lamella or equiaxed grain morphologies are obtained after ART annealing, as illustrated in Fig. 13A and B [12]. In general, the hot rolled MMnS inherits the morphology of martensite and possesses lamella morphology. In contrast, the cold-rolled MMnS shows equiaxed grain morphology because of the active recrystallization of martensite.

The microstructure design concepts of the AHSS 3. Gen. aim at obtaining a considerable amount of retained austenite (>20 vol%) in a martensitic/ferritic matrix. To achieve a sophisticated multiphase structure, complex thermal processing routes are employed. Q&P is a novel process to produce martensitic steel with a certain amount of retained austenite by controlling carbon partitioning [39, 47]. Austenite-reverted-transformation (ART) annealing brings new opportunities to produce an ultrafine-grained (UFG) duplex ferrite-austenite microstructure in MMnS [52]. In the latter case, the carbon and mainly manganese partitioning play an essential role in stabilizing austenite [53].

The optimum annealing temperature for the highest percentage of stable austenite phase depends on the chemical composition of the alloy. For cold rolled steel, the optimum temperature is estimated to be +20 [26]. Increasing the amount of austenite phase is necessary to obtain the best combination of yield stress, ultimate tensile strength and elongation. However, the stability of the austenite phase with increasing annealing

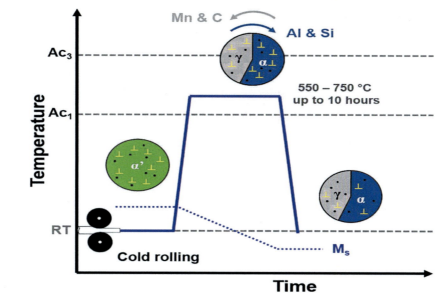

Fig. 12 Schematic illustration of austenite-reverted-transformation (ART) annealing profile of medium-Mn steel (MMnS) [46].

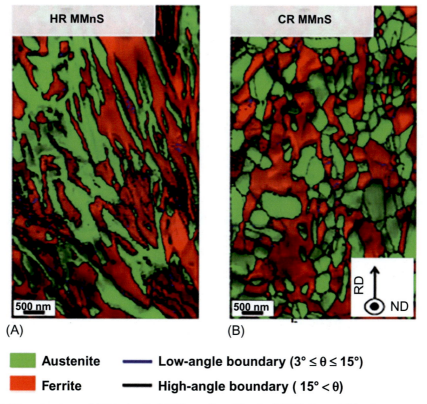

Fig. 13 Microstructure of (A) hot-rolled (HR) medium-Mn steel (MMnS), and (B) microstructure of a cold-rolled (CR) MMnS [11].

temperature, it decreases. It is possible to transform the austenite phase with low stability to martensite, during cooling to room temperature. Increasing the annealing temperature increases the amount of carbon and manganese elements that penetrate the austenite phase.

By increasing the annealing temperature under the same annealing time conditions, the yield stress and the ultimate tensile strength decrease and the ductility increases. Ductility decreases [27].

As the annealing time increases at a constant annealing temperature, the strength and ductility increase, which improves the mechanical properties by increasing the austenite phase percentage. It should be noted that based on the specified goal for the mechanical properties and the specified goals, the factors that improve the mechanical properties of medium manganese steels can be used alone or in combination of all three parameters (annealing temperature and time and chemical composition). For example, for industrial and mass production purposes, annealing for 6 h is not feasible and the chemical composition and heat treatment temperature should be considered as determinants of mechanical properties.

Improvement of mechanical properties of medium manganese steels by austenite return process from martensite phase has been done by strain partitioning in the remaining austenite phase with nanometer grain size. In these steels, both TRIP and TWIP effects occur during deformation and are transformed based on the strain percentage of the austenite structure. The design of these steels is such that the stability of the austenite phase is higher and by selecting the defective energy of the set in a certain range, both the trip and TWIP effects occur during deformation and the mechanical properties of the steel are improved. The nanometer austenite grain size determines the deformation mechanism of medium manganese steels. The large size of the nanometer austenite phase causes twinning and creates a complex and dense deformation structure inside the grain. The deformation created inside the grain leads to more stability of austenite and leads to the subsequent deformation of austenite. Smaller grains are less likely to have twins. Doing very little deformation in these small grains also leads to the formation of substructures, and this leads to martensitic transformation at low strains. As a result, smaller austenite grains are less stable than tripod transformation. This effect of austenite grain size in medium manganese steels with nanometer austenite grain size has led to increase in the fracture toughness and decrease in the strain localization by controlling the returned austenite grain size [24].

4.3.5 TRIP and TWIP effects

The typical tensile properties represented by the yield strength (YS), ultimate tensile strength (UTS) and total elongation (TEL) of the selected AHSS are listed in Table 3. It can be seen that the AHSS 3. Gen. offers an extraordinary combination of high strength and superior formability. Besides, the tensile properties of AHSS 3. Gen. cover a large spectrum, which is strongly dependent on the heat treatment. The activation of additional deformation mechanisms like TRIP and TWIP beside a dislocation glide provides a high strain-hardening capacity and promotes an excellent combination of high strength and good ductility. Fig. 6A shows the engineering stress—engineering strain curves of Q&P980 specimens. The evolution of austenite measured by in situ synchrotron X-ray diffraction during tensile tests is illustrated in Fig. 14B. The initial austenite fraction prior to the tensile test is around 0.10. During deformation, the austenite progressively transforms into α'-martensite and its fraction declines to 0.03–0.04 at the end of deformation. The TRIP effect contributes to the enhanced strain-hardening rate and improved ductility in martensitic steels. In MMnS, the TRIP effect also plays a crucial role in the enhancement of strain-hardening behavior and ductility. Besides, the TWIP and TRIP effects can occur progressively during the tensile tests, as illustrated in Fig. 15. During mechanical deformation, primary twins are generated and followed by secondary twins. The twin intersections are expected to be the nucleation sites for the following deformation-induced α'-martensite transformation. The succession of the TWIP and TRIP effects improves the strain hardening capacity of the steel. As it can be seen in Fig. 15B, the TWIP-TRIP MMnS exhibits an extraordinary combination of high strength and superior ductility [37].

4.3.6 TRIP-aided bainitic ferrite (TBF) steels

Since the 1970s, various high-, medium-, and low carbon transformation-induced plasticity (TRIP)-aided bainitic steels containing a high silicon content have been developed for the manufacture of various structural parts and components. These steels have a characteristic microstructure mainly consisting of carbide-free bainite (or bainitic ferrite) matrix and the metastable-retained austenite films, with a small amount of α'-martensite

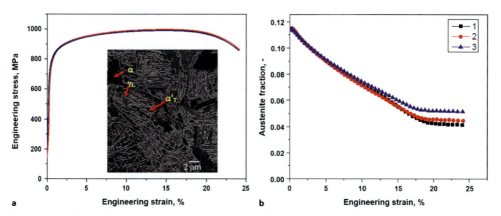

Fig. 14 Mechanical properties of a commercial quenching and partitioning (Q&P) steel: (A) Engineering stress–strain curve and (B) fraction of retained austenite as a function of strain. The test has been repeated with three samples, designated as *1*, *2*, and *3* in the figure [37].

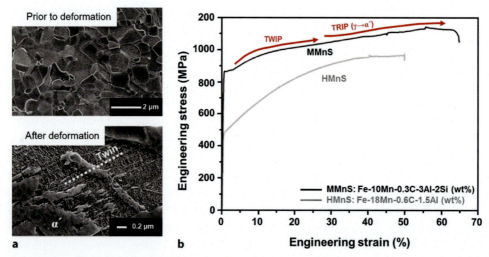

Fig. 15 Microstructure and mechanical behavior of a cold-rolled medium-Mn steel (MMnS): (A) microstructure prior to deformation (*top*) and after deformation (*bottom*) [37] and (B) engineering stress–strain curve [37].

and martensite–austenite constituents. TRIP-aided bainitic steels with a high-carbon content are known as very strong bainitic or nanostructured bainitic steels, and those with low- and medium-carbon contents are known as TRIP-aided bainitic ferrite (TBF) or quenching and partitioning steels. Because of their excellent ductility and formability, there are strong expectations that "lean alloy" TBF and quenching and partitioning steels will emerge as the third-generation advanced ultrahigh and high-strength steels (AHSS).

TBF steels can be produced by heat treatment using a multistep thermal processing route, austenitizing (or annealing in the γ regime) and then isothermal transforming (or austempering) at temperatures between the bainite-start (BS) and the martensite-start (MS) temperatures of the steels. Making TBF steels through heat treatment includes the following steps:
- Austenitizing.
- Isothermal transforming (or austempering) at temperatures between the bainite-start (BS) and the martensite-start (MS) temperatures of the steels.

In the final stage of cooling to room temperature, the microstructure will consist of carbon-rich austenite with a small amount of martensite. If the second stage of heat treatment of TBF steels is done at the temperature between the beginning and the end of martensite formation, part of the austenite will be converted to martensite and with the continuation of the isothermal process, the remaining austenite will be converted to benzite ferrite. If isothermal transformation is carried out at temperatures lower than the M_f temperature, most of the austenite first transforms to coarse α'-martensite containing a small amount of transition carbides. So in general it can be concluded that depending on the temperature of the isothermal process, the microstructure of these steels will change. The microstructure of TBF steels consists of a soft lath-structure matrix and a hard second phase (inter lath retained austenite films and blocky martensite–austenite constituents). Hence, deformation is mainly controlled by the following two factors, much like TRIP-aided steels with polygonal ferrite matrix.

(i) compressive long-range internal stress in the matrix which is resulted from untransformed retained austenite films and other secondary phases and

(ii) the strain-induced transformation of the retained austenite, which results in an increased martensite fraction and stress relaxation (or plastic relaxation).

The first factor contributes to large strain-hardening at an early stage and the resultant continuous yielding. On the other hand, the second factor brings on a relatively high strain-hardening rate in a large strain range and consequently suppresses the onset of diffuse necking, together with the first factor [54].

5. Conclusion

The third-generation advanced high-strength steels have been developed for an excellent balance of strength and formability. The upgrade of technologies of modern annealing lines with accurate online control of process parameters has stimulated new ideas for complex heat treatment schedules in order to create mixtures of different phases with ultrafine-grained morphologies and chemical gradients. The local enrichment of carbon and/or manganese in austenite contributes to the enhanced stability of austenite. Using the TRIP effect or/and the TWIP effect results in improved formability of the new steel concepts.

Automakers are always looking for ways to improve the structure of used steels. One of these methods is to reach the optimal grain size. For each of the DP, Q, P, and medium and high manganese steels and TBF steels, by controlling the process, it is possible to achieve nano grain size. Some of these steels are still in the laboratory stage and have not reached the production stage.

Several third-generation AHSS have been proposed and developed in the last few years, but only two classes currently are in series production through several steelmakers: Q&P and TBF steels. Nano steel has been only recently produced in coil scale.

References

[1] M. Kawasaki, Defect structure and properties of nanomaterials, In: J. Gubicza (Ed.), Microscopy and Microanalysis, Second and Extended Edition, Woodhead Publishing, Cambridge, UK, 2018. doi:10.1017/S1431927618000260.

[2] J. Ding, Z. Shang, J. Li, H. Wang, X. Zhang, Microstructure and tensile behavior of nanostructured gradient TWIP steel, Mater. Sci. Eng. A 785 (2020) 139346.

[3] X. Lu, J. Zhao, Z. Wang, B. Gan, J. Zhao, G. Kang, X. Zhang, Crystal plasticity finite element analysis of gradient nanostructured TWIP steel, Int. J. Plast. 130 (2020) 102703.

[4] H. Halfa, Recent trends in producing ultrafine grained steels, J. Miner. Mater. Charact. Eng. 2014 (2014).

[5] W. Morrison, Past and Future Development of HSLA Steels. HSLA Steels 2000, Xi'an, China, Beijing, 30 October-2 November 2000, 2000, pp. 11–19.
[6] Y. Tomita, K. Okabayashi, Effect of microstructure on strength and toughness of heat-treated low alloy structural steels, Metall. Trans. A. 17 (1986) 1203–1209.
[7] R. Lazarova, R.H. Petrov, V. Gaydarova, A. Davidkov, A. Alexeev, M. Manchev, Microstructure and mechanical properties of P265GH cast steel after modification with TiCN particles, Mater. Des. 32 (2011) 2734–2741.
[8] D. Rasouli, S. Khameneh Asl, A. Akbarzadeh, G. Daneshi, Optimization of mechanical properties of a micro alloyed steel, Mater. Des. 30 (2009) 2167–2172.
[9] A. Azushima, R. Kopp, A. Korhonen, D.Y. Yang, F. Micari, G.D. Lahoti, A. Yanagida, Severe plastic deformation (SPD) processes for metals, CIRP Ann. 57 (2) (2008) 716–735.
[10] R.Z. Valiev, T.G. Langdon, Principles of equal-channel angular pressing as a processing tool for grain refinement, Prog. Mater. Sci. 51 (7) (2006) 881–981.
[11] R.Z. Valiev, Y. Estrin, Z. Horita, T.G. Langdon, M.J. Zehetbauer, Y.T. Zhu, Producing bulk ultrafine-grained materials by severe plastic deformation, JOM 58 (4) (2006) 33–39.
[12] T.C. Lowe, R.Z. Valiev, The use of severe plastic deformation techniques in grain refinement, JOM 56 (10) (2004) 64–77.
[13] N. Niikura, M. Fujioka, A. Adachi, A. Matsukura, T. Yokota, Y. Shirota, New concepts for ultra refinement of grain size in super metal project, JMPT 2001 (117) (2001) 141–146.
[14] H. Beladi, G. Kelly, A. Shokouhi, P. Hodgson, The evolution of ultrafine ferrite formation through dynamic strain-induced transformation, Mater. Sci. Eng. A 371 (2004) 343–352.
[15] A. Abdollah-Zadeh, B. Eghbali, Mechanism of ferrite grain refinement during warm deformation of a low carbon Nb-microalloyed steel, Mater. Sci. Eng. A 457 (2007) 219–225.
[16] J. Jonas, S. Yue, Microstructural evolution during hot rolling, in: Proceedings of the International Symposium on Mathematical Modeling of Hot Rolling of Steel, Hamilton, 26–29 August 1990, 1990, pp. 99–118.
[17] B. Dutta, E.J. Palmiere, C.M. Sellars, Modelling the kinetics of strain induced precipitation in Nb microalloyed steels, Acta Mater. 49 (2001) 785–794.
[18] B. Dutta, E.J. Palmiere, Effect of prestrain and deformation temperature on the recrystallization behavior of steels microalloyed with niobium, Metall. Mater. Trans. A 34 (2003) 1237–1247.
[19] M. Alberto, H. Luiz, B. Oscar, Ultra grain refinement during the simulated thermo-mechanical processing of low carbon steel, J. Mater. Res. Technol. 1 (2012) 141–147.
[20] G. Speich, V. Demarest, R. Miller, Formation of austenite during intercritical annealing of dual-phase steels, Metall. Mater. Trans. A 12 (1981) 1419–1428.
[21] C. Sellars, J. Whiteman, Recrystallization and grain growth in hot rolling, Metal Sci. 13 (1979) 187–194.
[22] M. Tokizane, K. Ameyama, K. Takao, Ultra-fine austenite grain steel produced by thermomechanical processing, Scr. Metall. 22 (1988) 697–701.
[23] T. Furuhara, K. Kikumoto, H. Saito, T. Sekine, T. Ogawa, S. Morito, T. Maki, Phase transformation from fine-grained austenite, ISIJ Int 48 (2008) 1038–1045.
[24] J. Zhao, Z. Jiang, Thermomechanical processing of advanced high strength steels, Prog. Mater. Sci. 94 (January) (2018 May) 174–242.
[25] J. Sun, M. Wang, W. Hui, H. Dong, W. Cao, Enhanced work hardening behavior and mechanical properties in ultrafine-grained steels with large-fractioned metastable austenite, Scr. Mater. 63 (2010) 815–818.
[26] E. Emadoddin, A. Akbarzadeh, G.H. Daneshi, Effect of intercritical annealing on retained austenite characterization in textured TRIP-assisted steel sheet, Mater. Charact. 57 (4–5) (2006) 408–413.
[27] H. Luo, H. Dong, New ultrahigh-strength Mn-alloyed TRIP steels with improved formability manufactured by intercritical annealing, Mater. Sci. Eng. A 626 (2015) 207–212.
[28] M.Y. Demeri, Advanced High-Strength Steels: Science, Technology, and Applications, ASM International, 2013.
[29] World Steel Association, Steel in the Circular Economy, 2015.
[30] WorldAutoSteel Internal report, 2015.

[31] https://newsroom.posco.com/en/ahss-intensive-vehicles-make-major-debuts-at-the-2019-detroit-auto-show/.
[32] W. Bleck, K. Phiu-on, Grain refinement and mechanical properties in advanced high strength sheet steels, in: The Joint International Conference of HSLA Steels 2005 and ISUGS 2005, 2005, p. 50.
[33] E. Billur, Hot Stamping of Ultra High-Strength Steels. From a Technological and Business Perspective, Cham, 2019.
[34] Z.P. Xiong, A.G. Kostryzhev, N.E. Stanford, E.V. Pereloma, Microstructures and mechanical properties of dual phase steel produced by laboratory simulated strip casting, Mater. Des. 88 (2015) 537–549.
[35] J. Reed, Advanced High-Strength Steel Technologies in the 2015 Ford Edge, Great Designs in Steel, Livonia MI, 13 May, 2015.
[36] M.K. Singh, Application of steel in automotive industry, Int. J. Emerg. Technol. Adv. Eng. 6 (7) (2016) 246–253.
[37] W. Bleck, F. Brühl, Y. Ma, C. Sasse, Materials and processes for the third-generation advanced high-strength steels, BHM Berg-Huttenmann. Monatsh. 164 (11) (2019) 466–474.
[38] J.G. Speer, A.M. Streicher, D.K. Matlock, F. Rizzo, G. Krauss, Quenching and partitioning: a fundamentally new process to create high strength TRIP sheet microstructures, in: E.B. Damm, M.J. Merwin (Eds.), Austenite Formation and Decomposition, ISS/TMS, Warrendale, PA, 2003, pp. 505–522.
[39] J.G. Speer, D.K. Matlock, B.C. De Cooman, J.G. Schroth, Carbon partitioning into austenite after martensite transformation, Acta Mater. 51 (2003) 2611–2622.
[40] G. Gao, H. Zhang, X. Gui, P. Luo, Z. Tan, B. Bai, Enhanced ductility and toughness in an ultrahigh-strength Mn–Si–Cr–C steel: The great potential of ultrafine filmy retained austenite, Acta Mater. 76 (2014) 425–433.
[41] E.J. Seo, L. Cho, B.C. De Cooman, Application of quenching and partitioning (Q&P) processing to press hardening steel, Metall. Mater. Trans. A 45 (2014) 4022–4037.
[42] L. Cho, E.J. Seo, B.C. De Cooman, Near-Ac3 austenitized ultra-fine-grained quenching and partitioning (Q&P) steel, Scri. Mater. 123 (2016) 69–72.
[43] M.J. Santofimia, T. Nguyen-Minh, L. Zhao, R. Petrov, I. Sabirov, J. Sietsma, New low carbon Q&P steels containing film-like intercritical ferrite, Mater. Sci. Eng. A 527 (23) (2010) 6429–6439.
[44] E. Paravicini Bagliani, M.J. Santofimia, L. Zhao, J. Sietsma, E. Anelli, Mater. Sci. Eng. A 559 (2013) 486–495.
[45] E. De Moor, J.G. Speer, D.K. Matlock, C. Fojer, J. Penning, Effect of Si, Al and Mo alloying on tensile properties obtained by quenching and partitioning, Proc. MS&T (2009) 1554–1563.
[46] D.V. Edmonds, K. He, F.C. Rizzo, B.C. De Cooman, D.K. Matlock, J.G. Speer, Quenching and partitioning martensite—a novel steel heat treatment, Mater. Sci. Eng. A 438–440 (2006) 25–34.
[47] W. Bleck, X. Guo, Y. Ma, The TRIP effect and its application in cold formable sheet steels, Steel Res. Int. 88 (2017) 1700218.
[48] H. Liu, X. Jin, H. Dong, J. Shi, Martensitic microstructural transformations from the hot stamping, quenching and partitioning process, Mater. Charact. 62 (2011) 223–227.
[49] M.J. Santofimia, L. Zhao, R. Petrov, C. Kwakernaak, W.G. Sloof, J. Sietsma, Microstructural development during the quenching and partitioning process in a newly designed low-carbon steel, Acta Mater. 59 (2011) 6059–6068.
[50] L. Yang, L. Yu-peng, W. Chong, L. Shi-tong, C. Lu-bin, Phase stability of residual austenite in 60Si2Mn steels treated by quenching and partitioning, J. Iron. Steel Res. Int. 18 (2) (2011) 70–74.
[51] S. Morito, J. Nishikawa, T. Maki, Dislocation density within lath martensite in Fe-C and Fe-Ni alloys, ISIJ Int. 43 (2003) 1475–1477, https://doi.org/10.2355/isijinternational.43.1475.
[52] R.L. Miller, Ultrafine-grained microstructures and mechanical properties of alloy steels, Metall. Mater. Trans. B Process Metall. Mater. Process. Sci. 3 (1972) 905–912.
[53] S.-J. Lee, S. Lee, B.C. De Cooman, Mn partitioning during the intercritical annealing of ultrafine-grained 6% Mn transformation induced plasticity steel, Scr. Mater. 64 (2011) 649–652.
[54] K.I. Sugimoto, T. Hojo, J. Kobayashi, Critical assessment 29: TRIP-aided bainitic ferrite steels, Mater. Sci. Technol. 33 (17) (2017) 2005–2009.

SECTION C

Nanocoatings for automotive application

CHAPTER 16

Magnetic nanoparticles-based coatings

P. Poornima Vijayan[a], Archana Somadas Radhamany[a], Ansar Ereath Beeran[b], Maryam Jouyandeh[c], and Mohammad Reza Saeb[d]

[a]Department of Chemistry, Sree Narayana College for Women (affiliated to University of Kerala), Kollam, Kerala, India
[b]Department of Chemistry, MES Asmabi College, Thrissur, Kerala, India
[c]Center of Excellence in Electrochemistry, School of Chemistry, College of Science, University of Tehran, Tehran, Iran
[d]Department of Polymer Technology, Faculty of Chemistry, Gdańsk University of Technology, Gdańsk, Poland

Chapter outline

1. Introduction — 317
2. MNPs used to prevent corrosion in metal — 318
   2.1 MNPs as direct corrosion inhibitors — 319
   2.2 MNPs as active components in anticorrosion coating — 324
   2.3 Magnetic fluids for corrosion protection — 332
3. MNPs as antifouling component — 333
4. Smart coatings based on MNPs — 334
5. MNPs for electromagnetic absorbing coatings — 337
6. MNP coating for textiles — 338
7. Conclusions and future perspectives — 339
References — 340

1. Introduction

Magnetic nanoparticles (MNPs) have unique combination of superparamagnetic behavior along with other superior properties associated with nano-scale dimension [1, 2]. Mostly, nanostructured MNPs consist of a magnetic element coated at least with an organic entity. The magnetic part is mostly iron-oxide-based compounds and sometimes Mn, Co, Ni, Zn, Mg, and their oxides [3, 4]. It is possible to control the magnetic characteristics of MNPs to an extent by proper tuning of particle size, shape, composition, and structure. MNPs have emerged as promising nanomaterials for various analytical [5] and biomedical applications [6]. Apart from their popularity in these fields, MNPs have a crucial role in coating technology. Development of protective coatings with MNPs as active ingredients has been considered as one of the exciting applications of MNPs. Researchers have explored various protection strategies based on special features of MNPs. Overall, they have been used to protect metal and fabrics from corrosion, fouling, and radiation.

The magnetic nanoparticles have been used in their native form or as an active component in polymer nanocomposite for specific or integrated coating properties. The current chapter discusses such attempts in developing advanced protective coatings for metals and fabrics, which could have potential application in various fields like

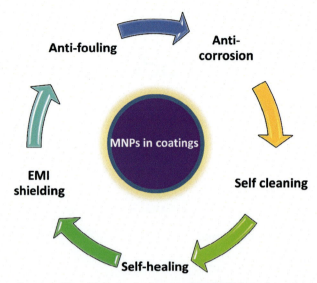

Fig. 1 The different functionality which can be imparted by MNP-based coatings.

marine, oil and gas, biomedical implants, defence, and many more. The MNP-based coating can offer multifunctionality depending on the nature of magnetic entity, nature of capping, size, and morphology of the particles. Fig. 1 shows those functionalities which MNPs can offer to a coating system for substrates like metal and fabrics.

Various fields where coatings with MNPs as functional components have significant role are schematically shown in Fig. 2.

To understand the reasons and molecular mechanisms behind the performance of MNP-based coatings, it is required to investigate and overview their role in different fields.

2. MNPs used to prevent corrosion in metal

Corrosion of metal structures is a serious concern in many industries including automobile, marine, civil, power plants, oil and gas, fertilizer, metallurgical industries, etc. [7]. Metallic corrosion causes huge economic loss and environmental contamination [8]. Among different approaches used to tackle corrosion, providing the metal surface with a protective coating is the first-line defence strategy. It is found that the MNPs could act as an active ingredient in such protective coatings [9]. Traditionally, corrosion inhibitors directly added to the electrolyte preferentially migrate to the anodic and/or cathodic sites to slow down the corrosion process. Herein, MNPs could be effectively applied rather than using toxic anticorrosion agents. The two cases, where MNPs could act as essential components to protect the metal structures from corrosion, are detailed in the following sections.

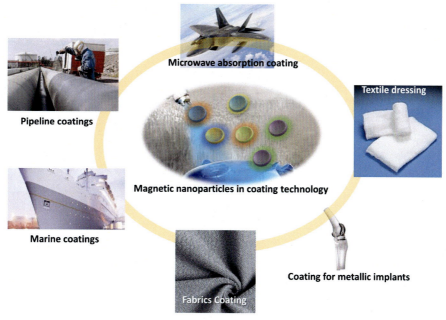

Fig. 2 The various applications of MNP-based coatings.

2.1 MNPs as direct corrosion inhibitors

MNPs found application as excellent corrosion inhibitors, which are incorporated into the electrolyte solution during pickling in order to reduce the degree of metal attack. Since they are highly active, MNPs get aggregate during the synthesis [10]. Proper functionalization provides monodispersed MNPs to facilities effective adhesion to the metal surface, which can result in a continuous and stable thin film over the metal substrate. Polymers are now used for the stabilization of corrosion inhibiting nanoparticles because they are cost-effective and stable corrosion inhibitors for metallic materials in acidic media. MNPs can be dispersed or chemically bonded to the polymer matrix to form a metal complex [11]. The role of polymers in such corrosion inhibitors is directly correlated to their microstructure, since they provide active centers of adsorption such as cyclic rings, hetero-atoms as oxygen and nitrogen. A major advantage of using suitably capped or stabilized MNPs is that they would eliminate the use of toxic corrosion inhibitors. Researchers have proposed several ideas to prepare MNPs as corrosion inhibitors such as magnetic nanogel and polymer-stabilized MNPs. The following sections discuss such strategies.

2.1.1 Magnetic nanogel hybrid as corrosion inhibitor

Iron-oxide-based MNPs are important class of corrosion inhibitors to protect metal from aggressive environments. Magnetite (Fe_3O_4) MNPs hybridized (coated) with cross-linked polymer nanogel are one of the widely studied corrosion inhibitor systems in acidic environments. Wherein, superparamagnetic Fe_3O_4 core is relatively low toxic, while inside encapsulated and the shell prevents the Fe_3O_4 core from oxidation and aggregation [12]. Researchers developed magnetic nanogel with controllable particle size below 200 nm via free radical polymerization technique, wherein the monomer and cross-linker adsorbed on the surface of the modified Fe_3O_4 particle through chelation between Fe_3O_4 and the oxygen and nitrogen atoms of monomers followed by polymerization onto the surface of the MNPs. Atta et al. [12] synthesized polyacrylamide (PAM) superparamagnetic Fe_3O_4 nanogels based on poly(2-acrylamido-2-methylpropane sulfonic acid) (PAMPS) and copolymers with acrylic acid (AA) or acrylamide (AM) via polymerization technique and obtained particles in the size range of 25–180 nm. They studied the corrosion inhibition ability of the PAM-based Fe_3O_4 nanogels to protect carbon steel in 1 M HCl using electrochemical impedance spectroscopy (EIS). The electrochemical parameters (Table 1) derived from EIS data were used to predict the mechanism of inhibitor action of nanogel wherein formation of self-assembled films over iron sheet by adsorption mechanism at carbon steel/hydrochloric acid solution interface was predicted.

Likewise, El-Mahdy et al. [14] developed cross-linked acrylamide-co-sodium acrylate (AM-co-AA-Na)/magnetite (Fe_3O_4) nanogel hybrid via chemical tailoring of magnetite.

Table 1 Electrochemical impedance spectroscopy parameters of carbon steel electrode in aqueous 1 M HCl solution in the absence and presence of various concentrations of the PAM-based MNP inhibitor nanogel films at 298 K [13].

Inhibitor name	Conc. ppm by weight	R_s (ohm cm^2)	R_{ct} (ohm cm^2)	C_{dl} ($\mu F\ cm^{-2}$)	η_1 (%)
PAMPS-Na magnetite nanogel	0	1.2	49	583.0	–
	50	2.8	101	558.0	50.7
	100	2.8	143	488.6	65.4
	150	2.2	163	212.6	69.9
	200	2.5	187	99.7	73.8
	250	1.5	271	74.0	81.9
AMPS/AA magnetite nanogel	50	1.8	99	221.4	50.5
	100	2.6	134	82.5	63.4
	150	6.9	190	40.0	74.2
	200	1.8	227	37.5	78.4
	250	3.7	483	26.5	89.9
AMPS/AM magnetite nanogel	50	2.2	163	488.6	69.9
	100	1.8	227	296.7	78.4
	150	1.5	271	221.4	81.9
	200	2.9	339	74.0	85.5
	250	9.1	553	45.4	91.1

The adduct (NMA) obtained by the reaction of polyoxyethylene 4-nonyl-2-propylene-phenol nonionic reactive surfactant with maleic anhydride (MA) formed as the core and magnetite as the shell at periphery hydrophilic carbonyl and ethoxy groups of NMA. Therefore, modified magnetite adsorbed the acrylamide (AM) and sodium acrylate monomer (AA-Na) monomers followed by polymerization onto the surface of the magnetic nanoparticles by a free radical polymerization in the presence of N-methylenebisacrylamide (MBA) cross-linker. The entire process involved in the generation of AM-co-AA-Na/Fe$_3$O$_4$ composites is shown in Fig. 3A. This special designing process resulted in AM-co-AA-Na/Fe$_3$O$_4$ composites with spherical polydisperse particles that have a diameter range of 5–35 nm (Fig. 3C). This molecular level tailoring of magnetite enabled them to act as a strong corrosion inhibitor at very low concentration for steel even in aggressive solution (1.0 M HCl solution). This was evident from potentiodynamic polarization curves for AM-co-AA-Na/Fe$_3$O$_4$ hybrid nanoparticles-coated steel in 1.0 M HCl (Fig. 4) with a shift in cathodic and anodic curves to lower current densities values due to the presence of magnetic hybrid nanoparticles.

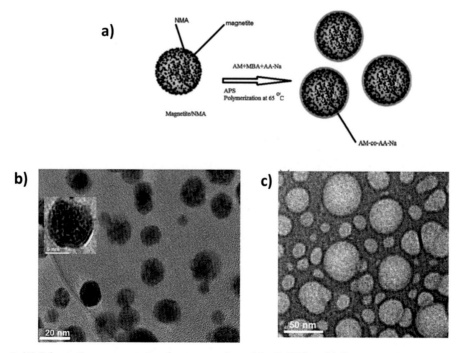

Fig. 3 (A) Schematic representation for preparation of Fe$_3$O$_4$/AM-co-AA-Na composite; transmission electron microscopy (TEM) images of (B) Fe$_3$O$_4$/NMA hybrid nanoparticles and (C) Fe$_3$O$_4$/AM-co-AA-Na nanocomposite [14].

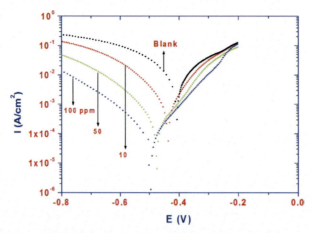

Fig. 4 Potentiodynamic polarization curves for steel in 1.0 M HCl without and with different concentrations of inhibitor [14].

2.1.2 Polymer-MNPs hybrids as corrosion inhibitors

Polymer-MNPs hybrids are used as corrosion inhibitors, where the polymer chains form complexes with metal ions and on the metal surface and these complexes occupy a large surface area thereby isolating the surface and protecting the metals from corrosive medium. MNPs hybridized (stabilized) with polyvinylpyrrolidone (PVP) having size of 5–20 nm were prepared with high inhibition efficiency for carbon steel corrosion in 1 M HCl solution. The electrochemical parameters derived from the EIS studies revealed the PVP stabilized MNPs adsorbed on the metal surface create a physical barrier to charge and mass transfer for metal dissolution. Recently, Fe_3O_4 magnetic nanoparticles hybridized with hyperbranched polyglycerol (HPG) (Fe_3O_4@SiO_2/HPG) (Fig. 5A) in

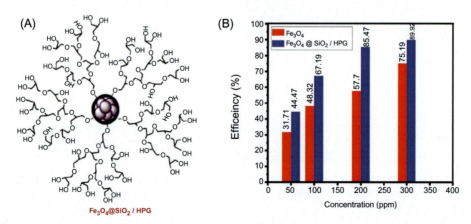

Fig. 5 (A) The structure of Fe_3O_4@SiO_2/HPG hybrid MNPs and (B) the relationship between inhibitor efficiency and inhibitors concentration [13].

the size range of 35–50 nm were found to effectively control the corrosion of mild steel in 1 M HCl [13]. The inhibitor efficiency was increased with inhibitor concentration and the maximum inhibitor efficiency was achieved at 300 ppm (Fig. 5B).

2.1.3 Green capping agent for MNP nanoparticle

As the researcher seeking green methods for developing corrosion inhibitors, the use of natural materials for stabilizing of MNPs got much attraction. Green MNP-based nanoparticle was developed by coating it with cheap, naturally occurring nontoxic biomaterials such as Myrrh gum [15] and rosin [16, 17]. Myrrh capped magnetite was developed with a diameter of 17–20 nm with a core-shell morphology (Fig. 6A). Magnetite/Myrrh nanohybrids inhibit both anodic and cathodic reactions by adsorption on the steel surface in 1 M HCl solution (Fig. 6B). Further, amidoxime obtained from rosin gum has been used as a capping agent to synthesis highly dispersed magnetite nanoparticles in the size range of 28–42 nm (Fig. 6C). While using magnetite-rosin ketone/amidoxime nanoparticles in 1 M HCl solution as corrosion inhibitor, a protective film gets adsorbed onto the steel substrate hinders the dissolution of steel in anodic sites (Fig. 6D).

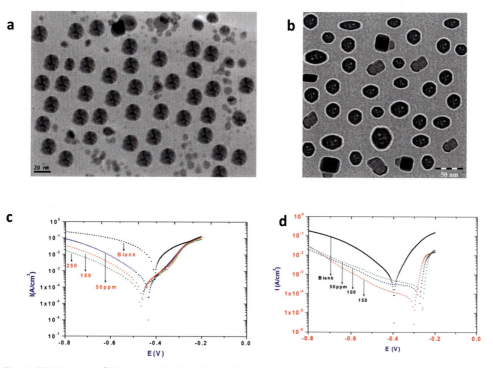

Fig. 6 TEM images of (A) magnetite/Myrrh and (B) magnetite-rosin ketone/amidoxime nanoparticles; Tafel polarization curves for steel in 1 M HCl solution containing different concentration of (C) magnetite/Myrrh nanohybrids (D) magnetite-RK/amidoxime hybrid nanoparticles [15, 17].

2.2 MNPs as active components in anticorrosion coating

2.2.1 MNP-metal hybrid thin films

Among different polymorphs of iron oxide, hematite (α-Fe_2O_3) has more stable anticorrosive properties and hence it has found active component in many corrosion protection coatings. Chandrappa et al. [18] made use of hematite (α-Fe_2O_3) nanoparticles for the fabrication of composite thin films with metal to make protective coating over mild steel. The α-Fe_2O_3 nanoparticles generated by electrochemical-thermal method were used to form composite thin film of Zn-Co-α-Fe_2O_3 over mild steel. An interesting observation was that the Zn-Co-α-Fe_2O_3 composite film formed with more uniform crystals with fine-grained structure than Zn-Co coatings (Figs. 5B and 7A). The figures showed that the irregular hexagonally shaped particles in Zn-Co film get converted to a compact surface upon the Zn-Co-α-Fe_2O_3 thin film. The corrosion performance of Zn-Co and Zn-Co-α-Fe_2O_3 thin films were evaluated using electrochemical impedance spectroscopic (EIS) studies. Corrosion resistance of Zn-Co-α-Fe_2O_3 composite coated (0.25–1.0 g/L) samples was higher when compared to bare Zn-Co film.

2.2.2 MNPs as nanofillers in anticorrosive polymer coating

Usually, polymer coatings are used to protect the metal substrate from corrosion. Epoxy [19, 20], acrylic [21], polyurethane [22], etc. are the most common polymeric resins used as major components in protective coating paints. These polymer-based coatings provide outstanding toughness, adhesion to metal substrates, and durability. However, the barrier properties of these coating have certain limitations like defects over coating surface, due to high cross-link density of epoxy network. As a remedial measure, nanofillers and modifiers are extensively used to improve the barrier performance of polymeric coatings [23, 24]. A well-established chemical interaction between the MNPs and the polymer matrix has been proved the generation effective protective coating for aggressive corrosion environment.

Nowadays, MNPs are used either as individual fillers or as hybrids in polymeric resin coatings [25]. Apart from imparting anticorrosion properties, these MNPs are used to tailor smart functionality to the coating. MNPs are capable to enhance hardness, impact resistance, adhesion strength, contact angle, etc. [26, 27]. Several mechanisms were proposed behind the anticorrosion action of MNPs which depends on the functionalization/capping of MNPs and the nature of magnetic parts in MNPs [28].

Various ferrite nanoparticles including nickel ferrite ($NiFe_2O_4$) [29], cobalt ferrite ($CoFe_2O_4$) [30], $ZnFe_2O_4$ [31], $NiLaFeO_4$ [32] are found as nontoxic anticorrosive agents suitable to incorporate into various polymeric resin coatings. It was widely accepted that the properly dispersed MNPs in polymeric coatings effectively fill the cavities in polymer coating, hence block the electrolyte pathways. Moreover, the release of inhibitive species like Co^{2+} and Ni^{2+} cations in the scratched area protects the metal surface by

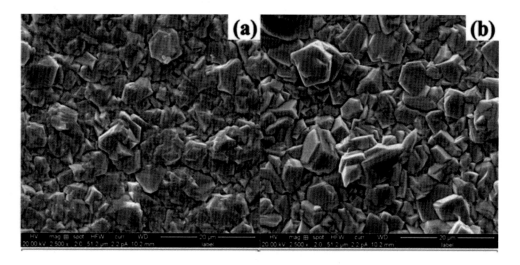
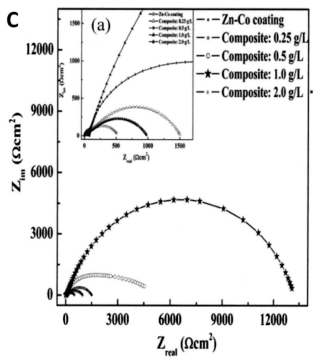

Fig. 7 Scanning electron microscope (SEM) images of (A) Zn-Co and (B) Zn-Co-α-Fe$_2$O$_3$ coated sample (composite: 1.0 g/L) and (C) impedance spectra of Zn-Co and Zn-Co-a-Fe$_2$O$_3$ coatings in 3.5 wt% NaCl solution [18].

forming a passive layer [25, 29]. Moreover, recent studies showed that some MNPs have the ability to accelerate the cure reaction of thermosetting resin which would be a major advantage during the coating process. Gharagozlou et al. [25] ensured proper dispersion of cobalt ferrite nanopigment in epoxy matrix by prior dispersion of the nanoparticles in a silica matrix ($NiFe_2O_4$-SiO_2) through which effective anticorrosion was enabled.

Zhan et al. [27] decorated graphene oxide sheets with Fe_3O_4 spherical particles followed by surface functionalization to generate a hybrid filler which imparts synergic effect on anticorrosion ability of epoxy coating (Fig. 8I). The bio-inspired surface functionalization has been done by self-polymerization between dopamine and secondary functional monomer (KH550). Moreover, the surface-functionalized hybrid filler effectively enhances the hydrophobicity of the epoxy coating surface. The passage of H_2O and O_2 can be effectively prevented and hence corrosion by epoxy coating modified by both GO-Fe_3O_4 hybrid and surface functionalized GO-Fe_3O_4 (Fig. 8II). The surface functionalization facilities uniform dispersion of the hybrid filler, which resulted in superior anticorrosion performance.

Recently, AA3105 aluminum has been effectively protected by dispersing $CoFe_2O_4$-nanopowder along with Na-montmorillonite (Na-MMT) into epoxy coating [30]. Soltani et al. prepared 70/30 wt% 80/20 wt% and 90/10 wt% of NaMMT/$CoFe_2O_4$ by doping method. From the EIS studies, it was found that corrosion protection properties of the 70/30 wt% of NaMMT/$CoFe_2O_4$ (EC80) coating have not significantly changed after 22 days. After 36 days of immersion, the EC80 coating still has the maximum resistance among all the other coating formulations (Fig. 9). A combined action of both Na-MMT and $CoFe_2O_4$ nanoparticles in forming protective layer on the AA3105 aluminum surface and on avoiding defects and holes on the coating surface contribute to the excellent corrosion behavior.

Network formation in polymer/MNPs hybrids for controlling anticorrosion property

Study on cure reaction of resins and their composites is the key to perceive the relation between cross-linking reaction and the final properties and the performance of the system [33–35]. This subject is more fundamental for thermoset polymers due to the complex nature and interaction between the thermoset chains and additional elements such as fillers, pigments, and nanoparticles [36–38]. *Cure Index* is a new dimensionless criterion that can simply classify the thermoset resins reinforced with fillers in terms of cure state from a qualitative point of view [39, 40]. *Poor, Good,* and *Excellent* cure are three labels of cure reaction that can be dedicated to the coatings by the use of the *Cure Index* [41]. The *Cure Index* is a fast-detecting criterion that can classify the cross-linking ability of nanocomposites with respect to the neat resin. It can be calculated as *Cure Index* $= \Delta T^* \times \Delta H^*$. The dimensionless cure heat release is defined as $\Delta H^* = \Delta H_{Comp}/\Delta H_{Ref}$, such that ΔH_{Comp} and ΔH_{Ref} are the enthalpies of the cure

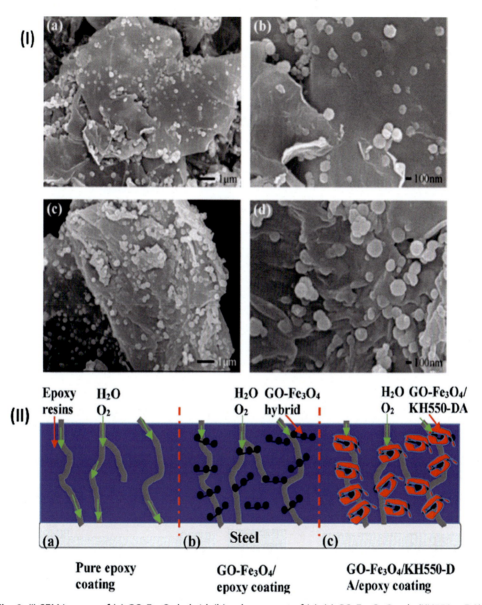

Fig. 8 (I) SEM images of (a) GO-Fe_3O_4 hybrid, (b) enlargement of (a), (c) GO-Fe_3O_4@poly (KH550 + DA) hybrid, and (d) enlargement of (c) (II) Anticorrosion mechanism of (a) pure epoxy coating, (b) GO-Fe_3O_4/epoxy coating, and (c) GO-Fe_3O_4@poly (KH550 + DA) hybrid/epoxy coating [27].

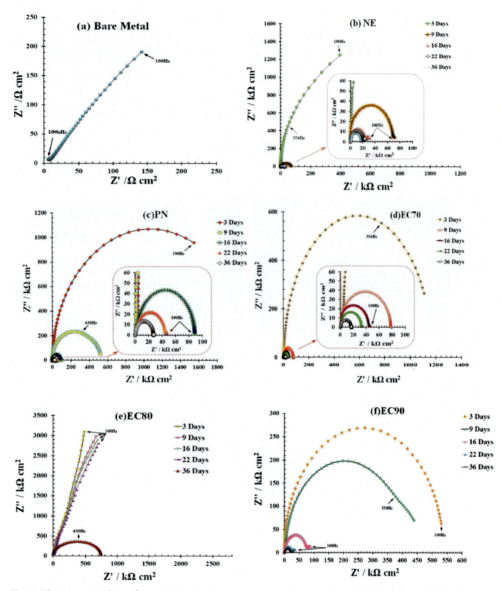

Fig. 9 The Nyquist plots of nanocomposite coatings containing various combinations of Na-MMT and $CoFe_2O_4$ nanoparticles after different immersion times in 3.5 wt% NaCl solution (NE, PN, EC70, EC80 and EC 90 are the codes for AA3105 aluminum that coated with neat epoxy, 1 wt% MMT, 1 wt% Co70 (70/30 wt% of NaMMT/$CoFe_2O_4$), 1 wt% Co80 (80/20 wt% NaMMT/$CoFe_2O_4$) and 1 wt% Co90 (90/10 wt% of NaMMT/$CoFe_2O_4$), respectively) [30].

of the composite and that of the neat resin, respectively. Likewise, dimensionless complete cure temperature with similar subscripts is defined as $\Delta T^* = \Delta T_{Comp}/\Delta T_{Ref}$. Among thermoset resin, epoxy resins have been widely applied in the industry due to their high performance as adhesive [42], flame-retardant coatings [43], and anticorrosion coatings [44]. Combination of inorganic nanoparticles and organic epoxy resin results in promising physicochemical properties. Diverse nanomaterials such as silica [45], carbon nanotubes and graphene oxide [46, 47], nanoclay, halloysite nanotubes, and layered double hydroxides [48–52], and iron and iron oxide nanoparticles [53, 54] have been incorporated into epoxy resin to enhance the cross-linking reaction and consequently ultimate properties. MNPs incorporated epoxy nanocomposites can improve the thermal, mechanical, flame retardancy, anticorrosion, and electrical properties of epoxy resins [55, 56].

The blank MNPs highly tend to agglomerate due to their high surface/volume ratio. Aggregation of MNPs hinders the curing reaction of epoxy matrix due to the steric hindrance. The improvement of ultimate properties of MNP/epoxy nanocomposites strongly depends on the interaction between epoxy and MNPs and its effect on epoxy network formation. Bulk and surface modifications of MNPs are effective techniques for preventing agglomeration of nanoparticles. Metal ions doping is the method of bulk modification of MNPs crystal structure and the treatment of the surface of MNPs by polymers and macromolecules is an efficient approach for the surface modification of MNPs. In addition, simultaneous bulk and surface modification of MNPs can synergistically affect the cross-linking reaction of epoxy. Recent studies that show the effect of bulk or surface modification of MNPs on cure reaction of epoxy resin are reported in Table 2. The simultaneous modification of the bulk and the surface of MNPs with the metal dopants and macromolecules were also reported.

It is clear that doping MNPs with Ni, Mn, Co, Zn, and Gd is responsible for a significant change in curing epoxy depending on the reactivity of surface of MNPs. In the crystal structure of Fe_3O_4, Fe^{2+} metal ion substitute by Ni^{2+} Mn^{2+}, Co^{2+}, Zn^{2+}, and Gd^{2+} and the reactivity of MNPs surface change depends on doping layer [77]. The Mn^{2+} and Zn^{2+} cations tend to locate in both surface and the bulk layer in the structure Fe_3O_4 nanoparticles and reduce the agglomeration of Fe_3O_4. The Ni^{2+} dopant has the tendency to locate in the top layer of Fe_3O_4 crystal which causes increment in the activity of MNPs surface and allowed them to agglomerate. By substituting Fe^{3+} with Gd^3 the cross-linking reaction of epoxy/MNPs can be improved. In the case of Fe_3O_4 nanoparticles doped with Co^{2+}, Fe^{2+} replace in bulk layers with CO^{2+} and reduce the tendency of MNPs to agglomerate. Though bulk modification of MNPs can effectively enhance cure potential of epoxy/MNPs nanocomposites, simultaneous bulk and surface treatment of MNPs results in dramatic facilitation of cure reaction.

As apparent final properties of epoxy coatings containing low and high amount of MNPs should be interpreted on account of *Cure Index* which helps in deep understanding of performance of polymeric coatings from molecular perspective. Jouyandeh et al. [78]

Table 2 Effect of bulk, surface, and bulk-surface modification of MNPs on cross-linking reaction of epoxy resin.

No.	Modifier	Cure Index	Functional group	Effect on epoxy curing	Refs.
Blank Fe$_3$O$_4$					
1	–	Poor	OH	Hinder the curing reaction of epoxy	[57]
Bulk modification					
2	Co-doping	Good at low heating rates	Co^{2+} cations as Lewis acid	Curing reactions between epoxy and amine curing agent was improved by the introduction of Fe$_3$O$_4$ nanoparticles doped with Co^{2+}	[57]
3	Mn-doping	Excellent	Mn^{2+} cations as Lewis acid	Catalyze the curing reaction of epoxy/amine and increase the order of autocatalytic reaction	[58, 59]
4	Ni-doping	Poor	Ni^{2+} cations as Lewis acid	Increase the agglomeration of Fe$_3$O$_4$ and decrease the curability of epoxy	[60, 61]
5	Zn-doping	Good cure, except for the high heating rate	Zn^{2+} cations as Lewis acid	Catalyze epoxy ring opening reaction and facilitate the autocatalytic reaction	[62, 63]
6	Gadolinium (Gd)–doping	Poor, Good, and Excellent depends on heating rate	Gd^{2+} and Gd^{3+} cations as Lewis acid	The curing reaction of epoxy system was improved in the case Gd-doped Fe$_3$O$_4$	[64]
Surface modification					
7	Chitosan and chitosan@imide@phenylalanine	–	OH, NH and NH$_2$	Modify cross-linking reaction of epoxy	[53]
8	2-Acrylamido–2-methylpropanesulfonic acid (AMPS)	–	NH and sulfonic acid	Competitive reactions caused by AMPS acid such as chain-growth homopolymerization of epoxide, intensified step-growth epoxide ring-opening with amine, and reaction of AMPS with amine groups of curing agent that possibly deactivated hardener results in no progress in curing level	[54]
9	β-Cyclodextrin	–	OH	Results in highly cured epoxy network	[65]

Bulk and surface modification

10	Ethylenediamine-tetraacetic acid (EDTA) and Co	Good	COOH and Co^{2+} cations as Lewis acid	Catalyzing effect of COOH of EDTA and Co^{2+} in the lattice of IONs lead in highly cured epoxy network	[66]
11	Polyethylene glycol (PEG) and Co	Excellent	OH and Co^{2+} cations as Lewis acid	Synergistic effect of OH and Co^{2+} facilitate the cure reaction of epoxy	[67]
12	PEG and Gd	Good	OH and Gd^{2+} and Gd^{3+} cations as Lewis acid	Replacement of Fe^{2+} by Gd^{2+} activated Fe_3O_4 surface and enhanced PEG grafting onto Fe_3O_4 which improve curing	[68]
13	PEG and Ni	Good	OH and Ni^{2+} cations as Lewis acid	Epoxy curing was completed by combined bulk (Ni) and surface modification (PEG) of MNPs	[69]
14	PEG and Zn	Good	OH and Zn^{2+} cations as Lewis acid	Significant catalytic effect on epoxide ring opening springing from Zn^{2+} cations along with hydroxyl groups of PEG	[70]
15	Polyvinyl chloride (PVC) and Co	From *Poor* to *Good* depending on heating rate	Cl^- and Co^{2+}	Bulk modification by Co significantly affected epoxy curing in support of surface functionalization with PVC	[71]
16	PVC and Ni	Poor	Cl^- and Ni^{2+}	Agglomeration of nanoparticles due to small number of PVC chains grafted on the surface of the Ni-doped MNPs decreased curability of epoxy system	[72]
17	Polyvinylpyrrolidone (PVP) and Mn	Good	Amide group and Mn^{2+}	Accelerating effect of PVP and Mn^{2+} on cross-linking between epoxy and amine	[73, 74]
18	PVP and Ni	Good	Amide group and Ni^{2+}	Enhancement of epoxy ring opening due to Ni^{2+} cations incorporation and PVP macromolecule	[75]
19	PVP and Zn	Good	Amide group and Zn^{2+}	Dipole-dipole interactions between amid group of PVP and the oxirane rings of epoxy and catalyzing effect of Zn^{2+} through Lewis acid action	[76]

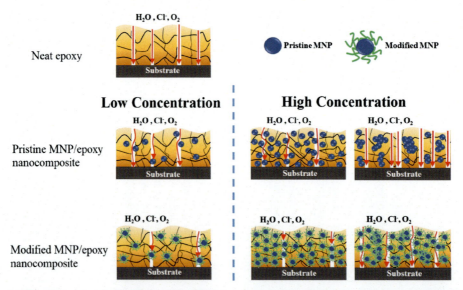

Fig. 10 Possible network formation and corrosion protection mechanisms of the epoxy/MNPs [78].

showed that sufficient amount of MNPs with appropriate functionality can improve cross-linking density of epoxy network and improve barrier property of epoxy coating by reduction of diffusion paths for anticorrosion and flame retardancy application. As indicated in Fig. 10, the well-dispersed modified MNPs in epoxy matrix showed the denser network and the highest corrosion barrier properties.

2.3 Magnetic fluids for corrosion protection

Magnetic fluids are nanomagnetic liquids that possess unusual properties making them applicable in technological and biomedical fields [79–81]. In magnetic fluids, magnetic and liquid state coexists and is a colloidal suspension of single-domain particles having dimension in between 5 and 20 nm dispersed in a carrier liquid and stabilized by a suitable organic surfactant [81]. Qiu et al. [82] reported a new strategy for the protection of commercially available strongest neodymium (NdFeB) magnets from corrosion. Using mussel-inspired chemistry and Michael's reaction, they prepared hydrophobic/superoleophilic nanostructured Fe_3O_4 magnetic fluid and used it for corrosion protective coating on NdFeB magnets. NdFeB magnets are excessively used in different fields but it is prone to corrosion in harsh environments. The corrosion protection using magnetic fluid is novel, in which the magnetic fluid coheres on the substrate surface, so that the liquid mantle is built to act as a barrier and isolate the substrate from external environment. During coating process, the imposed magnetic field will strongly attract Fe_3O_4 nanoparticles, densely aggregate on the surface, and acts as a barrier to prevent corrosion [83, 84]. In NaCl solution and acidic salt spray environment, magnetic fluid achieves the corrosion current density of three orders of magnitude lower than that of bare NdFeB. The magnetic fluid coating on NdFeB also exhibits excellent self-healing property [82].

3. MNPs as antifouling component

Apart from effective anticorrosion property of MNPs and MNP-based coatings, their antifouling properties were explored, which would further extend them as multifunctional coating ingredients especially in the marine environment. Inorganic nanoparticles are vital components in antifouling coating as they enhance the foul release phenomena [85, 86]. Among them, MNPs are also experimented for their antifouling ability in many applications [87–89]. Researchers are in continuous effort to find eco-friendly and low-cost antifouling coating alternatives for biocidal antifouling paints [90]. The antibacterial activity arises as MNPs induce the production of reactive oxygen species (ROS) which destruct the microbial cell wall [91]. Pallela et al. [92] studied the antibacterial efficiency of hematite prepared via green synthesis using *Sida cordifolia* plant extract. *S. cordifolia* mediated hematite with crystalline particle size of 20 nm has potent antifouling ability mainly arisen from the UV or visible light activated production of reactive oxygen species such as superoxide radical anions (O^{2-}) and hydroxyl radicals (OH^-). These reactive oxygen species lead to the desorption of membrane leading to the death of the bacteria. Apart from this, different interactions like electrostatic, dipole-dipole, hydrogen bond, hydrophobic, and van der Waals interactions between these MNPs and bacterial cell were reported to cause disruption of microbial cellular membrane. Fig. 11 shows the zone of inhibition for *S. cordifolia* mediated α-Fe_2O_3 nanoparticles against different bacteria.

MNPs incorporated polymer coatings/paints were reported to have the ability to effectively prevent fouling. MNPs incorporated silicone composites were used for making eco-friendly marine antifouling paints [93]. Apart from inducing the antifouling ability, uniform distribution of the spherical magnetite nanoparticles having diameter of 10–20 nm impart superhydrophobicity and corrosion resistance to the coating. PDMS/magnetite inhibits the settlement of spore, diatoms, algae, and bacteria settlements via lotus leaf effect and physical antiadhesion mechanism. Fig. 12 shows the complete inhibition of diatom growth on the PDMS/magnetite nanocomposite coatings.

Cobalt ferrite (CF) and graphene oxide have been simultaneously used to enhance the antibacterial activity of alkyd resin-based paint (magnetic graphene oxide (MGO) hybrid paint) to protect the galvanized iron (GI) substrate [94]. The antibacterial activity of the hybrid MGO film can be attributed to originated from the contribution of both GO and CF MNP. MGO paint damages the bacterial membrane by procuring ROS. The antibacterial efficacy of hybrid MGO film was found excellent with a mechanism of producing ROS and damaging bacterial membrane. The cell damage in *S. typhimurium* and *E. coli* after exposed to the MGO film was visualized via. SEM (Fig. 13). The interaction of MGO resulted in membrane rupture, membrane blabbing, chin breaking, and shape deformation on the morphology of these bacteria.

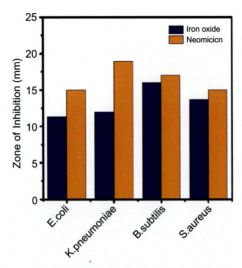

Fig. 11 Comparison of zone of inhibition for green synthesis of *S. cordifolia* mediated α-Fe_2O_3 (hematite) nanoparticles and standard antibiotic with 50 μg/mL against different bacteria [92].

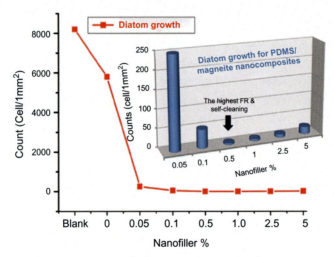

Fig. 12 Diatom growth measurements for the cured virgin and nano-magnetite filled PDMS composites (with different nanofiller percentages) after incubation for 7 days (error bars represent ±2 standard deviations from three replications) [93].

4. Smart coatings based on MNPs

MNPs are also found as a functional material in smart coatings including self-healing and self-cleaning coatings [95]. Self-healing is the ability in which the damaged coating repairs by itself without any manual intervention [96, 97]. Recently, it has been reported that magnetic nanomaterials have the ability to repair the defects of organic

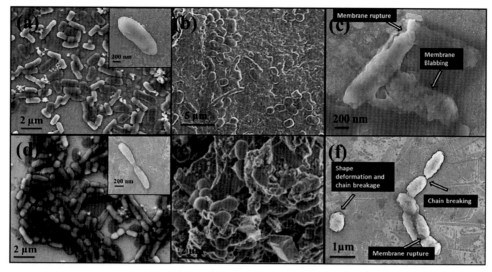

Fig. 13 SEM micrograph of bacterial strains with MGO film (A) untreated *S. typhimurium* (B) *S. typhimurium* with MGO film (C) morphological changes in *S. typhimurium* after treatment (D) untreated *E. coli* (E) *E. coli* with magnetic paint and (F) morphology changes of *E. coli* after treatment (inset in A, D shows single bacteria) [94].

coatings on the metal surface to make them durable and economically reasonable. Epoxy-based coatings with MNPs were developed as promising smart coatings [98, 99]. In epoxy/magnetite hybrid nanocomposite, oxygen and salt-sensitive magnetite nanoparticles healed the damaged coating at steel surface by the formation of passivation materials [98, 99]. In one of the studies, Myrrh gum capped magnetite nanoparticle not only acts as self-healing agent but also as anticorrosion and antibacterial agent. Atta et al. [100] tailored self-healing ability in epoxy coating by incorporating 2,4-dihydrazino-6-(methoxypolyethyleneglycol)-1,3,5-triazine (mPEGTH) decorated Fe_3O_4 nanoparticle. Along with the formation of passivation material, the reaction between mPEGTH (which acts as healing agent) with epoxy heals the damage in the coating and the reaction was proposed to be accelerated by the presence of Fe_3O_4. Proper healing on the damaged site reduces the diffusion paths for the corrosive ions and prevents further corrosion. The visualized self-healing process upon exposed to salt spray on epoxy/mPEGTH decorated Fe_3O_4 hybrid nanocomposite is shown in Fig. 14. A 3 wt% of mPEGTH decorated Fe_3O_4 nanoparticle was found to effectively resist the manually created scratch on the coated steel panel.

Another important smart functionality of a coating system is self-cleaning ability due to the superhydrophobicity of the coating surface. Self-cleaning coating is commonly used for surfaces like windshields of vehicles, entire body of vehicles, window

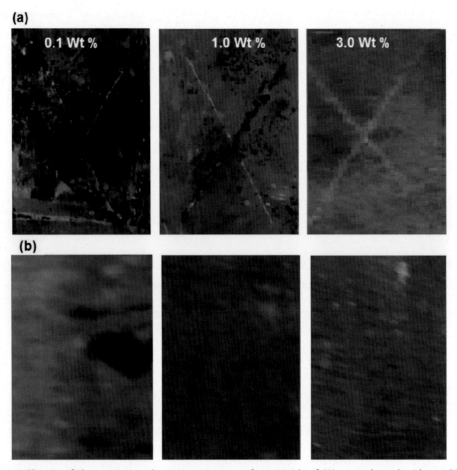

Fig. 14 Photos of the coatings salt spray resistance after 1500 h of (A) coated steel with modified epoxy Fe_3O_4 capped with mPEGTH and (B) cleaned steel after removal of modified epoxy coatings [100].

and door glasses, skyscrapers, solar cell panels, fabrics, sports shoes, metals, papers, sponges, woods, marbles, and the list is ceaseless. Many surfaces in day-to-day life eventually get contaminated due to the accumulation of dust/dirt or through air pollution. The self-cleaning characteristic, or lotus effect, of this coating makes it suitable for rendering a surface antimicrobial. Different methods are applied for making material surface superhydrophobic. Properly dispersed MNPs were found to impart self-cleaning ability to polymer coatings with good acid-base resistance. For instance, Alamri et al. [101] uniformly dispersed superhydrophobic silica-coated magnetite nanoparticles to provide self-cleaning ability to epoxy coating surface. They have tailored dendrimeric fibers morphology so as to spread out from the core-shell in radial manner, which effectively facilitated superhydrophobicity. They have measured

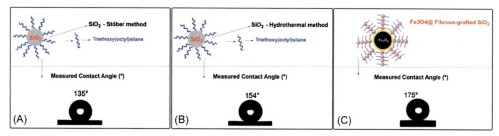

Fig. 15 The contact angles of superhydrophobic silica (A) and (B) prepared by the Stöber and hydrothermal processes, respectively, and those of magnetic superhydrophobic silica (C) [101].

contact angle of the coating surface on carbon steel so as to evaluate the superhyrdobhobicity and hence the self-cleaning ability. The epoxy coating without modification showed a contact angle of 65° due to the presence of plenty of hydroxyl functional groups. However, this contact angle value gets enhanced upon the dispersion of superhydrophobic silica nanoparticles prepared by different methods (Fig. 15A and B). It was found that fibrous grafted silica-coated magnetite nanoparticles have the ability to further enhance the contact angle value (Fig. 15C).

5. MNPs for electromagnetic absorbing coatings

Electromagnetic wave interference (EMI) pollution, the noise created when electromagnetic (EM) waves interfere with the input signal of the electronic devices, is a serious concern in the modern world [102]. The aftereffects of EMI pollution could be avoided to some extent by the use of EMI shielding materials. Electromagnetic interface (EMI) shielding materials help to shield gigahertz (GHz) EM radiation used in commercial, military, scientific, electronic, and communication systems. MNPs are effective in developing EMI shielding coatings for devices, commercial appliances, and structures used in those fields.

Iron-oxide-based MNPs act as EMI shielding material by absorbing EM due to their inherent magnetic dipoles [102]. It was found that nanoferrite can effectively absorb microwave radiations [103–105]. Nanoferrite has wideband absorption capacity which is not achieved with a pure ferrite microwave absorber [105]. Hexagonal plate-shaped nanoparticles of barium hexaferrite/nickel ferrite composite having size range of 85–95 nm showed maximum reflection loss of −27.17 at 11.79 GHz [106]. MNPs dispersed in polymer matrix especially in conducting polymers have excellent microwave and radar absorption property which has been explored in many fields [107–109]. Such nanocomposites were coated over many substrates to make efficient EMI shielding coatings that protect from microwave and radar. For instance, an efficient microwave absorbing coatings were developed by Açıkalın et al. [110] using Nickel-Zinc nanoferrite and nanosized polyaniline as a filler and epoxy as binder.

6. MNP coating for textiles

Nowadays, metallized textiles have become important due to their valuable features arise from combination of textiles and metals. Various methods have been used for the construction of metal-textiles coatings including electroplating, electroless plating, sputter coating, evaporation, airbrushing, vacuum deposition, and in situ synthesis of nanoparticles on textiles [111]. In addition, the magnetic-textile coatings can be prepared using magnetic fibers for textile gauges, transmitters, and actuators. They can be utilized in electronic circuits and intelligent clothing for monitoring the selected human physiological parameters [112]. Rubacha and Zięba [113] prepared a magnetic fiber using cellulose as matrix and powdered magnetic as modifier for preparation of a textile with magnetic coils and textile core. Longitudinal and cross-section views of the prepared magnetic fibers are shown in Fig. 16.

The main preparation methods of magnetic fabric are (1) Woven, knitted, and braided from a magnetic yarn; (2) Physical modification and magnetic-coating; and (3) 3D printing [114]. The performance of fabric is not be affected by magnetic powder, however magnetic elements have deterministic effect on wear ability of fabrics [115, 116]. The controllable ability of MNPs in textile increase the number of possible smart textile products as shown in Fig. 17.

Fig. 16 SEM images of (A) longitudinal and (B) cross-section of magnetic cellulose fibers [113].

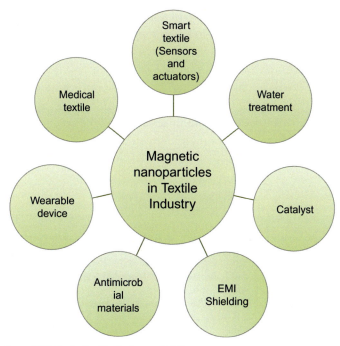

Fig. 17 Application of MNPs in textile industry [114].

7. Conclusions and future perspectives

The advantages of MNPs including large surface areas, size, and shape-dependent catalytic properties allow them for different applications in the environmental and medical fields. Hybridization of MNPs with polymer in coatings technology opens the door for various high-tech applications for health care, corrosion protection coatings, wearable devices, and shielding materials. The potential of MNPs in coatings technology is summarized as follows:

- MNPs have the potential to generate multifunctional coating for metals and fabrics.
- Proper tuning in functionalization of MNPs enabled the researchers to tailor functionalities like anticorrosion, self-cleaning, antifouling, etc.
- The reactivity of surface of MNPs can affect cross-linking reaction of thermoset coatings.
- Green synthesis, use of natural stabilizers/capping agents, etc. need to encourage for development MNPs for coating technology.
- EMI shielding coatings generated by dispersing potential MNPs in paints/resins would be the future coating technology for aircraft, ships, and military vehicles to make them imperceptible to radars.

- MNPs have promising potential application in textile industry for use in smart textiles, flexible sensors, textile wastewater treatment, medical textiles, antibacterial materials, and catalysts.
- When used in thermoset coatings, the state of cure in MNP incorporated coatings should be considered to find reasons for improving or deteriorating ultimate properties.

The future of magnetic nanoparticle applications involves the creation of multifunctional materials. Ultimately, safety and biocompatibility studies, in particular long-term toxicity studies, should be carried out.

References

[1] S.P. Gubin, Y.A. Koksharov, G. Khomutov, G.Y. Yurkov, Russ. Chem. Rev. 74 (2005) 489.
[2] T. Indira, P. Lakshmi, Int. J. Pharm. Sci. Nanotechnol. 3 (2010) 1035–1042.
[3] C.S. Kumar, F. Mohammad, Adv. Drug Deliv. Rev. 63 (2011) 789–808.
[4] R. Pareta, S. Sirivisoot, Calcium Phosphate-Coated Magnetic Nanoparticles for Treating Bone Diseases, Elsevier, Cambridge, UK, 2012, pp. 131–148.
[5] J.S. Beveridge, J.R. Stephens, M.E. Williams, Annu. Rev. Anal. Chem. 4 (2011) 251–273.
[6] V.F. Cardoso, A. Francesko, C. Ribeiro, M. Bañobre-López, P. Martins, S. Lanceros-Mendez, Adv. Healthc. Mater. 7 (2018) 1700845.
[7] M. Kermani, D. Harrop, SPE Prod. Facil. 11 (1996) 186–190.
[8] C. Hansson, Metall. Mater. Trans. A 42 (2011) 2952–2962.
[9] K. Jeyasubramanian, V. Benitha, V. Parkavi, Prog. Org. Coat. 132 (2019) 76–85.
[10] Q. Du, W. Zhang, H. Ma, J. Zheng, B. Zhou, Y. Li, Tetrahedron 68 (2012) 3577–3584.
[11] E.A. Khamis, A. Hamdy, R.E. Morsi, Egypt. J. Pet. 27 (2018) 919–926.
[12] A.M. Atta, O.E. El-Azabawy, H. Ismail, M. Hegazy, Corros. Sci. 53 (2011) 1680–1689.
[13] M. Amiri, M. Ghaffari, A. Mirzaee, G. Bahlakeh, M.R. Saeb, J. Mol. Liq. (2020) 113597.
[14] G.A. El-Mahdy, A.M. Atta, H.A. Al-Lohedan, J. Taiwan Inst. Chem. Eng. 45 (2014) 1947–1953.
[15] A.M. Atta, G.A. El-Mahdy, H.A. Al-Lohedan, S.A. Al-Hussain, Int. J. Electrochem. Sci. 9 (2014) 8446–8457.
[16] A.M. Atta, G.A. El-Mahdy, H.A. Al-Lohedan, S.A. Al-Hussain, Int. J. Mol. Sci. 15 (2014) 6974–6989.
[17] A.M. Atta, G.A. El-Mahdy, H.A. Al-Lohedan, S.A. Al-Hussain, Int. J. Electrochem. Sci. 10 (2015) 2621–2633.
[18] K. Chandrappa, T. Venkatesha, Mater. Corros. 65 (2014) 509–521.
[19] P. Vijayan, Y.M.H. El-Gawady, M.A.S. Al-Maadeed, Ind. Eng. Chem. Res. 55 (2016) 11186–11192.
[20] C. Chen, S. Qiu, M. Cui, S. Qin, G. Yan, H. Zhao, L. Wang, Q. Xue, Carbon 114 (2017) 356–366.
[21] D. Song, Z. Yin, F. Liu, H. Wan, J. Gao, D. Zhang, X. Li, Prog. Org. Coat. 110 (2017) 182–186.
[22] Y. González-García, S. González, R. Souto, Corros. Sci. 49 (2007) 3514–3526.
[23] N.H. Othman, M.C. Ismail, M. Mustapha, N. Sallih, K.E. Kee, R.A. Jaal, Prog. Org. Coat. 135 (2019) 82–99.
[24] M. Dong, Q. Li, H. Liu, C. Liu, E.K. Wujcik, Q. Shao, T. Ding, X. Mai, C. Shen, Z. Guo, Polymer 158 (2018) 381–390.
[25] M. Gharagozlou, B. Ramezanzadeh, Z. Baradaran, Appl. Surf. Sci. 377 (2016) 86–98.
[26] O.U. Rahman, M. Kashif, S. Ahmad, Prog. Org. Coat. 80 (2015) 77–86.
[27] Y. Zhan, J. Zhang, X. Wan, Z. Long, S. He, Y. He, Appl. Surf. Sci. 436 (2018) 756–767.
[28] A. Javidparvar, B. Ramezanzadeh, E. Ghasemi, J. Taiwan Inst. Chem. Eng. 61 (2016) 356–366.
[29] A.U. Chaudhry, V. Mittal, B. Mishra, Dyes Pigments 118 (2015) 18–26.
[30] N. Soltani, H. Salavati, A. Moghadasi, Surf. Interfaces 15 (2019) 89–99.

[31] A. Javidparvar, B. Ramezanzadeh, E. Ghasemi, Corrosion 72 (2016) 761–774.
[32] A.M. Asiri, M.A. Hussein, B.M. Abu-Zied, A.E.A. Hermas, Polym. Compos. 36 (2015) 1875–1883.
[33] M. Jouyandeh, O.M. Jazani, A.H. Navarchian, M. Shabanian, H. Vahabi, M.R. Saeb, Appl. Surf. Sci. 447 (2018) 152–164.
[34] M. Mauro, M.R. Acocella, C.E. Corcione, A. Maffezzoli, G. Guerra, Polymer 55 (2014) 5612–5615.
[35] M. Jouyandeh, O.M. Jazani, A.H. Navarchian, M. Shabanian, H. Vahabi, M.R. Saeb, Appl. Surf. Sci. 479 (2019) 1148–1160.
[36] M. Jouyandeh, F. Tikhani, N. Hampp, D. Akbarzadeh Yazdi, P. Zarrintaj, M. Reza Ganjali, M. Reza Saeb, Chem. Eng. J. 396 (2020) 125196.
[37] F. Tikhani, S. Moghari, M. Jouyandeh, F. Laoutid, H. Vahabi, M.R. Saeb, P. Dubois, Polymers 12 (2020) 644.
[38] M. Jouyandeh, F. Tikhani, M. Shabanian, F. Movahedi, S. Moghari, V. Akbari, X. Gabrion, P. Laheurte, H. Vahabi, M.R. Saeb, J. Alloys Compd. 829 (2020) 154547.
[39] M. Jouyandeh, S.M.R. Paran, A. Jannesari, M.R. Saeb, Prog. Org. Coat. 127 (2019) 429–434.
[40] M. Jouyandeh, S.M.R. Paran, A. Jannesari, D. Puglia, M.R. Saeb, Prog. Org. Coat. 131 (2019) 333–339.
[41] F. Tikhani, M. Jouyandeh, S.H. Jafari, S. Chabokrow, M. Ghahari, K. Gharanjig, F. Klein, N. Hampp, M.R. Ganjali, K. Formela, M.R. Saeb, Prog. Org. Coat. 135 (2019) 176–184.
[42] M. Jouyandeh, O. Moini Jazani, A.H. Navarchian, M.R. Saeb, J. Reinf. Plast. Compos. 35 (2016) 1685–1695.
[43] H. Vahabi, M. Jouyandeh, M. Cochez, R. Khalili, C. Vagner, M. Ferriol, E. Movahedifar, B. Ramezanzadeh, M. Rostami, Z. Ranjbar, B.S. Hadavand, M.R. Saeb, Prog. Org. Coat. 123 (2018) 160–167.
[44] F. Seidi, M. Jouyandeh, V. Akbari, S.M.R. Paran, S. Livi, F. Ducos, H. Vahabi, M.R. Ganjali, M.R. Saeb, Polym. Eng. Sci. 60 (2020) 1940–1957.
[45] S. Ghiyasi, M.G. Sari, M. Shabanian, M. Hajibeygi, P. Zarrintaj, M. Rallini, L. Torre, D. Puglia, H. Vahabi, M. Jouyandeh, F. Laoutid, S.M.R. Paran, M.R. Saeb, Prog. Org. Coat. 120 (2018) 100–109.
[46] M. Jouyandeh, E. Yarahmadi, K. Didehban, S. Ghiyasi, S.M.R. Paran, D. Puglia, J.A. Ali, A. Jannesari, M.R. Saeb, Z. Ranjbar, M.R. Ganjali, Prog. Org. Coat. 136 (2019) 105217.
[47] M.R. Saeb, F. Najafi, E. Bakhshandeh, H.A. Khonakdar, M. Mostafaiyan, F. Simon, C. Scheffler, E. Mäder, Chem. Eng. J. 259 (2015) 117–125.
[48] Z. Karami, M. Jouyandeh, J.A. Ali, M.R. Ganjali, M. Aghazadeh, M. Maadani, M. Rallini, F. Luzi, L. Torre, D. Puglia, V. Akbari, M.R. Saeb, Prog. Org. Coat. 136 (2019) 105228.
[49] Z. Karami, M. Jouyandeh, J.A. Ali, M.R. Ganjali, M. Aghazadeh, S.M.R. Paran, G. Naderi, D. Puglia, M.R. Saeb, Prog. Org. Coat. 136 (2019) 105218.
[50] Z. Karami, M. Jouyandeh, J.A. Ali, M.R. Ganjali, M. Aghazadeh, M. Maadani, M. Rallini, F. Luzi, L. Torre, D. Puglia, M.R. Saeb, Prog. Org. Coat. 136 (2019) 105264.
[51] X. Kornmann, H. Lindberg, L.A. Berglund, Polymer 42 (2001) 1303–1310.
[52] V. Akbari, F. Najafi, H. Vahabi, M. Jouyandeh, M. Badawi, S. Morisset, M.R. Ganjali, M.R. Saeb, Prog. Org. Coat. 135 (2019) 555–564.
[53] M. Jouyandeh, S.M.R. Paran, M. Shabanian, S. Ghiyasi, H. Vahabi, M. Badawi, K. Formela, D. Puglia, M.R. Saeb, Prog. Org. Coat. 123 (2018) 10–19.
[54] M. Jouyandeh, M. Shabanian, M. Khaleghi, S.M.R. Paran, S. Ghiyasi, H. Vahabi, K. Formela, D. Puglia, M.R. Saeb, Prog. Org. Coat. 125 (2018) 384–392.
[55] L. Wang, H. Qiu, C. Liang, P. Song, Y. Han, Y. Han, J. Gu, J. Kong, D. Pan, Z. Guo, Carbon 141 (2019) 506–514.
[56] Y. Huangfu, C. Liang, Y. Han, H. Qiu, P. Song, L. Wang, J. Kong, J. Gu, Compos. Sci. Technol. 169 (2019) 70–75.
[57] M. Jouyandeh, M.R. Ganjali, J.A. Ali, M. Aghazadeh, F.J. Stadler, M.R. Saeb, Prog. Org. Coat. 137 (2019) 105252.
[58] M. Jouyandeh, M.R. Ganjali, J.A. Ali, M. Aghazadeh, F.J. Stadler, M.R. Saeb, Prog. Org. Coat. 136 (2019) 105199.

[59] M. Jouyandeh, S.M.R. Paran, S.S.M. Khadem, M.R. Ganjali, V. Akbari, H. Vahabi, M.R. Saeb, Prog. Org. Coat. 140 (2020) 105505.
[60] M. Jouyandeh, M.R. Ganjali, J.A. Ali, M. Aghazadeh, F.J. Stadler, M.R. Saeb, Prog. Org. Coat. 136 (2019) 105198.
[61] M. Jouyandeh, Z. Karami, S.M.R. Paran, A.H. Mashhadzadeh, M.R. Ganjali, B. Bagheri, P. Zarrintaj, S. Habibzadeh, P. Vijayan, M.R. Saeb, J. Compos. Sci. 4 (2020) 102.
[62] M. Jouyandeh, Z. Karami, S.M. Hamad, M.R. Ganjali, V. Akbari, H. Vahabi, S.-J. Kim, P. Zarrintaj, M.R. Saeb, Prog. Org. Coat. 136 (2019) 105290.
[63] M. Jouyandeh, J.A. Ali, M. Aghazadeh, K. Formela, M.R. Saeb, Z. Ranjbar, M.R. Ganjali, Prog. Org. Coat. 136 (2019) 105246.
[64] M. Jouyandeh, P. Zarrintaj, M.R. Ganjali, J.A. Ali, I. Karimzadeh, M. Aghazadeh, M. Ghaffari, M.R. Saeb, Prog. Org. Coat. 136 (2019) 105245.
[65] M.R. Saeb, H. Rastin, M. Shabanian, M. Ghaffari, G. Bahlakeh, Prog. Org. Coat. 110 (2017) 172–181.
[66] M. Jouyandeh, M.R. Ganjali, J.A. Ali, M. Aghazadeh, I. Karimzadeh, K. Formela, X. Colom, J. Cañavate, M.R. Saeb, Prog. Org. Coat. 136 (2019) 105248.
[67] M. Jouyandeh, M.R. Ganjali, M. Aghazadeh, S. Habibzadeh, K. Formela, M.R. Saeb, Polymers 12 (2020) 1820.
[68] M. Jouyandeh, M.R. Ganjali, J.A. Ali, V. Akbari, Z. Karami, M. Aghazadeh, P. Zarrintaj, M.R. Saeb, Prog. Org. Coat. 137 (2019) 105283.
[69] M. Jouyandeh, Z. Karami, J.A. Ali, I. Karimzadeh, M. Aghazadeh, F. Laoutid, H. Vahabi, M.R. Saeb, M.R. Ganjali, P. Dubois, Prog. Org. Coat. 136 (2019) 105250.
[70] M. Jouyandeh, S.M. Hamad, I. Karimzadeh, M. Aghazadeh, Z. Karami, V. Akbari, F. Shammiry, K. Formela, M.R. Saeb, Z. Ranjbar, M.R. Ganjali, Prog. Org. Coat. 137 (2019) 105285.
[71] M. Jouyandeh, M.R. Ganjali, B.S. Hadavand, M. Aghazadeh, V. Akbari, F. Shammiry, M.R. Saeb, Prog. Org. Coat. 137 (2019) 105364.
[72] M. Jouyandeh, M.R. Ganjali, Z. Karami, M. Rezapour, B. Bagheri, P. Zarrintaj, A. Mouradzadegun, S. Habibzadeh, M.R. Saeb, J. Compos. Sci. 4 (2020) 107.
[73] M. Jouyandeh, J.A. Ali, V. Akbari, M. Aghazadeh, S.M.R. Paran, G. Naderi, M.R. Saeb, Z. Ranjbar, M.R. Ganjali, Prog. Org. Coat. 136 (2019) 105247.
[74] M. Jouyandeh, M.R. Ganjali, F. Seidi, H. Xiao, M.R. Saeb, J. Compos. Sci. 4 (2020) 55.
[75] M. Jouyandeh, M.R. Ganjali, J.A. Ali, M. Aghazadeh, M.R. Saeb, S.S. Ray, Prog. Org. Coat. 136 (2019) 105259.
[76] M. Jouyandeh, M.R. Ganjali, J.A. Ali, M. Aghazadeh, S.M.R. Paran, G. Naderi, M.R. Saeb, S. Thomas, Prog. Org. Coat. 136 (2019) 105227.
[77] R. Gargallo-Caballero, L. Martín-García, A. Quesada, C. Granados-Miralles, M. Foerster, L. Aballe, R. Bliem, G.S. Parkinson, P. Blaha, J.F. Marco, J. Chem. Phys. 144 (2016) 094704.
[78] M. Jouyandeh, N. Rahmati, E. Movahedifar, B.S. Hadavand, Z. Karami, M. Ghaffari, P. Taheri, E. Bakhshandeh, H. Vahabi, M.R. Ganjali, K. Formela, M.R. Saeb, Prog. Org. Coat. 133 (2019) 220–228.
[79] M. Shuai, A. Klittnick, Y. Shen, G.P. Smith, M.R. Tuchband, C. Zhu, R.G. Petschek, A. Mertelj, D. Lisjak, M. Čopič, Nat. Commun. 7 (2016) 1–8.
[80] K. Raj, B. Moskowitz, R. Casciari, J. Magn. Magn. Mater. 149 (1995) 174–180.
[81] I. Torres-Díaz, C. Rinaldi, Soft Matter 10 (2014) 8584–8602.
[82] Y. Ouyang, R. Qiu, Y. Xiao, Z. Shi, S. Hu, Y. Zhang, M. Chen, P. Wang, Chem. Eng. J. 368 (2019) 331–339.
[83] P. Irajizad, M. Hasnain, N. Farokhnia, S.M. Sajadi, H. Ghasemi, Nat. Commun. 7 (2016) 1–7.
[84] W. Wang, J.V. Timonen, A. Carlson, D.-M. Drotlef, C.T. Zhang, S. Kolle, A. Grinthal, T.-S. Wong, B. Hatton, S.H. Kang, Nature 559 (2018) 77–82.
[85] L. Zheng, Y. Lin, D. Wang, J. Chen, K. Yang, B. Zheng, W. Bai, R. Jian, Y. Xu, RSC Adv. 10 (2020) 24241.
[86] W.-Z. Shi, Y.-S. Liang, B. Lu, M. Chen, Y. Li, Z. Yang, Quim Nova 42 (2019) 638–641.

[87] S. Zinadini, A.A. Zinatizadeh, M. Rahimi, V. Vatanpour, H. Zangeneh, M. Beygzadeh, Desalination 349 (2014) 145–154.
[88] T. Zhao, K. Chen, H. Gu, J. Phys. Chem. B 117 (2013) 14129–14135.
[89] G. Luongo, P. Campagnolo, J.E. Perez, J. Kosel, T.K. Georgiou, A. Regoutz, D.J. Payne, M.M. Stevens, M.P. Ryan, A.E. Porter, I.E. Dunlop, ACS Appl. Mater. Interfaces 9 (2017) 40059–40069.
[90] S. Mohr, R. Berghahn, W. Mailahn, R. Schmiediche, M. Feibicke, R. Schmidt, Environ. Sci. Technol. 43 (2009) 6838–6843.
[91] S. Novak, D. Drobne, M. Golobič, J. Zupanc, T. Romih, A. Gianoncelli, M. Kiskinova, B. Kaulich, P. Pelicon, P. Vavpetič, L. Jeromel, N. Ogrinc, D. Makovec, Environ. Sci. Technol. 47 (2013) 5400–5408.
[92] P.N.V.K. Pallela, S. Ummey, L.K. Ruddaraju, S. Gadi, C.S. Cherukuri, S. Barla, S.V.N. Pammi, Heliyon 5 (2019), e02765.
[93] M.S. Selim, A. Elmarakbi, A.M. Azzam, M.A. Shenashen, A.M. EL-Saeed, S.A. El-Safty, Prog. Org. Coat. 116 (2018) 21–34.
[94] T. Arun, S.K. Verma, P.K. Panda, R.J. Joseyphus, E. Jha, A. Akbari-Fakhrabadi, P. Sengupta, D. Ray, V. Benitha, K. Jeyasubramanyan, Mater. Sci. Eng. C 104 (2019) 109932.
[95] P.P. Vijayan, D. Puglia, Emerg. Mater. (2019) 1–25.
[96] P.P. Vijayan, M. Al-Maadeed, Materials 12 (2019) 2754.
[97] M.A.S. Al-Maadeed, Sci. Rep. 6 (2016) 38812.
[98] A.M. Atta, A.M. El-Saeed, H.A. Al-Lohedan, M. Wahby, Molecules 22 (2017) 905.
[99] A.M. Atta, A.M. El-Saeed, G.M. El-Mahdy, H.A. Al-Lohedan, RSC Adv. 5 (2015) 101923–101931.
[100] A.M. Atta, A. El-Faham, H.A. Al-Lohedan, Z.A.A. Othman, M.M. Abdullah, A.O. Ezzat, Prog. Org. Coat. 121 (2018) 247–262.
[101] H. Alamri, A. Al-Shahrani, E. Bovero, T. Khaldi, G. Alabedi, W. Obaid, I. Al-Taie, A. Fihri, J. Colloid Interface Sci. 513 (2018) 349–356.
[102] V. Shukla, Nanoscale Adv. 1 (2019) 1640–1671.
[103] U. Lima, M. Nasar, R. Nasar, M. Rezende, J. Araújo, J. Magn. Magn. Mater. 320 (2008) 1666–1670.
[104] F. Mohammad, J. Siddiqui, K. Ali, H. Arshad, M. Mudsar, A. Ijaz, J. Mater. Sci. Mater. Electron. 30 (2019) 2278–2284.
[105] G. Shen, G. Cheng, Y. Cao, Z. Xu, Mater. Sci.-Pol. 28 (2010) 327–334.
[106] S. Tyagi, V. Pandey, S. Goel, A. Garg, Integr. Ferroelectr. 186 (2018) 25–31.
[107] K. Didehban, E. Yarahmadi, F. Nouri-Ahangarani, S.A. Mirmohammadi, N. Bahri-Laleh, J. Chin. Chem. Soc. 62 (2015) 826–831.
[108] H. Hosseini, H. Mahdavi, Appl. Organomet. Chem. 32 (2018) e4294.
[109] J. Azadmanjiri, P. Hojati-Talemi, G. Simon, K. Suzuki, C. Selomulya, Polym. Eng. Sci. 51 (2011) 247–253.
[110] E. Açıkalın, K. Çoban, A. Sayıntı, Prog. Org. Coat. 98 (2016) 2–5.
[111] I. Bibi, N. Nazar, M. Iqbal, S. Kamal, H. Nawaz, S. Nouren, Y. Safa, K. Jilani, M. Sultan, S. Ata, Adv. Powder Technol. 28 (2017) 2035–2043.
[112] M. Rubacha, J. Zięba, Fibres Text. East. Eur 14 (2006) 59.
[113] M. Rubacha, J. Zięba, Fibres Text. East. Eur. 15 (2007) 64.
[114] S. Shahidi, J. Ind. Text. (2019). 1528083719851852.
[115] F.F. Wu, G.J. Yu, Y. Chen, The Influence of the Magnetic Fiber Content on Fabric Wearability, Trans Tech Publ, Switzerland, 2013, pp. 144–147.
[116] A. Ehrmann, T. Blachowicz, Examination of Textiles With Mathematical and Physical Methods, Springer, Switzerland, 2017.

CHAPTER 17

Nano coatings for scratch resistance

Sahar Amiri
Petroleum and Chemical Engineering Faculty, Islamic Azad University, Tehran, Iran

Chapter outline

1. Background of polymeric coating — 345
2. Introduction to scratch process — 346
   - 2.1 Effects of viscoelastic on scratch resistance of polymers — 346
   - 2.2 Scratch resistance of polymers at nano-, micro-, and macro level — 347
   - 2.3 Scratch resistance at macro level — 347
3. Typical organic coatings on coil coated steel — 347
   - 3.1 Binder systems — 347
   - 3.2 Polymeric coating based on sol–gel method — 347
4. Introduction to sol–gel method — 348
   - 4.1 Addition of nano-particles in coatings — 349
5. Applications of sol–gel derived coating — 350
   - 5.1 Metallic nano-coating — 350
   - 5.2 Scratch resistance coatings based on waterborne polymers — 352
   - 5.3 Durability of organic coatings against mechanical wear — 353
   - 5.4 The effect of silica nanoparticles on scratch resistance of coatings — 355
   - 5.5 Scratch resistance of epoxy-zeolite coatings — 356
   - 5.6 Scratch resistance coating based on bio-based materials — 356
   - 5.7 Scratch resistance coating based on magnesium alloys — 359
   - 5.8 Evaluation of anti-scratch properties of graphene oxide — 359
   - 5.9 Anti-scratch and anti-corrosion hybrid nanocomposite coatings based on zirconia — 361
   - 5.10 Scratch resistance of acrylate coatings based on bio-polymers — 362
   - 5.11 Anti-reflection coatings with enhanced abrasion and scratch resistance properties — 364
6. Concluding remarks — 366
References — 366

1. Background of polymeric coating

Polymeric materials show unique functionalities such as toughness, brightness, and weather ability which is related to their flexible formulation and the adjustable structure which can be used in various industrial applications such as automotive [1, 2]. Polymeric coatings provide a new route to inhibit metal damages or scratches and improve their lifetime [3, 4].

Polymeric films showed specific and controllable physical and mechanical properties, recyclability, and commercial cost of laminates Polymeric films, applying coating is a commercial method to obtaining scratch resistance surfaces [5, 6]. Polymeric coating shows wide variety of applications such as multilayer polyelectrolyte, biodegradable polymeric capsules and containers [7], functionalized biomaterials [8], antimicrobial packaging [9], anti-reflective layers [10], flame retardants coatings [11], sound dampers [12], and photonic devices [13]. Polymeric coatings despite high cost, polymeric coating containing some nanoparticles could be sued as scratch resistance and anticorrosion coatings and laminates. Coated metal or steel sheets may show barrier properties against moisture to protect the steel or metal coating and improve corrosion resistance of substrate. Applying uniform coating on the substrate led to efficient resistance to corrosion and providing scratch resistance properties. Polymeric coating may be consisting of functional groups which improve the anticorrosion activity of the steel substrate [14]. Scratch and wear resistance is affected by a combination of many different factors such as hardness, elasticity and adhesion to substrate [15].

2. Introduction to scratch process

Surface scratch is an important parameter in many industries which limits the application of metals in various applications to obtain scratch resistance coatings for automotive and applications which led to efficient surface quality [1]. After occurring scratches on plastics, surface uniformity of polymer decreases and also leads to stress concentration, making products disposed to untimely damage [16]. Due to good performance and recyclability of polymers such as thermoplastic polyolefins (TPOs), they are used widely in automotive industry [17–19].

Scratch or corrosion damage may be caused by plastic flow scratch pattern and the fracture scratch and investigation of subsurface scratch damage is important [6]. Scratch visibility is a result of the nonuniform espousing of light from the surface and depends mainly on the size of the yield zone and the fragmentation from the scratch which include cracking, crazing, voiding, and deboning [20].

2.1 Effects of viscoelastic on scratch resistance of polymers

Structural parameters like shape and size, scratch velocity and rate, strain and strain rates have been considered impressive conditions for the fracture process [21]. Abrasion is often the cause of the loss of their optical and tribological performances and mechanical properties.

Scratch protective coatings can act as hard coatings which formed good adhesion with coated substrate but when delamination process starts and functionality of surface became led then fracture starts to growth. Scratch process in ductile manner is a result of contact pressure, the remaining depth of the track, and the height of the pile-up [22–26]. Hardness in serration determined contact pressure and determined the fracture toughness [27].

2.2 Scratch resistance of polymers at nano-, micro-, and macro level

By specification of effective factors in defect formation and propagation, scratch or corrosion process may reduce which is different in nano-, micro-, and macro scale. At nano scale, the innocence of surface harshness and the indenter tip radius on scratch resistance of polymer were determined under constant load [24]. Results indicated that roughness didn't effect on surface damage which can be used as coatings for aesthetic sense to be smaller than the surface it encounters. By increasing cross-link density at nano scratch testing, the higher modulus is obtained which leads to higher stresses.

2.3 Scratch resistance at macro level

Pencil hardness test is the routine method to investigate scratch coatings resistance behavior at macro-scale level and is sensitive to scratch speed, load on the surface and the fillers present in the material [25]. By applying scratch resistance coating, wear resistance improved and delayed the delamination process via formation of new strong interaction [25]. Performance of the obtained coatings is dependent on thickness and composition of polymeric films, although uniform coating coverage drastically improved anti-scratch activity of metals and decreased damage effect.

3. Typical organic coatings on coil coated steel

Organic coatings may consist of binders, solvents, and fillers which can be used in wide range of applications for UV-resistance, chemical resistance, and resistance to wear. Coatings may be formed as thermosetting resin or thermoplastic resin in which thermosetting resin reacts during curing process and formed new cross-links, but thermoplastic resin may be molten or to form a uniform coating [2].

3.1 Binder systems

Aliphatic or cycloaliphatic polyester coatings, PUR coatings may form new strong bonds led to higher resistance to corrosion, scratch, and UV light [2]. Smart coating with UV-protection, hydrophobic or hydrophilic coating, anticorrosion or anti-scratch activity obtained via incorporation of specific additives which dependent to, manufacturing method, chemical structure, shape, etc. [8]. Fillers may modify friction or bulk properties of polymers, surface roughness which led to stress distribution of matrix [11]. On the other hand, mechanical and chemical properties of surface properties may improve via formation of new bonding between substrate surface and polymer coating chains [2].

3.2 Polymeric coating based on sol–gel method

Using environmentally friendly coating with multi-properties and consisting of functional groups via sol–gel method is a promising way in which flexible coatings are obtained. Incorporation of corrosion inhibiting or scratch resistance nanoparticles such as functionalized metal oxide nanoparticles developed new coatings for using as nano- and micro-scale materials with specific applications.

Nanoscale filler composition, size, polydispersity, shape, porosity, crystallinity, functional level, and functionalization types are critical parameters for obtaining smart coatings with uniform coating coverage without any agglomeration.

4. Introduction to sol–gel method

Sol–gel method is a good idea to combine inorganic and organic parts to form a homogeneous hybrid structure based on new string interaction between polymer chains and inorganic components active sited [28, 29]. The main advantages of sol–gel techniques are low cost, low temperature, controllable chemical conditions, and flexibility, and led to formation of hybrid materials for various applications [28, 29].

Obtained metal oxides or coatings directly change with reaction condition, temperature, and pH of sol–gel route. In low pH, hydrolysis is the main reaction and after this step, condensation usually starts, but at high pH, condensation is faster. Under acidic conditions, hydrolysis is favored and condensation usually starts when hydrolysis is completed, while in alkaline conditions condensation is faster and often occurs on terminal silanols which can result in highly condensed species or chain-like structures in the sol and network-like gels (Fig. 1) [30–32].

Sol–gel-derived coatings can be applied on glasses, metals, or plastics which improve lifetime of coated substrate compared to traditional systems [30]. The use of sol–gel coatings enables the deposition of a barrier coat which extends durability and resistance to corrosion. Coatings may have been composed of pure inorganic coatings and

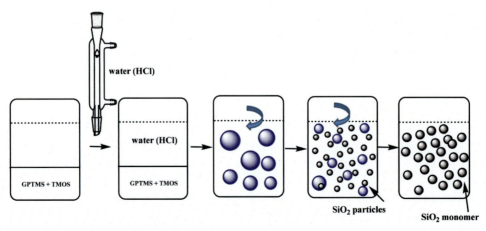

Fig. 1 Schematic diagram for Sol–gel process [30].

organic–inorganic hybrids. Pure inorganic coatings containing metal alkoxides such as SiO_2, ZrO_2, SiO_2-ZrO_2 may be brittle which by incorporation of flexible organic this limitation will be removed. Inorganic parts of hybrid nanocomposites determined scratch resistance and adhesion of coating to substrates and the organic parts are responsible for flexibility, density, and functional compatibility. The obtained coating showed good barrier to water and corrosion initiators. Direct addition of organic parts into inorganic sol–gel system may be without chemical bonding between both components formed new interactions between presence parts in the medium. Stainless steel and aluminum and its alloys were coated with silane-based sol–gel coatings which are good anti-corrosion coatings [33, 34], but delamination, brittleness and thickness restrictions are limitations for this method. Micro- and nanostructures, cracks, and defects affect mechanical strength of obtained coatings with unique synergistic properties which related to nature of precursors structure [35, 36]. The most commonly utilized sol–gel precursor is tetraethyl orthosilicate (TEOS), tetramethyl orthosilicate (TMOS), methyl triethoxysilane (MTES), methyl trimethoxysilane (MTMS), 3-glycidoxipropyltrimethoxysilane (GPTMS), vinyl trimethoxysilane (VTMS), and 3-aminopropyl trimethoxysilane (APS). The most interesting system is based on GPTMS and TEOS which epoxide ring of GPTMS opened, formed an organic chain, and in the presence of Lewis acids, an organic and inorganic network obtained with acceptable mechanical properties with controllable hardness and elastic properties [37].

4.1 Addition of nano-particles in coatings

Micro/nano particles or micro-capsules encapsulated active species via direct addition or in-situ in the organic–inorganic hybrid coatings or added to system after coating formation via sol–gel and improve the protective activity of coatings [38, 39]. Nanoparticles or nanocapsules such as graphite, graphene, clays, and layered hydroxides (LDHs) are incorporated into polymeric coatings for advantageous barrier applications to protect the material against moisture and oxygen to ingress and corrosion protection [40–46]. Corrosion inhibitor or healing nanocapsules and nanocontainers added into sol–gel obtained coatings which activated by occurring damaged and releasing the healing agent into the damaged area and helping to avoid the corrosion attack (Fig. 2) [37]. Silica (SiO_2), ceria (CeO_2), zinc oxide (ZnO), titania (TiO_2), zirconia (ZrO_2), and alumina (Al_2O_3) nanoparticles are used as corrosion protection additives [47–50].

Between the above nanoparticles, SiO_2 nanoparticles with low cost and availability is a good choice due to increase in the barrier properties of the coating and improved mechanical durability [51, 52]. Direct addition of healing agent or other nanoparticles into organic–inorganic coating have unavoidable disadvantage which caused indirect addition of nanoparticles into coating and improved mechanical properties of coating. Direct addition of nanoparticles led to chemical interaction formation between coating

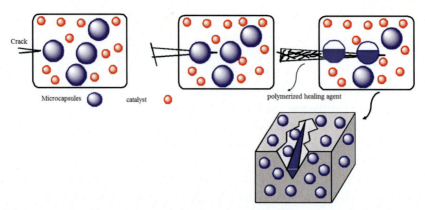

Fig. 2 Capsule-based healing process [37].

matrix chains and nanoparticles which limits activity of coating and barrier properties. First, it is quite difficult to control diffused of entrapped nanoparticles especially when they are poorly soluble within the coating matrix [53–57].

Controlled release of organic corrosion inhibitors from hybrid coatings nanocomposites led to protective coating with various functional groups and new cross-linking formed. Micro- or nano-containers containing corrosion inhibitor or scratch resistance nanoparticles incorporated into coating which release host material slowly with controlled release rate to damaged sites and protect the surface against corrosive parameters [53–57].

5. Applications of sol–gel derived coating

Chemically homogeneous and flexible coatings are formed via sol–gel method with good stability against chemical or mechanical attacks, oxidation control, and modified corrosion resistance of the coating using silicates or transition metal oxide nanoparticles [55–57]. Hybrid properties strongly depend on the nature and strength of the strong covalent bonding between the organic and inorganic components. Combination of rigidity of an inorganic phase and toughness of an organic phase make such systems interesting also from the aspect of mechanical behavior caused by crack-free hybrid coating which slows down corrosion [37, 58, 59].

5.1 Metallic nano-coating

Nanocoating containing metal oxides, or alloys such as Cadmium (Cd), Nickel (Ni), Tungsten (W), Zinc (Zn), Phosphorous (P), Cobalt (Co), Iron (Fe), and Copper (Cu), enhanced the properties obtained coating. The corrosion behavior of metallic nano-coating involves different factors such as composition of coating, structure size,

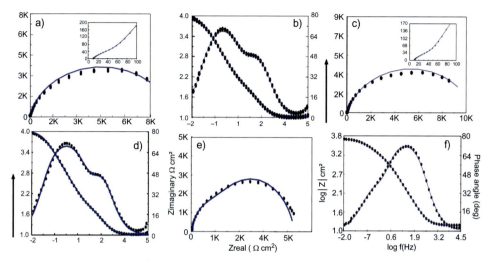

Fig. 3 Nyquist and bode plots obtained for pulse deposited Co-P coatings on MS substrate: (A and B) Co-P1, (C and D) Co-P2, and (E and F) Co-P3.

matrix of nanocomposite coatings. Coating containing metallic nanoparticles showed improvement in their properties such as protection against corrosive conditions. Zinc-based coating showed good weld ability which by increasing zinc thickness, protection increased [60–63].

Due to insufficient corrosion resistance of pure zinc, Zn-Ni alloy is used with good efficiency against adhesive wear resistance [64]. Nanocrystalline cobalt and its alloys, nano-crystalline Ni– alloy, nickel-copper alloy showed better hardness and scratch resistance under a corrosive environment [65, 66]. By increasing grain size of nano-crystalline Zn-Ni alloy in coating, corrosion resistance increased (13.31 wt% Ni to 17.62 wt% Ni for 26-nm grain size [67] and 37-nm grain [68], respectively). Fig. 3 indicated that addition of phosphorous to Co-Ni alloy coatings improved corrosion resistance in 3.5 wt% NaCl solution and 9% of the phosphorus (P) is optimum percent for this reason. Higher amount of P content led to higher film resistance (Rf), but indicated surface heterogeneity with higher P content improved the film resistance [69].

By increasing phosphorous in Ni-P nano-coating, amorphous structure is characteristic structure which is active phase in passivates in acidic and neutral solutions, so obtained coating showed better corrosion resistance in pH = 7 or less than 7 [70]. In another research EIS results showed that appropriate grain size of cobalt may improve corrosion potential value to the positive side with respect to the bigger nano-crystalline grain size [71]. Using zinc nanoparticles in coatings on electrogalvanized steel samples decreased corrosion rate compared to zinc micro-particles in NaOH solution which related to full and homogenous coverage of nanoparticles on substrate and enhanced the stability of the passive film [71]. Nano-crystalline structures showed higher grain

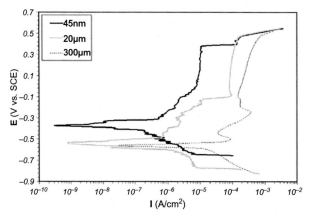

Fig. 4 Potentiodynamic polarization curves in 10 wt% sodium hydroxide aqueous solutions [72].

boundary density, so formed a stable and protective oxide layer and improved corrosion or scratch resistance [63, 68] and by increasing grain sizes of nano-crystalline iron from 20 to 500 μm, corrosion resistance improved (Fig. 4) [72].

Polymer nanocomposite coatings can have entrapped filler or pigments in their structure and formed organic–inorganic hybrid nanocomposites which led to specific properties such as stiffness, strength, conductivity, thermal resistance, and scratch [73, 74]. Polymeric coating containing ceramic alumina or silica nano-fillers without agglomeration improved scratch and wear resistance of covered substrate and create self-healing properties, which will result in increasing the scratch resistance. MWCNT, Al_2O_3, graphene oxide (GO), ZrO, and SiO_2 were added to various polymers via in situ polymerization or melt intercalation to achieve good dispersion and distribution which caused improvement of cohesive and adhesive properties of the nanostructure [73, 74]. Incorporation of Al_2O_3 nanoparticles into the polymer laminates or films improves the mechanical properties and increases corrosion resistance of the polymer via decrease in the corrosion current density of the nanocomposite coating (Fig. 5) and showed direct dependence to weight percent of nanoparticle in the matrix. The scratch resistance was two times higher than that of the polymer [75].

5.2 Scratch resistance coatings based on waterborne polymers

To reduce environmental disadvantage of volatile organic compounds(VOCs) plasticizers, waterborne coating with water as a solvent to disperse a resin are introduced and reinforced with nanoparticles such as Fe_3O_4, Fe_2O_3, and ZnO to investigate scratch behavior, UV resistance, and abrasion resistance of the coating as compared to the neat coating system [76–79]. Incorporation of 3.75% nano-silica into acrylic system, which is a good method to obtain scratch resistance automotive coating, improved the gloss retention about 99% from 96% after 1000 h accelerated weathering test (Fig. 6), and decreased

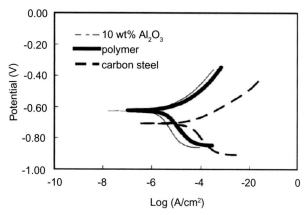

Fig. 5 Potentiodynamic curves for samples containing Al_2O_3 nanoparticles at 24 h immersion in 3 wt% NaCl solution [75].

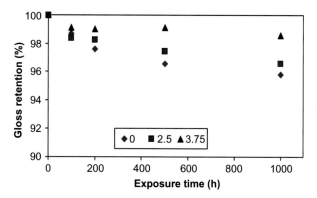

Fig. 6 Gloss retention of coatings containing nanosilica after various weathering [80].

both the roughness ratio and hardness [80]. Addition of nano-size Al_2O_3, Fe_2O_3, ZnO on the alkyd-based waterborne coating system improve corrosion resistance, scratch resistance and abrasion resistance of the alkyd-based waterborne coating was investigated which is due to the cross-link density of nano-particles (Fig. 7) [82].

5.3 Durability of organic coatings against mechanical wear

Wear and scratch are a result of motion or transfer layers which are usually adhered or interlocked mechanically to either of the counterparts. When two solid bodies moved, single asperities cause scratching to the worn surface and micro-scratches divided into micro-ploughing, micro-cutting and micro-cracking. If another scratch partially overlaps a previous scratch, the phenomenon can be called micro-fatigue (Fig. 8) [83].

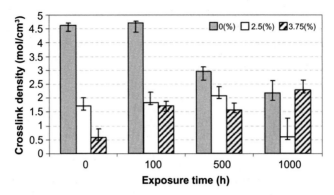

Fig. 7 Cross-link density of coating containing nano-silica after weathering [81].

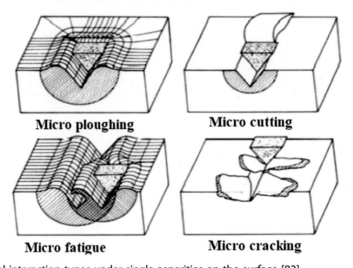

Fig. 8 Physical interaction types under single asperities on the surface [83].

The transition between micro-cutting and micro-ploughing is related to friction coefficient, operating conditions, and cohesion and elasticity influence on the transition between micro-cutting and micro-ploughing. Micro-cracking typically occurs on brittle materials when the cross-links between fillers and polymer matrix start cracking and propagate into detaching the fillers. Micro-cracking types are shown in Fig. 9 [83].

Hardness of polymers improved with molecular weight and crystallinity, on other hand, semi-crystalline polymers without spherulitic structures have shown higher wear resistance. By applying a coating on the surface of substrate increased interaction between polymer chains and surface, polymer molecules orientated and into slight increase of strength in the direction of sliding, improved the scratch and wear resistance. If the coating was brittle, orientation of polymer chains may lead to high amount of micro-cracking

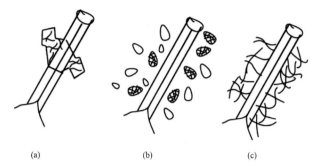

Fig. 9 Micro-cracking at (A) large filler, (B) around fillers, and (C) in the matrix [83].

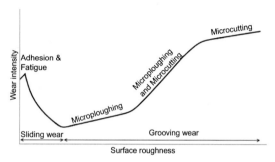

Fig. 10 Wear intensity, wear modes, and wear mechanisms as a function of the surface roughness of unlubricated sliding pairs [83].

and premature breakage. Crystallinity, fatigue properties, polymer chain length, cross-link density, interaction between chain and substrate, and cross-linking density influence on the wear properties [12].

The particle structure and size play an important role in the wear and scratch resistance properties of the coating. Fillers due to high surface particles, increase the surface roughness and surface quality which is showed in Fig. 10 [83].

Coatings strength may change via incorporation of soft and lubricating agent which increased flexibility of polymer chains and created a balance between reduction of friction and wear properties, but fillers may decrease toughness of polymer [84]. In case of nanoscale particles, the properties can differ significantly depending on the particle dispersion and amount of particles. [85].

5.4 The effect of silica nanoparticles on scratch resistance of coatings

Uniform distribution of nanoparticles and the amount of nanoparticles into polymeric coating improved strength, mechanical properties, and corrosion performance. Addition of 8 vol% silica nanoparticles to epoxy coating improved hardness, stiffness and scratch resistance and mechanical properties improved [86]. Increase of silica content in sol–gel coatings increase elastic modulus of polymer and hardness increased up to a volume fraction of about 0.58 and then decreased, which they associated it with revealed porosity inside the coating [87]. Hardness of obtained coatings was drastically dependent on silica amount and showed highest elastic modulus when 30 wt% of silica was added to hybrid coating [88]. Bauer et al. showed that silica and alumina particles with some modifications improved scratch resistance of polyacrylates with high modulus and heat resistance [38]. Silica nanoparticles without modification may decrease scratch resistance and with modification, matrix–filler interactions became weaker and radically affect the scratch resistance without drastically effect on Young's modulus [89].

5.5 Scratch resistance of epoxy-zeolite coatings

Nanotechnology and nanoparticles showed a critical role in corrosion protection and scratch resistance of coatings. Characteristic properties of surface coatings are dependent on nanoparticle structure and size. Coating containing nanoparticles is attractive for paint industries and automotive industries which led to multipurpose coatings [1–3]. Microporous zeolites are alumino-silicate minerals that can be used as adsorbents and catalysts for water purification due to Na^+, K^+, Ca^{2+}, Mg^{2+}, and others cations. Epoxy coatings can be used as corrosion inhibitors to assist the steel surface in resisting the attack by aggressive species such as chloride anions, but after crack formation poor resistance is obtained [90]. Incorporation of inorganic filler particles at nanometer decreases the porosity of epoxy coating and prevents diffusion path of harmful species and improves scratch resistance or corrosion protection properties via formation of new strong cross-links [91]. Zeolite due to porosity and mechanical strength showed good elastic modulus larger than 40 GPa. Indent hardness and elastic modulus of samples containing Zeolite before and after the dilute ammonia rinse shows in Fig. 11 (thicker than 500 nm) [28]. Zeolite-based coating showed good anti-reflective properties less than 100 nm. Results showed that at 5 nm into the polymeric coatings, hardness and modulus decrease which is due to surface properties in which the hardness and elastic modulus are above 1 and 10 GPa, respectively. At indent depth near to 40 nm, hardness curve became stable and plateau value is 1 GPa after rinsing with ammonia. By increasing plateau of the modulus curves at 30 nm depth, indicating an increase of film modulus from 25 to 35 GPa after the ammonia rinse (Fig. 11) [93].

5.6 Scratch resistance coating based on bio-based materials

Thermoplastic applications in automotive industries increasing drastically but show low resistance to mechanical damage. To overcome this limitation, polymer networks formed via using multifunctional compounds polymerization and radical (photo)-polymerization which led to heterogeneity at the micro- or nano scale. Bio-based UV-curable materials are good candidates for this application such as vegetable oils extracts, castor oil, tung oil,

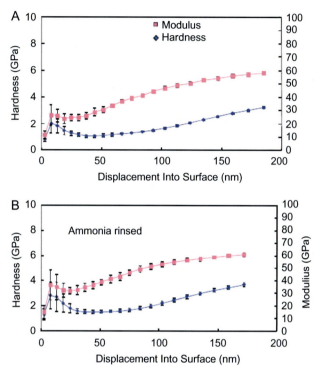

Fig. 11 Indent hardness and elastic modulus of sample with 4 h of humid baking before firing (A) and that after further ammonia wash (B) [92].

and a commercial epoxidized soybean oil [94]. Due to long alkyl chains, vegetable oils can't form stable networks for protective application, so networks became stable via photo-polymerizable. Photo-polymerizable bio-sourced acrylates coated on polycarbonate substrates were investigated for scratch resistance properties. Fig. 12 shows micro-scratch tests and results indicated that bio-based coatings did not show sufficient scratch resistance but improve the elastic recovery due to flexibility and thus notably limit the residual depth of the scratch. By increasing bio-based content, critical load under the same test condition increased from ~60% to ~175% for 12 and 13%, respectively [94]. The effect of DiPEPHA and PETA (Bio1 and 3, respectively) indicated an increase in elastic recovery which is related to saturated alkyl segments can't form new string interactions, so scratch resistance of the coatings did not improve and did nit effect indented depth. Cross-linking density of all coatings are similar but the different structure of acrylate-based caused variable scratch resistance of coated samples (Fig. 13) which is related to high elastic modulus and a wide mechanical relaxation.

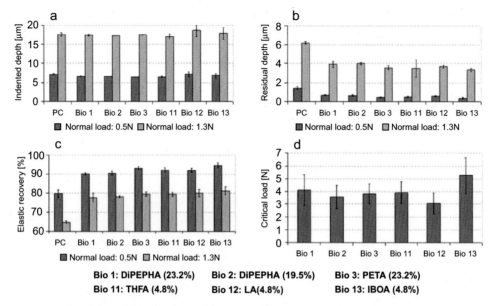

Fig. 12 Results of the micro-scratch tests for the bio-based 15 µm-thick coatings deposited on PC: indented depth (A), residual depth (B), elastic recovery (C), and critical load (D).

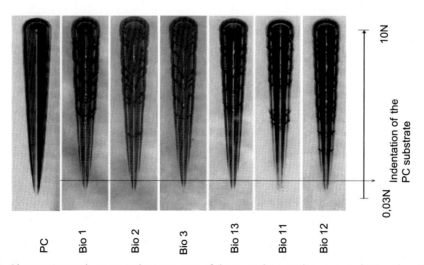

Fig. 13 Observation with an optical microscope of the scratches on the uncoated PC and on the bio-based 15 µm-thick coatings deposited on PC.

5.7 Scratch resistance coating based on magnesium alloys

Self-healing and scratch resistance coating containing magnesium/graphene oxide (GO) alloy with high strength to weight ratio and high damping capacity which are important for biomedical applications are applicable in various industrial applications [95–98]. GO formed strong hydrogen bonding with the polymer and improved thermal stability and decrease the oxygen permeability of obtained nanocomposites [99]. Multilayer membrane by alternative deposition of PEI and GO showed low permeation to oxygen and carbon dioxide due to layer structure of graphene [100]. Then, poly(ethylene imine) (PEI) and poly (acrylic acid) (PAA) were alternatively deposited on the Ce layer. Fig. 14 showed the buffering effect of the PEI/PAA multilayers in the presence of Ce(IV) after immersion of these samples in a 3.5 wt% NaCl solution. The sample with fewer bilayers of PEI/PAA (Ce(IV)/(PEI/PAA)5) showed a better corrosion resistance which absorbed the salt ions and the subsequent dissolving of the multilayers. More polyelectrolytes absorbed more salt ions during the same immersion time [101]. Because PEI and PAA are weak polyelectrolytes, the multilayers of PEI and PAA demonstrated an exponential growth and the cross-link density decreased with the growth of the multilayers due to the extrinsic charge compensation [102].

The decrease of thickness with the increase of multilayers of PEI/PAA confirmed the reaction between the polyelectrolytes and the cerium compounds during the coating preparation process. Because the cerium conversion layer also contributed to the corrosion resistance of magnesium alloy, the thicker layer provided a better protection to the magnesium substrate.

5.8 Evaluation of anti-scratch properties of graphene oxide

GO is a good candidate for anticorrosion coating due to specific liquid and gas barrier, chemical and thermal stabilities, mechanical stiffness and strength, and anticorrosion, antimicrobial, and antifouling features and can be introduced in organic coatings structure [103]. The development of GO-polymer-functionalized composite coatings is a potential method to effectively enhance anticorrosion properties and physicomechanical properties of the coatings used for industrial equipment anticorrosion protection [104]. Modified GO/polypropylene (PP) nanocomposites were synthesized, then mechanical and scratch resistance of these nanocomposites were determined (Fig. 15).

OTES act as coupling agent and increased reactive sites on the GO surface which caused more strong cross-links, so mechanical properties, elastic modulus, fracture toughness and scratch protection increased [36]. 1.0 wt% of GO and OTES led to improve hardness, elastic modulus and fracture toughness (286%, 127%, and 117%,

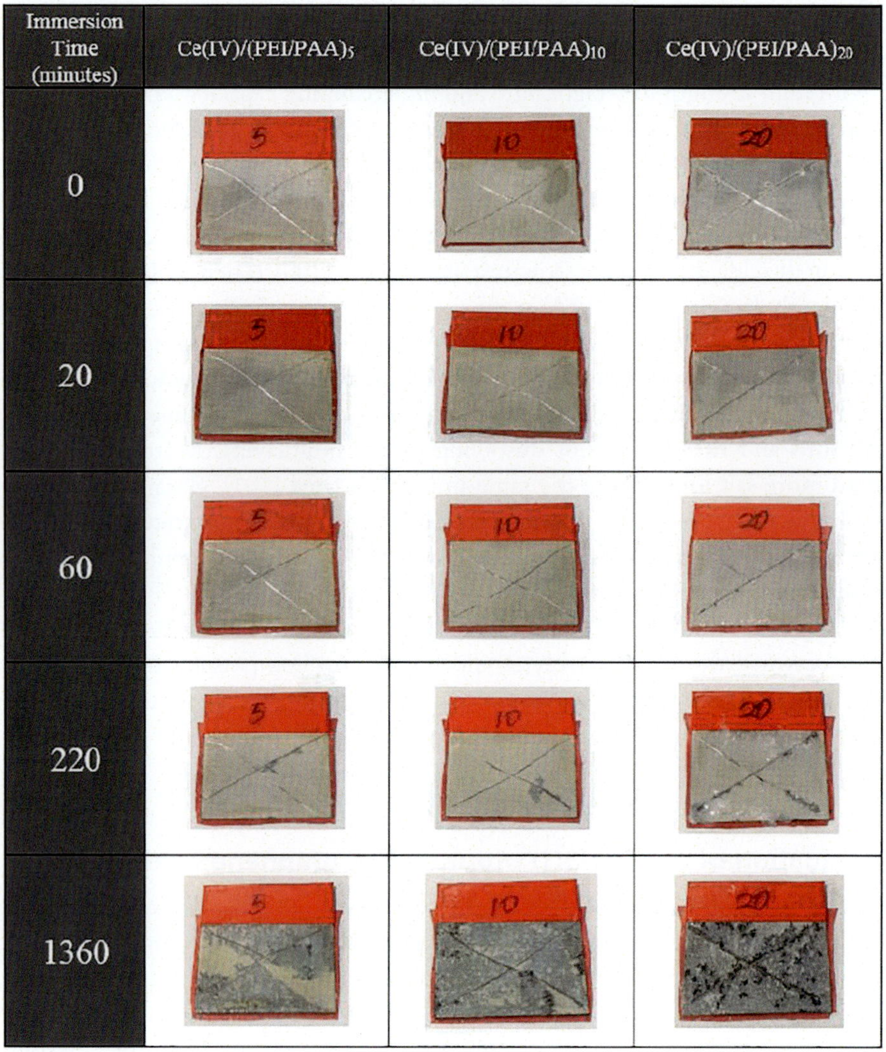

Immersion Time (minutes)	Ce(IV)/(PEI/PAA)$_5$	Ce(IV)/(PEI/PAA)$_{10}$	Ce(IV)/(PEI/PAA)$_{20}$
0			
20			
60			
220			
1360			

Fig. 14 Salt immersion tests of the multilayer coating.

respectively). These properties effect on scratch hardness and resistance of obtained nanocomposites and the scratch hardness increased 189% in the maximum amount [36]. SEM of GO/PP nanocomposites with and without OTES showed in Fig. 15, which in the presence of OTES, regular zigzag scratch was observed on the surface of GO/PP/OTES nanocomposites. By increasing GO amount, plastic damage decreased and tangential forces released and acting against the physical damage such as scratch.

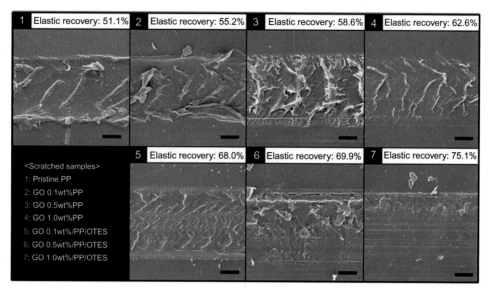

Fig. 15 The SEM micrographs of the scratch damage on the surface of several GO/PP nanocomposites [36].

Table 1 Scratch resistance of obtained coatings with different ZrO2 contents.

No.	ZrO$_2$ (g)	Scratch (g)	Surface gloss
a	0	650	111.1
b	0.0009	2750	154.3
c	0.0015	3950	148.1
d	0.0018	1200	133.5
e	0.0021	2150	143.8
f	0.003	2900	124.8
g	0.0042	2350	136.1
h	0.006	1900	135.2

5.9 Anti-scratch and anti-corrosion hybrid nanocomposite coatings based on zirconia

Metal chlorides such as zirconia led to improvement in chemical resistance and mechanical strength and also induced thermal-barrier coating to wear-resistant prosthesis. SiO_2/TiO_2/ZnO/ZrO_2 nanocomposites were reported which showed good scratch resistance. Scratch test results indicated that increasing the ZrO_2 weight ratio up to 0.0015, improved corrosion resistance of polyurethane (PU) matrix (Table 1). The hardness value increases by increasing the ZrO_2 weight ratio up to 0.0015 (Fig. 16). The nanostructured zirconia monolayer coating yielded good adhesion and less porosity, thus showing improvement. Incorporation of ZrO_2 nanoparticles into polyurethane

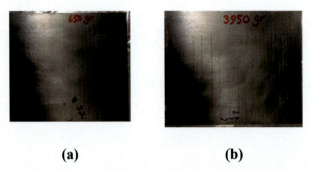

Fig. 16 Scratch test for (A) sample a and (B) sample c according to Table 1.

increased adhesion between PU chains, formed new hydrogen bonding and reduced cross-link density in the polymer to homogenous distribution of ZrO_2 nanoparticles in PU cross-links, so hardness improved [105]. The highest D value is seen which increases the level of the contact surface between the components ZrO_2 composite materials with high strength, hardness, and toughness. The very high level of contact between the resin matrix and the nanoparticles helps in the transfer of physical stress [105].

By good dispersion of ZrO_2 into PU matrix and formation of new interactions, mobility of the polymer chains restricted, which decrease crack growth and led to anti-scratch and anti-corrosion coatings. By further increasing the filler loading, the hardness effect of the polyurethane/ZrO_2 systems slowly increases or may even decrease. After 24 and 48 h immersion, uncoated samples start to corrode and the resistance decreased. By increasing immersion time, corrosion products accumulated on the surface and blockage of water and intrusive ion pores and surface resistance increased slightly [105].

5.10 Scratch resistance of acrylate coatings based on bio-polymers

Incorporation of modified SiO_2 nanoparticles with coupling agents into bio-based acrylate is a good method to improve hydrophobic properties of metals which is due to the presence of organosilanes and formation of stable interactions with resin. Pure epoxy coating without SiO_2 nanoparticles showed pencil hardness of 2B and epoxy coatings reinforced with specific SiO_2 nanoparticles showed pencil hardness of 6H which indicated good scratch resistance (Table 2). New and strong cross-linking and grafting processes led to movement restriction and scratch-resistance improved.

Silica nanoparticles also influenced the curing behavior (radical polymerization) of UV coatings due to synergistic effect of silica nanoparticles during the photopolymerization process (Table 3) and localization of silica nanoparticles on the surface of EPOLA composite, so mechanical properties improved.

Table 2 Variation in pencil hardness grade of modified surface cured by UV radiation.

Formulation	SiO$_2$	EPOLA	UV radiation (No. of passes)				
			1	2	3	4	5
F1	–	45	4B	3B	HB	F	H
F2	5	40	HB	F	F	2H	2H
F3	10	35	HB	F	H	2H	4H
F4	15	30	4F	F	3H	3H	5H
F5	20	25	2F	H	3H	3H	5H
F6	25	20	H	2H	4H	4H	6H
F7	30	15	H	2H	4H	5H	6H

Table 3 Scratch resistance of radiation cured materials using diamond tip with 90 degree.

Formulation	UV curing	EB curing
F1	0.1	0.1
F2	0.2	0.3
F3	0.3	0.4
F4	0.4	0.6
F5	0.5	0.6
F6	0.6	0.8
F7	0.8	0.9

In recent years, metal oxide nanoparticles are used to improved mechanical properties of coatings. Nano-alumina nanoparticles were used as additives in polyurethane coating to modify scratch properties of these coatings for industrial applications and investigated with laser scanning confocal microscopy [106]. For this reason, two kinds of fillers (A: non-modified alumina and B: modified alumina) were used with three different particle concentrations. When filler A percent is higher (2.5 and 5 wt%) scratch resistance increased (Fig. 17) which is due to formation of efficient cross-links, and change in scratch profile of the 1 wt% filler A system is also noticeable and the profile shape becomes wider and shallower shape similar to the profiles at higher loads. LSCM images showed interaction of light reflected and scattered by different layers of the coating and measured the relationship between damage and visibility which indicated that higher amount of nanoparticles formed higher cross-links led to higher scratch resistance.

Mousavi et al. studied the aluminum oxide coated glass. The scratch resistance of coatings containing nano-porous g-Al$_2$O$_3$ with grain sizes of approximately 30–80 nm showed hardness values ranging from 5 to 9H, which is within the desirable range for

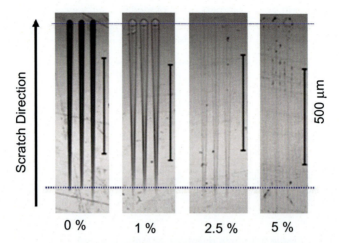

Fig. 17 LSCM images of P-scratch of an unfilled coating and the nanocomposite coatings containing filler A (The scale bar represents 500 μm. The *dashed lines* are for the visual guidelines for the onset force comparison).

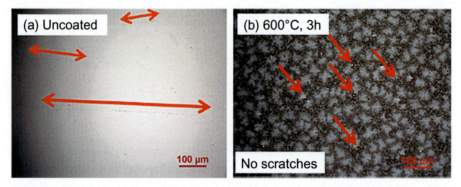

Fig. 18 Light micrographs of the samples after the Buehler scratch test: (A) uncoated glass with microscratches, (B) only test particles but no scratches on the coated glass were observed (annealed at 600 °C, 3 h). The *arrows* mark the traces of aluminum oxide particles from Buehler test.

industrial scratch-resistant coatings and indicated inter diffusion [107]. Fig. 18 showed Buehler scratch test which uncoated glass showed scratches and coated glass can be seen without scratches and introduced a new route to scratch-protective layers on optical elements, e.g., on solar cells, antireflective optical coatings, and displays [107].

5.11 Anti-reflection coatings with enhanced abrasion and scratch resistance properties

Multifunctional coatings containing nanoparticles showed specific optical and solar applications coated on glass or metals with high transmittance. Scratch resistance and

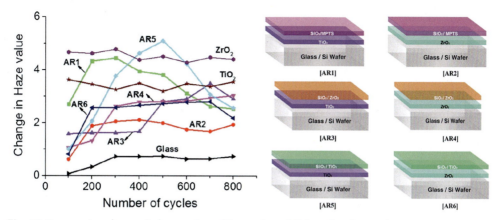

Fig. 19 Progressive change in haze value with number of Taber abrasion cycles.

anti-corrosion properties of obtained coatings incorporated with ZrO_2, TiO_2, or silane compounds may act as anti-reflection (AR), scratch and abrasion resistance properties. For this reason, low refractive nanocomposite layer as a top layer and a high refractive layer as a bottom layer showed a high transmission in the visible range with comparatively better abrasion and scratch resistance. Coated silicon wafers act as optical and photovoltaic coatings with high mechanical stability and excellent low reflectance property. Taber abrasion tests using CS10F wheels with a 500-g load on each wheel, and the substrates were allowed to be abraded cumulatively to 800 cycles, then the haze value through wear track area was measured. According to the Taber Abrader test results shown in Fig. 19, the AR coatings, AR2, AR4, and AR6, respectively, exhibit a low change in haze value, 3 even after 800 cycles. In the case of AR sample, a low change in haze value was also which is seen for TiO_2 and ZrO_2 layer alone coated. For AR1 and AR5 a great change in haze value when compared to other AR coating systems. Haze value was increased with number of abrading cycles from 100 to 500 cycles due to a low abrasive resistance of the top layer, and after 500 cycles haze value decrease which is related to a high abrasive resistance of bottom layer. Zirconia, with a low refractive composite SiO_2/MPTS and exhibits an excellent abrasion resistance compared to all the other AR systems. Presence of zirconia in the presence of and a low refractive composite SiO_2/MPTS (mercaptopropyl trimethoxysilane) and exhibits an excellent abrasion resistance compared to all the other AR systems. When silane-functionalized organic coupling agents such as TEOS (tetraethyl orthosilicate) improves the scratch resistance which is related to reaction of hydroxyl groups on ZrO_2 layer and formation of new cross-links.

6. Concluding remarks

Apply protective coating on the surface of metals and alloys to protect the surface from environmental corrosive conditions and improve lifetime of coated substrate. Incorporation of metal oxide as filler improved scratch resistance and corrosion protection of coatings and prevent crack initiation of propagation on the surface. Presence of inorganic and organic components led to chemical or physical interactions and also formed covalent cross-links with metals or alloys surface with inorganic/organic hybrid coating. Al_2O_3, TiO_2, SiO_2, ZrO_2, graphene oxide, zeolite, and magnesium may be incorporated into polymeric coating via direct or indirect addition methods which ensure long-term protection. In this chapter, the ability of various nanocontainers was investigated which indicated long-term corrosion protection and scratch resistance for metals.

References

[1] G.J.B.S. Decher, Multilayer Thin Films: Sequential Assembly of Nanocomposite Materials, second ed., John Wiley & Sons, 2012.
[2] S. Bekele, M.D. Esakov, Polyamide multilayer film, 2012. U.S. Patent No. 8,206,818.
[3] T. Puntous, S. Pavan, D. Delafosse, M. Jourlin, J. Rech, Ability of quality controllers to detect standard scratches on polished surfaces, Precis. Eng. 37 (2013) 924–928.
[4] M. Klecka, G. Subhash, Grain size dependence of scratch-induced damage in alumina ceramics, Wear 265 (2008) 612–619.
[5] K.H. Nitta, Y.W. Shin, H. Hashiguchi, S. Tanimoto, M. Terano, Morphology and mechanical properties in the binary blends of isotactic polypropylene and novel propylene-co-olefin random copolymers with isotactic propylene sequence. Ethylene–propylene copolymers, Polymer 46 (2005) 965–975.
[6] X. Zhang, A. Ajji, Oriented structure of PP/LLDPE multilayer and blends films, Polymer 46 (2005) 3385–3393.
[7] L.J. De Cock, S. De Koker, B.G. De Geest, J. Grooten, C. Vervaet, J.P. Remon, G.B. Sukhorukov, M.N. Antipina, Polymeric multilayer capsules in drug delivery, Angew. Chem. Int. Ed. 49 (2010) 6954–6973.
[8] P.H. Chua, K.G. Neoh, E.T. Kang, W. Wang, Surface functionalization of titanium with hyaluronic acid/chitosan polyelectrolyte multilayers and RGD for promoting osteoblast functions and inhibiting bacterial adhesion, Biomaterials 29 (2008) 1412–1421.
[9] J.H. Han, Antimicrobial Food Packaging. Novel Food Packaging Technique, Wood head Publishing Limited, Cambridge, England, 2003, pp. 50–70.
[10] J.A. Hiller, J.D. Mendelsohn, M.F. Rubner, Reversibly erasable nanoporous anti-reflection coatings from polyelectrolyte multilayers, Nat. Mater. 1 (2002) 59–63.
[11] Y.C. Li, S. Mannen, A.B. Morgan, S. Chang, Y.H. Yang, B. Condon, J.C. Grunlan, Intumescent all-polymer multilayer nanocoating capable of extinguishing flame on fabric, Adv. Mater. 23 (2011) 3926–3931.
[12] M.C.O. Chang, K.R. Hansford, Sound absorbing non-woven material, sound absorbing multilayer film, and laminates made thereof. U.S. Patent Application No. 14/368,220, 2012.
[13] T.C. Wang, R.E. Cohen, M.F. Rubner, Metallodielectric photonic structures based on polyelectrolyte multilayers, Adv. Mater. 14 (2002) 1534–1537.
[14] ASTM International, ASTM G 171-03, Standard test method for scratch hardness of materials Using a Diamond Stylus, Annual Book of ASTM Standards, 3:02, 2003.
[15] International Organization for Standardization, ISO 12137-1:1997—Determination of Mar Resistance—Part 1: Method Using a Curved Stylus, ISO Standards, 1997.

[16] C. Xiang, H.J. Sue, J. Chu, B. Coleman, Scratch behavior and material property relationship in polymers, J. Polym. Sci. B 39 (2001) 47–59.
[17] J. Chu, L. Rumao, B. Coleman, Scratch and mar resistance of filled polypropylene materials, Polym. Eng. Sci. 38 (1998) 1906–1914.
[18] J. Chu, C. Xiang, H.J. Sue, R. Hollis, Scratch resistance of mineral-filled polypropylene materials, Polym. Eng. Sci. 40 (2000) 944–955.
[19] J.L. Courter, E.A. Kamenetzky, Micro- and nano-indentation and scratching for evaluating the mar resistancle of automotive clearcoats, J. Eur. Coat. 7 (1999) 100–115.
[20] H.J. Sue, A.F. Yee, Study of fracture mechanisms of multiphase polymers using the double-notch four-point-bending method, J. Mater. Sci. 28 (1993) 2975–2980.
[21] J. Vincent, M. Pierre, Viscoelastic effects on the scratch resistance of polymers: relationship between mechanical properties and scratch properties at various temperatures, Prog. Org. Coat. 48 (2003) 322–331.
[22] C. Gauthier, S. Lafaye, R. Schirrer, Elastic recovery of a scratch in a polymeric surface: experiments and analysis, Tribol. Int. 34 (2001) 469–479.
[23] C. Gauthier, R. Schirrer, Time and Temperature Dependence of the Scratch Properties of PMMA Surfaces, vol. 30, Kluwer Academic Publishers, Dordrecht, 2000, pp. 2121–2130.
[24] C. Gauthier, R. Schirrer, The visco-elastic visco-plastic behavior of a scratch on a polymeric surface, in: World Tribology Conference, WTC No. 2, Vienna, 2001.
[25] J. Vincent, W.C. Oliver, Viscoelastic behavior of polymers films during scratch test: a quantitative analysis, in: Proceedings of the MRS Symposium, 2000, p. 594.
[26] J. Vincent, Understanding and quantification of elastic and plastic deformation during a scratch test, Wear 218 (1998) 8–14.
[27] J. Vincent, W.C. Oliver, On the robustness of scratch testing for thin films: the issue of tip geometry for critical load measurement, in: Proceedings of the MRS Symposium, 2000, p. 594.
[28] S. Amiri, A. Rahimi, Sol-gel technology and various hybrid nanocomposite coatings applications, Iran. Polym. J. 25 (2016) 559–577.
[29] A. Rahimi, Inorganic and organometallic polymers: a review, Iran. Polym. J. 13 (2004) 149–164.
[30] S. Amiri, S. Amiri, Cyclodextrins: Properties and Industrial Applications, first ed., Wiley, 2017.
[31] S. Kang, S. Hong, C.R. Choe, M. Park, S. Rim, J. Kim, Preparation and characterization of epoxy composites filled with functionalized nanosilica particles obtained via sol-gel process, Polymer 42 (2001) 879–887.
[32] A. Ghanbari, M.M. Attar, A study on the anticorrosion performance of epoxy nanocomposite coatings containing epoxy-silane treated nano-silica on mild steel substrate, J. Ind. Eng. Chem. 23 (2015) 145–153.
[33] PPG, Comparing PPG polysiloxane coatings and traditional coating systems, in: PPG Protective and Marine Coatings, 2011. https://www.cgedwards.com/ameron/PSX%20700SG%20Whitepaper%20Final.pdf.
[34] D. Wang, G.P. Bierwagen, Sol-gel coatings on metals for corrosion protection, Prog. Org. Coat. 64 (2009) 327–338.
[35] B.A. Latella, B.K. Gan, C.J. Barbe, D.J. Cassidy, Nanoindentation hardness, Young's modulus, and creep behaviour of organic-inorganic silica-based sol-gel thin films on copper, J. Mater. Res. 23 (2008) 2357–2365.
[36] F. Mammeri, E. Le Bourhis, L. Rozes, C. Sanchez, Mechanical properties of hybrid organic-inorganic materials, J. Mater. Chem. 15 (2005) 3787–3811.
[37] R.B. Figueira, C.J.R. Silva, E.V. Pereira, Organic-inorganic hybrid sol-gel coatings for metal corrosion protection: a review of recent progress, J. Coat. Technol. Res. 12 (2015) 1–35.
[38] F. Bauer, H. Glasel, U. Decker, H. Ernst, A. Freyer, E. Hartmann, V. Sauerland, R. Mehnert, Trialkoxysilane grafting onto nanoparticles for the preparation of clearcoat polyacrylate systems with excellent scratch performance, Prog. Org. Coat. 47 (2003) 147–153.
[39] N. Jalili, K. Laxminarayana, A review of atomic force microscopy imaging systems: application to molecular metrology and biological sciences, Mechatronics 14 (2004) 907–945.

[40] J. Yang, L. Bai, G. Feng, X. Yang, M. Lv, C. Zhang, H. Hu, X. Wang, Thermal reduced graphene based polyethylene vinyl alcohol nanocomposites: enhanced mechanical properties, gas barrier, water resistance, and thermal stability, Ind. Eng. Chem. Res. 52 (2013) 16745–16754.

[41] L. Zhang, F. Zhan, X. Yang, G. Long, W. Yingpeng, T. Zhang, Y. Huang, Y. Ma, A. Yu, Y. Chen, Porous 3D graphene-based bulk materials with exceptional high surface area and excellent conductivity for supercapacitors, Sci. Rep. 3 (2013), 1408.

[42] C. Lee, X. Wei, J.W. Kysar, J. Hone, Measurement of the elastic properties and intrinsic strength of monolayer graphene, Science 321 (2008) 385–388.

[43] A. Araujo, G.L. Botelho, M. Silva, A.V. Machado, UV stability of poly(lactic acid) nanocomposites, J. Mater. Sci. Eng. B 3 (2013) 75–83.

[44] L. Moyo, W.W. Focke, D. Hedenreich, F.J.W.J. Labuschagne, H.-J. Radusch, Properties of layered double hydroxide micro- and nanocomposites, Mater. Res. Bull. 48 (2013) 1218–1227.

[45] Z. Wang, E. Han, W. Ke, Influence of nano-LDHs on char formation and fire-resistant properties of flame-retardant coating, Prog. Org. Coat. 53 (2005) 29–37.

[46] C. Arunchandran, S. Ramya, R.P. George, U.K. Mudali, Self-healing corrosion resistive coatings based on inhibitor loaded TiO2 nanocontainers, J. Electrochem. Soc. 159 (2012) 552–559.

[47] D. Zhou, A.A. Keller, Role of morphology in the aggregation kinetics of ZnO nanoparticles, Water Res. 44 (2010) 2948–2956.

[48] S.M. Gupta, M. Tripathi, A review of TiO2 nanoparticles, Chin. Sci. Bull. 56 (2011) 1639–1657.

[49] G.X. Shen, Y.C. Chen, L. Lin, C.J. Lin, D. Scantlebury, Study on a hydrophobic nano-TiO2 coating and its properties for corrosion protection of metals, Electrochim. Acta 50 (2005) 5083–5089.

[50] S.A. Khorramie, M.A. Baghchesara, R. Lotfi, S.M. Dehagi, Synthesis of ZrO2 nanoparticle by combination of sol-gel auto-combustion method-irradiation technique, and preparation of Al-ZrO2 metal matrix composites, Int. J. Nano Dimension 2 (2012) 261–267.

[51] M.V. Esther, R. Frederik, W. Daming, S.M. David, E.S. Mark, V.H. John, R.J. Julian, Bioactivity in silica/poly(γ-glutamic acid) sol–gel hybrids through calcium chelation, Acta Biomater. 9 (2013) 7662–7671.

[52] X.F. Song, L. Gao, Fabrication of hollow hybrid microspheres coated with silica/titania via sol-gel process and enhanced photocatalytic activities, J. Phys. Chem. C 23 (2007) 8180–8187.

[53] S. Amiri, A. Rahimi, Preparation of supramolecular corrosion inhibitor nanocontainers for self-protective hybrid nanocomposite coatings, J. Polym. Res. 21 (2014) 556, https://doi.org/10.1007/s10965-014-0566-5. 2014.

[54] Amiri, S., Rahimi, A. (2014) Self-healing hybrid nanocomposite coatings containing encapsulated organic corrosion inhibitors nanocontainers, J. Polym. Res., 22:624, DOI: https://doi.org/10.1007/s10965-014-0624-z, 2014.

[55] S. Amiri, A. Rahimi, Synthesis and characterization of supramolecular corrosion inhibitor nanocontainers for anticorrosion hybrid nanocomposite coatings, J. Polym. Res. 22 (2015) 66, https://doi.org/10.1007/s10965-015-0699-1.

[56] S. Amiri, A. Rahimi, Anti-corrosion hybrid nanocomposite coatings with encapsulated organic corrosion inhibitors, J. Coat. Technol. Res. 12 (2015) 587–593.

[57] S. Amiri, A. Rahimi, Self-healing anticorrosion coating containing 2-mercaptobenzothiazole and 2-mercaptobenzimidazole nanocapsules, J. Polym. Res. 23 (2016) 83, https://doi.org/10.1007/s11998-014-9652-1.

[58] M.L. Zheludkevich, I.M. Salvado, M.G.S. Ferreira, Sol-gel coatings for corrosion protection of metals, J. Mater. Chem. 15 (2005) 5099–5111.

[59] C.J. Brinker, G.W. Scherer, Sol-Gel Science: The Physics and Chemistry of Sol Gel Processing, Academic Press, 1990.

[60] K.M.S. Youssef, C.C. Koch, P.S. Fedkiw, Improved corrosion behavior of nanocrystalline zinc produced by pulse-current electrodeposition, Corros. Sci. 46 (2004) 51–64.

[61] L. Wang, J. Zhang, Y. Gao, Q. Xue, L. Hu, T. Xu, Grain size effect in corrosion behavior of electrodeposited nanocrystalline Ni coatings in alkaline solution, Scr. Mater. 55 (2006) 657–660.

[62] L. Wang, Y. Lin, Z. Zeng, W. Liu, Q. Xue, L. Hu, J. Zhang, Electrochemical corrosion behavior of nanocrystalline Co coatings explained by higher grain boundary density, Electrochim. Acta 52 (2007) 4342–4350.

[63] S.K. Ghosh, G.K. Dey, R.O. Dusane, A.K. Grover, Improved pitting corrosion behaviour of electrodeposited nanocrystalline Ni-Cu alloys in 3.0 wt.% NaCl solution, J. Alloys Compd. 426 (2006) 235–243.

[64] K.R. Sriraman, H.W. Strauss, S. Brahimi, R.R. Chromik, J.A. Szpunar, J.H. Osborne, S. Yue, Tribological behavior of electrodeposited Zn, Zn–Ni, Cd and Cd–Ti coatings on low carbon steel substrates, Tribiol. Int. 56 (2012) 107–120.

[65] A. Aledresse, A. Alfantazi, A study on the corrosion behavior of nanostructured electrodeposited cobalt, J. Mater. Sci. 39 (2004) 1523–1526.

[66] M. Mirak, M. Alizadeh, M. Ghaffari, Characterization, mechanical properties and corrosion resistance of biocompatible Zn-HA/TiO2 nanocomposite coatings, J. Mech. Behav. Biomed. Mater. 62 (2016) 282–290.

[67] Z. Feng, Q. Li, J. Zhang, P. Yang, H. Song, M. An, Electrodeposition of anocrystalline Zn-Ni coatings with single gamma phase from an alkaline bath, Surf. Coat. Technol. 270 (2015) 47–56.

[68] S.H. Mosavat, M.H. Shariat, M.E. Bahrololoom, Study of corrosion performance of electrodeposited nanocrystalline Zn-Ni alloy coatings, Corros. Sci. 59 (2012) 81–87.

[69] V.E. Selvi, H. Seenivasan, K.S. Rajam, Electrochemical corrosion behavior of pulse and DC electrodeposited Co-P coatings, Surf. Coat. Technol. 206 (2012) 2199–2206.

[70] Z.O.U. Longfei, L.U.O. Shoufu, L.I. Pengxing, A study on the anodic polarization behaviours of electroless nickel coatings in acidic, alkaline and neutral solutions, Surf. Coat. Technol. 36 (1988) 455–462.

[71] H. Jung, A. Alfantazi, An electrochemical impedance spectroscopy and polarization study of nanocrystalline Co and Co-P alloy in 0.1 M H2SO4 solution, Electrochim. Acta 51 (2006) 1806–1814.

[72] V. Afshari, C. Dehghanian, Effects of grain size on the electrochemical corrosion behaviour of electrodeposited nanocrystalline Fe coatings in alkaline solution, Corros. Sci. 51 (2009) 1844–1849.

[73] V. Mittal, Polymer nanocomposites: Synthesis, microstructure, and properties, in: Optimization of PolymerNanocomposite Properties, WILEY-VCH Verlag GmbH & Co. KGaA, Weinheim, Germany, 2010, pp. 1–19. ISBN 9783527325214.

[74] M. Oliveira, A. Machado, Preparation of polymer-based nanocomposites by different routes, Nanocomp. Synth. Charact. Appl. 1–22 (2013).

[75] 128. Y. Wang, S. Lim, J.L. Luo, Z.H. Xu, Tribological and corrosion behaviors of Al2O3/polymer nanocomposite coatings, Wear 260 (2006) 976–983.

[76] O.U. Rahman, M. Kashif, S. Ahmad, Nanoferrite dispersed waterborne epoxy-acrylate: anticorrosive nanocomposite coatings, Prog. Org. Coat. 80 (2015) 77–86.

[77] S. Wang, H.M. Ang, O. Moses, Volatile organic compounds in indoor environment and photocatalytic oxidation: state of the art, Environ. Int. 33 (2007) 694–705.

[78] S.K. Dhoke, T.J. Mangal Sinha, A.S. Khanna, Effect of nano-Fe2O3 particles particles on the corrosion behavior of alkyd based waterborne coatings, J. Coat. Technol. Res. 6 (2009) 353–368.

[79] S.K. Dhoke, A.S. Khanna, T.J.M. Sinha, Effect of nano-ZnO particles on the corrosion behavior of alkyd-based waterborne coatings, Prog. Org. Coat. 64 (2009) 371–382.

[80] H. Yari, S. Moradian, N. Tahmasebi, The weathering performance of acrylic melamine automotive clearcoats containing hydrophobic nanosilica, J. Coat. Technol. Res. 11 (2014) 351–360.

[81] H. Yari, S. Moradian, N. Tahmasebi, M. Arefmanesh, The effect of weathering on tribological properties of an acrylic melamine automotive nanocomposite, Tribol. Lett. 46 (2012) 123–130.

[82] B. Ahmadi, M. Kassiriha, K. Khodabakhshi, E.R. Mafi, Effect of nano layered silicates on automotive polyurethane refinish clear coat, Prog. Org. Coat. 60 (2007) 99–104.

[83] K. Zum Gahr, Microstructure and Wear of Materials, Elsevier, Amsterdam, 1987.

[84] B.J. Briscoe, S.K. Sinha, Tribological applications of polymers and their composites: past, present and future prospects, in: K. Friedrich, A.K. Schlarb (Eds.), Tribology and Interface Engineering Series, Elsevier, 2008, pp. 1–14.

[85] A. Abdelbary, Wear of Polymers and Composites, Woodhead Publishing, Oxford, 2014.

[86] Cripps, D. Polyester Resins. n.d. NetComposites. Available from https://netcomposites.com/guide-tools/guide/resin-systems/polyester-resins/, 2019.

[87] J. Malzbender, J.M.J. Toonder, A.R. Balkenende, G. With, Measuring mechanical properties of coatings: a methodology applied to nano-particle-filled sol-gel coatings on glass, Mater. Sci. Eng. R 36 (2002) 7–103.

[88] J. Ballare, E. Jimenez-Pique, M. Anglada, S.A. Pellice, A.L. Cavalieri, Mechanical characterization of nano-reinforced silica based sol-gel hybrid coatings on AISI 316L stainless steel using nanoindentation techniques, Surf. Coat. Technol. 203 (2009) 3325–3331.

[89] J. Douce, J. Boilot, J. Biteau, L. Scodellaro, A. Jimenez, Effect of filler size and surface condition of nano-sized silica particles in polysiloxane coatings, Thin Solid Films 466 (2004) 114–122.

[90] F. Dietsche, Y. Thomann, R. Thomann, R. Mulhaupt, Translucent acrylic nanocomposites containing anisotropic laminated nanoparticles derived from intercalated layered silicates, J. Appl. Polym. Sci. 75 (2000) 396–405.

[91] R. Li, L. Chen, A paint containing nano titanium oxide and nano silver, and its preparation method. CN 10027622, 2005.

[92] E. Berasategui, S.J. Bull, T.F. Page, Mechanical modelling of multilayer optical coatings, Thin Solid Films 447 (2004) 26–32.

[93] W.C. Oliver, G.M. Pharr, An improved technique for determining hardness and elastic-modulus using load and displacement sensing indentation experiments, J. Mater. Res. 7 (1992) 1564–1583.

[94] P. Emeline, L. Sébastien, M. Michel, A. Justine, V. Vincent, F. Etienne, M. Françoise, Effect of bio-based monomers on the scratch resistance of acrylate photopolymerizable coatings, J. Polym. Sci. Polym. Phys. B53 (2015) 379–388.

[95] X. Cui, X. Lin, C. Liu, R. Yang, X. Zheng, M. Gong, Fabrication and corrosion resistance of a hydrophobic micro-arc oxidation coating on AZ31 Mg alloy, Corros. Sci. 90 (2015) 402–412.

[96] L. Wu, J. Dong, W. Ke, Potentiostatic deposition process of fluoride conversion film on AZ31 magnesium alloy in 0.1 M KF solution, Electrochim. Acta 105 (2013) 554–559.

[97] Y. Dou, S. Cai, X. Ye, G. Xu, K. Huang, X. Wang, M. Ren, 45S5 bioactive glass–ceramic coated AZ31 magnesium alloy with improved corrosion resistance, Surf. Coat. Technol. 228 (2013) 154–161.

[98] L. Zhang, J. Zhang, C. Chen, Y. Gu, Advances in microarc oxidation coated AZ31 Mg alloys for biomedical applications, Corros. Sci. 91 (2015) 7–28.

[99] R. Liu, S. Liang, X.Z. Tang, D. Yan, X. Li, Z.Z. Yu, Tough and highly stretchable graphene oxide/polyacrylamide nanocomposite hydrogels, J. Mater. Chem. 22 (2012) 14160–14167.

[100] Y.H. Yang, L. Bolling, M.A. Priolo, J.C. Grunlan, Super gas barrier and selectivity of graphene oxide-polymer multilayer thin films, Adv. Mater. 25 (2013) 503–508.

[101] N. Joseph, P. Ahmadiannamini, R. Hoogenboom, I.F.J. Vankelecom, Layer-by-layer preparation of polyelectrolyte multilayer membranes for separation, Polym. Chem. 5 (2014) 1817–1831.

[102] K.M. Holder, B.R. Spears, M.E. Huff, M. Priolo, E. Harth, J.C. Grunlan, Stretchable gas barrier achieved with partially hydrogen-bonded multilayer nanocoating, Macromol. Rapid Commun. 35 (2014) 960–964.

[103] Z. Lei, W. Haitang, W. Ming, Z. Zeyu, D.V. Dinh, T.X.B. Thi, H. Xiaohua, Preparation, characterization, and properties of graphene oxide/urushiol-formaldehyde polymer composite coating, J. Coat. Technol. Res. 15 (2018) 1343–1356.

[104] L.H. He, Y. Zhao, L.Y. Xing, P.G. Liu, Z.Y. Wang, Y.W. Zhang, X.F. Liu, Preparation of phosphonic acid functionalized graphene oxide-modified aluminum powder with enhanced anticorrosive properties, Appl. Surf. Sci. 411 (2017) 235–239.

[105] L.S. Karimi, S. Amiri, M.H. Zori, Synthesis and characterization of anti-scratch and anti-corrosion hybrid nanocomposite coatings based on sol-gel process, Silicon (2020), https://doi.org/10.1007/s12633-020-00721-w.

[106] S. Li-Piin, C. Jeffrey, M.F. Aaron, H. Haiqing, F. Bryce, B. Lukas, H.F. Raymond, Scratch behavior of nano-alumina/polyurethane coatings, J. Coat. Technol. Res. (2008), https://doi.org/10.1007/s11998-008-9110-z.

[107] S.H. Mousavi, M.H. Jilavi, M. Koch, E. Arzt, P.O. William, Development of a transparent scratch resistant coating through direct oxidation of Al-coated glass, Adv. Eng. Mater. (2016), https://doi.org/10.1002/adem.201600617.

CHAPTER 18

Self-healing nanocoatings

Andressa Trentin, Mayara Carla Uvida, Adriana de Araújo Almeida, Thiago Augusto Carneiro de Souza, and Peter Hammer

São Paulo State University (UNESP), Institute of Chemistry, Araraquara, SP, Brazil

Chapter outline

1. Introduction — 371
2. Inorganic corrosion inhibitors — 376
3. Organic corrosion inhibitors — 384
4. Conclusions — 396
Acknowledgments — 396
References — 396

1. Introduction

The urgent need to reduce greenhouse gas emissions presses the automotive industry to face the challenge of reducing vehicle weight. Although there is no single solution for the development of lightweight car bodies, the sector has succeeded in creating hybrid structures, considerably increasing the use of aluminum and magnesium alloys, due to their excellent strength-to-weight ratio [1, 2]. Efforts to replace steel as the main component of the vehicle body have been applied in the customization of materials for each part of the car, improving drivability, fuel economy, and cabin silence [3]. It is estimated that in the next 30 years the percentage of iron and steel alloys in automobiles will be reduced from presently around 75% to approximately 50% due to the advances in the development of aluminum, magnesium alloys, and reinforced carbon fiber composites [3, 4]. Fig. 1 gives an overview of the alloy components used for the car body.

However, working with multiple metals is not an easy task. In addition to issues such as cost, recyclability, thermal expansion, joining methods, and the supply chain, metal corrosion remains an important challenge in terms of durability and safety [3, 4]. The advances of polymeric coatings in the last century resulted in technologies capable of meeting current demands in the automotive sector, solving issues such as faster curing time, improved finishing, and environmental safety [1, 2]. Currently, the durability of these coating systems has been significantly extended by the constant development of paints and varnishes that passively act against mechanical, thermal, and chemical damage, namely by increasing adhesion to the metal substrate, improving thermal stability, and acting as an efficient barrier against corrosion. This was possible thanks to the evolution of the chemistry of polymers and solvents, the application methods, and the optimization

Fig. 1 Representative scheme of the main metal components of car body. *(Callout: Car body metallic components.)*

of the engineering production line [1, 2], thus leading to estimated revenue in the automotive industry of around EUR 79 billion worldwide in 2020, an increase of about 32% compared to the profits of 2012 [5].

The process chain of automotive industries has undergone many adaptations to save time and energy. In recent years, the wet-on-wet-on-wet (3W) process has been adopted by an increasing number of automakers. This implies that the curing of the primer layer is eliminated, and the clear coat is applied to the wet body, reducing the energy consumed by eliminating drying steps between the layers, shown in Fig. 2 [4]. In addition, the number of layers was reduced to four or five and, although there are variations between industries, the technologies are currently almost standardized worldwide. The lower part of Fig. 2 shows a cross-sectional view of a generic coating system consisting of inorganic surface pretreatment (phosphating), electrocoating (cathodic dip painting based on epoxy-polyurethanes) [6], primer (electrostatically deposited powders or liquid formulations based on UV-resistant polyesters, epoxies or acrylates) [7], base coating (typically composed of high solids, waterborne amines or powders containing pigments and particles for metallic effects), and the transparent topcoat of one or two components (acrylic resin bonded to reactive polyurethane or melamine crosslinker) [1, 2, 8]. Consequently, the development of water-based technologies, wet-on-wet deposition, and in particular, automation of cleaning, pretreatment, and painting by robots, reduced the total coating process to about 8 h, as well as the occupational risk during the application of paints [8].

Despite these advances, nowadays countless research activities are conducted to develop self-triggered smart coatings capable of regenerating themselves after failure, extending the safety, lifetime, and functionality of automobiles. A smart coating is based on a material that, if damaged, can repair or heal itself, slowing or even stoping corrosion

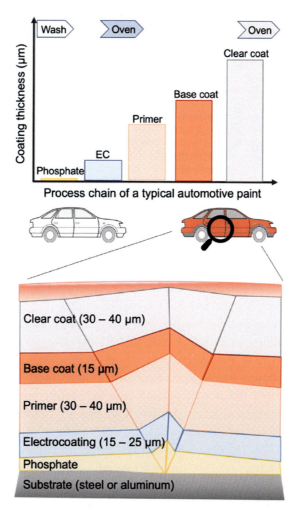

Fig. 2 Simplified 3W process chain of an automotive paint and representative cross-section of automotive coatings. *(Callout: Multilayer structure of automotive paint.)*

and, in the best-case scenario, restoring completely its functionality and aesthetic aspect. This ability, also known as self-healing, can occur either autonomously (intrinsic) or through external intervention (extrinsic) [9–12]. Consequently, these advanced coating systems can be designed to respond to either internal or external stimuli such as pH, chemical reactions, temperature, mechanical action, UV light, among others.

Since the work of White et al. in 2001 [13], the self-healing concept was introduced in material sciences and became the main trend of anti-corrosion applications. Among the strategies developed, stand out the polymers with shape memory property [14], reversible processes based on Diels-Alder (DA) reactions [15], dynamic supramolecular

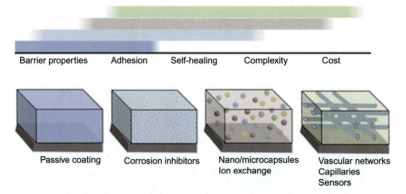

Fig. 3 Schematic representation of the main approaches adopted for developing protective coatings with self-healing ability and the effect of additives on the potential for regeneration and matrix compatibility. *(Callout: Strategies for self-healing coatings.)*

bonds or ionomers [16, 17], reversible covalent bonds [18], microcapsules containing monomer and catalyst [19], interconnected vascular networks [20], and corrosion inhibitors in form of organic [21] and inorganic compounds [22].

In view of the options described earlier, Fig. 3 displays a possible classification of protective coatings, extending from a passive system (left) toward smart devices with 3D vascular networks (right) [23]. Passive coatings, based on polymers or organic-inorganic materials, provide a physical barrier against corrosion, mechanical and/or thermal damage. An optimized formulation and nanostructure ensure strong adhesion to the metallic substrate along with an efficient barrier that remains intact until a natural failure by chemical agents or mechanical damage occurs [24, 25]. The addition of corrosion inhibitors, such as inorganic salts [22, 26], organic molecules [27], or carbon dots [28, 29], provides the possibility of active responsive repair under specific conditions such as pH gradient, presence of water, corrosive species, or formation of metallic corrosion products. When incorporated in an appropriate concentration range, typically 500–5000 ppm, they act the self-healing agents, without significantly impairing the structural integrity and adhesion of the polymeric or hybrid coating [22, 23, 26].

Another class of smart coatings are the nano/microcapsule systems, containing inorganic or organic healing agents [11, 13, 30], and ion exchange systems, in the form of lamellar double hydroxides (natural or synthetic clays) [31]. Coatings containing nano/microcapsules exhibit remarkable regenerative activity against corrosion and mechanical damage by liberating active agents through the crack-induced rupture. However, considering their complexity and cost, this can be an obstacle for large-scale applications. Also, depending on the loading and size of the capsules, other undesired effects can occur, namely, the loss of barrier property and adhesion to the substrate [23].

Finally, the integration of bio-inspired microvascular networks, capillaries, or sensors (for signaling changes in pH, color, fluorescence, or transmittance) based on pH

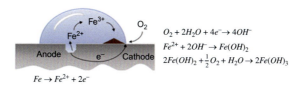

Fig. 4 Corrosion process on the surface of unprotected steel after contact with water. *(Callout: Metallic corrosion process.)*

indicators, 8-hydroxyquinoline or nanoengineered polyaniline films containing Ce(III) and Cr(VI), among others [32–34], act as reservoirs of healing substances. Although vascular networks, in analogy to human tissue, are able to liberate over several cycles healing agents (monomers and polymerization initiators), they represent complex systems from the perspective of development and large-scale application [11, 20, 35]. Furthermore, if the loading of the active component exceeds a critical value, the compatibility with the polymer matrix is affected, leading to the formation of percolation paths for corrosive agents and subsequently to corrosion reactions at the metal/coating interface. Nevertheless, in proper amounts, these systems can provide a synergistic effect between containers/capillaries and the passive barrier, leading to a significant increase in the coating service life [23].

In view of the possibility of achieving high-performance anticorrosive coatings with relatively simple systems, excellent cost-benefit, and compatibility with the matrix, studies using corrosion inhibitors have increased significantly in the last decades [9, 11, 36, 37]. A corrosion inhibitor can be defined as a substance that reduces the corrosion rate of a metal surface exposed to the corrosive environment by its chemical activity, thus extending the service life of metallic components [36–38]. More specifically, an inhibitor acts in the cathodic, anodic reaction, or both, depending on its activity in the partial oxidation and reduction reactions occurring during the corrosion process. Cathodic inhibitors reduce the current density of partial reduction reactions while anodic inhibitors reduce the current density of partial oxidation reactions, resulting in a potential shift [36, 37].

In Fig. 4 the illustration of steel corrosion is depicted. The corrosion of steel results from contact with humid, saline, or acidic environment since the species involved act as oxidizing agents in the redox process that corrodes unprotected metals [38, 39]. When steel is exposed to water and oxygen, iron is oxidized to Fe^{2+} (anodic reaction) and oxygen from the air is reduced to hydroxyl ions, reactions that occur separately but simultaneously. Then, Fe^{2+} and OH^- ions combine producing solid hydroxides such as $Fe(OH)_2$, which reacts further with oxygen and water to form $Fe(OH)_3$, the so-called rust (cathodic reactions) [38, 39]. During this process, a pH gradient is established, being more acid at the anodic sites (the pH drops from 6 to 2–3) and more alkaline at the cathodic sites (due to the OH^- formed at the oxygen reduction reaction (ORR)). The presence of effective inhibitors in the coatings prevents the progress of redox reactions in metals by (i) interacting with ions from anodic or cathodic sites, (ii) decreasing the ORR, and (iii) precipitating insoluble products in the form of a layer of low electrical conductivity [36, 37]. The mechanisms of inhibition vary according to the local pH, however, they can be classified as (i) adsorption between organic compounds and the metal substrates or products of the anodic reaction or (ii) precipitation of a protective film by the reaction of the inhibitor (usually an inorganic salt) with the dissolved metal or products of the cathodic reaction [36,37].

The changes of the local potential and current density provided by corrosion inhibitors can be detected and analyzed by Tafel diagrams [36, 40], electrochemical impedance spectroscopy graphs (EIS) [22], and localized electrochemical techniques, such as scanning vibrating electrode technique (SVET) [21], scanning ion-selective electrode technique (SIET) [41], scanning electrochemical microscopy (SECM) [14], localized electrochemical impedance spectroscopy (LEIS) [30], and scanning Kelvin probe (SKP) [42].

From this perspective, the chapter covers the main research advances of different classes of high-performance coatings modified with dispersed or encapsulated inorganic and organic corrosion inhibitors, with an emphasis on their self-healing ability, revealed by the Bode profile recovery features assessed by EIS and localized electrochemical techniques. Special attention is given to the healing mechanism and its effectiveness in feedback with the nature of the inhibitors and metallic components (steel, magnesium, and aluminum alloys) used in the automobile industry.

2. Inorganic corrosion inhibitors

Studies on nontoxic inorganic inhibitors, such as cerium, lithium, and molybdenum salts [22, 43–45] revealed a corrosion inhibition activity comparable to that of chromates, well known to affect human health and the environment by toxic Cr^{6+} species [46, 47]. Recent studies have shown that the incorporation of these salts improves the anticorrosive performance of acrylic [22, 26, 48, 49], epoxy [50–56], and polyurethane [45] coatings used to protect metal surfaces such as steel [26], aluminum alloys [22, 50, 57], and magnesium [51, 52, 55]. Although the inhibiting activity of cerium-based compounds is widely accepted in the scientific community, only few studies have demonstrated effective self-healing ability in terms of restoration of the electrochemical barrier properties. The curing effect can be easily verified using electrochemical impedance spectroscopy (EIS), where a partial or total recovery of the impedance modulus, and even more sensitively, of the phase angle is expected. Here, the Bode plots serve as indicators of the coating's anticorrosive performance. A low-frequency (1 mHz) impedance modulus of more than 100 MΩ cm^{-2} associated with a capacitive phase angle profile with values lower than −80 degrees over a wide frequency range, is generally considered excellent protection. When water and corrosive species (O_2, Cl^-) reach the substrate, the protection fails locally due to pitting corrosion, however, the activity of an effective inhibitor reduces the current density of partial oxidation/reduction reactions, resulting in a distinct response in the Bode and Nyquist plots.

Several cerium salts, such as nitrates [26, 48, 49, 54, 58, 59], phosphate [51–53], tartrates [50, 60], molybdates [61], and cinnamates [62], had their inhibitory activity evaluated in polymer coatings. The promising results of these compounds are associated with the cathodic protection provided by the Ce^{3+} and Ce^{4+} ions, which are depleted into metal-exposed regions and react with OH^- of the oxygen reduction reactions to form an insoluble oxides/hydroxide layer, which blocks the progress of corrosion reactions in the affected area [46].

The chemical role of cerium was investigated by Suegama et al. evaluating the effect of the addition of Ce(IV) salt in the form of cerium ammonium nitrate (CAN) on the polymerization of coatings based on silane bis-[triethoxysilyl] ethane (BTSE) [63]. The authors proposed that Ce^{4+} ions act by accelerating the redox polymerization through a combined mechanism of complex formation and free radical reactions. As a result, the Ce^{4+} modified coating presented approximately 2 orders of magnitude higher impedance modulus than that observed for the pristine coating in 0.1 mol L^{-1} NaCl solution [63]. Harb et al. studied the effects of CAN in organic-inorganic poly (methyl methacrylate) (PMMA)-silica coatings. The authors found that besides inhibition ability, Ce^{3+} and Ce^{4+} ions play an active role in the densification of the hybrid structure, inducing an increase in the degree of polycondensation of the silica phase and the polymerization efficiency of MMA, thus improving the barrier properties of the coatings [26]. The best results were obtained for intermediate Ce(IV) concentrations (700 ppm) of 2 μm-thick coatings deposited by dip-coating on 1010 carbon steel. EIS assays and potentiodynamic polarization curves in 0.6 M NaCl solution revealed a low-frequency (5 mHz) impedance modulus higher than 100 MΩ cm^2 after 304 days of exposure and current densities lower than 10^{-11} A cm^2, respectively.

Another study evaluated the effects of cerium oxide nanoparticles covalently linked to the PMMA matrix through the 2-hydroxyethylmethacrylate (HEMA) coupling agent [49]. The covalent conjugation between both phases was crucial to guarantee a homogeneous distribution of CeO_2/Ce_2O_3 nanoparticles and to suppress the formation of larger agglomerates. The coatings with a thickness of about 10 μm deposited on carbon steel 1020 exhibited a low-frequency impedance modulus higher than 100 GΩ cm^2 for more than 6 months of exposure to 0.6 M NaCl solution. The authors attributed the high anti-corrosion performance to a densely cross-linked structure acting as an effective barrier against corrosive species. The self-healing ability of the scratched coating with a molar proportion of 1Ce:2HEMA:25MMA was observed by EIS after 24 h exposure to saline medium showing one order of magnitude higher impedance modulus than the bare carbon steel. XPS analysis performed in the scratched zone confirmed the presence of elevated amounts of cerium oxide phases, suggesting the liberation of cerium ions

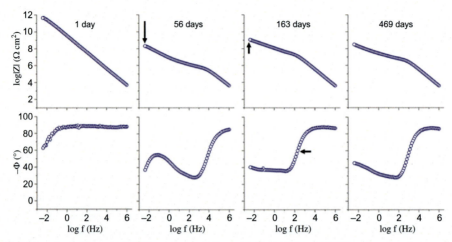

Fig. 5 Bode plots of smart PMMA-cerium oxide coatings on the AA7075 alloy showing a partial impedance and phase angle restoration after 163 days immersion in 0.6 M NaCl solution. *(Callout: Self-healing ability of cerium oxide-PMMA coating revealed by electrochemical impedance spectroscopy.)*

that act as a self-healing agent by the formation of an insoluble layer in the scratch track, thus inhibiting the progression of the corrosion process.

In an ongoing study, we have shown the self-healing ability of PMMA-cerium oxide coatings with a thickness of 15–20 μm on AA7075 aluminum alloy during long-term exposure to 0.6 M NaCl medium. Coatings prepared using LiOH as an oxidating agent to form ceria nanoparticles at a ratio LiOH:Ce(NO$_3$)$_3$ = 3:1 presented an initial low-frequency impedance modulus of 440 GΩ cm^2. After 56 days, the appearance of localized corrosion (pit) resulted in a sharp drop in the impedance modulus of 4 orders of magnitude, followed by a restoration of 1 order of magnitude after 163 days, remaining stable for another 306 days (Fig. 5, *top*). The self-healing effect is also visible in the phase angle plot (Fig. 5, *bottom*), showing a shift of the capacitive profile to lower frequencies. The healing ability was explained by the depletion of cerium cations from ceria nanoparticles and the subsequent formation of insoluble cerium species in the corroded zone [40, 49].

Another recent study evaluated the effect of different concentrations of Ce(IV) ions (500, 1000, 3000, and 5000 ppm) on the structure and anti-corrosion properties of PMMA-silica coatings with a thickness of 3–4 μm on AA7075 aluminum alloy [64]. The authors found that at intermediate cerium loadings (500–1000 ppm), the homogeneous dispersion of CAN and silica nodes favors the formation of a highly reticulated structure resulting in an excellent barrier, with an impedance modulus of up to 20 GΩ cm^2 and durability of more than 720 days in 0.6 M NaCl medium. For higher Ce concentrations, inclusions of larger ceria nanoparticles (20–40 nm) were found, however, not significantly affecting the barrier property. The coatings containing 500 and 5000 ppm of Ce(IV) presented active corrosion protection, showing a partial restoration

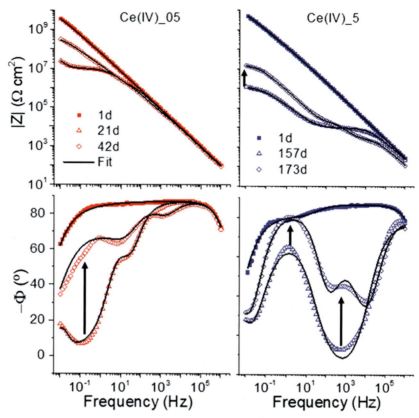

Fig. 6 Nyquist and Bode plots of PMMA-silica-Ce(IV) coatings on the AA7075 alloy containing (A) 500 ppm and (B) 5000 ppm of Ce(IV) exposed to 0.6 M NaCl solution showing partial impedance restoration due to self-healing in the region affected by corrosion. *(Reproduced with permission of Elsevier. Callout: Self-healing ability of Ce(IV) ions in PMMA-silica hybrid coatings.)*

of the impedance modulus and the phase angle in the low and intermediate frequency range, after 42 and 173 days of immersion, respectively (Fig. 6). According to the authors, the regeneration of PMMA-silica-Ce(IV) coatings is associated with Ce(IV) driven formation of insoluble oxides and hydroxides at intermetallic particles (Cu, Zn) of AA7075 at pH values close to 3.

Mosa et al. reported the self-healing property of bilayer siloxane-methacrylate coatings with an inner layer enriched by cerium nitrate, used as a precursor of Ce(III) ions [48]. Coatings with a thickness of 26 μm showed high corrosion protection on mild steel with a low-impedance modulus higher than 1 GΩ cm^2 after more than 290 days of immersion in 0.6 M NaCl. After 295 days, the authors observed an increase in the impedance modulus accompanied by a shift in the phase angle to lower frequencies. With the aid of scanning electron microscopy (SEM) coupled to energy-dispersive X-ray spectroscopy (EDS) and micro-Raman analysis, they detected elevated concentrations of cerium in the corrosion-affected zone, indicative of Ce depletion and activation of self-healing process by a precipitated oxide/hydroxide layer.

Recently, Castro et al. used a similar strategy to study an integrated system that combines anodization of the aluminum alloy AA2024 with the deposition of a silica bilayer coating based on 3-(glycidyloxypropyl)trimethoxysilane (GPTMS), tetraethoxysilane (TEOS) and colloidal silica precursors (Ludox® 40 wt% suspension in water), containing an inner layer enriched with cerium nitrate [57]. EIS measurements in 0.6 M NaCl solution confirmed the self-healing activity of the coating, showing an increment in the low-frequency impedance modulus from ~1 to ~100 MΩ cm^2 (100 mHz) between 24 and 48 h of exposure, remaining constant for 120 h (Fig. 7A). Accelerated corrosion tests performed in saline solution confirmed the performance of the protective coating system due to the absence of visual corrosion signs after 75 days of immersion (Fig. 7B).

The inhibitory effect of cerium tartrate as a precursor of Ce(III) ions in commercial 25 μm-thick epoxy coating was evaluated by Hu et al. for the protection of the AA2024 aluminum alloy [50]. By adding 15 wt% of the inhibitor, the coatings showed an improvement in anticorrosive performance paired with self-healing ability, as evidenced by an increase in the impedance modulus (10 mHz) from 0.1 to 1 GΩ cm^2 after 45 days of immersion in 0.6 M NaCl solution (Fig. 8A). In addition, the ability of cerium tartrate to heal two artificial defects was shown employing SVET (scanning vibrating electrode technique), a localized electrochemical technique. The SVET current maps showed after 3 days of immersion an increase in the cathodic and anodic current density in the defective zones, which disappeared after 5 days, indicating the inhibitory effect of cerium tartrate by the formation of insoluble precipitates of cerium (Fig. 8B).

Calado et al. evaluated the use of tris(bis(2-ethylhexyl) cerium phosphate) (Ce(DEHP)$_3$) as corrosion inhibitor in ~4 μm-thick epoxy silane hybrid coating for protection of the WE43 magnesium alloy [51]. The modified Ce(DEHP)$_3$ coating showed an anti-corrosion performance superior to the epoxy reference, with impedance modulus at 10 mHz higher than 200 MΩ cm^2 during 150 days of immersion in 0.5 M NaCl solution. After 100 days of immersion, a small restoration effect was observed in the Bode plots, increasing the low-frequency impedance to 700 MΩ cm^2. SVET maps of artificial defects in contact with 0.05 M NaCl showed a strong decrease in the cathodic current density compared to the cerium-free coating, an indicator of the active role of Ce(DEHP)$_3$ as an inhibitor of corrosion propagation. The healing activity was related to the combined action of cerium ions providing protection of cathodic sites, and of organophosphates reducing the corrosive activity at anodic sites due to the interaction with Mg^{2+} formed during the dissolution process. A similar corrosion inhibition activity of Ce(DEHP)$_3$ was observed for epoxy coatings on the AZ31 magnesium alloy [52] and carbon steel [53].

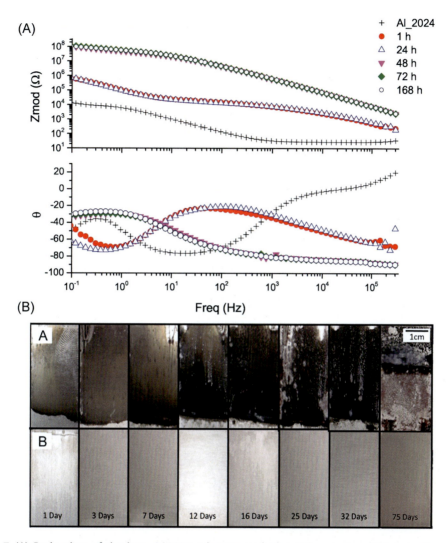

Fig. 7 (A) Bode plots of the bare AA2024 substrate and silane/silane-cerium bilayer coating as a function of immersion time in 0.6 M NaCl and (B) images of the AA2024 substrate (A), and silane-cerium-anodizing bilayer coating (B), taken during 75 days of immersion in saline solution [57]. *Reproduced with permission of Elsevier. (Callout: Self-healing ability of cerium in bilayer silica sol–gel coatings.)*

In recent studies, lithium salts, such as carbonate and oxalate, have been evaluated as self-healing agents in coatings for the protection of AA2024 and AA7075 aluminum alloys [20,22,45,65–68]. Similar to the cerium salts, promising results were obtained for lithium salts, which are associated with the diffusion of lithium ions toward the damaged region, inducing the formation of a protective layer of aluminum oxide/hydroxide

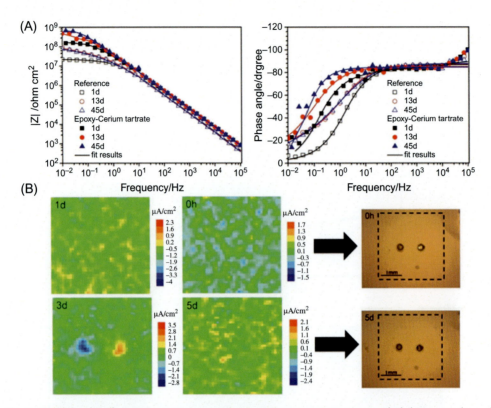

Fig. 8 (A) Bode plots for epoxy coatings containing 15% cerium tartrate recorded during 45 days of immersion in 0.6 M NaCl solution and (B) SVET maps of artificially damaged cerium tartrate containing epoxy coating after 0 h, 1 day, 3 days, and 5 days of immersion and the corresponding images after 0 h and 5 days (0.6 M NaCl) [50]. *(Reproduced with permission of Elsevier. Callout: Self-healing ability of cerium tartrate in epoxy coatings.)*

corrosion products. Visser et al. and Liu et al. studied the use of lithium carbonate and lithium oxalate as corrosion inhibitors in commercial polyurethane-based organic coatings [45,65,66]. The authors observed that under conditions of neutral salt spray, lithium species were leached from the coating into the artificial scratch to form a protective layer [45,65,66].

In another study, Dieleman et al. used electrospinning to produce water-responsive inhibition networks based on polyvinyl alcohol and corrosion inhibitors such as cerium chloride and lithium carbonate incorporated into epoxy coatings with thicknesses of 100–180 μm [20]. The inhibitory effect of lithium carbonate was evidenced by a slight increase in the low-frequency impedance modulus of the coating from ~100 kΩ cm^2 to 1 MΩ cm^2 remaining stable for up to 28 days.

More recently, Trentin et al. investigated the structural properties and the active corrosion inhibition induced by the addition of lithium carbonate in poly(methyl methacrylate) (PMMA)-silica coatings with thicknesses of 4–6 μm [22]. Structural analysis performed using nuclear magnetic resonance (NMR), small-angle X-ray scattering, and thermal analysis, revealed that the addition of increasing amounts of lithium carbonate (0, 500, 1000, and 2000 ppm) enhanced the crosslinking of the silica domains covalently bonded to PMMA matrix. EIS results revealed that the coatings have high anticorrosive performance with a low-frequency impedance modulus of up to 50 GΩ cm^2 after exposure to 0.6 M NaCl solution (Fig. 9A). The self-healing ability was identified for all

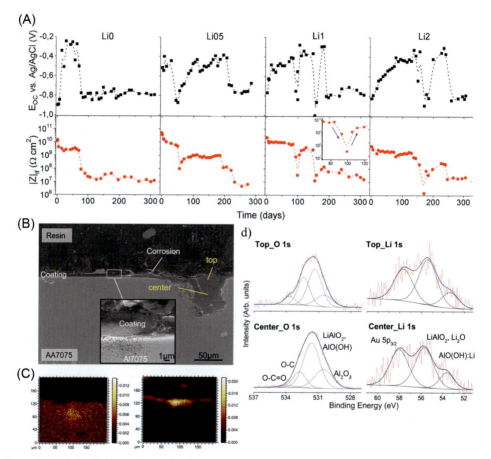

Fig. 9 (A) Time evolution of the open circuit potential, E_{OC}, and low-frequency impedance modulus, $|Z|_{5mHz}$, for lithium modified PMMA-silica coatings, immersed up to 310 days in NaCl 0.6 M solution (the *inset* shows in more detail the first recovery event of the impedance modulus after 99 days for the Li1 coating); (B) cross-sectional view of the pit obtained by SEM after 310 days of immersion in 0.6 M NaCl for the Li1 coating; (C) normalized cross-sectional ToF-SIMS map of AlOH$^+$ (*yellow, left*), and Li$^+$ (*yellow, right*) ions recorded in the cross-sectional area of the center and top and of the pit; and (D) deconvoluted XPS O 1 s and Li 1 s spectra, obtained at the top and center of the pit. (The Au signal comes from gold deposition to reduce SEM charging effects) [22]. *(Reproduced with permission of ACS. Callout: Self-healing ability of lithium in PMMA-silica hybrid coatings.)*

Li-loaded coatings, by monitoring the time-dependence of the open circuit potential (E_{OC}) and impedance modulus (Fig. 9A), recorded during 310 days of immersion. The occurrence of a local failure of the coating resulted in a sharp impedance drop, followed by a gradual recovery of about two orders of magnitude, observed twice for the Li1 coating (1000 ppm). A detailed cross-sectional analysis of the pitted region using SEM/EDS (Fig. 9B), time of flight secondary ion spectroscopy Tof-SIMS (Fig. 9C), and X-ray photoelectron spectroscopy (XPS) (Fig. 9D) evidenced that the regeneration process occurs through the depletion of lithium ions from the coating to the corrosion sites, which were restored by the formation of a layer composed of $LiAlO_2$ and/or Li containing AlO(OH).

Additionally, several other approaches for the controlled release of inorganic inhibitors have been explored, such as cerium molybdate in layered double hydroxides (LDHs) [31], Ce and mercaptobenzothiazole (MBT) in zeolites [69], montmorillonite (MMT) clays as cerium nanocontainers [70], and others reviewed in the reference [47]. Some of the most interesting results reported for high-performance coatings showing active corrosion protection by inorganic inhibitors are summarized in Table 1.

3. Organic corrosion inhibitors

Several classes of organic compounds are widely used as corrosion inhibitors for metal surfaces. This is due to the ability of molecules such as nitrogen, sulfur, oxygen, and π electrons to interact with the metallic substrates through heteroatoms in their structure [71]. The active protection against corrosion by adsorbed organic inhibitors is associated with their ability to reduce the rate of cathodic and anodic corrosion reactions by forming a thin hydrophobic film on the metal surface [47]. Using molecular agents, several strategies have been adopted to develop smart anticorrosive coatings with self-healing properties. One of the main approaches used is the microencapsulation of healing agents in different structures, such as silica [72], graphene [73], lamellar double hydroxides (LDH) [74], polymers [75], and zeolites [76]. The encapsulation of functional active materials is an interesting approach to store and protect them from aggressive environments until their controlled release in response to specific intrinsic and extrinsic stimuli, such as the variation of pH, ionic strength, temperature, ultrasonic treatment, alternating magnetic and electromagnetic field, and mechanical wear [77].

White et al. have pioneered the development of anticorrosive self-healing systems by encapsulating active agents in an epoxy matrix. They used polyurea-formaldehyde (PUF) microcapsules containing dicyclopentadiene monomers (DCPD), which were released after subjecting the coatings to intentional mechanical damage. The monomers polymerized by the action of a Grubbs catalyst in artificial scratches, resulting in a healing efficiency of approximately 60% [13].

Table 1 Results reported for coatings with active corrosion protection by inorganic agents in organic-inorganic and polymeric coatings: composition, substrate, coating thickness, low-frequency impedance modulus ($|Z_{lf}|$), coating lifetime, and electrolyte used.

Coating	Substrate	Thickness (μm)	$|Z_{lf}|$ (GΩ cm^2) Lifetime (days) Solution	Ref.
PMMA-MPTS-TEOS-Ce(IV)	1010 carbon steel	~2	~10 304 0.6 M NaCl	[26]
PMMA-MPTS-TEOS-Ce(III)	Mild steel	26	~10 362 0.6 M NaCl	[48]
PMMA-MPTS-TEOS-Ce(IV)	Al alloy AA7075	~3–4	~20 720 0.6 M NaCl	[64]
PMMA-HEMA-cerium oxide	1010 carbon steel	10	~290 345 0.6 M NaCl	[49]
PMMA-HEMA-cerium oxide	1020 carbon steel	15–20	~440 469 0.6 M NaCl	[This work]
GPTMS-TEOS-silica-Ce(III)	Al alloy AA20224	53	~0.3 168 h 0.6 M NaCl	[57]
Epoxy-Ce(III)	Al alloy AA20224	25	~1 45 0.6 M NaCl	[50]
Epoxy-Ce(III)	Mg alloy WE43	25	~1 373 0.6 M NaCl	[51]
Epoxy-APTES-Ce(III)	Mg alloy AZ31	~7	~10 600 0.6 M NaCl	[52]
Epoxy-Ce (III)	Mild steel	5–50	~100 454 0.05 M NaCl	[53]
Epoxy-MMT-Ce(III)	Carbon steel	50–70	~0.1 100 0.6 M NaCl	[54]
Epoxy-APTES-CeO$_2$	Mg alloy AZ31	~10	~30 29 0.05 M NaCl	[55]
Epoxy-PANI-CeO$_2$-GO	Mild steel	75	~4 70 0.6 M NaCl	[56]
PMMA-MPTS-TEOS-Li	Al alloy AA7075	4–6	~50 310 0.6 M NaCl	[22]

PMMA, poly (methyl methacrylate); *MPTS*, 3-(trimethoxysilyl) propyl methacrylate; *TEOS*, tetraethoxysilane; *HEMA*, 2-hydroxyethyl methacrylate; *GPTMS*, (3-glycidyloxypropyl)trimethoxysilane; *APTES*, aminopropyltriethoxysilane; *MMT*, montmorillonite; *PANI*, polyaniline; *GO*, graphene oxide, *Li*, lithium.

Currently, environmentally friendly, and low-cost self-healing coatings are the focus of research activities. Natural products, such as tung oil, linseed oil, and sunflower oil, are increasingly applied as healing agents [78]. Following this approach, Siva et al. evaluated the corrosion protection of scratched epoxy coatings containing urea formaldehyde (UF) microcapsules loaded with linseed oil (LO) and mercaptobenzothiazole (MBT) on the mild steel panels (Fig. 10A) [79]. Using optical microscopy the authors observed that the scratch (Fig. 10C) was gradually reduced after 45 days due to the filling provided by the microcapsules (Fig. 10D). The authors related the healing effect to the capacity of the LO containing MBT to dry out by oxidation in the presence of atmospheric oxygen, thus filling the defect. The time evolution of the impedance curves of pristine and microcapsules loaded coatings was studied using EIS in 0.5 M NaCl solution (Fig. 10E and F). Pure epoxy coatings showed during 15 days of immersion initally a passivation effect in the defective zone resulting in an increase in the coating impedance from 3.4 to 77.0 kΩ cm^2, however, after 5 days followed by an impedance drop at high frequencies

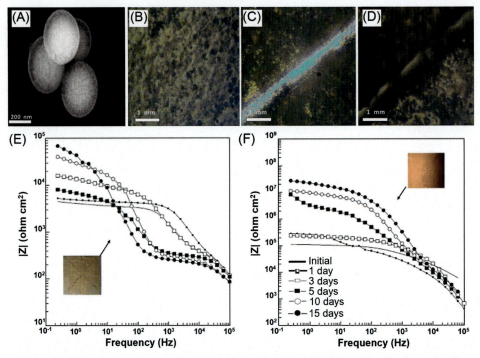

Fig. 10 (A) SEM image of urea formaldehyde microcapsules loaded with linseed oil and mercaptobenzothiazole in epoxy coatings; optical microscopies of the coating containing microcapsules (B) without defect, (C) after scratching, and (D) after 45 min healing; bode impedance plots of the damaged region obtained for (E) pure epoxy coating, and (F) for microcapsules modified coating [79]. *(Reproduced with permission of Elsevier. Callout: Self-healing ability of linseed oil and mercaptobenzothiazole in epoxy coatings.)*

(Fig. 10E). In contrast, the modified coating presented an initial resistance of 93.3 kΩ cm^2 and reached 23.7 MΩ cm^2 after 15 days of exposure (Fig. 10F). The increase of more than 2 orders of magnitude was attributed to formation of insoluble compounds by the release of the healing agent. Using SVET, the self-healing ability was confirmed by a strong suppression of the anodic current density in the defective zone.

Smart coatings based on graphene loaded with organic inhibitors have shown promising anti-corrosion properties [80,81]. Ye et al. coated Q235 carbon steel with ~50 μm-thick epoxy coatings containing graphene nanosheets with porous polyhedral oligomeric silsesquioxane (POSS) structures loaded with benzotriazole (BTA) inhibitor [80]. Using local electrochemical impedance spectroscopy (LEIS) the self-healing ability was observed for artificially scratched coating after 24 h of exposure to 0.6 M NaCl solution. The authors showed that the admittance values of the BTA composite coating significantly decreased in the scratch, accompanied by a gradual shrinking of the corrosion zone. Time-dependent EIS measurements of the intact coating showed a low-frequency (0.01 Hz) impedance modulus above 10 GΩ cm^2 after 90 days of immersion in saline solution. In addition, after 40 days, there was a small drop in impedance of about one order of magnitude, followed by partial restoration after 60 days. According to the authors, the reaction of released BTA with metal ions of the substrate contributed to the formation of a protective film that restored the defect and inhibited redox reactions at the interface.

Similar results were obtained by Liu et al. who synthesized 40 μm-thick coatings on carbon steel containing supramolecular β-cyclodextrin (βCD) nanocontainers supported on reduced graphene oxide (rGO) sheets [81]. The nanocontainers were loaded with the BTA inhibitor and dispersed in a commercial epoxy matrix (EP). The self-healing ability of scratched and intact coatings was revealed using the LEIS and EIS techniques, respectively. LEIS maps showed that for the rGO-βCD–BTA/EP coating the corrosion process in the scratched zone was inhibited during 20 h of exposure to 0.6 M NaCl solution (Fig. 11A–C). Time-dependent EIS measurements in saline solution showed for intact epoxy coating a high initial corrosion resistance (~1 GΩ cm^2), however, then a continuous decrease to 10 MΩ cm^2 was observed during the next 55 days (Fig. 11D). In contrast, the low-frequency impedance modulus of the rGO-βCD–BTA/EP coated sample remained above 200 MΩ cm^2 in the same period (Fig. 11F). According to the authors, the self-healing mechanism consists, initially, in the oxidation of the metal in the anodic zone reducing the local pH due to the hydrolysis of metal ions. In the cathodic reaction, oxygen is reduced to form hydroxyl ions, increasing local pH. Due to the higher BTA release rate, observed at pH = 10, the BTA molecules are gradually released and adsorbed on the metal, inhibiting the corrosion reaction [81].

Quantum carbon dots (CDs) are particles with nanometric dimensions which have been widely studied due to their unique photocatalytic and photoluminescence properties, biocompatibility, and ease of synthesis [82,83]. In addition to their small size and

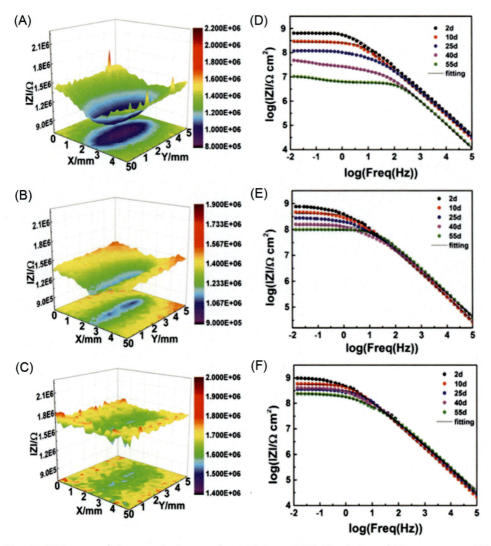

Fig. 11 LEIS maps of the scratched zone after 20 h in 0.6 M NaCl solution of (A) pure epoxy (EP), (B) rGO−βCD/EP, and (C) rGO−βCD−BTA/EP coatings on carbon steel; time evolution of the Bode impedance curves for intact coatings of (D) pure epoxy, (E) rGO−βCD/EP and (F) rGO−βCD−BTA/EP immersed in 0.6 M NaCl solution [81]. *(Reproduced with permission of ACS. Callout: Self-healing ability of graphene oxide supported β-cyclodextrin and benzotriazole in epoxy coatings.)*

high specific surface area, the possibility of modifying the surface with functional groups enables homogeneous dispersion and new functionalities when inserted in polymer or hybrid matrices. Consequently, the choice of specific functional groups on the surface allows CDs to act as self-healing agents after corrosive or mechanical damage due to the interfacial interaction (van der Waals forces, covalent bonds, and hydrogen bonds) between CD and the host matrix [28].

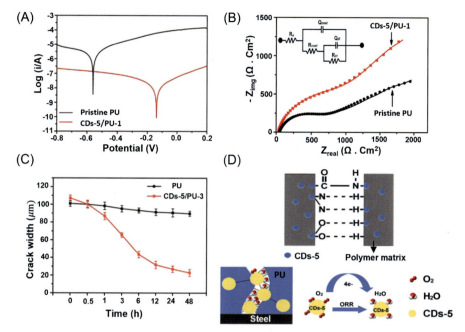

Fig. 12 (A) Potentiodynamic polarization curves and (B) EIS Nyquist plot obtained for polyurethane (PU) and carbon dots loaded polyurethane coating (CDs/PU) on stainless steel; (C) crack width time evolution probed by an Alpha step profilometer for CDs-5/PU-3 and PU self-healed at room temperature; and (D) illustration of O_2 and H_2O following a transmission path through CDs-5/PU and ORR process on the surface of a CD-5 (see text) [28]. *(Reproduced with permission of Wiley-VCH GmbH. Callout: Self-healing ability of carbon dots in polyurethane coating.)*

Zhu et al. dispersed 5 nm-sized carbon nanocrystallites in polyurethane matrix (CDs/PU) applied as 600 μm-thick coating for active corrosion protection of stainless steel [28]. The corrosion protection efficiency studied using potentiodynamic polarization curves and EIS (Nyquist plots) showed that the CD modified coating presented about two orders of magnitude lower current density (~10 nA) and higher corrosion resistance compared to the pristine sample (Fig. 12A and B). The self-healing ability of the artificially scratched CDs/PU coating was studied by recording the time dependence of scratch track width, which was compared to the scratched PU coating. The profilometry results showed for CDs/PU coating a gradual reduction of the defect width from ~100 to ~20 μm within 48 h, while for the pristine PU sample no significant changes were observed (Fig. 12C). According to the authors, the self-healing mechanism in the defect zone occurs due to the interactions between adjacent CDs and also between CDs and polymer chains. The formation of covalent C=O—NH bonds, hydrogen bonds (O—H—N, O—H—O, N—H—N, and N—H—O), and interaction by van der Waals forces leads to the filling of local pores and cracks, thus preventing the expansion and formation of macropores. Hence, the presence of CDs reduces the diffusion of water and corrosive substances. In addition, hydrophilic functional groups on the surface of CDs absorb water and oxygen. The O_2 molecules are reduced to H_2O through a four-electron pathway, thus suppressing the corrosion process (Fig. 12D) [28].

Lamellar double hydroxides (LDH) are potential candidates to replace chromate-based protection systems. LDH, also known as anionic clays, are characterized by the stacking of hydroxides layers with cations of different valences, and the electroneutrality of the system is achieved by the presence of anions and water molecules between the lamellae. In general, the structure of LDH is represented by the formula $[M^{II}_{1-x}M^{III}_x(OH)_2]^{x+}(A^{y-})_{x/y} \, zH_2O$, where M^{II} and M^{III} are trivalent metal cations, while A^{y-} represents the anions of the interlamellar domains [84]. The structural characteristics of LDH allow the intercalation of a variety of inhibitors between the lamellae to be released in a controlled manner through the anion exchange property. Specifically in anticorrosive coatings, LDHs act by releasing inhibitors and capturing corrosive species followed by the cathodic formation of a hydroxide film in the affected zone [85].

Zadeh et al. evaluated the synergy of inhibition of mercaptobenzothiazole (MBT) and Ce^{3+} ions present in the two-dimensional structure of LDH (LDH-MBT) and the three-dimensional zeolite structure of NaY (NaY-Ce) [69]. The nanosized systems were incorporated in the epoxy coatings used to protect the AA2024-T3 aluminum alloy. The synergy between the inhibitors (MBT and Ce^{3+}) was studied using EIS in 0.05 M NaCl solution. After 7 days of immersion, the authors observed a low-frequency impedance modulus of 2 MΩ cm^2 for both LDH-MBT and NaY-Ce sample, compared to 600 kΩ cm^2 found for inhibitor-free NaY coating. The best result was obtained for the combined activity of both inhibitors with an low-frequency impedance modulus of 10 MΩ cm^2. According to the authors, the MBT inhibition occurs through adsorption on aluminum and aluminum oxide surfaces, especially in copper-rich AA2024-T3 domains, while Ce^{3+} ions act suppressing the corrosion by the formation of insoluble hydroxides or oxide layers.

In a recent study, Nardeli et al. have investigated the self-healing ability of 60 μm-thick polyurethane (PU) coatings derived from vegetable oils for the protection of the AA1200 aluminum alloy [86]. They found that the healing functionality can be achieved by tuning the interactions between the flexible segments (polyester groups) and rigid segments (diisocyanate and diol groups) in the polymer. The anticorrosion property of the coatings was evaluated using EIS for different immersion times in a 0.6 M NaCl solution. For the coating containing a higher proportion of flexible segments (Coat-I), the authors observed an increase in the low-frequency modulus from 20 MΩ cm^2 (1 day) to 70 MΩ cm^2 after 37 days (Fig. 13A). In contrast, the coating with a smaller number of flexible segments (Coat-II) showed during the first days a slight decrease of the

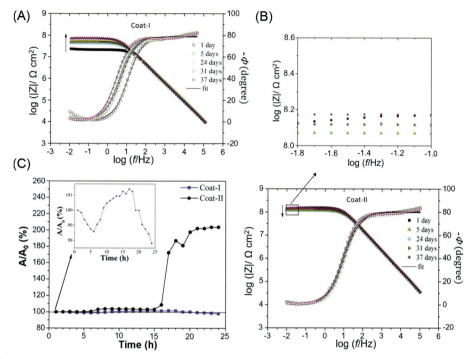

Fig. 13 Time dependence of Bode plots obtained for coatings (A) with higher number of flexible segments (Coat-I) and (B) with smaller amount of flexible segments Coat-II, after immersion in 0.6 M NaCl solution; (C) time dependence of the admittance (A) to initial admittance (A_0) ratio [86]. (Reproduced with permission of Elsevier. Callout: Self-healing ability of polyurethane coatings derived from vegetable oils.)

impedance modulus, which then increased to 110 MΩ cm² after 27 days (Fig. 13B). LEIS was used to investigate the self-healing property of the scratched coatings after immersion in 0.005 M NaCl. Fig. 13C presents the time dependence of the ratio of admittance (A) to initial admittance (A_0) for both coatings. While for the Coat-II a strong increase of the A/A_0 ratio was detected after 15 h immersion, for the Coat-I sample the values remained slightly below 100% after 20 h, indicating a healing effect. As suggested by the authors, the regeneration is associated with the formation of hydrogen bonds between the rigid and flexible segments, and due to the higher proportion of flexible segments present in the Coat-I sample.

Sulfur-containing compounds have been widely studied as curing agents in smart polymeric coatings due to the ability to form reversible covalent disulfide bonds (S—S) in the damaged zone of the coating [18,87]. Zadeh et al. evaluated this potential for 50–300 μm-thick tetrasulfide groups containing sol-gel epoxy-amine hybrids coatings on the AA2024-T3 aluminum alloy [18]. The tetrasulfide groups were inserted using (3-Aminopropyl) trimethoxysilane (APS) and Bis[3-(triethoxysilyl)propyl]-tetrasulfide (BS) in different proportions. After artificially damaging the coating, the extrinsic self-healing process was promoted by an annealing step for 2 h at 70 °C, showing a 100% surface healing efficiency, as illustrated by the scratch profiles (Fig. 14A). According to the authors, the recovery of the initial properties can be explained by the reversible dynamic nature of the S—S cross bonds of the tetrasulfide groups that restored the integrity of the polymer. For the sulfur-free coating a regeneration effect was not observed. As shown in Fig. 14B, the impedance modulus of the scratched coatings recovered to the value of the intact coating ($\sim$1 GΩ cm^2) after 1 h exposure to 0.5 M NaCl electrolyte, suggesting a complete restoration of the barrier properties of the coating. After healing, coatings with 25 μm wide scratches presented a high resistance and low capacitance even after 1 month of exposure to the 0.5 M NaCl solution (Fig. 14C). Regarding the long-term stability, the authors observed that only for a scratch width/coating thickness ratio smaller than 1 the resistance of the healed coating followed the time evolution of the intact film (Fig. 14D).

Magnesium alloys play a fundamental role in the automotive industry due to their low specific density, reducing the weight and thus the fuel consummation of vehicles. However, these materials are very susceptible to corrosion and have low wear resistance, which makes the development of self-healing coatings with high anti-corrosion efficiency essential [88,89].

Coatings prepared by plasma electrolytic oxidation (PEO) have been used as a strategy to protect Mg, Al, and Ti alloys, however, the presence of micropores and defects makes the oxide layer susceptible to corrosive processes. Here, the porous microstructure of the PEO layer can act as micro and nano reservoirs to store or transport corrosion inhibitors, providing active protection to the coatings. Using this strategy for pure Mg substrate, Yang et al. studied active corrosion protection of a porous PEO layer ($\sim$20 μm) loaded with sodium 3-methylsalicylate (3-MSA) inhibitor, which was sealed with a 13 μm-thick epoxy-silica upper layer [88]. The hybrid coating acted as a physical barrier preventing the diffusion of corrosive species and the PEO layer suppressed the corrosion reaction in case of overcoat failure. The bilayer coating presented a high corrosion protection with impedance modulus (10 mHz) higher than 1 GΩ cm^2, remaining stable after 30 days in contact with 0.6 M NaCl solution. To study the healing ability, SVET measurements were performed for coatings with and without the inhibitor, monitoring during 24 h the current density in the damaged area. For the inhibitor-free sample, the authors observed a strong increase in the current density, indicating a corrosive activity in the defect region, while the inhibitor containing coating showed after 6 h a significant reduction of the corrosive activity, accompanied by a decrease of the current density. The active protection was attributed to the release of the inhibitor molecules from the pores

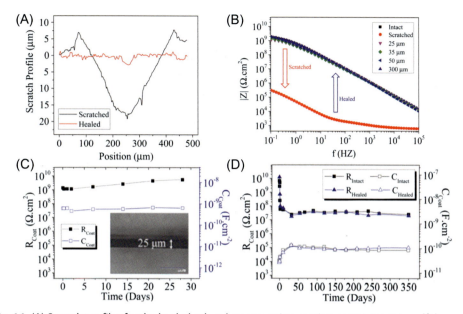

Fig. 14 (A) Scratch profiles for the healed sol–gel epoxy-amine coating containing tetrasulfide groups and the pristine sample (*black curve*); (B) bode impedance modulus plots of the intact, scratched, and healed coating with different scratch widths after 1 h exposure to 0.5 M NaCl solution; (C) evolution of coating resistance (R_{Coat}) and capacitance (C_{Coat}) for coatings with scratch width of 25 μm; (D) evolution of the resistance and capacitance of a 300-μm-thick intact and scratched (300-μm width) coating in saline solution [18]. *(Reproduced with permission of ACS. Callout: Self-healing ability of tetrasulfide groups containing sol–gel epoxy-amine coatings.)*

of the PEO coating toward the damaged area following two main mechanisms: physical/chemical adsorption and complex formation between cations of detrimental impurities and healing activity of inhibitory molecules.

Recently, Zhao et al. developed a superamphiphobic self-healing coating for efficient corrosion protection of the AZ31B magnesium alloy by combining a ceresin wax containing self-healing epoxy resin layer (SHEP) of ~130 μm thickness and a porous superamphiphobic topcoat (~13 μm) consisting of perfluorodecyl polysiloxane modified silica (PF-POS@silica) [89]. The authors used the synergistic effect of the two layers to obtain a coating that combines high corrosion resistance with self-healing ability. The elevated anticorrosive performance of the SHEP and SHEP/PF-POS@silica coating was confirmed by its high impedance modulus (1 mHz) of ~10 GΩ cm² (Fig. 15A). To assess the self-healing property, the samples were scratched and then heated at 65 °C for 30 min followed by 80 °C for 30 min. SEM images showed for the SHEP and SHEP/PF-POS@ coating the regeneration of artificial damage after the curing step

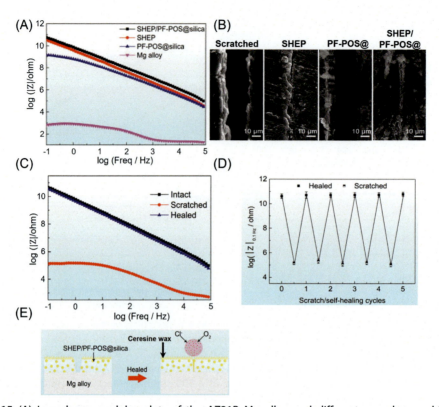

Fig. 15 (A) Impedance modulus plots of the AZ31B Mg alloy and different monolayer e bilayer coatings; (B) SEM images of the scratched zone before and after thermal healing of the coatings; (C) impedance modulus plot of the SHEP/PF-POS@ coating before and after self-healing; (D) variation of the low-frequency impedance modulus during five scratch/healing cycles; and (E) schematic illustration of self-healing mechanisms of the SHEP/PF-POS@silica coating [89]. *(Reproduced with permission of Elsevier. Callout: Self-healing ability of ceresin wax containing epoxy bi-layer coatings.)*

(Fig. 15B). EIS measurements performed in 0.6 M NaCl solution showed for the scratched double-layer coating a drop of the impedance modulus of about five orders of magnitude to ~100 kΩ cm^2, which was completely restored after the heat treatment (Fig. 15C). In addition, after five scratches/curing cycles, the low-frequency impedance modulus remained stable after each healing step (Fig. 15D). According to the authors, the self-healing ability is associated with the glass transition of the ceresin wax in the SHEP coating. The reversible transformation from the vitreous state to the rubbery state occurs due to the memory effect of the resin, while the released wax allows to further seal the defects (Fig. 15E).

Table 2 summarizes the most important results reported for self-healing coatings using organic agents.

Table 2 Reported active coatings loaded with organic corrosion inhibitors: composition, substrate, coating thickness, low-frequency impedance modulus ($|Z_{lf}|$), coating lifetime, and electrolyte solution.

Coating	Substrate	Thickness (µm)	$\|Z_{lf}\|$ (GΩ cm^2) Lifetime (days) Solution	Ref.
Epoxy-UF–LO–MBT	Mild steel	40	~0.023 15 0.6 M NaCl	[79]
Epoxy-GO-BTA	Q235 carbon steel	50	~1 90 0.6 M NaCl	[80]
Epoxy-GO – CD–BTA	Carbon steel	40	~0.1 55 0.6 M NaCl	[81]
Epoxy-NaY-Ce/LDH-MBT	Al alloy AA2024	50	~0.1 7 0.05 M NaCl	[69]
Polyester/tannin	Al alloy AA1200	40–44	~10 30 0.6 M NaCl	[90]
PU/PS/PLA 7	Al alloy AA 2024-T3	152	~0.1 83 0.6 M NaCl	[75]
Epoxy/HT-BZ 1.5 wt%	Carbon steel	35–45	~10–0.1 60 0.5 M NaCl	[74]
PU	Al alloy AA1200	59–62	~0.1 37 0.6 M NaCl	[86]
Epoxy-APS-BS	Al alloy AA2024-T3	50	~1 14 0.5 M NaCl	[18]
PEO-epoxy-3-MSA	Pure Mg	24	~1 32 0.6 M NaCl	[88]
SHEP/PF-POS@silica	Mg alloy AZ31B	142	~1 17 0.6 M NaCl	[89]

UF, urea formaldehyde; *LO*, linseed oil; *MBT*, mercaptobenzothiazole; *GO*, graphene oxide; *CD*, carbon dots; *LDH*, layered double hydroxide; *NaY*, NaY-zeolite; *BTA*, benzotriazole; *PLA*, polylactic acid; *PU*, polyurethane; *PS*, polysiloxane; *HT*, hydrotalcites; *BZ*, benzoate; *APS*, (3-aminopropyl)trimethoxysilane; *BS*, bis[3-(triethoxysilyl) propyl] tetrasulfide; *PEO*, plasma electrolytic oxidation; *3-MSA*, 3-methylsalicylate; *SHEP*, epoxy resin; *PF-POS@silica*, perfluorodecyl polysiloxane modified silica.

4. Conclusions

The most challenging tasks for the development of smart anticorrosive coatings for the automotive industry include (i) the selection of effective and environmental compliant systems that supply one or multiple healing agents to the damage site; (ii) compatibility of the additives with the coating matrix, preserving its integrity; (iii) utilization of simple, green, and economic preparation methods; and (iv) establishment of reproducible performance over several damage/curing cycles. The overview of strategies given in this chapter for developing smart protective coating on different metallic substrates shows that in the past decade significant advances have been made to meet most of these requirements, although the latter criterion remains the most challenging. Several organic and inorganic additives have been shown to be very efficient as healing agents in pure polymeric or hybrid matrices, although, some approaches are less favorable in terms of matrix compatibility, the complexity of preparation, and the need for external stimuli. Using the sol–gel chemistry and polymerization process, it is possible to produce less than 10 μm thick, adherent, and highly crosslinked hybrid coatings that combine a long-term high-performance barrier with efficient self-healing capability using inexpensive, green ionic or molecular agents. These characteristics make the modified acrylic/epoxy-silica hybrid coatings the material of choice for future smart coatings for the automotive industry.

Acknowledgments

The authors would like to acknowledge the financial support of funding agencies, namely Conselho Nacional de Desenvolvimento Científico e Tecnológico (CNPq) [grants 424133/2016-4, 142305/2020-0, 141078/2020-0], Coordenação de Aperfeiçoamento de Pessoal de Nível Superior (CAPES) [88887.341571/2019-00, 88887.495487/2020-00, Finance Code 001], and Fundação de Amparo à Pesquisa do Estado de São Paulo (FAPESP) [grants 2015/11907-2, 2015/09342-7, 2019/13871-6].

References

[1] M. Doerre, L. Hibbitts, G. Patrick, N.K. Akafuah, Advances in automotive conversion coatings during pretreatment of the body structure: a review, Coatings 8 (2018), https://doi.org/10.3390/coatings8110405.

[2] N.K. Akafuah, S. Poozesh, A. Salaimeh, G. Patrick, K. Lawler, K. Saito, Evolution of the automotive body coating process—a review, Coatings 6 (2016) 1–22, https://doi.org/10.3390/coatings6020024.

[3] B. Smith, A. Spulber, S. Modi, T. Fiorelli, Technology roadmaps: intelligent mobility technology, materials and manufacturing processes, and light duty vehicle propulsion, Cent. Automot. Res. (2017) 1–27, https://doi.org/10.1044/leader.ppl.22062017.20.

[4] K.W. Thomer, Materials and concepts in body construction, in: H.-J. Streitberger, K.-F. Dossel (Eds.), Automotive Paints Coatings, second ed., Wiley VCH, Munster and Wuppertal, 2008, pp. 13–59, https://doi.org/10.1002/9783527622375.ch2.

[5] D. Mohr, N. Muller, A. Krieg, P. Gao, H.-W. Kaas, A. Krieger, R. Hensley, The road to 2020 and beyond: what's driving the global automotive industry? Adv. Ind. 21 (2013).

[6] H.-J. Streitberger, Electrodeposition coatings, in: H.-J. Streitberger, K. Dossel (Eds.), Automotive Paints Coatings, second ed., Wiley VCH, Munster and Wuppertal, 2008, pp. 89–127, https://doi.org/10.1002/9783527622375.ch4.

[7] H. Wonnemann, Primer Surfacer, in: H.-J. Streitberger, K. Dossel (Eds.), Automotive Paints Coatings, second ed., Wiley VCH, Munster and Wuppertal, 2008, pp. 129–174, https://doi.org/10.1002/9783527622375.ch5.

[8] H.-J. Streitberger, Introduction, in: H.-J. Streitberger, K.-F. Dossel (Eds.), Automotive Paints Coatings, second ed., Wiley VCH, Munster and Wuppertal, 2008, pp. 1–11, https://doi.org/10.1002/9783527622375.ch1.

[9] M.F. Montemor, Functional and smart coatings for corrosion protection: a review of recent advances, Surf. Coat. Technol. 258 (2014) 17–37, https://doi.org/10.1016/j.surfcoat.2014.06.031.

[10] J. Baghdachi, Smart coatings, in: Smart Coatings II, ACS Symposium Series, Washington, DC, 2009, pp. 3–24, https://doi.org/10.1021/bk-2009-1002.ch001.

[11] B.J. Blaiszik, S.L.B. Kramer, S.C. Olugebefola, J.S. Moore, N.R. Sottos, S.R. White, Self-healing polymers and composites, Annu. Rev. Mat. Res. 40 (2010) 179–211, https://doi.org/10.1146/annurev-matsci-070909-104532.

[12] H.R. Fischer, S.J. García, Active protective coatings: sense and heal concepts for organic coatings, in: M.L. Zheludkevich, R.G. Buchheit, A.E. Hughes, J.M.C. Mol (Eds.), Active Protective Coatings New-Generation Coatings for Metals, Springer Science+Business Media B.V, Dordrecht, 2016, pp. 139–156, https://doi.org/10.1007/978-94-017-7540-3_7 139.

[13] S.R. White, N.R. Sottos, P.H. Geubelle, J.S. Moore, M.R. Kessler, S.R. Sriram, E.N. Brown, S. Viswanathan, Autonomic healing of polymer composites, Nature 409 (2001) 794–797, https://doi.org/10.1038/35057232.

[14] L. Wang, L. Deng, D. Zhang, H. Qian, C. Du, X. Li, J.M.C. Mol, H. Terryn, Shape memory composite (SMC) self-healing coatings for corrosion protection, Prog. Org. Coat. 97 (2016) 261–268, https://doi.org/10.1016/j.porgcoat.2016.04.041.

[15] J. Kötteritzsch, S. Stumpf, S. Hoeppener, J. Vitz, M.D. Hager, U.S. Schubert, One-component intrinsic self-healing coatings based on reversible crosslinking by Diels-Alder cycloadditions, Macromol. Chem. Phys. 214 (2013) 1636–1649, https://doi.org/10.1002/macp.201200712.

[16] D. Döhler, H. Peterlik, W.H. Binder, A dual crosslinked self-healing system: supramolecular and covalent network formation of four-arm star polymers, Polymer (Guildf). 69 (2015) 264–273, https://doi.org/10.1016/j.polymer.2015.01.073.

[17] A. Rekondo, R. Martin, A. Ruiz De Luzuriaga, G. Cabañero, H.J. Grande, I. Odriozola, Catalyst-free room-temperature self-healing elastomers based on aromatic disulfide metathesis, Mater. Horiz. 1 (2014) 237–240, https://doi.org/10.1039/C3MH00061C.

[18] M. Abdolah Zadeh, S. van der Zwaag, S.J. Garcia, Adhesion and long-term barrier restoration of intrinsic self-healing hybrid sol–gel coatings, ACS Appl. Mater. Interfaces 8 (2016) 4126–4136, https://doi.org/10.1021/acsami.5b11867.

[19] S.H. Cho, S.R. White, P.V. Braun, Self-healing polymer coatings, Adv. Mater. 21 (2009) 645–649, https://doi.org/10.1002/adma.200802008.

[20] C.D. Dieleman, P.J. Denissen, S.J. Garcia, Long-term active corrosion protection of damaged coated-AA2024-T3 by embedded electrospun inhibiting nanonetworks, Adv. Mater. Interfaces 5 (2018), https://doi.org/10.1002/admi.201800176.

[21] A. Lutz, O. van den Berg, J. Wielant, I. De Graeve, H. Terryn, A multiple-action self-healing coating, Front. Mater. 2 (2016) 1–12, https://doi.org/10.3389/fmats.2015.00073.

[22] A. Trentin, S.V. Harb, M.C. Uvida, S.H. Pulcinelli, C.V. Santilli, K. Marcoen, S. Pletincx, H. Terryn, T. Hauffman, P. Hammer, Dual role of lithium on the structure and self-healing ability of PMMA-silica coatings on AA7075 alloy, ACS Appl. Mater. Interfaces 11 (2019) 40629–40641, https://doi.org/10.1021/acsami.9b13839.

[23] A.E. Hughes, J.M.C. Mol, M.L. Zheludkevich, R.G. Buchheit, Introduction, in: A.E. Hughes, J.M.C. Mol, M.L. Zheludkevich, R.G. Buchheit (Eds.), Active Protective Coatings New-Generation Coatings for Metals, Springer Science+Business Media B.V, Dordrecht, 2016, pp. 1–13, https://doi.org/10.1007/978-94-017-7540-3.

[24] P. Hammer, F.C. Dos Santos, B.M. Cerrutti, S.H. Pulcinelli, C.V. Santilli, Highly corrosion resistant siloxane-polymethyl methacrylate hybrid coatings, J. Sol-Gel Sci. Technol. 63 (2012) 266–274, https://doi.org/10.1007/s10971-011-2672-8.

[25] F.C. Dos Santos, S.V. Harb, M.-J. Menu, V. Turq, S.H. Pulcinelli, C.V. Santilli, P. Hammer, On the structure of high performance anticorrosive PMMA–siloxane-silica hybrid coatings, RSC Adv. 5 (2015) 106754, https://doi.org/10.1039/C5RA20885H.

[26] S.V. Harb, F.C. Dos Santos, B.L. Caetano, S.H. Pulcinelli, C.V. Santilli, P. Hammer, Structural properties of cerium doped siloxane-PMMA hybrid coatings with high anticorrosive performance, RSC Adv. 5 (2015) 15414–15424, https://doi.org/10.1039/C4RA15974H.

[27] S. Habib, E. Fayyed, R.A. Shakoor, R. Kahraman, A. Abdullah, Improved self-healing performance of polymeric nanocomposites reinforced with talc nanoparticles (TNPs) and urea-formaldehyde microcapsules (UFMCs), Arab. J. Chem. 14 (2021) 102926, https://doi.org/10.1016/j.arabjc.2020.102926.

[28] C. Zhu, Y. Fu, C. Liu, Y. Liu, L. Hu, J. Liu, I. Bello, H. Li, N. Liu, S. Guo, H. Huang, Y. Lifshitz, S.T. Lee, Z. Kang, Carbon dots as fillers inducing healing/self-healing and anticorrosion properties in polymers, Adv. Mater. 29 (2017) 1–8, https://doi.org/10.1002/adma.201701399.

[29] D. Yang, Y. Ye, Y. Su, S. Liu, D. Gong, H. Zhao, Functionalization of citric acid-based carbon dots by imidazole toward novel green corrosion inhibitor for carbon steel, J. Clean. Prod. 229 (2019) 180–192, https://doi.org/10.1016/j.jclepro.2019.05.030.

[30] D. Snihirova, S.V. Lamaka, M.F. Montemor, "SMART" protective ability of water based epoxy coatings loaded with CaCO3 microbeads impregnated with corrosion inhibitors applied on AA2024 substrates, Electrochim. Acta 83 (2012) 439–447, https://doi.org/10.1016/j.electacta.2012.07.102.

[31] M.F. Montemor, D.V. Snihirova, M.G. Taryba, S.V. Lamaka, I.A. Kartsonakis, A.C. Balaskas, G.C. Kordas, J. Tedim, A. Kuznetsova, M.L. Zheludkevich, M.G.S. Ferreira, Evaluation of self-healing ability in protective coatings modified with combinations of layered double hydroxides and cerium molibdate nanocontainers filled with corrosion inhibitors, Electrochim. Acta 60 (2012) 31–40, https://doi.org/10.1016/j.electacta.2011.10.078.

[32] G.S. Frankel, Corrosion-sensing behavior of an acrylic-based coating system, Corrosion 55 (1999) 957–967.

[33] G. McAdam, P.J. Newman, I. McKenzie, C. Davis, B.R.W. Hinton, Fiber optic sensors for detection of corrosion within aircraft, Struct. Heal. Monit. 4 (2005) 47–56, https://doi.org/10.1177/1475921705049745.

[34] M. Kendig, M. Hon, L. Warren, "Smart" corrosion inhibiting coatings, Prog. Org. Coat. 47 (2003) 183–189, https://doi.org/10.1016/S0300-9440(03)00137-1.

[35] C.J. Hansen, W. Wu, K.S. Toohey, N.R. Sottos, S.R. White, J.A. Lewis, Self-healing materials with microvascular networks, Adv. Mater. 21 (2009) 1–5, https://doi.org/10.1038/nmat1934.

[36] L. Fedrizzi, F. Andreatta, Corrosion inhibitors, in: R.A.E. Hughes, J.M.C. Mol, M.L. Zheludkevich (Eds.), Active Protective Coatings New-Generation Coatings for Metals, Springer Science+Business Media B.V, Dordrecht, 2016, pp. 59–84, https://doi.org/10.1007/978-94-017-7540-3.

[37] O.M. Magnussen, Corrosion protection by inhibition, in: A.J. Bard (Ed.), Encyclopedia of Electrochemistry, Wiley VCH, Weinheim, 2007, pp. 435–459, https://doi.org/10.1002/9783527610426.bard040502.

[38] J. Alcántara, D. de la Fuente, B. Chico, J. Simancas, I. Díaz, M. Morcillo, Marine atmospheric corrosion of carbon steel: a review, Materials (Basel) 10 (2017) 1–65, https://doi.org/10.3390/ma10040406.

[39] E.D. Verink Jr., Simplified procedure for constructing pourbaix diagrams, in: R.W. Revie (Ed.), Uhlig's Corrosion Handbook, second ed., John Wiley & Sons, 1979, pp. 111–124, https://doi.org/10.1002/9780470872864.ch7.

[40] M. Schem, T. Schmidt, J. Gerwann, M. Wittmar, M. Veith, G.E. Thompson, I.S. Molchan, T. Hashimoto, P. Skeldon, A.R. Phani, S. Santucci, M.L. Zheludkevich, CeO2-filled sol-gel coatings for corrosion protection of AA2024-T3 aluminium alloy, Corros. Sci. 51 (2009) 2304–2315, https://doi.org/10.1016/j.corsci.2009.06.007.

[41] P. Visser, A. Lutz, J.M.C. Mol, H. Terryn, Study of the formation of a protective layer in a defect from lithium-leaching organic coatings, Prog. Org. Coat. 99 (2016) 80–90, https://doi.org/10.1016/j.porgcoat.2016.04.028.

[42] K.A. Yasakau, M.L. Zheludkevich, S.V. Lamaka, M.G.S. Ferreira, Mechanism of corrosion inhibition of AA2024 by rare-earth compounds, J. Phys. Chem. B 110 (2006) 5515–5528, https://doi.org/10.1021/jp0560664.

[43] H. Verbruggen, H. Terryn, I. De Graeve, Inhibitor evaluation in different simulated concrete pore solution for the protection of steel rebars, Construct. Build Mater. 124 (2016) 887–896, https://doi.org/10.1016/j.conbuildmat.2016.07.115.

[44] M.S. Vukasovich, J.P.G. Farr, Molybdate in corrosion inhibition—a review, Polyhedron 5 (1986) 551–559, https://doi.org/10.1016/S0277-5387(00)84963-3.

[45] P. Visser, Y. Liu, X. Zhou, T. Hashimoto, G.E. Thompson, S.B. Lyon, L.G.J. Van Der Ven, A.J.M.C. Mol, H.A. Terryn, The corrosion protection of AA2024-T3 aluminium alloy by leaching of lithium-containing salts from organic coatings, Faraday Discuss. 180 (2015) 511–526, https://doi.org/10.1039/c4fd00237g.

[46] A. Trentin, S.V. Harb, T.A.C. de Souza, M.C. Uvida, P. Hammer, Organic-Inorganic Hybrid Coatings for Active and Passive Corrosion Protection, first ed., IntechOpen, 2020, https://doi.org/10.5772/57353.

[47] F. Zhang, P. Ju, M. Pan, D. Zhang, Y. Huang, G. Li, X. Li, Self-healing mechanisms in smart protective coatings: a review, Corros. Sci. 144 (2018) 74–88, https://doi.org/10.1016/j.corsci.2018.08.005.

[48] J. Mosa, N.C. Rosero-Navarro, M. Aparicio, Active corrosion inhibition of mild steel by environmentally-friendly Ce-doped organic–inorganic sol–gel coatings, RSC Adv. 6 (2016) 39577–39586, https://doi.org/10.1039/C5RA26094A.

[49] S.V. Harb, A. Trentin, T.A.C. de Souza, M. Magnani, S.H. Pulcinelli, C.V. Santilli, P. Hammer, Effective corrosion protection by eco-friendly self-healing PMMA-cerium oxide coatings, Chem. Eng. J. 383 (2020), https://doi.org/10.1016/j.cej.2019.123219.

[50] T. Hu, H. Shi, S. Fan, F. Liu, E.-H. Han, Cerium tartrate as a pigment in epoxy coatings for corrosion protection of AA 2024-T3, Prog. Org. Coat. 105 (2017) 123–131, https://doi.org/10.1016/j.porgcoat.2017.01.002.

[51] L.M. Calado, M.G. Taryba, Y. Morozov, M.J. Carmezim, M.F. Montemor, Cerium phosphate-based inhibitor for smart corrosion protection of WE43 magnesium alloy, Electrochim. Acta 365 (2021) 137368, https://doi.org/10.1016/j.electacta.2020.137368.

[52] L.M. Calado, M.G. Taryba, Y. Morozov, M.J. Carmezim, M.F. Montemor, Novel smart and self-healing cerium phosphate-based corrosion inhibitor for AZ31 magnesium alloy, Corros. Sci. 170 (2020) 108648, https://doi.org/10.1016/j.corsci.2020.108648.

[53] Y. Morozov, L.M. Calado, R.A. Shakoor, R. Raj, R. Kahraman, M.G. Taryba, M.F. Montemor, Epoxy coatings modified with a new cerium phosphate inhibitor for smart corrosion protection of steel, Corros. Sci. 159 (2019) 108128, https://doi.org/10.1016/j.corsci.2019.108128.

[54] I. Danaee, E. Darmiani, G.R. Rashed, D. Zaarei, Self-healing and anticorrosive properties of Ce(III)/Ce(IV) in nanoclay-epoxy coatings, Iran. Polym. J. 23 (2014) 891–898, https://doi.org/10.1007/s13726-014-0288-x.

[55] L.M. Calado, M.G. Taryba, M.J. Carmezim, M.F. Montemor, Self-healing ceria-modified coating for corrosion protection of AZ31 magnesium alloy, Corros. Sci. 142 (2018) 12–21, https://doi.org/10.1016/j.corsci.2018.06.013.

[56] B. Ramezanzadeh, G. Bahlakeh, M. Ramezanzadeh, Polyaniline-cerium oxide (PAni-CeO2) coated graphene oxide for enhancement of epoxy coating corrosion protection performance on mild steel, Corros. Sci. 137 (2018) 111–126, https://doi.org/10.1016/j.corsci.2018.03.038.

[57] Y. Castro, E. Özmen, A. Durán, Integrated self-healing coating system for outstanding corrosion protection of AA2024, Surf. Coat. Technol. 387 (2020) 125521, https://doi.org/10.1016/j.surfcoat.2020.125521.

[58] P. Hammer, M.G. Schiavetto, F.C. dos Santos, A.V. Benedetti, S.H. Pulcinelli, C.V. Santilli, Improvement of the corrosion resistance of polysiloxane hybrid coatings by cerium doping, J. Non Cryst. Solids 356 (2010) 2606–2612, https://doi.org/10.1016/j.jnoncrysol.2010.05.013.

[59] P.S. Correa, C.F. Malfatti, D.S. Azambuja, Corrosion behavior study of AZ91 magnesium alloy coated with methyltriethoxysilane doped with cerium ions, Prog. Org. Coat. 72 (2011) 739–747, https://doi.org/10.1016/j.porgcoat.2011.08.005.

[60] T. Hu, H. Shi, T. Wei, F. Liu, S. Fan, E.H. Han, Cerium tartrate as a corrosion inhibitor for AA 2024-T3, Corros. Sci. 95 (2015) 152–161, https://doi.org/10.1016/j.corsci.2015.03.010.
[61] K.A. Yasakau, S. Kallip, M.L. Zheludkevich, M.G.S. Ferreira, Active corrosion protection of AA2024 by sol–gel coatings with cerium molybdate nanowires, Electrochim. Acta 112 (2013) 236–246, https://doi.org/10.1016/j.electacta.2013.08.126.
[62] H. Shi, E.-H. Han, S.V. Lamaka, M.L. Zheludkevich, F. Liu, M.G.S. Ferreira, Cerium cinnamate as an environmentally benign inhibitor pigment for epoxy coatings on AA 2024-T3, Prog. Org. Coat. 77 (2014) 765–773, https://doi.org/10.1016/j.porgcoat.2014.01.003.
[63] P.H. Suegama, H.G. de Melo, A.V. Benedetti, I.V. Aoki, Influence of cerium (IV) ions on the mechanism of organosilane polymerization and on the improvement of its barrier properties, Electrochim. Acta 54 (2009) 2655–2662, https://doi.org/10.1016/j.electacta.2008.11.007.
[64] A. Trentin, S. Harb V., M. Uvida C., K. Marcoen, S. Pulcinelli H., C. Santilli V., H. Terryn, T. Hauffman, P. Hammer, Effect of Ce(III) and Ce(IV) ions on the structure and active protection of PMMA-silica coatings on AA7075 alloy, Corros. Sci. 189 (2021) 109581, https://doi.org/10.1016/j.corsci.2021.109581.
[65] Y. Liu, P. Visser, X. Zhou, S.B. Lyon, T. Hashimoto, M. Curioni, A. Gholinia, G.E. Thompson, G. Smyth, S.R. Gibbon, D. Graham, J.M.C. Mol, H. Terryn, Protective film formation on AA2024-T3 aluminum alloy by leaching of lithium carbonate from an organic coating, J. Electrochem. Soc. 163 (2016) C45–C53, https://doi.org/10.1149/2.0021603jes.
[66] P. Visser, Y. Liu, H. Terryn, J.M.C. Mol, Lithium salts as leachable corrosion inhibitors and potential replacement for hexavalent chromium in organic coatings for the protection of aluminum alloys, J. Coat. Technol. Res. 13 (2016) 557–566, https://doi.org/10.1007/s11998-016-9784-6.
[67] M. Meeusen, P. Visser, L. Fernández Macía, A. Hubin, H. Terryn, J.M.C. Mol, The use of odd random phase electrochemical impedance spectroscopy to study lithium-based corrosion inhibition by active protective coatings, Electrochim. Acta 278 (2018) 363–373, https://doi.org/10.1016/j.electacta.2018.05.036.
[68] A. Hughes, J. Laird, C. Ryan, P. Visser, H. Terryn, A. Mol, Particle characterisation and depletion of Li2CO3 inhibitor in a polyurethane coating, Coatings (2017), https://doi.org/10.3390/coatings7070106.
[69] M. Abdolah Zadeh, J. Tedim, M. Zheludkevich, S. van der Zwaag, S.J. Garcia, Synergetic active corrosion protection of AA2024-T3 by 2D-anionic and 3D-cationic nanocontainers loaded with Ce and mercaptobenzothiazole, Corros. Sci. 135 (2018) 35–45, https://doi.org/10.1016/j.corsci.2018.02.018.
[70] T.T. Thai, A.T. Trinh, M.-G. Olivier, Hybrid sol–gel coatings doped with cerium nanocontainers for active corrosion protection of AA2024, Prog. Org. Coat. 138 (2020) 105428, https://doi.org/10.1016/j.porgcoat.2019.105428.
[71] A. Yabuki, Self-healing coatings for corrosion inhibition of metals, Mod. Appl. Sci. 9 (2015) 214, https://doi.org/10.5539/mas.v9n7p214.
[72] I. Recloux, M. Mouanga, M.E. Druart, Y. Paint, M.G. Olivier, Silica mesoporous thin films as containers for benzotriazole for corrosion protection of 2024 aluminium alloys, Appl. Surf. Sci. 346 (2015) 124–133, https://doi.org/10.1016/j.apsusc.2015.03.191.
[73] J. Li, Q. Feng, J. Cui, Q. Yuan, H. Qiu, S. Gao, J. Yang, Self-assembled graphene oxide microcapsules in Pickering emulsions for self-healing waterborne polyurethane coatings, Compos. Sci. Technol. 151 (2017) 282–290, https://doi.org/10.1016/j.compscitech.2017.07.031.
[74] D. Nguyen Thuy, H.T.T. Xuan, A. Nicolay, Y. Paint, M.G. Olivier, Corrosion protection of carbon steel by solvent free epoxy coating containing hydrotalcites intercalated with different organic corrosion inhibitors, Prog. Org. Coat. 101 (2016) 331–341, https://doi.org/10.1016/j.porgcoat.2016.08.021.
[75] M.M. Alrashed, S. Jana, M.D. Soucek, Corrosion performance of polyurethane hybrid coatings with encapsulated inhibitor, Prog. Org. Coat. 130 (2019) 235–243, https://doi.org/10.1016/j.porgcoat.2019.02.005.
[76] E.L. Ferrer, A.P. Rollon, H.D. Mendoza, U. Lafont, S.J. Garcia, Double-doped zeolites for corrosion protection of aluminium alloys, Microporous Mesoporous Mater. 188 (2014) 8–15, https://doi.org/10.1016/j.micromeso.2014.01.004.

[77] M.F. Haase, D.O. Grigoriev, H. Möhwald, D.G. Shchukin, Development of nanoparticle stabilized polymer nanocontainers with high content of the encapsulated active agent and their application in water-borne anticorrosive coatings, Adv. Mater. 24 (2012) 2429–2435, https://doi.org/10.1002/adma.201104687.

[78] Z. Baharom, N.B. Baba, R. Ramli, M.I. Idris, H.Z. Abdullah, Microencapsulation of natural self-healing agent as corrosion coating, AIP Conf. Proc. 2068 (2019), https://doi.org/10.1063/1.5089402.

[79] T. Siva, S. Sathiyanarayanan, Self healing coatings containing dual active agent loaded urea formaldehyde (UF) microcapsules, Prog. Org. Coat. 82 (2015) 57–67, https://doi.org/10.1016/j.porgcoat.2015.01.010.

[80] Y. Ye, H. Chen, Y. Zou, Y. Ye, H. Zhao, Corrosion protective mechanism of smart graphene-based self-healing coating on carbon steel, Corros. Sci. 174 (2020) 108825, https://doi.org/10.1016/j.corsci.2020.108825.

[81] C. Liu, H. Zhao, P. Hou, B. Qian, X. Wang, C. Guo, L. Wang, Efficient graphene/cyclodextrin-based nanocontainer: synthesis and host-guest inclusion for self-healing anticorrosion application, ACS Appl. Mater. Interfaces 10 (2018) 36229–36239, https://doi.org/10.1021/acsami.8b11108.

[82] P. Miao, K. Han, Y. Tang, B. Wang, T. Lin, W. Cheng, Recent advances in carbon nanodots: synthesis, properties and biomedical applications, Nanoscale 7 (2015) 1586–1595, https://doi.org/10.1039/c4nr05712k.

[83] J. Liu, Y. Liu, N. Liu, Y. Han, X. Zhang, H. Huang, Y. Lifshitz, S.T. Lee, J. Zhong, Z. Kang, Metal-free efficient photocatalyst for stable visible water splitting via a two-electron pathway, Science 347 (2015) 970–974, https://doi.org/10.1126/science.aaa3145.

[84] M. Richetta, P.G. Medaglia, A. Mattoccia, A. Varone, R. Pizzoferrato, et al., Layered double hydroxides: tailoring interlamellar nanospace for a vast field of applications, J. Mater. Sci. Eng. 06 (2017), https://doi.org/10.4172/2169-0022.1000360.

[85] J. Tedim, S.K. Poznyak, A. Kuznetsova, D. Raps, T. Hack, M.L. Zheludkevich, M.G.S. Ferreira, Enhancement of active corrosion protection via combination of inhibitor-loaded nanocontainers, ACS Appl. Mater. Interfaces 2 (2010) 1528–1535, https://doi.org/10.1021/am100174t.

[86] J.V. Nardeli, C.S. Fugivara, M. Taryba, M.F. Montemor, A.V. Benedetti, Self-healing ability based on hydrogen bonds in organic coatings for corrosion protection of AA1200, Corros. Sci. 177 (2020), https://doi.org/10.1016/j.corsci.2020.108984.

[87] M. Goyal, S. Kumar, I. Bahadur, C. Verma, E.E. Ebenso, Organic corrosion inhibitors for industrial cleaning of ferrous and non-ferrous metals in acidic solutions: a review, J. Mol. Liq. 256 (2018) 565–573, https://doi.org/10.1016/j.molliq.2018.02.045.

[88] J. Yang, C. Blawert, S.V. Lamaka, D. Snihirova, X. Lu, S. Di, M.L. Zheludkevich, Corrosion protection properties of inhibitor containing hybrid PEO-epoxy coating on magnesium, Corros. Sci. 140 (2018) 99–110, https://doi.org/10.1016/j.corsci.2018.06.014.

[89] X. Zhao, J. Wei, B. Li, S. Li, N. Tian, L. Jing, J. Zhang, A self-healing superamphiphobic coating for efficient corrosion protection of magnesium alloy, J. Colloid Interface Sci. 575 (2020) 140–149, https://doi.org/10.1016/j.jcis.2020.04.097.

[90] J.V. Nardeli, C.S. Fugivara, M. Taryba, E.R.P. Pinto, M.F. Montemor, A.V. Benedetti, Tannin: a natural corrosion inhibitor for aluminum alloys, Prog. Org. Coat. 135 (2019) 368–381, https://doi.org/10.1016/j.porgcoat.2019.05.035.

CHAPTER 19

Self-healing nanocoatings for automotive application

Abhinay Thakur and Ashish Kumar
Department of Chemistry, Faculty of Technology and Science, Lovely Professional University, Phagwara, Punjab, India

Chapter outline

1. Introduction — 404
   1.1 History — 405
2. Nanocoatings — 405
   2.1 Self-healing nanocoatings — 407
3. Types of nanocontainers-based self-healing coatings — 409
   3.1 Halloysite nanocontainers — 410
   3.2 Mesoporous silica nanocontainers — 411
4. The release of nanomaterials — 411
   4.1 There are two types of self-healing process — 412
5. Self-healing process investigation — 414
6. Impact on self-healing nanocoatings on various aspects — 416
   6.1 Environmental and health aspects — 416
7. Commercially available self-healing nanocoatings — 418
   7.1 Ceramic — 418
   7.2 Ceramic lite — 419
   7.3 Ceramic plus — 419
   7.4 Heal lite — 419
   7.5 Heal plus — 420
   7.6 Wheel and caliper — 420
   7.7 Topcoat — 420
   7.8 Hydro (marine) — 421
8. Applications in other field of the automobile industry — 421
   8.1 Lotus coatings — 421
   8.2 In windshields — 421
   8.3 Reduction in weight of vehicles — 421
   8.4 Catalytic converters — 422
9. Advantages and disadvantages — 422
   9.1 Advantages — 422
   9.2 Disadvantages — 422
10. Conclusion — 422
References — 424

1. Introduction

To be successful in a global market, products have to meet the most stringent local rules and regulations with respect to product properties and environmental concerns. For this reason, the automotive industry belongs to the most demanding customers in the world. Nanocoatings are valuable for anticorrosive surfaces in addition to other enhanced material functionalities. Some of the desired and achievable properties of this class of coatings on surfaces are dirt or water repellent, nonstick, hydrophobic, conductive, colored, transparent, decorative, anticorrosive, good adhesion, high chemical and temperature resistance, good diffusion barrier for certain metal ions, easy to clean/self-cleaning, scratchproof, insulating, antimicrobial, etc. [1–3]. In the area of corrosion inhibition, nanocoatings provide significant benefits that are valuable for aerospace, defense, medical, marine, oil industries, etc. [4–7].

The function of automotive coatings so far has been twofold—decoration and conservation. The former not only consists of providing a colored and smooth surface but also enhancing the shape of the car body via a viewing angle dependent on brightness and/or color. These optical effects mostly rely on tiny mirrors of metal flakes like aluminum or coated mica platelets that are more or less homogeneously dispersed within one coating layer; their surface normal is predominantly orthogonal to the substrates surface [8]. The particles' lateral dimensions are typically in the order of several hundred microns. The role of the coatings in the conservation of a metallic car body mainly consists of the protection of the substrate against electrochemical degradation (= corrosion) either actively or via providing a barrier layer as well as a best possible limitation of the coatings defects produced in the wake of dramatic mechanical perturbations like impacting gravel stones. On the contrary, mounted plastic parts or body panels in most cases have to be protected against chemical degradation triggered by exposure to UV light, physical erosion by swelling solvents like fuel or water depending on the nature of the polymers and the morphology of the bulk as well as against catastrophic failure of the part upon mechanical impacts (= brittle crash behavior). The means of achieving the latter are similar to those that are used in the design of stone chip-resistant coatings for car bodies. According to the report, self-healing materials using microencapsulation systems will generate revenues of $1.1 billion in revenues in 2022 [9,10]. Although the automotive sector already uses self-healing aftermarket coatings but has begun to sample much higher performing self-healing materials that can comply with the industry's demanding coatings requirements. The rapidly growing use of relatively fragile composites in the automotive sector is a strong driver for the development of self-healing composites. There are similar trends in the construction sector, with self-healing concrete undergoing commercialization. Although it will take until 2019 to take off, n-tech analysts expect sales of self-healing materials for construction to reach $475 million by 2022 [11,12]. This chapter describes the potential use of nanoscaled particles for self-healing purposes for novel

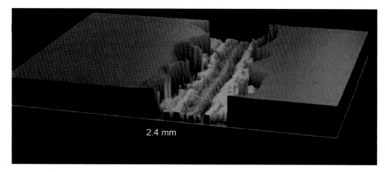

Fig. 1 Fracture scratch caused by macro-scratch tester equipped with 460 mm radius diamond tip at 24 N force. *(Adapted from C.M. Seubert, M.E. Nichols, Scaling behavior in the scratching of automotive clearcoats, J. Coat. Technol. Res. 4 (2007) 21–30. https://doi.org/10.1007/s11998-007-9006-3.)*

protective as well as decorative automotive coatings. The focus lies on the types of self-healing processes, their diffusion barrier properties as well as their role in active corrosion inhibition [13]. An example of a larger, fractured scratch produced using the same macro-scratch tester is shown in Fig. 1. These fracture scratches are caused by more severe contact damage from larger asperities and higher forces, such as keys, tree branches, grocery carts, and anything else that is translated across the surface with a large amount of force. Here, a jagged scratch profile was present after loading. Much of the profile's shape was a result of material removal.

1.1 History

Historians claim that some kind of lime mortar from the time of the Romans was the first self-healing material. While several articles, reports, and presentations were published after the mid-80s worldwide, the first popular one was published in the 21st century [14,15]. Self-healing/self-repair paints and coatings were developed in technical universities primarily in the United States, Canada, Europe, and Australia. These paints and coatings have often found commercial use in different units of the automotive industry in the years since 2010.

2. Nanocoatings

The modification of matter on an atomic and molecular scale is known as nanotechnology. Nanotechnology deals with materials, devices, and other structures that have at least one dimension between 1 and 100 nm. Nanotechnology has been around for 30 years. Nanoscience and nanotechnology can be applied to a wide range of small objects and to all other science areas, including chemistry, physics, material science, and engineering. Convergence of scientific developments such as the development of the scanning tunneling microscope in 1981 and the discovery of fullerenes in 1985 led to the advent of

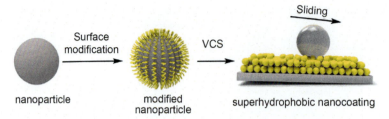

Fig. 2 Schematic illustration of the fabrication process of superhydrophobic nanocoatings. *(Adapted from Z. Wang, X. Chen, Y. Gong, B. Zhang, H. Li, Superhydrophobic nanocoatings prepared by a novel vacuum cold spray process, Surf. Coat. Technol. 325 (2017) 52–57. https://doi.org/10.1016/j.surfcoat.2017.06.044.)*

nanotechnology in the 1980s [16,17]. Today's scientists and engineers are experimenting with a number of methods to make materials at the nanoscale to take advantage of their improved properties, such as increased power, lighter weight, better light spectrum control, and better chemical reactivity. Fig. 2 shows a schematic illustration of the fabrication process of the superhydrophobic nanocoatings on the substrate.

Coatings have been used in many aspects of society for decades. Coatings' primary purpose is to cover and decorate objects, and their use has grown in scope as social and industrial growth has progressed. Vehicle efficiency and cost-effectiveness are already being improved by nano-enhanced materials, and this impact can only grow in the coming years as harder, stronger, and lighter nanomaterials become commercially available [18]. Natural resins and wax were used by the ancient Egyptians to create coatings, and artists used lacquers made from dried oils to cover their paintings. Although polymeric coatings have historically been used to protect different surfaces, there are several other important applications for this form of coating. Ancient Egyptian scientists created a very fine coating technology that resembled nanotechnology [19]. Nowadays, there is a strong market for thinner, but more efficient organic coatings for corrosion safety of metallic substrates used in the automotive industry, such as galvanized steel. The nanocoatings industry is expected to expand from $2 billion in 2012 to over $6.8 billion in 2020, according to optimistic estimates [20–22]. High transparency, modern functionalities, and high-quality performance are becoming increasingly important criteria in the coatings industry. However, protection from ice, pollutants, UV, fire, heat, bacteria, marine life, contact, and corrosion is the most important feature of nanostructured coatings. These factors cost the global industry billions of dollars in repairs, damage, and downtime each year, and they can be a major public health risk. Direct corrosion costs, for example, account for 3%–4% of a country's GDP globally [23,24]. Nanotechnology is best applied to coatings in automobiles. Protective and decorative finishes are applied to regular vehicle parts. Glass has already been coated to make it heat-reflective. Water- and dust-repellent coatings have also been added to automobiles. However, there is still a large area where nanocoatings could make their mark. Ceramic hard coatings boost part wear and friction properties, as well as the ability to detect even fractional gas concentrations in vehicle interiors. Another idea is self-healing, in which materials can refine their original shape under the influence of temperature. Furthermore, electro-chromic coatings foreshadow a huge boon for future automobiles [25,26].

Nanocomposite coatings, nanoscale multilayer coatings, superlattice coatings, nano-graded coatings, and other design models for nanocoatings serve a purpose. The potential for improving coating performance for particular applications such as environmentally friendly anti-corrosion coatings for the automotive and aerospace industries has sparked researchers' interest in nanostructured coatings over the last decade.

Advantages of nanocoatings include:
- Reduced costs in a variety of applications.
- Better functionality than conventional coatings (transparency, improved barrier properties, erosion resistance, and spectral control) (UV, IR).
- Coatings are made using less electricity.
- Excellent coating properties.
- Thin and light: Saves money on packaging, transportation, and storage.
- Nontoxic: This product is safe for the environment.
- Compatibility with various surfaces.
- Longevity and resistance have been improved.
- Corrosion resistance in harsh environments.
- Cost-effectiveness.
- Reduced prep time, application time, and coat count.
- Longer lifespan.
- Improved processing.

2.1 Self-healing nanocoatings

The self-healing material was first proposed in the 1980s, and it remains a hot spot of science to this day. A self-healing smart coating is a substrate that changes its behavior in response to external stimuli such as light, heat, mechanical initiation, permeability, pH, temperature, and hostile ions, and can repair the damage without external interference [27]. The two types of self-healing coatings are intrinsic and external, depending on whether the coating is implanted with an external repair agent. Since the coating involves a complex molecular structure, the intrinsic self-healing coating does not require the implantation of an external repair agent. Reversible reactions, such as the Diels-Alder reaction, hydrogen bonding reaction, and disulfide bonding reaction, restore the coating's original role and structure in response to an external stimulus. Although the intrinsic self-healing coating can self-heal several times, the realization of reversible reactions is limited by particular functional groups, making it difficult to use in industrial production. The external self-healing coating, on the other side, will release the carried healing agent

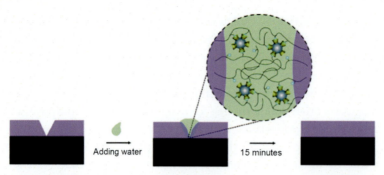

Fig. 3 Self-healing mechanism of β-CD-TiO$_2$/P(HEMA-co-BA). *(Adapted from L. Peng, M. Lin, S. Zhang, L. Li, Q. Fu, J. Hou, A Self-Healing Coating with UV-Shielding Property (2019).)*

and fill the damaged area of the coating, allowing it to self-heal. In the beginning, inorganic compounds like chromate were doped directly in the coating. However, since these inorganic corrosion inhibitor ions are poisonous, their leakage can result in significant environmental emissions [28]. Direct contact between the repairing agent and the coating substrate, on the other hand, will cause a chemical reaction that will affect the repairing agent's and coating's output. As a result, microcapsules, hollow fibers, and microvascular networks can be used to encapsulate the repair agent. As the local environment changes, the micro/nano container reacts by releasing the encapsulated active material and actively repairing the coating's structure and efficiency. This approach prevents the repair agent from leaking prematurely, selectively repairs the damaged area of the coating, and significantly enhances the coating's overall efficiency and service life. When the coating -CD-TiO$_2$/P(HEMA-co-BA) is mechanically compromised, the inclusion interaction between host and guest molecules breaks down, resulting in a large number of free host and guest molecules in the wound cross-section as shown in Fig. 3 [29,30]. The cross-section extends as it comes into contact with moisture, allowing the two sides to touch. The free host and guest molecules on the cross-section can become entangled with one another in this case, and the polymer molecular chains can become entangled with one another. The coating eventually completes self-healing with the aid of chain entanglement and host–guest interaction.

As a result, the smart coating method, which is more receptive, durable, and efficient than recent coating techniques, is a reliable tool for resolving issues like the ones described earlier. New manufacturing principles for the manufacture of functional coatings from a nanoscale layer loaded with active components on a classical layer with a solid base material layer and an integrated host structure have been developed recently in surface science and technology. The smart coating is currently being produced, and nanocoating has recognized it. Nanocoating is the application of nanotechnology toward corrosion protection techniques seeing as nanoscale materials have unique electrical, physical and physicochemical characteristics. Nanocoating is formulated by adding nanoparticles to the coating, which can thus improve the properties of the coating [31].

Self-healing nanocoatings improve the lifespan of the working material by reducing wear and friction while also reducing energy dissipation as heat, improving the vehicle's performance. Improved solvent, fuel, and gas barriers, as well as flame resistance, stiffness, and other mechanical properties, are all advantages of nanocoatings. This results in higher tool efficiency (longer tool life, higher cycle speeds, and less workpiece finishing), lower production costs, better product quality (due to smoother surfaces, better dimensional stability, higher degrees of metal deformation, and fewer manufacturing steps), and lower lubricant consumption [32].

Desirable functional properties for the automotive self-healing nanocoatings industry afforded by nanomaterials include:
- Scratch resistance
- Anti-fingerprint
- Self-cleaning
- Thermal barrier
- Flame retardance
- Chemical resistance
- UV resistance
- Self-healing
- Abrasion resistance.

Various applications include:
- Hydrophobic and oleophobic anti-fingerprint coatings in automotive displays
- Anti-fingerprint mirror and interior surface coatings
- Scratch-proof coatings
- Wear-resistant nanocoatings for engines
- Lubricant additives
- Self-cleaning coatings on glass
- Anti-corrosion engine coatings
- Anti-bacterial interior trim and upholstery coatings.

3. Types of nanocontainers-based self-healing coatings

Thin tubular nanocontainers are more appealing than microcapsules because they have superior aero- and hydrodynamic properties, as well as improved processability. The ability to design nanocontainers with fine-tuned properties, such as varying hydrophobicity, is allowed by the modification of the inner and outer surfaces of nanocontainers with well-defined functionality [33,34]. Immobilization, storage, and controlled release of inhibitors embedded in self-healing anticorrosive coatings have been accomplished using halloysite nanotubes, mesoporous silica containers, hydroxyapatite microparticles, and layered double hydroxides as carriers and reservoirs. The organic or inorganic inhibitors are loaded into the target nanocontainer through absorption through the porous nanocontainers' structure, encapsulation using emulsion polymerization, or ion exchange with counter positive/negative ions in the corresponding nanocontainers in the first step [35]. The inhibitor-loaded nanocontainers are then coated with pH-sensitive polyelectrolyte multilayers in the second stage. Finally, the inhibitor-loaded nanocontainers are distributed in a polymeric matrix material that is organic or inorganic.

3.1 Halloysite nanocontainers

Nanocontainers can be made from different tubular materials such as carbon, polymeric, metal, and metal oxide nanotubes. Polymeric nanotubes can be templated by molecular sieves or cylindrical nanopores to form tubular structures [36,37]. Metal and metal oxide nanotubes are synthesized by employing polymeric or inorganic nanorods as scaffold template. The shortcoming of these types of nanotubes is the employment of a template that needs to be prepared separately and requires extra post-synthesis removal steps, which is time-consuming and a costly process.

Halloysite is defined as a two-layered aluminosilicate with a hollow tubular structure in the submicrometer range. The adjacent alumina and silica layers create a packing disorder causing them to curve. The size of halloysite particles varies within 1–15 μm in length and 10–150 nm in inner diameter depending on the deposits. Thus, a variety of active agents such as drugs, corrosion inhibitors, and marine biocides can be entrapped within the halloysite inner lumen as well as within void spaces in the multilayered aluminosilicate shell [38,39]. The entrapped and stored active agents are retained and released in a controlled manner for a specific application. Both hydrophilic and hydrophobic agents can be entrapped after suitable pretreatment and conditioning of the halloysite. A hybrid sol-gel film doped with halloysite nanotubes for controlled release of entrapped corrosion inhibitor on aluminum substrate has been proposed. Initially, the halloysite nanocontainers were doped with 5 wt% 2-mercaptobenzothiazole inhibitor. To prevent the leakage of the doped inhibitor, the surface of the inhibitor-doped halloysite nanocontainers was coated with several alternating polyelectrolyte multilayers of (poly(styrene sulfonate)/poly(allylamine hydrochloride)). This step was essential to close the edges of the nanocontainers as well as to enable the controlled release of the doped 2-mercaptobenzothiazole inhibitor to the surrounding environment upon pH change. In a separate step, the organically modified hybrid sol-gel was prepared by using zirconium (IV) n-propoxide (TPOZ) and 3-glycidoxypropyltrimethoxysilane (GPTMS) precursors. Finally, the inhibitor-doped halloysite nanocontainers suspension was incorporated into the sol-gel solution before the dip coating of the AA2024-T3 samples [40]. The prepared sol-gel films with the halloysite nanocontainers provided long-term corrosion protection in comparison with the undoped sol-gel film.

3.2 Mesoporous silica nanocontainers

Monodisperse, mesoporous silica nanoparticles could be filled with any organic corrosion inhibitor and coated in a combination gel-sol coating for resistance to metal corrosion. Mesoporous silica nanocontainers (MSNs) have several fascinating characteristics: (1) higher surface area (2) small pore size distribution, wide pore volume and high inhibitor loading power. The silica nanocontainer-filled inhibitor can be incorporated in the sol-gel hybrid coating with no further encapsulation (i.e., without external polyelectrolyte multilayer coating) ensuring the cause release mechanism is focused entirely on the corrosion process, offering a self-healing effect [41–43].

4. The release of nanomaterials

Various studies have also been conducted on the impending release of nanomaterials from coatings. Particles smaller than 100 nm were released during the abrasion, sanding, and aging processes. The nanoparticles that had been added, on the other hand, remained securely embedded in the binder matrix. Abrasion measurements, peripheral-longitudinal surface grinding (DREMEL), the effect of aging processes, and the influence of climatic influences on surface weathering were all used in the experiments to model the daily use of coated surfaces (parquet floor) [44]. The release of TiO_2 nanomaterials from coatings that had been applied to various surfaces was investigated, among other things. The impact of various forms of stress (e.g., climatic stress, mechanical stress) was also investigated. Particle size distributions of 15–616 nm were measured, with a maximum of 630 particles per cm^3 of air. There was no evidence of isolated TiO_2 nanoparticles being released [45]. There was no discernible difference in particle size distribution between coating systems containing nanomaterials. Experts agree that mechanical treatment, rather than chemical or thermal degradation of the matrix material, is the only way to release isolated nanoobjects from coatings [46]. However, it is still unclear if nanomaterial-containing matrix material undergoes any processes after being released into the atmosphere, which could lead to further degradation and, eventually, the release of isolated nanomaterials. It appears that nanomaterials (e.g., photochemically degradable materials) could be released during the weathering of their carrier matrix. Weathering processes have been shown to release small quantities of synthetic TiO_2 particles ranging in size from 20 to 300 nm, as well as nanosilver particles smaller than 15 nm, from facade paint. Storm water drains can allow these particles to enter the atmosphere. Dry coatings do not leach out nano-TiO_2. Normal wear and tear, however, can cause it to be released into the breathable air alongside the binder. In Fig. 4, an illustration of the intrinsic self-healing nano polymer system with reversible chemical bonds can be observed.

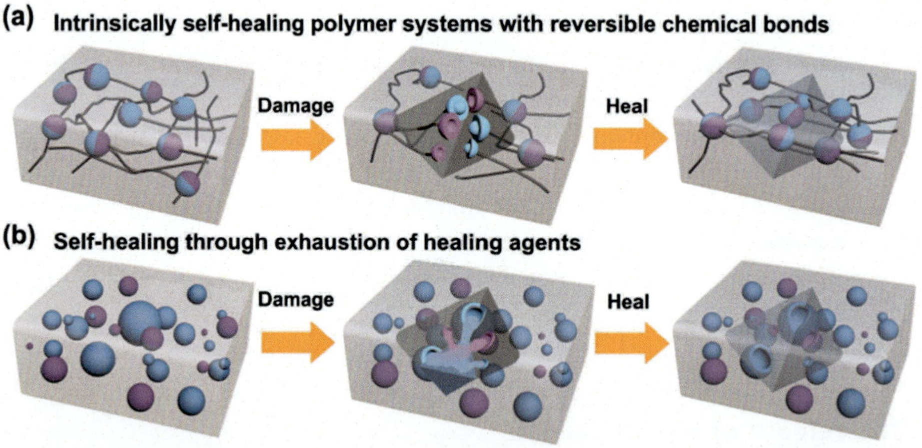

Fig. 4 Schematic illustration of intrinsically self-healing polymer systems with reversible chemical bonds (A) and self-healing through exhaustion of healing agents (B). (A) The self-healing of damages is realized by the reformation of chemical bonds at the fractured surfaces through the supramolecular assembly. (B) The self-healing is realized through exhaustion of liquid agents pre-embedded in the polymer matrix to fill in the ruptured space. (Adapted from D. Chen, D. Wang, Y. Yang, Q. Huang, S. Zhu, Z. Zheng, Self-healing materials for next-generation energy harvesting and storage devices, Adv. Energy Mater. 7 (2017) 27–31. https://doi.org/10.1002/aenm.201700890.)

4.1 There are two types of self-healing process

Non-autonomic, in which the healing process is triggered by external stimuli rather than being completely self-contained and requiring no external input; and autonomic, in which the healing process is triggered by external stimuli. Encapsulated healing agents, microvascular healing networks, and intrinsic healing are three main mechanisms for promoting self-healing in materials [47].

4.1.1 Encapsulated healing

Encapsulated healing is perhaps the most popular method for integrating self-healing properties into polymers and polymer composites. During the manufacturing stage, this mechanism integrates micro- or nano-capsules of healing agent into the material's structure. Catalysts would be spread in the material matrix, along with healing agent microcapsules, in materials that use this process. When a crack appears, the microcapsule ruptures, allowing the healing agent to enter the crack. When the released agent comes into contact with the catalyst, it undergoes a chemical reaction that causes it to harden and fill the gap. Professor Scott R. White and his colleagues showed one of the earliest practical examples of this in a polymer epoxy material in 2001. White et al. produced an epoxy composite by combining 100 parts Epon Resin 828, a bisphenol-A and epichlorohydrin mixture, with 12 parts diethylenetriamine in the sample (DETA) [48–50].

This mechanism, however, is constrained by its nature. Each healing agent capsule is microscopic to ensure that the material's overall integrity is not jeopardized. Self-healing mechanisms that restore damage to a material with lower fracture resistance than its non-healing counterparts would seem counterintuitive. The amount of healing agent contained in each capsule is limited due to the capsule's size constraints. As a result, the amount of damage it can effectively fix is limited [51].

4.1.2 Microvascular healing systems

Materials scientists took cues from Mother Nature's book to solve the embedded capsules approach's single-use limitations. "Healing in biological systems is achieved by a pervasive vascular network that provides the requisite biochemical components," wrote Dr. Kathleen Toohey in a seminal 2007 study in the field of self-healing materials [22,52]. Toohey was the lead researcher on a study that integrated a similar vascular network into an epoxy resin's substrate. By exploiting capillarity, or the ability of liquids in capillary tubes to flow independently of external forces, such a material contains a network of microtubes that enable the healing agent to flow to the location of injury. When a crack appears, the vascular network's surface tension changes, causing the healing agent to pump to the point of damage, where it interacts with embedded catalyst particles, hardening and sealing the crack. Microvascular healing materials can solve the problem of repeated localized damage, which restricts the use of microencapsulated healing, with this design. It also has a lot of potential for a variety of materials used in transportation, construction, and architecture [53,54]. Carbon-fiber reinforced plastic (CFRP), for example, is a composite material commonly used in transportation due to its high stiffness and strength at a low density and weight. As a result, it's becoming more popular in everything from car chassis and roof frames to train frames and the fuselage of Boeing's 787 Dreamliner plane. CFRPs with self-healing properties have been formed by using capillary networks to transport healing agents, and the healed material often does not display a noticeable reduction in bend power. G Williams points this out in a 2007 paper titled "A self-healing carbon fiber reinforced polymer for aerospace applications."

4.1.3 Intrinsic healing

In the field of self-healing materials, both encapsulation mechanisms and microvascular structures hold promise. However, necessitates the creation of extrinsic healing material, which necessitates the use of a separate healing agent. Some materials scientists and researchers aim to create intrinsic self-healing polymers that regenerate themselves using complex chemical bonds within the material. Extrinsic healing materials' lifetime is restricted by the amount of latent healing agent present in the material, while intrinsic healing materials could potentially have almost limitless potential for repair. It's not easy to achieve intrinsic healing, particularly for materials that would be useful in the automotive industry. According to Yu Yanagisawa in a 2018 research paper, the typical problems are that "these healable materials are typically soft and deformable." Cross-linking with dynamic covalent bonds has also resulted in the creation of certain healable materials with high mechanical robustness [55]. In most cases, however, heating to high temperatures (120°C or higher) to reorganize their cross-linked networks is needed to repair the broken portions." In this field, research is still ongoing, especially in the development of durable materials that heal themselves. Progress has been made, with one study achieving low elastic modulus polymers that can heal at temperatures as high as 70°C. There have also been a few other materials reported that have inherent self-healing properties. Self-healing ceramics can repair themselves due to a reaction between oxygen molecules that penetrate a crack and silicon carbide in the ceramic. This reaction produces silicon dioxide, which combines with another ceramic ingredient, alumina, to form a substance that fills the gap and hardens.

Adding trace amounts of manganese oxide to alumina grains allowed the ceramic to heal in under 60 s at 1000°C, according to research published in 2018. This was noted as the operating temperature of aircraft engines that could use this ceramic as a turbine material. However, the majority of these materials are not yet ready to withstand the rigors of transportation. Similarly, these non-autonomous healing processes render them inadequate for transportation applications, where materials that heal in real time in response to damage during service are the most valuable.

5. Self-healing process investigation

Encapsulation of moisture or oxygen-reactive substances, like metal oxide by-products or synthetic oils, adheres to the surface of nanocontainer to ambient oxygen as well as moisture, creating an impenetrable film in the degraded area of coating. The pH change on the metal substrate surface during the corrosion phase is the second essential cause studied for self-healing corrosion resistant coatings. The corrosion phenomenon is correlated by a regional pH that decreases within anodized area and increases within the cathodic zone. Consequently, an encapsulated inhibitor may be released for corrosion mitigation by a nanocontainer shell that is responsive to one or either of the acidic and alkaline regions. Polyelectrolyte capsules, nano-sized capsules produced by interfacial polymerization, hydroxyapatite and porous bio-inspired nanovalves with low acidic or basic shell functional groups have been documented in the literature for reversible and (or) irreversible shell permeability changes in a broad pH spectrum. The following are several basic external stimuli used in capsules and nanocontainers for self-healing coatings:
- Electromagnetic irradiation, under which container shell must have sensitive elements such as metal nanoparticles which are resistant to visible light dyes, IR light and TiO_2 nanoparticles for UV radiation.
- Ultrasonic treatment

- Temperature
- Ionic strength

The electrochemical potential of the corroded metal surface.
- Despite their specific characteristics, each proposed micro container-based self-healing device should satisfy the following criteria:
- All reagents and newly formed polymeric materials must be consistent with the polymer matrix and should have strong adhesion.
- Both the reagents and the container shell must remain stable to degradation and several chemical reactions including self-polymerization for a longer period.
- The chemical reaction between a healing agent and hardener resulting in the formation of a healing film must be completed within a temperature range of 10–35°C at a relatively short time.

The efficacy of self-healing coatings in corrosive conditions can be examined by utilizing a variety of techniques, like vibrating electrode scanning (SVET), electrochemical impedance spectroscopy (EIS) and ion-selective electrode scanning (SIET). EIS can be used to calculate the degradation of coating and corrosion kinetics even though the exposure period for coating in a corrosive media is not prolonged [56]. This strategy is generally aided by SVET or localized impedance spectroscopy (LEIS). SVET containing non-intrusive scanning and vibration study of the probe. In particular, the electrical field produced in a plane just above the electrochemically active layer is measured using the regional electrochemical behavior of coating can be measured and evaluated in clear text. This approach is especially helpful for monitoring the activity of corrosion inhibitors in self-healing coatings. The ion-electrode scanning methodology is used to trace regional pH changes to the coating surface. Electron-probe microanalysis (EPMA) and X-ray photoelectron spectroscopy (XPS) has been used to understand the mechanism of self-healing. XPS allows the analysis of the chemical structure of the targeted compound. EPMA provides information on the movement of ions to the scratch area. Microscopic techniques are often used to assess the ability of the coating to heal itself. For this reason, the coating is removed, submerged in a corrosive solution and the cutting position is examined at fixed time intervals using optical microscopy, atomic force microscopy (AFM), electron scanning microscope (SEM). The scanning electrochemical microscope (SECM) is often used to test the self-healing capabilities of the coatings. Using this approach, it is not only possible to obtain microscopic data through surfaces damaged via corrosion, but also to determine local variations in an electrochemical reaction. In addition, the discharge of metal ions at the anodized site and absorption of oxygen at the cathode site could be controlled. The ability to monitor both the insulation and the activity of surfaces is a major advantage of SECM. In Fig. 5, the diagrammatic representation of the automatic mechanically triggered self-healing mechanism is being shown.

Additionally, the self-healing mechanism can be measured by healing efficiency estimation. Equations used to calculate it has been shown below. Crack healing efficiency ($η$)

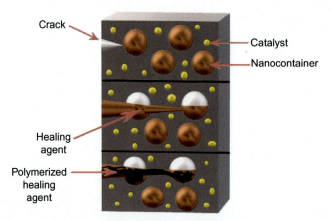

Fig. 5 Diagrammatic representation of the automatic mechanically triggered self-healing mechanism.

is assessed as the capacity to restore fracture durability under monotonous conditions and described as the ratio of fracture durability of healed KI and virgin KI of virgin materials.

$$\eta = \frac{K_{Ic(healed)}}{K_{IC(virgin)}}$$

A tapered double-cantilever beam (TDCB) test may also be used to detect fracture strength values. To determine the fracture strength of the TDCB specimen, the critical fracture load (Pc) must be calculated and geometric terms must be understood. In that case, healing efficiency can easily be found.

$$\eta = \frac{P_{c(healed)}}{P_{C(virgin)}}$$

6. Impact on self-healing nanocoatings on various aspects

6.1 Environmental and health aspects

Nanocoatings are primarily used in the automotive industry to protect equipment, engines, and other essentials. They contribute to resource management by extending the life cycle and maintenance cycles of machinery and vehicles when applied to metal and other surfaces. Nanocoating systems have the potential to reduce emissions even further. Nanocoatings result in thinner layers of coating [57]. Nano-based lacquer, for example, needs just a tenth of the volume of traditional lacquer. This helps to save raw materials. As a result, the trend toward lighter-weight product design can have environmental benefits during use, particularly in the transportation sector. Not only in the automotive industry but also in the aerospace and rail industries, strong positive effects are anticipated. Another possible advantage to the environment may be the reduction or replacement of solvents and harmful chemicals in paints and lacquers (e.g., chromium compounds). Surfaces that are self-cleaning or easy to clean can reduce the need for cleaning. Self-cleaning or easy-to-clean surfaces, especially in industrial cleaning, may save energy and cleaning agents while also extending the life of coated items. An inorganic-organic binder is used in UV-curable coatings. They will dry in seconds under UV light, reducing the amount of energy used during the drying process. UV curable coatings often typically have no or few toxic solvents. Nanomaterial-containing coatings can have a range of environmental benefits. However, no accurate quantitative data on the real potential for environmental benefits is currently available. Descriptions of the environmental benefits of nanocoatings usually do not include any analysis or evaluation of the raw material and energy savings that can be made during production or any evaluation of the materials' fate and behavior in their end-of-life phase (waste phase).

6.1.1 Environmental impacts

Since the release of isolated nanomaterials is relatively restricted, the environmental risks presented by such coatings containing nanomaterials that are securely embedded in a carrier matrix are currently considered to be negligible [58]. There is currently no evidence that nano-based surface coatings pose a threat to the environment. However, nanomaterials found in coatings may be released as a result of aging processes. The formation of free oxygen radicals by photo-catalytically active TiO_2 has a toxic effect on aquatic species, according to laboratory studies using simulated solar radiation. Nanosilver, which is another nanomaterial used in coatings, is primarily applied for its antibacterial effect. The release of silver ions prevents microorganisms from evolving. Colloidal silver is listed as "Very harmful to water" because it is poisonous to marine life. Nanosilver is harmful to marine organisms in several studies. The released silver ions were primarily responsible for the toxic effect of nanoscale silver that was discovered during these studies. The release of silver ions is dependent on the nanosilver's stability, form, and coating, as well as the surrounding environmental conditions. Some reports show the nanoparticles themselves have an additional impact. When considering long-term consequences, this must be taken into account. Other nanomaterials that may be present in self-healing coatings, such as nanoscale zinc oxide, iron oxide, and carbon black, have ecotoxicological evidence. However, as with nanoscale TiO_2 and silver, there isn't enough knowledge about potential side effects to draw any conclusions, particularly when it comes to long-term exposure [59]. There is no existing information on the environmental concentrations of nanomaterials used in coatings. However, newly developed coating products containing nanomaterials should be designed in such a way that throughout their entire life cycle the release of nanomaterials into the environment is avoided to the maximum extent possible.

6.1.2 Health impacts

The intended use of substances coated with nanomaterials that are securely embedded in a matrix may not pose a quantifiable nano-specific health risk to the customer. Inhalation of dust during the manufacture and further processing of nanomaterials, or when grinding, cutting, drilling, or milling nanocoatings, poses the greatest health risk. Spray painting, if done incorrectly, can endanger the user's health. Aerosol sprays produce aerosols (very small liquid droplets) ranging in size from a few tens of nanometers to around 100 μm. Droplets smaller than 10 μm can penetrate the lungs and possibly enter the pulmonary alveoli [60–62]. Alveolar collapse may be caused by inhaling surface-active compounds from impregnating sprays. As a result, inhaling spray mist from such materials should be avoided at all costs. The directions for safe use must be strictly followed. They should not be used in enclosed spaces under any conditions. On the chronic inhalation toxicity of synthetic nanomaterials, only a few studies (TiO_2 and carbon black) are visible. Inflammatory responses and tumors have been observed in rats in these studies. However, whether primary genotoxic effects or the results of overloading and inflammation are responsible for the carcinogenicity of such nanomaterials is still being debated. It's also uncertain if adverse effects are likely in the low-dose range that matters to the environment. As a result, the Federal Ministry of the Environment (BMUB) and BASF, as an industrial partner, have started a large-scale chronic in-vivo inhalation study on rats. Various concentrations of nanoscale cerium dioxide (Nano-CeO_2) are investigated during the project. The project will take four years to complete. The project adheres to the Organization for Economic Cooperation and Development's testing guidelines (OECD). As independent professional authorities, the Federal Environment Agency (UBA), Federal Institute for Risk Assessment (BfR), and Federal Institute for Occupational Safety and Health (BAuA) will review and analyze the study's findings (see also BMUB Press Release No. 066/12 dated 15 May 2012). Dermal absorption of nanomaterials can occur during the manufacture and use of nanomaterial-containing lacquers and paints, as well as during the degradation of matrix material in current nanocoatings [10,60–70].

7. Commercially available self-healing nanocoatings

7.1 Ceramic

Ceramic was developed for painted automotive surfaces to provide extreme toughness and safety. Ceramic has outstanding heat resistance, a highly glossy finish, and excels in hydrophobics as an ultra-high solids coating (sheeting water). Ceramic provides excellent chemical resistance as well as improved protection from light washing swirl marks.

Since the formula penetrates deeper into the established paint structure than anything else available, FEYNLAB's patented nanotechnology and bonding chemistry provide an unrivaled degree of safety.

Features
- Ridiculously glossy
- Extremely durable
- Outstanding hydrophobic sheeting properties

7.2 Ceramic lite

Ceramic lite is a low-cost coating that outperforms conventional waxes and sealants in terms of performance. It's the ideal entry-level coating for keeping your vehicle cleaner for longer due to its impressive longevity, shiny finish, and hydrophobic properties. Ceramic lite is the only way to get this legacy technology on the market. Ceramic lite is a long-lasting, slick, and water-repellent paint coating that adds a high degree of gloss to every paint job. The term "LITE" refers to a coating with a lower solids content than FEYNLAB's conventional offerings, but it still packs a punch in terms of toughness and efficiency.

Features
- Ridiculously glossy
- Extreme slickness
- Outstanding hydrophobic sheeting and self cleaning properties

7.3 Ceramic plus

Ceramic Plus has the same thickness and protection as Ceramic, but it also has a 40% self-healing capability for light surface scratches and micro marring. Although some deeper scratches may not be fully healed, they will be much less visible and almost undetectable on lighter-colored paints. Ceramic Plus has the same heat resistance, gloss finish, hydrophobic properties, and simple cleaning properties as Ceramic.

Features
- Ridiculously glossy
- Durable long lasting protection
- Outstanding UV resistance

7.4 Heal lite

Heal lite is the ideal remedy for dark-colored vehicles that are difficult to keep clean. When the panel is heated by the sun, washed caused stains and other light scratches can disappear easily. Heal lite will restore up to 60% of finer scratches and blemishes. Heal lite is a one-step ultra-durable coating with excellent UV protection, extreme hydrophobic properties, and up to 60% of Heal plus' healing capability. Heal lite was created with darker colors in mind, especially nonmetallic vehicles, which show defects like spider web swirl marks and other micro defects more clearly. In addition to the impressive healing capabilities, Heal lite has all the UV and chemical protection, gloss, and hydrophobic properties that FEYNLAB coatings are known for.

Features
- Healable
- Ridiculously glossy
- Superior hydrophobic sheeting and self cleaning properties

7.5 Heal plus

Heal Plus is the perfect way to ensure that your vehicle looks great and is well covered for years to come. In terms of automotive surface care, Heal Plus is the thickest and most advanced nano-coating available. Heal Plus' ceramic backbone provides a highly durable protective layer with super gloss and intense slickness, in addition to its innovative self-healing properties. Heal Plus coats painted surfaces with a super thick, self-healing, strong, chemical, and UV resistant coating that keeps them looking great for a long time.

Features
- Ridiculously glossy
- Extreme slickness
- Outstanding UV protection
- Excellent chemical resistance
- Durable long lasting protection
- Superior hydrophobic sheeting and self cleaning properties
- Industry leading self-healing ceramic nano technology

7.6 Wheel and caliper

The Wheel and Caliper were created to protect vehicle wheels from the harsh conditions they are exposed to, as well as make wheel cleaning a breeze! This dense and long-lasting ceramic nano-coating was created to withstand extreme temperatures and a continuous stream of pollution.

Product Benefits

With extremely high solids content (70%), WHEEL AND CALIPER will stand up to the daily beating delivered by road contamination and brake dust.

7.7 Topcoat

The topcoat is the most hydrophobic coating. Its sole aim is to give any of the ceramic coatings excellent water sheeting and self-cleaning properties. Furthermore, Topcoat's super-low free surface energy makes the surface feel silky smooth to the touch.

Product Benefits

TOPCOAT will help keep the exterior vehicle surfaces cleaner for longer due to its super high hydrophobic properties.

7.8 Hydro (marine)

Hydro Marine Coating is made to withstand the rigors of watercraft. Both painted and gel-coat finishes are compatible with this hardcore UV coating safety. HYDRO employs smart nano particles that disperse UV rays with a high degree of effectiveness, reducing the likelihood of UV rays touching the paint surface by around 1%. This solves the age-old problem of UV-induced yellow, chalky, and powdery gel-coat oxidation.

Product Benefits

Hydro produces hydrophobic surfaces that allow water and contaminants to escape easily. This reduces the time it takes to rinse down after use, making cleanup a breeze. When applied below the water line, the hydro's hydrophobic properties aid in reducing water drag. The propellers benefit from the application because it reduces fuel consumption. Hyrdo has a similar appearance to ceramic; it is extremely glossy, crisp, and vivid.

8. Applications in other field of the automobile industry

8.1 Lotus coatings

The lotus effect refers to the very high water repellence exhibited by the leaves of the lotus flower. Some nanotechnologists have created treatments, coatings, paints, and other surfaces that, like lotus leaves, can remain dry and clean themselves. Lotus coatings are made up of hydrophobic silsesquioxanes nanoparticles with adhesion promoters and low surface energy groups. When these coatings are added to the tyres, the water condenses into tiny droplets that fall off the surface without getting wet. These coatings also prevent dust from adhering to the vehicle's surface, which helps to prevent corrosion.

8.2 In windshields

Another advanced application of nanotechnology in automobiles is the development of nanoparticle-based mirrors and side panels. As a result, they filter the sun's rays, smoke, and other toxins in the air. The same technology allows radio and phone signals, as well as sound waves, to freely reach cars, allowing passengers to remain aware of their surroundings. This technology produces mirrors with increased stiffness and clarity. Fog formation on windshields can be completely removed by adding nanoparticle coatings to them.

8.3 Reduction in weight of vehicles

The weight of the vehicle's engine and transmission system accounts for a significant portion of the vehicle's weight in modern automobiles. As a result, vehicles need more fuel to propel those devices forward. Nonetheless, with the introduction of alloys, engines were made lighter, but not in a way that made them more fuel-efficient. Engines and parts were made much lighter thanks to nanotechnology, eliminating the need to use more fuel just to move the vehicle forward. Carbon Nanotubes are used in this case, which is 100 times lighter than steel but 100 times stronger. Additionally, by reducing the weight of the chassis, the weight of the vehicle can be decreased.

8.4 Catalytic converters

Nanomaterials are used in catalytic converters made of nanotechnology to prevent platinum from aggregating, requiring less platinum per converter. The porous silica layer is used to coat platinum nanoparticles, which is a significant advancement. The silica coating acts as an energy shield, keeping the platinum in place even at very high temperatures, preventing aggregation, and preserving catalytic activity. Experiments show that uncoated platinum started to accumulate at temperatures as low as 350°C, while silica-coated platinum retained its catalytic capacity (around 750°C) at far higher temperatures.

9. Advantages and disadvantages

9.1 Advantages

- Self-healing nanomaterial coatings on vehicles can prevent corrosive materials from corroding.
- They lower the rate of fuel consumption by reducing frictional, tear, and wear losses.
- They are very small and light in weight, but they are much tougher than steel, iron, and other metals.
- There are plenty of raw materials available for manufacturing, and they are long-lasting.

9.2 Disadvantages

- Nanotechnology is currently very costly due to its high manufacturing costs, and certain sectors, such as steel and diamond production, may be adversely affected.
- The environmental impact of nanotechnology is still uncertain.

10. Conclusion

Interest in nanocoatings has increased because of the potential to synthesize materials with unique physical, chemical, and mechanical properties. The integration of nanomaterials into thin films, coatings, and surfaces results in new functionalities, entirely novel properties, and the potential for multi-functional and smart coatings. The use of nanomaterials also improves wear, corrosion-wear, fatigue, and corrosion-resistant coating efficiency. As compared to conventional coatings, nanocoatings show a substantial increase in outdoor toughness as well as greatly improved hardness and versatility. Self-healing nanocoatings are a rapidly emerging technology, and it is a foregone conclusion that we will hear about them more often in the future. New techniques that take advantage of nanoscale effects can be used to produce tailor-made coatings with dramatically improved properties in the field of surface coatings. Nanotechnology's ultimate effect in the field of coatings will be determined by its ability to guide the assembly of hierarchical systems containing nanostructures. The use of nanocoatings improves water and corrosive ion permeation barrier properties. This protective coating has a wide range of applications in the automotive, aerospace, marine, and manufacturing industries. A cost-effective non-epoxy-based corrosion-resistant coating that can withstand temperature changes, strong acid, water, and/or road salt is in demand. The surface area of nanoparticle materials is extremely large. When functionalized, this surface can deliver high levels of organic corrosion inhibitors. As a result, optimized nanoparticles are an excellent vehicle for delivering the required amount of active corrosion inhibitors. The key issues with micro/nano container-based self-healing coatings are reduced load ability and inconsistent micro/nano container dispersion. It's usual to increase the volume of a container to increase the load capacity, but this compromises the container's compatibility with the coating. Increasing the number of micro/nanocontainers is also an efficient process, but it introduces more defects, reducing the coating's corrosion resistance. The self-healing nanocoating will continue to be a hot topic in the future, with the following development trend:

- **Initiate self-healing method is much better**. To begin, there is a pressing need to find cost-effective, environmentally friendly, and long-lasting healing agents to enhance the industrial application of self-healing coatings. Second, optimizing container loading ability and dispersion by coating or surface modification is also needed. Finally, it is fashionable to increase the coating's efficiency by reducing the container size microns to nanometers.
- **Integrate the self-healing system's initial components**. To boost the coating's anticorrosion efficiency, a coating system with external and intrinsic healing mechanisms, or a micro/nano container and microvascular network, should be added to the coating substrate at the same time.
- **Develop coatings that are multi-responsive or multi-functional**. The shift from single-response to multi-response coatings appears to be promising. Most research reports on self-healing nanocoatings currently focus solely on their short-term application impact. Such articles do not include long-term sulfide stress cracking, cyclic tests, or even field tests. As a result, more rigorous long-term or accelerated corrosion tests are needed to assess coating efficiency.

References

[1] Y. Xi, Z. Xie, Corrosion Effects of Magnesium Chloride and Sodium Chloride on Automobile Components, Colorado Department of Transportation, 2002, pp. 1–91.
[2] J. Zhang, C. Wu, Corrosion and protection of magnesium alloys – a review of the patent literature, Recent Patents Corros. Sci. 2 (2010) 55–68.
[3] T. Dhanabalan, K. Subha, R. Shanthi, A. Sathish, Factors influencing consumers' car purchasing decision in indian automobile industry, Int. J. Mech. Eng. Technol. 9 (2018) 53–63.
[4] S. Bashir, A. Thakur, H. Lgaz, I.-M. Chung, A. Kumar, Computational and experimental studies on phenylephrine as anti-corrosion substance of mild steel in acidic medium, J. Mol. Liq. 293 (2019) 111539, https://doi.org/10.1016/j.molliq.2019.111539.
[5] S. Bashir, A. Thakur, H. Lgaz, I.-M. Chung, A. Kumar, Corrosion inhibition performance of acarbose on mild steel corrosion in acidic medium: an experimental and computational study, Arab. J. Sci. Eng. 6 (2020), https://doi.org/10.1007/s13369-020-04514-6.
[6] S. Bashir, A. Thakur, H. Lgaz, I.M. Chung, A. Kumar, Corrosion inhibition efficiency of bronopol on aluminium in 0.5 M HCl solution: insights from experimental and quantum chemical studies, Surf. Interfaces 20 (2020), https://doi.org/10.1016/j.surfin.2020.100542, 100542.
[7] G. Parveen, S. Bashir, A. Thakur, S.K. Saha, P. Banerjee, A. Kumar, Experimental and computational studies of imidazolium based ionic liquid 1-methyl-3-propylimidazolium iodide on mild steel corrosion in acidic solution, Mater. Res. Express. 7 (2020), https://doi.org/10.1088/2053-1591/ab5c6a, 016510.
[8] Y. Song, K.P. Meyers, J. Gerringer, R.K. Ramakrishnan, M. Humood, S. Qin, A.A. Polycarpou, S. Nazarenko, J.C. Grunlan, Fast self-healing of polyelectrolyte multilayer nanocoating and restoration of super oxygen barrier, Macromol. Rapid Commun. 38 (2017) 1–7, https://doi.org/10.1002/marc.201700064.
[9] Y. Song, K.P. Meyers, J. Gerringer, R.K. Ramakrishnan, M. Humood, S. Qin, A.A. Polycarpou, S. Nazarenko, J.C. Grunlan, Fast self-healing of polyelectrolyte multilayer nanocoating and restoration of super oxygen barrier, Macromol. Rapid Commun. 38 (2017), https://doi.org/10.1002/marc.201700064.
[10] F. Paquin, J. Rivnay, A. Salleo, N. Stingelin, C. Silva, Multi-phase semicrystalline microstructures drive exciton dissociation in neat plastic semiconductors, J. Mater. Chem. C 3 (2015) 10715–10722, https://doi.org/10.1039/b000000x.
[11] S. Das, S. Kumar, S.K. Samal, S. Mohanty, S.K. Nayak, A review on superhydrophobic polymer nanocoatings: recent development and applications, Ind. Eng. Chem. Res. 57 (2018) 2727–2745, https://doi.org/10.1021/acs.iecr.7b04887.
[12] M.S. Ahmmad, M.B.H. Hassan, M.A. Kalam, Comparative corrosion characteristics of automotive materials in Jatropha biodiesel, Int. J. Green Energy 15 (2018) 393–399, https://doi.org/10.1080/15435075.2018.1464925.
[13] D. Jiang, X. Xia, J. Hou, X. Zhang, Z. Dong, Enhanced corrosion barrier of microarc-oxidized Mg alloy by self-healing superhydrophobic silica coating, Ind. Eng. Chem. Res. 58 (2019) 165–178, https://doi.org/10.1021/acs.iecr.8b04060.
[14] S. Chen, X. Li, Y. Li, J. Sun, C.E.T. Al, Intumescent flame-retardant and coatings on cotton fabric, ACS Nano (2015) 4070–4076.
[15] M. Serhan, M. Sprowls, D. Jackemeyer, M. Long, I.D. Perez, W. Maret, N. Tao, E. Forzani, Total iron measurement in human serum with a smartphone, AIChE Annu. Meet. Conf. Proc. 2019 (2019), https://doi.org/10.1039/x0xx00000x.
[16] R. Bahaskaran, N. Palaniswamy, N.S. Rengaswamy, A review of differing approaches used to estimate the cost of corrosion, anti-corrosion method and materials, J. Appl. Sci. Res. 52 (2013) 29–41.
[17] M.J. Anjum, J. Zhao, H. Ali, M. Tabish, H. Murtaza, G. Yasin, M.U. Malik, W.Q. Khan, A review on self-healing coatings applied to Mg alloys and their electrochemical evaluation techniques, Int. J. Electrochem. Sci. 15 (2020), https://doi.org/10.20964/2020.04.36.
[18] Y. Ma, H. Zhang, Development of the Sharing Economy in China: Challenges and Lessons, 2019, https://doi.org/10.1007/978-981-13-8102-7_20.

[19] D.Y. Zhu, M.Z. Rong, M.Q. Zhang, Self-healing polymeric materials based on microencapsulated healing agents: from design to preparation, Prog. Polym. Sci. 49–50 (2015) 175–220, https://doi.org/10.1016/j.progpolymsci.2015.07.002.

[20] L. Vakhitova, Fire retardant nanocoating for wood protection 16 L.N. Vakhitova, in: Nanotechnology in Eco-efficient Construction, Elsevier Ltd., 2019, pp. 361–383.

[21] J. Telegdi, Multifunctional Smart Layers with Self-Cleaning, Self-Healing, and Slow-Release Activities, Elsevier Inc., 2019, https://doi.org/10.1016/B978-0-12-849870-5.00012-4.

[22] Y. Huang, L. Deng, P. Ju, L. Huang, H. Qian, D. Zhang, X. Li, H.A. Terryn, J.M.C. Mol, Triple-action self-healing protective coatings based on shape memory polymers containing dual-function microspheres, ACS Appl. Mater. Interfaces 10 (2018) 23369–23379, https://doi.org/10.1021/acsami.8b06985.

[23] A.S.H. Makhlouf, Current and Advanced Coating Technologies for Industrial Applications, Woodhead Publishing Limited, 2011, https://doi.org/10.1533/9780857094902.1.3.

[24] I.L. Hia, V. Vahedi, P. Pasbakhsh, Self-healing polymer composites: prospects, challenges, and applications, Polym. Rev. 56 (2016) 225–261, https://doi.org/10.1080/15583724.2015.1106555.

[25] A.S. Hamdy, I. Doench, H. Möhwald, Smart self-healing anti-corrosion vanadia coating for magnesium alloys, Prog. Org. Coat. 72 (2011) 387–393, https://doi.org/10.1016/j.porgcoat.2011.05.011.

[26] D.H. Abdeen, M. El Hachach, M. Koc, M.A. Atieh, A review on the corrosion behaviour of nanocoatings on metallic substrates, Materials (Basel) 12 (2019) 210, https://doi.org/10.3390/ma12020210.

[27] D.V. Andreeva, D. Fix, H. Möhwald, D.G. Shchukin, Self-healing anticorrosion coatings based on pH-sensitive polyelectrolyte/inhibitor sandwichlike nanostructures, Adv. Mater. 20 (2008) 2789–2794, https://doi.org/10.1002/adma.200800705.

[28] M.F. Montemor, D.V. Snihirova, M.G. Taryba, S.V. Lamaka, I.A. Kartsonakis, A.C. Balaskas, G.C. Kordas, J. Tedim, A. Kuznetsova, M.L. Zheludkevich, M.G.S. Ferreira, Evaluation of self-healing ability in protective coatings modified with combinations of layered double hydroxides and cerium molibdate nanocontainers filled with corrosion inhibitors, Electrochim. Acta 60 (2012) 31–40, https://doi.org/10.1016/j.electacta.2011.10.078.

[29] B.W. Soutter, Nanotechnology in the Automotive Industry Nano - Enhanced Adhesives Nanoparticle Fillers for Tyres, 2012, pp. 1–4.

[30] A.K. Naskar, J.K. Keum, R.G. Boeman, Polymer matrix nanocomposites for automotive structural components, Nat. Nanotechnol. 11 (2016) 1026–1030, https://doi.org/10.1038/nnano.2016.262.

[31] OECD International Futures Programme, Small sizes that matter: opportunities and risks of Nanotechnologies, Allianz 46 (2007).

[32] H. Kantamneni, Advanced materials manufacturing & characterization avant-garde nanotechnology applications in automotive industry, Adv. Mater. Manuf. Charact. 3 (2013) 195–198.

[33] M.M. Parvez, Application of Nanomaterials in Lubricants, Vol. 2, 2020, pp. 107–113.

[34] R. Vera, C. Matteo, P. Candido, V. Francesca, Current and future nanotech applications in the oil industry, Am. J. Appl. Sci. 9 (2012) 784–793. http://www.magforce.de.

[35] H. Presting, U. König, Future nanotechnology developments for automotive applications, Mater. Sci. Eng. C 23 (2003) 737–741, https://doi.org/10.1016/j.msec.2003.09.120.

[36] Harilaos Vasiliadis, Bax & Willems Consulting Venturing, TRANSPORT Nanotechnology in automotive tyres, ObservatoryNANO (2011) 4. https://nanopinion.archiv.zsi.at/sites/default/files/briefing_no.23_nanotechnology_in_automotive_tyres.pdf.

[37] I.J. Gomez, B. Arnaiz, M. Cacioppo, F. Arcudi, M. Prato, Nitrogen-doped carbon nanodots for bioimaging and delivery of paclitaxel, J. Mater. Chem. B 6 (2018), https://doi.org/10.1039/x0xx00000x.

[38] H. Agentur, Nanotechnologies in Automobiles, 2008, pp. 1–56.

[39] M. Werner, V. Igel, W. Wondrak, Nanotechnology and nanoelectronics for automotive applications, in: Nano-Micro Interface Bridg. Micro Nano Worlds, second ed., Vol. 2, 2015, pp. 459–472, https://doi.org/10.1002/9783527679195.ch22.

[40] Y.C. Lu, S. Pilla, The Use of Nano Composites in Automotive Applications, 2015, https://doi.org/10.4271/pt-172.

[41] L.H. Xi, Review on applications of superamphiphiles, Xiandai Huagong/Modern Chem. Ind. 39 (2019) 1479–1482, https://doi.org/10.16606/j.cnki.issn0253-4320.2019.12.007.

[42] S. Veličković, B. Stojanović, L. Ivanović, S. Miladinović, S. Milojević, Application of nanocomposites in the automotive industry, Mobil. Veh. Mech. 45 (2019) 51–64, https://doi.org/10.24874/mvm.2019.45.03.05.
[43] M.N. Nadagouda, D.D. Dionysiou, D.A. Lytle, T.F. Speth, S.M. Mukhopadhyay, Nanomaterials synthesis, applications, and toxicity 2012, J. Nanotechnol. 2013 (2013) 2–4, https://doi.org/10.1155/2013/978541.
[44] P. Louda, Applications of thin coatings in automotive industry, J. Achiev. Mater. Manuf. Eng. 24 (2007) 51–56. http://jamme.acmsse.h2.pl/papers_vol24_1/24105.pdf.
[45] B. Aïssa, D. Therriault, E. Haddad, W. Jamroz, Self-healing materials systems: overview of major approaches and recent developed technologies, Adv. Mater. Sci. Eng. 2012 (2012), https://doi.org/10.1155/2012/854203.
[46] A.H. Makhlouf, I. Tiginyanu, Nanocoatings and Ultra-Thin Films, 2011.
[47] K. Müller, E. Bugnicourt, M. Latorre, M. Jorda, Y.E. Sanz, J.M. Lagaron, O. Miesbauer, A. Bianchin, S. Hankin, U. Bölz, G. Pérez, M. Jesdinszki, M. Lindner, Z. Scheuerer, S. Castelló, M. Schmid, Review on the processing and properties of polymer nanocomposites and nanocoatings and their applications in the packaging, automotive and solar energy fields, Nanomaterials 7 (2017) 74, https://doi.org/10.3390/nano7040074.
[48] S. Mohan, A. Mohan, Wear, Friction and Prevention of Tribo-Surfaces By Coatings/Nanocoatings, Elsevier Ltd, 2014, https://doi.org/10.1016/B978-0-85709-211-3.00001-7.
[49] M.C. Coelho, G. Torrão, N. Emami, J. Grácio, Nanotechnology in automotive industry: research strategy and trends for the future-small objects, big impacts, J. Nanosci. Nanotechnol. 12 (2012) 6621–6630, https://doi.org/10.1166/jnn.2012.4573.
[50] S.B. Ulaeto, J.K. Pancrecious, K.K. Ajekwene, G.M. Mathew, T.P.D. Rajan, Advanced Nanocoatings for Anticorrosion, Elsevier Inc., 2020, https://doi.org/10.1016/b978-0-12-819359-4.00025-8.
[51] V.S. Saji, R. Cook, Corrosion Protection and Control Using Nanomaterials, 2012, https://doi.org/10.1533/9780857095800.
[52] A.S. Malani, A.D. Chaudhari, R.U. Sambhe, A review on applications of nanotechnology in automotive industry, Int. J. Mech. Mechatronics Eng. 10 (2015) 36–40, https://doi.org/10.5281/zenodo.1110867.
[53] D.G. Shchukin, D.O. Grigoriev, Corrosion Protection and Control Using Nanomaterials, Woodhead Publishing, 2012. http://www.sciencedirect.com/science/article/pii/B9781845699499500101.
[54] C. Krishnamoorthy, R. Chidambaram, Nanostructured Thin Films and Nanocoatings, Elsevier Inc., 2018, https://doi.org/10.1016/B978-0-323-51254-1.00017-8.
[55] N.B. Dahotre, S. Nayak, Nanocoatings for engine application, Surf. Coat. Technol. 194 (2005) 58–67, https://doi.org/10.1016/j.surfcoat.2004.05.006.
[56] D. Brabazon, E. Pellicer, F. Zivic, J. Sort, M.D. Baró, N. Grujovic, K.L. Choy, Commercialization of Nanotechnologies – A Case Study Approach, Springer, 2017, pp. 1–315, https://doi.org/10.1007/978-3-319-56979-6.
[57] M. Shafique, X. Luo, Nanotechnology in transportation vehicles: an overview of its applications, environmental, health and safety concerns, Materials (Basel). 12 (2019) 11–17, https://doi.org/10.3390/ma12152493.
[58] M. Aliofkhazraei, A.S.H. Makhlouf, Handbook of Nanoelectrochemistry: Electrochemical Synthesis Methods, Properties, and Characterization Techniques, 2016, pp. 1–1451, https://doi.org/10.1007/978-3-319-15266-0.
[59] A. Stankiewicz, Self-Healing Nanocoatings for Protection Against Steel Corrosion, Elsevier Ltd, 2019, https://doi.org/10.1016/b978-0-08-102641-0.00014-1.
[60] Y. Si, H. Zhu, L. Chen, T. Jiang, Z. Guo, A multifunctional transparent superhydrophobic gel nanocoating with self-healing properties, Chem. Commun. 51 (2015) 16794–16797, https://doi.org/10.1039/c5cc06977g.
[61] A. Hikasa, T. Sekino, Y. Hayashi, R. Rajagopalan, K. Niihara, Preparation and corrosion studies of self-healing multi-layered nano coatings of silica and swelling clay, Mater. Res. Innov. 8 (2004) 84–88, https://doi.org/10.1080/14328917.2004.11784835.

[62] W.S. Miller, L. Zhuang, J. Bottema, A.J. Wittebrood, P. De Smet, A. Haszler, A. Vieregge, Recent development in aluminium alloys for the automotive industry, Mater. Sci. Eng. A 280 (2000) 37–49, https://doi.org/10.1016/S0921-5093(99)00653-X.

[63] Z. Wang, X. Chen, Y. Gong, B. Zhang, H. Li, Superhydrophobic nanocoatings prepared by a novel vacuum cold spray process, Surf. Coat. Technol. 325 (2017) 52–57, https://doi.org/10.1016/j.surfcoat.2017.06.044.

[64] D. Chen, D. Wang, Y. Yang, Q. Huang, S. Zhu, Z. Zheng, Self-healing materials for next-generation energy harvesting and storage devices, Adv. Energy Mater. 7 (2017) 27–31, https://doi.org/10.1002/aenm.201700890.

[65] M. Mohseni, B. Ramezanzadeh, H. Yari, M. Moazzami, The role of nanotechnology in automotive industries, in: New Advances in Vehicular Technology and Automotive Engineering, 2012, pp. 3–54, https://doi.org/10.5772/49939.

[66] J. Mathew, J. Joy, S.C. George, Potential applications of nanotechnology in transportation: a review, J. King Saud Univ. Sci. 31 (2019) 586–594, https://doi.org/10.1016/j.jksus.2018.03.015.

[67] M.G.S. Ferreira, M.L. Zheludkevich, J. Tedim, K.A. Yasakau, Self-Healing Nanocoatings for Corrosion Control, Woodhead Publishing Limited, 2012, https://doi.org/10.1533/9780857095800.2.213.

[68] P. Dhaundiyal, S. Bashir, V. Sharma, A. Kumar, An investigation on mitigation of corrosion of mildsteel by *origanum vulgare* in acidic medium, J. Chem. Inf. Model 33 (2019) 159–168, https://doi.org/10.1017/CBO9781107415324.004.

[69] A. Kumar, S. Bashir, Ethambutol: a new and effective corrosion inhibitor of mildsteel in acidic medium, Russ. J. Appl. Chem. 89 (2016) 1158–1163, https://doi.org/10.1134/S1070427216070168.

[70] A. Singh, T. Pramanik, A. Kumar, M. Gupta, Phenobarbital: a new and effective corrosion inhibitor for mild steel in 1 M HCl solution, Asian J. Chem. 25 (2013) 9808–9812, https://doi.org/10.14233/ajchem.2013.15414.

CHAPTER 20

Conductive nanopaints: A remarkable coating

Maria Nayane de Queiroz[a], Antônia Millena de Oliveira Lima[a], Manuel Edgardo Gomez Winkler[a], Vanessa Hafemann Fragal[a], Adley Forti Rubira[a], Thiago Sequinel[b], Lucas da Silva Ribeiro[c], Francisco Nunes de Souza Neto[c], Emerson Rodrigues Camargo[c], Mauricio Zimmer Ferreira Arlindo[d], Christiane Saraiva Ogrodowski[d], and Luiz Fernando Gorup[d,e,f]

[a]Department of Chemistry, UEM—State University of Maringa Avenida Colombo, Maringá, Paraná, Brazil
[b]Faculty of Exact Sciences and Technology (FACET), Federal University of Grande Dourados, Dourados, Mato Grosso do Sul, Brazil
[c]LIEC—Interdisciplinary Laboratory of Electrochemistry and Ceramics, Departament of Chemistry, UFSCar-Federal University of São Carlos, São Carlos, São Paulo, Brazil
[d]School of Chemistry and Food Science, Federal University of Rio Grande, Rio Grande, Rio Grande do Sul, Brazil
[e]Materials Engineering, Federal University of Pelotas, Campus Porto, Pelotas, Brazil
[f]Institute of Chemistry, Federal University of Alfenas, Alfenas, Minas Gerais, Brazil

Chapter outline

1. Introduction — 429
2. Conductive coating—Market value — 431
3. Types, characteristics, and use of conductive nanopaints — 432
4. Recent development of conductive nanopaints — 437
5. Commercial conductive nanopaints — 439
6. Application of conductive nanopaints in the automotive industry — 441
7. General conclusions and future perspectives — 443
References — 444
Web References — 449

1. Introduction

The automotive industry is undergoing a major transformation to produce cars with greater functionalities [1], such as lighter, more efficient, and environmentally friendly vehicles [2–4]. The global demand for automobiles has increased at a steady rate [5]. For instance, the production of cars in the U.S. increased from 5.6 million vehicles in 2009 to 11.3 million in 2017 [6], while world production increased from 69 million to 97 million vehicles in the same period. However, the Coronavirus pandemic (COVID-19), caused by the SARS-CoV-2 virus, affected all countries and economies [7], thus reducing global production to 79 million vehicles in 2020.

In this context, the automotive paint industry plays a special and important role in the automotive industry due to protection from corrosion [8,9], heat differences, bumps, and UV degradation [10]. Paints are emulsions of solid pigments in a solvent medium used for esthetic or protective coatings [11–13]. The development of better and more efficient painting processes has brought about dramatic changes over the last century.

In the early 1900s, cars were manually painted with varnishes that, after drying, needed to be sanded and polished to produce shiny surfaces. The varnish was reapplied and refinished to establish several coating layers, taking up to 40 days for the entire process [14]. The Ford Motor Company started using nitrocellulose lacquers in paints for automotive assembly in the 1920s. These paints had a substantially shorter drying time than older varnishes [15]. Between 1930 and 1940, automakers started using stoving enamel paints (alkaline resins). These paints provided brighter shines and much faster drying times [16]. Acrylic stoving improved the durability of enamel in the 1960s [17].

In the late 1970s, a new type of wet-on-wet finish was developed and introduced to further improve the coating's appearance and durability that consisted of a thin basecoat and a thicker clearcoat [18]. In the late 1980s, car manufacturers started using urethane and polyurethane paints in their vehicles [19]. The application of these types of paints resulted in durable and high-gloss finishes [20].

The technological advancement resulted in a new type of paint in the late 1980s, called conductive paints. These paints currently use polymeric composites with conductive fillers of nanometric sizes, most of them are metals or carbon-based additives. Among the polymers, epoxies are the most used due to their mechanical properties, durability, and resistance to the weather [21]. Acrylic is another type of polymer that has been gaining popularity due to its low cost, climate resistance, good conductivity, and resistance to humidity [22]. These conductive nanopaints are used in many industrial sectors, ranging from automotive and aerospace to electronics and pharmaceuticals.

In recent years, extensive studies have been done on resin-based conductive coatings [23,24]. Carbon steel is one example of an additional material that preserves activity and exhibits "metal-like" characteristics [25,26]. The life cycle of coated steel with conductive coatings would be prolonged to 10–20 years or more in automobiles [27]. However, due to its cost, only specific parts of the automobile that need more protection or greater thermal conductivity receive this type of special conductive nanopaints. Conductive nanocoatings are used to dissipate static charges on ground straps, slip rings, flexible circuits, contact points, and commutator segments. Non-conductive substrates, such as polymers, ceramics, or composite materials, can be painted with conductive coatings, as well as less conductive metallic substrates.

The use of nanoparticles decreased the amount of paints necessary to preserve the original properties, which is a great advantage over traditional paints [28]. An example is the replacement of carbon black by multiwall carbon nanotubes (MWCNTs) in epoxy matrices, resulting in excellent nanopaints for the automotive industry with enhanced spraying processes [29], which improves film cohesion and durability [30]. In this sense, this chapter will address the application of different nanoparticles as fillers in conductive nanopaints focused on the automotive industry. These nanocoatings have improved properties that are not present in traditional paintings, such as scratch resistance, self-healing, self-cleaning, and electrical and thermal conductivity.

2. Conductive coating—Market value

The global market for conductive coatings was estimated at US$ 15.6 billion in 2019 and US$ 13.7 billion in 2020. This 13.8% drop in 2020 was due to the global economic recession led by COVID-19. Since the outbreak of the virus in December 2019, the disease has spread to almost every country around the world. Consequently, the World Health Organization (WHO) has declared it as a public health emergency. The temporary shutdown of manufacturing units and sales channels resulted in a drop in revenue in this market [31]. Even in a difficult situation, the market is likely to recover soon, reaching an estimate of US$ 21.3 billion in 2025 with a CAGR (compound annual growth rate) of 4% to 6% from 2022 to 2025 [32]. In the coming years, conductive coatings may be one of the drivers of this growing market due to the growth of the electric vehicle market, the increase in electronic technology, and expansions in the aerospace and defense industries (Fig. 1) [33].

The demand for electrically conductive coatings is increasing in important exporting countries with large electronics manufacturing units, such as China, Vietnam, and Taiwan. The United States also accounts for a significant share of the global conductive coatings market in terms of value, estimated to US$4 billion in 2021, attributed to the

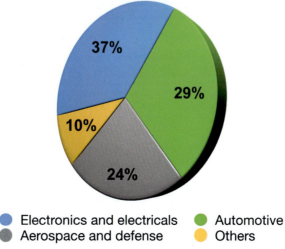

Fig. 1 Representation of electrically conductive coating market by global application in 2019.

increasing employment rate and rising disposable income. China is the second-largest economy and is expected to reach a projected market size of US$ 5.2 billion by 2021. Japan and Canada also present a forecasting growth of 1.8% and 6.2%, respectively, in the projection period. In Europe, Germany is expected to grow with a CAGR of 4.4% (Market Data, 2021).

Conductive nanocoatings are commonly classified into epoxies, polyesters, acrylics, and polyurethanes. The material used in these coatings consists of different nanoparticles from metals such as copper, silver, aluminum, molybdenum, or carbon-based such as graphite or carbon black (Fig. 2) [34].

Silver nanopaints was the largest share in the conductive coatings market in 2020, as silver offers the best protection and resistance to corrosion and can be applied with a very thin thickness. Silver is often used in board-level shielding and mission-critical applications. Silver nanopaints attracted attention due to their electrical, optical, and thermal properties deposited in the material applied. Among other fillers, silver has the highest electrical conductivity and is considered a good candidate for use in highly conductive, transparent, and flexible electronic coatings [35].

3. Types, characteristics, and use of conductive nanopaints

The addition of nanoparticles changes and improves the properties of paints, forming a new material described as a nanostructured paint or coating. These nanopaints are suspensions containing metallic or non-metallic nanoparticles used for many applications in industry for protection and decoration [36,37]. Fig. 3 shows the structural difference between a conventional paint and a nanopaint [38].

Nanoparticles can play various roles to enhance paint properties, such as scratch resistance and UV protection, which are determined by the type of nanoparticle added to the paint composition [37]. Titanium dioxide (TiO_2) nanoparticles, for example, are used as a biocide [39] and UV absorber to protect wood [40]. Silicon dioxide (SiO_2) nanoparticles enhance scratch resistance and prevent corrosion of a film deposited on the substrate during the curing process, particularly in automotive electrocoating [40–43]. In addition, SiO_2 can improve the efficiency of the solar still [42].

The process used to prepare nanopaints is similar to that used in the synthesis of nanocomposites. First, the nanoparticles are added to a matrix, usually polymeric paint, followed by solvent addition. At this time, the nanopaint is considered a nanofluid. The nanopaint remains as suspension until its application on the substrate. After solvent evaporation, a thin layer of a nanostructured film is formed on the substrate. [44]. The nanopaint can be prepared by various methods. The sonication method (Fig. 4A), mostly applied on a laboratory scale, has many disadvantages for industrial applications, such as the rapid heat generated, the dependence on the viscosity of the solvent, and potential

Trends and forecast for the global electrically conductive coating market (2014 to 2025)

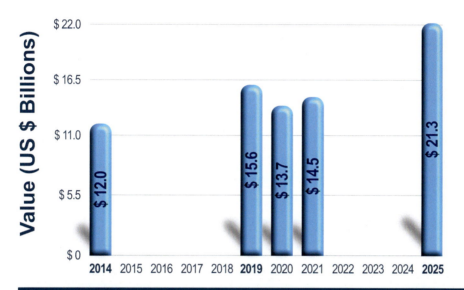

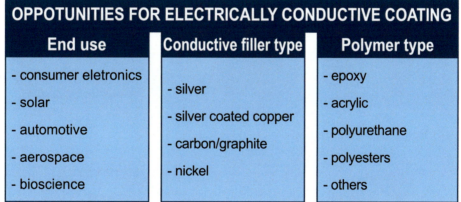

Fig. 2 Graphic trends and forecast for the global electrically conductive coating market from 2014 to 2025. Box: The global conductive nanocoatings market segments based on material and use.

mechanical failures. However, this method is the most efficient for low viscosity solvents, such as water-based paint [45].

The roll mills method, on the other hand, is an inexpensive technique widely used for industrial applications. Fig. 4B shows the roll mills' equipment. Each roll runs at different velocities. The gap between rolls controls the dispersion, preventing agglomeration.

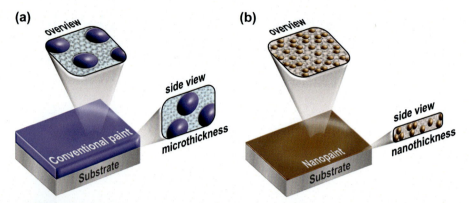

Fig. 3 Difference in structural between a conventional paint and a nanopaint and the thickness of the films formed after application.

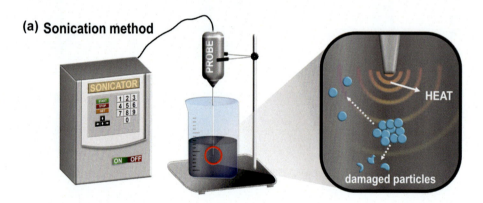

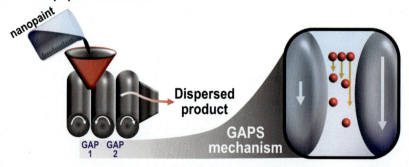

Fig. 4 Scheme of two usual methods used to disperse nanofluid in the synthesis of nanopaint with controllable size nanoparticles: (A) is the sonication method with the sound wave propagation, and (B) is three roll equipment, and in detail the gap between rollers.

The main advantage of this technique is the adjustment of the gap between rollers [45]. This method, commonly used for dispersing paints and cosmetics, is industrially known as a calendaring process [46]. The addition of conductive nanoparticles in paints can change the thermal and electrical properties. The composition of the nanopaints results from the incorporation of a conductive agent or the use of a conductive polymeric resin. Sometimes, special nanopaints are prepared combining conductive additives with conductive resins. However, the use of conductive polymer is limited due to its cost, poor solubility, and loss of conductivity with environmental exposure. Therefore, the conductive pigment is the most appropriate due to the low cost in relation to the conductive polymer [47–49].

The size and morphology of nanoparticles (spheres, wires, or lamellar forms) affect thermal and electrical conductivity. The importance of using nanoparticles instead of larger particles is a key factor for a good application of paint or coating. Although larger particles alone conduct heat better than small particles, they tend to precipitate quickly, resulting in non-homogeneous and unstable suspensions.

In some special cases, the conductive polymer can act as a conductive element [50]. However, metallic or non-metallic nanoparticles have been preferred as conductive agents to replace polymers. The main examples are copper nanoparticles [51], silver nanoparticles [52], graphene [53], and carbon nanotubes [37].

Among metallic nanoparticles, silver and copper are the most commonly used. These materials stand out for presenting high electrical conductivity. In addition, silver exhibits high thermal conductivity. However, sometimes copper nanoparticles can oxidize, thus decreasing conductive efficiency and the lifetime of the material [54].

On the other hand, nanostructured carbons are an excellent alternative material for conductive nanopaints. Some examples of carbon-based materials are graphite, graphene and graphene oxide, carbon black, single-wall carbon nanotubes (SWCNTs) [45] and multiple-wall carbon nanotubes (MWCNTs). Graphitic carbon is characterized by overlapping hexagonal honeycomb carbon sheets of Csp^3-Csp^3 and Csp^2-Csp^2 bonds, as shown in Fig. 5A [44]. From the exfoliation of graphite, the overlapped carbon planes can be separated to obtain a carbon monolayer, called graphene (Fig. 5B) [44,48]. Graphene oxide is formed when oxidized functional groups are inserted into the graphene layer. The properties of graphene oxide can be adjusted by the degree of oxidation (Fig. 5C) [48]. Moreover, two graphene allotropes have occupied a prominent position as fillers in conductive nanopaints. These graphene allotropes called carbon nanotubes (CNTs) can be interpreted as graphene sheets rolled up to form a cylinder. Based on the number of layers they are either SWCNTs or MWCNTs (Fig. 5D and E) [48]. Disordered carbon allotropes also exhibit similar thermal and electrical properties (Fig. 5F). Carbon black is fine powders produced by the incomplete combustion of hydrocarbons. The simplicity of obtaining and the low cost of the process make them a very attractive option to substitute more expensive forms of carbon [52].

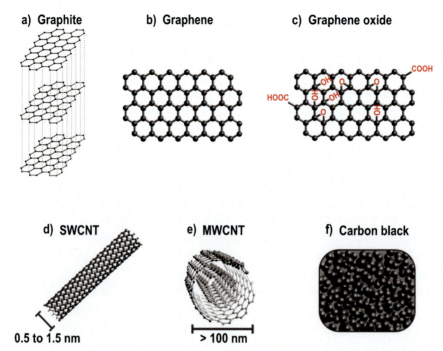

Fig. 5 Nanostructured carbons (A–E), and amorphous carbon (F) applied in conductive nanopaints.

Graphene is a special material for conductive applications. Pure graphene exhibits an electrical conductivity (10^8 S/m) [55–57] of one order higher than silver or copper (10^7 S/m). The carbon atoms in the graphene layer make three covalent bonds and have a free electron responsible for the electrical and thermal conductivities. However, graphene is expensive for applications in large areas [55]. Carbon nanotubes (CNT), also have many properties that make them a good conductive pigment for nanopaints. Pure CNT have high electrical conductivity, similar to silver and copper (10^6 to 10^7 S/m). Despite their excellent properties, CNT are also very expensive for applications in large areas [56].

Graphene and CNT present the percolation phenomenon that enhances the electrical conductivity of composite material. This phenomenon occurs due to the packing of particles in concentrations above a critical value. Both materials present good thermal conductive and can be applied in radiators and microelectronics [58]. In addition, the combination of metallic and non-metallic nanoparticles can improve the electrical conductivity of the nanopaint. The addition of carbon black into silver nanopaints increased overall conductivity and scratch resistance [52]. Although carbon black particles are lighter than metallic nanoparticles, their presence in nanopaints makes their application on transparent surfaces unfeasible. On the other hand, the replacement of metal nanoparticles by carbon derivates contributes to the reduction of the thickness of the substrate [59].

4. Recent development of conductive nanopaints

Thermally conductive coatings are used to increase heat dissipation. Many automotive vehicle locations need heat dissipation to increase the overall efficiency, such as heat exchangers, radiators, electronic circuits, etc. There are solid and liquid thermal conductive coatings. Thermally conductive coating in liquids is enhanced by loading the base fluid into a liquid with carbon-based fillers such as graphite, graphene, or CNT. In this way, graphene nanoplatelets (GnP) have been used to enhance paints used in the automotive industry. For example, GnP fillers spray-coated on the aluminum foil of larger size always exhibited higher thermal conductivity than smaller GnP, regardless of the filler concentration [58]. Moreover, colloidal dispersions of GnP with lower viscosity showed higher thermal conductivity values. Thermally conductive GnP-loaded paint presents a higher thermal conductivity (1.6 W/mK) than the base paint (0.63 W/mK).

Carbon black exhibits good electrical, thermal, and physical-chemical properties, which facilitates the production of conductive nanopaint. Carbon black pigments have applications in electrical energy storage (EES), such as lithium-ion batteries [60] and supercapacitors [61]. Several biomass-based materials are used as precursors to produce carbon black, such as corn stalk, cow dung, rice straw, etc. Interestingly, cow dung is a good material to produce carbon black due to a large number of functional groups that contain oxygen and nitrogen.

Bhakare et al. [53] developed a carbon black pigment from graphitic carbon for conductive nanopaints using cow dung precursors with alkyd resin (used in electronic circuits and high-performance supercapacitors) and turpentine oil. The conductive paint was prepared using the sonication method and the main advantage of this method was the shorter drying time. The conductive nanopaints presented high electrical conductivity (resistance 8.61×10^{-7} Ω sq^{-1}) and good energy storage (capacitance 216 F g^{-1}). Thus, according to the experimental results, the conductive nanopaint could be used for conductive nanocoating technologies, such as energy storage devices and electronic circuits, which can enhance with the nanopaint in the automotive industry.

Another application of conductive nanopaints in EES is in ion-lithium batteries for cell phone batteries, pumped hydroelectric storage, and electrical vehicles [43]. Among the samples synthesized by simple-layer and multi-layer methods, the hybrid filler systems presented better electrical conductivity (4.92 S m^{-1}) than the simple systems (7.42×10^{-4} S m^{-1}) due to the reduction of distance (<2–3 nm) between the filler conductive particles (Fig. 6).

Conductive nanopaints with graphene have received much attention due to their superior electrical conductivity. However, the use of graphene is limited due to the high cost of mass production. Graphene nanosheets are applied in several electronic devices, such as field-effect transistors, storage devices, and thermoelectric devices.

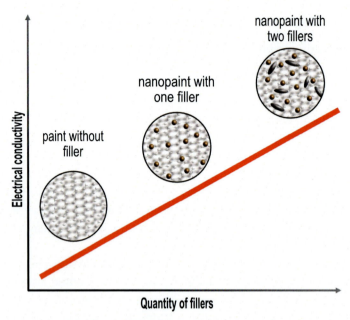

Fig. 6 Influence of reducing the distance between nanoparticles and filling in electrical conductivity.

Krishnamoorthy et al. [62] developed electrically conductive nanopaints using graphene as a pigment. The graphene was prepared using the mechanochemical route. To prepare the graphene nanopaint, it was used a ball milling approach using graphene and alkyd resin, and other components. The preparation by the ball milling process provided an efficient mixture, which presented a formation of conductive pathways for the electrons. These conductive pathways facilitate the electron flux throughout the interconnected graphene. The current-voltage curve showed a non-linear behavior with high symmetric nature. The material with graphene paint had a resistance of ∼1.77 kW. This resistance value is lower than that of other paints, such as acrylic paint, and it is associated with the difference between the insulating binder and the graphene. Thus, graphene paint synthesized by a cost-effective route presented good properties for conductive nanopaint technologies.

As already mentioned, the combination of metallic and non-metallic nanoparticles can improve many conductivity properties of nanopaints. For instance, Leong and Chung [52] investigated conductive nanopaint composed of different amounts of carbon black and silver nanoparticles. The results demonstrated that the sample without carbon black presents a higher sheet resistance (1.47 Ω) than a sample mixture of 0.0071% w/w of carbon black and 0.7071% of silver particles (0.66 Ω). The reduction in resistance caused an increase in electrical conductivity due to carbon black increasing the connectivity between the silver particles (Fig. 7). Furthermore, the scratch resistance did not change significantly. The replacement was promising because it decreased resistivity and improved scratch resistance.

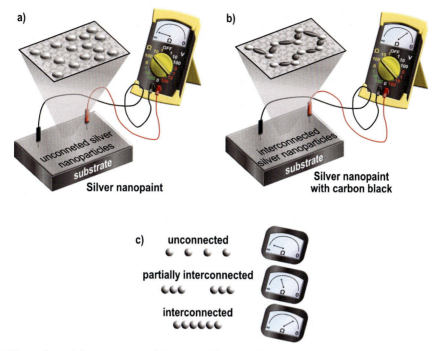

Fig. 7 Illustration of the resistance of the material using silver nanopaint. (A) Material resistance with silver paint without carbon-black filling; (B) resistance of the material with silver paint with carbon-black fillers and (C) relationship of connectivity and material resistance.

An overview in the field of conductive nanopaints is presented in Table 1. This table is including the type of nanomaterial used to enhance the properties of nanopaints.

5. Commercial conductive nanopaints

Among metals, silver nanoparticles are one of the most investigated materials due to their high electrical conductivity, oxidation resistance, and antimicrobial properties against fungi, bacteria, and virose [72–78]. Different Ag-based conductive nanopaints are commercially available, each optimized for: specific substrates (such as ITO, glass, PET, PA, PC, PC/ABS),[a] the main solvent (DGME, TGME, PM, EG, TD, and TPM),[b] viscosity,[c]

[a] Substrates: *ITO*, indium tin oxide; *PET*, polyethylene terephthalate; *PA*, polyacrylate; *PC*, polycarbonate; *PC/ABS*, polycarbonate/acrylonitrile butadiene styrene blends.
[b] Solvents: *DGME*, diethylene glycol methyl ether; *TGME*, triethylene glycol methyl ether; *PM*, propylene glycol methyl ether; *EG*, ethylene glycol; *TD*, tetradecane; *TPM*, tripropylene glycol methyl ether.
[c] Typical viscosity at 25°C: from 5 to 37 cP.

Table 1 Nanoparticles and properties of the nanopaints.

Nanoparticles	Properties	References
Carbon black	Electrical conductivity and scratch resistance	[52]
Copper	Electrical conductivity	[51]
Graphite and graphene nanoparticles	The electrical conductivity and anticorrosion	[43]
Graphene	Thermal conductivity	[58]
Graphene nanosheets	Electrical conductivity	[62]
Graphitic carbon	Electrical conductivity	[53]
Multiwall Carbon Nanotubes	Electrical conductivity	[37]
Silver	Electrical and thermal conductivity	[63]
Titanium oxide	Improve the surface appearance and structure	[64]
Zinc ferrite	Anticorrosion	[65]
Graphite	Electrical conductivity	[47]
Silver	Electrical conductivity	[66]
Zinc oxide	Color fading of the paint	[49]
Magnetic materials	Electromagnetism and heating	[36]
Graphite and Graphene	Anticorrosion	[57]
Graphene	Electrical conductivity	[67]
Silver	Electrical conductivity	[68]
Carbon nanotube	Electrical conductivity	[69]
Graphene	Thermal conductivity	[70]
Silver	Electrical conductivity	[71]

electrical resistivity, printing technology (inkjet or aerosol) and sintering methods (thermal, laser, NIR, UV, photonic, and pressure) [79–81]. The sintering methods are an essential step exclusive to metallic nanoparticles, in which the deposited nanoparticles connect each other, resulting in conductive trails. Fig. 8 illustrates the application (printing and solvent removal) and the process of sintering a metallic conductive nanopaint on microchip.

Copper nanopaints have emerged as a cheaper alternative and are also available on the market according to consumer needs [80,81]. It is important to follow the sintering protocol strictly according to the manufacturer's recommendations to avoid the formation of cupric oxide increasing the electrical resistivity—literature data show that it raises from 10^{-6} Ω cm for Cu to 10^7 Ω cm for CuO [51]. In addition, the technology adopted for sintering directly affects the resistivity of the final material. For example, when adopting laser sintering, the resistivity value obtained is around 6.5 Ω cm, while for thermal sintering is around 10^{-4} Ω cm. Although Cu paint is cheaper than silver paint, its susceptibility to oxidation is the biggest problem, the lifetime of the Cu paint is just around 4 months [51], very short compared to the lifetime of an automobile.

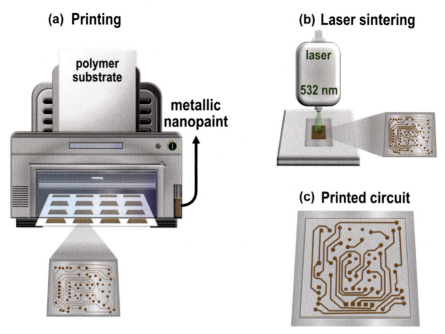

Fig. 8 Illustration of steps for made a circuit with metallic conductive nanopaint (A) printing, (B) laser sintering, and (C) printed circuit.

SWCNTs and graphene paints are already available commercially, while the other forms of carbon are still being studied [53,55]. In contrast to metallic nanoparticles paints, which use exclusively organic solvents, carbon-derived nanopaints are found with solvent-based and water-based paints [80].

The process of using carbon paints is very similar to that of metallic nanoparticles, consisting of two steps: printing and drying/curing. The latter is responsible for improving the adhesion of the carbon particles to the substrate and preventing them from detaching from the surface. To this end, polymers binders, such as Nafion and acrylic, are used in the paints formulations [82]. In addition to the electrical and thermal properties, the result with the use of SWCNTs paints is a transparent surface, which makes it ideal for screen printing devices [83].

6. Application of conductive nanopaints in the automotive industry

As stated above, nanopaints have several electrical and thermal properties, which make them suitable for application in the automotive industry. Based on this fact, this topic describes the advantages of using nanopaints and their relevance in the automotive area.

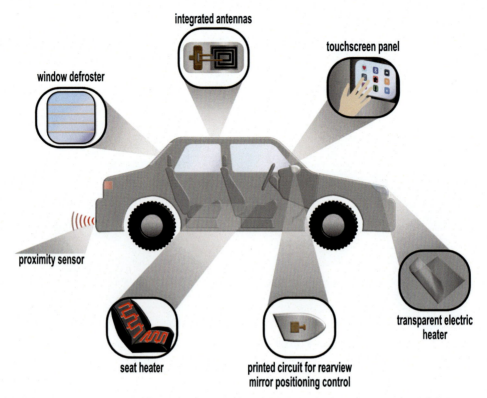

Fig. 9 Conductive nanopaints for automotive applications: integrated antennas; printed defrosters on rear-windows; seat heater; electronic circuits; printed circuit for a rearview mirror; battery monitoring/heating and touchscreen panel.

Conductive nanopaints are applied in several automotive devices, as demonstrated in Fig. 9 and highlighted below.

- *Antennas integrated*

The production of integrated antennas is possible due to conductive nanopaints. A miniaturized antenna internalized in the parts of the car allows the best aerodynamic performance and is flexible and adaptable to different shapes. Besides that, it is also possible to include the radio frequency identification (RFID) technology in this system to improve the functionalization of antennas [84].

- *Printed defrosters on rear-windows*

Defrosters are the conductive paint system that allows the elimination of ice and condensed water formed on the car's glass. In addition, this technology protects the perception sensors used in safety driving. The main vantage of conventional defrosters is transparency, which helps to maintain the initial transparency of the glasses. This property makes conductive paint an actual application in the automotive industry [85].

- *Seat heater*

Another application of conductive nanopaint in the automotive industry is in heated seats. Among the several materials applied in heated seats, the conductive paint, with carbon nanotube and silver in the composition, presents itself as an alternative material for this application. This heated seat shows excellent properties, such as fast heating consuming less energy [86].

- *Touchscreen panel*

Touchscreens are displays that allow communication between the user and a computer through touch. In the automotive industry, they were first inserted in the multimedia center replacing the physical buttons. As these displays have recently become more affordable, they have become very popular on car dashboards, controlling all the electrical devices (such as air conditioning, rearview mirror positions, seat heater, etc.), as well as supporting multimedia control and global position system (GPS) navigation, all in one display.

- *Electronic circuits*

The electrical system has several circuits, such as starting circuit, charging circuit, ignition circuit, injection circuit, and electronic circuit. Electronic circuits are devices that have interconnections between various electronic components and their electrical conductivity can be enhanced using nanopaint.

- *Battery monitoring/heating*

Battery heating/monitoring allows efficient work and the battery capacity is dependent on temperature. Thus, the increase in temperature and, consequently, in pressure, indicates a bad function. One way to solve this problem is to use printed arrays of temperature sensors that monitor the variation of temperature. These printed arrays can be fabricated using conductive nanopaints [87].

In summary, the use of conductive nanopaints in vehicles enhances several properties, such as durability, comfort, safety, aerodynamic performance, and economy, among others. As a result, improving these and other properties can help improve the cost-benefit of next-generation vehicles.

7. General conclusions and future perspectives

Our work highlights the importance of the technological development of conductive nanopaints over the years. Conductive nanopaints have a huge diversity of applications due to their versatile type of coating presenting many suitable properties and useful applications, such as space heating, solar cells, electromagnetic interference shielding, and microelectronics. Conductive nanocoating is commonly produced using epoxies, polyesters, acrylics, and polyurethanes, whereas conductive nanopaints are prepared using two basic constituents: a pigment and a binder. The pigments in conductive nanopaints usually are nanoparticles of metals and carbon such as copper, silver, graphite, carbon black, and graphene. Thus, manufacturers of paints are interested in using the inherent advantages of conductive nanopaints to create better products.

Conductive nanopaints with the specifically cited nanoparticles have properties such as electric conductivity, inhibition of corrosion, and scratch resistance that depend directly on the type of nanomaterial used. Electrically conductive coatings are generally applied to non-conductive substrates and less conductive metallic substrates. It is worth mentioning that the thickness of conductive nanopaints is reduced in comparison with traditional paints.

The challenges of the market of conductive nanopaints are the processing that involves high operational and investment costs. Thus, there is still much space for advances in the development of conductive nanopaints materials, such as reduction of costs and production of nanoparticles on a large scale. Therefore, conductive nanopaints have become a fertile field for advances in terms of cost reduction and manufacturing limitations.

We believe that the future is promising from a scientific and technological point of view for the development of nanocoating. Even with the declines in 2020 in the global market of electrically conductive coatings due to the global economic recession led by COVID-19. However, the forecast is for the market to recover this year. News and significant advances made in the fundamental understanding of the nanomaterial's properties will provide new opportunities for the technological development of conductive nanopaints.

References

[1] N. El Hajj, S. Seif, N. Zgheib, Recycling of poly(propylene)-based car bumpers as carrier resin for short glass fiber composites, J. Mater. Cycles Waste Manage. 23 (2021) 288–300, https://doi.org/10.1007/s10163-020-01128-w.

[2] D. Loukatos, E. Petrongonas, K. Manes, I.-V. Kyrtopoulos, V. Dimou, K.G. Arvanitis, A synergy of innovative technologies towards implementing an autonomous DIY electric vehicle for harvester-assisting purposes, Machines 9 (4) (2021) 82, https://doi.org/10.3390/machines9040082.

[3] A. Peppas, K. Kollias, D.A. Dragatogiannis, C.A. Charitidis, Sustainability analysis of aluminium hot forming and quenching technology for lightweight vehicles manufacturing, Int. J. Thermofluids 10 (2021) 100082, https://doi.org/10.1016/j.ijft.2021.100082.

[4] X. Shu, Y. Guo, W. Yang, K. Wei, G. Zhu, Life-cycle assessment of the environmental impact of the batteries used in pure electric passenger cars, Energy Rep. 7 (2021) 2302–2315, https://doi.org/10.1016/j.egyr.2021.04.038.

[5] G. Salihoglu, N.K. Salihoglu, A review on paint sludge from automotive industries: generation, characteristics and management, J. Environ. Manage. 169 (2016) 223–235, https://doi.org/10.1016/j.jenvman.2015.12.039.

[6] N. Shigeta, S.E. Hosseini, Sustainable development of the automobile industry in the United States, Europe, and Japan with special focus on the vehicles' power sources, Energies 14 (1) (2021) 78, https://doi.org/10.3390/en14010078.

[7] A. Belhadi, S. Kamble, C.J.C. Jabbour, A. Gunasekaran, N.O. Ndubisi, M. Venkatesh, Manufacturing and service supply chain resilience to the COVID-19 outbreak: lessons learned from the automobile

and airline industries, Technol. Forecast. Soc. Change 163 (2021) 120447, https://doi.org/10.1016/j.techfore.2020.120447.
[8] R.R. Abakah, F. Huang, Q. Hu, Y. Wang, L. Jing, Comparative study of corrosion properties of different graphene nanoplate/epoxy composite coatings for enhanced surface barrier protection, Coatings 11 (3) (2021) 285, https://doi.org/10.3390/coatings11030285.
[9] T. Ge, W. Zhao, X. Wu, Y. Wu, S. Lu, X. Ci, Y. He, Design alternate epoxy-reduced graphene oxide/epoxy-zinc multilayer coatings for achieving long-term corrosion resistance for Cu, Mater. Des. 186 (2020) 108299, https://doi.org/10.1016/j.matdes.2019.108299.
[10] A. Hottenrott, L. Waidner, M. Grunow, Robust car sequencing for automotive assembly, Eur. J. Oper. Res. 291 (3) (2021) 983–994, https://doi.org/10.1016/j.ejor.2020.10.004.
[11] C. Costa, B. Medronho, A. Filipe, I. Mira, B. Lindman, H. Edlund, M. Norgren, Emulsion formation and stabilization by biomolecules: the leading role of cellulose, Polymers 11 (10) (2019) 1570, https://doi.org/10.3390/polym11101570.
[12] M.R. Khan, N. Ahmad, M. Ouladsmane, M. Azam, Heavy metals in acrylic color paints intended for the school children use: a potential threat to the children of early age, Molecules 26 (8) (2021) 2375, https://doi.org/10.3390/molecules26082375.
[13] R. Pal, New generalized viscosity model for non-colloidal suspensions and emulsions, Fluids 5 (3) (2020) 150, https://doi.org/10.3390/fluids5030150.
[14] N.K. Akafuah, S. Poozesh, A. Salaimeh, G. Patrick, K. Lawler, K. Saito, Evolution of the automotive body coating process—a review, Coatings 6 (2) (2016) 24, https://doi.org/10.3390/coatings6020024.
[15] H.H. Nelson, High-temperature application of nitrocellulose lacquers, Ind. Eng. Chem. 31 (1) (1939) 70–75, https://doi.org/10.1021/ie50349a015.
[16] H. Standeven, The development of decorative gloss paints in Britain and the United States C. 1910–1960, J. Am. Inst. Conserv. 45 (1) (2006) 51–65, https://doi.org/10.1179/019713606806082210.
[17] A. Hofland, alkyd resins: from down and out to alive and kicking, Prog. Organ. Coatings 73 (4) (2012) 274–282, https://doi.org/10.1016/j.porgcoat.2011.01.014.
[18] C.C. Okoye, O.D. Onukwuli, M.I. Ejimofor, C.F. Okey-Onyesolu, Modelling and optimal viscometry formulation evaluation of a modified green based self-healing automotive paint, Int. J. Adv. Eng. Manage. Sci. 7 (3) (2021) 39.
[19] E. Koh, N.-K. Kim, J. Shin, Y.-W. Kim, Polyurethane microcapsules for self-healing paint coatings, RSC Adv. 4 (31) (2014) 16214–16223, https://doi.org/10.1039/C4RA00213J.
[20] Y.-G. June, K.I. Jung, D.G. Lee, S. Jeong, T.-H. Lee, Y.I. Park, S.M. Noh, H.W. Jung, Influence of functional group content in hydroxyl-functionalized urethane methacrylate oligomers on the crosslinking features of clearcoats, J. Coatings Technol. Res. 18 (2021) 229–237, https://doi.org/10.1007/s11998-020-00398-1.
[21] M. Barletta, A. Gisario, Electrostatic spray painting of carbon fibre-reinforced epoxy composites, Prog. Organ. Coatings 64 (4) (2009) 339–349, https://doi.org/10.1016/j.porgcoat.2008.07.020.
[22] K. Hu, S. Liu, J. Lei, C. Zhou, Dispersion and resistivity optimization of conductive carbon black in environment-friendly conductive coating based on acrylic resin, Polym. Compos. 36 (3) (2015) 467–474, https://doi.org/10.1002/pc.22961.
[23] D. Andre, S.-J. Kim, P. Lamp, S.F. Lux, F. Maglia, O. Paschos, B. Stiaszny, Future generations of cathode materials: an automotive industry perspective, J. Mater. Chem. A 3 (2015) 6709–6732, https://doi.org/10.1039/C5TA00361J.
[24] K. Friedrich, Polymer composites for tribological applications, Adv. Ind. Eng. Polym. Res. 1 (2018) 3–39, https://doi.org/10.1016/j.aiepr.2018.05.001.
[25] A. Das, H.T. Hayvaci, M.K. Tiwari, I.S. Bayer, D. Erricolo, C.M. Megaridis, Superhydrophobic and conductive carbon nanofiber/PTFE composite coatings for EMI shielding, J. Colloid Interface Sci. 353 (2011) 311–315, https://doi.org/10.1016/j.jcis.2010.09.017.
[26] Y. Qian, Y. Li, S. Jungwirth, N. Seely, Y. Fang, X. Shi, The application of anti-corrosion coating for preserving the value of equipment asset in chloride-laden environments: a review, Int. J. Electrochem. Sci. 10 (2015) 10756–10780.
[27] P.P. Chung, J. Wang, Y. Durandet, Deposition processes and properties of coatings on steel fasteners—a review, Friction 7 (2019) 389–416, https://doi.org/10.1007/s40544-019-0304-4.

[28] S.B. Ulaeto, J.K. Pancrecious, K.K. Ajekwene, G.M. Mathew, T.P.D. Rajan, Chapter 25: Advanced nanocoatings for anticorrosion, in: S. Rajendran, T.A.N.H. Nguyen, S. Kakooei, M. Yeganeh, Y. Li (Eds.), Corrosion Protection at the Nanoscale, Elsevier, 2020, pp. 499–510.

[29] S. Laurenzi, M. Clausi, F. Zaccardi, U. Curt, M. Gabriella Santonicola, Spray coating process of MWCNT/epoxy nanocomposite films for aerospace applications: effects of process parameters on surface electrical properties, Acta Astronaut. 159 (2019) 429–439, https://doi.org/10.1016/j.actaastro.2019.01.043.

[30] H. Bahramnia, H.M. Semnani, A. Habibolahzadeh, H. Abdoos, Epoxy/polyurethane hybrid nanocomposite coatings reinforced with MWCNTs and SiO_2 nanoparticles: processing, mechanical properties and wear behavior, Surf. Coatings Technol. 415 (2021) 127121, https://doi.org/10.1016/j.surfcoat.2021.127121.

[31] Wicz Market, Report: Conductive Coatings Market Growth Volume, Share 2021, Opportunities and Challenges, Revenues, Industry Key Factors, Main Competitive Landscape and COVID-19 Impact Forecast By 2027. Available online: https://www.wicz.com/story/44167921/conductive-coatings-market-growth-volume-share-2021-opportunities-and-challenges-revenues-industry-key-factors-main-competitive-landscape-and-covid19 (Accessed on 30 April, 2021).

[32] Lucintel, Report published by Lucintel: Opportunities and Competitive Analysis of the Electrically Conductive Coatings Market, July 2020. Available online: https://www.lucintel.com/electrically-conductive-coatings-market.aspx (Accessed on 30 April, 2021).

[33] Mordor, Electrically Conductive Coating Market – Growth, Trends, COVID-19 Impact, And Forecasts (2021–2026), Mordor Inteligence, 2021. Available online: https://www.mordorintelligence.com/industry-reports/electrically-conductive-coating-market. (Accessed 30 April 2021).

[34] Lucitel, Electrically Conductive Coatings Market Report: Trends, Forecast and Competitive Analysis, Lucitel Insigth of Matter, 2021. Available online: https://www.lucintel.com/electrically-conductive-coatings-market.aspx. (Accessed 30 April 2021).

[35] Industry ARC, Conductive Coatings Market – By Resin, By Pigment, By Application and By Geography Analysis – Forecast 2020–2025, 2021, Available online: https://www.industryarc.com/Research/Conductive-Coatings-Market-Research-501523. (Accessed 30 April 2021).

[36] N. Wang, M. Prodanović, H. Daigle, Nanopaint application for flow assurance with electromagnetic pig, J. Petrol. Sci. Eng. 180 (2019) 320–329, https://doi.org/10.1016/j.petrol.2019.05.028.

[37] Á. Yedra, G. Gutiérrez-Somavilla, C. Manteca-Martínez, M. González-Barriuso, L. Soriano, Conductive paints development through nanotechnology, Prog. Organ. Coatings 95 (2016) 85–90, https://doi.org/10.1016/j.porgcoat.2016.03.001.

[38] J. Mathew, J. Joy, S.C. George, Potential applications of anotechnology in transportation: a review, J. King Saud Univ. Sci. 31 (4) (2019) 586–594, https://doi.org/10.1016/j.jksus.2018.03.015.

[39] J.-P. Kaiser, S. Zuin, P. Wick, Is nanotechnology revolutionizing the paint and lacquer industry? A critical opinion, Sci. Total Environ. 442 (2013) 282–289, https://doi.org/10.1016/j.scitotenv.2012.10.009.

[40] A.T. Saber, K.A. Jensen, N.R. Jacobsen, R. Birkedal, L. Mikkelsen, P. Møller, S. Loft, H. Wallin, U. Vogel, Inflammatory and genotoxic effects of nanoparticles designed for inclusion in paints and lacquers, Nanotoxicology 6 (5) (2012) 453–471, https://doi.org/10.3109/17435390.2011.587900.

[41] M. Bakhtiary-Noodeh, S. Moradian, Z. Ranjbar, Edge protection improvement of automotive electrocoatings in the presence of silica nanoparticles, Surf. Coatings Technol. 317 (2017) 134–147, https://doi.org/10.1016/j.surfcoat.2017.03.004.

[42] R. Sathyamurthy, A.E. Kabeel, M. Balasubramanian, M. Devarajan, S.W. Sharshir, A. Muthu Manokar, Experimental study on enhancing the yield from stepped solar still coated using fumed silica nanoparticle in black paint, Mater. Lett. 272 (2020) 127873, https://doi.org/10.1016/j.matlet.2020.127873.

[43] Y. Tong, S. Bohm, M. Song, The capability of graphene on improving the electrical conductivity and anti-corrosion properties of polyurethane coatings, Appl. Surf. Sci. 424 (2017) 72–81, https://doi.org/10.1016/j.apsusc.2017.02.081.

[44] S. Schincariol, M. Fonseca, V. Neto, Development of a Nanopaint for polymeric auto components, in: A.W. Jackson, M. (Eds.), Micro and Nanomanufacturing, Springer, Cham, 2018, pp. 157–201.

[45] P.-C. Ma, N.A. Siddiqui, G. Marom, J.-K. Kim, Dispersion and functionalization of carbon nanotubes for polymer-based nanocomposites: a review, Compos. Part A: Appl. Sci. Manuf. 41 (10) (2010) 1345–1367, https://doi.org/10.1016/j.compositesa.2010.07.003.

[46] E.T. Thostenson, T.-W. Chou, Processing-structure-multi-functional property relationship in carbon nanotube/epoxy composites, Carbon 44 (14) (2006) 3022–3029, https://doi.org/10.1016/j.carbon.2006.05.014.

[47] S. Azim, A. Syed, K. Satheesh, K. Ramu, S. Ramu, G. Venkatachari, Studies on graphite based conductive paint coatings, Prog. Organ. Coatings 55 (2006) 1–4, https://doi.org/10.1016/j.porgcoat.2005.09.001.

[48] K. Krishnamoorthy, M.P. Pazhamalai, J.H. Lim, K.H. Choi, S.-J. Kim, Mechanochemical reinforcement of graphene sheets into alkyd resin matrix for the development of electrically conductive paints, ChemNanoMat 4 (6) (2018) 568–574, https://doi.org/10.1002/cnma.201700391.

[49] H.E. Yong, K. Krishnamoorthy, K.T. Hyun, S.J. Kim, Preparation of ZnO nanopaint for marine antifouling applications, J. Ind. Eng. Chem. 29 (2015) 39–42, https://doi.org/10.1016/j.jiec.2015.04.020.

[50] Y. Nishimura, T. Otsuka, M. Murashima, K. Maruyama, M. Suezaki, Electrically Conductive Paint Composition, Sekisui Chemical Co., Ltd., Japan, 1999.

[51] Y. Hokita, M. Kanzaki, T. Sugiyama, R. Arakawa, H. Kawasaki, High-concentration synthesis of sub-10-nm copper nanoparticles for application to conductive nanoinks, ACS Appl. Mater. Interfaces 7 (34) (2015) 19382–19389, https://doi.org/10.1021/acsami.5b05542.

[52] C.-K. Leong, D.D.L. Chung, Improving the electrical and mechanical behavior of electrically conductive paint by partial replacement of silver by carbon black, J. Electron. Mater. 35 (1) (2006) 118–122, https://doi.org/10.1007/s11664-006-0193-y.

[53] M.A. Bhakare, P.H. Wadekar, R.V. Khose, M.P. Bondarde, S. Some, Eco-friendly biowaste-derived graphitic carbon as black pigment for conductive paint, Prog. Organ. Coatings 147 (2020) 105872, https://doi.org/10.1016/j.porgcoat.2020.105872.

[54] J. Flynt, What is Conductive Paint? Pros, Cons, and Applications 2020, 2021, Available from https://3dinsider.com/conductive-paint/. (Accessed 17 May 2021).

[55] K. Krishnamoorthy, K. Jeyasubramanian, M. Premanathan, G. Subbiah, H.S. Shin, S.J. Kim, Graphene oxide nanopaint, Carbon 72 (2014) 328–337, https://doi.org/10.1016/j.carbon.2014.02.013.

[56] Y. Wang, Electrical conductivity of carbon nanotube- and graphene-based nanocomposites, in: G.J. Weng, S.A. Meguid (Eds.), Micromechanics and Nanomechanics of Composite Solids, Springer International Publishing, 2018, pp. 123–156.

[57] M. Nascimento Silva, E. Kassab, O.G. Pandoli, J.L. de Oliveira, J.P. Quintela, I.S. Bott, Corrosion behaviour of an epoxy paint reinforced with carbon nanoparticles, Corros. Eng. Sci. Technol. 55 (8) (2020) 603–608, https://doi.org/10.1080/1478422X.2020.1767322.

[58] S. Ligati, A. Ohayon-Lavi, J. Keyes, G. Ziskind, O. Regev, Enhancing thermal conductivity in graphene-loaded paint: effects of phase change, rheology and filler size, Int. J. Therm. Sci. 153 (2020) 106381, https://doi.org/10.1016/j.ijthermalsci.2020.106381.

[59] D. Coetzee, M. Venkataraman, J. Militky, M. Petru, Influence of nanoparticles on thermal and electrical conductivity of composites, Polymers 12 (4) (2020) 742, https://doi.org/10.3390/polym12040742.

[60] L.S. Roselin, R.-S. Juang, C.-T. Hsieh, S. Sagadevan, A. Umar, R. Selvin, H.H. Hegazy, Recent advances and perspectives of carbon-based nanostructures as anode materials for Li-ion batteries, Materials 12 (8) (2019) 1229, https://doi.org/10.3390/ma12081229.

[61] D. Bhattacharjya, Y. Jong-Sung, Activated carbon made from cow dung as electrode material for electrochemical double layer capacitor, J. Power Sources 262 (2014) 224–231, https://doi.org/10.1016/j.jpowsour.2014.03.143.

[62] K. Krishnamoorthy, M.P. Pazhamalai, J.H. Lim, K.H. Choi, S.-J. Kim, Mechanochemical reinforcement of graphene sheets into alkyd resin matrix for the development of electrically conductive paints, ChemNanoMat 4 (2018) 568–574, https://doi.org/10.1002/cnma.201700391.

[63] R.V.K. Rao, K. Venkata Abhinav, P.S. Karthik, S.P. Singh, Conductive silver inks and their applications in printed and flexible electronics, RSC Adv. 5 (2015) 77760–77790, https://doi.org/10.1039/C5RA12013F.

[64] S.Z.A. Sakinah, W.H. Azmi, J. Alias, Characterization of TiO_2 nanopaint for automotive application, IOP Conf. Ser. Mater. Sci. Eng. 863 (2020) 012053.

[65] K. Nechvílová, A. Kalendová, Influencing the anticorrosion efficiency of pigments based on zinc ferrite by conductive polymers, Koroze a Ochrana Materialu 62 (3) (2018) 83–86, https://doi.org/10.1515/kom-2018-0012.

[66] C.Y. Lai, C.F. Cheong, J.S. Mandeep, H.B. Abdullah, N. Amin, K.W. Lai, Synthesis and characterization of silver nanoparticles and silver inks: review on the past and recent technology roadmaps, J. Mater. Eng. Perform. 23 (10) (2014) 3541–3550, https://doi.org/10.1007/s11665-014-1166-6.

[67] M.H. Overgaard, M. Kühnel, R. Hvidsten, S.V. Petersen, T. Vosch, K. Nørgaard, B.W. Laursen, Highly conductive semitransparent graphene circuits screen-printed from water-based graphene oxide ink, Adv. Mater. Technol. 2 (2017) 1700011, https://doi.org/10.1002/admt.201700011.

[68] J. Li, X. Zhang, X. Liu, Q. Liang, G. Liao, Z. Tang, T. Shi, Conductivity and foldability enhancement of Ag patterns formed by PVAc modified Ag complex inks with low-temperature and rapid sintering, Mater. Des. 185 (2020) 108255, https://doi.org/10.1016/j.matdes.2019.108255.

[69] N.H.A. Aziz, H. Jaafar, R.M. Sidek, S. Shafie, M.N. Hamidon, Raman study on dispersion of carbon nanotube in organic solvent as the preparation of conductive nano-ink, in: Paper read at 2019 IEEE Regional Symposium on Micro and Nanoelectronics (RSM), 21–23 August 2019, 2019.

[70] L. Peng, Y. Sun, G. Xiaobin, P. Liu, B. Liang, B. Wei, Thermal conductivity enhancement utilizing the synergistic effect of carbon nanocoating and graphene addition in palmitic acid/halloysite FSPCM, Appl. Clay Sci. 206 (2021) 106068, https://doi.org/10.1016/j.clay.2021.106068.

[71] F. Meng, J. Huang, Fabrication of conformal array patch antenna using silver nanoink printing and flash light sintering, AIP Adv. 8 (8) (2018), https://doi.org/10.1063/1.5039951, 085118.

[72] R.A. Fernandes, A.A. Berretta, E.C. Torres, A.F.M. Buszinski, G.L. Fernandes, C.C. Mendes-Gouvêa, F.N. de Souza-Neto, L.F. Gorup, E.R. de Camargo, D.B. Barbosa, Antimicrobial potential and cytotoxicity of silver nanoparticles phytosynthesized by pomegranate peel extract, Antibiotics 7 (3) (2018) 51, https://doi.org/10.3390/antibiotics7030051.

[73] V.H. Fragal, T.S.P. Cellet, G.M. Pereira, E.H. Fragal, M.A. Costa, C.V. Nakamura, T. Asefa, A.F. Rubira, R. Silva, Covalently-layers of PVA and PAA and in situ formed Ag nanoparticles as versatile antimicrobial surfaces, Int. J. Biol. Macromol. 91 (2016) 329–337, https://doi.org/10.1016/j.ijbiomac.2016.05.056.

[74] L.F. Gorup, F.N. de Souza Neto, A.M. Kubo, J.A.S. Souza, R.A. Fernandes, G.L. Fernandes, D.R. Monteiro, D.B. Barbosa, E.R. Camargo, Nanostructured functional materials: silver nanoparticles in polymer for the generation of antimicrobial characteristics, in: E. Longo, L.P.F.A. Gewerbestrasse (Eds.), Recent Advances in Complex Functional Materials, Springer, Switzerland, 2017.

[75] D.R. Monteiro, L.F. Gorup, S. Silva, M. Negri, E.R. de Camargo, R. Oliveira, D.B. Barbosa, M. Henriques, Silver colloidal nanoparticles: antifungal effect against adhered cells and biofilms of *Candida albicans* and *Candida glabrata*, Biofouling 27 (7) (2011) 711–719, https://doi.org/10.1080/08927014.2011.599101.

[76] D.R. Monteiro, L.F. Gorup, A.S. Takamiya, E.R. de Camargo, A.C.R. Filho, D.B. Barbosa, Silver distribution and release from an antimicrobial denture base resin containing silver colloidal nanoparticles, J. Prosthodont. 21 (2012) 7–15, https://doi.org/10.1111/j.1532-849X.2011.00772.x.

[77] J.A.S. Souza, D.B. Barbosa, A.A. Berretta, J.G. do Amaral, L.F. Gorup, F.N. de Souza Neto, R.A. Fernandes, G.L. Fernandes, E.R. Camargo, A.M. Agostinho, A.C.B. Delbem, Green synthesis of silver nanoparticles combined to calcium glycerophosphate: antimicrobial and antibiofilm activities, Future Microbiol. 13 (3) (2018) 345–357, https://doi.org/10.2217/fmb-2017-0173.

[78] S. Neto, F. de Nunes, R.L. Sala, R.A. Fernandes, T.P.O. Xavier, S.A. Cruz, C.M. Paranhos, D.R. Monteiro, D.B. Barbosa, A.C.B. Delbem, E.R. de Camargo, Effect of synthetic colloidal nanoparticles in acrylic resin of dental use, Eur. Polym. J. 112 (2019) 531–538, https://doi.org/10.1016/j.eurpolymj.2018.10.009.

[79] D.H. Abdeen, M. El Hachach, M. Koc, M.A. Atieh, A review on the corrosion behaviour of nanocoatings on metallic substrates, Materials (Basel, Switzerland) 12 (2) (2019) 210, https://doi.org/10.3390/ma12020210.

[80] G. Barroso, Q. Li, R.K. Bordia, G. Motz, Polymeric and ceramic silicon-based coatings – a review, J. Mater. Chem. A 7 (5) (2019) 1936–1963, https://doi.org/10.1039/C8TA09054H.

[81] O.F. Ngasoh, V.C. Anye, B. Agyei-Tuffour, O.K. Oyewole, P.A. Onwualu, W.O. Soboyejo, Corrosion behavior of 5-hydroxytryptophan (HTP)/epoxy and clay particle-reinforced epoxy composite steel coatings, Cogent Eng. 7 (2020) 1797982, https://doi.org/10.1080/23311916.2020.1797982.

[82] K. Chatterjee, J. Tabor, T.K. Ghosh, Electrically conductive coatings for fiber-based e-textiles, Fibers 7 (6) (2019) 51, https://doi.org/10.3390/fib7060051.

[83] L. Cai, C. Wang, Carbon nanotube flexible and stretchable electronics, Nanoscale Res. Lett. 10 (2015) 1013, https://doi.org/10.1186/s11671-015-1013-1.

[84] Y. Li, D. Lu, C.P. Wong, Conductive nano-inks, in: Y. Li, D. Lu, C.P. Wong (Eds.), Electrical Conductive Adhesives with Nanotechnologies, Springer US, Boston, MA, 2010, pp. 303–360.

Web References

[85] M. Dyson, Y. Yamamoto, Report: Marchés d'encre conductive 2020–2030: Prévisions, Technologies, Joueurs, PV, 5G, automobile, électronique de puissance, blindage EMI, électronique In-Mold, E-textiles, patchs de peau, capteurs imprimés, électronique hybride flexible, RFID, métallisation 3D, chauffage, maille métallique hybride/imprimée, et bien d'autres, 2021, Available online: https://www.idtechex.com/fr/research-report/conductive-ink-markets-2020-2030-forecasts-technologies-players/734. (Accessed 30 April 2021).

[86] S.-H. Yue, Y.-B. Jung, J.-B. Kim, United States Patent: Carbon nanotube sheet heater, January 12, 2016, Assignee: LG Hausys, Ltd., Seoul, 2016. Available online: https://patents.google.com/patent/US9237606B2/en. (Accessed 30 April 2021).

[87] M. Dyson, L. Gear, J. Edmondson, R. Das, Report: Printed and Flexible Electronics for Automotive Applications 2021-2031: Technologies and Markets, IDTechEx, 2021. Available online: http://www.idtechex.com/en/research-report/printed-and-flexible-electronics-for-automotive-applications-2021-2031-technologies-and-markets/806. (Accessed 30 April 2021).

SECTION D

Nanodevices for energy conversion and storage in the automotive application

CHAPTER 21

Battery-supercapacitor hybrid systems: An introduction

Anuj Kumar[a], Shumaila Ibraheem[b], Ram K. Gupta[c], Tuan Anh Nguyen[d], and Ghulam Yasin[b]

[a]Department of Chemistry, GLA University, Mathura, Uttar Pradesh, India
[b]Institute for Advanced Study, Shenzhen University, Shenzhen, Guangdong, China
[c]Department of Chemistry, Pittsburg State University, Pittsburg, KS, United States
[d]Institute for Tropical Technology, Vietnam Academy of Science and Technology, Hanoi, Vietnam

Chapter outline

1. Introduction 453
2. Combination of battery and supercapacitor 454
3. Hybridization of battery and supercapacitor 455
4. Conclusion 457
References 457

1. Introduction

To enhance the performance of materials and devices used in modern cars, various nanotechnology-based approaches and techniques have been developed, such as nanoalloys, nanocomposites, nanocoatings, nanodevices, nanocatalysts, and nanosensors. Thus, nanotechnology is the key to this important development (Fig. 1).

Not the electric motor, the electric power source is the most important part in electric vehicles (EV). Currently, it takes about 30 min (with fast charging) up to nearly a full day to recharge the batteries of battery electric vehicles (BEVs). Regarding the 800 Volt charging system, the KIA EV6 only needs 18 min to go from 10% to 80%, managing to charge for a 100-km trip in just 4 min and 30 s. Regarding the battery used in hybrid supercars, the Lamborghini Sian Roadster uses supercapacitors that can store more power (10–100 times) with fast charging time. The combination/hybridization of the battery and supercapacitor may exhibit superior features as compared to the pure battery or supercapacitor. This new hybrid approach can overcome the limits of a single component and even achieve super-fast charging time, more efficiency, and a high mileage range on a single charge with improved safety.

Battery-supercapacitor hybrid (BSH) refers to the combination/hybridization of battery and supercapacitor.

Fig. 1 Nanotechnology in the automotive industry.

2. Combination of battery and supercapacitor

Rising energy costs, coupled with increased alertness and recognition of global warming, have promoted the development of the green technology field that is booming today. The fields of electronics, automobiles, and electrical engineering depend to a large extent on the search for clean, sustainable, and transportation energy. Hybrid electric vehicles have become a potential solution to some of the world's energy problems [1]. Although a hybrid vehicle can save fuel, it has its own costs. Batteries are composed of highly reactive materials, which are very expensive, heavy, and challenging to replace [2]. In order for hybrid electric vehicles to become a suitable solution, these issues must be solved. The emergence of new energy storage capacitors and lighter-weight rechargeable batteries with higher energy density has led to new developments in the field of clean energy [3].

Rechargeable batteries, such as Li-ion batteries (LIBs), are ideal energy resources because they reduce replacement costs as well as serious environmental concerns. For example, modern hybrid electric vehicles (HEV) use rechargeable batteries with gasoline engines to power these vehicles [4]. In these systems, batteries act as the main energy source, whereas gasoline is used as a backup to reduce fuel consumption. The problem with these systems is that there is no buffer between the battery itself and the load.

Without a buffer, the battery could easily be damaged, and its life would be greatly reduced [5]. The preferred operation of the battery is to charge it constantly with the smallest average current. It changes the energy generated and received from the load. Due to the low energy density of most rechargeable batteries, their lifespan is shortened due to constant fluctuations in demand. To address this problem, an energy storage system consisting of a high-performing battery-type electrode and a fast-rate capacitive electrode, called a battery supercapacitor hybrid (BSH), offers the merits of both rechargeable batteries and supercapacitors. In this hybrid system, a supercapacitor acts as a buffer, having high power density and it can handle erratic oscillations in demand without sustaining any damage [6].

BSH is expected to display higher energy and power density for future multifunctional electronic devices, hybrid transport vehicles, and industrial equipment [7]. BSH systems can reduce the bandwidth effect loss of modern portable electronic devices with large load curve fluctuations. The traditional hybrid supercapacitor battery needs a large supercapacitor in parallel to reduce the maximum discharge current of the battery, which is enough to reduce the impact of the rapid-acting capacity on the battery. However, larger supercapacitors will greatly reduce the total available energy density, which is one of the most critical limitations of portable electronic power supplies. By isolating the battery from the supercapacitor, and by using a supercapacitor smaller than the supercapacitor used in parallel, the proposed constant-current architecture effectively reduces the impact of the rate capacity factor [8].

Moreover, in the case of a hybrid battery-supercapacitor system, the charging/discharging frequency of the battery can be reduced by integrating the supercapacitor. For instance, a BSH system is used to compensate for the fluctuation of wind energy, where the battery is responsible for balancing the P-steady to avoid charging, which can extend the battery life in the hybrid system [9].

3. Hybridization of battery and supercapacitor

With growing concerns regarding environmental pollution due to the heavy use of fossil fuels, the modern world is using renewable resources to produce energy. Solar, wind, and waves are some of the renewable resources which are being used to generate green energy. Since the outcome of the energy using these resources is very much dependent on conditions, such as solar energy can be only produced during a sunny day or a windmill can be only installed in an open remote area, there is a need for devices that can store the generated energy and can be used at night or transported to the community. Among various portable energy storage devices, batteries and supercapacitors are very popular due to their several advantages. A battery can deliver high energy density, while a supercapacitor can provide high power density. These devices also vary in their long-term performance. For example, supercapacitors can be used for many years while batteries have a limited life. Recent technology demands energy devices that are capable of providing both high energy and power density as well as a long life. The hybridized battery and supercapacitor could meet such demand.

Generally, the creation of more advanced energy storage devices not only depends on the micro/nanostructure design of the electrode materials but also mainly depends on the design of the device configuration. In the design structure of the BSH system, the hierarchical electrode showed high areal capacity, good rate performance, and excellent durability [10, 11]. Therefore, these hybrid battery supercapacitor energy storage systems can provide high energy and power densities as well as the ability to store energy for a shorter time at a lower cost than traditional supercapacitors. Although their electrochemical properties and behavior are different, batteries and supercapacitors have the same configuration, including the anode, cathode, electrolyte, separator, and current collector. As a result, the hybrid systems directly exhibit high performance in terms of significant energy and power as well as long cycling life due to the good incorporation between the battery and supercapacitor, respectively [11].

Amatucci et al. first reported a BSH device that was based on Li-ion, where activated carbon was used as the cathode and $Li_4Ti_5O_{12}$ was used as the anode [12]. The device was very stable up to 5000 cycles of charge-discharge studies and provide an energy density of over 20 Wh/kg which bridged the gap between Li-ion batteries and supercapacitors. Aqueous and non-aqueous electrolytes are being used in BSH devices. However, non-aqueous (organic and ionic) electrolytes are mainly used in Li-ion based hybrid devices. Recent research is also focusing on using Na-ion based hybrid devices due to the low cost of sodium and similar electrochemical properties compared to lithium. The technology for Na-ion based hybrid devices is similar to what being used for Li-ion based devices, therefore, the commercialization of Na-ion based devices is not a constrain. Fig. 2 shows the energy and power density of various devices and expected values for the battery-supercapacitor hybrid device [10]. BSH devices based on Na-ion also use organic and ionic electrolytes. The choice of electrolyte is very important as it may affect the overall performance of any device. Redox-active electrolytes are currently receiving attention due to their capability to enhance the electrochemical properties by participating in the redox process at the electrode-electrolyte interface.

Flexible/wearable devices are the future of the electronic industry and need high-performance and flexible energy devices to power them. BSHs are the potential devices for such applications. These devices can be fabricated on flexible electrodes such as carbon nanotubes, graphene, and textiles. Apart from various types of carbon, several other materials, such as conducting polymers, transition metal oxide, sulfide, and phosphides, are being used in flexible devices. For example, a combination of CoO@polypyrrole as a cathode and activated carbon as an anode was used in a flexible device [13]. The synergism between CoO and polypyrrole provided a device with over 3 times increased in areal capacity, high energy density, and exceptional cycleability. Such devices

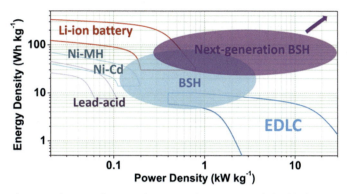

Fig. 2 Energy and power density of various batterer, electrochemical double-layer capacitors (EDLC), and BSHs. *(Adapted with permission from reference W. Zuo, R. Li, C. Zhou, Y. Li, J. Xia, J. Liu, Battery-supercapacitor hybrid devices: recent progress and future prospects, Adv. Sci. 4(7) (2017), 1600539. The article is published using CC-BY license.)*

can use liquid electrolytes, gel electrolytes, or even solid-state electrolytes. Solid-state or quasi-solid-state electrolytes can also function as separators in these devices.

4. Conclusion

In summary, the battery-supercapacitor hybrid power system has a number of merits like high power density and high energy density as well as long cycling life. Therefore, the hybrid battery-supercapacitor systems can meet the performance requirements of the energy storage system for long-drive electric automobiles and various other energy applications.

References

[1] H. Presting, U. König, Future nanotechnology developments for automotive applications, Mater. Sci. Eng. C 23 (6–8) (2003) 737–741.
[2] M. Farhadi, O. Mohammed, Energy storage technologies for high-power applications, IEEE Trans. Ind. Appl. 52 (3) (2015) 1953–1961.
[3] S. Hajiaghasi, A. Salemnia, M. Hamzeh, Hybrid energy storage system for microgrids applications: A review, J. Energy Storage 21 (2019) 543–570.
[4] H. Wang, C. Zhu, D. Chao, Q. Yan, H.J. Fan, Nonaqueous hybrid lithium-ion and sodium-ion capacitors, Adv. Mater. 29 (46) (2017) 1702093.
[5] S. Park, Y. Kim, N. Chang, Hybrid energy storage systems and battery management for electric vehicles, in: 2013 50th ACM/EDAC/IEEE Design Automation Conference (DAC), IEEE, 2013, pp. 1–6.
[6] I.J. Cohen, J.P. Kelley, D.A. Wetz, J. Heinzel, Evaluation of a hybrid energy storage module for pulsed power applications, IEEE Trans. Plasma Sci. 42 (10) (2014) 2948–2955.
[7] D. Shin, Y. Kim, J. Seo, N. Chang, Y. Wang, M. Pedram, Battery-supercapacitor hybrid system for high-rate pulsed load applications, in: 2011 Design, Automation & Test in Europe, IEEE, 2011, pp. 1–4.

[8] J. Ding, W. Hu, E. Paek, D. Mitlin, Review of hybrid ion capacitors: from aqueous to lithium to sodium, Chem. Rev. 118 (14) (2018) 6457–6498.
[9] P. Thounthong, S. Rael, B. Davat, Energy management of fuel cell/battery/supercapacitor hybrid power source for vehicle applications, J. Power Sources 193 (1) (2009) 376–385.
[10] W. Zuo, R. Li, C. Zhou, Y. Li, J. Xia, J. Liu, Battery-supercapacitor hybrid devices: recent progress and future prospects, Adv. Sci. 4 (7) (2017) 1600539.
[11] M. Zheng, H. Tang, L. Li, Q. Hu, L. Zhang, H. Xue, H. Pang, Hierarchically nanostructured transition metal oxides for lithium-ion batteries, Adv. Sci. 5 (3) (2018) 1700592.
[12] G.G. Amatucci, F. Badway, A. Du Pasquier, T. Zheng, An Asymmetric Hybrid Nonaqueous Energy Storage Cell, J. Electrochem. Soc. 148 (2001) A930.
[13] C. Zhou, Y. Zhang, Y. Li, J. Liu, Construction of High-Capacitance 3D CoO@Polypyrrole Nanowire Array Electrode for Aqueous Asymmetric Supercapacitor, Nano Lett. 13 (2013) 2078–2085.

CHAPTER 22

Supercapacitors: An introduction

Narendra Lakal[a], Sumit Dubal[b], and P.E. Lokhande[a]
[a]Sinhgad Institute of Technology, Lonavala, India
[b]Symboisis Skills and Professional University, Pune, India

Chapter outline

1. Introduction 459
2. Supercapacitor component development 460
   2.1 Electrode development 460
   2.2 Separator development 461
   2.3 Development of solid polymer electrolyte 462
3. Structural supercapacitor performance parameters 462
4. Conclusion 463
   References 463

1. Introduction

Limited fossil fuel sources generate a need for renewable energy sources such as wind, solar, geothermal, and tidal energy. But due to the fluctuating nature of these sources, it requires energy storage technology to store such generated renewable energy and use it for various applications. Nowadays, various advanced energy storage technologies, like batteries, supercapacitors, fuel cells, etc., are used to store such renewable energy. Among them, the supercapacitor gained the attention of researchers in various fields, such as automotive, electronics, and mechanical, due to its superior properties such as high-power density, fast charge-discharge, wide operating temperature range, and excellent cycle life. Based on charge storage mechanism, supercapacitors are grouped into three categories: electrochemical double-layer capacitor (EDLC), pseudo capacitor and hybrid capacitor.

In the case of EDLC, charge is stored electrostatically where no transfer of ions between two electrodes instead of the formation of a double layer of opposite charged ions on respective electrodes. Carbon-based materials such as graphene, activated carbon, carbon aerogels, and carbon nanotubes demonstrate the EDLC type charge storage mechanism. Due to carbon-based materials exhibiting large surface area, they showed higher power density and superior cycle life, but suffered from poor energy density. On the other hand, in the case of the pseudocapacitor, reversible redox reactions are responsible for the charge storage mechanism. The transition metal oxides/hydroxides conductive polymer demonstrates a pseudocapacitive charge storage mechanism. Due to reversible faradaic reaction, energy density for the pseudocapacitor is on the higher side but suffers from poor stability. To mitigate the disadvantages of EDLC and supercapacitor, a hybrid supercapacitor is fabricated by combining EDLC and a pseudocapacitor. Nowadays, transition metal oxides/hydroxides such as NiO, Co_3O_4, MnO, Fe_2O_3, Ni $(OH)_2$, CuO have attracted enormous attention from the pseudocapacitive group, while graphene and carbon nanotubes from the EDLC group. Nanocomposites of EDLC and pseudocapacitor exhibited excellent specific capacitance, higher energy density and long cycle life.

The increasing cost and pollution from declining fossil fuel sources has prompted the scientific community in the automotive field to consider alternative ways, such as electric vehicles, the use of biofuels, etc. Another strategy converts the entire automotive body into energy storage by using the assembly of multifunctional structural laminates that will provide high load-bearing capacity with minimal weight to volume fraction of the composite laminate and, simultaneously, it will act as energy storage. Composite materials such as carbon fiber have structural and electrochemical functions in a single material, which will be important in applications such as hybrid electric vehicles and the aerospace field. This book chapter will provide the development of supercapacitors in the automotive sector in terms of electrode materials and composite structural materials.

2. Supercapacitor component development

Supercapacitors are made up of two electrodes separated by a separator which allows ions to pass through it but restricts the electrons [1]. Electrolytic capacitors use electrolytes as a third component. The electrolyte can be liquid as in the case of conventional electrolytic supercapacitors or it can be solid as in the case of structural supercapacitors (SC). SC's use solid-state electrolyte (glass fiber-GF) to act as separator [2]. Materials for each of these three components play a pivotal role in SCs, numerous studies explored the same.

2.1 Electrode development

For structural supercapacitors to be useful in automobile applications, the electrodes used should provide excellent mechanical as well as electrical properties. Carbon fiber (CF) is one such candidate material. For SC [3–5]. But the CF laminates if used as electrodes do not provide good Mechanical properties due to impropergraphitization and low surface area [6, 7]. One of the alternatives considered for improving these properties is CF surface treatment using acids and bases [8–10]. The concentration of these acids and bases is one of the important governing parameters for surface quality. Other alternatives explored were oxygen plasma and the large pulse electron beam technique. It is found that this treatment significantly affects the surface area and adhesion between matrix and separator [11, 12].

The next alternative approach to improve the surface area and specific capacitance is the growing and sizing of carbon nanotubes (CNT) on CF [13, 14]. Out of the two

Table 1 BET surface area and CSP of CFs treated by different processes [25].

Sl. No	Samples(carbon fiber)	BET surface area(m² g⁻¹)	Specific capacitance(F g⁻¹)
1	As received	0.035	0.057
2	Heat treated	0.068	1.453
3	KOH treated	37.263	29.451
4	CNT-sizing	38.562	27.291
5	CNT growth	41.027	30.783
6	CAG treated	165.252	156.462

BET, Brunauer–Emmett–Teller.

methods, growing CNT was found to be more effective than sizing CNT [15, 16]. Multiple studies adopted different approaches for growing CNT on CF. the two methods reported are: conventional chemical vapor deposition (CVD technique) [15–17] and a nonconventional method of microwave irradiation [18–20].

One more way to improve the properties of CF is by supporting CF with nanostructured metal oxides. In one study CuO nanowires were embedded on woven CF by hydrothermal process. The improvement in the mechanical and chemical properties was found to be significant [21]. Another study used spinel $NiCo_2O_4$ nanostructures on CF cloth to improve the properties [22]. One more similar study used NiCo layered double hydroxide nanorods to improve the properties of electro-spun CF [23]. Among these attempts, the best approach was the use of carbon aerogel (CAG) on the surface of CF [24]. As indicated by Table 1 the use of the CAG method improved the BET surface area and the specific capacitance significantly compared to the heat-treated, KOH treated, and CNT growing and sizing methods. [25].

Thus, the use of CF for electrodes in structural SC's is not sufficient to get expected electrical and mechanical properties. To improve both the properties, the key parameters are the surface area and the specific capacitance as demonstrated by the CF electrodes. The research is going on in the direction of surface treatment of the CF to improve the two parameters and to make the CF electrodes provide the desired properties to be used in structural SC's.

2.2 Separator development

The requirement for a separator in a capacitor is high ionic conductivity while demonstrating electronic resistivity [26, 27]. Conventional materials like porous polypropylene and cellulose membranes have good ionic conductivity but they are poor in mechanical properties required for structural SC's [28, 29]. The key parameter required for achieving good mechanical properties for the separator is inter facial adhesion (IFA) of the separator with the electrodes. This IFA governs the interlaminar shear strength (ILSS). Another characteristic of the separator which affects both the electrical and mechanical properties of SC's is the thickness of the separator. A study on woven CF-epoxy capacitors revealed that with decreasing thickness, the capacitance and dielectric strength increases whereas ILSS decreases [30]. Researchers tried to strike balance between the electrical and mechanical properties of the separator by using woven glass fiber (WGF) material [31, 32]. Surface modification of separators may also provide high capacitance as well as high IFA values.

Thus, for the separator, the electrical and mechanical properties are inversely proportional to each other. With the increase in the thickness of the separator, Mechanical properties improve whereas electrical properties deteriorate. Hence striking a balance between these two properties is the key challenge for structural SC's.

2.3 Development of solid polymer electrolyte

Developing solid polymer electrolyte (SPE) is the most difficult challenge for SCS's as structural polymers are resistive to both electrons as well as ions. The requirement for SC's is that the polymer needs to be conducive for ions to be considered as solid electrolytes [33, 34]. However, with the increase in ionic conductivity, the mechanical properties of SPE deteriorate [35]. One of the alternatives explored for improving ionic conductivity in structural polymers is the addition of ionic salts. But the addition of the ionic salts adversely affected the mechanical properties [36–38]. The next alternative approach is the addition of ionic liquids (IL) with polymeric resins. IL improves the ionic conductivity and simultaneously provides thermal stability to the resins [39–42]. Shirsova explored the combination of epoxy-based SPE with IL and Li salts. The outcome was good mechanical properties with better electrical properties [43]. Another study experimented with silicon-based porous material with IL and demonstrated high energy density and mechanical strength [44].

Thus, for getting SPE's, ionic salts and IL are mixed with solid polymers to improve the ionic conductivity. Out of the two approaches, IL proves to be more promising as it provides an improvement in the mechanical properties also with the improvement in the ionic conductivity of the solid polymers.

3. Structural supercapacitor performance parameters

Structural SC performance parameters mainly include mechanical and electrical properties. Mechanical properties of SC are governed by the surface area of the electrodes. BET method is used to measure the surface area and pore size distribution for the supercapacitor surface [45]. Apart from the electrode surface area, mechanical properties also include various mechanical strengths of the supercapacitor [46, 47].

Electrochemical performance parameters supercapacitor includes specific capacitance C_{sp}, electrical contact resistance between the electrodes R_p, power density P, energy density E, and series resistance R_s [48, 49]. Specific capacitance depends upon the surface area of the electrode, pore size distribution, and the ionic strength of the electrolyte [50]. CV scan rate affects the energy density and power density of the capacitor and increasing scan rate deteriorates both the densities of the capacitor.

As the structural supercapacitor needs to have good electrochemical and mechanical properties together, the ultimate performance parameter is multifunctionality. It can be measured as the electrical conductivity with respect to the mechanical analysis [2]. As suggested by O'Brian, for a capacitor to be considered multifunctional, the combination of mechanical and electrical efficiencies (which is called multifunctional efficiency) need to be greater than unity [51].

4. Conclusion

The structural supercapacitor is an ambitious plan to expect the supercapacitor, which is primarily the power source for various applications, to act as a structural member also, thus making it multifunctional. It is not an easy task to get structural functionalities from the supercapacitor as the mechanical and electrical properties of the components of the supercapacitor are inversely related. The various approaches investigated to get multifunctionality are from the material point of view. The common research direction is to use mechanical components like a polymer matrix and use various methods to get electrical conductivity to use it as a solid electrolyte, or use electrical components like CF electrodes and improve their mechanical properties by methods like surface treatment. Mechanical design point of view, studies in this field are rare where the SC components are designed with different design approaches like honeycomb structure to improve the mechanical properties without affecting the electrochemical properties.

With these kinds of future directions, significant improvements can be achieved to overcome the major limitations of the electric propulsion technology, i.e., range and the specific energy density, and the relative contribution of the power pack to the mass of the vehicle. This could lead to a cleaner and cooler future for all of us.

References

[1] J. Zhao, C. Lai, Y. Dai, J. Xie, Pore structure control of mesoporous carbon as supercapacitor material, Mater. Lett. 61 (2007) 4639–4642. https://doi.org/10.1016/j.matlet.2007.02.071.

[2] B.K. Deka, A. Hazarika, J. Kim, Y.-B. Park, H.W. Park, Recent development and challenges of multifunctional structural supercapacitors for automotive industries, Int. J. Energy Res. 41 (2017) 1397–1411. https://doi.org/10.1002/er.3707.

[3] P.P. Andonoglou, A.D. Jannakoudakis, P.D. Jannakoudakis, E. Theodoridou, Preparation and electrocatalytic activity of rhodium modified pitch-based carbon fiber electrodes, Electrochim. Acta 44 (1998) 1455–1465. https://doi.org/10.1016/S0013-4686(98)00269-2.

[4] X.J. Zhang, H.Y. Li, Y.H. Tian, Activated carbon Fiber for super-capacitor electrode, Adv. Mater. Res. 97–101 (2010) 510–513. https://doi.org/10.4028/www.scientific.net/AMR.97-101.510.

[5] F. Ke, J. Tang, S. Guang, H. Xu, Controlling the morphology and property of carbon fiber/polyaniline composites for supercapacitor electrode materials by surface functionalization, RSC Adv. 6 (2016) 14712–14719. https://doi.org/10.1039/C5RA22208G.

[6] M. Rzepka, E. Bauer, G. Reichenauer, T. Schliermann, B. Bernhardt, K. Bohmhammel, et al., Hydrogen storage capacity of catalytically grown carbon nanofibers, J. Phys. Chem. B 109 (2005) 14979–14989. https://doi.org/10.1021/jp051371a.

[7] Y.-H. Li, Q.-Y. Li, H.-Q. Wang, Y.-G. Huang, X.-H. Zhang, Q. Wu, et al., Synthesis and electrochemical properties of nickel–manganese oxide on MWCNTs/CFP substrate as a supercapacitor electrode, Appl. Energy 153 (2015) 78–86. https://doi.org/10.1016/j.apenergy.2014.09.055.

[8] F. Severini, L. Formaro, M. Pegoraro, L. Posca, Chemical modification of carbon fiber surfaces, Carbon 40 (2002) 735–741. https://doi.org/10.1016/S0008-6223(01)00180-4.

[9] M. Ishifune, R. Suzuki, Y. Mima, K. Uchida, N. Yamashita, S. Kashimura, Novel electrochemical surface modification method of carbon fiber and its utilization to the preparation of functional electrode, Electrochim. Acta 51 (2005) 14–22. https://doi.org/10.1016/j.electacta.2005.04.002.

[10] P. Ekabutr, P. Sangsanoh, P. Rattanarat, C.W. Monroe, O. Chailapakul, P. Supaphol, Development of a disposable electrode modified with carbonized, graphene-loaded nanofiber for the detection of dopamine in human serum, J. Appl. Polym. Sci. 131 (2014). https://doi.org/10.1002/app.40858.

[11] K. Okajima, K. Ohta, M. Sudoh, Capacitance behavior of activated carbon fibers with oxygen-plasma treatment, Electrochim. Acta 50 (2005) 2227–2231. https://doi.org/10.1016/j.electacta.2004.10.005.

[12] B.K. Deka, K. Kong, Y.-B. Park, H.W. Park, Large pulsed electron beam (LPEB)-processed woven carbon fiber/ZnO nanorod/polyester resin composites, Compos. Sci. Technol. 102 (2014) 106–112. https://doi.org/10.1016/j.compscitech.2014.07.026.

[13] H. Qian, H. Diao, M. Houllé, J. Amadou, N. Shirshova, E.S. Greenhalgh, Carbon fibre modifications for composite structural power devices, Proceedings of the 15th European Conference on Composite Materials., 2012, pp. 1–5. http://www.escm.eu.org/eccm15/data/assets/1884.pdf.

[14] E. Saito, V.E. Caetano, E.F. Antunes, A.O. Lobo, F.R. Marciano, Trava-AiroldiVJ, et al., Vertically aligned carbon nanotubes/carbon Fiber composites for electrochemical applications, Mater. Sci. Forum 802 (2014) 192–196. https://doi.org/10.4028/www.scientific.net/MSF.802.192.

[15] H. Qian, A. Bismarck, E.S. Greenhalgh, M.S.P. Shaffer, Carbon nanotube grafted carbon fibres: a study of wetting and fibre fragmentation, Compos. A: Appl. Sci. Manuf. 41 (2010) 1107–1114. https://doi.org/10.1016/j.compositesa.2010.04.004.

[16] R.J. Sager, P.J. Klein, D.C. Lagoudas, Q. Zhang, J. Liu, L. Dai, et al., Effect of carbon nanotubes on the interfacial shear strength of T650 carbon fiber in an epoxy matrix, Compos. Sci. Technol. 69 (2009) 898–904. https://doi.org/10.1016/j.compscitech.2008.12.021.

[17] R. Samsur, J.S. RangariVK, L. Zhang, Z.Y. Cheng, Fabrication of carbon nanotubes grown woven carbon fiber/epoxy composites and their electrical and mechanical properties, J. Appl. Phys. 113 (2013) 214903. https://doi.org/10.1063/1.4808105.

[18] Z. Liu, J. Wang, V. Kushvaha, S. Poyraz, H. Tippur, S. Park, et al., Poptube approach for ultrafast carbon nanotube growth, Chem. Commun. 47 (2011) 9912–9914. https://doi.org/10.1039/C1CC13359D.

[19] Y.-L. Hsin, C.-F. Lin, Y.-C. Liang, K.C. Hwang, J.-C. Horng, J.A. Ho, et al., Microwave arcing induced formation and growth mechanisms of Core/Shell metal/carbon nanoparticles in organic solutions, Adv. Funct. Mater. 18 (2008) 2048–2056. https://doi.org/10.1002/adfm.200701407.

[20] A.S. Anisimov, A.G. Nasibulin, H. Jiang, P. Launois, J. Cambedouzou, S.D. Shandakov, et al., Mechanistic investigations of single-walled carbon nanotube synthesis by ferrocene vapor decomposition in carbon monoxide, Carbon 48 (2010) 380–388. https://doi.org/10.1016/j.carbon.2009.09.040.

[21] B.K. Deka, A. Hazarika, J. Kim, Y.-B. Park, H.W. Park, Multifunctional CuO nanowire embodied structural supercapacitor based on woven carbon fiber/ionic liquid–polyester resin, Compos. A: Appl. Sci. Manuf. 87 (2016) 256–262. https://doi.org/10.1016/j.compositesa.2016.05.007.

[22] N. Padmanathan, S. Selladurai, Controlled growth of spinel NiCo2O4 nanostructures on carbon cloth as a superior electrode for supercapacitors, RSC Adv. 4 (2014) 8341–8349. https://doi.org/10.1039/C3RA46399K.

[23] F. Lai, Y. Huang, Y.-E. Miao, T. Liu, Controllable preparation of multi-dimensional hybrid materials of nickel-cobalt layered double hydroxide nanorods/nanosheets on electrospun carbon nanofibers for high-performance supercapacitors, Electrochim. Acta 174 (2015) 456–463. https://doi.org/10.1016/j.electacta.2015.06.031.

[24] H. Qian, A.R. Kucernak, E.S. Greenhalgh, A. Bismarck, M.S.P. Shaffer, Multifunctional structural supercapacitor composites based on carbon aerogel modified high performance carbon Fiber fabric, ACS Appl. Mater. Interfaces 5 (2013) 6113–6122. https://doi.org/10.1021/am400947j.
[25] N. Shirshova, H. Qian, M. Houllé, J.H.G. Steinke, F.Q.P.V. KucernakARJ, et al., Multifunctional structural energy storage composite supercapacitors, Faraday Discuss. 172 (2014) 81–103. https://doi.org/10.1039/C4FD00055B.
[26] D. Pech, M. Brunet, H. Durou, P. Huang, V. Mochalin, Y. Gogotsi, et al., Ultrahigh-power micrometre-sized supercapacitors based on onion-like carbon, Nat. Nanotechnol. 5 (2010) 651–654. https://doi.org/10.1038/nnano.2010.162.
[27] A. Lewandowski, M. Galinski, Practical and theoretical limits for electrochemical double-layer capacitors, J. Power Sources 173 (2007) 822–828. https://doi.org/10.1016/j.jpowsour.2007.05.062.
[28] X.-Z. Sun, X. Zhang, B. Huang, Y.-W. Ma, Effects of separator on the electrochemical performance of electrical double-layer capacitor and hybrid battery-supercapacitor, Acta Phys.-Chim. Sin. 30 (2014) 485–491. https://doi.org/10.3866/PKU.WHXB201401131.
[29] J.R. Nair, A. Chiappone, C. Gerbaldi, V.S. Ijeri, E. Zeno, R. Bongiovanni, et al., Novel cellulose reinforcement for polymer electrolyte membranes with outstanding mechanical properties, Electrochim. Acta 57 (2011) 104–111. https://doi.org/10.1016/j.electacta.2011.03.124.
[30] T. Carlson, D. Ordéus, M. Wysocki, L.E. Asp, Structural capacitor materials made from carbon fibre epoxy composites, Compos. Sci. Technol. 70 (2010) 1135–1140. https://doi.org/10.1016/j.compscitech.2010.02.028.
[31] K.M. Kim, Y.-G. Lee, D.O. Shin, J.M. Ko, Supercapacitive properties of activated carbon electrode in potassium-polyacrylate hydrogel electrolytes, J. Appl. Electrochem. 46 (2016) 567–573. https://doi.org/10.1007/s10800-016-0927-3.
[32] N. Shirshova, H. Qian, M.S.P. Shaffer, J.H.G. Steinke, E.S. Greenhalgh, P.T. Curtis, et al., Structural composite supercapacitors, Compos. A: Appl. Sci. Manuf. 46 (2013) 96–107. https://doi.org/10.1016/j.compositesa.2012.10.007.
[33] R.C. Agrawal, G.P. Pandey, Solid polymer electrolytes: materials designing and all-solid-state battery applications: an overview, J. Phys. D. Appl. Phys. 41 (2008) 223001. https://doi.org/10.1088/0022-3727/41/22/223001.
[34] C. Meng, C. Liu, L. Chen, C. Hu, S. Fan, Highly flexible and all-solid-state Paperlike polymer supercapacitors, Nano Lett. 10 (2010) 4025–4031. https://doi.org/10.1021/nl1019672.
[35] L.E. Asp, Multifunctional composite materials for energy storage in structural load paths, Plast., Rubber Compos. 42 (2013) 144–149. https://doi.org/10.1179/1743289811Y.0000000043.
[36] J.-H. Shin, W.A. Henderson, S. Passerini, PEO-based polymer electrolytes with ionic liquids and their use in Lithium metal-polymer electrolyte batteries, J. Electrochem. Soc. 152 (2005) A978. https://doi.org/10.1149/1.1890701.
[37] D.K. Pradhan, R.N.P. Choudhary, B.K. Samantaray, Studies of dielectric and electrical properties of plasticized polymer nanocomposite electrolytes, Mater. Chem. Phys. 115 (2009) 557–561. https://doi.org/10.1016/j.matchemphys.2009.01.008.
[38] A. Javaid, K. Ho, A. Bismarck, M. Shaffer, J. Steinke, E. Greenhalgh, Multifunctional structural supercapacitors for electrical energy storage applications, J. Compos. Mater. 48 (2014) 1409–1416. https://doi.org/10.1177/0021998313487239.
[39] P. Kurzweil, M. Chwistek, Electrochemical stability of organic electrolytes in supercapacitors: spectroscopy and gas analysis of decomposition products, J. Power Sources 176 (2008) 555–567. https://doi.org/10.1016/j.jpowsour.2007.08.070.
[40] M.F. DR, N. Tachikawa, M. Forsyth, J.M. Pringle, P.C. Howlett, G.D. Elliott, et al., Energy applications of ionic liquids, Energy Environ. Sci. 7 (2014) 232–250. https://doi.org/10.1039/C3EE42099J.
[41] A. Balducci, R. Dugas, P.L. Taberna, P. Simon, D. Plée, M. Mastragostino, et al., High temperature carbon–carbon supercapacitor using ionic liquid as electrolyte, J. Power Sources 165 (2007) 922–927. https://doi.org/10.1016/j.jpowsour.2006.12.048.
[42] C. Merlet, B. Rotenberg, P.A. Madden, P.-L. Taberna, P. Simon, Y. Gogotsi, et al., On the molecular origin of supercapacitance in nanoporous carbon electrodes, Nat. Mater. 11 (2012) 306–310. https://doi.org/10.1038/nmat3260.

[43] N. Shirshova, A. Bismarck, S. Carreyette, Q.P.V. Fontana, E.S. Greenhalgh, P. Jacobsson, et al., Structural supercapacitor electrolytes based on bicontinuous ionic liquid–epoxy resin systems, J. Mater. Chem. A 1 (2013) 15300–15309. https://doi.org/10.1039/C3TA13163G.

[44] A.S. Westover, J.W. Tian, S. Bernath, L. Oakes, R. Edwards, F.N. Shabab, et al., A multifunctional load-bearing solid-state supercapacitor, Nano Lett. 14 (2014) 3197–3202. https://doi.org/10.1021/nl500531r.

[45] M.J. Bleda-Martínez, J.A. Maciá-Agulló, D. Lozano-Castelló, E. Morallón, D. Cazorla-Amorós, A. Linares-Solano, Role of surface chemistry on electric double layer capacitance of carbon materials, Carbon 43 (2005) 2677–2684. https://doi.org/10.1016/j.carbon.2005.05.027.

[46] I. De Baere, W. Van Paepegem, M. Quaresimin, J. Degrieck, On the tension–tension fatigue behaviour of a carbon reinforced thermoplastic part I: limitations of the ASTM D3039/D3479 standard, Polym. Test. 30 (2011) 625–632. https://doi.org/10.1016/j.polymertesting.2011.05.004.

[47] Y. Liang, H. Wang, X. Gu, In-plane shear response of unidirectional fiber reinforced and fabric reinforced carbon/epoxy composites, Polym. Test. 32 (2013) 594–601. https://doi.org/10.1016/j.polymertesting.2013.01.015.

[48] R. Kötz, M. Carlen, Principles and applications of electrochemical capacitors, Electrochim. Acta 45 (2000) 2483–2498. https://doi.org/10.1016/S0013-4686(00)00354-6.

[49] A.G. Pandolfo, A.F. Hollenkamp, Carbon properties and their role in supercapacitors, J. Power Sources 157 (2006) 11–27. https://doi.org/10.1016/j.jpowsour.2006.02.065.

[50] C. Kim, Electrochemical characterization of electrospun activated carbon nanofibres as an electrode in supercapacitors, J. Power Sources 142 (2005) 382–388. https://doi.org/10.1016/j.jpowsour.2004.11.013.

[51] D.J. O'Brien, D.M. Baechle, E.D. Wetzel, Design and performance of multifunctional structural composite capacitors, J. Compos. Mater. 45 (2011) 2797–2809. https://doi.org/10.1177/0021998311412207.

CHAPTER 23

Nanomaterials based solar cells

Ritesh Jaiswal[a], Anil Kumar[a], and Anshul Yadav[b]
[a]Kamla Nehru Institute of Technology, Sultanpur, Uttar Pradesh, India
[b]CSIR—Central Salt and Marine Chemicals Research Institute, Bhavnagar, Gujarat, India

Chapter outline

1. Introduction — 467
2. Working of solar cell — 468
3. Classification of solar cell — 471
   3.1 First-generation (wafer-based) solar cell — 471
   3.2 Second-generation (thin film) solar cells — 472
   3.3 Third-generation solar cells — 475
4. Nanotechnology in solar cells — 479
   4.1 Different forms of nanomaterials based solar cell — 479
5. Summary — 481
Acknowledgments — 482
References — 482

1. Introduction

Because of global warming and the widespread use of nonrenewable energy sources, including oil, natural gas, coal, etc., there has been a surge in interest in developing renewable, efficient, and clean energy sources. Different renewable energy sources like wind, wave, solar panels, etc., are being investigated for their ability to meet demand at a large scale [1]. Solar photovoltaic (PV) technology, which harnesses the sun's energy, has long been regarded as one of the most available, inexhaustible, renewable, and long-term energy sources. The cost and reliability of solar energy production, on the other hand, is the most significant impediment to moving toward an environmentally friendly and sustainable technology [2]. The third-generation PV technology has a solar cell (SC) that uses the photovoltaic effect to turn incident solar radiation into electricity directly. A solar cell can convert 20% of solar radiation from the sun into electricity [3].

Nanomaterials have recently emerged as new solar cell assembly building blocks. Academics, industry, and government officials are all buzzing about the potential of nanomaterials in solar cell applications. Nanotechnology's unique contribution to different solar energy sources is being developed, as is the desire to produce an economic and high-efficiency solar cell. Nanomaterials provide a flexible and exciting material platform for solar energy conversion, opening up new possibilities. Nanostructured materials like graphene, quantum dots, fullerene and carbon nanotubes, etc., are becoming increasingly important in solar cell applications [4]. As a result, nanostructured materials have been shown to improve solar cell performance by increasing the trapped solar radiation. The shape, size, and structure of nanomaterials significantly affect solar energy conversion efficiency [5].

Over the last 30 years, numerous studies on nanomaterials-based solar cell applications have been performed. These included metal oxide solar cells, dye-sensitized solar cells, quantum dot solar cells, and polymer nanocomposites solar cells. Graphene, Graphene derivatives, carbon nanotubes, and fullerene are nanocarbon materials that have recently been extensively studied in solar cells. The working theory of solar cells, their classification, and the various types of nanomaterials used for high-efficiency solar cells production are covered in this chapter. An illustration in Fig. 1 shows the overview of the chapter.

2. Working of solar cell

A dye-sensitized solar cell (DSSC) consist of mechanical support, which has a coating of transparent conductive oxides, a semiconductor film (TiO_2), absorbed sensitizer on the semiconductor surface, an electrolyte with redox mediator, and a counter electrode is having the capability of generating redox mediator (palatine). The schematic of the

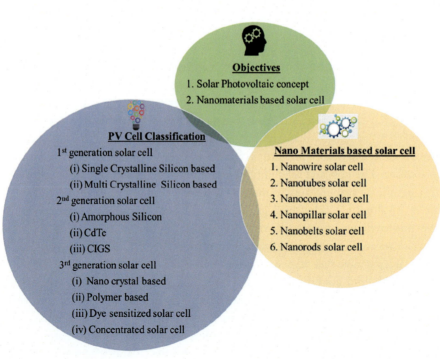

Fig. 1 Overview of the chapter.

dye-sensitized solar cell is shown in Fig. 2 [6]. Titanium oxide is widely chosen for semiconductors for image electrodes because of its various advantages. TiO_2 is an easily accessible, nontoxic, and low-cost material. Since the beginning, Ruthenium complexes, such as [Ru(4,40-dicarboxylic-2,20-bipyridine ligand)3], have been used and are still the most common. Finally, triiodide/iodine is the most popular redox pair.

The operating principle of DSSCs is explained in Fig. 3 [6]. In the first step, the photon is absorbed by the sensitizer. It causes the sensitizer to excite and inject an electron into the semiconductor's bandgap, which oxidizes the sensitizer. The electron injected passed through the semiconductor network and then through the external load to the

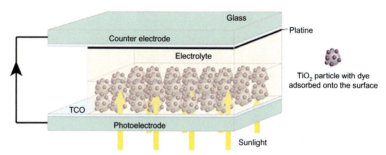

Fig. 2 A schematic of the dye-sensitized solar cell. *(Reprinted with permission from reference M.K. Nazeeruddin, E. Baranoff, M. Grätzel, Dye-sensitized solar cells: a brief overview, Sol. Energy 85 (2011) 1172–1178, https://doi.org/10.1016/j.solener.2011.01.018. Copyright 2011 Elsevier.)*

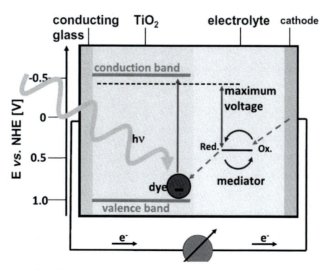

Fig. 3 Working principle of a dye-sensitized solar cell. *(Reprinted with permission from reference M.K. Nazeeruddin, E. Baranoff, M. Grätzel, Dye-sensitized solar cells: a brief overview, Sol. Energy 85 (2011) 1172–1178, https://doi.org/10.1016/j.solener.2011.01.018. Copyright 2011 Elsevier.)*

electrode counter. This process reduces the redox mediator and allows the sensitizer to regenerate.

When the device is illuminated, it functions as a stable and regenerative PV energy conversion system.

$$S_{(adsorbed)} + h\nu \rightarrow S^*_{(adsorbed)} \tag{1}$$

$$S^*_{(adsorbed)} \rightarrow S^+_{(adsorbed)} + S^-_{(injected)} \tag{2}$$

$$I_3^- + 2 \cdot e^-_{(cathode)} \rightarrow 3I^-_{(cathode)} \tag{3}$$

$$S^+_{(adsorbed)} + \frac{3}{2}I^- \rightarrow S_{(adsorbed)} + \frac{1}{2}I_3^- \tag{4}$$

There are a few unfavorable reactions that result in a decrease in cell production. The electrons injected recombine with the oxidized sensitizer (Eq. 5) or the oxidized redox pair at the TiO_2 surface (Eq. 6) are what they are.

$$S^+_{(adsorbed)} + e^-_{(TiO_2)} \rightarrow S_{(adsorbed)} \tag{5}$$

$$I_3^- + 2 \cdot e^-_{(TiO_2)} \rightarrow 3I^-_{(anode)} \tag{6}$$

The dye-sensitized solar cell's total efficiency is determined by optimization and compatibility of DSSC especially dye spectral response and film of semiconductor [7]. Surface area and semiconductor thickness are the critical factors of DSSC. They help in dye loading improvement of optical density. This results in effective light harvesting. The "external quantum efficiency" (EQE), also called incident monochromatic photon-to-current conversion efficiency (IPCE), is a crucial system characteristic. The light-harvesting efficiency (LHE) of sensitizers is compared using devices with similar architecture. LHE is defined as the ratio of electrons produced by light and the number of incident photons [8]:

$$IPCE(\lambda) = \frac{Photocurrent\ density}{Wavelength \times Photon\ flux} = LHE(\lambda) \times \varphi_{inj} \times \eta_{coll} \tag{7}$$

where $LHE(\lambda)$ is light-harvesting efficiency at wavelength λ, η_{coll} represents the efficiency of electron collection, and φ_{inj} is the quantum yield for electron injection.

The open-circuit potential (V_{OC}), photocurrent density (J_{SC}), the cell's fill factor (ff), and the strength of the incident light (IS) decide the dye-sensitized solar cell's overall conversion efficiency (η) (Eq. 8).

$$\eta_{global} = \frac{J_{SC} \cdot V_{OC} \cdot ff}{I_s} \tag{8}$$

The open-circuit photovoltage is calculated from the difference between the Fermi levels of solid under illumination. The distance between the redox pair and conduction band edge, on the other hand, is smaller for various sensitizers than the experimentally observed open-circuit potential (V_{OC}). This is due to competition for electrons between electron transport and charge recombination pathways. Understanding the mechanism of and rates of these reactions is crucial for developing successful sensitizers and, as a result, system improvement [9]. The ratio of the maximum power of device P_{max} and theoretical maximum power defines the fill factor *ff*. The fill factor is J_{SC} in short-circuit current and V_{OC} in open-circuit voltage. The range of fill factor values lies between 0 and 1. The fill factor represents losses, including electrical and electrochemical, which occur during the activity of DSSC.

3. Classification of solar cell

Depending on the materials used and the production process used, solar cells are categorized into three generations. The first-generation solar cells use single crystal silicon (Si) and bulk polycrystalline silicon (Si). These cells have solar conversion efficiencies ranging from 12% to 16%, depending on the manufacturing process and wafer quality. They dominate the solar cell market. The thin-film solar cell includes amorphous silicon (a-Si), cadmium telluride, and cadmium indium selenide, etc. These are the solar cell of the second generation that are less expensive. These solar cells have a lower conversion of solar energy than silicon wafers. The third-generation solar cell is evolving currently and includes dye-sensitized SC, polymer SC cell, quantum dots SC, perovskite SC. The third-generation solar cell produces high-performance devices at a low cost of solar cell production [10].

3.1 First-generation (wafer-based) solar cell

The first-generation solar cells were made of silicon wafers. This technology is the oldest and most used because of its high efficiency. The basic structure of silicon solar cells is shown in Fig. 4 [11]. The silicon wafer-based (first generation) technology is categorized into single and multicrystalline solar cells.

3.1.1 Single crystalline silicon solar cell

Monocrystalline silicon solar cell is the oldest and most popular solar photovoltaic technology. These are made up of a thin film of a silicon wafer or pure silicon. Monocrystalline silicon is made up of precisely arranged atoms in ordered crystal structures. Single-crystalline silicon wafers are produced in a languid and precise manner. As a result, with a conversion rate of 25%, they are one of the most costly forms of solar cells. c-Si has a low manufacturing cost of $0.38/W, in addition to module reliability and stability. The monocrystalline solar cell module has a 4%–8% more power output than the other silicon solar cell module [12].

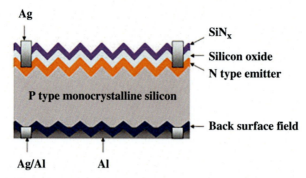

Fig. 4 Silicon solar cell structure. *(Reprinted with permission from reference C.P. Liu, M.W. Chang, C.L. Chuang, Effect of rapid thermal oxidation on structure and photoelectronic properties of silicon oxide in monocrystalline silicon solar cells, Curr. Appl. Phys. 14 (2014) 653–658, https://doi.org/10.1016/j.cap.2014.02.017. Copyright 2014 Elsevier.)*

3.1.2 Multicrystalline silicon solar cell

Poly/multicrystalline cells are made up of several monocrystalline silicon grains. The polycrystalline is manufactured by casting silicon into ingots. These Ingots are cut down into very thin wafers and then assembled to complete cells. These cells are less costly to manufacture than single crystalline cells due to the simple manufacturing process. Polycrystalline Si solar cells are more cost-effective to produce. These cells are produced by cooling the molten silicon into the graphite mold. They are slightly less costly to produce than monocrystalline silicon solar panels, but they are inefficient (12%–14%) [13].

3.2 Second-generation (thin film) solar cells

The second-generation solar cell, also called a thin-film solar cell, is cost-efficient than the first-generation silicon wafer-based solar cells. The light-absorbing layers in silicon wafer solar cells can be up to 350 m thick, whereas light-absorbing layers in thin-film solar cells are usually on the order of 1 m thick. The following are the classifications for thin-film solar cells:

3.2.1 Amorphous silicon (a-Si) solar cell

During the mid-1980s, research in low-temperature supporting materials led to the development of hydrogenated amorphous silicon deposited at 200°C. The deposition of a-Si:H is done by the plasma-enhanced chemical vapor deposition (PECVD) technique. A-Si solar cell has the advantage of producing PV electricity at a low cost, including a high value of optical absorption coefficient. The amorphous silicon solar cell does not significantly share in the global market of photovoltaic technology due to its low efficiency of 6%. The reason behind the modest stable efficiency is the "Staebler–Wronski effect," which is based on the degradation of the initial module efficiency to the stabilized module efficiency. The other reason for the low efficiency of amorphous silicon solar

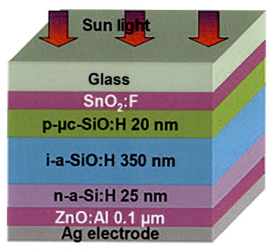

Fig. 5 Amorphous silicon solar cell device structure. *(Reprinted with permission from reference J. Sritharathikhun, A. Moollakorn, S. Kittisontirak, A. Limmanee, K. Sriprapha, High quality hydrogenated amorphous silicon oxide film and its application in thin film silicon solar cells, Curr. Appl. Phys. 11 (2011) S17–S20, https://doi.org/10.1016/j.cap.2010.11.100. Copyright 2011 Elsevier.)*

cells is a manufacturing problem with a broad substrate like transparent conductive oxide layer and non-uniformity in silicon film [14]. The structural configuration of the amorphous solar cell is shown in Fig. 5 [15].

3.2.2 Cadmium telluride (CdTe) solar cell

Cadmium telluride (CdTe) solar cell is a kind of thin-film solar cell. It is both cost-effective and commercially viable. CdTe has a high value of optical absorption coefficient with good chemical stability and bandgap of 1.5 eV. The properties of CdTe make it the most attractive material for thin-film solar cell design. CdTe is a crystalline compound semiconductor with a direct bandgap that makes light absorption easier and increases efficiency. A p-n junction diode is usually made by sandwiching cadmium sulfide layers between them. The configuration of the CdTe solar cell is shown in Fig. 6 [16]. There are three stages in the manufacturing process: First, polycrystalline materials are used to make CdTe-based solar cells, with glass as the substrate. The second stage is deposition, which involves multiple coating of CdTe solar cells using various cost-effective methods. As previously mentioned, CdTe has a 1.45 eV direct bandgap and a high absorption coefficient of over 5×10^{15}/cm [17]. As a result, its performance usually ranges between 9% and 11%. The cadmium component of solar cells, on the other hand, raises environmental concerns. Cadmium is a heavy metal that can accumulate in humans, animals, and plants, making it potentially toxic. The disposal and recycling of hazardous Cd-based materials can be very costly and harmful to our environment and community. As a result, the key problem with this CdTe technology is environmental hazards associated with the use of cadmium, and hence its supply is restricted [10].

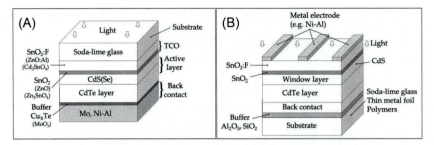

Fig. 6 CdTe solar cell configurations: (A) superstrate and (B) substrate. *(Reprinted from reference A. Bosio, S. Pasini, N. Romeo, The history of photovoltaics with emphasis on CdTe solar cells and modules, Coatings 10 (2020) Article No. 344, https://doi.org/10.3390/coatings10040344.)*

3.2.3 Copper indium gallium di-selenide (CIGS) solar cells

Copper indium gallium selenide solar cell is a thin-film technology that converts sunlight into electricity. It is also known as CIGS cell, CIS cell, or CI(G)S cell. A copper, indium, gallium, and selenium film is deposited on plastic or glass backing with electrodes on the front and back to absorb current. The schematic of the CIGS solar cell is shown in Fig. 7 [18]. Since the material absorbs sunlight so well and has such a large absorption coefficient, it needs a thinner film than other semiconductor materials. CIGS layers may be deposited on many flexible substrates because they are thin enough to be versatile. However, since both of these technologies depend on high-temperature deposition techniques, cells deposited on glass produce the best results, despite advances in

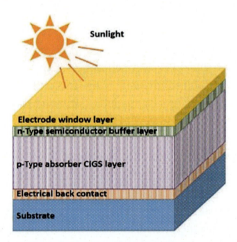

Fig. 7 Schematic of CIGS solar cell. *(Reprinted with permission from reference N. Mufti, T. Amrillah, A. Taufiq, S. Aripriharta, M. Diantoro, Zulhadjri, H. Nur, Review of CIGS-based solar cells manufacturing by structural engineering, Sol. Energy 207 (2020) 1146–1157, https://doi.org/10.1016/j.solener.2020.07.065. Copyright 2020 Elsevier.)*

low-temperature CIGS cell deposition erasing much of the performance difference. CIGS outperformed polysilicon at the cell stage. The performance of CIGS is still lower due to less mature upscaling. Polycrystalline thin film is the most popular form of CIGS.

The best efficiency obtained by the CIGS solar cell is 19.9% by adjusting the CIGS surface and making it look like CIS [19]. CIGS cell is produced by different methods like evaporation, sputtering, printing, electrochemical coating, printing, and electron beam deposition technique. Sputtering may be a single, double or multistep reactive process. Sputtering requires the deposition of materials and later interaction with selenium. Evaporation, on the other hand, is similar to sputtering. It can be used in one, two, or several stages of processing.

3.3 Third-generation solar cells

The third-generation PV solar cells are a promising new technology that is yet thoroughly investigated commercially. The following are the most popular third-generation solar cell types:

3.3.1 Nanocrystal based solar cells

Organic cell based on nanocrystal is a more economical alternative to the conventional inorganic solar cell. A cost-effective alternative of colloidal quantum dots has been used for developing hybrid solar cells due to its excellent solution processing performance, compatibility with conjugated polymer, and well-suited optical property. Nanocrystal exhibit few same properties as bulk inorganic semiconductors. But they are easy to work with than polymer. The photoactive layer is inserted between two electrodes with separate work functions in these solar cells, which work on electrochemical principles. The most critical factors in a solar cell's efficiency are the generation of charge. The primary excitations must be separated into the charge carriers, and they should be transferred to appropriate electrodes, with recombination losses minimized, to collect charge carriers efficiently [20].

3.3.2 Polymer-based solar cells

Low-cost electronics and photovoltaic (PV) applications may benefit from conjugated polymers. The efficiency of solar power conversion of 5% has recently been achieved with polymer-based solar cells. Fig. 8 shows the basic configuration of polymer-based solar cells [21]. Screen printing, inkjet printing, doctor balding, and spray deposition are used to deposit organic on the PV module. Throughout high roll-to-roll processing is needed for these techniques, which bring the cost of polymer-based PV almost closer to electricity produced by the grid. Furthermore, since all of these deposition processes occur at low temperatures, they can be used to fabricate devices on flexible plastic substrates. Lightweight and flexibility and the inherent economics of high-throughput

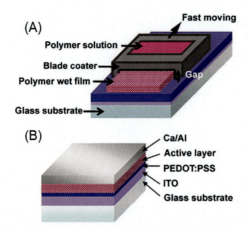

Fig. 8 Configuration of the polymer-based solar cell. *(Reprinted with permission from reference Y.H. Chang, S.R. Tseng, C.Y. Chen, H.F. Meng, E.C. Chen, S.F. Horng, C.S. Hsu, Polymer solar cell by blade coating, Org. Electron. 10 (2009) 741–746, https://doi.org/10.1016/j.orgel.2009.03.001. Copyright 2009 Elsevier.)*

processing are said to lower PV panel installation costs, resulting in a price reduction. New niche markets, such as portable power generation and PV in building design, are also generated by flexible PV. A mixture of C60, poly [2-methoxy-5-(2′-ethylhexyloxy)-p-phenylene vinylene] (PPV), and other derivatives was used to produce the first polymer-based solar cell with high power conversion efficiency [17]. Researchers were able to achieve 3.0% efficiency for PPV-style PSCs after significantly optimizing the parameters. PSCs' unique properties have opened the door to a new possibility for stretchable solar products developments like fabrics and textiles. Polarizing organic photovoltaic (ZOPVs) is a modern recycling concept developed to increase liquid crystal display function. The functioning of liquid crystal display increased when it is used with a polarizer, proper light condition, solar panel, and a photovoltaic system [22].

3.3.3 Dye-sensitized solar cells

A dye-sensitized solar cell (DSSCs) is classified under a thin-film solar cell. It has been the subject of extensive research for more than two due to its low cost, nontoxicity, simple preparation technique, and ease of production. In 1972, the first solar cell used was chlorophyll-sensitized zinc oxide (ZnO) electrode. Photons were converted into electricity for the first time by injecting electrons from excited dye molecules into semiconductors with a large bandgap. Since most of the research has been performed on ZnO crystal, the poor result is shown by the dye-sensitized solar cell for a monolayer of dye molecules absorbing just 1% of incident solar radiation.

The performance of the fine oxide powder electrode was improved by absorbing more dye across the electrodes. This process increases the light-harvesting efficiency (LHE). In 1991, a significant solar conversion efficiency of 7% was invented for DSSCs

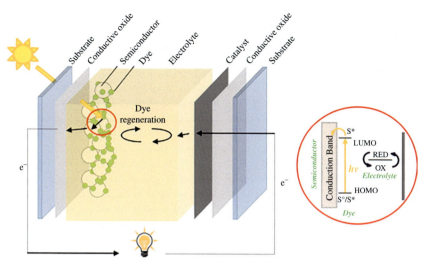

Fig. 9 A schematic diagram for the dye-sensitized solar cell. *(Reprinted with permission from reference M.L. Parisi, S. Maranghi, R. Basosi, The evolution of the dye sensitized solar cells from Grätzel prototype to up-scaled solar applications: a life cycle assessment approach, Renew. Sustain. Energy Rev., 39 (2014) 124–138. https://doi.org/10.1016/j.rser.2014.07.079, Copyright 2014 Elsevier.)*

for nonporous titanium dioxide (TiO_2) electrodes with a roughness factor of about 1000 [23]. However, some challenges persist for DSSCs, like stability issues due to the degradation of dye molecules. The low stability is due to the low optical absorption of the sensitizer. As a result, the poor conversion efficiency is achieved. The DSSSCs solar cells' lifetime is reduced as the dye molecules degrade when it is exposed to infrared and ultraviolet radiation. Thus, a coating is made on the surface of dye molecules to reduce degradation. However, it increases manufacturing costs and lowers DSSCs solar cell conversion efficiency [24,25] (Fig. 9).

3.3.4 Concentrated solar cells

Since the 1970s, concentrating photovoltaic (CPV) technology has been used for solar cell development. It's the most recent breakthrough in research and development for solar cells. The main focus of concentrated solar cells is to collect a high amount of the sun's energy into a small region of PV solar cells. The CPV technology is based on optics. It works by using large mirrors and lens arrangements to concentrate sun rays on a small solar cell area. As the sun's energy is converted into a small region, energy is produced in bulk. The heat energy is utilized by an integrated power generator to run the heat engine. Depending on the intensity of the lens system, the CPV can be categorized into low, medium, and high CPV solar cells. The highest solar efficiency is achieved by this kind of solar cell which is approximately 40%. CPV solar cells have many advantages, which include fast response time, no moving parts, scalability across a wide range of sizes [26] (Table 1).

Table 1 Comparison of different types of solar cells [6,9,10,27–29].

Cell type		Efficiency (%)	High-temperature performance	Size	Cost
Crystalline silicon	Monocrystalline	14–17.5	Not Good	For the same amount of powerless volume is used	2 × costlier than thin film
	Polycrystalline	12–14	Not Good	For the same amount of power, less volume is used	2 × costlier than thin film
Thin film	Amorphous silicon	4–8	Good	Product design range is wide	0.5 × costlier than silicon cells
	CIGS	10–12	Good	Product design range is wide	0.5 × costlier than silicon cells
	CdTe	9–11	Good	Product design range is wide	0.5 × costlier than silicon cells
Third generation	Nanocrystal	7–8	Excellent thermal stability	Product design range is wide	0.5 × costlier than silicon cells
	Dye-sensitized	10	Not Good	Product design range is wide	0.5 × costlier than silicon cells
	Polymer	3–10	Not Good	Product design range is wide	0.5 × costlier than silicon cells
	Concentrated	40	Excellent thermal stability	Product design range is wide	0.5 × costlier than silicon cells
Perovskite		31	Excellent thermal stability	Product design range is wide	0.5 × costlier than silicon cells

4. Nanotechnology in solar cells

The nanotechnology-based solar cell absorbs both outdoor and indoor light and converts it into electricity. Plastic is made up of nanoscale titanium particles coated with photovoltaic dyes that produce electricity when exposed to light. Improvements like this will make it possible to produce low-cost solar cells that are as efficient as or more efficient than current technology. Since one of the most significant disadvantages of traditional solar cells is their high production cost, this innovative technology can significantly impact our everyday lives. Even if this new technology were only capable of providing enough energy to low-power devices, the societal consequences would be enormous.

Wireless capabilities can aid in environmental protection, minimize soldier carrying loads, provide power to rural areas, and have various commercial applications. New solar cell technology may also have military consequences. According to Konarka's executive vice president, Daniel McGahn, "In today's world, a typical field soldier holds 1.5 pounds of batteries. A special operations soldier has more time off and must bear 140 pounds of gear, including 60–70 pounds of batteries". Soldier mobility can be significantly improved by using nanotechnology to make highly efficient and low-cost solar cells. Less expensive solar cells may also help rural areas and third-world countries get electricity. The electricity grid cannot be connected to these areas because the energy demand is insufficient, and the areas are too far apart. In this setting, solar energy, on the other hand, thrives. By making solar energy inexpensive, it can be used for electricity, hot water, and cooking. Nanotechnology in the solar cell has raised millions of people's living standards across the globe [5].

4.1 Different forms of nanomaterials based solar cell

4.1.1 Nanowires solar cell

Nanowire solar cells are advantageous because a dense absorption is allowed to absorb on the entire solar spectrum. It also allows fast diffusion between the two active compounds for a small distance. The solar cell is made up of vertical n-type ZnO nanowires that are encased in a film of p-type Cu_2O nanoparticles. The solution processing techniques are used to develop n-type ZnO nanowires solar cells. A blocking layer of TiO_2 is deposited on the ZnO nanowires to get the working solar cell. It is done before Cu_2O film could expand. Since there is no TiO2, a shunt pathway is formed, bypassing the ZnO nanowires. To bypass the shunt pathways, TiO_2 is used as it increases the shunt resistance. Despite this, the efficiency of converting this kind of solar cell is very low, approximately 0.053%. It is due to the presence of shunt pathways and parasitic phases of Cu_2O films [30].

4.1.2 Nanotubes solar cell

Carbon allotropes with a cylinder-shaped nanostructure are known as carbon nanotubes (CNTs). Nanotubes have a length-to-diameter ratio of up to 132,000,000:1, far exceeding that of any other material. Carbon nanotubes have a unique property useful in nanotechnology, electronics, optics, and other technological fields. The use of carbon nanotubes in solar panel photovoltaic can increase efficiency up to 80%. The photovoltaic effect in solar panels occurs when an electric voltage is created or, more significantly, electric current flow in the closed circuit. The photovoltaic effect is almost close to the photoelectric effect in many ways. When the solar panel is exposed to the sun, i.e., stream of photon incident on it, the electron is emitted and transferred between different energy bands of atoms. The transition of the energy state of the electron takes place from valence bond to conduction band. This process occurs only within the material of the solar panel. The evolution of electrons occurs due to the electron transfer, resulting in a voltage build-up between the two electrodes. In thin-film solar panels, double-walled carbon nanotubes are used as charge carrier collecting and photon generation sites. A semitransparent thin film of nanotubes is coated on the n-type crystalline substrate. A high-density p-n junction is formed between n-Si and nanotubes. It facilitates the extraction of electrons and holes (via n-Si through nanotubes) and charges separation. The efficiency of solar PV cells can be improved up to 7%–8% by nanotubes solar photovoltaic technology [31].

4.1.3 Nanocones solar cell
Nanocones are nanomaterials that can be used to improve light absorption and, hence, PV efficiency. These materials have excellent properties due to their high refractive index. Each cone of nanocones solar cell is made of a material having insulating properties inside and conducting properties outside. Under the microscope, these materials show the mass of bullets standing with having a flat base at the top end. Nanocones, similar to other topological insulators, use oscillation induced by changes in electron concentration caused by photons hitting the material. They allow material to have superior light absorption properties, making it suitable for a broad range of photovoltaic applications, including optical fibers, waveguides, and even lenses. Each cone consists of a dielectric core and metal shell coating—a material made from them will have superior light absorption properties, making it suitable for a wide range of PV applications, from optical fibers to waveguides. According to the researchers, Nanocones in thin-film solar cells can enhance light absorption by 15% in visible and UV ranges [32].

4.1.4 Nanopillar solar cell
In the field of nanotechnology, nanopillars are a new technology. Nanopillars are pillar-shaped nanostructures with a diameter of about 10 nm that can be arranged in lattice-like arrays [33]. Nanopillars are very good at absorbing light because of their tapered ends. Nanopillar-coated solar collector surfaces are three times more effective than nanowire solar cells. As compared to standard semiconductive materials, nanopillars need less material to build a solar cell. They also hold up well during the solar panel manufacturing

process. Because of this durability, solar panels can be made from less costly materials and using less expensive processes. Researchers are looking into putting dopants in the bottoms of nanopillars to increase the number of times photons bounce around the pillars and hence the amount of light captured. Nanopillars in solar panels can make them more flexible and effective at capturing light. Manufacturers have more options on how their solar panels are shaped, which reduces costs by reducing how delicately the panels must be treated. Scientists have yet to mass-produce nanopillars, although they are more efficient and less costly than traditional materials. This is a significant drawback to using nanopillars in the manufacturing process [34].

4.1.5 Nanobelt solar cell

CdSe nanobelts were covered with transparent graphene, or carbon nanotube (CNT) membranes, at specific locations. This process has generated exciting Schottky junction solar cells using different connections and configurations of single and multiple nanobelts assemblies. Firstly, CdSe is deposited, and graphene is moved, and finally, a contact of Ag paste is formed. Nanobelts solar cell manufacturing is based on chemical vapor deposition (CVD), a low-cost manufacturing process. In addition, these are flexible thin-film solar PV cells. Nanobelts solar cell achieved a conversion efficiency of 0.1% at an open-circuit voltage of 0.5 V and short circuit current density (Jsc) of 0.94 mA/cm^2. It's worth mentioning that the fill factor's low value (0.1%) has a big impact on the conversion performance (less than 23.7%). This parameter must be improved to check the solar cell function as an alternative to conventional ones and increase conversion efficiency sensibly [35].

4.1.6 Nanorods solar cell

Nanorod/wire frameworks have sparked interest in implementation and integration into devices due to their benefits. A well-oriented nanorods/wires (DSC) with dense arrays must fabricate nanorod/wire solar cells. Nanowires are nanorods having a higher aspect ratio than nanorods. Nanorods are the same as nanowires for the main purpose of simplification. Nanowires can be made using the same fabrication techniques that are used to make nanorods. The fabrication of nanorod structures is divided into two stages: catalyst or seed deposition and growth. The most common and reliable method for rising silicon nanorods is vapor-liquid-solid (VLS). At the same time, VLS and solution-based chemical synthesis are also viable options for DSC materials [36].

5. Summary

Solar energy has evolved as one of the most demanding sources of energy. It has several benefits over the conventional source of energy like petroleum and fossil fuels. It is a promising and consistent solution for meeting the high-energy demand. Thin-film solar cell modules are rapidly entering the market, providing incentives for these potentially lower-cost approaches to develop themselves. Several different thin-film technologies based on amorphous silicon, polycrystalline, or mixed phases, etc., are now visible or are on the verge of being so. Nanotechnology appears to be a viable option for fabricating commonly used, economically cheaper, and high-performance solar cells. It is primarily since these processes are performed at low temperatures, which significantly decreases the energy bill for photovoltaic cell manufacturing. Even though traditional solar cells have higher conversion efficiency than nanomaterial-based solar cells, they remain more appealing due to their lower production costs and potential for widespread use in daily life.

Acknowledgments

The manuscript has been assigned CSIR-CSMCRI PRIS no. 167/2021.

References

[1] R. Singh, A. Kumar, A. Yadav, Performance analysis of the solar photovoltaic thermal system using phase change material, in: IOP Conference Series: Materials Science and Engineering, 2019, https://doi.org/10.1088/1757-899X/577/1/012166.
[2] C.C. Raj, R. Prasanth, A critical review of recent developments in nanomaterials for photoelectrodes in dye sensitized solar cells, J. Power Sources 317 (2016) 120–132, https://doi.org/10.1016/j.jpowsour.2016.03.016.
[3] B. O'Regan, M. Grätzel, A low-cost, high-efficiency solar cell based on dye-sensitized colloidal TiO_2 films, Nature 353 (6346) (1991) 737–740, https://doi.org/10.1038/353737a0.
[4] V. Sugathan, E. John, K. Sudhakar, Recent improvements in dye sensitized solar cells: a review, Renew. Sust. Energ. Rev. 52 (2015) 54–64, https://doi.org/10.1016/j.rser.2015.07.076.
[5] F. Ghasemzadeh, M.E. Shayan, Nanotechnology in the service of solar energy systems. Nanotechnol. Environ. 59 (2020). https://doi.org/10.5772/intechopen.93014.
[6] M.K. Nazeeruddin, E. Baranoff, M. Grätzel, Dye-sensitized solar cells: a brief overview, Sol. Energy 85 (2011) 1172–1178, https://doi.org/10.1016/j.solener.2011.01.018.
[7] D. Cahen, G. Hodes, M. Grätzel, J.F. Guillemoles, I. Riess, Nature of photovoltaic action in dye-sensitized solar cells, J. Phys. Chem. B 104 (2000) 2053–2059, https://doi.org/10.1021/jp993187t.
[8] J. Gong, S. Krishnan, Chapter 2—mathematical modeling of dye-sensitized solar cells, in: M. Soroush, K.K.S. Lau (Eds.), Dye. Sol. Cells, Academic Press, 2019, pp. 51–81, https://doi.org/10.1016/B978-0-12-814541-8.00002-1.
[9] P. Würfel, U. Würfel, Physics of Solar Cells: From Basic Principles to Advanced Concepts, John Wiley & Sons, 2016.
[10] A. Sharma, S.K.K. Jain, Sharma, Solar cells: in research and applications—a review, Mater. Sci. Appl. 29 (2017) 762–770.
[11] C.P. Liu, M.W. Chang, C.L. Chuang, ffect of rapid thermal oxidation on structure and photoelectronic properties of silicon oxide in monocrystalline silicon solar cells, Curr. Appl. Phys. 14 (5) (2014) 653–658, https://doi.org/10.1016/j.cap.2014.02.017.
[12] S. Sundaram, D. Benson, T.K. Mallick, Chapter 2—overview of the PV industry and different technologies, in: S. Sundaram, D. Benson, T.K. Mallick (Eds.), Sol. Photovolt. Technol. Prod., Academic Press, 2016, pp. 7–22, https://doi.org/10.1016/B978-0-12-802953-4.00002-0.

[13] S.A. Kalogirou, Chapter nine—photovoltaic systems, in: S.A. Kalogirou (Ed.), Solar Energy Engineering, Academic Press, Boston, 2009, pp. 469–519, https://doi.org/10.1016/B978-0-12-374501-9.00009-1.
[14] A.G. Aberle, Thin-film solar cells, Thin Solid Films 517 (2009) 4706–4710, https://doi.org/10.1016/j.tsf.2009.03.056.
[15] J. Sritharathikhun, A. Moollakorn, S. Kittisontirak, A. Limmanee, K. Sriprapha, High quality hydrogenated amorphous silicon oxide film and its application in thin film silicon solar cells, Curr. Appl. Phys. 11 (1) (2011) S17–S20, https://doi.org/10.1016/j.cap.2010.11.100.
[16] A. Bosio, S. Pasini, N. Romeo, The history of photovoltaics with emphasis on CdTe solar cells and modules, Coatings 10 (4) (2020), https://doi.org/10.3390/coatings10040344, 344.
[17] K.M. Elsabawy, W.F. El-Hawary, M.S. Refat, Advanced synthesis of titanium-doped-tellerium-cadmium mixtures for high performance solar cell applications as one of renewable source of energy, Int. J. Chem. Sci. 10 (2012) 1869–1879.
[18] Mufti N., T. Amrillah, A. Taufiq, S. Aripriharta, M. Diantoro, Zulhadjri, H. Nur, Review of CIGS-based solar cells manufacturing by structural engineering, Solar Energy 207 (1) (2020) 1146–1157, https://doi.org/10.1016/j.solener.2020.07.065.
[19] M.A. Green, Y. Hishikawa, W. Warta, E.D. Dunlop, D.H. Levi, J. Hohl-Ebinger, A.W.H. Ho-Baillie, Solar cell efficiency tables (version 50), Prog. Photovolt. Res. Appl. 25 (2017) 668–676, https://doi.org/10.1002/pip.2909.
[20] S. Kumar, G.D. Scholes, Colloidal nanocrystal solar cells, Microchim. Acta 160 (2008) 315–325, https://doi.org/10.1007/s00604-007-0806-z.
[21] Y.H. Chang, S.R. Tseng, C.Y. Chen, H.F. Meng, E.C. Chen, S.F. Horng, C.S. Hsu, Polymer solar cell by blade coating, Org. Electron. 10 (5) (2009) 741–746, https://doi.org/10.1016/j.orgel.2009.03.001.
[22] P.E. Henderson, United States Patent (19), (1993).
[23] K. Sharma, V. Sharma, S.S. Sharma, Dye-sensitized solar cells: fundamentals and current Status, Nanoscale Res. Lett. 13 (2018), https://doi.org/10.1186/s11671-018-2760-6.
[24] M. Bertolli, Solar Cell Materials, 2014, https://doi.org/10.1002/9781118695784.
[25] M.L. Parisi, S. Maranghi, R. Basosi, The evolution of the dye sensitized solar cells from Grätzel prototype to up-scaled solar applications: a life cycle assessment approach, Renew. Sustain. Energy Rev. 39 (2014) 124–138, https://doi.org/10.1016/j.rser.2014.07.079.
[26] W.T. Xie, Y.J. Dai, R.Z. Wang, K. Sumathy, Concentrated solar energy applications using Fresnel lenses: a review, Renew. Sust. Energ. Rev. 15 (2011) 2588–2606, https://doi.org/10.1016/j.rser.2011.03.031.
[27] B. Srinivas, S. Balaji, M. Nagendra Babu, Y.S. Reddy, Review on present and advance materials for solar cells, Int. J. Eng. Res. 3 (2015) 178–182.
[28] M. Kouhnavard, S. Ikeda, N.A. Ludin, N.B. Ahmad Khairudin, B.V. Ghaffari, M.A. Mat-Teridi, M.A. Ibrahim, S. Sepeai, K. Sopian, A review of semiconductor materials as sensitizers for quantum dot-sensitized solar cells, Renew. Sust. Energ. Rev. 37 (2014) 397–407, https://doi.org/10.1016/j.rser.2014.05.023.
[29] L.J. Brennan, M.T. Byrne, M. Bari, Y.K. Gun'ko, Carbon nanomaterials for dye-sensitized solar cell applications: a bright future, Adv. Energy Mater. 1 (2011) 472–485, https://doi.org/10.1002/aenm.201100136.
[30] T. Fix, Oxide and Ferroelectric Solar Cells, Elsevier Inc., 2019, https://doi.org/10.1016/B978-0-12-814501-2.00002-5.
[31] G. Conibeer, A. Willoughby, Solar Cell Materials: Developing Technologies, 2014, https://doi.org/10.1002/9781118695784.
[32] Z. Yue, B. Cai, L. Wang, X. Wang, M. Gu, Intrinsically core-shell plasmonic dielectric nanostructures with ultrahigh refractive index, Sci. Adv. 2 (2016) 2–3, https://doi.org/10.1126/sciadv.1501536.
[33] Z. Fan, R. Kapadia, P.W. Leu, X. Zhang, Y.L. Chueh, K. Takei, K. Yu, A. Jamshidi, A.A. Rathore, D.J. Ruebusch, M. Wu, A. Javey, Ordered arrays of dual-diameter nanopillars for maximized optical absorption, Nano Lett. 10 (2010) 3823–3827, https://doi.org/10.1021/nl1010788.
[34] R.Z. Tala-ighil, Handbook of Nanoelectrochemistry, 2016, https://doi.org/10.1007/978-3-319-15207-3.

[35] L. Zhang, L. Fan, Z. Li, E. Shi, X. Li, H. Li, C. Ji, Y. Jia, J. Wei, K. Wang, H. Zhu, D. Wu, A. Cao, Graphene-CdSe nanobelt solar cells with tunable configurations, Nano Res. 4 (2011) 891–900, https://doi.org/10.1007/s12274-011-0145-6.

[36] Y. Hames, Z. Alpaslan, A. Kösemen, S.E. San, Y. Yerli, Electrochemically grown ZnO nanorods for hybrid solar cell applications, Sol. Energy 84 (2010) 426–431, https://doi.org/10.1016/j.solener.2009.12.013.

CHAPTER 24

Two dimensional MXenes for highly stable and efficient perovskite solar cells

Sahil Gasso, Manreet Kaur Sohal, Navdeep Kaur, and Aman Mahajan
Department of Physics, Guru Nanak Dev University, Amritsar, Punjab, India

Chapter outline

1. Introduction — 485
2. Perovskite solar cells — 488
3. Issues with PSCs and their solutions — 491
   - 3.1. Recombination reactions — 491
   - 3.2. Stability — 492
   - 3.3. High cost — 492
4. Introduction to 2D materials — 493
   - 4.1. MXenes — 493
5. Mxene additive in PSCs — 495
   - 5.1. Recombination reactions — 495
   - 5.2. Doping of transition metal in Mxenes — 498
   - 5.3. Nucleation and growth of perovskite layer — 498
   - 5.4. Cost-effective HTM and electrodes — 500
   - 5.5. Encapsulation of PSCs — 501
6. Future scope — 502
7. Conclusion — 503
- Acknowledgement — 503
- Conflict of interest — 503
- References — 504

1. Introduction

The era of electronics needs abundant energy resources to power up innumerable devices, ranging from miniaturized gadgets to gigantic machines. Traditionally, this energy has been produced from nonrenewable energy resources like coal, crude oil, and natural gas. Since these resources are limited in amount, their overuse in a relatively short time has led to a global energy crisis. Growing population and industrialization have further fueled the existing problem due to the emission of greenhouse gases that leads to drastic variation in the climate [1]. Thus, there is an urgent need to replace current energy resources with some environmentally friendly resources to meet the energy demand. In this respect, renewable energy resources, mainly solar energy, have gained considerable attention to power up the urban as well as remote areas at a comparatively lower cost using photovoltaic (PV) devices (Fig. 1) [2].

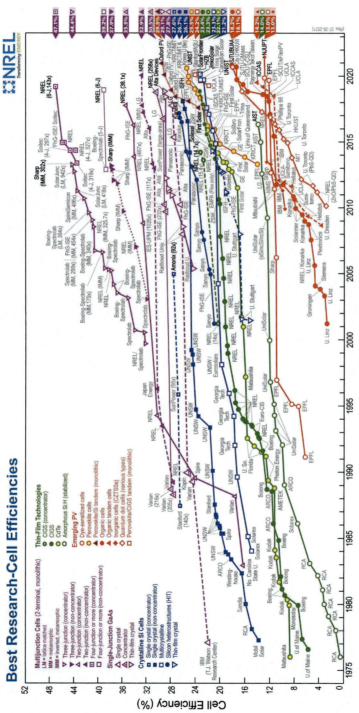

Fig. 1 Evolution of power conversion efficiency of different solar cells from 1975 to 2020.

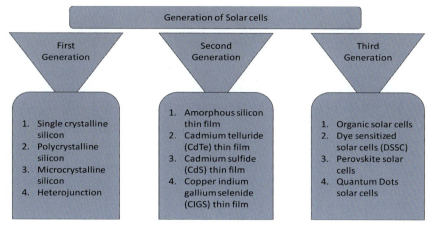

Fig. 2 Three generations of solar cells with their sub categories.

The PV industry has observed a research boom after the practical fabrication of the first silicon solar cell by Russel Ohl in 1946. The PV technology has evolved dramatically with the development of various organic and inorganic materials used in these devices and has been classified into three generations (Fig. 2). First-generation solar cells employ wafer-based single-junction crystalline Si material with single-crystalline, polycrystalline, microcrystalline, and heterojunction with an intrinsic thin layer (HIT), as active layers and have achieved a maximum power conversion efficiency (PCE) of 28% [3]. However, they are highly expensive because of high fabrication cost of Si and involve wastage of a large amount of solar energy in the form of heat. Thus second-generation solar cells emerged, which involve thin films of amorphous Si, semiconductor compounds, such as cadmium sulfide (CdS) and cadmium telluride (CdTe), single-junction gallium arsenide (GaAs), and copper indium gallium selenide (CIGS) [4]. But the complex manufacturing process, high cost, and toxic nature of these materials restrict their commercialization. Then comes the third-generation solar cells that comprise organic solar cells (OSCs), dye sensitized solar cells (DSSCs), perovskite solar cells (PSCs), and quantum dot solar cells (QDSSCs) with maximum PCE of 15.6%, 13.3%, and 25.2%, respectively [5, 6]. Among them, PSCs are relatively new in the solar cell space and utilize perovskite materials, namely methylammonium lead iodide ($MAPbI_3$) and methylammonium lead bromide ($MAPbBr_3$), as photosensitizers. However, PSCs show lesser stability and durability along with a comparatively higher recombination rate and costs. To overcome these issues, worldwide research efforts have been made to explore 2D MXenes as potential candidates in PSCs. This chapter gives the basic overview of PSCs, their structure and components, and presents the issues faced in the development of efficient PSCs. In addition, the potential use of MXene to enhance the PSCs' efficiency and stability has also been discussed in detail.

2. Perovskite solar cells

A typical PSC constitutes a perovskite absorber layer, an electron transport layer (ETL), a hole transport layer (HTL), and charge collecting electrodes. The schematic of the working mechanism of PSCs is shown in Fig. 3. The incident sunlight absorbed by the perovskite material results in the generation of excitons that are then dissociated into electrons and holes at the interface of the HTL and ETL, followed by their respective diffusion toward anode and cathode. The transportation of these photogenerated charge carriers in the external circuit produces electrical energy. The performance of PSCs significantly depends upon the configuration of different components, the morphology of each functional layer, and interfacial characteristics.

The perovskite absorber layer is made of a perovskite material with an ABX3 structure, where A and B are cations of variable size with A > B and X is an anion. The B cation and X halide are involved in forming corner-sharing BX6 octahedron, and the A is in the octahedral cage and coordinates 12-fold X in the unit cell (Fig. 4). The bandgap of the standard perovskite material $MAPbI_3$ can be easily engineered by altering its composition. The bandgap of $MAPbI_3$ has been observed to reduce to a desired value matching the Shockley–Queisser limit, by replacing MA+ with FA+, as

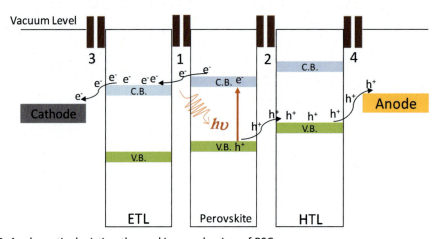

Fig. 3 A schematic depicting the working mechanism of PSCs.

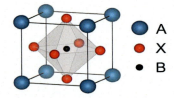

Fig. 4 Standard ABX3 structure of perovskite.

it leads to the lattice expansion, which improves the symmetry from the tetragonal phase of MAPbI$_3$ to the cubic phase of FAPbI$_3$ [7]. Although this structure produces comparable PCE to standard PSCs, FAPbI$_3$ undergoes instability of α-phase. Further, the mixed anion and mixed cation structures, i.e., (FAPbI$_3$)$_{0.85}$(MAPbBr$_3$)$_{0.15}$, have been explored and found to potentially improve phase stability in PSCs [8]. The unique compositional engineering of perovskite materials provides them the ability to exhibit tunable bandgaps, broader absorption of sunlight spectra from the UV–Vis-NIR region, high absorption coefficient, a larger distance between electrons and holes (>1 nm) leading to their longer lifetimes (~273 ns) at respective interlayers and faster charge transfer processes. In general, the stability parameter of perovskite materials is identified from the octahedral (μ) and Goldschmidt (t) tolerance factor [9]. Here, μ is obtained from the ratio of the radii of cation B (R_B) to anion X (R_x), i.e., (R_B/R_X), and t is obtained from $(R_A + R_X)/\sqrt{2}(R_B + R_X)$, where R_A is the ionic radii of cation A. The structure of perovskite depends upon the variation of these two parameters and generally their values vary from $0.44 < \mu < 0.90$ and $0.81 < t < 0.9$. The most interesting perovskite materials are organic-inorganic mixed halides because of their favorable transformation from being an insulator to metal by increasing the number of inorganic layers. Here, A is an organic cation with radii in the range of 0.19–0.22 nm, e.g., methyl ammonium ($CH_3NH_3^+$), ethyl ammonium ($CH_3CH_2NH_3^+$), and formamidium ($NH_2CH=NH_2^+$); and X is generally I, Br, and Cl with their radii ranging from 0.196 to 0.216 nm, whereas Pb and Sn based inorganic compounds with ionic radii of 0.119 and 0.110 nm, respectively, are usually used as B cation due to their well-matched energy bandgaps.

ETL's main function is to block recombination reactions occurring at the ETL/perovskite and perovskite/HTL interfaces; hence it is also named as the blocking layer. Moreover, ETL improves the crystallinity of the perovskite material and aids to reduce the dark current and series resistance in the device. Various materials involving both organic and inorganic, possessing high electron mobility and compatible energy band levels with perovskite materials for efficient charge transportation, and reduced energy losses are suitable as ETL in PSCs. Among the different ETLs studied so far, such as SnO_2, In_2O_3, TiO_2, ZnO, spiro-OMeTAD, C60 fullerenes and its derivatives, the most used is the mesoporous anatase-TiO_2 owing to its excellent hole-blockage ability, stability, and well-matched conduction band edge [10].

Also, a wide range of materials constituting small molecules, polymers, and inorganic materials have been investigated as HTLs to serve the purpose of collection of photogenerated holes and their transportation to metal electrodes. A typical HTL must have high internal hole mobility, thermal and chemical stability, and a compatible HOMO energy level with the concerned perovskite material. Various investigations into HTL materials, such as NiO, polyaniline (PANI), CuI, poly (triarylamine) (PTAA), CuSCN, and poly (3-hexylthiophene-2,5-diyl) (P3HT), suggest the superiority of spiro-OMeTAD as HTL in PSCs [11, 12].

Further, the photogenerated charge carriers are collected by the charge collecting electrodes placed at the outer edges of HTL and ETL to complete the electrical circuit. The most suitable candidates for electrodes are the ones whose difference between the Fermi level and valence band maximum of HTL or conduction band minimum of ETL is small, to reduce the carrier injection barrier at the interface. Generally, for the extraction of photogenerated electrons, transparent conducting electrodes, such as FTO and ITO, have been used and for the collection of photogenerated holes, metal electrodes, such as Au, Ag, Pt, Ni, and Ti, have been widely studied [13]. The commonly used electrodes at the outer edges of HTL and ETL are FTO and Au metal electrodes owing to their favorable chemical stability and compatible energy band.

PSCs are classified based on the type of charge transport layer, i.e., HTL or ETL, used to face the incident sunlight and hence are named as n-i-p (regular) and p-i-n (inverted) structures. These structures are further categorized into mesoscopic and planar structures that solely depend upon the type of layer used, whether mesoporous or planar. Additionally, fabrication of HTL- and ETL-free PSCs have also been explored. The different types of configurations of PSCs are shown in Fig. .5. The first PSC was fabricated in a mesoscopic n-i-p configuration, where a mesoporous perovskite material was used as an absorbing layer instead of the standard TiO_2 semiconductor. In a planar n-i-p configuration, the perovskite layer is usually sandwiched between HTL on the electrode side and ETL toward the cathode. In general, planar geometry based PSCs provide higher PV performance in comparison to mesoscopic n-i-p structure based PSCs. However, they undergo uneven hysteresis behavior that has been investigated and controlled by using inorganic HTMs or poly (3,4-ethylenedioxythiophene)-poly (styrenesulfonate) (PEDOT: PSS) instead of spiro-OMeTAD in PSC design. The highest obtained PCE in the case of mesoscopic and planar n-i-p device configuration is 21.6% and 20.7%, respectively [14]. Notably, in inverted p-i-n device architecture, the perovskite layer acts as both a light absorber and an HTM.

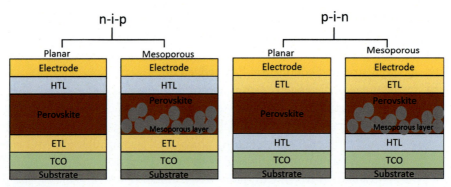

Fig. 5 Different types of PSCs with planar and mesoporous absorber layer in n-i-p and p-i-n configurations.

Till 2021, the PSCs with inverted configuration have shown a maximum of 18% PCE along with low temperature processing and negligible hysteresis [15]. Further, the ETL-free PSCs have also been explored, where a perovskite absorber layer is directly deposited over the transparent conducting oxide substrate. The performance of ETL-free PSCs depends on the technique used for deposition of the layers and has shown a maximum of 15.1% PCE [16]. Also, HTL-free PSCs have been studied as the commonly used HTMs, such as fullerenes, are highly expensive. The organic lead halide materials show an ambipolar nature along with longer charge transfer lifetimes, hence they have been used as a hole conductor and a light absorber in HTL-free PSC.

3. Issues with PSCs and their solutions

Although PSCs show great potential for commercialization, there are several issues with the available perovskite materials that need to be addressed. Mainly, PSCs have high recombination rate, low stability and high cost, which hinder their practical applications.

3.1 Recombination reactions

A large number of interfacial layers leads to higher variation in the work function (WF) values. The mismatched WF results in recombination of photogenerated charge carriers at various interfaces and induces significant current losses in PSCs. The interfacial charge transfer depends upon the difference in the bandgap of materials, their thickness, synthesis methods, and defects of different layers. Since the perovskite layer is sandwiched between ETM and HTM, the energy level alignment between the adjacent layers affects the charge carrier extraction. Besides the synthesis method used for fabricating these layers, the thickness and temperature also affect the band alignment. To reduce the recombination reactions, the interfacial band alignment is done in such a way to ease the movement of charge carriers, and thus, HTM and ETM are carefully selected depending on their WF tunability [17, 18]. Further, the electronic traps formed in the pristine perovskite materials readily react with the ambient environment and thereby, act as centers for degradation of perovskite. Mainly, single cation/halide induces point defects, which intensify the perovskite degradation. Also, the incorporation of mixed organic ions increase the formation of interfacial vacancies, which easily absorb water and oxygen molecules present in the ambient atmosphere. However, mixed organic-inorganic perovskite are less prone to form vacancies, and thus are more stable against degradation [19]. Additionally, the incorporation of monovalent halogen anions, like Cl^-, Br^-, and I^-, fill these vacancies and enhance the perovskite stability. Nevertheless, the optimum amount of these anions should be added as excessive addition of halogen blue shifts the bandgap resulting in decreased PCE of PSCs.

3.2 Stability

The instability of the perovskite absorber layer is another key factor responsible for the reduced durability and performance of PSCs. The perovskite absorber layer gets decomposed under UV-light illumination, high temperatures, humid environment, and oxygen exposure. Since the HTL shows a good adhesive property with the perovskite material, the perovskite layer on exposure to the ambient environment degrades because of the interaction of H2O molecules at the perovskite/HTL interface, thereby degrading the material further [20]. Lee et al. proposed Li-TFSI (lithium bis(trifluoromethanesulfonyl)imide) as an additive in the perovskite material to decrease the adhesion up to 60% and increase the stability of the PSCs [21]. The adhesion free layer act as a capping layer over the perovskite to enhance the stability and efficiency of PSCs.

3.3 High cost

The use of expensive HTL and electrodes raises the overall cost of PSCs. The best alternative to replacing the HTL and electrodes is to utilize carbon-based electrodes because of their properties like lower work function, lesser recombination rate, along with its low cost. However, there exists an energy level mismatch between the perovskite film and carbon, which impedes the charge carrier transfer and thus affects the performance of the cell [22]. Therefore the choice of HTL and electrode are the bottleneck for device application. Although the organic HTL reduces the trap defects and charge recombinations, the organic material has high cost and results in instability issues due to its hydrophilic nature that can deteriorate the quality of the perovskite.

Hence, the presence of a large number of interlayers in PSCs creates interfacial defects or vacancies due to inappropriate adhesion of layers. These defects provide a resistive path for interfacial charge transfer processes and contribute to the recombination of photogenerated electrons and thus become an obstacle for emerging PSCs.

Various strategies based on optimized solution processing techniques of perovskite materials, variations in stoichiometry, structure, and composition have been adopted to improve the stability, durability, and PV performance of PSCs. For instance, the crystalline and uniform perovskite absorber layer with larger grain domains has shown an improvement in charge carrier lifetime, which helps to reduce nonradiative recombination reactions, i.e., Shockley-Read-Hall, and enhance the charge carrier extraction at the perovskite/HTL or ETL interface in PSCs. Also, the incorporation of alkali metals into the perovskite material enlarges its grain size and lowers the defect state density, which enhances the charge transfer processes and hence PCE of PSCs. Further, it has been observed that the addition of a carefully designed chemical linker, such as 4-imidazoleacetic acid hydrochloride (ImAcHCl), at SnO_2 ETL and perovskite interface not only increases the grain size and crystallinity of perovskite material but also improves their stability under high humidity conditions [23]. Moreover, the nonstoichiometric in situ method of using an excessive precursor of MAI in the coating solution to prepare MAPbI$_3$ creates large number of unreactive grain boundaries of MAI in the perovskite absorber layer, which aids in the enhancement of charge carrier lifetime from 18.6 to 24.9 ns [24]. Nevertheless, this approach is useful for reducing recombination reactions at interfaces in PSCs, but at the expense of stability of perovskite layer. To overcome this issue, another approach known as additive engineering has been adopted, where the chemical additives having strong electron donating functional groups are used in the precursor solution for the formation of perovskite layer. As an example, –SH functionalized (3-mercaptopropyl)trimethoxysilane (MPTS) with electron-donating properties has been added in the PbI$_2$ precursor solution to obtain perovskite layer. Here, the MPTS aids in increasing the charge carrier lifetime, reducing the defect concentration and improving the stability of perovskite material in PSCs [25]. Furthermore, the engineering at HTL/perovskite interface effectively reduces the defects via ion exchange reactions and tunes bandgap of the perovskite material. This helps to reduce the recombination reactions and enhance the light harvesting ability of the perovskite material.

4. Introduction to 2D materials

The potential impact of graphene on the present state of technology has further fueled the exploration of other 2D materials. Like graphene, other 2D materials, such as transition metal dichalcogenides, graphitic carbon nitrides, and MXenes, are also made of one or few atomic thick layers of constituting atoms with periodic structures. Due to the outstanding physical and chemical properties of these materials, they have found applications in energy storage [26], catalysis, solar cells [27], sensing [28], and optoelectronics [29]. Since these materials have strong in-plane bonding and weak out-of-plane bonding, they can be easily stacked on top of one another resulting in flexible heterogeneous structures. Mainly, 2D materials exhibit a large surface area, a greater number of exposed active sites, easy surface functionalization in conjunction with excellent charge transfer property, and high conductivity [30]. Also, as the properties of these materials can be easily tuned through dimensionality reduction, structural and compositional variations, they show remarkable performance when used in PV applications. Among these materials, MXene shows remarkable performance when used in PSCs.

4.1 MXenes

The novel class of transition metal carbide/nitride (MXene) 2D materials has been introduced by the group of Yury Gogosti and Michel Barsoum at Drexel University, USA, in 2011. The general configuration of MXene is $M_{n+1}X_nT_X$, where $n = 1$, 2, or 3, M corresponds to an early transition metal element from group III–VII, X corresponds to carbide/nitride, and T corresponds to functional termination groups (—O, –OH, —F). The name "MXenes" comes from the MAX phase of their parent ternary carbides and nitrides and represents enough similarities with graphene [31, 32]. The parent MAX phase is a layered structure of carbides and nitrides having general formula of $M_{n+1}AX_n$, where $n = 1$, 2, or 3, M is an early transition metal element from group III–VII, A represents elements from the group XIII and XIV of the periodic table, and X represents carbides or nitrides as shown in Fig. 6. With variation of n, MAX family can have three atomic structures M_2AX, M_3AX_2, M_4AX_3 and subsequently, MXene can have corresponding M_2XT_X, $M_3X_2T_X$, $M_4X_3T_X$ structures, which are used for different applications [33].

Various methods have been investigated to synthesize MXenes, such as chemical etching and transformation, and bottom-up approach. The chemical etching method is considered the most suitable method if MXene has to be etched from its corresponding MAX phase. Different chemical etching reagents, such as HF, LiF, have been used to detach the A layer from the MAX phase, which results in the formation of —F, —OH, and —O functionalized MX surfaces [34]. Further, the chemical transformation method has been adopted to synthesize MXenes from different precursors other than MAX [35]. Here, the ammoniation process is employed to replace C atoms of transition metal carbides with N atoms available from NH_3 atmosphere. Moreover, a few members of MXene family have also been prepared via bottom up techniques such as chemical vapor deposition and salt template growth [36].

As of 2021, 70 members of layered MXene have been discovered and showed strong bonding due to the combination of metallic, covalent, and ionic bonding. The nontoxic nature of MXenes and properties, like high stability, high electrical and thermal conductivity, mobility, large surface area, high transparency, high melting point, excellent oxidation resistance, compositional variability, ability to host a broad range of intercalants,

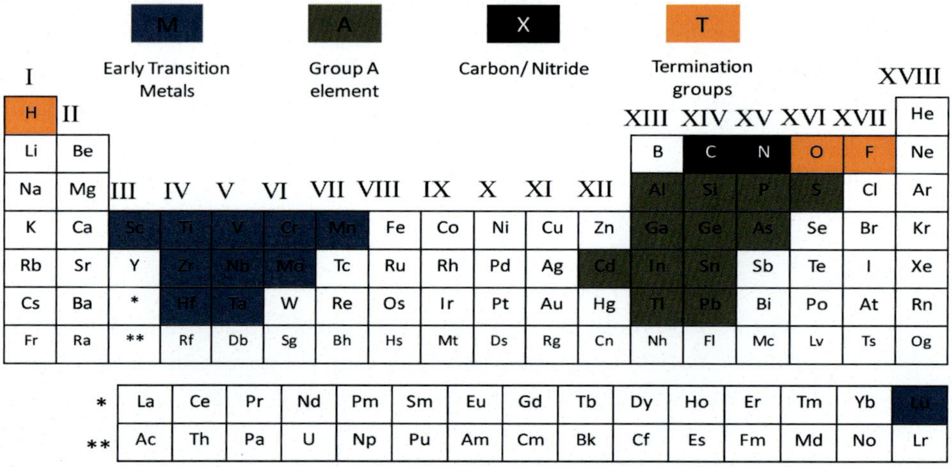

Fig. 6 Periodic table showing various elements forming MAX and MXene family.

and presence of functional ligands, open the gate for a large number of applications, like conducting thin films, sensors, piezoelectric devices, supercapacitors, PVs, lithium-ion batteries, topological insulators, and electrocatalysts [37].

5. MXene additive in PSCs

Interestingly, MXene, especially Ti_3C_2, when incorporated into the perovskite layer, is capable of improving the uniformity, crystallinity, and enlarging the grain size of the perovskite layer along with reduction of surface defects, to maximize the photon absorption via improving the light scattering and reflection ability of perovskite materials and charge transportation toward the electrodes. Thus, to materialize the fabrication of low-cost, highly stable and efficient PSCs, a low-cost material like MXene is required. The introduction of MXenes into PSCs enhances work function tunability, increases nucleation and grain size, and can replace high-cost HTM and electrodes.

5.1 Recombination reactions

MXenes exhibit excellent WF tunability that arises even with the slightest change in the number of —OH, —F, and —H terminations present on their surface, which indirectly depends on the type of transition metals, carbides and nitrides. The variation in the amount of termination groups induces a change in the electrostatic potential on the surface of MXenes, which alters its electronic structure and changes the WF [38]. In addition, termination groups change the thermal, mechanical, and chemical stabilities of MXenes. Hu et al. have shown through Bader charge analysis and thermodynamic calculations that the termination groups in $Ti_3C_2T_X$ MXenes help to improve its stability in the order $Ti_3C_2O_2 > Ti_3C_2F_2 > Ti_3C_2(OH)_2 > Ti_3C_2H_2 > Ti_3C_2$ owing to their specific chemical and structural features [39]. Ashton et al. further verified that the chemisorbed termination groups form strong covalent bonds with the MXene surface and hence affect the charge transportation [40]. Also, among different termination groups, —O draws most of the electrons from the Ti_3C_2 layer, making it neutral unlike others. Thus stable functionalized MXene broadens the range of chemical and physical properties and is beneficial WF tunability of perovskite material. Liu et al. performed first principle calculations based on density functional theory (DFT) and suggested that surface chemistry of termination groups, originating from surface dipoles effect, also affects the Fermi level of MXenes [41]. The —O terminations increase the WF of MXene and inject holes spontaneously, whereas -OH decreases the WF and spontaneously transfer electrons. However, the change in WF by —F terminations depends on the etching material and can show either trend. The theoretical WF calculations of these termination groups provide an estimation of their use in charge transfer processes in PSCs. Huang et al. used MXene incorporated 3D SnO_2 nanoparticles and 0D TiO_2 quantum dots as an ETL for the fabrication of planar multidimensional conductive heterojunction structure (MDCN) of PSCs [42]. The addition of an optimal concentration of MXene resulted in crystalline perovskite layer, which improved optical properties, effective charge extraction and transportation in PSCs, and led to a PCE of 19.14%. MDCN tuned the WF with the perovskite layer for easy electron transportation and reduced the recombination rate at the ETL/perovskite interface. Also, the MXene reduced the degradation of the perovskite layer under 30%–40% humidity. Yang et al. used MXene nanosheets in SnO_2 based ETL in a planar perovskite structure [43]. The increase in the Ti—O bonds at the surface of Ti_3C_2 MXenes improved electron transfer, reduced recombination at the ETL/perovskite interface and enhanced PCE from 5% to 17.17% after UV-ozone treatment. Further, Agresti et al. finely tuned the WF of the perovskite absorber layer and TiO_2 ETL by the addition of MXene flakes [44]. The introduction of MXene into TiO_2 ETL decreased the WF from 3.91 to 3.85 eV. Further, DFT studies showed that the termination groups strongly influence the density of state and WF of MXene, as —OH and —O terminated MXene had WF 1.6 and 6.25 eV, respectively. Also, charge transfer at the $MAPbI_3/Ti_3C_2(OH)_2$ interface showed a reduction in WF values due to the interface diploes. The electron and hole transportation with varying WF of HTM and electrodes are shown in Fig. 7. Previous reports on the additives incorporated into perovskite materials based PSCs along with their corresponding WFs and PCEs are tabulated in Table 1. It can be observed that additives, especially MXene, significantly alters the WF of perovskite material and enhance the efficiency.

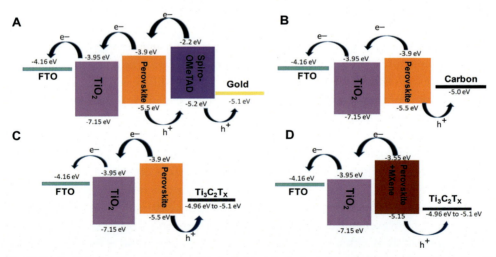

Fig. 7 Tuning of work function to accelerate hole and electron transportation. (A) HTM and gold electrode, (B) carbon electrode, (C) MXene electrode, and (D) MXene incorporation in perovskite layer and MXene electrode.

Table 1 List of different additives incorporated in perovskites along with work function and power conversion efficiency (PCE).

S. No.	Perovskite	Additive	Work function (eV) Perovskite	Work function (eV) Additive	Power conversion efficiency	Reference
1.	MAPbI$_3$	Formamide	−4.36	−4.63	15.21%	[45]
		Acetamide		−4.65	15.57%	
		Urea		−4.61	15.07%	
2.	FAMAPbI$_3$Br$_3$	Poly(methylmethacrylate) (PMMA)	—	—	21.6%	[46]
3.	MAPbI$_3$	Graphdiyne (GD)	—	—	21.01%	[47]
4.	MAPbI$_3$	Graphitic carbide nitride (g–C$_3$N$_4$)	—	—	19.49%	[48]
5.	MAPbI$_3$	MXene (Ti$_3$C$_2$T$_x$)	−3.82	−4.12	19.12%	[49]
6.	MAPbI$_3$	MXene (Ti$_3$C$_2$T$_x$)		—	17.41%	[50]
7.	FAMACsPbI$_3$Br$_3$	MXene (Ti$_3$C$_2$T$_x$)	−3.90	−3.55	16.3%	[44]

MAPbI$_3$, methylammonium lead iodide; *FAMAPbI$_3$Br$_3$*, formamidinium methylammonium lead iodide bromide; *FAMACsPbI$_3$Br$_3$*, formamidinium methylammonium cesium lead iodide bromide.

5.2 Doping of transition metal in MXenes

The surface termination groups on MXene also change the optical and dielectric properties of MXene, which is beneficial for PV applications. For instance, Hart et al. reported that the replacement of Ti with Mo from the surface of MXene successfully has been observed to change the metallic behavior of MXene to the semiconductor [39]. Also, excess addition of –OH terminations showed the semiconductor-like conductivity of MXene, although it is considered a metallic material. These studies suggest that MXene exhibits outstanding metal-to-semiconductor and semiconductor-to-metal transition by replacing transition metals and adding some transition dopants. Wang et al. decorated transition magneto metals (Fe, Co, Ni) in Ti_3C_2 and studied their electronic properties through DFT [51]. The introduction of transition metals significantly promoted the separation of photoexcited charges and suppressed the recombination rate, which effectively increased the overall photon utilization rate. Further, these elements promoted the absorption coefficient of MXene to all three ultraviolet, visible, and near-infrared regions. The Co and Ni composites widened the spectral range by 50% in the visible and ultraviolet region, whereas Fe increased it by more than 100% especially in the infrared region. The remarkable enhancement in the optical absorbance is attributed to the change in electronic structure and specific surface area induces due to transition magnetic metals decoration. This broadening of spectral range opened the gates for practical applications of MXene.

5.3 Nucleation and growth of perovskite layer

The degradation in the performance of PSCs arises due to the poor perovskite film quality with small grain sizes forming through a quick reaction crystallization process. The crystallization occurs due to the annealing effects on the morphology and grain size. The crystallinity of perovskite strongly affects the perovskite interface, which influences the device performance. MXene as an additive improves the uniformity, crystallinity, and homogeneity of perovskite. The incorporation of MXene in the perovskite absorber layer during the film growth enlarges its grain size and reduces surface defects that improve the charge transportation toward the electrode in PSCs. The large grain size of perovskite reduces the number of grain boundaries and promotes charge transfer through the material resulting in higher efficiency. The termination groups also retard the crystallization rate and hence increase the crystal size of organic-inorganic halide perovskite. Lan et al. deposited different concentrations of MAI on PbI_2 film to obtain high-quality perovskite layer in ambient conditions via a two-step deposition method and achieved PCE of 11.48% with an average grain size of 1.4 μm [52]. The addition of MXene in the perovskite precursor slowed the crystallization process and increase the grain size as shown in Fig. 8. The increased grain size of perovskite material decreased the boundary defects, which resulted in higher conductivity and mobility through the

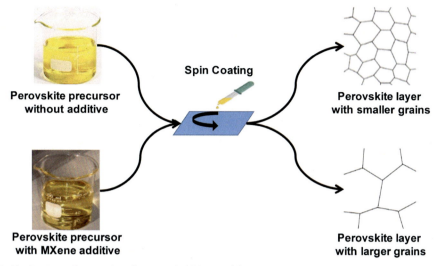

Fig. 8 Nucleation of perovskite layer with MXene additive.

grain boundaries. The nucleation in this process occurred when the precursor solution (perovskite material) with smaller grains is deposited on the larger surface area (MXenes flakes). Moreover, the interactions of —F termination group attached on the surface of MXene with methylammonium (CH_3NH_3) facilitates the protonation of F with H atom of CH_3NH_3. Additionally, –OH group interacts with MAI via van der Waals interaction. All these interactions suppress the number of nucleus generating around the MXene, thus leading to the retardation of the nucleation process. As compared with the pristine, the nucleus generates randomly on the ETL layer. While fabricating the perovskite layer with MXene as an additive, the rate of crystal growth is lower, which results in larger-sized crystals of perovskite. So, the addition of MXene in perovskite precursor solution provides heterogeneous nucleation sites to efficiently transfer the charge carriers, improve the morphology, enlarge the grain size, and enhance the PCE of PSCs. Huang et al. showed that fabrication of PSCs by low-temperature annealing in both air and N_2 atmosphere increases the grain size and thereby reduces the grain boundaries at the interfaces [43]. Average grain size of 658 nm has been achieved using MDCN PSCs, which further decreased with an increase in the amount of MXene. Agresti et al. incorporated MXene in both ETL and perovskite layers not only to reduce the charge recombination rate at the interface but also to reduce the passivation of the mesoporous TiO_2. Here, the combined effect of MXene with perovskite slowed down the crystal growth as compared to the pristine film and resulted in a uniform, defect-free, and highly crystalline perovskite layer with larger grain boundaries [44]. Ma et al. showed that incorporation of MXene in the perovskite layer can accelerate the charge transfer processes resulting in 12% enhancement in PCE. The increase in particle size due to the incorporation of MXenes highly influenced the scattering behavior of incident light. The improvement in incident photon-to-electron conversion efficiency (IPCE) depends upon the open-circuit voltage (V_{OC}) and fill factor (FF), which directly depend on the perovskite/HTL interface. The lesser number of grain boundaries facilitate the transportation and collection of charge carriers, which improve the FF and the larger grain size increases the V_{OC}. Moreover, the surface passivation of perovskite by MXene results in improvement in hole selectivity and reduced recombinations at perovskite/HTL interface. Contrarily, excess addition of MXene can lead to agglomeration on the surface of film that acts as a carrier recombination centers.

5.4 Cost-effective HTM and electrodes

The stability of PSCs mainly depends on the interface of spiro-OMeTAD HTL with $CH_3NH_3PbI_3$ perovskite absorber layer. This interface facilitates the iodine migration from the perovskite layer into HTL and results in direct contact of a metal electrode with the active layer due to incomplete surface coverage, which further promotes the charge recombinations. Moreover, the HTL, ETL, and Au electrodes in PSCs raise the fabrication cost of PSCs. Different strategies are adopted to overcome these factors such as introduction of various cost-effective materials (2D materials) as both HTL and electrodes, inorganic HTL, and capping of the active layer. All these materials lower the price and improve the stability of PSCs, however at the expense of PCE. In this respect, Sheng et al. used carbon-based electrode materials (c-PSCs) without HTL and achieved PCE of 14.5% [53]. Carbon-based material effectively replaces the Au electrode due to its water-resistance ability, stability, and ion migration inertness. However, the achieved PCE of c-PSCs is comparatively lower than the existing spiro-OMeTAD HTL and Au electrode-based PSCs (23.7%) [54]. The low PCE in c-PSCs mainly originates from mismatching of energy levels of different layers of perovskite at the interfacial layer with carbon, which increases charge carrier recombination rate and lowers the hole selectivity. Nevertheless, the device structure and working principle of c-PSCs are completely different from those of HTM based PSCs [55]. In a PSC, MAI based perovskite layer absorbs light and consequently, transfers charge carriers toward the electrode, but the carbon electrode is incapable of reflecting the incident light for secondary absorption in the perovskite layer [55]. In contrast, the mesoscopic PSC effectively reflects the incident light to the perovskite layer for secondary absorption resulting in the transfer of more charge carriers and thereby increase the PCE as shown in Fig. 9. Later, Mi et al. used a mixed carbon electrode by incorporating carbon nanotubes (CNTs 1D) and Ti_3C_2 MXenes in carbon paste [56]. The mixed carbon in comparison to pure carbon-based PSCs showed point-to-point contact and provided a multidimensional charge transfer path with an effective increase in the charge carriers. The addition of CNTs and Ti_3C_2 MXenes enhanced the surface contact and promoted the electrode conductivity in inorganic perovskite, where the multidimensional charge transfer path effectively

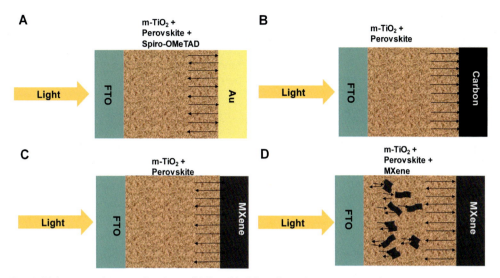

Fig. 9 Light scattering mechanism of (A) gold, (B) carbon, (C) MXene, and (D) MXene additive in perovskite layer and MXene electrode.

extracted the charge carriers and transported them. The interfaces in the multidimensional electrodes somehow resisted the charges and resulted in a lower PCE of 7.09% for inorganic PSCs. The different types of electrode materials used in PSCs are summarized in Table 2. Similarly, MXenes having same energy level as that of carbon material are effectively used as noble metal-free electrodes and HTM in mesoporous PSCs by Cao et al. [54] An increment of 27% in PCE of MXene based PSCs over that of carbon based electrode has been achieved because of smooth interfacial contacts. Enhancement of 80% in IPCE in the visible range (400 to 700 nm) has also been observed, which is beneficial for the large production of PSCs because of easy preparation of MXenes. Moreover, it also showed good reproducibility and long term stability of about 360 h in ambient atmosphere at 30% humidity due to the 300 μm thick encapsulating Ti_3C_2 layer that protected perovskite from air and water.

5.5 Encapsulation of PSCs

All inorganic perovskite materials exhibit higher stability, but lower efficiency as compared to mixed inorganic-organic perovskite materials. The main factors that affect the stability in PSCs are light irradiation, thermal attack, and moisture [64]. The direct contact of the metallic electrode with the active perovskite layer promotes the charge recombination, thus reduces the PCE of PSCs. The encapsulation of perovskite material with MXene layer forms a protective layer around it, which increases the stability of organic-inorganic mixed halide perovskite. Ma et al. showed that the termination groups retard the crystallization rate and increase the crystal size of perovskite material, which in turn decreases the grain boundary defects and lowers the humidity effect [50].

Table 2 List of different perovskite and electrode materials used in PSCs with their work functions and power conversion efficiencies.

Perovskite	Electrode	Work function (eV)	Power conversion efficiency	Reference
FAMACsPbI$_3$Br$_3$	Au	−5.10	15.90%	[44]
MAPbI$_3$	Carbon	−5.00	13.70%	[57]
MAPbI$_3$	Multilayer graphene carbon	−5.00	11.50%	[58]
	Single layer graphene carbon	−4.80	6.70%	[58]
MAPbI$_3$	Multiwalled carbon nanotubes	−4.46	10.70%	[59]
	B-doped multiwalled carbon nanotubes	−4.55	15.23%	[59]
MAPbI$_3$	B-doped graphite	−4.81	13.60%	[60]
MAPbI$_3$	B/P-doped carbon	−5.01 (C) −4.81 (PC) −5.14 (BC) −5.12 (BPC)	3.72% (C) 5.39% (PC) 5.20% (BC) 6.78% (BPC)	[61]
MAPbI$_3$	Oxygen-enriched black carbon	−5.00	15.70%	[62]
MAPbI$_3$	MoS$_2$	−5.40	13.30%	[63]
MAPbI$_3$	MXenes	−4.96	13.83%	[54]

FAMACsPbI$_3$Br$_3$, formamidinium methylammonium cesium lead iodide bromide; *MAPbI$_3$*, methylammonium lead iodide; *C*, carbon electrode; *PC*, phosphorus-doped carbon electrode; *BC*, boron-doped carbon electrode; *BPC*, boron phosphorus codoped carbon electrode.

This retardation improves the charge transportation and hence increases the current as well as FF. MXene sheets passivate the electron-hole recombinations increasing the V_{OC} thereby. Also, termination groups passivate the grains through protonation with H of MAI and hence improve the stability of perovskite against humidity. Additionally, Cao et al. showed that more than 300 μm thickness of MXene results in efficient encapsulation of the perovskite layer [54].

6. Future scope

Metal-doped MXenes have been investigated at different interfacial layers and showed an improvement in the PV performance of PSCs via modification of their optical, structural, and electronic properties. Recently, it has been theoretically observed that decorating MXenes with 2D transition metal dichalcogenides and transition metals aids the charge carrier extraction at the metal-semiconductor junction. Their inclusion induces a fine spacing between the Ti- and C-layer of MXenes, which results in the expansion of band structure and increased specific surface area that in turn would increase the amount of perovskite absorber layer. Further, not only the chemical methods but also the cost-effective physical methods, such as ion implantation techniques and sputtering, for incorporating transition metals in PSCs need to be explored, as they effectively alter the WF. Thus metal ion-implanted MXenes would pave a way to their utilization as an ETL, or an interfacial layer between ETL and perovskite, and electrode in PSCs for easy charge carrier transportation and reduction in the recombination reactions [65]. Hence the incorporation of TMDs and transition metals based MXenes with varied methods at different interfaces of PSCs would benefit their light scattering and harvesting ability as well as improve their charge transfer processes, which would result in higher PCE and stability.

7. Conclusion

In summary, this chapter presents the effect of incorporation of MXenes at varied interfaces on the PV performance and stability of PSCs. MXenes as an additive in PSCs induce a WF tunability of perovskite, which results in the easy mobility of charge carriers. Moreover, the nucleation process leads to an increase in grain size of perovskite, thus reducing the grain boundaries. This helps to improve the charge transfer processes at the interfaces along with the reduction in recombination reaction rate and hence enhance the PCE of PSCs. Moreover, MXenes are concluded to be the best replacement for highly expensive organic spiro-OMeTAD HTM and Au metal electrodes owing to their well-matched WF. Further, the encapsulation of MXene layer over the perovskite film protects it from humid environment and helps to work under ambient conditions. However, the termination groups of MXenes (—O, –OH, —F) exhibit different WFs. As a higher concentration of —O termination exhibit higher WF and results in mismatching of bandgap and less charge transportation, so it is difficult to modulate the amount of termination groups and tune the WF within the perovskite material. Thus it is essential to work on the synthesis process of termination groups enriched MXene.

Acknowledgement

The authors highly acknowledge the Science and Engineering Research Board (SERB), Department of Science and Technology, New Delhi, India for providing financial assistance under the sanctioned project no. CRG/2020/003741. One of the authors, S.G., greatly acknowledges RUSA 2.0 component 4 for providing junior research fellowship.

Conflict of interest

There are no conflicts to declare.

References

[1] Change NGC, Global Land-Ocean Temperature Index, Available from https://climate.nasa.gov/vital-signs/global-temperature/.

[2] Data SIR, Massive Growth Since 2000 SEts the Stage for the Solar+ Decade, Available from https://www.seia.org/solar-industry-research-data#:~:text=Thanks%20to%20strong%20federal%20policiespower%20nearly%2018%20million%20homes.

[3] C. Battaglia, A. Cuevas, S. De Wolf, High-efficiency crystalline silicon solar cells: status and perspectives, Energ. Environ. Sci. 9 (5) (2016) 1552–1576, https://doi.org/10.1039/C5EE03380B.

[4] I.M. Dharmadasa, A.A. Ojo, Unravelling complex nature of CdS/CdTe based thin film solar cells, J. Mater. Sci. Mater. Electron. 28 (22) (2017) 16598–16617, https://doi.org/10.1007/s10854-017-7615-x.

[5] B. Kippelen, J.-L. Brédas, Organic photovoltaics, Energ. Environ. Sci. 2 (3) (2009) 251–261, https://doi.org/10.1039/B812502N.

[6] I. Hussain, H.P. Tran, J. Jaksik, J. Moore, N. Islam, M.J. Uddin, Functional materials, device architecture, and flexibility of perovskite solar cell, Emerg. Mater. 1 (3) (2018) 133–154, https://doi.org/10.1007/s42247-018-0013-1.

[7] F.F. Targhi, Y.S. Jalili, F. Kanjouri, MAPbI3 and FAPbI3 perovskites as solar cells: case study on structural, electrical and optical properties, Results Phys. 10 (2018) 616–627. https://doi.org/10.1016/j.rinp.2018.07.007.

[8] Y. Reyna, M. Salado, S. Kazim, A. Pérez-Tomas, S. Ahmad, M. Lira-Cantu, Performance and stability of mixed FAPbI3(0.85)MAPbBr3(0.15) halide perovskite solar cells under outdoor conditions and the effect of low light irradiation, Nano Energy 30 (2016) 570–579. https://doi.org/10.1016/j.nanoen.2016.10.053.

[9] N.-G. Park, Perovskite solar cells: an emerging photovoltaic technology, Mater. Today 18 (2) (2015) 65–72. https://doi.org/10.1016/j.mattod.2014.07.007.

[10] Y. Yang, M.T. Hoang, D. Yao, N.D. Pham, V.T. Tiong, X. Wang, et al., Spiro-OMeTAD or CuSCN as a preferable hole transport material for carbon-based planar perovskite solar cells, J. Mater. Chem. A 8 (25) (2020) 12723–12734.

[11] P. Kaur, M.-S. Shin, J.-S. Park, G. Verma, S.S. Sekhon, Supramolecular modification of carbon nanofibers with poly(diallyl dimethylammonium) chloride and Triton X-100 for electrochemical application, Int. J. Hydrog. Energy 43 (13) (2018) 6575–6585. https://doi.org/10.1016/j.ijhydene.2018.02.075.

[12] D.K. Jarwal, A. Kumar, A.K. Mishra, S. Ratan, R.K. Upadhyay, C. Kumar, et al., Fabrication and TCAD validation of ambient air-processed ZnO NRs/CH3NH3PbI3/spiro-OMeTAD solar cells, Superlattice. Microst. 143 (2020) 106540.

[13] I. Jeon, A. Shawky, S. Seo, Y. Qian, A. Anisimov, E.I. Kauppinen, et al., Carbon nanotubes to outperform metal electrodes in perovskite solar cells via dopant engineering and hole-selectivity enhancement, J. Mater. Chem. A 8 (22) (2020) 11141–11147.

[14] M. Saliba, T. Matsui, K. Domanski, J.-Y. Seo, A. Ummadisingu, S.M. Zakeeruddin, et al., Incorporation of rubidium cations into perovskite solar cells improves photovoltaic performance, Science 354 (6309) (2016) 206–209.

[15] L. Etgar, P. Gao, Z. Xue, Q. Peng, A.K. Chandiran, B. Liu, et al., Mesoscopic CH3NH3PbI3/TiO2 heterojunction solar cells, J. Am. Chem. Soc. 134 (42) (2012) 17396–17399.

[16] L. Huang, J. Xu, X. Sun, Y. Du, H. Cai, J. Ni, et al., Toward revealing the critical role of perovskite coverage in highly efficient electron-transport layer-free perovskite solar cells: an energy band and equivalent circuit model perspective, ACS Appl. Mater. Interfaces 8 (15) (2016) 9811–9820.

[17] S. Wang, T. Sakurai, W. Wen, Y. Qi, Energy level alignment at interfaces in metal halide perovskite solar cells, Adv. Mater. Interfaces 5 (22) (2018) 1800260. https://doi.org/10.1002/admi.201800260.

[18] A. Fakharuddin, L. Schmidt-Mende, G. Garcia-Belmonte, R. Jose, I. Mora-Sero, Interfaces in perovskite solar cells, Adv. Energy Mater. 7 (22) (2017) 1700623. https://doi.org/10.1002/aenm.201700623.

[19] M.I. Saidaminov, J. Kim, A. Jain, R. Quintero-Bermudez, H. Tan, G. Long, et al., Suppression of atomic vacancies via incorporation of isovalent small ions to increase the stability of halide perovskite solar cells in ambient air, Nat. Energy 3 (8) (2018) 648–654, https://doi.org/10.1038/s41560-018-0192-2.

[20] S. Yang, W. Fu, Z. Zhang, H. Chen, C.-Z. Li, Recent advances in perovskite solar cells: efficiency, stability and lead-free perovskite, J. Mater. Chem. A 5 (23) (2017) 11462–11482, https://doi.org/10.1039/C7TA00366H.

[21] I. Lee, J.H. Yun, H.J. Son, T.-S. Kim, Accelerated degradation due to weakened adhesion from Li-TFSI additives in perovskite solar cells, ACS Appl. Mater. Interfaces 9 (8) (2017) 7029–7035, https://doi.org/10.1021/acsami.6b14089.

[22] H. Chen, S. Yang, Carbon-based perovskite solar cells without hole transport materials: the front runner to the market? Adv. Mater. 29 (24) (2017) 1603994. https://doi.org/10.1002/adma.201603994.

[23] J. Chen, X. Zhao, S.G. Kim, N.G. Park, Multifunctional chemical linker imidazoleacetic acid hydrochloride for 21% efficient and stable planar perovskite solar cells, Adv. Mater. 31 (39) (2019), https://doi.org/10.1002/adma.201902902, e1902902.

[24] Y.-M. Xie, B. Yu, C. Ma, X. Xu, Y. Cheng, S. Yuan, et al., Direct observation of cation-exchange in liquid-to-solid phase transformation in FA 1 − x MA x PbI 3 based perovskite solar cells, J. Mater. Chem. A 6 (19) (2018) 9081–9088.

[25] H. Ye, Dual Layer-Encapsulated Halide Perovskite with Enhanced Stability for Bioimaging: Drexel University, 2020.

[26] L. Lin, W. Lei, S. Zhang, Y. Liu, G.G. Wallace, J. Chen, Two-dimensional transition metal dichalcogenides in supercapacitors and secondary batteries, Energy Storage Mater. 19 (2019) 408–423. https://doi.org/10.1016/j.ensm.2019.02.023.

[27] S. Das, D. Pandey, J. Thomas, T. Roy, The role of graphene and other 2D materials in solar photovoltaics, Adv. Mater. 31 (1) (2019) 1802722. https://doi.org/10.1002/adma.201802722.

[28] S. Yang, C. Jiang, S.-H. Wei, Gas sensing in 2D materials, Appl. Phys. Rev. 4 (2) (2017) 021304, https://doi.org/10.1063/1.4983310.

[29] F. Yang, S. Cheng, X. Zhang, X. Ren, R. Li, H. Dong, et al., 2D organic materials for optoelectronic applications, Adv. Mater. 30 (2) (2018) 1702415. https://doi.org/10.1002/adma.201702415.

[30] H. Kim, H.N. Alshareef, MXetronics: MXene-enabled electronic and photonic devices, ACS Mater. Lett. 2 (1) (2020) 55–70, https://doi.org/10.1021/acsmaterialslett.9b00419.

[31] C. Wang, S. Chen, H. Xie, S. Wei, C. Wu, L. Song, Atomic Sn4 + decorated into vanadium carbide MXene interlayers for superior Lithium storage, Adv. Energy Mater. 9 (4) (2019) 1802977. https://doi.org/10.1002/aenm.201802977.

[32] G. Gao, A.P. O'Mullane, A. Du, 2D MXenes: a new family of promising catalysts for the hydrogen evolution reaction, ACS Catal. 7 (1) (2017) 494–500, https://doi.org/10.1021/acscatal.6b02754.

[33] R. Khaledialidusti, M. Khazaei, S. Khazaei, K. Ohno, High-throughput computational discovery of ternary-layered MAX phases and prediction of their exfoliation for formation of 2D MXenes, Nanoscale (2021).

[34] M.A. Hope, A.C. Forse, K.J. Griffith, M.R. Lukatskaya, M. Ghidiu, Y. Gogotsi, et al., NMR reveals the surface functionalisation of Ti3C2 MXene, Phys. Chem. Chem. Phys. 18 (7) (2016) 5099–5102, https://doi.org/10.1039/C6CP00330C.

[35] I.R. Shein, A.L. Ivanovskii, Graphene-like titanium carbides and nitrides tin + 1Cn, tin + 1Nn (n = 1, 2, and 3) from de-intercalated MAX phases: first-principles probing of their structural, electronic properties and relative stability, Comput. Mater. Sci. 65 (2012) 104–114. https://doi.org/10.1016/j.commatsci.2012.07.011.

[36] X. Xiao, P. Urbankowski, K. Hantanasirisakul, Y. Yang, S. Sasaki, L. Yang, et al., Scalable synthesis of ultrathin Mn3N2 exhibiting room-temperature Antiferromagnetism, Adv. Funct. Mater. 29 (17) (2019) 1809001. https://doi.org/10.1002/adfm.201809001.

[37] X. Zhan, C. Si, J. Zhou, Z. Sun, MXene and MXene-based composites: synthesis, properties and environment-related applications, Nanoscale Horiz. 5 (2) (2020) 235–258.

[38] M. Khazaei, M. Arai, T. Sasaki, A. Ranjbar, Y. Liang, S. Yunoki, OH-terminated two-dimensional transition metal carbides and nitrides as ultralow work function materials, Phys. Rev. B 92 (7) (2015) 075411, https://doi.org/10.1103/PhysRevB.92.075411.

[39] J.L. Hart, K. Hantanasirisakul, A.C. Lang, B. Anasori, D. Pinto, Y. Pivak, et al., Control of MXenes' electronic properties through termination and intercalation, Nat. Commun. 10 (1) (2019) 522, https://doi.org/10.1038/s41467-018-08169-8.

[40] M. Ashton, K. Mathew, R.G. Hennig, S.B. Sinnott, Predicted surface composition and thermodynamic stability of MXenes in solution, J. Phys. Chem. C 120 (6) (2016) 3550–3556, https://doi.org/10.1021/acs.jpcc.5b11887.

[41] Y. Liu, H. Xiao, W.A. Goddard, Schottky-barrier-free contacts with two-dimensional semiconductors by surface-engineered MXenes, J. Am. Chem. Soc. 138 (49) (2016) 15853–15856, https://doi.org/10.1021/jacs.6b10834.

[42] L. Huang, X. Zhou, R. Xue, P. Xu, S. Wang, C. Xu, et al., Low-temperature growing anatase TiO_2/SnO_2 multi-dimensional heterojunctions at MXene conductive network for high-efficient perovskite solar cells, Nanomicro Lett. 12 (1) (2020) 44, https://doi.org/10.1007/s40820-020-0379-5.

[43] L. Yang, Y. Dall'Agnese, K. Hantanasirisakul, C.E. Shuck, K. Maleski, M. Alhabeb, et al., SnO_2–Ti_3C_2 MXene electron transport layers for perovskite solar cells, J. Mater. Chem. A 7 (10) (2019) 5635–5642, https://doi.org/10.1039/C8TA12140K.

[44] A. Agresti, A. Pazniak, S. Pescetelli, A. Di Vito, D. Rossi, A. Pecchia, et al., Nat. Mater. 18 (11) (2019) 1228–1234, https://doi.org/10.1038/s41563-019-0478-1.

[45] S. Liu, S. Li, J. Wu, Q. Wang, Y. Ming, D. Zhang, et al., Amide Additives induced a fermi level shift to improve the performance of hole-conductor-free, printable mesoscopic perovskite solar cells, J. Phys. Chem. Lett. 10 (21) (2019) 6865–6872, https://doi.org/10.1021/acs.jpclett.9b02463.

[46] D. Bi, C. Yi, J. Luo, J.-D. Décoppet, F. Zhang, M. Zakeeruddin Shaik, et al., Polymer-templated nucleation and crystal growth of perovskite films for solar cells with efficiency greater than 21%, Nat. Energy 1 (10) (2016) 16142, https://doi.org/10.1038/nenergy.2016.142.

[47] J. Li, T. Jiu, S. Chen, L. Liu, Q. Yao, F. Bi, et al., graphdiyne as a host active material for perovskite solar cell application, Nano Lett. 18 (11) (2018) 6941–6947, https://doi.org/10.1021/acs.nanolett.8b02863.

[48] L.-L. Jiang, Z.-K. Wang, M. Li, C.-C. Zhang, Q.-Q. Ye, K.-H. Hu, et al., Passivated perovskite crystallization via g-C_3N_4 for high-performance solar cells, Adv. Funct. Mater. 28 (7) (2018) 1705875. https://doi.org/10.1002/adfm.201705875.

[49] Y. Zhao, X. Zhang, X. Han, C. Hou, H. Wang, J. Qi, et al., Tuning the reactivity of PbI_2 film via monolayer $Ti_3C_2T_x$ MXene for two-step-processed $CH_3NH_3PbI_3$ solar cells, Chem. Eng. J. 127912 (2020). https://doi.org/10.1016/j.cej.2020.127912.

[50] Z. Guo, L. Gao, Z. Xu, S. Teo, C. Zhang, Y. Kamata, et al., High electrical conductivity 2D MXene serves as additive of perovskite for efficient solar cells, Small 14 (47) (2018) 1802738. https://doi.org/10.1002/smll.201802738.

[51] X. Wang, S. Huang, L. Deng, H. Luo, C. Li, Y. Xu, et al., Enhanced optical absorption of Fe-, co- and Ni- decorated Ti_3C_2 MXene: a first-principles investigation, Physica E Low Dimens. Syst. Nanostruct. 127 (2021) 114565. https://doi.org/10.1016/j.physe.2020.114565.

[52] C. Lan, S. Zhao, C. Zhang, W. Liu, S. Hayase, T. Ma, Concentration gradient-controlled growth of large-grain $CH_3NH_3PbI_3$ films and enhanced photovoltaic performance of solar cells under ambient conditions, CrystEngComm 18 (48) (2016) 9243–9251, https://doi.org/10.1039/C6CE02151D.

[53] Y. Sheng, Y. Hu, A. Mei, P. Jiang, X. Hou, M. Duan, et al., Enhanced electronic properties in $CH_3NH_3PbI_3$ via LiCl mixing for hole-conductor-free printable perovskite solar cells, J. Mater. Chem. A 4 (42) (2016) 16731–16736, https://doi.org/10.1039/C6TA08021A.

[54] J. Cao, F. Meng, L. Gao, S. Yang, Y. Yan, N. Wang, et al., Alternative electrodes for HTMs and noble-metal-free perovskite solar cells: 2D MXenes electrodes, RSC Adv. 9 (59) (2019) 34152–34157, https://doi.org/10.1039/C9RA06091J.

[55] H. Chen, S. Yang, Methods and strategies for achieving high-performance carbon-based perovskite solar cells without hole transport materials, J. Mater. Chem. A 7 (26) (2019) 15476–15490, https://doi.org/10.1039/C9TA04707G.

[56] L. Mi, Y. Zhang, T. Chen, E. Xu, Y. Jiang, Carbon electrode engineering for high efficiency all-inorganic perovskite solar cells, RSC Adv. 10 (21) (2020) 12298–12303, https://doi.org/10.1039/D0RA00288G.

[57] T. Liu, L. Liu, M. Hu, Y. Yang, L. Zhang, A. Mei, et al., Critical parameters in TiO2/ZrO2/carbon-based mesoscopic perovskite solar cell, J. Power Sources 293 (2015) 533–538. https://doi.org/10.1016/j.jpowsour.2015.05.106.

[58] K. Yan, Z. Wei, J. Li, H. Chen, Y. Yi, X. Zheng, et al., High-performance graphene-based hole conductor-free perovskite solar cells: Schottky junction enhanced hole extraction and Electron blocking, Small 11 (19) (2015) 2269–2274. https://doi.org/10.1002/smll.201403348.

[59] X. Zheng, H. Chen, Q. Li, Y. Yang, Z. Wei, Y. Bai, et al., Boron doping of multiwalled carbon nanotubes significantly enhances hole extraction in carbon-based perovskite solar cells, Nano Lett. 17 (4) (2017) 2496–2505, https://doi.org/10.1021/acs.nanolett.7b00200.

[60] M. Duan, C. Tian, Y. Hu, A. Mei, Y. Rong, Y. Xiong, et al., Boron-doped graphite for high work function carbon electrode in printable hole-conductor-free mesoscopic perovskite solar cells, ACS Appl. Mater. Interfaces 9 (37) (2017) 31721–31727, https://doi.org/10.1021/acsami.7b05689.

[61] M. Chen, R.-H. Zha, Z.-Y. Yuan, Q.-S. Jing, Z.-Y. Huang, X.-K. Yang, et al., Boron and phosphorus co-doped carbon counter electrode for efficient hole-conductor-free perovskite solar cell, Chem. Eng. J. 313 (2017) 791–800. https://doi.org/10.1016/j.cej.2016.12.050.

[62] C. Tian, A. Mei, S. Zhang, H. Tian, S. Liu, F. Qin, et al., Oxygen management in carbon electrode for high-performance printable perovskite solar cells, Nano Energy 53 (2018) 160–167. https://doi.org/10.1016/j.nanoen.2018.08.050.

[63] A. Capasso, F. Matteocci, L. Najafi, M. Prato, J. Buha, L. Cinà, et al., Few-layer MoS2 flakes as active buffer layer for stable perovskite solar cells, Adv. Energy Mater. 6 (16) (2016) 1600920. https://doi.org/10.1002/aenm.201600920.

[64] H. Tsai, R. Asadpour, J.-C. Blancon, C.C. Stoumpos, O. Durand, J.W. Strzalka, et al., Light-induced lattice expansion leads to high-efficiency perovskite solar cells, Science 360 (6384) (2018) 67, https://doi.org/10.1126/science.aap8671.

[65] J.T.-W. Wang, Z. Wang, S. Pathak, W. Zhang, deQuilettes DW, Wisnivesky-Rocca-Rivarola F, et al. Efficient perovskite solar cells by metal ion doping, Energy Environ. Sci. 9 (9) (2016) 2892–2901, https://doi.org/10.1039/C6EE01969B.

SECTION E

Nanocatalysts for automotive application

CHAPTER 25

Nanocatalysts for exhaust emissions reduction

Ramesh Ch. Deka[a], Sudakhina Saikia[a], Nishant Biswakarma[a], Nand Kishor Gour[a], and Ajanta Deka[b]

[a]Department of Chemical Sciences, Tezpur University, Tezpur, Assam, India
[b]Department of Physics, Girijananda Chowdhury Institute of Management and Technology, Guwahati, Assam, India

Chapter outline

1. Introduction — 511
2. Methods — 513
   2.1 Three-way catalysis — 513
   2.2 Selective catalytic reduction — 516
3. Theoretical studies — 519
4. Conclusion and future scope — 522
References — 522

1. Introduction

With the rapid upsurge in the number of vehicles, automotive exhaust emissions have begun to impart a serious concern upon environmental pollution. These emissions have undoubtedly impacted the population worldwide, as the World Health Organization (WHO) reports about 4.2 million deaths per year due to air pollution [1, 2]. The incomplete combustion of fuels in engines generates massive amounts of hazardous exhaust gases consisting of carbon monoxide (CO), unburned hydrocarbons (HC), oxides of nitrogen (NO_x), and other particulate matter (PM) [3, 4]. These harmful pollutants have serious adverse effects on human health, as they cause respiratory diseases, cancers, chronic diseases, etc. Automobile exhausts are also responsible for environmental problems, such as global warming, ozone film depletion, acid rain, and photochemical smog [5, 6]. The first observation of photochemical smog dates back to the 1970s, which is formed due to the reaction occurring between two primary exhaust pollutants: HC and NO_x, in presence of sunlight [7]:

$$HC + NO_x + \text{sunlight}(h\nu) \rightarrow O_3 + HNO_3 + \text{other products} \qquad (1)$$

All these exhaust pollutants can greatly affect human health and are harmful to plants [8]. CO is an extremely toxic gas that is colorless and odorless. It reacts with blood hemoglobin to form carboxy-hemoglobin which reduces oxygen at the tissue level, thus causing hypoxia [9]. The human body relies mostly on oxygen for the cells to function. The vital organs such as the heart, brain, and lungs require large amounts of oxygen to function normally. So, any amount of exposure to CO results in symptoms such as nausea, vomiting, fatigue, lethargy, dizziness, etc. [10]. Nitrogen oxides which are referred to as NO_x are composed of nitric oxide (NO) and nitrogen dioxide (NO_2) [11], with NO (80%–95%) being the main constituent [12]. They are one of the major air pollutants that are considered to be causing severe human health as well as environmental problems. NO_x is largely accountable for the formation of ozone at the ground-level, which is harmful to the human respiratory system [13]. The ozone affects vegetation adversely and is also the major cause of the formation of photochemical smog in metropolitan cities [14]. NO and NO_2 can combine with the mixture of ammonia to form an acidic solution and some other particles. These small particles when enter the human system have the ability to cause premature death. In addition to this, they are responsible for acid rain and haze. The emission of CO_2 gas also contributes to changing the carbon cycle thus, altering the climate by the effect popularly known as "Green House Effect." CO_2 is a greenhouse gas that contributes to the annual increase in the global temperature. Another exhaust product, PM is a complex mixture of hydrocarbons, sulfur compounds, and other species [15, 16]. It largely consists of carbonaceous material known as soot [16]. The health effects of inhaling airborne particulate matter are observed to be serious in all living beings and include asthma, lung cancer, and premature death.

United States (in the mid-1960s) and Europe (in 1970s) recommended safety and clean environment rules to enforce restrictions on limits of exhaust gases. Emission rule in Europe is based on European Union Research Organization (EURO) whereas Super Ultra Low Emission Vehicles (SULEV) and Zero-Emission Vehicles (ZEV) standards are used in California. Taking EURO norms as a reference, India has also implemented Bharat norms in the early 2000s. Fig. 1 shows the progression of European emission standards for light commercial diesel vehicles. As it can be seen from Fig. 1, exhaust emissions

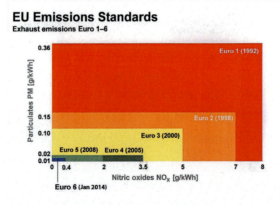

Fig. 1 Progression of European emission standards for light commercial diesel vehicles.

per test kilometer were reduced significantly as we move from EURO-I to EURO-IV. EURO-I was implemented in the year 1992. Since then, European countries have been able to regulate exhaust gases emission.

For the past two decades, strict norms and regulations for limiting the emission of these harmful pollutants from vehicles have been continuously imposed [17]. Therefore, the development of advanced catalyst technologies for reducing exhaust emissions is extremely desirable. So far, various controlling techniques of automotive exhausts have been introduced, such as three-way catalysts (TWC), diesel particulate filters (DPF), diesel oxidation catalyst (DOC), NO_x storage reduction (NSR), continuously regenerating trap (CRT), and selective catalytic reduction (SCR) [18]. Among these techniques, TWC catalysts have been widely utilized for the effective reduction of all the three major pollutants (CO, HC, and NO_x) in exhaust systems. Although NO_x from gasoline engines is effectively reduced by TWC under low oxygen conditions, this technology cannot be applied to lean-burn gasoline and diesel engines in presence of excess oxygen [2, 19]. In such cases, the SCR technology exhibits an impressive reduction of NO_x, with a durable performance at a reasonable cost [20]. Both these methods have become a vital enabling technology for the reduction of emissions from buses, trucks, passenger cars, and other automobiles.

In recent years, remarkable progress has been achieved by the implementation of nanometer-scale systems due to their unique physicochemical properties compared to bulk counterparts, such as large surface-to-volume ratio, high surface area, and quantum effects [21,22]. A comprehensive review, focusing on the effect of size and shape of nanoparticles (NPs) on chemisorption and catalytic reactivity was published by Cuenya and Behafarid in 2015 [23]. In this chapter, we mainly discuss the recent advances in the development of nanocatalysts for the TWC and SCR techniques of exhaust emission reduction. Moreover, we have also discussed some theoretical aspects in regard to this study.

2. Methods
2.1 Three-way catalysis

The three-way catalysis has been an emission control approach since the 1970s, which derived its name for its simultaneous catalytic conversion of CO, HC, and NO_x, into less hazardous CO_2, H_2O, N_2, and O_2 gases under low oxygen exhaust conditions with near stoichiometric combustion [5, 24]. Fig. 2 represents the three-way catalytic converter for the removal of automobile exhaust emissions. The basic reactions involved in the TWC technology are shown below [25]:

Oxidation reactions:

$$2CO + O_2 \rightarrow 2CO_2 \qquad (2)$$

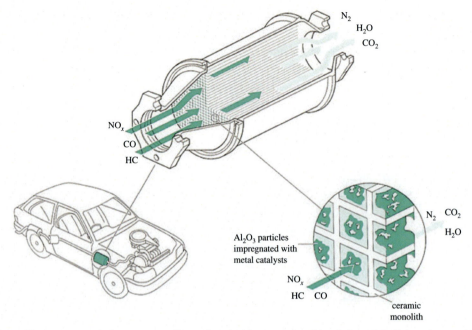

Fig. 2 Three-way catalytic converter. *(Reproduced with permission from S. Dey, N.S. Mehta, Automobile pollution control using catalysis, Resources, Environment and Sustainability 2 (2020) 100006.)*

$$\text{HC (hydrocarbons)} + O_2 \rightarrow CO_2 + H_2O \qquad (3)$$

Reduction reaction:

$$NO_x + 2CO \rightarrow N_2 + CO_2 \qquad (4)$$

Typically, the catalytic material in TWC is composed of platinum group metals (PGMs), also known as noble metals, such as Pt, Pd, Rh, Ir, Au, and Ag [2, 4]. Pt and Pd–based catalysts act as oxidation catalysts for CO and HC oxidation, whereas Rh promotes the reduction of NO_x to N_2 [26].

The preference for noble metals can be attributed to their higher specific activity for hydrogen oxidation, high resistance to deactivation and thermal sintering, superior cold-start performance, and low deactivation by sulfur in the fuel at temperatures below 500°C [27, 28]. The noble metals are usually well dispersed together with a washcoat (alumina modified with small amounts of barium or lanthanum and *ca.* 10%–20% ceria and zirconia) on a ceramic honeycomb-like structure or metallic monolith [29]. However, due to the significantly high cost of noble metals, tremendous efforts have been made by researchers to develop inexpensive, stable, and easily available nonnoble metal/metal oxide-based catalysts. In this regard, various reports on transition and nonnoble metal-based catalysts have been found to show superior catalytic activity for exhaust control like noble metals [30–35]. Moreover, precious metals in the form of nanoparticles have been found to reduce metal consumption owing to the decrease in the size of the catalyst [36, 37]. The size and morphology of catalyst largely affect the catalytic activity, and it has been observed that nanosized catalysts exhibit better catalytic activity owing to a larger specific surface area which is readily accessible for adsorption of oxygen and exhaust gases for conversion reactions [38]. Metals supported over metal oxides are also recognized as efficient catalytic systems for enhancing exhaust reduction with lower metal loading [39, 40]. In addition, catalyst support stabilizes the metal species from sintering and facilitates better dispersion of nanoparticles [41].

Ceria-based catalysts are widely utilized in TWC technology due to their ability for better dispersion of metals and higher oxygen storage capacity [42]. The generation of oxygen vacancy defects by Ce^{4+}/Ce^{3+} redox cycle facilitates the adsorption and activation of O_2 for CO oxidation [4]. However, the catalytic activity of ceria is reduced at high temperatures as a result of sintering, owing to its poor thermal stability. Therefore, the dispersion of appropriate metal ions into the ceria lattice improves its thermal stability thereby enhancing the structural and redox properties [43]. Bera et al. studied the catalytic performance of Pt/CeO_2 prepared by solution combustion method with different Pt loadings for CO oxidation. They observed that 1% Pt/CeO_2 catalyst exhibited better catalytic activity compared to Pt metal particles. It was found that the Pt ions are substituted for Ce^{4+} ions in the catalyst, thereby creating oxide ion vacancies. These oxide ion vacancies lead to a strong Pt^{2+}–CeO_2 interaction and enhance the catalytic activity of CO oxidation [44]. Preparation of ultra-small Pd particles (1.5–2.0 nm) supported over ceria was reported by Tereshchenko et al. [45]. The synthesized Pd/CeO_2 catalyst exhibited 100% CO conversion at ~50°C, unlike Pd-based catalysts which show higher conversion at higher temperatures.

The introduction of zirconia (ZrO_2) into ceria lattice improves the thermal stability and oxygen storage capacity (OSC) of the latter by creating a high concentration of defects. Various reports are devoted to the investigation of catalytic activity using different modifications of CeO_2-ZrO_2 catalysts for TWC. González-Velasco and co-workers investigated the catalytic activity of a $Pd/Ce_{0.68}Zr_{0.32}O_2$ catalyst using complex gas mixtures that simulates the nature of automotive exhaust gases [46]. It was observed that the catalysts have a surface area of $103 m^2 g^{-1}$ and show good conversion of the exhaust gases with temperature rise. Furthermore, they studied the effects of prereduction and aging treatments in cycled redox feed streams at temperature 1173 K under different compositions of oxidizing and reduction streams with a cooling environment. Yang et al. reported the synthesis of a series of Zr^{4+} doped Mn_3O_4 TWCs and found that the activities of the zirconium-doped samples were higher than that of pure Mn_3O_4 [47]. The highest conversion of CO (at 228°C), NO (at 334°C), and C_2H_6 (at 400°C) was achieved for a 20% Zr/Mn molar ratio. Yang and co-workers synthesized nickel-ceria catalysts by a facile deposition-precipitation method and investigated the structure-dependent catalytic performance of the catalysts for CO and NO removal [48]. The Ni species were distributed on two types of supports: CeO_2 nanorods with {110} exposed facets and CeO_2 nanocubes with {100} exposed facets. They observed that the shape of the ceria support has a significant effect on the type of distribution of Ni species. Experimental studies suggested that the CeO_2 exposed {110} facets possessed more active sites to strongly combine nickel species to form a more stable system (Ni−Ce−O solid solution, NiO species strongly or less weakly interacted with the surface) than CeO_2 exposed {100} facets. Therefore, a better catalytic activity was observed for Ni-CeO_2 catalyst with rod shape support.

2.2 Selective catalytic reduction

Selective catalytic reduction (SCR) of oxides of nitrogen (NO_x) has been an effective technology for the abatement of NO_x from automobile exhausts [49, 50]. In this technique, the NO_x containing exhaust gas passes over a catalyst and gets transformed into environmentally friendly N_2 and H_2O in the presence of a reductant, such as ammonia (NH_3). The key reactions in SCR are as follows [3, 51–53]:

$$4NO + 4NH_3 + O_2 \rightarrow 4N_2 + 6H_2O \tag{5}$$

$$2NO_2 + 4NH_3 + O_2 \rightarrow 3N_2 + 6H_2O \tag{6}$$

$$2NO + 2NO_2 + 4NH_3 \rightarrow 4N_2 + 6H_2O \tag{7}$$

$$6NO_2 + 8NH_3 \rightarrow 7N_2 + 12H_2O \tag{8}$$

Although the SCR technology is commonly utilized for stationary applications, its use in automobiles is comparatively new. Since the storage and handling of gaseous NH_3 are impractical in automobiles, an ammonia alternate, such as urea (NH_2CONH_2) is commonly utilized. In this technique, NH_3 is generated in situ in the vehicle by injecting a solution of urea and water into the exhaust upstream [53]. The water vapor hydrolyses urea into NH_3 which participates in selective catalytic reduction of NO_x:

$$NH_2(CO)NH_2 + H_2O \rightarrow 2NH_3 + H_2O \tag{9}$$

Shimizu and Satsuma have specified AdBlue (32.5% aqueous urea solution) as the standard precursor of NH_3 for automobiles [3, 54]. It should be noted herein that sophisticated urea injection control strategies are required to ensure proper operation without ammonia "slip." In this context, CO has attracted significant attention as a reductant in SCR as it is already present in vehicle exhaust [55].

So far, various studies have been conducted on NH_3-SCR catalysts for NO_x reduction. Noble metals (Pt, Pd, Rh, and Au) supported on ZrO_2, TiO_2, Al_2O_3, and SiO_2 supports were used as SCR catalysts in the early 1970s due to high activity at low temperatures. However, the SCR system consumed large amounts of reductant using these catalysts, as part of NH_3 gets oxidized by the noble metal catalyst [56]. Among the noble metals, Rhodium exhibits better activity for NO_x reduction than Pt and Pd in the trace presence of O_2 or SO_2 [57, 58]. Pt-based SCR catalysts are also reported to have shown superior activity for NO reduction. Eyring and co-workers investigated the NO elimination activity of a series of Pt/Al_2O_3, Cu/Al_2O_3, and $Pt-Cu/Al_2O_3$ catalysts [59]. They observed that the Pt/Al_2O_3 catalyst exhibited the highest CO oxidation efficiency and NO_x removal efficiency among the three catalysts. Efficient NO reduction using Au supported on metal oxides was reported by Ueda and Haruta [60].

In recent years, nonnoble metals, metal oxide-based NH_3-SCR catalysts have been extensively studied for SCR reactions [61–65]. The most commercially applied NH_3-SCR catalyst for vehicular applications is V_2O_5 dispersed on TiO_2 and promoted by WO_3 (V_2O_5-WO_3/TiO_2), which shows high activity and N_2 selectivity at 300–400°C [66]. Wang et al. studied the performance of NH_3-SCR with V_2O_5-WO_3/TiO_2 using different vanadia loadings (1, 3, and 5 wt%) [67]. A particle size of 10–30 nm was observed from TEM images for all the V_2O_5-WO_3/TiO_2 samples. They found that catalytic activity of the samples depends mainly on vanadium content and the maximum NO conversion is achieved with 1 wt% vanadium. Nevertheless, the synthesized V_2O_5-WO_3/TiO_2 catalyst operates only at high temperatures of 300–500°C. Ceria-based NH_3-SCR catalysts have attracted attention for the high OSC, excellent redox property, and better activity at low temperatures [68]. Bruckner and co-workers reported the low-temperature NH_3-SCR using V_2O_5 dispersed over CeO_2-TiO_2 support [69]. They observed that the synthesized catalyst shows ~100% NO conversion and N_2 selectivity at temperatures below 200°C. Based on their in situ spectroscopic investigation and the commonly accepted Mars-van Krevelen mechanism for NH_3-SCR, they reported a tentative mechanism for low temperature (LT) NH_3-SCR on $V/Ce_{1-x}Ti_xO_2$ catalysts (Scheme 1). The oxygen vacancies created near V^{4+} upon reduction of V^{5+} by ammonia are restored by lattice oxygen in the immediate vicinity. Subsequently, other vacancies created at higher distances are filled up by gas-phase oxygen. This observation of uptake of oxygen into deeper layers of catalyst lattice might be the reason for the higher catalytic activity of ceria-based oxides.

Although SCR technology is now in its third commercial generation approaching 95%–96% NO_x reduction efficiency, improvement of ammonia adsorption, reliability, device design, and catalyst development are still in progress [24]. Zhao et al. synthesized a novel Zr doped $CeVO_4$ [$Ce_{1-x}Zr_xVO_4$] as a low temperature (150–375°C) catalyst for NH_3-SCR [70]. As shown in Scheme 2, the introduction of proper amount of Zr into $CeVO_4$ leads to an increase in active oxygen species/oxygen vacancy due to the enhancement of electron interaction through the redox reactions of $V^{4+} + Zr^{4+} \leftrightarrow V^{5+} + Zr^{3+}$ and $Ce^{4+} + Zr^{3+} \leftrightarrow Ce^{3+} + Zr^{4+}$. Moreover, the Brønsted and Lewis acid sites on the catalysts are also beneficial to NH_3-SCR performance.

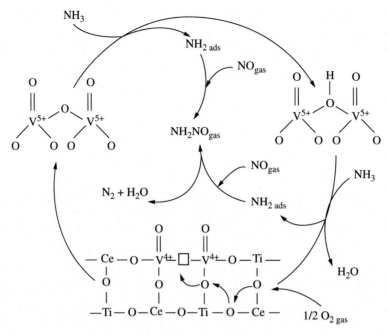

Scheme 1 Tentative mechanism of LT-NH$_3$-SCR on V/Ce$_{1-x}$Ti$_x$O$_2$ catalysts. *(Reproduced with permission from T.H. Vuong, Jo. Radnik, J. Rabeah, U. Bentrup, M. Schneider, H. Atia, U. Armbruster, W. Grünert, A. Brückner, Efficient VOx/Ce$_{1-x}$Ti$_x$O$_2$ catalysts for low-temperature NH$_3$-SCR: Reaction mechanism and active sites assessed by in situ/operando spectroscopy, ACS Catal. 7 (2017) 1693–1705.)*

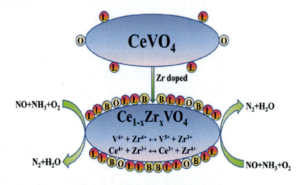

Scheme 2 The relationships of active oxygen species/oxygen vacancy, acid sites and catalytic performance over Ce$_{1-x}$Zr$_x$VO$_4$ catalysts. *(Reproduced with permission from X. Zhao, L. Huang, H. Li, H. Hu, X. Hu, L. Shi, D. Zhang, Promotional effects of zirconium doped CeVO$_4$ for the low-temperature selective catalytic reduction of NO$_x$ with NH$_3$, Appl. Catal. B. 183 (2016) 269–281.)*

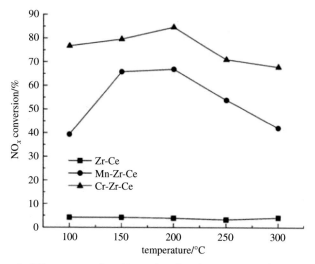

Fig. 3 SCR activity of different samples. *(Reproduced with permission from W. Li, H. Liu, Y. Chen, Promotion of transition metal oxides on the NH$_3$-SCR performance of ZrO$_2$-CeO$_2$ catalyst, Front Environ Sci Eng. 11 (2017).)*

Promotion of transition metal oxides (MnO$_x$ and CrO$_x$) on the low temperature (100–300°C) NH$_3$-SCR activity over ZrO$_2$-CeO$_2$ catalyst was investigated by Chen and co-workers [71]. They observed that the particle size for both Cr-Zr-Ce and Mn-Zr-Ce catalysts was ~5 nm unlike Zr-Ce (15 nm). Also, the addition of Cr and Mn into Zr-Ce composite led to high specific surface area and more surface oxygen species. Therefore, the Cr-Zr-Ce and Mn-Zr-Ce catalysts were found to be much more active than ZrO$_2$-CeO$_2$ binary oxide as shown in Fig. 3.

Recently, Xu et al. investigated the NH$_3$-SCR of NO$_x$ using Sb-containing CeO$_2$-ZrO$_2$ catalyst and found that SbCeZr performed better than SbZr and SbCe, and showed the highest activity with 80% NO conversion within the temperature range 202–422°C as shown in Fig. 4 [72]. They reported that the presence of Sb in the metal oxides resulted in a difference in acid distribution and redox property, which is beneficial to NH$_3$ adsorption and NO oxidation.

3. Theoretical studies

Numerous theoretical studies [73–76] have been performed for CO oxidation as well as NO reduction on the metal surface to find the mechanistic nature of the reaction. The oxidation of CO on metal surfaces is a typical example in catalysis that has been investigated comprehensively. CO oxidation is based on two distinguished mechanisms, the Langmuir-Hinshelwood (LH) mechanism and Eley-Rideal (ER) mechanism [77–79]. Some of the theoretical studies showed CO oxidation following mostly Langmuir–Hinshelwood

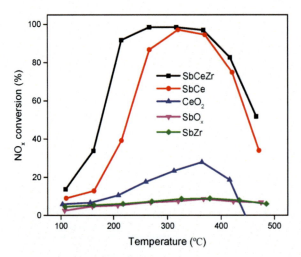

Fig. 4 NO$_x$ conversions over the catalysts (reaction conditions: 500 ppm NO, 500 ppm NH$_3$, 5% O$_2$, with Ar balance). *(Reproduced with permission from Q. Xu, D. Liu, C. Wang, W. Zhan, Y. Guo, L. Wang, Q. Ke, M.N. Ha. Catalysts 10 (2020) 1154.)*

(LH) mechanism [80, 81]. The elementary steps for CO oxidation on surfaces using LH mechanism have been proposed as (i) CO adsorption on surfaces (ii) O$_2$ adsorption and dissociation into O atoms on surfaces; and (iii) Combination of CO + O to form CO$_2$ on surface. However, the mechanism may vary with respect to different reaction conditions. In contrary to the above findings, some of the studies [82, 83] showed CO oxidation proceeding via Eley Rideal (ER) mechanism. The Eley Rideal mechanism for CO oxidation undergoes as follows: (i) O$_2$ adsorption on surfaces (ii) Combination of O + CO (in the gas phase) to form CO$_2$.

Moreover, a lot of theoretical studies [84–86] have been performed on noble metal surfaces for CO oxidation. Since Haruta et al. [87, 88] reported that the small gold nanoparticles supported on the 3d transition metal oxides exhibit remarkable catalytic effect toward CO oxidation at low temperatures, attention shifted toward Au nanocatalysts for CO oxidation reaction. An et al. [77] investigated the reaction mechanism of CO oxidation using unsupported single-walled helical gold nanotubes as well as structural and electronic properties of adsorbates and adsorbents. It was established that the CO oxidation on the Au-nanotube is most likely to proceed through LH mechanism and its high activity was attributed to the presence of under coordinated Au sites in the helical geometry. Chen et al. [89] applied the first-principles study to investigate the complex graphene-based gold catalyst and found that the reaction for CO oxidation would start with the LH process to release a CO$_2$ molecule, then to be followed by the ER process. Gao et al. [90] did Ab initio calculations to study the CO oxidation on six endohedral gold-cage clusters and found Nb@Au$_{13}$, Zr@Au$_{14}$, and Sc@Au$_{15}$

to be very effective nanocatalysts toward CO oxidation. Recently, a novel trimolecular LH mechanism, also termed as CO self-promoting oxidation, has been proposed by Liu et al. [91]. They found that the protruded triangular Au_3 site acts as the main active site and the co-adsorbed CO molecule at a triangular Au_3 active site acts as an electrophilic agent which can serve as a promoter for the scission of the O−O bond. Spendelow et al. [92] considered the mechanism of CO oxidation on well-ordered and disordered Pt(111) in aqueous NaOH solutions and proposed that the improved oxidation of CO in alkaline media is credited to the higher affinity of the Pt(111) surface for the adsorption of OH at low potentials. Duan et al. [93] performed density functional theory calculations to model CO oxidation on the Pd(111) surface using both LH and ER mechanisms. They observed that threefold oxygen atoms in the surface readily react with CO molecules following an Eley–Rideal mechanism. Studies have also been done on nonnoble metal surfaces in order to make a cost-effective catalyst. Theoretical investigation for CO oxidation on Ni_1/FeO_x was done by Liang et al. [94]. They reported that on comparing Ni_1/FeO_x with noble-metal catalysts Pt_1/FeO_x and Ir_1/FeO_x, its activity for CO oxidation is found to be comparable to that of Pt_1/FeO_x and higher than that of Ir_1/FeO_x. Lu et al. [95] explained the direct CO oxidation using lattice oxygen of SnO_2 (110) surface.

The reduction of NO on metals is a fascinating route since NO can be either reduced (e.g., N_2, N_2O) or oxidized (e.g., NO_2). Rh catalysts are believed to exhibit high catalytic activity toward NO_x reduction as it facilitates NO dissociation quite profoundly. Thus, density functional theory (DFT) plays an important role in understanding the behavior of NO molecules adsorbed on the Rh surface. Some theoretical studies [96, 97] have shown that more than half of the NO molecules adsorbed onto the Rh(111) surface readily dissociate into N and O atoms at 100 K, whereas NO dissociation does not take place on the Pt(111) and Pd(111) surfaces. Many experimental and theoretical studies have shown that NO dissociation is accelerated on the step and edge sites of the metal surfaces which indicates that small Rh particles are expected to display more reactivity toward facile NO dissociation [98, 99]. Efforts have been made by experimentalists to develop new series of catalysts which should not only be cost-effective but also catalytically active as noble metal surfaces. Zhao et al. [100] performed DFT study on the surface of vanadium-titanium-based selective catalytic reduction (SCR) denitrification catalysts. Lai et al. [101] did extensive work on the SCR of NO with NH_3 by Supported $V_2O_5-WO_3/TiO_2$ catalysts.

Recently, a lot of efforts have been directed toward studying the reaction mechanism of NO catalytic oxidation as it is an important process to mitigate NO_x concentration. The reaction mechanisms of NO oxidation over different catalysts obeyed either ER mechanism or LH mechanism or the combination of ER and LH mechanisms. Olsson et al. [102] proposed an ER model for NO oxidation over Pt/Al_2O_3 and found Pt-O + NO → Pt-NO_2 as the rate-determining step. Li et al. [103] studied NO oxidation on Pt/TiO$_2$ catalyst and proposed that the main pathway for NO$_2$ formation is through the direct oxidation of gaseous NO to NO$_2$. Bhatia et al. [104] investigated the kinetics of NO oxidation on Pt/Al$_2$O$_3$ and suggested NO oxidation pathway to follow the LH mechanism. Zhong et al. [105] proposed the possible reaction pathways of NO oxidation over cerium modified Cr/Ti-PILC catalyst. They found that the strong interaction between O$_2^-$ and NO resulted in the formation of free nitrates and the formation of nitrates was clearly improved via superoxide ions. From the above discussion, it can be proposed that both ER and LH mechanisms are efficient toward the reduction of exhaust gases.

4. Conclusion and future scope

In this chapter, we have described different nanotechnology techniques for exhaust emission reduction. Among the various exhaust after-treatment methods, the three-way catalytic converter (TWC), selective catalytic reduction (SCR) methods have been found to be very effective in reducing exhaust gas emissions (CO, NO$_x$, CH) to a large extent. Moreover, an attempt has been made to review the recent advances in metal/metal oxide-based nanocatalysts for automobile exhaust emission reduction. The development in metal oxide-supported metal nanocatalysts has shown significant scope to replace the expensive noble metal catalysts. However, further research on the development of highly stable, low-cost, high oxygen storage capacity, and wide working temperature-window catalysts using nonnoble metals is required to meet the future challenges of exhaust emissions. Furthermore, a large reduction in NO$_x$ emissions can be achieved by utilizing modern combustion techniques along with modified injection timings to reach its peak position in performance. Since the catalytic treatment of exhaust pollutants is considered as environmentally benign, it is highly preferable to investigate nanosized metal/metal oxide-based catalysts for the efficient removal of automotive exhaust emissions. Theoretical studies suggest that both Langmuir-Hinshelwood (LH) and Eley-Rideal (ER) mechanisms are preferred paths for CO oxidation and NO reduction depending upon the catalysts and the reaction conditions. Some recent studies have shown nonnoble metals to be as efficient as noble ones for the reduction of exhaust gases. The mentioned techniques and theoretical aspects would help develop a set of novel nanocatalysts with better performance in reducing exhaust emissions and improving engine performance in future.

References

[1] A.J. Cohen, M. Brauer, R. Burnett, H.R. Anderson, J. Frostad, K. Estep, M.H. Forouzanfar, Estimates and 25-year trends of the global burden of disease attributable to ambient air pollution: an analysis of data from the global burden of diseases study 2015, Lancet 389 (10082) (2017) 1907–1918.
[2] S. Dey, N.S. Mehta, Automobile pollution control using catalysis, Resour. Environ. Sustain. 2 (2020) 1–13. 100006.

[3] P. Brijesh, S. Sreedhara, Exhaust emissions and its control methods in compression ignition engines: a review, Int. J. Automot. Technol. 14 (2) (2013) 195–206.
[4] B. Tudu, R. Bortamuly, P. Saikia, Nanosized metal/metal oxides for auto-exhaust purification, in: Advanced Heterogeneous Catalysts Volume 1: Applications at the Nano-Scale, ACS Symposium Series, 2020, pp. 373–401.
[5] G. Gerasimov, M. Pogosbekian, Recent advances in nanostructured catalysts for vehicle exhaust gas treatment, Nanotechnol. Environ. Sci. (2018) 39–70.
[6] T. Zhu, X. Zhang, W. Bian, Y. Han, T. Liu, H. Liu, DeNO$_x$ of nanocatalyst of selective catalytic reduction using active carbon loading MnO$_x$-Cu at low temperature, Catalysts 10 (1) (2020) 1–17. 135.
[7] J. Wang, H. Chen, Z. Hu, M. Yao, Y. Li, A review on the Pd-based three-way catalyst, Catal. Rev. 57 (1) (2015) 79–144.
[8] E. Sher, Handbook of Air Pollution from Internal Combustion Engines: Pollutant Formation and Control, Academic Press, 1998.
[9] G.D. Kelen, J.S. Stapczynski, in: J.E. Tintinalli (Ed.), Emergency Medicine: A Comprehensive Study Guide, Mcgraw-hill, New York, 1985, pp. 218–221.
[10] M. Goldstein, Carbon monoxide poisoning, J. Emerg. Nurs. 34 (6) (2008) 538–542.
[11] F.S. Mirhashemi, H. Sadrnia, NO$_X$ emissions of compression ignition engines fueled with various biodiesel blends: a review, J. Energy Inst. 93 (1) (2020) 129–151.
[12] S.K. Hoekman, C. Robbins, Review of the effects of biodiesel on NO$_x$ emissions, Fuel Process. Technol. 96 (2012) 237–249.
[13] M. Kampa, E. Castanas, Human health effects of air pollution, Environ. Pollut. 151 (2) (2008) 362–367.
[14] Y. Deng, J. Li, Y. Li, R. Wu, S. Xie, Characteristics of volatile organic compounds, NO$_2$, and effects on ozone formation at a site with high ozone level in Chengdu, J. Environ. Sci. 75 (2019) 334–345.
[15] Y. Wang, H. Liu, C.F.F. Lee, Particulate matter emission characteristics of diesel engines with biodiesel or biodiesel blending: a review, Renew. Sust. Energ. Rev. 64 (2016) 569–581.
[16] L. Wei, P. Geng, A review on natural gas/diesel dual fuel combustion, emissions and performance, Fuel Process. Technol. 142 (2016) 264–278.
[17] E. Meloni, V. Palma, Most recent advances in diesel engine catalytic soot abatement: structured catalysts and alternative approaches, Catalysts 10 (7) (2020) 745.
[18] N. Biswakarma, P.J. Sarma, S.D. Baruah, N.K. Gour, R.C. Deka, Catalytic oxidation of NO on [Au–M]$^-$ (M = Pd and Pt) bimetallic dimers: an insight from density functional theory approach, J. Phys. Chem. C 124 (5) (2020) 3059–3068.
[19] J. Li, H. Chang, L. Ma, J. Hao, R.T. Yang, Low-temperature selective catalytic reduction of NO$_x$ with NH$_3$ over metal oxide and zeolite catalysts—a review, Catal. Today 175 (1) (2011) 147–156.
[20] S. Chaturvedi, P.N. Dave, N.K. Shah, Applications of nano-catalyst in new era, J. Saudi Chem. Soc. 16 (3) (2012) 307–325.
[21] C. Ruehl, J.D. Smith, Y. Ma, J.E. Shields, M. Burnitzki, W. Sobieralski, R. Ianni, D.J. Chernich, M.-C.O. Chang, J.F. Collins, S. Yoon, D. Quiros, S. Hu, H. Dwyer, Emissions during and real-world frequency of heavy-duty diesel particulate filter regeneration, Environ. Sci. Technol. 52 (10) (2018) 5868–5874.
[22] J.C. Summers, L.L. Hegedus, Effects of platinum and palladium impregnation on the performance and durability of automobile exhaust oxidizing catalysts, J. Catal. 51 (2) (1978) 185–192.
[23] B.R. Cuenya, F. Behafarid, Nanocatalysis: size-and shape-dependent chemisorption and catalytic reactivity, Surf. Sci. Rep. 70 (2) (2015) 135–187.
[24] T. Johnson, Vehicular emissions in review, SAE Int. J. Engines 6 (2) (2013) 699–715.
[25] C. Huang, W. Shan, Z. Lian, Y. Zhang, H. He, Recent advances in three-way catalysts of natural gas vehicles, Cat. Sci. Technol. 10 (19) (2020) 6407–6419.
[26] B.M.V. Twigg, Haren Gandhi 1941-2010: contributions to the development and implementation of catalytic emissions control systems, Platin. Met. Rev. 55 (1) (2011) 43–53.
[27] B.E. Milton, Control technologies in spark-ignition engines, in: Handbook of Air Pollution From Internal Combustion Engines: Pollutant Formation and Control, Academic Press, New York, 1998, pp. 189–258.
[28] J. Waikar, H. Pawar, P. More, Review on CO oxidation by noble and nonnoble metal based catalyst, Catal. Green Chem. Eng. 2 (1) (2019).
[29] A. Crucq, Frennet, A. (Eds.),, Catalysis and Automotive Pollution Control, Elsevier, 1987.

[30] T.J. Huang, D.H. Tsai, CO oxidation behavior of copper and copper oxides, Catal. Lett. 87 (3) (2003) 173–178.
[31] D. Jampaiah, P. Venkataswamy, V.E. Coyle, B.M. Reddy, S.K. Bhargava, Low-temperature CO oxidation over manganese, cobalt, and nickel doped CeO_2 nanorods, RSC Adv. 6 (84) (2016) 80541–80548.
[32] S. Royer, D. Duprez, Catalytic oxidation of carbon monoxide over transition metal oxides, ChemCatChem 3 (1) (2011) 24–65.
[33] X. Wang, W. Wu, Z. Chen, R. Wang, Bauxite-supported transition metal oxides: promising low-temperature and SO_2-tolerant catalysts for selective catalytic reduction of NO_x, Sci. Rep. 5 (1) (2015) 1–6.
[34] P.G. Smirniotis, D.A. Peña, B.S. Uphade, Low-temperature selective catalytic reduction (SCR) of NO with NH_3 by using Mn, Cr, and cu oxides supported on Hombikat TiO_2, Angew. Chem. Int. Ed. 40 (13) (2001) 2479–2482.
[35] F. Liu, H. He, L. Xie, XAFS study on the specific deoxidationbehavior of iron titanate catalyst for the selective catalytic reduction of NO_x with NH_3, ChemCatChem 5 (12) (2013) 3760–3769.
[36] A.M. Gololobov, I.E. Bekk, G.O. Bragina, V.I. Zaikovskii, A.B. Ayupov, N.S. Telegina, V.I. Bukhtiyarov, A.Y. Stakheev, Platinum nanoparticle size effect on specific catalytic activity in n-alkane deep oxidation: dependence on the chain length of the paraffin, Kinet. Catal. 50 (6) (2009) 830–836.
[37] F. Behafarid, L.K. Ono, S. Mostafa, J.R. Croy, G. Shafai, S. Hong, T.S. Rahman, S.R. Bare, B.R. Cuenya, Electronic properties and charge transfer phenomena in Pt nanoparticles on γ-Al_2O_3: size, shape, support, and adsorbate effects, Phys. Chem. Chem. Phys. 14 (33) (2012) 11766–11779.
[38] Z.Y. Zhou, N. Tian, J.T. Li, I. Broadwell, S.G. Sun, Nanomaterials of high surface energy with exceptional properties in catalysis and energy storage, Chem. Soc. Rev. 40 (7) (2011) 4167–4185.
[39] X.Y. Liu, A. Wang, T. Zhang, C.Y. Mou, Catalysis by gold: new insights into the support effect, Nano Today 8 (4) (2013) 403–416.
[40] T. Zheng, H.E. Junjun, Z. Yunkun, X.I.A. Wenzheng, H. Jieli, Precious metal-support interaction in automotive exhaust catalysts, J. Rare Earths 32 (2) (2014) 97–107.
[41] B.M. Reddy, J.L.G. Fierro, Metal Oxides, Chemistry and Applications, CRC, Boca Raton, 2006.
[42] A. Trovarelli, J. Llorca, Ceria catalysts at nanoscale: how do crystal shapes shape catalysis? ACS Catal. 7 (7) (2017) 4716–4735.
[43] B.M. Reddy, L. Katta, G. Thrimurthulu, Novel nanocrystalline $Ce_{1-x}La_xO_{2-\delta}$ (x = 0.2) solid solutions: structural characteristics and catalytic performance, Chem. Mater. 22 (2) (2010) 467–475.
[44] P. Bera, A. Gayen, M.S. Hegde, N.P. Lalla, L. Spadaro, F. Frusteri, F. Arena, Promoting effect of CeO_2 in combustion synthesized Pt/CeO_2 catalyst for CO oxidation, J. Phys. Chem. B 107 (25) (2003) 6122–6130.
[45] A. Tereshchenko, V. Polyakov, A. Guda, T. Lastovina, Y. Pimonova, A. Bulgakov, A. Tarasov, L. Kustov, V. Butova, A. Trigub, A. Soldatov, Ultra-small Pd nanoparticles on ceria as an advanced catalyst for CO oxidation, Catalysts 9 (4) (2019) 385. 1–14.
[46] M.P. González-Marcos, B. Pereda-Ayo, A. Aranzabal, J.A. González-Marcos, J.R. González-Velasco, On the effect of reduction and ageing on the TWC activity of $Pd/Ce_{0.68}Zr_{0.32}O_2$ under simulated automotive exhausts, Catal. Today 180 (1) (2012) 88–95.
[47] G. Zhao, J. Li, W. Zhu, X. Ma, Y. Guo, Z. Liu, Y. Yang, Mn_3O_4 doped with highly dispersed Zr species: a new nonnoble metal oxide with enhanced activity for three-way catalysis, New J. Chem. 40 (12) (2016) 10108–10115.
[48] K. Tang, W. Liu, J. Li, J. Guo, J. Zhang, S. Wang, S. Niu, Y. Yang, The effect of exposed facets of ceria to the nickel species in nickel-ceria catalysts and their performance in a NO + CO reaction, ACS Appl. Mater. Interfaces 7 (48) (2015) 26839–26849.
[49] J. Wang, H. Zhao, G. Haller, Y. Li, Recent advances in the selective catalytic reduction of NO_x with NH_3 on Cu-Chabazite catalysts, Appl. Catal. B Environ. 202 (2017) 346–354.
[50] P. Forzatti, I. Nova, E. Tronconi, Enhanced NH_3 selective catalytic reduction for NO_x abatement, Angew. Chem. 121 (44) (2009) 8516–8518.

[51] M. Rodkin, S.J. Tauster, X. Wei, T. Neubauer, Purification of Automotive Exhaust Through Catalysis, ACS Symposium Series, 2008.
[52] J. Kašpar, P. Fornasiero, N. Hickey, Automotive catalytic converters: current status and some perspectives, Catal. Today 77 (4) (2003) 419–449.
[53] J.K. Hochmuth, K. Wassermann, R.J. Farrauto, Car exhaust cleaning, in: Comprehensive Inorganic Chemistry II, Elsevier, 2013, pp. 505–523.
[54] K.I. Shimizu, A. Satsuma, Hydrogen assisted urea-SCR and NH_3-SCR with silver–alumina as highly active and SO_2-tolerant de-NO_x catalysis, Appl. Catal. B Environ. 77 (1–2) (2007) 202–205.
[55] K. Liu, A.I. Rykov, J. Wang, T. Zhang, Recent advances in the application of Mößbauer spectroscopy in heterogeneous catalysis, Adv. Catal. 58 (2015) 1–142.
[56] X. Gou, K. Zhang, L.S. Liu, W.Y. Liu, Z.F. Wang, G. Yang, J.X. Wu, E.Y. Wang, Study on noble metal catalyst for selective catalytic reduction of NO_x at low temperature, in: Applied Mechanics and Materials, vol. 448, Trans Tech Publications Ltd., 2014, pp. 885–889.
[57] T. Nakatsuji, T. Yamaguchi, N. Sato, H. Ohno, A selective NO_x reduction on Rh-based catalysts in lean conditions using CO as a main reductant, Appl. Catal. B Environ. 85 (1–2) (2008) 61–70.
[58] M. Jabłońska, R. Palkovits, It is no laughing matter: nitrous oxide formation in diesel engines and advances in its abatement over rhodium-based catalysts, Cat. Sci. Technol. 6 (21) (2016) 7671–7687.
[59] M. Kang, D.J. Kim, E.D. Park, J.M. Kim, J.E. Yie, S.H. Kim, L.H. Weeks, E.M. Eyring, Two-stage catalyst system for selective catalytic reduction of NO_x by NH_3 at low temperatures, Appl. Catal. B Environ. 68 (1–2) (2006) 21–27.
[60] A. Ueda, M. Haruta, Reduction of nitrogen monoxide with propene over au/Al_2O_3 mixed mechanically with Mn_2O_3, Appl. Catal. B Environ. 18 (1–2) (1998) 115–121.
[61] Y. Shan, X. Shi, J. Du, Y. Yu, H. He, Cu-exchanged RTH-type zeolites for NH_3-selective catalytic reduction of NO_x: Cu distribution and hydrothermal stability, Cat. Sci. Technol. 9 (1) (2019) 106–115.
[62] H. Zhang, N. Li, L. Li, A. Wang, X. Wang, T. Zhang, Selective catalytic reduction of NO with CH_4 over in-Fe/sulfated zirconia catalysts, Catal. Lett. 141 (10) (2011) 1491–1497.
[63] Y. Dong, X. Wang, H. Liu, H. Wang, New insights into the relationships between performance and physicochemical properties of FeO_x–NbO_x mixed oxide catalysts for the NH_3-SCR reactions, Waste Dispos. Sustain. Energy 3 (2021) 97–106.
[64] X. Zhao, L. Huang, S. Namuangruk, H. Hu, X. Hu, L. Shi, D. Zhang, Morphology-dependent performance of Zr–$CeVO_4$/TiO_2 for selective catalytic reduction of NO with NH_3, Cat. Sci. Technol. 6 (14) (2016) 5543–5553.
[65] A. Dankeaw, F. Gualandris, R.H. Silva, K. Norrman, M.G. Sørensen, K.K. Hansen, B. Ksapabutr, V. Esposito, D. Marani, Amorphous saturated cerium–tungsten–titanium oxide nanofiber catalysts for NO_x selective catalytic reaction, New J. Chem. 42 (12) (2018) 9501–9509.
[66] S.S.R. Putluru, L. Schill, A. Godiksen, R. Poreddy, S. Mossin, A.D. Jensen, R. Fehrmann, Promoted V_2O_5/TiO_2 catalysts for selective catalytic reduction of NO with NH_3 at low temperatures, Appl. Catal. B Environ. 183 (2016) 282–290.
[67] C. Wang, S. Yang, H. Chang, Y. Peng, J. Li, Dispersion of tungsten oxide on SCR performance of $V_2O_5WO_3$/TiO_2: acidity, surface species and catalytic activity, Chem. Eng. J. 225 (2013) 520–527.
[68] G. Qi, R.T. Yang, A superior catalyst for low-temperature NO reduction with NH_3, Chem. Commun. 7 (2003) 848–849.
[69] T.H. Vuong, J. Radnik, J. Rabeah, U. Bentrup, M. Schneider, H. Atia, U. Armbruster, W. Grünert, A. Brückner, Efficient VO_x/$Ce_{1-x}Ti_xO_2$ catalysts for low-temperature NH_3-SCR: reaction mechanism and active sites assessed by in situ/operando spectroscopy, ACS Catal. 7 (3) (2017) 1693–1705.
[70] X. Zhao, L. Huang, H. Li, H. Hu, X. Hu, L. Shi, D. Zhang, Promotional effects of zirconium doped $CeVO_4$ for the low-temperature selective catalytic reduction of NO_x with NH_3, Appl. Catal. B Environ. 183 (2016) 269–281.
[71] L. Weiman, L. Haidi, C. Yunfa, Promotion of transition metal oxides on the NH_3-SCR performance of ZrO_2-CeO_2 catalyst, 2nd, 11, Springer, 2017, pp. 1–9. 6.
[72] Q. Xu, D. Liu, C. Wang, W. Zhan, Y. Guo, Y. Guo, L. Wang, Q. Ke, M.N. Ha, Sb-containing metal oxide catalysts for the selective catalytic reduction of NO_x with NH_3, Catalysts 10 (10) (2020) 1154. 1–17.

[73] I.N. Remediakis, N. Lopez, J.K. Nørskov, CO oxidation on gold nanoparticles: theoretical studies, Appl. Catal. A Gen. 291 (1–2) (2005) 13–20.
[74] C. Dupont, Y. Jugnet, D. Loffreda, Theoretical evidence of PtSn alloy efficiency for CO oxidation, J. Am. Chem. Soc. 128 (28) (2006) 9129–9136.
[75] J. Jin, N. Sun, W. Hu, H. Yuan, H. Wang, P. Hu, Insight into room-temperature catalytic oxidation of nitric oxide by Cr_2O_3: a DFT study, ACS Catal. 8 (6) (2018) 5415–5424.
[76] D. Mei, J. Du, M. Neurock, First-principles-based kinetic Monte Carlo simulation of nitric oxide reduction over platinum nanoparticles under lean-burn conditions, Ind. Eng. Chem. Res. 49 (21) (2010) 10364–10373.
[77] W. An, Y. Pei, X.C. Zeng, CO oxidation catalyzed by single-walled helical gold nanotube, Nano Lett. 8 (1) (2008) 195–202.
[78] S. Nigam, C. Majumder, CO oxidation by BN−fullerene cage: effect of impurity on the chemical reactivity, ACS Nano 2 (7) (2008) 1422–1428.
[79] D. Tang, C. Hu, DFT insight into CO oxidation catalyzed by gold nanoclusters: charge effect and multi-state reactivity, J. Phys. Chem. Lett. 2 (23) (2011) 2972–2977.
[80] A. Tripathi, R. Thapa, Promoting reactivity of graphene based catalysts to achieve LH mechanism for CO oxidation, Catal. Today 370 (2020) 142–150.
[81] F. Li, Y. Li, X.C. Zeng, Z. Chen, Exploration of high-performance single-atom catalysts on support M_1/FeO_x for CO oxidation via computational study, ACS Catal. 5 (2) (2015) 544–552.
[82] G. Xu, R. Wang, F. Yang, D. Ma, Z. Yang, Z. Lu, CO oxidation on single Pd atom embedded defect-graphene via a new termolecular Eley-Rideal mechanism, Carbon 118 (2017) 35–42.
[83] J.T. Hirvi, T.J.J. Kinnunen, M. Suvanto, T.A. Pakkanen, J.K. Nørskov, CO oxidation on PdO surfaces, J. Chem. Phys. 133 (8) (2010), 084704.
[84] S. Karmakar, C. Chowdhury, A. Datta, Noble-metal-supported GeS monolayer as promising single-atom catalyst for CO oxidation, J. Phys. Chem. C 122 (26) (2018) 14488–14498.
[85] Z.P. Liu, P. Hu, CO oxidation and NO reduction on metal surfaces: density functional theory investigations, Top. Catal. 28 (1) (2004) 71–78.
[86] L.M. Molina, B. Hammer, Theoretical study of CO oxidation on au nanoparticles supported by MgO (100), Phys. Rev. B 69 (15) (2004) 155424.
[87] M. Haruta, T. Kobayashi, H. Sano, N. Yamada, Novel gold catalysts for the oxidation of carbon monoxide at a temperature far below 0 °C, Chem. Lett. 16 (2) (1987) 405–408.
[88] M. Haruta, N. Yamada, T. Kobayashi, S. Iijima, Gold catalysts prepared by coprecipitation for low-temperature oxidation of hydrogen and of carbon monoxide, J. Catal. 115 (2) (1989) 301–309.
[89] G. Chen, S.J. Li, Y. Su, V. Wang, H. Mizuseki, Y. Kawazoe, Improved stability and catalytic properties of Au_{16} cluster supported on graphane, J. Phys. Chem. C 115 (41) (2011) 20168–20174.
[90] Y. Gao, N. Shao, S. Bulusu, X.C. Zeng, Effective CO oxidation on endohedral gold-cage nanoclusters, J. Phys. Chem. C 112 (22) (2008) 8234–8238.
[91] C. Liu, Y. Tan, S. Lin, H. Li, X. Wu, L. Li, X.C. Zeng, CO self-promoting oxidation on nanosized gold clusters: triangular Au_3 active site and CO induced O–O scission, J. Am. Chem. Soc. 135 (7) (2013) 2583–2595.
[92] J.S. Spendelow, J.D. Goodpaster, P.J.A. Kenis, A. Wieckowski, Mechanism of CO oxidation on Pt (111) in alkaline media, J. Phys. Chem. B 110 (19) (2006) 9545–9555.
[93] Z. Duan, G. Henkelman, CO oxidation on the Pd (111) surface, ACS Catal. 4 (10) (2014) 3435–3443.
[94] J.X. Liang, X.F. Yang, A. Wang, T. Zhang, J. Li, Theoretical investigations of nonnoble metal single-atom catalysis: Ni1/FeOx for CO oxidation, Cat. Sci. Technol. 6 (18) (2016) 6886–6892.
[95] Z. Lu, D. Ma, L. Yang, X. Wang, G. Xu, Z. Yang, Direct CO oxidation by lattice oxygen on the SnO_2 (110) surface: a DFT study, Phys. Chem. Chem. Phys. 16 (24) (2014) 12488–12494.
[96] R.J. Gorte, L.D. Schmidt, J.L. Gland, Binding states and decomposition of NO on single crystal planes of Pt, Surf. Sci. 109 (2) (1981) 367–380.
[97] C.S. Gopinath, F. Zaera, A molecular beam study of the kinetics of the catalytic reduction of NO by CO on Rh (111) single-crystal surfaces, J. Catal. 186 (2) (1999) 387–404.
[98] R.D. Ramsier, Q. Gao, H.N. Waltenburg, K.W. Lee, O.W. Nooij, L. Lefferts, J.T. Yates Jr., NO adsorption and thermal behavior on Pd surfaces. A detailed comparative study, Surf. Sci. 320 (3) (1994) 209–237.

[99] J. Wintterlin, S. Völkening, T.V.W. Janssens, T. Zambelli, G. Ertl, Atomic and macroscopic reaction rates of a surface-catalyzed reaction, Science 278 (5345) (1997) 1931–1934.
[100] Z. Zhao, E. Li, Y. Qin, X. Liu, Y. Zou, H. Wu, T. Zhu, Density functional theory (DFT) studies of vanadium-titanium based selective catalytic reduction (SCR) catalysts, J. Environ. Sci. 90 (2020) 119–137.
[101] J.K. Lai, I.E. Wachs, A perspective on the selective catalytic reduction (SCR) of NO with NH_3 by supported V_2O_5–WO_3/TiO_2 catalysts, ACS Catal. 8 (7) (2018) 6537–6551.
[102] L. Olsson, B. Westerberg, H. Persson, E. Fridell, M. Skoglundh, B. Andersson, A kinetic study of oxygen adsorption/desorption and NO oxidation over Pt/Al_2O_3 catalysts, J. Phys. Chem. B 103 (47) (1999) 10433–10439.
[103] L. Li, J. Cheng, Z. Hao, Catalytic oxidation of NO over TiO_2 supported platinum clusters. II: mechanism study by in situ FTIR spectra, Catal. Today 158 (3–4) (2010) 361–369.
[104] D. Bhatia, R.W. McCabe, M.P. Harold, V. Balakotaiah, Experimental and kinetic study of NO oxidation on model Pt catalysts, J. Catal. 266 (1) (2009) 106–119.
[105] L. Zhong, W. Cai, Q. Zhong, Evaluation of cerium modification over Cr/Ti-PILC for NO catalytic oxidation and their mechanism study, RSC Adv. 4 (82) (2014) 43529–43537.

CHAPTER 26

Automobile exhaust nanocatalysts

Kevin V. Alex[a], K. Kamakshi[b], J.P.B. Silva[c], S. Sathish[a], and K.C. Sekhar[a]
[a]Department of Physics, School of Basic and Applied Sciences, Central University of Tamil Nadu, Thiruvarur, Tamil Nadu, India
[b]Department of Science and Humanities, Indian Institute of Information Technology Tiruchirappalli, Tiruchirappalli, Tamil Nadu, India
[c]Centre of Physics of Minho and Porto Universities (CF-UM-UP), Campus de Gualtar, Braga, Portugal

Chapter outline

1. Introduction — 529
   1.1 Exhaust system of vehicles — 533
2. Catalytic convertor — 534
   2.1 Different types of catalytic convertors in an internal combustion engine — 535
3. Nanocatalysts — 537
   3.1 Platinum group metals (PGMs) — 537
   3.2 Nonplatinum group metals — 542
   3.3 Metal oxide-based catalysts — 544
   3.4 Carbon nanostructures — 546
   3.5 Perovskite-based nanocatalysts — 547
4. Wash-coat compositions and oxygen storage components (OSC) — 548
5. Catalytic convertors in diesel and learn-burn gasoline engines — 551
   5.1 Selective catalytic reduction (SCR) — 551
   5.2 Lean NO_x trap (LNT) — 553
6. Conclusion — 555
7. Future prospects — 556
Acknowledgments — 556
References — 556

1. Introduction

The automotive industry has become an integral part of human life since its inception and has become one of the important drivers of economic growth worldwide. The invention of automobiles has revolutionized the various dimensions of mobility, transportation, and motorsport entertainment and led to a highly interconnected world as seen today. However, its implications on the environmental and ecological imbalance become a flashpoint of huge social concern for a sustainable future. The lack of inefficient, cost-effective, and eco-friendly alternative energy sources increases the dependence on nonrenewable energy sources, which in turn proliferates petroleum fuel consumption and harmful gas emissions. The deleterious gas emissions from automobiles are similar to the effect of "slow poisoning" and thus remains catastrophic to the entire life on our planet [1–3]. The vehicular gas emissions can be broadly categorized into two such as

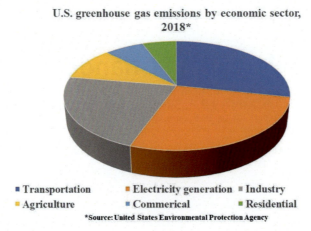

Fig. 1 The economic sector-wise greenhouse gas emissions in the United States, 2018 [6].

(a) greenhouse gases and (b) air pollutant emissions. Carbon dioxide (CO_2), nitrous oxide (N_2O), methane (CH_4), etc., are the greenhouse gas emissions from automobiles that cause serious climatic permutations due to the greenhouse effect and global warming [4, 5]. The transportation sector primarily constitutes for the total greenhouse gas emissions (28.2% as of 2018 as shown in Fig. 1) [6].

Air pollutant emissions are constituted mainly by CO_2, hydrocarbons (HC), nitrogen oxides (NO_x), and particulate matter result in the formation of smog and can cause serious health issues such as respiratory diseases, cancer, and cardiovascular disorders. Among all, CO_2 itself contributes to 56% of the emitted greenhouse gases (GHG) and in 2016 the concentration of GHG reached to value of $CO_2 > 400$ ppm [7]. The Paris agreement long-term goals are ambitious and for the EU it is expected a "40% reduction compared to 1990 levels" by 2030 [8]. To meet EU goals, it is of utmost importance to develop cost-effective materials and low-carbon energy technologies for CO_2 emissions mitigation.

CO_2 is the principal contributor to the exhaust emission from vehicles. The small quantity of emissions of CO and HC are also quickly converted into CO_2 in the atmosphere. The amount of CO_2 produced from fuel is proportional to the amount of carbon present on it and almost 99% of the carbon in the fuel is emitted in the form of CO_2 only [4, 5]. According to the datasheets of the United States Environmental Protection Agency (EPA), the CO_2 emission from a gallon of gasoline is 8887 g/gallon while that from diesel is 10,180 g/gallon. Even though diesel engines are highly fuel efficient than their respective gasoline engines with similar features, the CO_2 emissions in diesel engines are about 15% higher than that in gasoline engines. This is a matter of extreme concern as the majority of commercial and heavy-duty vehicles use diesel engines. A normal passenger vehicle emits about 404 g of CO_2 per mile and it varies from vehicle

Table 1 Comparison of GWP and emission percentage of various greenhouse gases [9].

Greenhouse gases	Global warming potential (GWP)	Percentage of emission from a passenger vehicle
Carbon dioxide (CO_2)	1	95%–99%
Methane (CH_4)	25	
Nitrous oxide (N_2O)	298	1%–5%
Hydrofluorocarbon (HFC)—A/c refrigerant	1430	

to vehicle as it depends on the fuel economy (mileage) and the carbon content in the fuel. The EPA and automobile manufactures estimated the fuel economy and gas emissions of a vehicle using a set of standard laboratory tests and found that the average annual CO_2 emissions of a typical passenger are estimated to be about 4.6 metric tons. Even though the quantity of CO_2 emission is much larger compared to other greenhouse gases, they possess a higher global warming potential (GWP) than CO_2 as given in Table 1 [9].

A boom has been observed in the global production and sale of conventional internal combustion engine (ICE) vehicles in the past few years according to the statistics of International Organization of Motor Vehicle Manufacturers, as shown in Fig. 2 [10]. The statistics of vehicle production in major countries and global year-wise vehicle production is illustrated in Fig. 2A and Fig. 2B, respectively. For instance, the global production of automobiles is increased from 77 million in 2010 to 91 million in 2019 [10]. As per the statistics of the United States EPA in 2018, the greenhouse gas emissions from the transportation sector contributed to 28.2% of the total U.S. greenhouse gas emissions [6]. Furthermore, according to the European Environmental Agency (EEA), the emissions from the transport sector in 2016 contribute to 26.96% of the overall emission of the European Union (EU) [11]. Therefore, it is imperative to take the steps to reduce the emissions for sustainable development.

One of the immediate strategies adopted by various governments and regulatory agencies is the promotion of public transport facilities as the reduction of a single passenger vehicle can lead to the reduction of 4.6 metric tons of CO_2 emission annually [9]. But the pandemic situations like COVID-19 etc. forces us to depend on personal or private vehicles to ensure safe day-to-day commutation and thus to prevent the spread of disease. According to the survey conducted by Euromonitor International, 13% of the respondents indicated the permanent usage of their vehicles for daily commuting due to health concerns [12]. The introduction of vehicle emission regulations by the governments also helped to decrease the rate of poisonous emissions. The U.S. vehicle emission standards, Euro emission standards of the European Union, and Bharat Stage Emission Standards (BSES) implemented by the Government of India (GoI) greatly aided the reduction of vehicular emissions [1]. The introduction of natural gas-based [liquified natural gas (LNG) or compressed natural gas (CNG)] vehicles also promoted the reduction of gas

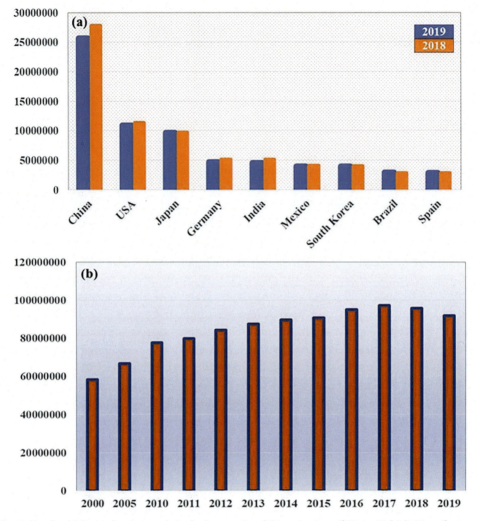

Fig. 2 Total vehicle production statistics by International Organization of Motor Vehicle Manufacturers (A) nation-wise (B) year-wise [10].

emissions to inappreciable extent as natural gas is sulfur-free. These alternative fuels possess a low carbon to hydrogen ratio than gasoline and hence emit a lesser amount of CO_2. The studies by United States EPA show that a light-duty CNG vehicle can reduce the CO and NO_x emission by 97% and 60%, respectively, when compared with their gasoline counterparts without three-way catalysts [9]. But all these strategies are not sufficient to effectively regulate exhaust gas emissions completely. Other strategies such as the deployment of more electric vehicles (EVs) or hybrid electric vehicles (HEVs) may help for the reduction of GHGs. An EV operates completely on electricity while HEVs use electricity as well as gasoline for their operation. EVs is exhausting gas emissions-free and only contribute to the hydrofluorocarbon (HFC) refringent emission. But the HEVs have both exhaust emissions (depending on the gasoline fuel efficiency) as well as refringent emissions [5]. The GoI has put forward the vision of having 100% EVs on the roads as of 2030 to steeply reduce vehicular exhaust emissions. But the possibility for the achievement of 100% conversion of conventional ICE vehicles to EVs seems to be less probable. As per the current statistics, the global sale of EVs (2.1 million in 2019) is less than 3% of the overall global vehicle sales (90 million in 2019), which is a matter of serious concern [6, 13]. Even though the financial savings and environmental benefits from EVs are far superior to the ICEs, the dependence on EVs for long routes is still a constraint as a result of insufficient recharging infrastructure and long charging time. Also, the initial investment and ownership cost of EVs or HEVs are higher when compared with the conventional ICE vehicles [5]. Therefore, a permanent and practical solution to the reduction of exhaust gas emissions has to be formulated for a sustainable future. In view of this, nanocatalysts are seen to be an amicable solution for the reduction of exhaust emissions in motor vehicles.

1.1 Exhaust system of vehicles

The combustion of fuels in the engines results in the generation of various gases of which some are certainly toxic. The exhaust system functions as a channel to carry away the gases from the engine chamber, reduce the emission of poisonous gases in to the atmosphere, decrease the engine noises and improve fuel efficiency. The major components of the exhaust system constitute the exhaust manifolds, oxygen sensors, catalytic convertors, exhaust pipes, resonator, muffler, and tailpipe as represented in Fig. 3. The exhaust manifold is directly connected with the engine chamber in order to collect the gases produced as a result of the combustion process and deliver them into the exhaust pipe. The size and structure of the manifold depend on the engine size and are usually made up of aluminum, cast iron, or stainless steel. The automobiles with fuel injection technology consist of an oxygen (O_2) sensor which measures the amount of O_2 present in the exhaust. This information is used by the automobile computer system to control the fuel injection and improving fuel efficiency. Usually, there are two O_2 sensors in vehicles with four-engine cylinders in which one is situated on the exhaust manifold and the other on the exhaust pipe after the catalytic convertor [5, 14].

The catalytic convertor which is located between the manifold and muffler helps in the reduction of toxic emissions from the exhaust system into the atmosphere. The resonator functions like a mini-muffler which is basically an empty echo chamber in which the exhaust noise gets reduced to an acceptable level and is located as a separate component before the muffler. The exhaust noise resonates in the resonator and the annoying

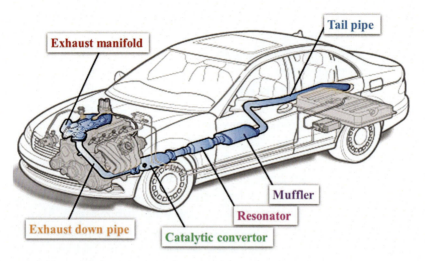

Fig. 3 Architecture of the exhaust system of vehicles [1].

higher tones cancel out each other. In addition to the resonator, the muffler decreases the noise level of the engine in accordance with government regulations. There are two variants of mufflers of which one uses a baffle chamber to decrease the noise level while the other type uses a sound-absorbing material to reduce the backpressure. The exhaust pipes connect the various components of the exhaust system and protect the heat-sensitive vehicle components. It is normally made out of stainless steel to minimize the corrosion issue and the interconnections are made via welds, clamps, or gaskets. The tailpipe is the last component of the exhaust system which comes after the muffler and delivers the exhaust gases directly into the atmosphere [5, 14]. As the catalytic convertor plays a vital role in the conversion of toxic gases in the exhaust emissions, we will be focussing on the features of the catalytic convertor in the following sections.

2. Catalytic convertor

It is the main part of the automobile exhaust emission control system which effectively converts highly toxic gases produced in the engine into less toxic pollutants. It consists of an efficient catalyst such as metal nanoparticles, metal oxides, carbon nanostructures, etc., which helps in the conversion of toxic gases via a series of chemical reactions. These catalysts result in the conversion of CO, HC, NO_x into CO_2, water vapor, and nitrogen, respectively, which are lesser harmful to the environment. The catalytic convertor was invented in 1930 by a French engineer named Eugene Houdry and was first introduced in 1975 in the USA as part of the regulations of the EPA [1].

In this chapter, we focus on the various types of catalytic convertor mechanisms currently used for the effective conversion of exhaust gases from automobiles. The current status of various nanostructure-based catalysts used in these control systems will be discussed in detail.

2.1 Different types of catalytic convertors in an internal combustion engine

Catalytic convertors have been continuously modified in order to meet the updated vehicle emission norms and regulations. Here, we are discussing the different catalytic convertor technologies used in internal combustion engines for the effective reduction of exhaust emissions [1].

Two-way catalyst: The first used catalyst convertor in automobiles was the two-way catalyst as shown in Fig. 4 [1]. It consists of an oxidation catalyst that converts the highly toxic CO and unburnt HC into less toxic CO_2 and water vapor as follows [15]:

Oxidation of carbon monoxide:

$$2CO + O_2 \rightarrow 2CO_2 \tag{1}$$

Oxidation of hydrocarbons:

$$C_xH_y + O_2 \rightarrow CO_2 + H_2O \tag{2}$$

But this emission control system becomes obsolete in the 1980s as it was insufficient to control the NO_x toxic emissions. The two-way catalytic system was replaced with the more efficient three-way catalyst system and later on the four-way catalyst system in conventional gasoline engines [1]. This system is still used in diesel or lean-burn gasoline engines for the oxidation of toxic CO and unburnt HC. The diesel oxidation catalyst (DOC), which is the main component in the diesel exhaust emission control system, is basically a two-way catalyst intended for the removal of CO and HC [2, 3].

Three-way catalyst (TWC): It is an emission control system (as in Fig. 5) found in conventional gasoline engines and helps in the emission reduction of NO_x, CO, and HC under low-oxygen exhaust conditions [1]. The oxidation catalyst performs the oxidation of CO and HC while the reduction catalyst executes the NO_x reduction process. The TWC simultaneously performs three catalytic reactions described as follows [15]:

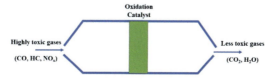

Fig. 4 Schematic of two-way catalytic convertor.

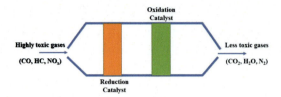

Fig. 5 Schematic of three-way catalytic convertor.

Reduction of NO_x with CO and HC:

$$CO + NO_x \rightarrow CO_2 + N_2 \qquad (3)$$

$$C_xH_y + NO_x \rightarrow CO_2 + H_2O + N_2 \qquad (4)$$

Oxidation of carbon monoxide:

$$2CO + O_2 \rightarrow 2CO_2 \qquad (5)$$

Oxidation of hydrocarbons:

$$C_xH_y + O_2 \rightarrow CO_2 + H_2O \qquad (6)$$

Four-way catalyst: This catalytic convertor system, as shown in Fig. 6, has an additional filter apart from the TWC system, to remove the particulate matter from the exhaust emissions [1].

The catalyst in the catalytic convertor typically consists of the active catalyst, wash coat with oxygen storage component, and monolith substrate, as shown schematically in Fig. 7. The catalytic material is normally comprised of platinum group metals such as Pt, Pd, or Rh, which help in the oxidation of CO as well as HC and reduction of NO_x. Usually, the gas conversion efficiency, light-off temperature (T_{50}), and CO/NO_x crossover are the parameters that determine the catalytic efficiency. The T_{50} value of a catalyst is the minimum temperature at which the 50% conversion efficiency is attained and it depends on the chemical stability, heat capacity, and geometric surface area of the catalyst. A lower T_{50} value and higher value of CO/NO_x crossover suggest a higher catalytic activity. The active catalyst materials are dispersed over a highly porous wash-coat support which provides a high surface area for the dispersion of the active

Fig. 6 Schematic of four-way catalytic convertor.

Fig. 7 Structure of the catalyst in the catalytic convertor.

catalyst and acts as a carrier/support for the active catalyst. The wash-coat is mainly composed of alumina-based materials because of its large surface area and excellent thermal stability [15–17].

In order to maintain a stoichiometric air to fuel ratio and to reduce the negative interactions between active catalysts at high operation temperatures, oxygen storage components (OSCs) are dispersed over the wash-coat support. The physical separation achieved via the OSCs help in the improvement of the catalytic efficiency and durability as it prevents alloy formation or other negative interactions between the active catalysts. The incorporation of OSCs is found to enhance the catalytic performance of the active catalyst as there exists a positive interaction between them. The wash-coat support is further attached to a monolith substrate with a honeycomb structure, usually made of cordierite or metallic foil substrates [15–17]. In the following section, we discuss the various nanocatalysts used as active catalysts in the catalytic convertor.

3. Nanocatalysts
3.1 Platinum group metals (PGMs)

The PGMs are generally used in catalytic convertors due to their high melting point, excellent corrosion resistance, high heat resistance, and superior catalytic properties than other catalytic materials [18]. The catalytic convertors are generally high-cost automobile components due to the use of PGMs as the catalyst. The amount of PGMs is optimized considering the cost factor without compromising the catalytic efficiency. The catalyst used in the early 1990s was made of Pd/Rh composites, as Pd was cheaper than Pt. Even though Rh is the cheaper PGM, it exhibits an excellent NO_x reduction efficiency and hence becomes unavoidable to regulate exhaust gas emissions. But, as the demand as well as price of Pd steeply increased with years, Pt/Rh composite-based catalysts were formulated and optimized in accordance with the governmental emission standards. However, the Pd/Rh-based catalysts are considered as the most suitable catalyst and have dominated the automobile market since 2002. The price increase of Rh paved the way for the fabrication of low-cost Rh formulations with outstanding NO_x reduction activity. The Pt or Pd catalysts are located at the bottom portion and perform the oxidation functions while the upper layer Rh acts as the reduction catalyst which reduces NO_x into N_2 before they diffuse into the lower Pt/Pd catalyst layer for the oxidation process [16]. Various strategies have been reported to improve the conversion efficiency and to reduce the cost factor of the PGM-based catalysts. The morphology and composition of the metal nanoparticles over the catalyst support have a significant effect on the catalytic properties.

a. Morphological tuning

The morphology can be tuned via various parameters in physical and chemical processing methods during the synthesis process and thus gas conversion efficiency of the nanocatalyst can be effectively tuned. Beniya et al. [19] illustrated that the CO oxidation activity of the Pt-based nanocatalysts can be enhanced by tuning its cluster size. They have fabricated Pt nanoclusters (Pt_n) over Al_2O_3 and TiO_2 support and found that the cluster size (n) of the Pt nanocatalyst has a strong influence over the CO oxidation process. The influence of the support oxide material improves as the size of the Pt nanocatalyst decreases and thus it leads to the alteration in the chemical states of Pt atoms in contact with the support layer. They have shown that there exists a strong correlation between the CO oxidation activity and the ratio of neutral to cationic Pt atoms (N_n/N_c). Since the cationic Pt atoms exhibit lower oxygen affinity, this results in a lower CO oxidation activity at small cluster sizes as shown in Fig. 8. Using temperature-programmed reduction (TPR) experiments, they have observed that the CO conversion efficiency of Pt_n/Al_2O_3 and Pt_n/TiO_2 catalyst becomes optimum at a cluster size of $n = 24$ and $n = 30$ respectively. Furthermore, they have theoretically predicted the ratio of (N_n/N_c) using a modified bond-additivity model (BAM) and found it to be in well agreement with the experimental results.

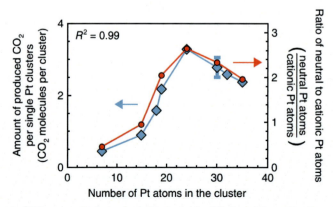

Fig. 8 The amount of CO_2 generated per single Pt cluster (*red line*) and the ratio of neutral to cationic Pt atoms (*blue line*) as a function of the number of Pt atoms in the cluster [19].

The hydrothermal treatment of the nanocatalyst can also affect the surface morphology of the nanocatalyst. Nie et al. [20] demonstrated that the steam treatment of Pt dispersed on CeO_2 at 750°C lead to the formation of active surface lattice oxygen near Pt and thus improved the low-temperature CO conversion efficiency with excellent hydrothermal stability. The T_{100} value of Pt/CeO_2 catalyst was found to be decreased from 320°C to 148°C after steam treatment and the CO conversion efficiency was maintained at 95% at 145°C even for 310 h confirming its excellent stability. Moreover, the CO oxidation activity of steam-treated Pt/CeO_2 catalyst was not deterred by other pollutants such as NO_x or HC.

b. Effect of PGM loadings and wash-coat composition

Since Pd is more abundant and cheaper than Pt or Rh catalysts, various groups have studied the effect of Pd loading [15]. Moreover, Pd shows better resistance to thermal sintering in comparison with Pt and Rh catalysts. Theis et al. [21] evaluated the effect of Pd loading over Al_2O_3, ZrO_2, and TiO_2 catalyst support using five different gas species such as CO, C_3H_6, NO, C_2H_4, C_3H_8. They found that the T_{90} value for all the gases decreased as Pd loading on Al_2O_3 support increased from 0.14 to 2 wt%, but there was no significant decrease in the T_{90} value as Pd loading further increased from 2 to 4 wt%. Moreover, they have studied the effect of Pd loading from 2 to 4 wt% over ZrO_2 and TiO_2 supports. The T_{90} value for the Pt/ZrO_2 catalyst exhibited a slight decrease while that for the Pt/TiO_2 catalyst showed a significant drop in the T_{90} value for all the gases, particularly for C_3H_8.

Similarly, Di Monte et al. [22] reported that the NO reduction reaction rate at 473 K of $Pd/Ce_{0.6}Zr_{0.4}O_2/Al_2O_3$ catalyst enhanced from 2.8 to 22 as Pd loading increased from 0.7 to 2.8 wt% as shown in Fig. 9. Moreover, the light-off temperature is found to be shifted toward a lower temperature as Pd loading increased. The increase in the number of active

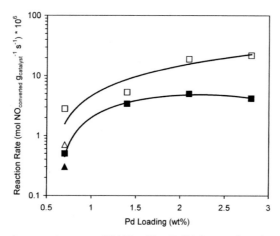

Fig. 9 Plot of NO reduction reaction rate of $Pd/Ce_{0.6}Zr_{0.4}O_2/Al_2O_3$ as a function of Pd loading (wt%) at 433 K (*blackened symbols*) and 473 K (*open symbols*) [22].

Table 2 Comparison of OSC and catalytic parameters of different Rh loadings under fresh as well as HTA conditions [23].

Rh loading over ZrCeYLaO$_2$ (wt%)	OSC at 500°C (mmol [O] per sample)	T$_{50}$ (°C)			Conversion efficiency at 400°C (%)		
		HC	CO	NO$_x$	HC	CO	NO$_x$
0.01% (fresh)	1.8	–	305	324	33	88	94
0.01% (HTA)	0.75	–	–	–	23	38	6
0.1% (fresh)	2.1	269	254	263	79	90	98
0.1% (HTA)	0.9	–	351	351	36	76	96
1% (fresh)	2.75	208	197	206	86	86	98
1% (HTA)	1.25	307	295	289	88	75	98

metal sites and enhancement in the promotional effect of Ce$_{0.6}$Zr$_{0.4}$O$_2$ (which acts as the OSC) with Pd loading leads to a higher NO conversion efficiency.

Even though Rh is the costliest among PGMs, it is highly necessary for the NO$_x$ reduction and thus it led to the fabrication of low Rh formulations. Alikin et al. [23] studied the effect of Rh loading (ranging from 0.01 to 1 wt%) on the catalytic activity and oxygen storage function by using CeO$_2$, CeZrO$_2$, ZrCeYLaO$_2$ as the support. They have concluded that the ZrCeYLaO$_2$ is the most thermally stable system and the negative effect of Rh loading follows the order CeO$_2$ > CeZrO$_2$ > ZrCeYLaO$_2$. The catalytic activity and oxygen storage capacity (OSC) of freshly prepared as well as hydrothermally aged (HTA) Rh/ZrCeYLaO$_2$ catalysts of different Rh loadings are compared in Table 2. The 1% Rh/ZrCeYLaO$_2$ catalyst exhibited an OSC of nearly twice that of 0.01 wt% Rh/ZrCeYLaO$_2$ as well as lower T$_{50}$ value and higher conversion efficiency, both under freshly prepared and hydrothermally aged conditions. Moreover, they observed that the low Rh loading of ≤0.1 wt% leads to the loss of catalytic efficiency as a result of strong Rh-ceria interactions at higher temperature aging.

c. Effect of alloy composition

Cooper et al. [24] studied the catalytic performance of different (Pt:Pd:Rh) compositions under Euro 5 vehicle standards as well as bench test conditions and are tabulated in Table 3. They have reported that a higher Rh loading leads to a better conversion efficiency rather than the Pt or Pd loadings. The HC, nonmethane HC as well as NO$_x$ gas emissions, and other bench test parameters are found to be optimum for the 0:25:7.5 catalyst composition while the CO emission is lower for 25:0:7.5 composition. Thus, it can also be concluded that the effect of Pd loading on the catalytic efficiency is slightly better than that of the Pt loading under most driving conditions and this difference becomes more visible under adverse driving conditions.

So, in general, even though Rh is more expensive among the PGMs, the Rh loadings have a significant effect on the catalytic performance compared to that of Pt or Pd.

Table 3 Comparison of various catalytic parameters of different Pt:Pd:Rh compositions [24].

Catalyst composition (Pt:Pd:Rh)	HC emission (g/km)	Nonmethane HC emission (g/km)	CO emission (g/km)	NO$_x$ (g/km)	HC T$_{50}$ (°C)	NO$_x$ T$_{50}$ (°C)	Efficiency at CO/NO$_x$ crossover (%)
0:15:5	0.070	0.054	0.263	0.051	377	375	94.7
0:25:5	0.068	0.052	0.281	0.044	370	374	96.6
0:35:5	0.066	0.051	0.264	0.044	365	371	97.4
15:0:5	0.065	0.05	0.249	0.037	375	375	97.8
25:0:5	0.065	0.05	0.228	0.035	380	381	97.4
35:0:5	0.069	0.055	0.297	0.045	379	376	95.6
0:25:2	0.071	0.056	0.289	0.047	376	379	97.4
0:25:7.5	0.056	0.042	0.261	0.027	354	356	99.0
25:0:2	0.081	0.062	0.304	0.051	383	386	91.7
25:0:7.5	0.063	0.051	0.242	0.03	362	360	99.4
Euro 5/6 standards	0.1	0.068	1.0	0.06	–	–	–

Moreover, the catalytic activity of Pd is slightly superior to that of Pt under most of the driving conditions.

3.2 Nonplatinum group metals

The high cost of the PGM-based catalysts forced researchers to focus on other low-cost catalysts. Noble metals such as Au, Ag, etc., and base metals such as Ni, Co, Fe, Cu, etc. can be employed as a substitute to the PGM-based catalysts. The bimodal pore structure present in these metals is favorable for the chemisorption of exhaust gases [1]. Bimodal pore structure consists of two-pore size distributions in which the large pores aid in the rapid molecular transportation while small pores provide a large surface area for better diffusion efficiency and supported metal dispersion [25]. Afanasev et al. [26] reported a low-temperature CO oxidation for Ag nanoparticles supported on fumed silica. They have demonstrated that the proper pretreatment of the fumed silica support as well as metal precursor and the Ag loading play a crucial role in the catalytic properties. The 4% Ag over preshaped silica exhibited a complete CO conversion at 30°C. Monodisperse Au nanoparticles of 6 nm size over graphitized porous carbon support achieved a complete CO conversion at −45°C as reported by Peng and co-workers [27]. Lou et al. [28] reported that Au single-atom catalysts supported over Co_3O_4 (Au_1/Co_3O_4) exhibits excellent catalytic activity and stability toward CO oxidation. The 0.067 wt% Au_1/Co_3O_4 exhibited 100% conversion of CO into CO_2 even at a lower temperature of −84°C as seen in Fig. 10 and also excellent stability without any degradation in the catalytic activity, even after storing it for 1 year under atmospheric conditions. Portillo-Velez et al. [29] demonstrated that the addition of Au nanoparticles over the transition-metal oxide catalysts such as MnO_x, FeO_x and CoO_x supported on TiO_2 improved the CO oxidation as a result of the formation of active sites at the

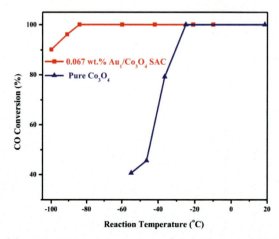

Fig. 10 Comparison of the CO conversion percentage of pure Co_3O_4 and 0.067 wt% Au_1/Co_3O_4 [28].

Table 4 Comparison of the T_{100} and CO oxidation reaction rates of various noble metal catalysts.

Catalyst	T_{100} (°C) for CO oxidation	Reaction rate (mole CO mole Au^{-1} s^{-1})	T_{50} (°C) for soot combustion	References
Ag/silica	30	–	–	[26]
Au/graphite	−45	–	–	[27]
Au/Co$_3$O$_4$	−84	–	–	[28]
Au/TiO$_2$	–	0.012	438	[29]
Au-MnO$_x$/TiO$_2$	–	0.008	353	[29]
Au-FeO$_x$/TiO$_2$	–	0.018	431	[29]
Au-CoO$_x$/TiO$_2$	–	0.041	359	[29]

Au-transition metal-oxide interface. The Au-CoO$_x$/TiO$_2$ catalyst exhibited a higher reaction rate for CO oxidation at −5 °C compared to Au-MnO$_x$/TiO$_2$, Au-FeO$_x$/TiO$_2$, and Au/TiO$_2$ as shown in Table 4. It is attributed to the increased number of interfacial active sites as a result of the better interaction between Au nanoparticles and CoO$_x$ oxide. Furthermore, they have also studied the soot (carbon particles formed as a result of the incomplete combustion of HC) combustion of these catalysts as shown in Table 4 and found that its T_{50} value is minimum for the Au-MnO$_x$/TiO$_2$ catalyst which is due to its lower heat of oxygen chemisorption and better surface lattice oxygen mobility. The oxygen can be easily extracted from the catalyst surface as the oxygen chemisorption heat reduces and thus the catalytic activity is enhanced.

Copper catalyst supported on Al$_2$O$_3$ and CeO$_2$ is reported to exhibit superior CO and CH$_4$ oxidation compared to other base metals [1]. Lykaki et al. [30] studied the effect of Cu incorporation on various ceria nanostructures for CO oxidation. The enhanced CO oxidation activity of Cu/CeO$_2$ nanorods than bare ceria structures is due to the formation of active sites as a result of copper-ceria interactions and adsorption of CO on Cu$^+$ species. The Cu/CeO$_2$ nanorods exhibited a lower T_{50} value of 72°C as well as a higher CO conversion of 25% (at 60°C) in comparison with Cu/CeO$_2$ nanocubes and nanopolyhedra. Similarly, Singh et al. [31] reported the effect of Cu loading (at.%) on the CO oxidation activity of Cu/Co$_3$O$_4$ catalyst as shown in Fig. 11 and found that 3% Cu/Co$_3$O$_4$ exhibited 99% CO conversion under H$_2$ rich conditions at around 170°C with a T_{50} value of 79.6°C. Lin and co-workers [32] theoretically studied the CO conversion of metal (Cu, Ag, Au, Pt, Rh, Pd, Fe, Co, and Ir) doped boron nitride nanosheets (BNNS) and observed that the catalytic activity is optimum for Co-BNNS and is

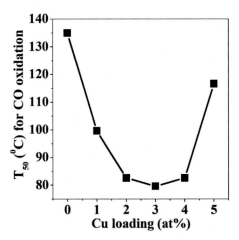

Fig. 11 T_{50} value of CO oxidation as a function of Cu loading (at.%) for Cu/Co$_3$O$_4$ catalyst [31].

attributed to the strong interaction between the cobalt 3d and oxygen 2p orbitals which activates the adsorbed oxygen atoms/molecules.

Bimetallic catalysts or alloys are found to exhibit remarkable catalytic properties due to the synergic effect existing between individual metal catalysts. Wang and co-workers [33] have fabricated Au—Ag nanoparticles over mesoporous support for CO oxidation and reported that the composition Au:Ag = 3:1 exhibited higher catalytic activity. Furthermore, the reduction (hydrogen) treatment of the foresaid catalyst composition between 550 and 600 °C leads to the complete conversion of CO at room temperature. The reduction treatment of the catalyst inhibited the particle enlargement and phase segregation caused by the formation of AgBr and benefitted in the re-alloying of Au with Ag (by removing Br$^-$ ions). Zhang et al. [34] theoretically studied various Au-based bimetallic nanoclusters (Ag-X; where X = Ag, Cu, Pd, Pt, Rh, and Ru) dispersed over CeO$_2$ support for CO oxidation and found out that the catalytic activity is optimum for Au-Cu/CeO$_2$ catalyst as a result of the preferential binding of O$_2$ to Cu rich sites.

3.3 Metal oxide-based catalysts

Base metal oxides such as CuO, Cu$_2$O, Co$_3$O$_4$, Fe$_2$O$_3$, V$_2$O$_5$, MnO, Cr$_2$O$_3$ etc exhibit good catalytic activity due to the highly porous structure as well as due to metal defects/oxygen vacancies. Copper oxide-based catalysts supported on CeO$_2$ or Al$_2$O$_3$ wash-coat are reported to be effective in CO and CH$_4$ oxidation [1]. Hossain et al. [35] compared the CO oxidation activity of Cu-O-Ce solid solutions and CuO/CeO$_2$ nanorods at different Cu content (wt%) as shown in Table 5.

The morphology of the metal oxide catalysts has a crucial impact on their catalytic properties. Feng et al. [36] fabricated CuO nanowires (NWs) over a Cu substrate and

Table 5 Comparison of the T_{50} and CO conversion efficiency of Cu-O-Ce solid solutions and CuO/CeO$_2$nanorods [35].

Cu content (wt%)	Cu-O-Ce solid solutions		CuO/CeO$_2$ nanorods	
	T_{50} (°C)	Maximum CO conversion (%)	T_{50} (°C)	Maximum CO conversion (%)
2	140	88	–	–
4	115	88	–	–
6	100	88	115	90
8	109	87	102	94
10	124	87	88	100
12	124	86	–	–

plasma-treated the as grown CuO NWs under Ar as well as H$_2$. The H$_2$ RF plasma-treated CuO NWs exhibited a higher CO conversion efficiency of 85% at 140°C in the fuel-lean condition in comparison to that of as-grown CuO NWs (24%) and Ar RF plasma-treated CuO NWs (29%). The enhanced activity of H$_2$ RF plasma-treated CuO NWs is attributed to the reduction of Cu (II) into more active Cu (I). Wang et al. [37] fabricated 1 wt% CuO$_x$ nanoclusters deposited on ceria nanospheres (1CuCe-NS) as well as nanorods (1CuCe-NR). It is observed that the 1CuCe-NS catalyst exhibited higher CO conversion with $T_{100} = 122$°C in comparison to that of 1CuCe-NR ($T_{100} = 194$°C). Moreover, the CO conversion of 1CuCe-NS and 1CuCe-NR at 105°C for 24 h is found to be 94% and 10%, respectively.

Another strategy to improve the catalytic activity of metal oxide-based catalysts is their incorporation with metal nanoparticles. Zhao et al. [38] reported that stoichiometric clusters of titania nanocluster anions $[(TiO_2)_nO_m^-]$ exhibited good CO oxidation activity when incorporated with noble metals. The surface lattice oxygen (O^{2-}) and molecularly adsorbed oxygen (O_2^- or O_2^{2-}) present at the metal-TiO$_2$ interface participate in the CO oxidation process. Therrien and co-workers [39] fabricated single Pt atoms on Cu$_2$O support for low-temperature CO oxidation. They have observed that no high-temperature CO or CO$_2$ desorption takes place without Pt coverage (Pt = 0%) and high-temperature desorption becomes evident when Pt is added to the surface of the support layer. It is also observed that as the Pt coverage increases, the peak temperature shifts to the lower side.

The gas conversion efficiency of metal-oxide-based catalysts can also be improved via their integration with another suitable metal-oxide resulting in the formation of metal-oxide composite structure. Liu and co-workers synthesized Cu$_x$—Mn composite oxide catalysts ($x = 0$ to 0.2) and tested them for CO oxidation as well as NO removal [40]. The Cu$_{0.075}$Mn oxide catalyst exhibited the superior CO oxidation of 100% at 65°C and further increase in Cu loading led to the deterrence of the catalytic activity. But the NO conversion was found to be higher for $Cu_{0.15}Mn$ oxide catalyst exhibiting 100% conversion at 275°C. The high activity in the metal oxide composite catalyst is attributed to the formation of $Cu_{1.5}Mn_{1.5}O_4$ spinel phase in the catalyst structure as a result of the synergistic effect between the binary metal oxides. AlKetbi and co-workers [41] fabricated a novel nanocatalyst composing of Ce-La-xCu-O mixed metal oxides ($x = 3$ to 20 at.%) for CO oxidation. They have shown that as the Cu content in the mixed oxide catalyst increases, the CO conversion efficiency is improved while T_{50} is found to be reduced and this is attributed to the increase in oxygen vacancy with the addition of Cu content.

3.4 Carbon nanostructures

Carbon nanostructures such as carbon nanotubes (CNTs), graphene, and their derivatives are promising low-cost catalytic materials due to their ability to form intermediate electronic configurations between sp^2 and sp^3. Despite their unique properties, they have not been used in vehicle emission control systems. The high surface area, lattice defects, oxygenated functional groups, and thermal stability of graphene make them favorable to be employed as wash-coat support for the dispersion of metal nanoparticles [2, 3]. Lin et al. [42] studied the effect of calcination temperature on oxidation of HC for nitrogen–doped carbon nanotubes encapsulated with cobalt nanoparticles (Co@GCN) as shown in Fig. 12. The ethylbenzene oxidation activity is found to be maximum at the calcination temperature of 800°C and is attributed to the synergistic effect between GCN and metallic Co.

Krishnan and co-workers [43] theoretically predicated that a single Pt-atom can be stably supported on penta-graphene and exhibits excellent CO oxidation. Similarly,

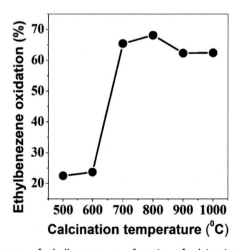

Fig. 12 Conversion percentage of ethylbenzene as a function of calcination temperature for nitrogen-doped carbon nanotubes encapsulated with cobalt nanoparticles (Co@GCN) [42].

Song et al. [44], using DFT, reported that Cu-embedded graphene shows good CO oxidation as a result of the enhanced interaction between Cu-3d, O_2-$2\pi^*$ and CO-$2\pi^*$ orbitals. Jiang and co-workers [45], via first-principle calculations, showed that Ni-embedded divacancy graphene (Ni-DG) exhibited good CO oxidation even under humid conditions at low temperatures. The higher adsorption energy of Ni-DG for CO than that for O_2, H_2O, and N_2 molecules makes it is a suitable candidate in metal-free catalysis.

3.5 Perovskite-based nanocatalysts

Perovskite metal-oxide structures of the type ABO_3 where "A" is a rare earth metal while "B" is a transition metal, exhibits good catalytic activity due to its variable oxygen stoichiometry. Only a few works are reported regarding perovskite-based catalysts since it becomes catalytically active for CO oxidation only at certain optimum conditions. Au/LaMnO$_3$ catalyst exhibited a CO conversion efficiency of 60% at 50°C [1]. Abdolrahmani et al. [46] fabricated $LaMn_{1-x}Cu_xO_{3\pm\delta}$ perovskite oxides (with $x=0$ to 1) by two different methods—Pechini and sol-gel technique for CO oxidation and methane combustion. The optimum amount of copper in the perovskite catalyst structure was found to be at $x=0.2$ and 0.4 as shown in Table 6. Furthermore, the catalysts prepared via Pechini method exhibited better activity than that synthesized by sol-gel process.

Seyfai et al. [47] proposed a novel perovskite composition $La_{0.8}Sr_{0.2}Co_{0.8}Cu_{0.2}O_3$ and it achieved a complete CO oxidation at 355°C and good stability up to 600°C. Similarly, Okejiri and co-workers [48] fabricated a new high entropy perovskite nanoparticle catalyst $Ru_{0.13}/Ba_{0.3}Sr_{0.3}Bi_{0.4}(Zr_{0.2}Hf_{0.2}Ti_{0.2}Fe_{0.27})O_3$. A high entropy material (HEM) is produced by five or more components into a single crystal phase to achieve a unique amalgamation of properties. The novel perovskite composition

Table 6 Comparison of the T_{50} and T_{90} value for CH_4 and CO oxidation of $LaMn_{1-x}Cu_xO_{3\pm\delta}$ perovskite oxides prepared via Pechini as well as sol-gel method [46].

Preparation method	X	T_{50} (°C)		T_{90} (°C)	
		CH_4	CO	CH_4	CO
Pechini	0	240	490	290	590
	0.2	220	470	270	570
	0.4	225	510	270	600
	0.6	245	530	300	640
	0.8	255	550	310	680
Sol-gel	0	250	520	300	610
	0.2	240	500	290	600
	0.4	235	530	285	620
	0.6	250	540	320	650
	0.8	260	560	330	700

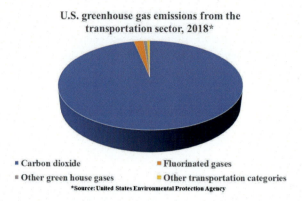

Fig. 13 The contribution of various greenhouse gas emissions in 2018 from the transportation sector of the United States [6].

exhibited a complete CO oxidation at 118°C while pure $BaRuO_3$ did not achieve a 100% CO conversion even at 300°C.

Even though the presently available active nanocatalysts are effective for the reduction of highly toxic vehicular emissions such as CO, HC, and NO_x, it is still unable to control the CO_2 emissions, which is the major contributor causing the global warming and related environmental issues. According to the U.S. EPA data in 2018, as shown in Fig. 13, CO_2 contributes to 96.7% of the overall greenhouse gas emissions from the transportation sector [6]. So, the ongoing researches are focused on effectively controlling the huge quantity of CO_2 emissions which possess a massive threat to the entire living world. Furthermore, fuel and lubricant additives such as sulfur, phosphorous, zinc, calcium, etc. can cause deleterious effects on the performance of the nanocatalyst. This can be effectively reduced by using alternative multipurpose fuel additives and lubricants having less impact on the catalytic performance or by the incorporation of certain novel materials into the catalyst which inhibits unnecessary reactions without affecting the catalytic activity [16].

4. Wash-coat compositions and oxygen storage components (OSC)

As discussed earlier, the wash-coat is basically composed of alumina (Al_2O_3) because of its high surface area and thermal stability. Ceria (CeO_2) based materials are dispersed over or incorporated into the alumina as OSCs and helps to maintain the stoichiometric air to fuel ratio in order to achieve the maximum catalytic efficiency. Since ceria can easily switch its oxidation states through the formation or annihilation of oxygen vacancies, it can store oxygen from the exhaust gases under stoichiometric excess conditions while releasing oxygen under deficient conditions and thus able to maintain the stoichiometric ratio. Ceria also helps to hinder the sintering of PGMs at high temperatures via the formation of Pt-O-Ce bond under oxygen presence as shown in Fig. 14 [15–17].

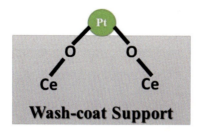

Fig. 14 Inhibition of sintering on Pt catalyst on ceria incorporated wash-coat support.

According to Sanchez and Gazquez [49], the oxygen vacancies formed at the surface of CeO_2 act as anchoring sites for the metal atoms which results in a "nesting effect." However, the diffusion of metal atoms into the CeO_2 structure is highly hindered by the cationic sublattice. The PGM-CeO_2 interface provides sufficient active sites for the catalytic reaction. Holles and co-workers [50] reported that the incorporation of ceria (CeO_2) into the Pd/Al_2O_3 catalyst enhances the NO reduction rate and this was further supported by Ciuparu et al. [51]. These works established the importance of Pd–CeO_2 interface in NO dissociation and stated that the improved interaction between Pd and CeO_2 significantly helps in the enhancement of oxygen storage capacity (OSC).

The previously used ceria (CeO_2) OSCs were found to be incompatible with the updated emission regulations as the catalytic convertor was positioned close to the engine manifold in order to decrease the catalyst light-off time. The TWC with pure CeO_2 is found to be thermally unstable at operation temperatures above 800°C. Hence, this led to the sintering of PGMs and loss of surface area of OSCs and support materials [15–17]. Various mixed oxide composites particularly formed with CeO_2, are employed as efficient OSCs in order to resolve the problems due to thermal instability. High thermal stability along with better TWC performance can be attained by using CeO_2-ZrO_2 (CZ) OSCs. It is reported that the CZ OSCs comprising of a ceria content ranging from 40% to 60% exhibit a higher OSC [15]. Madier and co-workers [52] studied the OSC property, redox property, and oxygen exchange process of $Ce_xZr_{1-x}O_2$ composites and reported that the $Ce_{0.63}Zr_{0.37}O_2$ composite has shown the highest OSC of 219 μmol O/g at 400°C, larger CO uptake, and highest O_2 exchange reactivity. Priya et al. [53] studied the OSC property of $Ce_xZr_{1-x}O_2$ composition; $0.4 \leq x \leq 0.8$ prepared via sol-gel synthesis and the $Ce_{0.6}Zr_{0.4}O_2$ composition has exhibited an optimum OSC of 0.1469 g/μmol of CeO_2 as shown in Fig. 15. They also reported that the optimum Ce/Zr ratio for a better OSC is determined to be about 1.5.

Similarly, Di Monte et al. [54] compared the thermal stability of nanostructured $Ce_mZ_{1-m}O_2$ (CZ)/Al_2O_3 and CeO_2-Al_2O_3 at 1100 °C and reported the formation of $CeAlO_3$ in CeO_2/Al_2O_3 while no such secondary phase formation in the CZ/Al_2O_3 composites. The zirconia (ZrO_2) acts as a stabilizer for the ceria structure at very

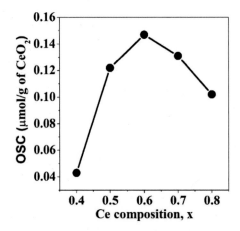

Fig. 15 Oxygen storage capacity (OSC) as a function of Ce composition (x) [53].

Table 7 Comparison of the OSC function of different compositions of CeO_2-ZrO_2 and $CeO2$-ZrO_2/Al_2O_3 nanostructure catalysts [54].

OSC composition	OSC at 400°C (/mL $O_2 g^{-1}_{solid\ solution}$)		OSC at 600°C (/mL $O_2 g^{-1}_{solid\ solution}$)	
	500°C, 5 h	1100°C, 5 h	500°C, 5 h	1100°C, 5 h
$Ce_{0.2}Zr_{0.8}O_2$	0.3	0	3.4	0.1
$Ce_{0.6}Zr_{0.4}O_2$	0.9	0	5.8	0.6
CeO_2/Al_2O_3	3.2	0.3	13.8	1.6
$Ce_{0.2}Zr_{0.8}O_2/Al_2O_3$	0.2	0.1	5.8	3.5
$Ce_{0.6}Zr_{0.4}O_2/Al_2O_3$	1.7	0.5	16.9	8.7

high temperatures with a higher surface area and excellent OSC properties even at thermal aging of 1100°C under atmospheric conditions. The $Ce_{0.6}Zr_{0.4}O_2/Al_2O_3$ nanostructure thermally aged at 500°C for 5 h exhibited a maximum OSC as shown in Table 7.

Granger et al. [55] studied the thermal stability of $Pd/Zr_xCe_{1-x}O_2$ at different Zr/Ce ratios and observed that the $Pd/Zr_{0.25}Ce_{0.75}O_2$ exhibited superior thermal stability than Pd/CeO_2 even though the latter exhibited a slightly lower T_{50} value for NO conversion (110°C) than the former one (130°C). Similarly, Fernandes et al. [56] showcased the higher OSC capacity of CZ even under extreme sintering conditions of 1200°C for 72 h without much affecting the catalytic performance of Pd/Rh commercial catalyst. Lan et al. [57] found that preparing a physical mixture of $Ce_{0.8}Zr_{0.2}O_2 + Ce_{0.2}Zr_{0.8}O_2$ as support for Pd-based TWCs exhibited a higher OSC activity regardless of the thermal conditions. The enhanced OSC of CZ can be attributed to the presence of catalytically active defective sites formed as a result of the introduction of ZrO_2 into the CeO_2

structure. The incorporation of Zr atoms in CeO_2 lattice aids in the lowering of the energy barrier for Ce^{4+} reduction and enhances the mobility of lattice oxygen. The enhanced catalytic activity as a result of the CZ addition is attributed to the ample interface, a large number of interfacial boundaries with different phase compositions, excellent structural stability, strong interaction between Pd and ceria-zirconia, and creation of more defective active sites [17].

Certain works have studied the benefits of multiatom doping into the ceria structure for improving the OSC property. Priya et al. [58] incorporated Mn^{4+} along with Zr^{4+} into the ceria lattice for enhanced OSC and $Ce_{0.6}Zr_{0.2}Al_{0.2}O_2$ showcased an increase of 22% in OSC when compared with that of $Ce_{0.6}Zr_{0.4}O_2$. Similarly, the same group studied the incorporation of Zr^{4+} and Al^{3+} ions into the ceria structure for improving the OSC property [59]. The $Ce_{0.6}Zr_{0.2}Al_{0.26}O_2$ composite exhibited an OSC of 56% higher than that of $Ce_{0.6}Zr_{0.4}O_2$ due to the enhanced structural stability and surface defects. In addition to this, various rare earth and alkaline rare earth metals are employed as promoters to improve the thermal stability, OSC property, and thus the TWC functionality. Zr-Ce-Pr-O and Zr-Ce-Nd-O mixed oxide composites exhibited excellent oxygen mobility at lower temperatures [17]. According to Wu et al. [60], the incorporation of Pr and Nd oxide promoters into the CZ structure enhanced its OSC property as a result of the improved thermal stability, higher Ce^{3+}/Ce^{4+} and hence creation of more catalytically active crystal defects. The introduction of promoters into the CZ structure does not form any phase segregation. Wang and co-workers [61] reported that the addition of about 5% PrO_x promoter to the Pt-Pd-Rh/CZ catalyst enhances its light-off performance toward NO reduction and hence improves the overall catalytic performance.

5. Catalytic convertors in diesel and learn-burn gasoline engines

As the catalytic convertors in conventional gasoline vehicles are ineffective in oxygen-rich emission conditions as in the case of diesel or lean-burn gasoline engines, three-way catalysts cannot be employed for efficient NO_x reduction in them. The two-way catalyst is used for the conversion of CO and HC in diesel or lean-burn gasoline engines. Diesel oxidation catalyst (DOC) typically consists of Pt and Pd nanoparticles dispersed over a wash-coat which is adhered to a honeycomb monolith substrate [2, 3].

5.1 Selective catalytic reduction (SCR)

The SCR system can be employed for efficient NO_x reduction in oxygen-rich conditions. Ammonia (NH_3) is used as the reducing agent in SCR for NO_x conversion. The commonly used SCR is a urea-based catalytic system (NH_3-SCR) in which the ammonia gas is introduced into the exhaust gas from the onboard urea under hydrothermal conditions. The ammonia in turn reacts with the NO_x (where $x = 1, 2$) adsorbed on the catalyst surface and is reduced into N_2. The gas emissions from diesel engines compose of nitrogen oxides mainly in the form of nitrogen monoxide (NO) and a minor quantity of nitrogen dioxide (NO_2) [2, 3]. The basic reactions involved in the SCR process are as described as follows [62];

$$4NH_3 + 4NO + O_2 \rightarrow 4N_2 + 6H_2O \tag{7}$$

But in newer diesel engines with advanced technologies, the percentage of NO_2 becomes more in the NO_x emissions and hence Eqs. (7) modify as (8). Reaction (8) occurring in a 1:1 reaction mixture of NO and NO_2 is faster than Eq. (7).

$$4NH_3 + 2NO + 2NO_2 + O_2 \rightarrow 4N_2 + 6H_2O \tag{8}$$

If the NO_2 quantity exceeds 50% of the total NO_x emission, then the SCR process modifies as follows,

$$8NH_3 + 6NO_2 \rightarrow 7N_2 + 12H_2O \tag{9}$$

The most commercially used SCR system is vanadium oxide (V_2O_5) dispersed over TiO_2 substrate and promoted by WO_3 (V_2O_5-WO_3/TiO_2) and exhibits a very good NO_x reduction (deNO_x) activity and N_2 selectivity at 300–400°C. Another used vanadium-based catalyst is V_2O_5-MoO_3/TiO_2 also shows a high selectivity and NO_x activity in the SCR process. Cheng and co-workers [63] incorporated CeO_2—TiO_2 composite structure with different Ce/Ti ratios into the V_2O_5-WO_3/TiO_2 structure. The catalysts with Ce/Ti < 1 exhibited better NO conversion than that for the catalysts with Ce/Ti ≥1 as shown in Fig. 16. The SCR activity of V_2O_5-WO_3/TiO_2 catalyst was

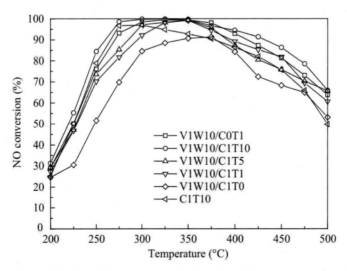

Fig. 16 NO conversion of CeO_2—TiO_2 incorporated V_2O_5- WO_3/TiO_2 catalyst as a function of reaction temperature at different Ce/Ti ratios [63].

found to be optimum at a Ce/Ti ratio of 1:10 (labeled as V1W10/C0T1 where C0T1 represents the ratio Ce/Ti = 1:10). Furthermore, the SCR activity of the catalyst with Ce/Ti = 0:10 degraded in the presence of SO_2. Ramis and co-workers [64] investigated the adsorption and transformation of ammonia on V_2O_5, V_2O_5/TiO_2, V_2O_5-WO_3/TiO_2, and CuO/TiO_2 catalyst systems. In the case of all the mentioned catalysts, the ammonia is primarily converted into NH_2 species or its dimeric form hydrazine (N_2H_4). The conversion of ammonia into NH_4^+ ions is exhibited by the V_2O_5 based catalysts, but not by the CuO/TiO_2 catalyst. Despite the high catalytic efficiency of vanadium-based catalysts, they may react with SO_2 to form ammonia bisulfate (ABM) which hinders the catalytically active sites and thus degrade the SCR catalyst efficiency [2]. Hence various researches are ongoing to tackle these issues effectively.

Mn-based metal oxides are observed to exhibit an excellent lower temperature SCR activity. Kijlstra et al. [65] studied the adsorption ability of MnO_x/Al_2O_3 catalyst using NO and NH_3 gases for the application in low-temperature SCR of NO. They have investigated regarding the surface complexes present during the SCR of NO. The NO adsorption results in the formation of nitrosylic species via the Mn^{3+} sites in the absence of oxygen. But these nitrosylic species are unstable under oxygen-rich conditions. At 423 K, the NO complexes such as nitrites/nitrates become sufficiently stable under oxygen-rich conditions, and their formation is also enhanced in the presence of O_2. The formation of reactive surface oxygen species in the presence of O_2 followed by NO oxidation is responsible for higher NO coverages. Similarly, upon NH_3 adsorption, the ammonia ions and coordinated NH_3 are formed and a certain amount of adsorbed NH_3 is converted into NH_2 species. It is observed that the increase in Mn loading prefers the conversion of NH_3 into NH_2 species rather than the formation of NH_4^+ ions. Similarly, Qi and co-workers [66] studied the effect of MnO_x doping on the SCR activity of MnO_x–CeO_2 mixed oxide catalyst, and the NO conversion was found to be optimum at 0.4 wt% of MnO_x as shown in Fig. 17.

Metal exchanged zeolites with narrow pores such as mordenite (MOR), ferrierite (FER), beta zeolite (BEA), zeolite Socony Mobil-5 (ZSM-5), etc. are seen as promising candidates for NH_3-SCR applications [2]. The NO_2 is formed at the metal centers of metal-exchanged zeolites and the SCR reaction occurs at the zeolite framework. The produced NO_2 in the SCR process is immediately adsorbed and hence not shown up as gas-phase NO_2. Ahn et al. [67] fabricated a copper exchanged LTA zeolite catalyst and observed that the hydrothermal aging of the catalyst improves the NO conversion due to the formation of catalytically active Cu^{2+} ions as shown in Fig. 18.

5.2 Lean NO_x trap (LNT)

The LNT system, also known as NO_x storage-reduction catalyst system, is the currently used control mechanism for the reduction of NO_x from the engine exhaust gases and it

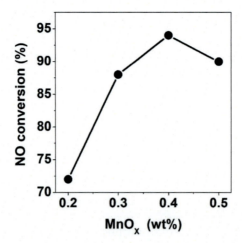

Fig. 17 NO conversion as a function of MnO$_x$ loading for MnO$_x$-CeO$_2$ mixed oxide catalyst [66].

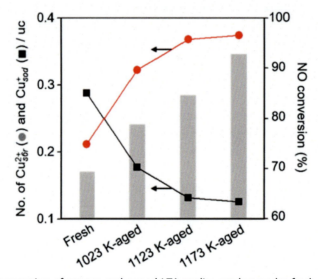

Fig. 18 The NO conversion of copper exchanged LTA zeolite catalyst under fresh as well as various hydrothermal conditions [67].

employs a complicated application of urea for NO$_x$ reduction. The LNT catalyst operates in both fuel-rich and fuel-lean environment conditions. Under the fuel-lean condition, the Pt/Rh catalyst oxidizes NO into NO$_2$ which is followed by the adsorption of NO and NO$_2$ on the surface of alkaline/alkaline earth metal storage components leading to the formation of its nitrates/nitrites. In fuel-rich environment, the NO$_x$ released from the alkaline/alkaline earth metal-based nitrates/nitrites are converted in N$_2$ by the Pt/Rh catalyst in the presence of either H_2, CO, or HC and thus resulting in the recovery of the catalyst [2, 3].

The $Pt/Ba/Al_2O_3$ catalyst is a typical LNT catalyst that exhibits a better NO_x storage-reduction efficiency and catalyst stability [2, 3]. Zhang and co-workers [68] have found that manganese-containing LNT catalysts exhibit higher activity for NO oxidation to NO_2 than the $Pt/Ba/Al_2O_3$ catalyst as it reduced the inhibiting effects of H_2O and CO_2 on NO_x conversion. Jeong et al. [69] studied the NO_x storage-reduction catalysis of Pt-BaO/Mg-Al mixed oxide catalyst by varying the Mg/Al ratio and found to exhibit enhanced performance than the Al_2O_3 based catalyst. The NO_x storage activity and thermal stability was found to be optimum at Mg/Al ratio of 4:6. Luo et al. [70] evaluated the NO oxidation and NO_x storage properties of $Pt/K/Al_2O_3$ catalyst with various potassium (K) loading. The NO_x storage results exhibited an optimum K loading of around 10% for the best performance at high temperatures. Since the majority of KNO_3 stayed as a surface layer at lower K loadings, the strong interaction between KNO_3 and Al_2O_3 promotes KNO_3 decomposition and hence deteriorates the high-temperature performance. At K loadings higher than 10%, the reduced performance is not caused by NO_x diffusion limitations as in the case of barium-based LNTs, but rather from the blocking of Pt sites by K species, which adversely affects the NO oxidation. The thermal aging at 800°C severely deactivates the $Pt/K/Al_2O_3$ catalysts due to Pt sintering. However, in the presence of potassium, some Pt remains in a dispersed and oxidized form. These Pt species interact strongly with K and, therefore, do not sinter. After a reduction treatment, these Pt species remain finely dispersed, contributing to a partial recovery of NO_x storage performance.

6. Conclusion

This chapter discusses the various exhaust emission control technologies such as two-/three-/four-way catalysis, selective catalytic reduction, and Lean-NO_x trap employed for the conversion of toxic exhaust gases into lesser toxic gases. The CO and hydrocarbons in the exhaust emission are oxidized into CO_2 and water vapor respectively, while NO_x is reduced into N_2. The nanocatalyst in the catalytic convertor is the active catalytic component in the exhaust system and its efficiency can be enhanced via different strategies such as morphology tuning, alloy/composite formation, and thermal treatments. Even though the platinum group metals exhibit high efficiency for the gas conversion, their cost of production and less availability forced the scientific community to look for other alternatives which can compete with a platinum metal-based nanocatalysts. Thus, various other nanostructures based on metal nanoparticles, metal oxides, and carbon nanostructures are rigorously under investigation to develop a suitable substitute for platinum-based catalysts.

7. Future prospects

Recently, 2D materials especially graphene-based catalysts are foreseen as effective catalysts due to their 2D layered structure, high surface area, excellent electron conductivity, and chemical stability. Couple of theoretical works have reported the superior catalytic activity of metal nanoparticles embedded graphene heterostructures, but their experimentation is still in the progress.

The presently used catalysts are effective in controlling toxic exhaust emissions such as CO, hydrocarbons, and NO_x, but they are incompetent to tackle the huge CO_2 emission, which is a real threat to the entire humanity. So, it is the need of time to effectively control the CO_2 emission. CO_2 capture and storage (CCS) is seen as a highly potential technology to control CO_2 emission [71, 72]. In this technology, CO_2 is captured and stored by a nanomaterial and later disposed of at a collection site. Even though this is currently employed in industries, its miniaturized version is yet to be installed in vehicles.

Acknowledgments

The author K.V.A. acknowledges DST, Govt. of India for the Inspire fellowship (IF170601). The Author K.C.S. and K.K. acknowledges the DST-SERB, Govt of India for the financial support through grant no. ECR/2017/000068 and ECR/2017/002537, respectively. J.P.B.S. is grateful for the financial support by the Portuguese Foundation for Science and Technology (FCT) in the framework of the Strategic Funding UID/FIS/04650/2019.

References

[1] S. Deya, N.S. Mehta, Automobile pollution control using catalysis, Environ. Dev. Sustain. 2 (2020) 100006. 1–13 https://doi.org/10.1016/j.resenv.2020.100006.
[2] G. Gerasimov, M. Pogosbekian, Recent advances in nanostructured catalysts for vehicle exhaust gas treatment, in: C.M. Hussain, A.K. Mishra (Eds.), Nanotechnology in Environmental Science, 2018, pp. 39–69, https://doi.org/10.1002/9783527808854.ch3.
[3] G. Gerasimov, M. Pogosbekian, Nanostructured catalysts in vehicle exhaust control systems, in: Handbook of Ecomaterials, 2019, pp. 1679–1700, https://doi.org/10.1007/978-3-319-68255-6_120.
[4] R.M. Heck, R.J. Farrauto, Automobile exhaust catalysts, Appl. Catal. A Gen. 221 (2001) 443–457, https://doi.org/10.1016/S0926-860X(01)00818-3.
[5] D.A. Crolla (Ed.), Automotive Engineering - Powertrain, Chassis System and Vehicle Body, 1st Edition, Butterworth-Heinemann, United Kingdom, 2009 (Section 3) https://www.elsevier.com/books/automotive-engineering/ribbens/978-1-85617-577-7.
[6] United States Environmental Protection Agency (EPA), Greenhouse Gas Inventory Data Explorer. https://cfpub.epa.gov/ghgdata/inventoryexplorer/.
[7] Z. Zoundi, CO_2 emissions, renewable energy and the environmental Kuznets curve, a panel cointegrationapproach, Renew. Sust. Energ. Rev. 72 (2017) 1067–1075, https://doi.org/10.1016/j.rser.2016.10.018.
[8] International Energy Agency (IEA), Co_2 Emissions From Fuel Combustion 2019 Highlights. https://webstore.iea.org/co2-emissions-from-fuel-combustion-2019-highlights.
[9] United States Environmental Protection Agency (EPA), Greenhouse Gas Emissions From A Typical Passenger Vehicle. https://www.epa.gov/greenvehicles/greenhouse-gas-emissions-typical-passenger-vehicle.

[10] International Organization of Motor Vehicle Manufactures, Productionstatistics, 2019. https://www.oica.net/category/production-statistics/2019-statistics/.

[11] European Environmental Agency(EEA), Greenhouse Gas Emissions From Transport in Europe. https://www.eea.europa.eu/data-and-maps/indicators/transport-emissions-of-greenhouse-gases/transport-emissions-of-greenhouse-gases-12.

[12] J. Liuima, Coronavirus increases demand for autonomous vehicles, Euromonitor International, 2020. https://blog.euromonitor.com/coronavirus-increases-demand-for-autonomous-vehicles/.

[13] International Energy Agency (IEA), Global EV outlook, 2020. https://www.iea.org/reports/global-ev-outlook-2020.

[14] A.K.M. Mohiuddin, A. Rahamn, M. Dzaidin, Optimaldesignofautomobileexhaustsystemusing GT-power, Int. J. Mech. Mater. Eng. 2 (1) (2007) 40–47.

[15] S. Rood, S. Eslava, A. Manigrasso, C. Bannister, Recent advances in gasoline three-waycatalyst formulation: a review, Proc. Inst. Mech. Eng., D 234 (4) (2019) 936–949, https://doi.org/10.1177/2F0954407019859822.

[16] N.R. Collins, M.V. Twigg, Three-way catalyst emissions control technologies for spark-ignitionengines—recent trends and future developments, Top. Catal. 42 (2007) 323–332, https://doi.org/10.1007/s11244-007-0199-6.

[17] J. Wang, H. Chen, Z. Hu, M. Yao, Y. Li, A review on the Pd-based three-way catalyst, Catal. Rev. 57 (1) (2015) 79–144, https://doi.org/10.1080/01614940.2014.977059.

[18] K. Nose, T.H. Okabe, Platinum group metals production, in: Treatise on Process Metallurgy: Industrial Processes, vol. 3, 2014, pp. 1071–1097, https://doi.org/10.1016/B978-0-08-096988-6.00018-3.

[19] A. Beniya, S. Higashi, N. Ohba, R. Jinnouchi, H. Hirata, Y. Watanabe, CO oxidation activity of non-reducible oxide-supported mass-selected few-atom Pt single-clusters, Nat. Commun. 11 (2020) 1888. 1–10 https://doi.org/10.1038/s41467-020-15850-4.

[20] L. Nie, D. Mei, H. Xiong, B. Peng, Z. Ren, X.I.P. Hernandez, A. DeLaRiva, M. Wang, M.H. Engelhard, L. Kovarik, A.K. Datye, Y. Wang, Activation of surface lattice oxygen in single-atom Pt/CeO_2 for low-temperature CO oxidation, Science 358 (2017) 1419–1423, https://doi.org/10.1126/science.aao2109.

[21] J.R. Theis, A. Getsoian, C. Lambert, The development of low temperature three-way catalysts for high efficiency gasoline engines of the future, SAE Int. J. Fuels Lubr. 10 (2) (2017) 583–592, https://doi.org/10.4271/2017-01-0918.

[22] R. Di Monte, P. Fornasiero, J. Kašpar, P. Rumori, G. Gubitosa, M. Graziani, Pd/$Ce_{0.6}Zr_{0.4}O_2$/Al_2O_3 as advanced materials for three-way catalystsPart 1. Catalyst characterisation, thermal stability and catalytic activity inthe reduction of NO by CO, Appl. Catal. B Environ. 24 (2000) 157–167, https://doi.org/10.1016/S0926-3373(99)00102-2.

[23] E.A. Alikin, A.A. Vedyagin, High temperature interaction of rhodium with oxygen storagecomponent in three-way catalysts, Top. Catal. 59 (2016) 1033–1038, https://doi.org/10.1007/s11244-016-0585-z.

[24] J. Cooper, J. Beecham, A study of platinum group metals in three-way autocatalysts, Platin. Met. Rev. 57 (4) (2013) 281–288. https://doi.org/10.1595/147106713X671457.

[25] J. Wang, H. Li, D. Li, J.P. den Breejenc, B. Hou, Influence of the bimodal pore structure on the COhydrogenation activity and selectivity of cobaltcatalysts, RSC Adv. 5 (2015) 65358. 1–7 https://doi.org/10.1039/C5RA10507B.

[26] D.S. Afanasev, O.A. Yakovina, N.I. Kuznetsova, A.S. Lisitsyn, High activity in CO oxidation of ag nanoparticles supported on fumed silica, Catal. Commun. 22 (2012) 43–47, https://doi.org/10.1016/j.catcom.2012.02.014.

[27] S. Peng, Y. Lee, C. Wang, H. Yin, S. Dai, S. Sun, A facile synthesis of monodisperse Au nanoparticles and their catalysis of CO oxidation, Nano Res. 1 (2008) 229–234, https://doi.org/10.1007/s12274-008-8026-3.

[28] Y. Lou, Y. Cai, W. Hu, L. Wang, Q. Dai, W. Zhan, Y. Guo, P. Hu, X. Cao, J. Liu, Y. Guo, Identification of active area as active center for CO oxidation over single Au atom catalyst, ACS Catal. 10 (2020) 6094–6101. https://doi.org/10.1021/acscatal.0c01303.

[29] N.S. Portillo-Vélez, R. Zanella, Comparative study of transition metal (Mn, Fe or co) catalysts supported ontitania: effect of Au nanoparticles addition toward CO oxidation and sootcombustion reactions, Chem. Eng. J. 385 (2020) 123848, https://doi.org/10.1016/j.cej.2019.123848.

[30] M. Lykaki, E. Pachatouridou, S.A.C. Carabineiro, E. Iliopoulou, C. Andriopoulou, N. Kallithrakas-Kontos, S. Boghosian, M. Konsolakis, Ceria nanoparticles shape effects on the structural defects and surface chemistry: Implications in CO oxidation by Cu/CeO$_2$ catalysts, Appl. Catal. B Environ. 230 (2018) 18–28, https://doi.org/10.1016/j.apcatb.2018.02.035.

[31] S.A. Singh, S. Mukherjee, G. Madras, Role of CO$_2$ methanation into the kinetics of preferential CO oxidation onCu/Co$_3$O$_4$, Mol. Catal. 466 (2019) 167–180, https://doi.org/10.1016/j.mcat.2019.01.020.

[32] S. Lin, X. Ye, R.S. Johnson, H. Guo, First-principles investigations of metal (Cu, Ag, Au, Pt, Rh, Pd, Fe, Co,and Ir) doped hexagonal boron nitride nanosheets: Stability and catalysis of CO oxidation, J. Phys. Chem. C 117 (33) (2013) 17319–17326, https://doi.org/10.1021/jp4055445.

[33] A. Wang, C. Chang, C. Mou, Evolution of catalytic activity of Au-ag bimetallic nanoparticles on mesoporous supportfor CO oxidation, J. Phys. Chem. B 109 (2005) 18860–18867, https://doi.org/10.1021/jp051530q.

[34] L. Zhang, H.Y. Kim, G. Henkelman, CO oxidation at the Au — cu Interface of bimetallic NanoclustersSupported on CeO$_2$(111), J. Phys. Chem. Lett. 4 (1) (2013) 216–221, https://doi.org/10.1021/jz301778b.

[35] S.T. Hossain, E. Azeeva, K. Zhang, E.T. Zell, D.T. Bernard, S. Balaz, R. Wang, A comparative study of CO oxidation over Cu-O-Ce solid solutions and CuO/CeO$_2$nanorods catalysts, Appl. Surf. Sci. 455 (2018) 132–143, https://doi.org/10.1016/j.apsusc.2018.05.101.

[36] Y. Feng, X. Zheng, Plasma-enhanced catalytic CuOnanowires for CO oxidation, Nano Lett. 10 (11) (2010) 4762–4766, https://doi.org/10.1021/nl1034545.

[37] W. Wang, W. Yu, P. Du, H. Xu, Z. Jin, R. Si, C. Ma, S. Shi, C. Jia, C. Yan, Crystal plane effect of ceria on supported copper oxide cluster catalyst for CO oxidation: importance of metal–support interaction, ACS Catal. 7 (2) (2017) 1313–1329, https://doi.org/10.1021/acscatal.6b03234.

[38] Y. Zhao, M. Wang, Y. Zhang, X. Ding, S. He, Activity of atomically precise titaniananoparticles in CO oxidation, 58 (24) (2019) 8002–8006, https://doi.org/10.1002/anie.201902008.

[39] A.J. Therrien, A.J.R. Hensley, M.D. Marcinkowski, R. Zhang, F.R. Lucci, B. Coughlin, A.C. Schilling, J. McEwen, E.C.H. Sykes, An atomic-scale view of single-site Pt catalysisfor low-temperature CO oxidation, Nat. Cata. 1 (2018) 192–198, https://doi.org/10.1038/s41929-018-0028-2.

[40] T. Liu, Y. Yao, L. Wei, Z. Shi, L. Han, H. Yuan, B. Li, L. Dong, F. Wang, C. Sun, Preparation and evaluation of copper — manganese oxide as a high-efficiency catalyst for CO oxidation and NO reduction by CO, J. Phys. Chem. C 121 (23) (2017) 12757–12770, https://doi.org/10.1021/acs.jpcc.7b02052.

[41] M. AlKetbi, K. Polychronopoulou, M.A. Jaoude, M.A. Vasiliades, V. Sebastian, S.J. Hinder, M.A. Baker, A.F. Zedan, A.M. Efstathiou, Cu-Ce-La-O$_x$ as efficient CO oxidation catalysts: effect of Cu content, Appl. Surf. Sci. 505 (2020) 144474, https://doi.org/10.1016/j.apsusc.2019.144474.

[42] X. Lin, Z. Nie, L. Zhang, S. Mei, Y. Chen, B. Zhang, R. Zhuc, Z. Liu, Nitrogen-doped carbon nanotubes encapsulatecobalt nanoparticles as efficient catalysts foraerobic and solvent-free selective oxidation ofhydrocarbons, Green Chem. 19 (2017) 2164–2173, https://doi.org/10.1039/C7GC00469A.

[43] R. Krishnan, S. Wu, H. Chen, Single Pt atom supported on penta-graphene Asan efficient catalyst for CO oxidation, Phys. Chem. Chem. 21 (2019) 12201–12208, https://doi.org/10.1039/C9CP02306B.

[44] E.H. Song, Z. Wen, Q. Jiang, CO catalytic oxidation on copper-embedded graphene, J. Phys. Chem. C 115 (9) (2011) 3678–3683, https://doi.org/10.1021/jp108978c.

[45] Q. Jiang, J. Zhang, H. Huang, Y. Wua, Z. Ao, A novel single-atom catalyst for CO oxidation inhumid environmental conditions: Ni-embeddeddivacancy graphene, J. Mater. Chem. A 8 (2020) 287–295, https://doi.org/10.1039/C9TA08525D.

[46] M. Abdolrahmani, M. Parvari, M. Habibpoor, Effect of copper substitution and preparation methods onthe LaMnO$_{3\pm\delta}$ structure and catalysis of methane combustion and CO oxidation, Chin. J. Catal. 31 (2010) 394–403, https://doi.org/10.1016/S1872-2067(09)60059-0.

[47] B. Seyfi, M. Baghalha, H. Kazemian, Modified LaCoO$_3$nano-perovskite catalysts for the environmental application ofautomotive CO oxidation, Chem. Eng. J. 148 (2009) 306–311, https://doi.org/10.1016/S1872-2067(09)60059-0.

[48] F. Okejiri, Z. Zhang, J. Liu, M. Liu, S. Yang, S. Dai, Room-temperature synthesis of high-entropy perovskite oxide nanoparticle catalysts via ultrasonication-based method, 13 (1) (2020) 111–115, https://doi.org/10.1002/cssc.201902705.

[49] M.G. Sanchez, J.L. Gazquez, Oxygen vacancy model in strong metal-supportinteraction, J. Catal. 104 (1) (1984) 120–135, https://doi.org/10.1016/0021-9517(87)90342-3.

[50] J.H. Holles, R.J. Davis, T.M. Murray, J.M. Howe, Effects of Pd particle size andceria loading on NO reduction with CO, J. Catal. 195 (1) (2000) 193–206, https://doi.org/10.1006/jcat.2000.2985.

[51] D. Ciuparu, A. Bensalem, L. Pfefferle, Pd-Ce interactions and adsorption properties of palladium: CO and NO TPD studies over Pd-Ce/Al2O3 catalysts, Appl. Catal. B Environ. 26 (4) (2000) 241–255, https://doi.org/10.1016/S0926-3373(00)00130-2.

[52] Y. Madier, C.C. Descorme, A.M. Le Govic, Oxygenmobility in CeO_2 and $Ce_xZr(1-x)O_2$ compounds: study byCO transient oxidation and $^{18}O/^{16}O$ isotopic exchange, J. Phys. Chem. B 103 (1999) 10999–11006, https://doi.org/10.1021/jp991270a.

[53] N.S. Priya, C. Somayaji, S. Kanagaraj, Optimization of Ceria-Zirconia solid solution based on OSC-measurement by cyclic heating process, Procedia Eng. 64 (2013) 1235–1241, https://doi.org/10.1016/j.proeng.2013.09.203.

[54] R. Di Monte, P. Fornasiero, J. Kašpar, M. Graziani, J.M. Gatica, S. Bernal, A. Gómez-Herrero, Stabilisation of nanostructured $Ce_{0.2}Zr_{0.8}O_2$ solid solution byimpregnation on Al_2O_3: a suitable method for the production of thermally stable oxygen storage/release promoters for three-way catalysts, Chem. Commun. 2000 (2000) 2167–2168, https://doi.org/10.1039/B006674P.

[55] P. Granger, J.F. Lamonier, N. Sergent, A.A. Aboukais, L. Leclercq, G. Leclercq, Investigation of the intrinsic activity of $Zr_xCe_{1-x}O_2$ mixed oxides in the CO + NOreactions: influence of Pd incorporation, Top. Catal. 16 (2001) 89–94, https://doi.org/10.1023/A:1016634915247.

[56] D.M. Fernandes, C.F. Scofield, A.A. Neto, M.J.B. Cardoso, F.M.Z. Zotin, Theinfluence of temperature on the deactivation of commercial Pd/Rh automotivecatalysts, Process Saf. Environ. Prot. 87 (5) (2009) 315–322, https://doi.org/10.1016/j.psep.2009.05.002.

[57] L. Lan, S. Chen, Y. Cao, M. Zhao, M. Gong, Y. Chen, Preparation of ceria-zirconiaby modified coprecipitation method and its supportedPd-only three-way catalyst, J. Colloid Interface Sci. 450 (2015) 404–416, https://doi.org/10.1016/j.jcis.2015.03.042.

[58] N.S. Priya, C. Somayaji, S. Kanagaraj, Characterization and optimization of $Ce_{0.6}Zr_{0.4-x}Mn_xO_2 (x \leq 0.4)$, J. Nanopart. Res. 16 (2014) 2661, https://doi.org/10.1007/s11051-014-2661-2.

[59] N.S. Priya, C. Somayaji, S. Kanagaraj, Optimization of $Ce_{0.6}Zr_{0.4-x}Al_{1.3x}O_2$ solid solution basedon oxygen storage capacity, J. Nanopart. Res. 16 (2014) 2214, https://doi.org/10.1007/s11051-013-2214-0.

[60] X.D. Wu, X.D. Wu, Q. Liang, J. Fan, D. Weng, Z. Xie, S.Q. Wei, Structure andoxygen storage capacity of Pr/Nd doped CeO_2-ZrO_2 mixed oxides, Solid StateSciences 9 (7) (2007) 636–643, https://doi.org/10.1016/j.solidstatesciences.2007.04.016.

[61] W.D. Wang, P.Y. Lin, M. Meng, Y.L. Fu, T.D. Hu, Y.N. Xie, T. Liu, Promoter of$(Ce-Zr)O_2$ solid solution modified by praseodymia in three-way catalysts, J. Rare Earths 21 (2003) 430–435.

[62] S. Brandenberger, O. Kröcher, A. Tissler, R. Althoff, The state of the art in selective catalytic reduction of NO_x by ammonia using metal-exchanged zeolite catalysts, Catal. Rev. 50 (4) (2008) 492–531, https://doi.org/10.1080/01614940802480122.

[63] K. Cheng, J. Liu, T. Zhang, J. Li, Z. Zhao, Y. Wei, G. Jiang, A. Duan, Effect of Ce doping of TiO_2 support on NH3-SCR activity over V2O5–WO3/CeO2–TiO2 catalyst, J. Environ. Sci. 26 (10) (2014) 2106–2113, https://doi.org/10.1016/j.jes.2014.08.010.

[64] G. Ramis, L. Yi, G. Busca, Ammonia activation over catalysts for the selective catalyticreduction of NO, and the selective catalytic oxidation of NH_3. AnFT-IR study, Catal. Today 28 (4) (1996) 373–380, https://doi.org/10.1016/S0920-5861(96)00050-8.

[65] W. SjoerdKijlstra, D.S. Brands, E.K. Poels, A. Bliek, Mechanism of the selective catalytic reduction of NOby NH_3 over MnO_x/Al_2O_3, J. Catal. 171 (1) (1997) 208–218, https://doi.org/10.1006/jcat.1997.1788.

[66] G. Qi, R.T. Yang, R. Chang, MnO_x-CeO_2 mixed oxides prepared by co-precipitation for selective catalytic reduction of NO with NH_3 at low temperatures, Appl. Catal. B: Environ. 51 (2) (2004) 93–106, https://doi.org/10.1016/j.apcatb.2004.01.023.

[67] N.H. Ahn, T. Ryu, Y. Kang, H. Kim, J. Shin, I. Nam, S.B. Hong, The origin of an unexpected increase in NH_3−SCR activity of agedCu-LTA catalysts, ACS Catal. 7 (10) (2017) 6781–6785, https://doi.org/10.1021/acscatal.7b02852.
[68] Z.S. Zhang, B.B. Chen, X.K. Wang, L. Xu, C. Au, C. Shi, M. Crocker, NO_x storage andreduction properties of model manganese-based lean NOx trap catalysts, Appl. Catal. B165 (2015) 232–244, https://doi.org/10.1016/j.apcatb.2014.10.001.
[69] S. Jeong, S. Youn, D.H. Kim, Effect of mg/Al ratios on the NO_x storage activity overPt-BaO/mg–Al mixed oxides, Catal. Today 231 (2014) 155–163, https://doi.org/10.1016/j.cattod.2013.12.047.
[70] J. Luo, F. Gao, D.H. Kim, C.H.F. Peden, Effects of potassium loading and thermal aging onK/Pt/Al_2O_3 high-temperature lean NO_x trap catalysts, Catal. Today 231 (2014) 164–172, https://doi.org/10.1016/j.cattod.2013.12.020.
[71] M.N. Anwar, A. Fayyaz, N.F. Sohaila, M.F. Khokhar, M. Baqar, W.D. Khan, K. Rasool, M. Rehan, A.S. Nizami, CO2 capture and storage: a way forward for sustainable environment, J. Environ. Manag. 226 (2018) 131–144, https://doi.org/10.1016/j.jenvman.2018.08.009.
[72] B. Dhrumil, Gohil, A. Pesyridis, J.R. Serrano, Overview of clean automotive thermal propulsion options for India to 2030, Appl. Sci. 10 (10) (2020) 3604, https://doi.org/10.3390/app10103604.

CHAPTER 27

Nanofuel additives

Luis A. Gallego-Villada[a], Edwin A. Alarcón[a], and Gustavo P. Romanelli[b,c]

[a]Chemical Engineering Department, Environmental Catalysis Research Group, Universidad de Antioquia, Medellín, Colombia
[b]Center for Research and Development in Applied Sciences "Dr. Jorge J. Ronco" (CINDECA-CCT La Plata-CONICET), National University of La Plata, La Plata, Argentina
[c]Course of Organic Chemistry, CISAV, Faculty of Agricultural and Forest Sciences, National University of La Plata, La Plata, Argentina

Chapter outline

1. Introduction — 561
2. Synthesis of nanoparticles as nanofuel additives — 562
   2.1 Metal-based nanoparticles — 562
   2.2 Carbon-based nanoparticles — 567
3. Properties of nanofuel additives in blends with fuels — 568
4. Application of nanofuel additives: Combustion performance — 569
5. Future aspects — 573
References — 573

1. Introduction

Transportation is mainly dependent on fossil-based fuels (>99.9%) such as diesel, gasoline fuel, compressed natural gas, and liquefied petroleum gas [1]; however, the availability of fossil-based fuels in the immediate future is a concern because their reserves will eventually be depleted [1, 2]. In addition, many harmful exhaust gases are released into the atmosphere while burning fossil fuels in internal combustion engines; for instance, 2.9 kg of greenhouse gas emissions is released into the atmosphere with the burning of 1 L diesel fuel [3]. The development of nanomaterials has led to nanoparticles (NPs) with applications in fuel additives. Fuel reformulation reduces exhaust emissions and improves combustion and fuel economy of combustion engines [4]. Additives are chemical compounds used to enhance the properties of the fuel by blending it with diesel, gasoline, and other alternative fuels in amounts ranging from 20 to 500 ppm. Nanoparticle additives, below 100 nm in size, have been successfully prepared and evaluated in the combustion engine. They are classified according to their composition: metal-based nanoparticles and carbon-based nanoparticles. Nanofuels are defined as common fuels whose nanoparticles are added as additives to improve their combustion characteristics [4]. Good reviews on the performance and emission characteristics of these additives in combustion engines fuelled with diesel, biodiesel, and blends have been published [4–8]. The potential of nanofuel additives, when blended with diesel-E waste plastics pyrolysis oil fuel blends has been reported. The effect of several nanoparticles on the enrichment in the performance characteristics and drop in the emission of CI engine tested with diesel and biodiesel fuel blends has been discussed [5]. According to their elements and properties, nanoparticles are classified into different types, the main ones being metal-based and carbon-based NPs [4] which are studied in detail below.

2. Synthesis of nanoparticles as nanofuel additives
2.1 Metal-based nanoparticles

Recently, novel synthesis approaches for metallic nanoparticles have been an interesting area in nanoscience and nanotechnology, in which two different fundamental principles of synthesis, top-down and bottom-up methods, have been investigated [9, 10]. Noble metals such as gold, silver, and platinum are good candidates for metal NP synthesis because of the easy reducibility of their salts [10, 11]. Top-down methods include synthesis procedures such as mechanical milling (ball milling, mechanochemical method), etching, and sputtering, while the bottom-up approach includes solid-state methods (physical and chemical vapor deposition), liquid state methods (sol-gel processes, chemical reduction, hydrothermal method, and solvothermal method), gas-phase methods (spray pyrolysis, laser ablation, flame pyrolysis), biological methods (bacteria, fungus, yeast, algae and plant extract) and other methods such as atomic/molecular condensation, precipitation/co-precipitation [9, 10]. The main methods that have been employed for the synthesis of nanoparticles used as nanoadditives in fuels are illustrated as follows.

2.1.1 Top-down methods
Bulk material is converted into small nanosized particles. The preparation of nanoparticles is based on the size reduction of the starting material by different physical and chemical treatments [12]. The major problem associated with this approach is the change in the surface chemistry and physicochemical properties of nanoparticles [13].

Ball milling: the main goal is the reduction in particle size, which requires a high amount of energy in the milling. The process success is affected by process variables and properties of the milling powder [10, 14], and this method is widely preferred for metal nanoparticle synthesis [15]. The procedure consists of adding bulk powder into a container along with several heavy balls and subsequently, high mechanical energy is applied to the material by a high-speed rotating ball [10, 15].

2.1.2 Bottom-up methods
This synthesis approach is based on the formation of nanoparticles from smaller molecules such as the joining of atoms, molecules, or small particles [10]. In this method, nanostructured building blocks of the nanoparticles are first formed and then assembled to produce the final nanoparticle [16].

2.1.2.1 Sol-gel method

The synthesis of nanoparticles by this method involves (i) mixing of preformed colloidal metal (oxide) with a sol containing the matrix-forming species followed by gel formation, (ii) direct mixing of metal and metal oxide or nanoparticles within a prehydrolyzed silica sol, or (iii) complexation of metal with silicone and reduction of metal before hydrolysis [17]. The method consists of a network formation using colloidal suspension (sol) and gelatin to form the network in a continuous liquid phase (gel). The ions of metal alkoxides (organometallic precursors) and aloxysilanes are used as a precursor for the synthesis of colloids, tetramethoxysilane, and tetraethoxysilane being the most commonly employed to form silica gel [10]. The sol-gel formation involves four main steps: hydrolysis, condensation, particle growth, and agglomeration.

Hydrolysis. In this stage, the addition of water leads to a replacement of [OR] group with [OH$^-$] group. Hydrolysis occurs by the attack of oxygen on silicon atoms in silica gel and it can be accelerated by adding a catalyst such as HCl and NH_3. The process continues until all alkoxy groups are replaced by hydroxyl groups, and the subsequent condensation, which involves silanol group (Si-OH), produces siloxane bonds (Si-O-Si) and alcohol and water [15].

Condensation. Polymerization to form siloxane bonds occurs by either water- or alcohol-producing condensation reaction, whose main products are monomers, dimers, cyclic tetramers, and high order rings. The pH, reagent concentration, and H_2O/Si molar ratio (in silica gels) are factors that affect the hydrolysis rate and that allow varying the structure and properties of sol-gel derived inorganic networks [15].

Growth and agglomeration. As the number of siloxane bonds increases, the molecules aggregate in the solution, forming a network, and a gel is formed upon drying. On the other hand, the water and alcohol are eliminated, and the network shrinks [15].

There are reports about nanoparticle synthesis by the sol-gel combustion process, which refers to an additional stage in the procedure of the sol-gel process described above. In this method, the obtained gel undergoes a strong self-propagation combustion process after a specified period, and then, a product composed of porous nanoparticles is formed [18–20].

Hydrothermal method

This procedure is based on the reaction of aqueous solutions, vapors, and/or fluids with solid materials at high temperature and pressure. It exploits the solubility of almost all inorganic substances in water at elevated temperatures and pressures and subsequent crystallization of the dissolved material from the fluid [21]. In this way, water at high temperatures plays an essential role in the transformation of precursor material because the vapor pressure is much higher and the water structure changes. In this method, the cations precipitate in polymeric hydroxide form and then these hydroxides get dehydrated and accelerate the formation of metal oxide crystal structure; the second metal cation formed is beneficial for controlling the particle formation process by preventing the formation of complex hydroxide when the base is added to the metal salt solution [10]. The addition of an oxidizer can suppress agglomeration between primary particles compared to the situation when particles are formed in the absence of an oxidizer [21]. The main advantages of the method are the desired size and shape of nanoparticles [21], well-crystallized powder [10], and high crystallinity of the synthesized nanocrystals [22]. However, there are some drawbacks due to the difficulty for process control and the limited reliability and reproducibility [10, 21].

Co-precipitation method
Precipitation/co-precipitation is a standard synthesis technique to prepare pure/multinary metal oxide nanoparticles [23]. The precipitation of oxides from both aqueous and nonaqueous solutions is less straightforward than the precipitation of metals. The reactions for the oxide synthesis can generally be classified into two categories: (i) reactions that produce an oxide directly and (ii) reactions that produce a precursor that must be subject to additional processes such as drying, calcination, among others. However, in both cases, monodispersed oxide nanoparticles, such as those of metals, frequently require a capping ligand or other surface-bound stabilizers to prevent particle agglomeration. The products of co-precipitation reactions, specifically those carried out at or near room temperature, are usually amorphous [17].

In a typical co-precipitation reaction, a salt precursor, generally a nitrate, chloride, or oxychloride, is dissolved in an aqueous solution and then, the corresponding hydroxides are precipitated by the addition of a base such as ammonium hydroxide or sodium hydroxide. After washing the produced ammonium or sodium salt, the hydroxides are calcined resulting in metal oxide powders [23]. When more than one precursor salt is used in the starting solution, multinary metal oxides can be obtained by co-precipitation of the corresponding hydroxides [24]. However, the particle size control is difficult if the precipitation rate is not controlled (slow); therefore, it has been reported that nanosized metal oxide powder is obtained with a modified co-precipitation method by controlling the pH at a constant value [25].

The co-precipitation of metal cations as carbonates, bicarbonates, or oxalates, followed by their subsequent calcination and decomposition, is a common method for producing crystalline nanoparticle oxides. However, the calcination procedure leads to agglomeration or at high temperatures, to aggregation and sintering. The advantage is that hydroxide, carbonate, and oxalate precursors tend to decompose at relatively low temperatures ($<400°C$) due to their high surface areas, hence minimizing agglomeration and aggregation [17].

Sonication-assisted co-precipitation method

Power ultrasound influences chemical reactivity through an effect known as cavitation [26]. The ultrasound (sonochemical) technique utilizes the principle of acoustic cavitation, which involves the formation, growth, and violent implosions of microcavities inside liquids irradiated with ultrasound, where these implosion regions are known as hotspots. The result of cavitation is the simple creation of supercritical conditions of very high pressures and temperatures of about 1000 atm and 5000 K, respectively, along with very fast heating and cooling rates. The supercritical conditions generated by ultrasound energy are on the nanosecond scale, ensuring safety to the sonochemical processes [27]. The extremely rapid cooling rates strongly favor the formation of amorphous products [17].

Microwave-assisted method

Recently, microwave-based techniques have been preferred to thermal heating for nanoparticle preparation. Microwave frequency in the range 300–300 GHz is applied, which orientates the polar molecule such as H_2O with the electric field. The reorientation of dipolar molecules with an alternating electric field causes molecular friction and loss of energy by heat [10]. The main advantages of this method are the simple and rapid volumetric heating and the consequent significant increase in reaction rate; in addition, homogeneous heating during all the processes can considerably speed up the reaction rate in comparison with conventional heating [10]. However, the short crystallization time and homogeneous nucleation because of uniform heat of the microwave oven are a drawback [28].

2.1.3 Examples of metal-based nanoparticles

Table 1 shows a state-of-the-art different methods for the synthesis of nanoparticles that have been investigated as nanofuel additives. Examples of synthesis procedures for each method reported in Table 1.

Silicon by ball milling: Microsized silicon particles are milled in a planetary ball mill for 45 h at 300 rpm speed, using a tungsten carbide vial and balls for milling. A 20:1 ball to material weight ratio was used [29].

TiO_2 by sol-gel method: Initially, the precursor material, titanium tetra isopropoxide, is treated with ethanol, hydrochloric acid, and deionized water. The mixture is stirred for 30 min until the pH level of 1.5 is reached. Deionized water is added to the obtained solution and stirred for 2 h at room temperature until the pH level of 6 is achieved. Subsequent titrations of the above mixture with deionized water turn it into a clady gel (pH 8), which is dried and calcined at 150°C for 12 h. Finally, TiO_2 nanoparticles are collected after the dried sample has been heated at 300°C for 2 h [33].

Table 1 Methods for the synthesis of nanoparticles used as nanofuel additives.

Method	Nanoparticle	Reference
Ball milling	Si	[29]
	Magnalium (Al—Mg)	[30]
Sol-gel	MnO	[31]
	TiO_2	[32, 33]
	CuO	[31, 34]
	CeO_2, CeO_2: Gd	[35]
	Al_2O_3	[36–38]
	Co_3O_4	[30]
Sol-gel combustion	CeO_2	[18–20]
	ZrO_2, TiO_2	[20]
	CuO	[39]
Hydrothermal	CeO_2, $Ce_{0.5}Co_{0.5}$	[40]
	Co_3O_4	[41]
	TiO_2	[41]
Precipitation/co-precipitation	Mn_2O_3	[42]
	CeO_2	[43, 44]
	$Ce_{0.7}Zr_{0.3}O_2$	[45]
	NiO	[46, 47]
	ZnO	[48]
	Co_3O_4	[49]
Sonication	Fe_3O_4	[50]
Microwave-assisted	Co_3O_4	[51]

CeO_2 by sol-gel combustion method: A required quantity of deionized water, an adequate amount of glycine, and cerium nitrate hexahydrate are dissolved in a beaker. The concentration of cerium nitrate hexahydrate is 1 M and glycine concentration is about 0.2 M. The obtained solution is kept under constant stirring at 70°C for approximately 2 h 30 min, and then, the obtained product is transferred to an electric mantle at 100°C. While heating at this temperature, excess water is dehydrated, and then, a transparent viscous gel is obtained. This gel undergoes a strong self-propagation combustion process after a specified period and then, a yellow product composed of porous foam nanoparticles is formed. The gases such as H_2O vapor, CO_2, and N_2 are exhausted in the form of brown fumes throughout the chemical reaction. The produced nanoparticles are annealed at 650°C for 2 h [18].

$Ce_{0.5}Co_{0.5}$ by the hydrothermal method: Cerium (III) nitrate hexahydrate and cobalt (II) nitrate hexahydrate in appropriate amounts are dissolved in water and mixed with concentrated sodium hydroxide solution. The solution is then transferred to an autoclave and gradually heated to 393 K and kept at that temperature for 24 h. Subsequently, the precipitates are collected by centrifugation and washed with distilled water and ethanol. The obtained materials are dried at 353 K overnight and calcined at 823 K for 4 h [40].

Co_3O_4 by co-precipitation method: An appropriate amount of cobalt (II) chloride hexahydrate is used as the precursor for this process, dissolved in deionized water with constant stirring for 20 min. Subsequently, an amount of carbonate solution (1 M) is added to the above solution. The mixtures are stirred for 5 h at 60°C. After 5 h, light purple color precipitates are obtained by centrifugation at 30,000 rpm. The precipitates are washed three times with deionized water and alcohol, respectively, and dried in an oven at 80°C for 12 h. The obtained product is calcined at 500°C for 3 h [49].

Fe_3O_4 by sonication method: Initially, an amount of ferric nitrate salt is dissolved in distilled water and then, an appropriate amount of sodium hydroxide is dissolved in distilled water. Both solutions are mixed and sonicated for 1 h (at 20 kHz) for uniform distribution, and the final homogeneous product is calcined up to 350°C [50].

Co_3O_4 by the microwave-assisted method: An amount of cobalt chloride is added to deionized water, followed by the addition of ethanol. The reaction mixture is stirred at room temperature for 1 h. Later, an appropriate amount of oxalic acid dihydrate is added to the reaction mixture, which is then transferred to a flask and refluxed in a microwave oven for 60 min at 50°C. After microwave treatment, light pink precipitates are collected by centrifugation and washed several times with deionized water at room temperature. Then, the obtained product is dried in a vacuum at 60°C for 6 h, ground, and calcined at 500°C for 2 h, hence the final product is obtained [51].

2.2 Carbon-based nanoparticles

2.2.1 Single-walled carbon nanotubes (SWCNTs)

SWCNTs were discovered by Iijima in 1991 while using an arc discharge evaporation method for producing fullerenes. The current methods for producing SWCNTs are plasma arc discharge (PAD), pulsed laser vaporization (PLV), and chemical vapor deposition (CVD). The latter method is the most popular because of the continuous, scalable, and controllable production [52]. In the CVD method, carbon nanotubes are produced from the carbon-containing source (liquid, gaseous, and solid hydrocarbons), as it decomposes at high temperature (500–1200°C) and passes over a metal catalyst (Fe, Co, Ni, Cu, Au, Ag, Pt, and Pd) deposited on a special powder carrier (e.g., Al_2O_3) or on a solid substrate, such as silicon wafers or glass in a tubular reactor [53]. The influence of the catalyst delivery speed on the synthesis output and activation conditions was studied based on the aerosol CVD approach using carbon monoxide as a carbon feedstock and ferrocene as catalyst precursor; the adjustment of the ferrocene injection strategy (injector flow rate) to cause a ninefold improvement in the synthesis yield while preserving the SWCNTs properties [54]. Optimization studies were performed to maximize the yield and minimize the diameter of SWCNTs, considering the complex dependence and interaction effects between the process parameters. The optimum condition for higher

SWCNTs yield with minimum diameter was obtained at a lower methane and hydrogen flow while keeping the furnace temperature and argon flow at the highest level and thiophene at the lowest level; ferrocene was used as catalyst [55]. SWCNTs were obtained by CVD using Co and Mo acetates as catalyst and ethanol as a carbon source [56]. SWCNTs were synthesized by the floating catalyst chemical vapor deposition (FC-CVD) in a horizontal two-zone reactor; the technique was used with ferrocene, methane, and hydrogen at 1000°C [57]. SWCNTs, with an average 1 nm diameter, were obtained in a stream surrounded by rich premixed laminar H_2/air flames using a feedstock containing ethanol and ferrocene [52]. Ru nanoparticles were supported on SWCNTs by microwave-assisted synthesis since they are potential materials as nanofuel additives [58].

2.2.2 Multiwalled carbon nanotubes (MWCNTs)

MWCNTs were obtained by aerosol-assisted CVD method using ferrocene in toluene and different carrier gases (N_2, Ar, He, 5% H_2–95% Ar, 3% H_2O-97% Ar). This process involves the atomization of a liquid precursor solution into aerosol droplets that are distributed throughout a gaseous medium. The generated aerosol is subsequently transported into a heated reaction zone where it undergoes rapid evaporation and/or decomposition (pyrolysis), causing the growth of carbon nanomaterial. The results also confirm that the use of different temperatures and atmospheres for the synthesis of carbon nanotubes allows controlling their parameters (e.g., diameter, number of walls) [53]. The low-cost red soil was directly used as a natural catalyst to obtain MWCNTs at 400°C in the presence of ethylene [59]. Chromium oxide and iron oxide solid solution were used as a catalyst for MWCNTs using natural gas as a carbon source at 950°C [60]. CVD synthesis using Fe—Ni over activated carbon as catalyst and acetylene as carbon source was carried out; the variables were studied using the statistical design of an experiment (DOE) involving the Taguchi method to control de diameter of MWCNTs under an optimal parameter combination [61]. Thiophene-based Ni-coordination polymer is a catalyst precursor and promoter for MWCNTs synthesis in CVD using propane and acetylene as carbon source; high yield and quality of MWCNTs were obtained by the CVD method [62]. Nanocomposite materials are also potential nanofluids. As an example, nanocomposite TiO_2-MWCNTs have been obtained by using a solution method previously synthesized MWCNTs [63].

3. Properties of nanofuel additives in blends with fuels

Numerous investigations have reported the influence of nanofuel additives on the physicochemical properties of fuel blends such as flash point, density, viscosity, cetane number, calorific value, among others [4, 64]. The addition of nanoparticles to fuel (diesel/biodiesel) blends improves the calorific value and cetane number and reduces the sulfur content in the fuel. The use of Al_2O_3, CNT, CeO_2, Al, Ag, and graphene nanoparticles with biodiesel lowered the flashpoint values, while they increased the viscosity and density values [65–67]. In addition, an increase in the calorific value in the fuel blend with the addition of CNT, graphene, and Al nanoparticles was reported [67]. Properties such as flash point, viscosity, density, and calorific value are directly affected by ZnO nanoparticles, while the addition of Mn, Mg, and Al nanoparticles reduces the flash point values [68–70]. Higher cetane indices were obtained for nanoaluminum and nanosilicon in the water-diesel emulsion as compared to diesel, which ensures better combustion quality [71, 72]. A detailed discussion about the properties of nanoadditives in fuel blends used in diesel engine applications is shown in reference [64].

4. Application of nanofuel additives: Combustion performance

Combustion performance has been discussed in detail in reference [73]. An update is shown in Table 2, which summarizes the performance of different nanoadditives in combustion engines. The first entry shows the application of a nanoadditive in diesel emulsion with water as fuel, which improves the energy and exergy performance parameters, and the sustainability of the engine without modification of diesel engines [74]. Entry 2 lists the different fuels (diesel, Mahua biodiesel) and modifications of common rail direct injection (CRDI) engine with ZnO nanoparticles mixed with diethyl ether and cetyl trimethyl ammonium bromide (CTAB); they enhance engine combustion, improve the performance characteristics, and lower emissions [48]. Manganese oxide or cobalt oxide nanoparticles (entry 3) decrease NOx emissions. However, the reduction level is lower compared to the urea-SCR system. Moreover, the results revealed that the pollutant reduction efficiency of the urea-SCR system was appreciably enhanced due to nanoparticle addition [75]. Al_2O_3 nanoparticles have been evaluated in a mixture of waste cooking oil (WCO) biodiesel - diesel as fuel, increasing BTE (Brake thermal efficiency) and decreasing BSFC (Brake specific fuel consumption). The results also showed that the blend of WCO biodiesel has similar characteristics to diesel [76]. Entry 5 shows the effect of NiO nanoadditives in *Azadirachta indica* biodiesel-diesel blend, prepared with different concentrations (25, 50, 75, and 100 ppm), on engine performance and emission characteristics. The results revealed that the presence of 75 ppm of NiO in fuel not only shows the best performance in BTE and BSFC but also lowers harmful emissions of CO, HC, and NOx [47]. Silicon nanoparticles (entry 6), synthesized by mall milling micron-sized silicon particles, were tested as nanoadditive in diesel fuel with three different concentrations (0.25, 0.5, and 0.75 wt%); the engine analysis showed that Si 0.5 wt% is better than the other tested nanofuel additives because it provides a better heat release rate and combustion chamber pressure during combustion [29]. Entries 7 and 8 show the positive effect of Al—Mg (synthesized by ball mill process) and Co_3O_4 (synthesized by the sol-gel

Table 2 Performance of nanoadditives in the combustion of diesel + biodiesel blends.

Entry	Nanoadditive	Fuel	Engine	Performance	Reference
1	Al_2O_3 and CNT, 25 ppm each one	5% Water in diesel emulsion	4-Cylinder 4-stroke Tata indigo variable speed diesel	BTE ↑1.49%, BSFC ↓5.61%	[74]
2	ZnO, 30 ppm	Diesel	4-Cylinder 4-stroke	BTE ↑7.3%–18.7%, smoke ↓20.6%, CO ↓13.2%, HC ↓10.1% NOx ↓5.7%	[48]
3	25 ppm Mn_2O_3 or 50 ppm Co_3O_4	20% Biodiesel, 80% diesel	4-Cylinder 4-stroke	CO ↓, NOx ↓	[75]
4	Al_2O_3	15% Biodiesel, 85% diesel	1-Cylinder 4-stroke	BTE ↑, BSFC ↑	[76]
5	NiO, 75 ppm	25% Biodiesel, 75% diesel	Direct injection-CI variable compression ratio	BTE ↑2.9%, BSFC ↓1.8%, CO ↓18.14%, HC ↓6.86% NOx ↓6.2%	[47]
6	Si, 0.5 wt%	Diesel	Twin cylinder direct injection diesel - Simpsons S217	BTE ↑8.93%, BSFC ↓14.56%, CO ↓28.57%, NOx ↓27.3%, HC ↑20.63%	[29]
7	Magnalium (Al—Mg)	Jatropha biodiesel	Single cylinder, air cooled, direct injection	BTE ↑, BSFC ↓3%, HC ↓70%, CO ↓41%, NOx ↓30%	[30]
8	Cobalt oxide (Co_3O_4)	Jatropha biodiesel	Single cylinder, air cooled, direct injection	BTE ↑, BSFC ↓2%, HC ↓60%, CO ↓50%, NOx ↓45%	[30]
9	CeO_2, 35 ppm	Diesel	Single cylinder, 4-stroke	BTE ↑6%, HC ↓45%, NOx ↓30%	[44]
10	Co_3O_4	*Calophyllum inophyllum* biodiesel	Single cylinder, 4-stroke, water cooled	BTE ↑7%, BSFC ↓4%, CO ↓30%, HC ↓80%, NOx ↑	[41]

Table 2 Performance of nanoadditives in the combustion of diesel + biodiesel blends—cont'd

Entry	Nanoadditive	Fuel	Engine	Performance	Reference
11	TiO_2	*Calophyllum inophyllum* biodiesel	Single cylinder, 4-stroke, water cooled	BTE ↑, BSFC ↓2%, CO ↑25%, HC ↓70%, NOx ↑	[41]
12	CNT, 100 ppm	30% Biodiesel, 70% diesel	Single cylinder, model: Yanmar TF 120 M	BTE ↑5.17%, BSFC ↓3.5%, CO ↓8.5%, HC ↑, NOx ↓3.92%	[77]
13	TiO_2, 100 ppm	30% Biodiesel, 70% diesel	Single cylinder, model: Yanmar TF 120 M	BTE ↑5.49%, BSFC ↓4.1%, CO ↓12.46%, HC ↓8.63%, NOx ↓1.84%	[77]
14	TiO_2, 60 ppm	20% Biodiesel, 80% diesel	Single cylinder, 4-stroke	BTE ↑, BSFC ↓, HC ↓, CO ↑, NOx ↓	[78]
15	CeO_2, 60 ppm	20% Biodiesel, 80% diesel	Single cylinder, 4-stroke	BTE ↑, BSFC ↓, HC ↓, CO ↓, NOx ↓	[78]
16	TiO_2 and CeO_2, 60 ppm each one	20% Biodiesel, 80% diesel	Single cylinder, 4-stroke	BTE ↑, BSFC ↓, HC ↓, CO ↓, NOx ↓	[78]
17	CeO_2 and CNT, 20 ppm each one	20% Biodiesel, 80% diesel	Single cylinder, 4-stroke, air cooled, direct injection	BTE ↑, BSFC ↓, NOx ↓40.6%, CO ↓, HC ↓	[79]
18	CNT, 100 ppm	20% Biodiesel from waste cooking oil, 80% diesel	Single cylinder, 4-stroke, air cooled	BTE ↑8%, BSFC ↓7%, CO ↓22%, NOx ↓22%, HC ↓28%	[80]
19	Graphene, 100 ppm	20% Biodiesel from waste cooking oil, 80% diesel	Single cylinder, 4-stroke, air cooled	BTE ↑19%, BSFC ↓18.2%, CO ↓47%, NOx ↓44%, HC ↓52%	[80]
20	Al_2O_3, 100 ppm	Diesel-methanol	Single cylinder, 4-stroke, water cooled, direct injection	BTE ↑3.6%, BSFC ↓3.7%, CO ↓83.3%, HC ↓40.9%, NOx ↑14.4%	[81]

Continued

Table 2 Performance of nanoadditives in the combustion of diesel + biodiesel blends—cont'd

Entry	Nanoadditive	Fuel	Engine	Performance	Reference
21	TiO_2, 90 ppm	10% Water in diesel emulsion	Single cylinder, 4-stroke, water cooled, direct injection	BTE ↑5.65%, CO ↓30%, HC ↓28.68%, NOx ↑16.26%	[82]

↓: decrease, ↑: increase, BSFC: brake specific fuel consumption, BTE: brake thermal efficiency.

process) nanoparticles, respectively, on the engine performance (increasing BTE and decreasing BSFC) and emission characteristics (HC, CO, NOx) of Jatropha biodiesel. Co_3O_4 acts as an oxygen buffer that improves the combustion and reduces the emissions. On the other hand, Al—Mg particles are highly energetic materials that reduce energy consumption and improve the thermal efficiency because the additive releases energy during combustion, in addition to the liquid fuel [30]. CeO_2 nanoparticles (entry 9), which were synthesized by precipitation method, have exceptional catalytic activity due to their oxygen buffering capability, especially in the nanosized form, and when they are used as an additive in diesel fuel they produce simultaneous reduction and oxidation of nitrogen dioxide and HC emissions, respectively. Furthermore, the reduction in the emissions is proportional to the dosing level of nanoparticles in the diesel, and an optimum dosing level of 35 ppm of CeO_2 was observed [44]. Co_3O_4 (entry 10) and TiO_2 (entry 11) nanoparticles were prepared by hydrothermal process and tested as fuel additives in biodiesel, showing that additives improve the engine performance and reduce HC emissions. However, a drawback of their use is the increase of NOx emissions, since the presence of fuel-bound oxygen promotes better combustion, resulting in higher cylinder temperature and thus, the oxidation of nitrogen [41]. Entries 12 and 13 correspond to CNT and TiO_2 nanoparticles, respectively, which were tested as fuel additives in biodiesel derived from palm-sesame oil, showing improvements in combustion performance and lower emissions of pollutants such as CO and NOx. However, HC emissions are mainly dependent on fuel atomization, fuel properties, and engine operating conditions, which increased when CNT nanoparticles were added to biodiesel due to the presence of carbon in their structure, while TiO_2 generated a decrease in HC emission that can be attributed to the improved burning of fuel because of the large reactive surface of nanoparticles [77]. Entries 14–16 show the positive effect on the engine performance and emissions when TiO_2 CeO_2 and a mixture of them, are used as nanofuel additives in palm oil-derived biodiesel. However, with titanium dioxide nanoparticles, a small increase in the carbon monoxide emissions is obtained due to incomplete combustion [78]. A mixture of CeO_2 and MWCNTs (entry 17) was evaluated as nanofuel additives in

20% *Calophyllum inophyllum* biodiesel +80% diesel, showing improved combustion performance and lower emissions of NOx, CO, and unburnt HC. On the other hand, the increase of nanoparticle dosage does not show a significant difference in the performance and emission characteristics of the engine, so it is concluded that this nanoparticle composition is the most suitable [79].

Another application of nanoadditives is in the engine radiator nanocoolants [83], improving the thermal conductivity and heat transfer coefficient. They also have major effects on internal combustion engines, such as energy savings, reduction in vehicular emissions due to lower fuel consumption, minimizing global warming. A complete review was reported by Hatami et al. [83]. Nanoparticles of TiO_2, Al_2O_3, graphene nanoplatelets (GNP), CNT, and Cu/CuO are usually used. Other hybrid nanoparticles are composed of combinations of the mentioned materials, e.g., 75% of Al_2O_3 and 25% CeO_2 increased the convective heat transfer coefficient with respect to the mass flow rate.

5. Future aspects

Nanotechnology will continue offering new nanofuel additives. Most of the evaluated nanomaterials use commercially available nanofuel additives. Great efforts should be made in evaluating newer formulations of nanofuel additives, apart from metal oxide and carbon-based nanoparticles. Moreover, significant investigations must be carried out to trap the possible unburnt nanoparticles from the engine exhaust to safeguard the global environment.

References

[1] Ü. Ağbulut, M. Karagöz, S. Sarıdemir, A. Öztürk, Impact of various metal-oxide based nanoparticles and biodiesel blends on the combustion, performance, emission, vibration and noise characteristics of a CI engine, Fuel 270 (February) (2020) 117521, https://doi.org/10.1016/j.fuel.2020.117521.

[2] Ü. Ağbulut, S. Sarıdemir, M. Karagöz, Experimental investigation of fusel oil (isoamyl alcohol) and diesel blends in a CI engine, Fuel 267 (November) (2020) 117042, https://doi.org/10.1016/j.fuel.2020.117042.

[3] Ü. Ağbulut, S. Sarıdemir, A general view to converting fossil fuels to cleaner energy source by adding nanoparticles, Int. J. Ambient Energy 0 (0) (2018) 1–6, https://doi.org/10.1080/01430750.2018.1563822.

[4] M. Tomar, N. Kumar, Influence of nanoadditives on the performance and emission characteristics of a CI engine fuelled with diesel, biodiesel, and blends—a review, Energy Sources A: Recover. Util. Environ. Eff. 42 (23) (2020) 2944–2961, https://doi.org/10.1080/15567036.2019.1623347.

[5] S. Deepankumar, K.L. Senthil Kumar, L.S. Gokul, B. Saravanan, The result of nano-additives in diesel engine using diesel-e waste plastics pyrolysis oil fuel blends: a review, IOP Conf. Ser. Mater. Sci. Eng. 764 (1) (2020) 012042, https://doi.org/10.1088/1757-899X/764/1/012042.

[6] S. Vellaiyan, K.S. Amirthagadeswaran, The role of water-in-diesel emulsion and its additives on diesel engine performance and emission levels: a retrospective review, Alexandria Eng. J. 55 (3) (Sep. 2016) 2463–2472, https://doi.org/10.1016/j.aej.2016.07.021.

[7] A. Londhekar, S. Shelke, N. Patil, R. Rajput, K. Phadkale, A review on: effect of nano-fuel additives on diesel engine, J. Emerg. Technolofies Innov. Res. 6 (3) (2019) 115–120.

[8] E. Jiaqiang, et al., Effect of different technologies on combustion and emissions of the diesel engine fueled with biodiesel: a review, Renew. Sustain. Energy Rev. 80 (2017) 620–647, https://doi.org/10.1016/j.rser.2017.05.250.

[9] J. Singh, T. Dutta, K.-H. Kim, M. Rawat, P. Samddar, P. Kumar, Green' synthesis of metals and their oxide nanoparticles: applications for environmental remediation, J. Nanobiotechnology 16 (1) (2018) 84, https://doi.org/10.1186/s12951-018-0408-4.

[10] P.G. Jamkhande, N.W. Ghule, A.H. Bamer, M.G. Kalaskar, Metal nanoparticles synthesis: an overview on methods of preparation, advantages and disadvantages, and applications, J. Drug Deliv. Sci. Technol. 53 (Oct. 2019) 101174, https://doi.org/10.1016/j.jddst.2019.101174.

[11] G. Schmid, Synthesis of metal nanoparticles, in: Encyclopedia of Inorganic Chemistry, Chichester, UK, John Wiley & Sons, Ltd, 2006.

[12] M.A. Meyers, A. Mishra, D.J. Benson, Mechanical properties of nanocrystalline materials, Prog. Mater. Sci. 51 (4) (May 2006) 427–556, https://doi.org/10.1016/j.pmatsci.2005.08.003.

[13] M.N. Nadagouda, T.F. Speth, R.S. Varma, Microwave-assisted green synthesis of silver nanostructures, Acc. Chem. Res. 44 (7) (Jul. 2011) 469–478, https://doi.org/10.1021/ar1001457.

[14] M. Ullah, M.E. Ali, S.B.A. Hamid, Surfactant-assisted ball milling: a novel route to novel materials with controlled nanostructure—a review, Rev. Adv. Mater. Sci. 37 (1–2) (2014) 1–14.

[15] N. Rajput, Methods of preparation of nanoparticles—a review, Int. J. Adv. Eng. Technol. 7 (4) (2015) 1806–1811.

[16] P. Mukherjee, et al., Fungus-mediated synthesis of silver nanoparticles and their immobilization in the mycelial matrix: a novel biological approach to nanoparticle synthesis, Nano Lett. 1 (10) (Oct. 2001) 515–519, https://doi.org/10.1021/nl0155274.

[17] B.L. Cushing, V.L. Kolesnichenko, C.J. O'Connor, Recent advances in the liquid-phase syntheses of inorganic nanoparticles, Chem. Rev. 104 (9) (2004) 3893–3946, https://doi.org/10.1021/cr030027b.

[18] B. Dhinesh, M. Annamalai, A study on performance, combustion and emission behaviour of diesel engine powered by novel nano nerium oleander biofuel, J. Clean. Prod. 196 (Sep. 2018) 74–83, https://doi.org/10.1016/j.jclepro.2018.06.002.

[19] M. Annamalai, et al., An assessment on performance, combustion and emission behavior of a diesel engine powered by ceria nanoparticle blended emulsified biofuel, Energy Convers. Manag. 123 (Sep. 2016) 372–380, https://doi.org/10.1016/j.enconman.2016.06.062.

[20] S. Janakiraman, T. Lakshmanan, V. Chandran, L. Subramani, Comparative behavior of various nano additives in a DIESEL engine powered by novel Garcinia gummi-gutta biodiesel, J. Clean. Prod. 245 (Feb. 2020) 118940, https://doi.org/10.1016/j.jclepro.2019.118940.

[21] A. Tavakoli, M. Sohrabi, A. Kargari, A review of methods for synthesis of nanostructured metals with emphasis on iron compounds, Chem. Pap. 61 (3) (2007) 151–170, https://doi.org/10.2478/s11696-007-0014-7.

[22] C. Burda, X. Chen, R. Narayanan, M.A. El-Sayed, Chemistry and properties of nanocrystals of different shapes, Chem. Rec. 36 (105) (2005) 1025–1102, https://doi.org/10.1002/chin.200527215.

[23] S.H. Soytaş, O. Oğuz, Y.Z. Menceloğlu, Polymer nanocomposites with decorated metal oxides, in: Polymer Composites with Functionalized Nanoparticles, Elsevier, 2019, pp. 287–323.

[24] M. Behrens, Coprecipitation: an excellent tool for the synthesis of supported metal catalysts—from the understanding of the well known recipes to new materials, Catal. Today 246 (May 2015) 46–54, https://doi.org/10.1016/j.cattod.2014.07.050.

[25] H. Ke, et al., Factors controlling pure-phase multiferroic BiFeO3 powders synthesized by chemical co-precipitation, J. Alloys Compd. 509 (5) (Feb. 2011) 2192–2197, https://doi.org/10.1016/j.jallcom.2010.09.213.

[26] D.N. Srivastava, N. Perkas, A. Zaban, A. Gedanken, Sonochemistry as a tool for preparation of porous metal oxides, Pure Appl. Chem. 74 (9) (2002) 1509–1517, https://doi.org/10.1351/pac200274091509.

[27] S. Sivasankaran, S. Sankaranarayanan, S. Ramakrishnan, A novel Sonochemical synthesis of metal oxides based Bhasmas, Mater. Sci. Forum 754 (Apr. 2013) 89–97, https://doi.org/10.4028/www.scientific.net/MSF.754.89.

[28] A. Ali, et al., Synthesis, characterization, applications, and challenges of iron oxide nanoparticles, Nanotechnol. Sci. Appl. 9 (2016) 49–67, https://doi.org/10.2147/NSA.S99986.

[29] S.A. Anand Kumar, V. Raja, R.P. Dhivakar Raviram, G. Sakthinathan, Impact of nano-silicon fuel additive on combustion, performance and emission of a twin cylinder CI engine, MATEC Web Conf. 172 (2018), https://doi.org/10.1051/matecconf/201817202007, 02007.

[30] D. Ganesh, G. Gowrishankar, Effect of nano-fuel additive on emission reduction in a biodiesel fuelled CI engine, in: 2011 International Conference on Electrical and Control Engineering, 2014, pp. 3453–3459, https://doi.org/10.1109/ICECENG.2011.6058240.

[31] M.A. Lenin, M.R. Swaminathan, G. Kumaresan, Performance and emission characteristics of a DI diesel engine with a nanofuel additive, Fuel 109 (Jul. 2013) 362–365, https://doi.org/10.1016/j.fuel.2013.03.042.

[32] D. Yuvarajan, M. Dinesh Babu, N. Beem Kumar, P. Amith Kishore, Experimental investigation on the influence of titanium dioxide nanofluid on emission pattern of biodiesel in a diesel engine, Atmos. Pollut. Res. 9 (1) (Jan. 2018) 47–52, https://doi.org/10.1016/j.apr.2017.06.003.

[33] H. Venu, L. Subramani, V.D. Raju, Emission reduction in a DI diesel engine using exhaust gas recirculation (EGR) of palm biodiesel blended with TiO_2 nano additives, Renew. Energy 140 (Sep. 2019) 245–263, https://doi.org/10.1016/j.renene.2019.03.078.

[34] V. Perumal, M. Ilangkumaran, The influence of copper oxide nano particle added pongamia methyl ester biodiesel on the performance, combustion and emission of a diesel engine, Fuel 232 (March) (2018) 791–802, https://doi.org/10.1016/j.fuel.2018.04.129.

[35] K. Dhanasekar, M. Sridaran, M. Arivanandhan, R. Jayavel, A facile preparation, performance and emission analysis of pongamia oil based novel biodiesel in diesel engine with CeO_2: Gd nanoparticles, Fuel 255 (2019) 115756, https://doi.org/10.1016/j.fuel.2019.115756.

[36] V. Dhana Raju, P.S. Kishore, K. Nanthagopal, B. Ashok, An experimental study on the effect of nanoparticles with novel tamarind seed methyl ester for diesel engine applications, Energy Convers. Manag. 164 (March) (2018) 655–666, https://doi.org/10.1016/j.enconman.2018.03.032.

[37] V. Dhana Raju, P.S. Kishore, M. Harun Kumar, S. Rami Reddy, Experimental investigation of alumina oxide nanoparticles effects on the performance and emission characteristics of tamarind seed biodiesel fuelled diesel engine, Mater. Today Proc. 18 (2019) 1229–1242, https://doi.org/10.1016/j.matpr.2019.06.585.

[38] M. Tomar, N. Kumar, Effect of multi-walled carbon nanotubes and alumina nano-additives in a light duty diesel engine fuelled with *schleichera oleosa* biodiesel blends, Sustain. Energy Technol. Assessments 42 (August) (2020) 100833, https://doi.org/10.1016/j.seta.2020.100833.

[39] V. Chandrasekaran, M. Arthanarisamy, P. Nachiappan, S. Dhanakotti, B. Moorthy, The role of nano additives for biodiesel and diesel blended transportation fuels, Transp. Res. Part D Transp. Environ. 46 (2016) 145–156, https://doi.org/10.1016/j.trd.2016.03.015.

[40] S. Akram, et al., Impact of cerium oxide and cerium composite oxide as nano additives on the gaseous exhaust emission profile of waste cooking oil based biodiesel at full engine load conditions, Renew. Energy 143 (Dec. 2019) 898–905, https://doi.org/10.1016/j.renene.2019.05.025.

[41] L. Jeryrajkumar, G. Anbarasu, T. Elangovan, Effects on nano additives on performance and emission characteristics of Calophyllim inophyllum biodiesel, Int. J. Chem Tech Res. 9 (04) (2016) 210–219.

[42] M. Amirabedi, S. Jafarmadar, S. Khalilarya, Experimental investigation the effect of Mn_2O_3 nanoparticle on the performance and emission of SI gasoline fueled with mixture of ethanol and gasoline, Appl. Therm. Eng. 149 (November) (2018) 512–519. 2019 https://doi.org/10.1016/j.applthermaleng.2018.12.058.

[43] S.A. Sheriff, et al., Emission reduction in CI engine using biofuel reformulation strategies through nano additives for atmospheric air quality improvement, Renew. Energy 147 (Mar. 2020) 2295–2308, https://doi.org/10.1016/j.renene.2019.10.041.

[44] A.C. Sajeevan, V. Sajith, Diesel engine emission reduction using catalytic nanoparticles: an experimental investigation, J. Eng. 2013 (2013) 1–9, https://doi.org/10.1155/2013/589382.

[45] P. Fu, X. Bai, W. Yi, Z. Li, Y. Li, L. Wang, Assessment on performance, combustion and emission characteristics of diesel engine fuelled with corn stalk pyrolysis bio-oil/diesel emulsions with $Ce0.7Zr0.3O2$ nanoadditive, Fuel Process. Technol. 167 (2017) 474–483, https://doi.org/10.1016/j.fuproc.2017.07.032.

[46] C. Srinidhi, A. Madhusudhan, S.V. Channapattana, Effect of NiO nanoparticles on performance and emission characteristics at various injection timings using biodiesel-diesel blends, Fuel 235 (2019) 185–193, https://doi.org/10.1016/j.fuel.2018.07.067.

[47] S. Campli, M. Acharya, S.V. Channapattana, A.A. Pawar, S.V. Gawali, J. Hole, The effect of nickel oxide nano-additives in *Azadirachta indica* biodiesel-diesel blend on engine performance and emission characteristics by varying compression ratio, Environ. Prog. Sustain. Energy (2020), https://doi.org/10.1002/ep.13514.

[48] M.E.M. Soudagar, et al., Study of diesel engine characteristics by adding nanosized zinc oxide and diethyl ether additives in Mahua biodiesel–diesel fuel blend, Sci. Rep. 10 (1) (2020) 15326, https://doi.org/10.1038/s41598-020-72150-z.

[49] M.R.S.A. Janjua, Synthesis of Co_3O_4 Nano aggregates by co-precipitation method and its catalytic and fuel additive applications, Open Chem. 17 (1) (Oct. 2019) 865–873, https://doi.org/10.1515/chem-2019-0100.

[50] S. Kumar, P. Dinesha, I. Bran, Influence of nanoparticles on the performance and emission characteristics of a biodiesel fuelled engine: an experimental analysis, Energy 140 (Dec. 2017) 98–105, https://doi.org/10.1016/j.energy.2017.08.079.

[51] S. Jamil, M.R.S.A. Janjua, S.R. Khan, Synthesis and structural investigation of polyhedron Co_3O_4 nanoparticles: catalytic application and as fuel additive, Mater. Chem. Phys. 216 (Sep. 2018) 82–92, https://doi.org/10.1016/j.matchemphys.2018.05.051.

[52] C. Zhang, et al., Synthesis of single-walled carbon nanotubes in rich hydrogen/air flames, Mater. Chem. Phys. 254 (6) (2020) 123479, https://doi.org/10.1016/j.matchemphys.2020.123479.

[53] P. Mierczynski, et al., Effect of the AACVD based synthesis atmosphere on the structural properties of multi-walled carbon nanotubes, Arab. J. Chem. 13 (1) (2020) 835–850, https://doi.org/10.1016/j.arabjc.2017.08.001.

[54] E.M. Khabushev, J.V. Kolodiazhnaia, D.V. Krasnikov, A.G. Nasibulin, Activation of catalyst particles for single-walled carbon nanotube synthesis, Chem. Eng. J. (October) (2020) 127475, https://doi.org/10.1016/j.cej.2020.127475.

[55] A. Kaushal, R. Alexander, P.T. Rao, J. Prakash, K. Dasgupta, Artificial neural network, Pareto optimization, and Taguchi analysis for the synthesis of single-walled carbon nanotubes, Carbon Trends 2 (2021) 100016, https://doi.org/10.1016/j.cartre.2020.100016.

[56] S. Inoue, S. Lojindarat, T. Kawamoto, Y. Matsumura, T. Charinpanitkul, Spontaneous and controlled-diameter synthesis of single-walled and few-walled carbon nanotubes, Chem. Phys. Lett. 699 (2018) 88–92, https://doi.org/10.1016/j.cplett.2018.03.054.

[57] M.D. Yadav, A.W. Patwardhan, J.B. Joshi, K. Dasgupta, Selective synthesis of metallic and semiconducting single-walled carbon nanotube by floating catalyst chemical vapour deposition, Diam. Relat. Mater. 97 (May) (2019) 107432, https://doi.org/10.1016/j.diamond.2019.05.017.

[58] T. Hemraj-Benny, N. Tobar, N. Carrero, R. Sumner, L. Pimentel, G. Emeran, Microwave-assisted synthesis of single-walled carbon nanotube-supported ruthenium nanoparticles for the catalytic degradation of Congo red dye, Mater. Chem. Phys. 216 (February) (2018) 72–81, https://doi.org/10.1016/j.matchemphys.2018.05.081.

[59] X. Yuan, et al., Low-cost synthesis of multi-walled carbon nanotubes using red soil as catalyst, Diam. Relat. Mater. 112 (September) (2020) 108241. 2021 https://doi.org/10.1016/j.diamond.2021.108241.

[60] M.D. Lima, R. Bonadiman, M.J. de Andrade, J. Toniolo, C.P. Bergmann, Synthesis of multi-walled carbon nanotubes by catalytic chemical vapor deposition using $Cr_2-xFe_xO_3$ as catalyst, Diam. Relat. Mater. 15 (10) (2006) 1708–1713, https://doi.org/10.1016/j.diamond.2006.02.009.

[61] T.C. Egbosiuba, et al., Taguchi optimization design of diameter-controlled synthesis of multi walled carbon nanotubes for the adsorption of Pb (II) and Ni(II) from chemical industry wastewater, Chemosphere 266 (2021) 128937, https://doi.org/10.1016/j.chemosphere.2020.128937.

[62] H.Y. Lin, J. Luan, Y. Tian, Q.Q. Liu, X.L. Wang, Thiophene-based Ni-coordination polymer as a catalyst precursor and promoter for multi-walled carbon nanotubes synthesis in CVD, J. Solid State Chem. 293 (September) (2021) 121782, https://doi.org/10.1016/j.jssc.2020.121782.

[63] M.B. Askari, Z. Tavakoli Banizi, M. Seifi, S. Bagheri Dehaghi, P. Veisi, Synthesis of TiO2 nanoparticles and decorated multi-wall carbon nanotube (MWCNT) with anatase TiO2 nanoparticles and study of optical properties and structural characterization of TiO2/MWCNT nanocomposite, Optik (Stuttg) 149 (2017) 447–454, https://doi.org/10.1016/j.ijleo.2017.09.078.

[64] M.E.M. Soudagar, N.-N. Nik-Ghazali, M. Abul Kalam, I.A. Badruddin, N.R. Banapurmath, N. Akram, The effect of nano-additives in diesel-biodiesel fuel blends: a comprehensive review on stability, engine performance and emission characteristics, Energy Convers. Manag. 178 (Dec. 2018) 146–177, https://doi.org/10.1016/j.enconman.2018.10.019.

[65] P. Arockiasamy, R.B. Anand, Performance, combustion and emission characteristics of a D.I. diesel engine Fuelled with nanoparticle blended Jatropha biodiesel, Period. Polytech. Mech. Eng. 59 (2) (2015) 88–93, https://doi.org/10.3311/PPme.7766.

[66] J. Sadhik Basha, R.B. Anand, The influence of nano additive blended biodiesel fuels on the working characteristics of a diesel engine, J. Brazilian Soc. Mech. Sci. Eng. 35 (3) (2013) 257–264, https://doi.org/10.1007/s40430-013-0023-0.

[67] G.R. Kannan, R. Karvembu, R. Anand, Effect of metal based additive on performance emission and combustion characteristics of diesel engine fuelled with biodiesel, Appl. Energy 88 (11) (Nov. 2011) 3694–3703, https://doi.org/10.1016/j.apenergy.2011.04.043.

[68] C.S. Aalam, C.G. Saravanan, Effects of nano metal oxide blended Mahua biodiesel on CRDI diesel engine, Ain Shams Eng. J. 8 (4) (Dec. 2017) 689–696, https://doi.org/10.1016/j.asej.2015.09.013.

[69] A. Keskin, M. Gürü, D. Altıparmak, Biodiesel production from tall oil with synthesized Mn and Ni based additives: effects of the additives on fuel consumption and emissions, Fuel 86 (7–8) (May 2007) 1139–1143, https://doi.org/10.1016/j.fuel.2006.10.021.

[70] M. Gürü, A. Koca, Ö. Can, C. Çınar, F. Şahin, Biodiesel production from waste chicken fat based sources and evaluation with mg based additive in a diesel engine, Renew. Energy 35 (3) (Mar. 2010) 637–643, https://doi.org/10.1016/j.renene.2009.08.011.

[71] R.N. Mehta, M. Chakraborty, P.A. Parikh, Impact of hydrogen generated by splitting water with nano-silicon and nano-aluminum on diesel engine performance, Int. J. Hydrog. Energy 39 (15) (May 2014) 8098–8105, https://doi.org/10.1016/j.ijhydene.2014.03.149.

[72] M. Norhafana, et al., A review of the performance and emissions of nano additives in diesel fuelled compression ignition-engines, IOP Conf. Ser. Mater. Sci. Eng. 469 (1) (2019) 012035, https://doi.org/10.1088/1757-899X/469/1/012035.

[73] M. Hatami, M. Hasanpour, D. Jing, Recent developments of nanoparticles additives to the consumables liquids in internal combustion engines: part I: Nano-fuels, J. Mol. Liq. 318 (Nov. 2020) 114250, https://doi.org/10.1016/j.molliq.2020.114250.

[74] R.K. Rai, R.R. Sahoo, Impact of different shape based hybrid nano additives in emulsion fuel for exergetic, energetic, and sustainability analysis of diesel engine, Energy 214 (Jan. 2021) 119086, https://doi.org/10.1016/j.energy.2020.119086.

[75] M. Mehregan, M. Moghiman, Experimental investigation of the distinct effects of nanoparticles addition and urea-SCR after-treatment system on NOx emissions in a blended-biodiesel fueled internal combustion engine, Fuel 262 (August) (2020) 116609, https://doi.org/10.1016/j.fuel.2019.116609. 2019.

[76] J. Prasad Pradhan, B. Singh, Experimental investigation on performance of a CI engine using waste cooking oil biodiesel blends with alcohol and nanoparticle additives as fuel, Mater. Today Proc. 24 (2020) 1332–1339, https://doi.org/10.1016/j.matpr.2020.04.449.

[77] M.A. Mujtaba, et al., Comparative study of nanoparticles and alcoholic fuel additives-biodiesel-diesel blend for performance and emission improvements, Fuel 279 (Nov. 2020) 118434, https://doi.org/10.1016/j.fuel.2020.118434.

[78] S.N.K. Reddy, M.M. Wani, Engine performance and emission studies by application of nanoparticles as additive in biodiesel diesel blends, Mater. Today Proc. 30 (3–4) (Nov. 2020) 175–180, https://doi.org/10.1016/j.matpr.2020.09.832.

[79] C.H. Sree Harsha, T. Suganthan, S. Srihari, Performance and emission characteristics of diesel engine using biodiesel-diesel-nanoparticle blends-an experimental study, Mater. Today Proc. 24 (2020) 1355–1364, https://doi.org/10.1016/j.matpr.2020.04.453.

[80] M.S. Gad, B.M. Kamel, I. Anjum Badruddin, Improving the diesel engine performance, emissions and combustion characteristics using biodiesel with carbon nanomaterials, Fuel 288 (Mar. 2021) 119665, https://doi.org/10.1016/j.fuel.2020.119665.

[81] J. Wei, et al., Impact of aluminium oxide nanoparticles as an additive in diesel-methanol blends on a modern DI diesel engine, Appl. Therm. Eng. 185 (93) (2021) 116372, https://doi.org/10.1016/j.applthermaleng.2020.116372.

[82] R. Vigneswaran, D. Balasubramanian, B.D.S. Sastha, Performance, emission and combustion characteristics of unmodified diesel engine with titanium dioxide (TiO2) nano particle along with water-in-diesel emulsion fuel, Fuel 285 (Feb. 2021) 119115, https://doi.org/10.1016/j.fuel.2020.119115.

[83] M. Hatami, M. Hasanpour, D. Jing, Recent developments of nanoparticles additives to the consumables liquids in internal combustion engines: part III: Nano-coolants, J. Mol. Liq. 319 (Dec. 2020) 114131, https://doi.org/10.1016/j.molliq.2020.114131.

CHAPTER 28

Nanocatalysts for fuel cells

Elisangela Pacheco da Silva[a], Vanessa Hafemann Fragal[a], Rafael Silva[a], Alexandre Henrique Pinto[b], Thiago Sequinel[c], Matheus Ferrer[g], Mario Lucio Moreira[h], Emerson Rodrigues Camargo[d], Ana Paula Michels Barbosa[e], Carlos Alberto Severo Felipe[e], Ramesh Katla[e], and Luiz Fernando Gorup[d,e,f]

[a]Department of Chemistry, UEM—State University of Maringa Avenida Colombo, Maringá, Paraná, Brazil
[b]Department of Chemistry & Biochemistry, Manhattan College, Riverdale, NY, United States
[c]Faculty of Exact Sciences and Technology (FACET), Federal University of Grande Dourados, Dourados, Mato Grosso do Sul, Brazil
[d]LIEC—Interdisciplinary Laboratory of Electrochemistry and Ceramics, Department of Chemistry, UFSCar-Federal University of São Carlos, São Carlos, São Paulo, Brazil
[e]School of Chemistry and Food Science, Federal University of Rio Grande, Rio Grande, Rio Grande do Sul, Brazil
[f]Institute of Chemistry, Federal University of Alfenas, Alfenas, Minas Gerais, Brazil
[g]CCAF—Advanced Crystal Growth and Photonics, CDTEC—Technological Development Center, Federal University of Pelotas, UFPEL, Pelotas, RS, Brazil
[h]Department of Physics, Federal University of Pelotas, Pelotas, RS, Brazil

Chapter outline

1. Introduction 579
2. Fuel cell technology 581
3. Nanocatalyst for fuel cell—Market in value 583
4. Types, characteristics, and synthesis of nanocatalysts 585
5. Recent development of nanocatalysts for fuel cells 591
   5.1. Nanostructured Pt-based catalysts 592
   5.2. Pt-free nanocatalyst 594
6. Advantages and challenges of nanocatalysts for fuel cells 596
7. General conclusions and future perspectives 599
References 599

1. Introduction

The growing concerns regarding the production and distribution of vehicles involving topics such as sustainability and the environment have gained more prominence over the years. Such a subject has caused certain governmental pressures on the products and processes used in current technologies considering the used raw materials and the by-products generated during the entire production and use cycle of automobiles. The growing number of vehicles in recent decades due to the growth of the global economy highlights the importance of solutions to the various problems linked to the automotive sector as a whole.

Among the problems, the consumption of fossil fuels and the emission of toxic gases generated in combustion engines have always been subjects that have been much discussed and constantly reassessed over the years [1, 2]. Observing the evolution of automobiles in the last decades, it is notable the decrease in relative fuel consumption per kilometer in popular vehicles. This can be attributed to several reasons, such as improving the efficiency of the engines concerning the cycle for carrying out work, decreasing the weight of the vehicles, the evolution of the technology of the materials used, and the design of the project. Following this evolutionary process in the automotive sector, the emission reduction of pollutants in the exhaust stage of the fuel-burning cycle resulted from incomplete fuel burning was also significant, reaching values close to zero [3]. The incomplete fuel burning has as main by-products carbon monoxide (CO), hydrocarbons and nitrogen oxides (NOx), toxic gases that contribute to the greenhouse effect, acidification of the environment, and smog.

As much as the current scenario shows a transitory moment for new energy production technologies in the automotive sector, such as research on the efficiency of internal combustion engines, efficiency in catalytic conversion, and the development of electric vehicles continue to be of great importance. Vehicles with internal combustion will still be used for many years until new technologies, like fuel cells, become viable compared to current technologies [4].

Currently, the main focus is dedicated to reduce the cost of fuel cell production to become viable commercially. The major barrier is the material used as catalysts: the platinum group metals (PGMs). Platinum (Pt) is a noble metal that has been shown the efficient catalytic activity, for both catalytic converter of internal combustion cycles and fuel cell catalysts. In addition to the catalytic efficiency, some other factors are important in the selection of the materials to be used in the catalytic converters of the vehicles and fuel cell vehicles. For example, the catalytic material must maintain its activity at different temperatures, besides having thermal stability and resistance to poisoning caused by the by-products of the fuel.

Several efforts have been made to improve the performance of Pt electrocatalysts for oxygen reduction reaction (ORR), such as the optimization of the size and shape of Pt particles. Also, Pt with 3D transition metals is linking creating composite nanostructures. However, the catalyst surface is maintained as Pt, because this metal is probably the only element that is active and stable under Proton Exchange Membrane Fuel Cell (PEMFC—the most used fuel cell to the automotive industry) conditions. Therefore, the design of advanced Pt and Pt-free electrocatalysts generally involves the manipulation of subsurface nanoscale architectures to induce changes in surface geometry and/or electronic structures to alter the adsorption and catalytic properties of the nanomaterials.

Thus, nanotechnology holds the promise of new solutions for fuel cell catalysts application. Companies are using Pt nanoparticles and other nanomaterials to improve the performance and reduce the cost of the catalysts. The application of nanotechnology in the automotive sector is considered to be highly successful because of the great efficiency in fulfilling its objective. Although we are dealing with precious metals (low availability and high cost) they are essential components for automobilist technology.

In this context, even though the catalytic processes already show excellent efficiency in the internal combustion cycles and are considered a success in fuel cell application, the subject attracts a lot of attention from researchers. The main objective of recent research deals with finding alternatives that involve processing to recycle or reduce the number of rare metals, either by increasing the catalytic efficiency or by substituting the rare metals with cheaper and more abundant metals [5, 6]. Thus, a general presentation on the recent researches, including synthesis strategies of catalysts used in the automotive sector and new nanocatalyst Pt-free is worthwhile, even if some methods are still in study stages.

2. Fuel cell technology

Climate change and energy safety concerns are the main subjects discussed in the automotive industry at the moment. Hydrocarbon fuels used for transport are one of the main responsible for emissions of large amounts of "greenhouse gases," like CO_2, to the atmosphere. The transportation sector contributes 14% to the total global CO_2 emissions [7]. Accordingly, the current fuel used will need to be substituted by low-carbon or no-carbon alternatives to reduce greenhouse gas emissions and avoid dangerous climate change. In this context, there are huge efforts dedicated to developing electric vehicles (EV) and fuel cell vehicles (FCVs) as an alternative to fossil fuel. Electric vehicles, like Nissan LEA, Toyota Corolla Cross, Peugeot e208, Fiat 500e, and Jeep are the reality in the whole world. However, the main issue of this type of car is the necessity of a long time to be rechargeable, making logistics difficult [8].

Hydrogen as a fuel has remarkable progress in the last decade, being used for transportation, as well as the stationary and portable fuel cell. Fuel cells and hydrogen have received a lot of attention from researchers as an alternative to batteries used in electric cars. At the moment, the industries that commercialize FCV are Toyota (Mirai), Hyundai (NEXO), and Honda (Clarity).

Fuel cell technology is expected to play an essential role in the future of the automotive industry since it can reduce carbon dioxide (CO_2) emissions. A fuel cell is an electrochemical device, which converts the chemical energy of fuel into electricity, which means a fuel cell operates like a battery. The advantage of the fuel cell is related to the capacity to generate energy rather than storing it, making it possible to continue to do as long as a fuel supply is maintained [9].

The concept of fuel cell had been demonstrated in the early nineteenth century and there is some controversy about the scientist behind the discovery. It is reported that Willian Grove (1838s) conducted an experiment in which electric current was produced from the electrochemical reaction between hydrogen and oxygen over a platinum catalyst. After that, a lot of scientists performed various fuel cell experiments until Francis Bacon (1939s) developed the first fuel cell to be used in submarines of the Royal Navy. Subsequently, the patent of Bacon's work was acquired to be used in Apollo super craft (1950s). At the same time, Harry Ihrig developed the first fuel-cell-powered vehicle including a tractor for Allis-Chalmer. The 1970s were marked by a burst of companies, such as General Motors and Shell, demonstrating fuel cell electric vehicles. The development continued in the following decades driven by the intense demand for sources of renewable energy until arrives in today's technology [10].

Based on its application, the fuel cell industry is divided into three segments: portable, transportation, and stationary, where the last one corresponds to 70% of the market in terms of volume. However, due to environmental concerns, the automobilist fuel cell industry has made a huge effort to promote fuel cell technology in automobiles. The advances in fuel cells industry have been categorized, based on type, into Proton Exchange Membrane Fuel Cell (PEMFC), Molten Carbonate Fuel Cell (MOFC), Phosphoric Acid Fuel Cell (PAFC), Direct Methanol Fuel Cell (DMFC), Solid Oxide Fuel Cell (SOFC), Alkaline Fuel Cell (AFC). The classification is based on the type of electrochemical reactions in the cell, the temperature range of operation, the type of electrolyte, fuel, and so on, according to Fig. 1. Among all of them, PEMFC has

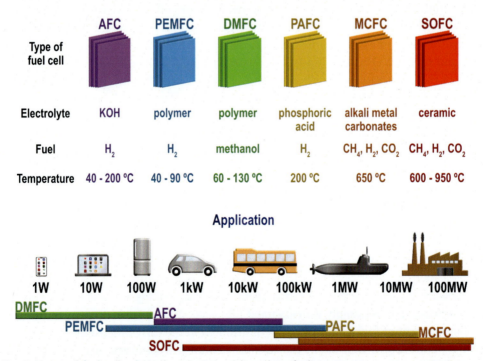

Fig. 1 Types of fuel cells, their characteristics (electrolyte, fuel, temperature range operation, and output power), and applications.

emerged as the most promising option for a future transport application, due to its high efficiency, low emissions, low operating temperature (80°C), fast startup time, and favorable power-to-weight ratio [11].

PEMFC is a type of fuel cell that uses hydrogen (H_2) with oxygen (O_2) to produce electricity while producing only water as a by-product when used pure hydrogen as fuel. The heart of a PEMFC is the membrane electrode assembly (MEA), which is a compound of a membrane, the catalyst layers, and gas diffusion layers (GDLs). A single fuel cell contains a membrane electrode assembly (MEA) and two flow-field plates distributing about 0.5 and 1 V voltage, which are too low in many applications, as shown in Fig. 1. Thus, it is common that the individual cells are stacked to increase the voltage while increasing the surface area of the cells, consequently, increases the current. The power output of a given fuel cell stack will depend on the type of application [12].

The principle of operations of a fuel cell is based on H_2 passing through the anode and O_2 through the cathode, Fig. 2. The conversion of chemical energy into electric energy occurs through two semi-reactions known as hydrogen oxidation reaction (HOR) and, oxygen reduction reaction (ORR), which occurs at the anode and cathode, respectively. At the anode, the fuel (H_2) is split by catalyst into hydrogen ions and electrons, and the electrons flow out of the cell to create current. At the cathode, oxygen molecules associate with electrons and hydrogen ions to produce water, this is the sluggish step of the reaction. In both electrodes, nanocatalysts are used to enhance the rates of oxygen reduction and hydrogen oxidation reactions. However, for ORR reaction more amount of catalyst is necessary, due to the high energy required to break the strong bond between O—O [13].

Pt supported in carbon is the most employed as nanocatalyst, due to the better selectivity, stability, activity, and resistance to poison than other metals. Pt catalyst is able to absorb species with strength enough that allows chemical bonds to be cleaved, but not so strong that the catalyst is blocked by the product, limiting the reaction. This is a required property for a good heterogeneous catalyst [14].

3. Nanocatalyst for fuel cell—Market in value

Nanocatalysts are becoming prevalent in the market owing to their ability properties on a nanometric scale that goes far beyond the larger surface area to accelerate the rate of reactions [15]. Some properties appear only when the catalyst is on a nanometric scale. Nanotechnology provides custom design specifications to catalysts to facilitate faultless selectivity, efficiency, durability in a catalyst-based reaction. Nanocatalysts also exhibit excellent power electro-catalytic such as nickel (Ni), platinum (Pt), and the Pt alloy are most prominently used as catalysts for the hydrogen fuel cell [16].

An increase in focus on environmental conservation and a rise in awareness about reducing carbon footprints are the key factor driving the fuel cell market. The global fuel

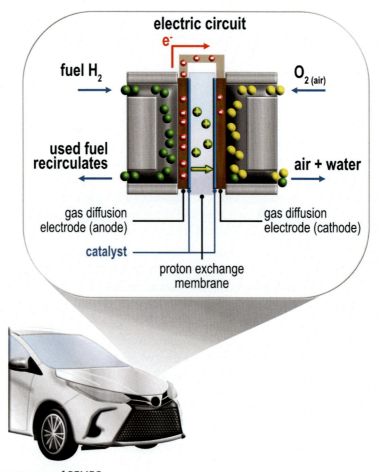

Fig. 2 Compartments of PEMFC.

cell market size was estimated at USD 2.6 billion in 2020. The automotive industry increasing demand for electric vehicles is driving the growth of the Hydrogen Fuel Cell Catalyst Market mainly in the market of PEMFC, which accounted for over 67.7% fuel cell market in 2020 [17]. Consequently, the growing global market of fuel cells pushes the nanocatalysts market [15].

According to the report, the global market for nanocatalysts was valued at USD 1.7 billion in 2020 and is projected to be $ 2.5 billion by 2027. Auto segment and Refinery & Petrochemical are responsible for record at 55% in the compound annual growth rate (CAGR) and will reach $ 957.3 million by the end of the analysis period [18]. Various nations across the globe including small and medium companies have been significantly impacted by the pandemic of COVID 19 resulting in various problems such as no cash flows, and fluctuations in demand in nanocatalysts manufacturers. However, the industry can have a fast development due to the mass vaccination, heating of the economy, pressure environment segment, and growth dynamics. The market is anticipated to grow with a healthy growth rate of more than 6.28% over the forecast period 2021–26 [19].

The nanocatalysts market is segmented into three parts such as nanomaterials, type, and application, where each part can also be divided into other classifications. For instance, depending on nanomaterials the nanocatalyst market is divided into carbon materials, metal nanoparticle catalysts, nanometer metal oxide catalysts, nanometer semiconductor photocatalyst particles, and others. Based on chemical properties, the market can be segmented into four main types like nanoparticle catalysts, nonporous catalysts, nanocrystalline catalysts, and supramolecular catalysts. Last, based on the applications of the nanocatalysts, the market can be divided into gas & fossil fuels, biomass, fuel cells, water & wastewater treatment, oil, and others, Fig. 3 [20].

As above mentioned, Pt-based nanocatalysts have been played an important role in the fuel cell application, increasing the demand for this noble metal. Pt holds the largest share in the nanocatalyst market while many emerging catalysts are in the nascent stage. This catalyst is used in various fuel cell applications such as methanol reformate, hydrogen, fuel cell, etc. Hydrogen fuel cells emerged as the leading player in the market in terms of their consumption in recent years [21], inducing an increase in consumption of this metal. For example, the global platinum market size was USD 502.8 million in 2020 and is projected to reach USD 563.6 million by 2026 with a CAGR of 1.9% during 2021–26 [22].

4. Types, characteristics, and synthesis of nanocatalysts

The big challenge in the large-scale implementation of fuel cells is catalysts. The catalyst is the substance that is used to increase the rate of reaction, without getting consumed and without changing its permanent physical or chemical states. Catalyst is necessarily recovered from the reaction mixture and has remained chemically unchanged. The use of catalyst species decreases the energy activation required for a reaction.

However, when the catalyst has at least one nanoscale dimension is called a nanocatalyst. This technology field has rapidly grown and has high added value. Nanomaterials are used as nanocatalysts for a variety of catalysis homogeneous and heterogeneous applications. The nanocatalysts are anchored in support to reduce their cost and to provide an increase in the active area of catalysis. Essential carbon supports for multiple 1D-3D substrates that are used in nanocatalysts with a single atom and nanoparticles are shown in Fig. 4A.

The great advances in the methods of synthesis and characterization in the nanoscale particles allow more and more precise control of the sizes and shapes of the particles. Controlling the size and type of exposed face according to the particle morphology obtained can often generate considerable changes in the catalytic activity of the material [23].

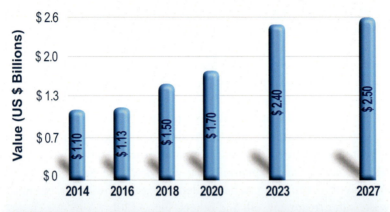

Fig. 3 Data and projections for the global nanocatalysts market from 2014 to 2027. Box: type, nanomaterials, and application for nanocatalysts.

The search for understanding about particle control always seeks to optimize properties and, in the case of precious metals, to decrease the quantities needed either by increasing catalytic efficiency or improving dispersion in the substrates used in the converters.

Obtaining an efficient electrocatalyst also depends on some conditions, such as a high presence of active centers and an area composed of pores that facilitate the transport of reagents and products, in addition to stability and electrical conductivity (Fig. 4B). Many studies highlight the electrocatalytic efficiency for fuel cells based on noble metals, such as Pt [24]. Thus, a synthesis of Pt nanostructures with controlled size and shape is one of the main factors of highly active Pt catalysts for many industrial applications. Pt is also the main catalyst in PEMFCs (Fig. 5); catalyzes the HOR or alcohol at the anode and the ORR at the cathode.

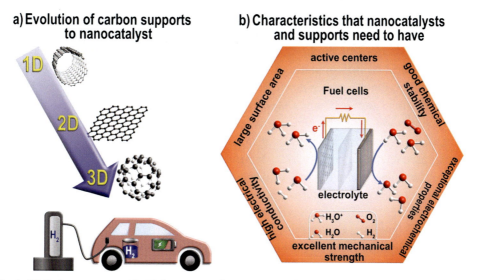

Fig. 4 (A) Carbon supports 1D-3D for nanocatalyst materials of one single atom and nanoparticles for fuel cells (B) characteristics synergistically obtained by nanocatalysts and support to catalyze PEMFC reactions.

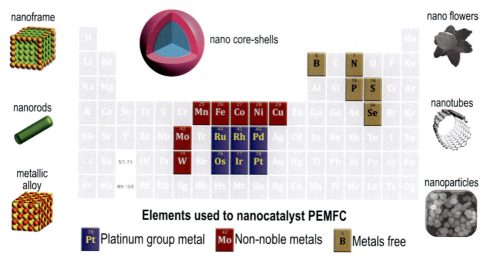

Fig. 5 Elements and structures used in nanocatalyst for PEMFC.

There is a large number of methods used for the production and/or processing of Pt nanoparticles. Usually, studies on the production of Pt particles with controlled sizes and shapes explore bottom-up methods such as solvothermal, sol-gel, electrochemical deposition, or physical deposition methods [25]. Each method has its particular reaction environment and processing strategy, which makes it possible to obtain particles with often different structural and electronic characteristics. It is important to comment that in addition to the method itself, chemical agents are often used to prevent growth or agglomeration, either by charge interactions or steric impediments [26]. Recent methods have been explored techniques that involve depositions of single-atoms or clusters of atoms to optimize the property and the ease of dispersion of catalyst agents on the substrate [27].

However, there are several issues against the use of pure Pt as fuel cell nanocatalysts. Firstly, Pt surfaces leading to a severe decrease in the catalytic performance caused by self-poisoned. Species as CO intermediates (originated from small organic fuel oxidation) have strong adsorption with Pt. Secondly, using Pt as cathodic catalysts in DMFC, methanol molecules crossover from anode to cathode may lower the ORR performance because of the mixed potentials formed from the simultaneous methanol oxidation and ORR. Thirdly, the limited reserve in nature of Pt makes the prices of nanocatalyst inviable, becomes one of the major barriers for the wide commercialization of fuel cells.

One of the strategies to minimize the cost and improve the stability and activity of the catalysts is the noble metal alloys with other lower-cost materials. Thus, the replacement of these metals by electrocatalysts obtained by nonnoble metals or even nonmetals has been investigated in different studies (Fig. 5). Also, nonprecious metal single-atom catalysts (NPM-SACs), particularly the transition metals such as Fe, Co, Ni, Cu, Zn, combined with the nonmetal dopants (N, P, S, B, O) show desirable catalytic activities for ORR (cathodic part of PMEFC) become a promising candidate to replace precious-metal catalysts. Many experimental techniques have been presented for NPM-SACs preparation with 1D, 2D, 3D. Good examples are ball milling, electrochemical deposition, physical and chemical vapor deposition, wet-chemistry pyrolysis, and thermal and microwave pyrolysis [28].

Heterogeneous nanomaterials such as metallic alloy, intermetallic ordered, core-shell, hollow interiors, cage-bell structures, stellated/dendritic morphologies, dimeric, or composite films also have been widely studied for catalysis (some examples are demonstrated in Fig. 5). The use of metal alloys was very important during the history of the development of catalyst technologies in automotive applications [29, 30]. The combination of two or more metals aims to supply catalytic deficiencies of a single metal to a certain type of reaction or to reduce costs from the use of cheaper metals so that the efficiency does not change considerably. The incorporation of Pd to replace a certain amount of Pt and Rh was a solution to help with the cost without changing the catalytic efficiency [31].

Heterogeneous nanomaterials could be synthesized by solid, solution, or gaseous state. The solid-state method does not obtain nanocrystalline compounds with high surface areas and requires high-temperature heating for long periods. Thus, the preferred method by many researchers to obtain heterogeneous nanomaterials is the solution-based method, because this method is more powerful and versatile. In the solution-based method, the nucleation and growth process of the nanomaterials can be easily controlled by adjusting the reaction parameters: temperature and time of the reaction, the concentration of reactants, the mole ratio between precursors, surfactants, and type of template.

A good example of the solution-based method is demonstrated with the synthesis of hollow nanostructures, this type of structure can be prepared using template-based methods as showed in Fig. 6. Broadly defining, these methods can be divided into (1) hard template methods, (2) soft template methods, (3) sacrificial template methods, and (4) template-free methods [32].

In the hard template methods, the active shell is grown on the surface of the template, then the template is selectively removed, for instance, by thermal or chemical methods, leaving the active shell unchanged and with a hollow interior. The word hard relates to the fact that in these methods the templates are solid materials, for instance, nanoparticles with well-defined size and morphology. In the soft template methods, generally, the templates are soft-matter materials, like micelles, emulsion droplets, or polymers. In the sacrificial template methods, the template is inherently consumed for being a participant in a chemical reaction leading to the formation of the hollow active shell material. In general, sacrificial templates are less resource-consuming, since they do not require an additional step to remove the template. In comparison, template-free methods produce hollow structures based on differences in the growth mechanism of the active shell material. For instance, the inside-out Ostwald ripening is a common mechanism in which the material in the bulk of the particle grows at a lower rate in comparison to the material on the surface of the particle. Consequently, the material in the bulk tends to dissolve and deposit onto the material on the surface. This diffusion of materials from the bulk to the surface of particles ultimately may lead to the formation of a hollow interior [32].

Therefore, the search for the nanocatalyst for fuel cells purpose remains a very explored subject. In this sense, the Pt-M (M = Pd, Cu, Au) remains a very attractive system to be studied by a wide variety of deposition strategies since, in addition to composition, size, shape, and structural organization also generates variations in properties [33–35]. It is worth mentioning that, in addition to bimetallic systems, alloys with more than two components are also alternatives that have been explored for the automotive sector, including some strategies for the use of nonprecious metals [36]. The use of nonprecious metal oxides as catalysts are also strategies to overcome the problem of easy oxidation of these metals [37, 38].

Fig. 7 summarizes the types of nanocatalysts platinum group metals (PGM) and PGM-free, to catalyze ORR in PEMFC. Also, are demonstrated the nanocatalysts morphology, supports, synthesis, combinations, order, most used structures, and characteristics.

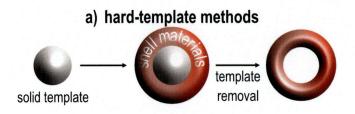
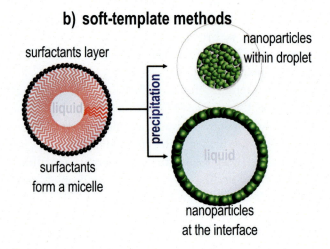
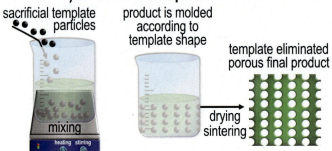

Fig. 6 Synthetic methods for nanostructured materials (A) hard-template, (B) soft-template, (C) sacrificial template, and (D) template-free method.

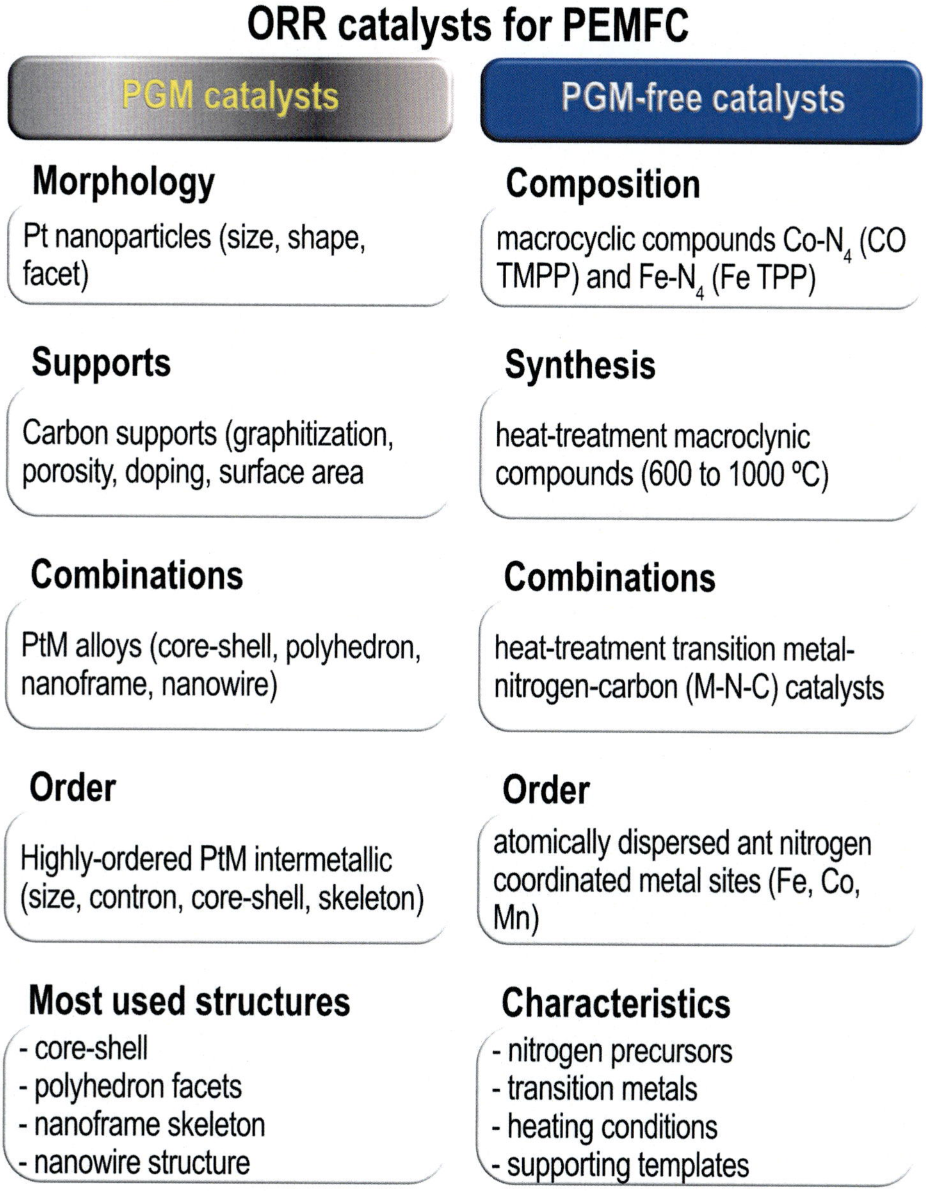

ORR catalysts for PEMFC

PGM catalysts

Morphology
Pt nanoparticles (size, shape, facet)

Supports
Carbon supports (graphitization, porosity, doping, surface area)

Combinations
PtM alloys (core-shell, polyhedron, nanoframe, nanowire)

Order
Highly-ordered PtM intermetallic (size, contron, core-shell, skeleton)

Most used structures
- core-shell
- polyhedron facets
- nanoframe skeleton
- nanowire structure

PGM-free catalysts

Composition
macrocyclic compounds Co-N_4 (CO TMPP) and Fe-N_4 (Fe TPP)

Synthesis
heat-treatment macroclynic compounds (600 to 1000 °C)

Combinations
heat-treatment transition metal-nitrogen-carbon (M-N-C) catalysts

Order
atomically dispersed ant nitrogen coordinated metal sites (Fe, Co, Mn)

Characteristics
- nitrogen precursors
- transition metals
- heating conditions
- supporting templates

Fig. 7 Types and characteristics of nanocatalysts PGM and PGM-free group to catalyze ORR in PEMFC.

5. Recent development of nanocatalysts for fuel cells

Despite research efforts in the use of nanostructures of alloys and composites, most of the catalysts used in PMEFC are still single-component Pt particles supported on high surface area carbon. Pt/C with many particle sizes and shapes have been extensively studied for the ORR catalytic activity. The extensive studies in history have not prevented the continued use of this simple system, where new insights and nanostructures with substantial catalytic improvement have been developed.

The recent advances in nanostructured Pt-based catalysts heavily rely on more efficient use of Pt. In other words, the Pt-based catalyst should present a higher activity and use a relatively lower amount of Pt [39]. In this sense, this section presents the state-of-the-art in the development of the Pt-based and Pt-free catalysts. These advances are based on innovative ways to tune the morphology and composition of these catalysts, leading to an improvement of their performance in comparison to the commercially available catalysts. Although none of the catalysts described here have been implemented in the market yet, the potential shown by them makes them promising alternatives for future applications in fuel cells for different types of reactions.

5.1 Nanostructured Pt-based catalysts

Here, we will discuss various efforts to develop alloys based on Pt and composite nanostructures for electrocatalytic applications. Especially how to adapt the surface structures to increase the catalytic activity and stability of the ORR. Although our emphasis is placed on high surface area nanocatalysts, we also try to link structure-property relationships of nanocatalysts located in extended surface model catalysts. The morphology is an important factor regarding catalyst activity since it can control the exposed facets of the Pt nanostructures, which consequently, impacts the number of active sites available for catalysis. Another mechanism from which morphology can control catalytic activity relates to the fact that it can improve Pt-based catalysts stability by decreasing the aggregation rate of these nanostructures [40].

The catalyst support also plays an important role in the performance of the hollow nanostructure Pt-based fuel cells. Recently, Yue and co-workers prepared a methanol oxidation fuel cell having as supporting double-shelled C/TiO_2 hollow spheres. Then, the Pt layer was deposited onto the surface of the support by homogeneously dispersing the support in water and adding $H_2PtCl_6 \cdot 6H_2O$, followed by the addition of formic acid [41]. The reaction mixture was heated at 80°C for 20 min, leading to Pt reduction and deposition onto the double-shelled C/TiO_2 hollow spheres support. This fuel cell presented a higher catalytic performance for methanol oxidation reaction, having a current density about 2.5 times higher than the commercial Pt supported on carbon.

Metal alloying is a viable way to reduce the quantity of Pt present in the catalyst. Due to the scarcity and high prices of Pt, it is desirable that the alloying could be made with earth-abundant metals like Ni, Cu, Co, and Fe [42].

A recent example from binary alloyed hollow catalyst was published by Chen and co-workers. In this paper, they prepared hollow octahedral PtCu nanostructures by using the Kirkendall effect, which is a type of sacrificial template method [43]. The ratio between Pt and Cu in the hollow octahedral PtCu nanostructures could be regulated by varying the time of the solvothermal treatment using dimethylformamide as a solvent, carried out at 115°C, and the catalysts were supported on carbon. The hollow octahedral PtCu supported on carbon showed to work as bifunctional catalysts. This is because they were able to catalyze the ORR and methanol reduction reaction, with higher durability and specific activity compared to the commercial Pt supported on carbon.

The effect of material processing history and the impact on its final morphology can be noticed when the same alloy presents the ability to catalyze other reactions simply by varying its production method, support, and morphology. For instance, when the PtCu alloyed catalysts had a hollow dodecahedron frame morphology catalyze the ORR, like the PtCu hollow octahedral nanostructures prepared by Chen et al. [43]. However, unlike the hollow octahedral nanostructures, the hollow dodecahedron nanostructures could not catalyze the methanol reduction reaction. Instead, they could catalyze the hydrogen evolution reaction (HER) [44]. The authors attributed the successful preparation of the hollow dodecahedron frames to using diglycolamine as a solvent in the solvothermal process. Since this molecule has an amino and a hydroxyl group, it can easily chelate to metallic cations controlling its particle growth rate. To further assist in the morphology control, cetyltrimethylammonium chloride was used during the solvothermal process as well [44]. The PtCu hollow dodecahedron nanostructures presented better specific activity than the commercial material Pt black, both for HER and ORR [44].

Although binary metal alloys are more common, it is possible to make ternary alloys catalysts. For example, Wu and co-workers made an alloyed hollow catalyst containing PtNiCu, with accessible surface area [45]. Initially, Pt^{+2}, Ni^{+2}, and Cu^{+2} were dissolved in oleylamine, which was used as a solvent and reducing agent. The surfactant dimethyldioctadecylammonium chloride was used as a capping agent. This mixture was heated at 200°C for 4 h under an N_2 atmosphere. The resulting material was dispersed in carbon black as support and underwent an acid and annealing treatment. The annealing treatment was necessary to increase the Pt content on the surface of the catalyst. Interestingly, this ternary alloyed hollow catalyst was demonstrated to be multifunctional. Because it was active for the HER, ORR, and methanol oxidation reaction, and presented better catalytic activity for all these reactions that the commercial Pt supported on carbon catalysts.

For being made up of two different materials, the core-shell structures, which are usually represented by the notation core material@shell material, have the possibility to tune how the relative position of the energy levels of the two materials will be located. Additionally, other factors to consider are the precursors and synthetic methods that will be used for the shell growth onto the core, and the final shell thickness. All these factors can influence the electronic, optical, and catalytic properties of core-shell nanostructures [46]. Consequently, by combining the same two constituent materials, it is possible to produce a wide array of nanomaterials with different properties, simply by varying slightly the properties of the core or shell material, such as particle size and crystalline phase.

A recent example from the literature is the paper published by Zou and co-workers [47]. In this paper, they prepared two types of Ru@Pt core-shell nanostructures. In the first nanostructure, the Ru core was crystalline, whereas in the second nanostructure the Ru core was amorphous. For each type of Ru@Pt core-shell, the authors also varied the thickness of the Pt shell. Comparing Ru@Pt core-shell types with the same Pt shell thickness, the Ru@Pt with crystalline and amorphous core were very different. For instance, the Ru@Pt core-shell with crystalline Ru core presented higher activity toward formic acid and ethanol oxidation reactions. This research reviewed important structural aspects of the core-shell nanomaterials used for small molecules oxidation for proton exchange membrane fuel cells.

Although Pt is usually deemed as the shell component in core-shell nanostructure, there are cases where Pt can be located in the core and still be the catalytic active material. For instance, Sun and co-workers encapsulated Pt with a layer of hexagonal boron nitride (h-BN) to produce the core-shell Pt@h-BN [48]. Since h-BN is a two-dimensional material with strong in-plane bonds, so it is robust enough to protect the Pt from losing its catalytic activity. However, as the bond between different h-BN layers is relatively weak, the h-BN end being flexible enough to allow the reactions to happen underneath the h-BN layers. These Pt@h-BN core-shell nanostructures were used as catalysts for oxygen reduction reactions and hydrogen oxidation reactions. For both reactions, it was shown that the h-BN shell does not have a negative impact on the Pt activity. Additionally, for the ORR, the Pt@h-BN/C presented mass activity higher than the one presented by the Pt/C catalyst.

Interestingly, there are core-shell nanostructures where the Pt can be found both in the core and the shell. Yuan and co-workers prepared core-shell nanostructures made of a bimetallic PtBi core and an ultrathin shell also of Pt (PtBi@Pt) [49]. The PtBi@Pt core-shell nanostructures were used as catalysts in the methanol oxidation and ethanol oxidation reactions. For both reactions, the PtBi@Pt core-shell nanostructures presented higher mass activity compared to the commercial Pt/C.

More complex configurations are possible for core-shell structures than simply being made by a core and a shell material. For example, Hu and co-workers prepared core-shell nanostructures made of Co_3O_4-Co in the core and 3-dimensional graphene (3-DG) as the shell (Co_3O_4-Co@3-DG) [50]. Then, reacting Co_3O_4-Co@3-DG with H_2PtCl_6 under stirring led to the formation of a Co_3O_4-Pt core shell@3-DG. The Co_3O_4-Pt core shell@3-DG were used as catalysts in the ORR and presented better specific activity than the commercial Pt/C catalysts.

5.2 Pt-free nanocatalyst

As mentioned before, Pt-based nanocatalyst is the most material used in automotive fuel cells. However, their widespread applications are limited because of high-cost technology and the low abundance of Pt metal. Hence, exploring low-cost nonnoble metal with efficient catalytic activity for the fuel cells is one of the current challenges [51].

Intense efforts have been made by researchers, in the whole world to overcome the challenge to replace Pt-based nanocatalyst. The main focus is on the kinetic sluggishness of ORR, which is due to its high overpotential required. Thus, promoting higher consumption of platinum. A variety of materials, including transition metals and their oxides, palladium, carbons, and others have been studied as electrocatalysts for ORR or HER.

Recently, our research group has used casein to produce nanoporous heteroatom-doped carbon conjugated with Cu^{2+} or Ni^{2+} to improve the ORR. The use of transition metal intends to create more electrocatalytically active sites in the nanoporous carbon, to reduce the overpotential and increase the current density of the reaction [52].

Carbon conjugated with N and transition metals (M = Co and/or Fe) has also been employed as a promising nanocatalyst to reduce fuel cell cost. For example, metal-organic polymer supramolecule was used as a strategy for dispersing efficiently and stabilize Fe ions through chelating metal-oxygen bonds with sodium alginate (SA). For use in the cathode, the new nanocatalyst demonstrated high performance, even acidic and alkaline environment [52]. Proietti and co-workers utilized iron (II) acetate (FeAc), 1,10-phenanthroline (Phen), and ZIF-8 (Z8) to prepare Fe/N/C-catalysts using a low-energy ball milling following by pyrolysis approach to improving the ORR performance [53]. In another work, Ahluwalia and co-workers also used a metal-organic framework (MOF) to produce Fe-N-C nanocatalyst with better catalyst activity. Although good results have been achieved, great advances to power density are necessary [54].

Wu et al. used an approach in which polyaniline is used as a precursor generating a carbon-nitrogen template. Also, the researchers incorporating iron and cobalt in the carbon precursor to employ the final material as a nanocatalyst for ORR. The material presented a high activity with remarkable performance stability compared to other nonprecious metal catalysts [55]. Carbon doped with Co and N have been produced as nonnoble metal catalysts. The nanocatalyst demonstrated remarkable activity for ORR in both acidic and alkaline media, showing an excellent performance to replace Pt-based catalyst [56]. Some other works have been demonstrated the efficiency of these types of nanocatalysts [57–60].

As an alternative for Pt-free nanocatalyst, Yang and co-workers synthesized MoO_2 nanowires doped with N and Co. The nanomaterial was employed as a nanocatalyst with high electrochemical performance for ORR and HER [61]. Kumar and co-workers prepared nanoparticles of Co_3Fe alloy on reduced graphene oxide (RGO) sheets [62]. FeNi alloy supported on nitrogen-doped graphene was prepared by Sirirak and co-workers [63]. All these researches have proved that the Pt-free nanomaterials are efficient nanocatalysts for ORR in PEMFC cathode.

Another metal widely employed as a nanocatalyst for ORR is palladium (Pd). Pd presents similar physicochemical characteristics, besides being cheaper than Pt. It is broadly considered to be a new material to replace Pt nanocatalyst. However, there still is a gap in ORR catalytic ability between them. To overcome this gap, new synthesis methods have been published, including alloy formation with other metallic elements [64, 65]. For instance, Ni@Pd core-shell nanostructure supported on multiwalled carbon nanotubes developed by Hosseini and co-workers demonstrated a good catalytic activity toward methanol oxidation in alkaline solutions [66]. Pd nanoparticles supported on tungsten (Pd/W@WCWO$_3$) have emerged as a new class of catalyst. Tungsten as support showed high stability for low-temperature fuel cells [67]. Another nanocomposite formed by NiOx@Pd nanocatalysts was reported by Bhalothia and co-workers. The strategy used by the authors was to deposit Ir-oxide clusters on the surface of Ni@Pd NCs to further improve the ORR performance, as activity and stability [68].

It is noted that all researches cited above have as main goals to produce a new catalyst Pt-free with high activity and stability to replace commercial Pt-base catalyst. All of them demonstrated a huge potential as nanocatalysts to be used in automotive fuel cell applications. However, despite progress, these new catalysts still suffer from the low activity and insufficient stability, which limit them reach the fuel cell performance of Pt/C-based catalysts.

An overview in the field of nanocatalyst for fuel cells is presented in Table 1. The table is including the characteristics and efficiency of nanomaterial used as nanocatalyst for fuel cells.

6. Advantages and challenges of nanocatalysts for fuel cells

The growing interest in FCVs, encouraged by the governments, leading to a boost in the demand for nanocatalysts in the automotive sectors, mainly one based on platinum. Pt-based nanocatalyst has a key role in a fuel cell. They have some advantages that make them huge useful; (i) activity catalytic at lower temperatures; (ii) good resistance to corrosion in concentrated electrolytes and up to moderate temperatures and; (iii) the ability to chemisorb hydrocarbons dissociative at low temperatures [14].

To comprehend the importance of nanocatalyst for fuel cell applications, abstract research on Google Scholar was made with the words "nanocatalyst for fuel cell" from 2015 to 2020. Fig. 8A demonstrates the overall trend of researches about all types of nanocatalysts. It is possible to notice that there is a perceptible increase in the total number of articles in the last 6 years. Additionally, a search of the number of articles was realized with the term "Platinum nanocatalyst for fuel cell" in the same years, and then, the percentage of papers that involve Pt-based nanocatalyst was calculated, as shown in Fig. 8B. The calculus was made concerning the total number of articles published in this area per year, from 2015 to 2020. Analyzing Fig. 8B, it is possible to see a tendency to decrease Platinum nanocatalysis researches, which leads to the belief that there is an expressive growth of interest in new nanocatalyst Pt-free.

Table 1 Details and performance of different nanomaterials used as nanocatalysts for PEMFC.

Nanocatalyst	Mode	Shape	Size/thicknesses (nm)	Specifc activity (mA cm^{-2}) $I_{0.9v}$	Number of cycles	Efficiency last cycle (%)	Refs.
Au@Pt/C[a]	Alloy	Spheric	6	2.95	–	–	[69]
Pd/rGO	–	Spheric	16	3	1K	82	[70]
Pt	–	Thin film	–	5.7	10K	95	[71]
Pt/C	–	Spheric	8	0.2	2K	61	[72]
Pt/HSC[b]	–	Spheric	4.3	1.3	30K	81	[73]
Pt/ITO	Coated	Spheric	10/20	0.7	0.2	82	[74]
Pt/TiO$_2$	Coated	Spheric	2/30	1.1	5K	88	[75]
Pt$_1$Au$_1$	Alloy	Spheric	120	0.44	5K	80	[76]
Pt$_5$Gd	Alloy	Thin film	2	10.4	10K	86	[71]
PtAu/C[a]	Alloy	Spheric	10	4.92	–	–	[69]
PtCo/C	Alloy	Spheric	4.8	2.4	10K	80	[77]
PtCo/HSC[b]	Alloy	Spheric	4.5	0.7	30K	106	[73]
PtCu	Alloy	Spheric	2	3.98	10K	91	[78]
PtCu	Alloy	Spheric	40	1.24	30K	85	[79]
PtCuAg	Alloy	Spheric	50	3.85	200K	120	[80]
Pt-free	–	spheric	5	1	–	–	[81]
Pt-free	–	Spheric	10	1.2	–	–	[82]
PtNi/C	Alloy	Octahedral	6	1.0	2K	76	[72]
PtNi@Pt	Alloy	Dendritic	5	3.66	10K	45	[83]
PtRu/C	Alloy	Spheric	5–200	1.4	10K	64	[84]
PtRu/W	Alloy	Spheric	5–200	1.3	10K	83	[84]
PtSmCo	Alloy	Spheric	45	5	10K	85	[85]
Pt$_3$Y	Alloy	Thin film	27	13.4	10K	80	[86]

[a]DMFC, direct methanol fuel cell.
[b]High surface area carbon (HSC) supported Pt and PtCo nanoparticle catalysts.

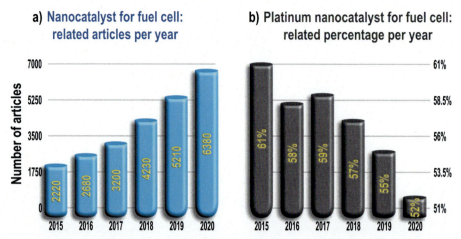

Fig. 8 (A) Number of articles about nanocatalyst for the fuel cell. (B) Percentage of the researches including Pt-based nanocatalyst for the fuel cell. Both data are based on abstract research at Google Scholar from 2015 to 2020, using the keywords *nanocatalyst for a fuel cell* and *Pt-based nanocatalyst for fuel cell*.

Even with the growing interest in new materials, platinum remains the first option for the industries being widely used in fuel cell electrodes. Companies like Umicore, IRD fuel cells, Tanaka, and N.E. ChemCat (NECC), are some of the companies responsible for electrocatalyst commercialization. Pt on carbon black (Pt/C), Pt-Co alloy on carbon black, and Pt/Ru on carbon black are the most common catalysts commercialized for fuel cell application. Current fuel cell technology uses a lot of nanocatalyst per unit of vehicles. It is reported that around 30–60 g ($\sim$0.4 mg/cm^2) of platinum nanocatalyst are required for a fuel cell vehicle to operate properly [8]. Among all components of the fuel cell, nanocatalyst corresponds to about 46% of the total cost of production, making the fuel cell highly cost for the automotive market.

Compared to the first fuel cell nanocatalyst, enormous advances technological have been reached in the last years. At the turn of the century, some cars industries, such as General Motors (Advanced HydroGen3), USA; Daimler (Mercedes Benz B-Class F-CELL), Germany; and Toyota (Mirai), Japan, thrown research programs to bring this technology to maturity of a mass-market product [87]. The last innovation was made by Toyota, with its innovative Mirai fuel cell vehicle (2014). This vehicle is considered the world's first mass-produced dedicated fuel cell electric vehicle. In the nanocatalysis field, the main innovation by Toyota was to increase the ORR activity via optimization of the composition of the PtCo catalyst. Compared to the previous generation of FCVs, the traditional hollow carbon support of Pt was replaced by a solid core, which enhanced the platinum availability to occur the reactions [88]. Thus, fewer nanocatalyst materials are used to decrease significantly the cost of production. Even with the price reduction and improvements, the cost of technology is still much higher than a conventional car. Therefore, it is still necessary to further reduce the amount of Pt in the catalyst or eliminate its use to increase the viability in the market.

7. General conclusions and future perspectives

Impressive progress has been reported during the last few years on the development of nanocatalysts for fuel cell technology. Many advances have been reported as (i) the incorporation of nonprecious metal and/or a second noble metal in the Pt structures; (ii) the development of new structures, like hollow nanostructures and core-shell; (iii) development of new morphology, like nanocages, nanowires, nanofibers and so forth; and (iv) synthesis of new nanocatalysts Pt-free. However, several challenges also remain. The new catalysts do not reach the efficiency of commercial nanocatalysts, they still have low activity and insufficient stability to be used commercially. Advancement in the research field is stuck by the high cost and durability of the nanocatalysts. The price of automobiles is much higher than that of a conventional car. In addition, the scarcity of Pt will soon be a big problem if fuel cell vehicles become popular. Finally, the fast-growing nanocatalyst field remains open many opportunities for discoveries ahead. Much work is made to generate commercially viable nanocatalysts also is expected that researches continue for the coming years. The efforts of researchers aim mainly at solving the problems of cost, durability, and find a new Pt-free nanocatalyst, these relationship needs to be solved before fuel cells can be made available for mass consumption.

References

[1] P. Geng, E. Cao, Q. Tan, L. Wei, Effects of alternative fuels on the combustion characteristics and emission products from diesel engines: a review, Renew. Sustain. Energy Rev. (2017), https://doi.org/10.1016/j.rser.2016.12.080.

[2] D. Shindell, C.J. Smith, Climate and air-quality benefits of a realistic phase-out of fossil fuels, Nature 573 (7774) (2019) 408–411, https://doi.org/10.1038/s41586-019-1554-z.

[3] H.S. Gandhi, G.W. Graham, R.W. McCabe, Automotive exhaust catalysis, J. Catal. 216 (2003) 433–442, https://doi.org/10.1016/S0021-9517(02)00067-2.

[4] D. Sperling, D. Gordon, Two Billion Cars: Driving Toward Sustainability, Oxford University Press, New York, 2010.

[5] T. Hirakawa, Y. Shimokawa, W. Tokuzumi, T. Sato, M. Tsushida, H. Yoshida, J. Ohyama, M. Machida, Multicomponent 3d transition-metal nanoparticles as catalysts free of Pd, Pt, or Rh for automotive three-way catalytic converters, ACS Appl. Nano Mater. 3 (9) (2020) 9097–9107, https://doi.org/10.1021/acsanm.0c01769.

[6] E. Vasile, A. Ciocanea, V. Ionescu, I. Lepadatu, C. Diac, S.N. Stamatin, Making precious metals cheap: a sonoelectrochemical—hydrodynamic cavitation method to recycle platinum group metals from spent automotive catalysts, Ultrason. Sonochem. 72 (April) (2021) 105404, https://doi.org/10.1016/j.ultsonch.2020.105404.

[7] R. Hannappel, The impact of global warming on the automotive industry, AIP Conf. Proc. 1871 (August) (2017) 1–7, https://doi.org/10.1063/1.4996530.

[8] B.G. Pollet, S.S. Kocha, I. Staffell, Current status of automotive fuel cells for sustainable transport, Curr. Opin. Electrochem. 16 (2019) 90–95, https://doi.org/10.1016/j.coelec.2019.04.021.
[9] K.I. Ozoemena, S. Chen, Nanomaterials for Fuel Cell Catalysis, Springer, Ontario, Canada, 2016.
[10] J.M. Andújar, F. Segura, Fuel cells: history and updating. A walk along two centuries, Renew. Sustain. Energy Rev. 13 (9) (2009) 2309–2322, https://doi.org/10.1016/j.rser.2009.03.015.
[11] G.P. Panayiotou, S.A. Kalogirou, S.A. Tassou, PEM fuel cells for energy production in solar hydrogen systems, Recent Pat. Mech. Eng. (2010), https://doi.org/10.2174/2212797611003030226.
[12] EERE, Office of Energy Efficiency & Renewable Energy, Report the Hydrogen and Fuel Cell Technologies Office (HFTO) Focuses on Research, Development, and Demonstration of Hydrogen and Fuel Cell Technologies Across Multiple Sectors Enabling Innovation, A Strong Domestic Economy, and a Clean, Equitable Energy Future, 2021, Available online https://www1.eere.energy.gov/hydrogenandfuelcells/mypp/pdfs/fuel_cells.pdf. (Accessed 30 April 2021).
[13] M. Ehsani, Y. Gao, S. Longo, K.M. Ebrahimi, Modern Electric, Hybrid Electric, and Fuel Cell Vehicles, CRC Press, Boca Raton, FL, 2018.
[14] O.T. Holton, J.W. Stevenson, The role of platinum in proton exchange membrane fuel cells, Platin. Met. Rev. 57 (4) (2013) 259–271.
[15] Market Research, Report: Nanocatalysts Market—Global Industry Analysis, Size, Share, Growth, Trends and Forecast 2018–2026, 2021, Available online https://www.transparencymarketresearch.com/nanocatalysts-market.html. (Accessed 30 April 2021).
[16] Grand View Research, Catalyst Market Size, Share & Trends Analysis Report By Raw Material (Chemical Compounds, Zeolites, Metals), By Product (Heterogeneous, Homogeneous), By Application, By Region, and Segment Forecasts, 2020–2027, 2021, Available online https://www.grandviewresearch.com/industry-analysis/catalyst-market. (Accessed 30 April 2021).
[17] Fortune, Global fuel cell catalyst market 2020 future estimations with top key players, production development and opportunities to 2025, 2021. Reports. Available online: https://www.fortunebusinessinsights.com/industry-reports/fuel-cell-catalyst-market-101335. (Accessed 30 April 2021).
[18] ReportLinker, ReportLinker Report: Global Nanocatalysts Industry, 2021, Available online: https://www.globenewswire.com/news-release/2020/07/23/2066572/0/en/Global-Nanocatalysts-Industry.html. (Accessed 30 April 2021).
[19] 360 Market Updates, Report Nanocatalysts in 2019: Global Market Analysis, Trends and Forecasts (2016–2024) With Profiles on 25 Players, ResearchAndMarkets.com, 2021. Available online: https://www.oilandgas360.com/nanocatalysts-in-2019-global-market-analysis-trends-and-forecasts-2016-2024-with-profiles-on-25-players-researchandmarkets-com/. (Accessed 30 April 2021).
[20] Alliedmarket, Nanocatalysts Market by Nanomaterials (Carbon Materials, Metal & Oxides, and Others), Type (Nanoparticle Catalysts, Nanoporous Catalysts, Nanocrystalline Catalysts, and Supermolecular Catalysts), and Application (Biomass, Oil, Gas & Fossil Fuels, Fuel Cells, Water & Wastewater Treatment, and Others): Global Opportunity Analysis and Industry Forecast, 2019–2026, 2021, Available online https://www.alliedmarketresearch.com/nanocatalysts-market-A06016. (Accessed 30 April 2021).
[21] The Source, New Catalyst Resolves Hydrogen Fuel Cell Cost, Longevity Issues, 2021, Available online: https://source.wustl.edu/2020/12/new-catalyst-resolves-hydrogen-fuel-cell-cost-longevity-issues/. (Accessed 30 April 2021).
[22] Cision, Platinum market size USD 563.6 million by 2026, 2021. Valuates Reports. Available online https://www.prnewswire.com/in/news-releases/platinum-market-size-usd-563-6-million-by-2026-valuates-reports-856458603.html. (Accessed 30 April 2021).
[23] P. Panagiotopoulou, X.E. Verykios, Metal–support interactions of Ru-based catalysts under conditions of CO and CO_2 hydrogenation, Catalysis 32 (2020) 1–23. The Royal Society of Chemistry https://doi.org/10.1039/9781788019477-00001.
[24] R. Vedarajan, J. Prithi, N. Rajalakshmi, M. Naushad, R. Saravanan, B.T. Kumar, 6—Advanced nanocatalysts for fuel-cell technologies, Materials Today, in: Nanomaterials for Sustainable Energy and Environmental Remediation Raju, Elsevier, 2020, pp. 165–191, https://doi.org/10.1016/B978-0-12-819355-6.00006-6.
[25] A. Chen, P. Holt-Hindle, Platinum-based nanostructured materials: synthesis, properties, and applications, Chem. Rev. 110 (6) (2010) 3767–3804, https://doi.org/10.1021/cr9003902.

[26] L.M. Rossi, J.L. Fiorio, M.A.S. Garcia, C.P. Ferraz, The role and fate of capping ligands in colloidally prepared metal nanoparticle catalysts, Dalton Trans. 47 (17) (2018) 5889–5915, https://doi.org/10.1039/C7DT04728B.

[27] A. Beniya, S. Higashi, Towards dense single-atom catalysts for future automotive applications, Nat. Catal. 2 (7) (2019) 590–602, https://doi.org/10.1038/s41929-019-0282-y.

[28] J. Liu, H. Zhang, M. Qiu, Z. Peng, M.K.H. Leung, W.-F. Lin, J. Xuan, A review of non-precious metal single atom confined nanomaterials in different structural dimensions (1D—3D) as highly active oxygen redox reaction electrocatalysts, J. Mater. Chem. A 8 (5) (2020) 2222–2245, https://doi.org/10.1039/C9TA11852G.

[29] J.S. Hepburn, K.S. Patel, M.G. Meneghel, H.S. Gandhi, Development of Pd-Only Three Way Catalyst Technology, SAE International, 1994, https://doi.org/10.4271/941058 UI - 941058.

[30] J. Wang, H. Chen, Z. Hu, M. Yao, Y. Li, A review on the Pd-based three-way catalyst, Catal. Rev. 57 (1) (2015) 79–144, https://doi.org/10.1080/01614940.2014.977059.

[31] R.J. Farrauto, M. Deeba, S. Alerasool, Gasoline automobile catalysis and its historical journey to cleaner air, Nat. Catal. 2 (7) (2019) 603–613, https://doi.org/10.1038/s41929-019-0312-9.

[32] X.W. Lou, L.A. Archer, Z. Yang, Hollow micro-/nanostructures: synthesis and applications, Adv. Mater. 20 (21) (2008) 3987–4019, https://doi.org/10.1002/adma.200800854.

[33] E.D. Goodman, S. Dai, A.-C. Yang, C.J. Wrasman, A. Gallo, S.R. Bare, A.S. Hoffman, et al., Uniform Pt/Pd bimetallic nanocrystals demonstrate platinum effect on palladium methane combustion activity and stability, ACS Catal. 7 (7) (2017) 4372–4380, https://doi.org/10.1021/acscatal.7b00393.

[34] G. Sharma, A. Kumar, S. Sharma, M. Naushad, R.P. Dwivedi, Z.A. ALOthman, G.T. Mola, Novel development of nanoparticles to bimetallic nanoparticles and their composites: a review, J. King Saud Univ. Sci. 31 (2) (2019) 257–269, https://doi.org/10.1016/j.jksus.2017.06.012.

[35] V. Ulrich, B. Moroz, P. Pyrjaev, I. Sinev, A. Bukhtiyarov, E. Gerasimov, V. Bukhtiyarov, B.R. Cuenya, W. Grünert, Three-way catalysis with bimetallic supported Pd—Au catalysts: gold as a poison and as a promotor, Appl. Catal. B Environ. 282 (2021) 119614, https://doi.org/10.1016/j.apcatb.2020.119614.

[36] H. Asakura, M. Kirihara, K. Fujita, S. Hosokawa, S. Kikkawa, K. Teramura, T. Tanaka, Fe-modified CuNi alloy catalyst as a nonprecious metal catalyst for three-way catalysis, Ind. Eng. Chem. Res. 59 (45) (2020) 19907–19917, https://doi.org/10.1021/acs.iecr.0c03389.

[37] L. Ma, C.Y. Seo, X. Chen, K. Sun, J.W. Schwank, Indium-doped Co_3O_4 nanorods for catalytic oxidation of CO and C_3H_6 towards diesel exhaust, Appl. Catal. B Environ. 222 (2018) 44–58, https://doi.org/10.1016/j.apcatb.2017.10.001.

[38] K. Ueda, C.A. Ang, Y. Ito, J. Ohyama, A. Satsuma, $NiFe_2O_4$ as an active component of a platinum group metal-free automotive three-way catalyst, Catal. Sci. Technol. 6 (15) (2016) 5797–5800, https://doi.org/10.1039/C6CY00795C.

[39] X. Ren, Q. Lv, L. Liu, B. Liu, Y. Wang, A. Liu, W. Gang, Current progress of Pt and Pt-based electrocatalysts used for fuel cells, Sustain. Energy Fuels 4 (2020) 15–30, https://doi.org/10.1039/C9SE00460B.

[40] A. Hoque, F.M. Hassan, A.M. Jauhar, G. Jiang, M. Pritzker, J.-Y. Choi, S. Knights, S. Ye, Z. Chen, Web-like 3D architecture of Pt nanowires and sulfur-doped carbon nanotube with superior electrocatalytic performance, ACS Sustain. Chem. Eng. 6 (2018) 93–98, https://doi.org/10.1021/acssuschemeng.7b03580.

[41] X. Yue, P. Yuguang, W. Zhang, T. Zhang, W. Gao, Ultrafine Pt nanoparticles supported on double-shelled C/TiO_2 hollow spheres material as highly efficient methanol oxidation catalysts, J. Energy Chem. 49 (2020) 275–282, https://doi.org/10.1016/j.jechem.2020.02.045.

[42] N.V. Long, Y. Yang, C.M. Thi, N. Van Minh, Y. Cao, M. Nogami, The development of mixture, alloy, and core-shell nanocatalysts with nanomaterial supports for energy conversion in low-temperature fuel cells, Nano Energy 2 (5) (2013) 636–676, https://doi.org/10.1016/j.nanoen.2013.06.001.

[43] G. Chen, X. Yang, Z. Xie, F. Zhao, Z. Zhou, Q. Yuan, Hollow PtCu octahedral nanoalloys : efficient bifunctional electrocatalysts towards oxygen reduction reaction and methanol oxidation reaction by regulating near-surface composition, J. Colloid Interface Sci. 562 (2020) 244–251, https://doi.org/10.1016/j.jcis.2019.12.020.

[44] H.-J. Niu, H.-Y. Chen, G.-L. Wen, J.-J. Feng, Q.-L. Zhang, A.-J. Wang, One-pot solvothermal synthesis of three-dimensional hollow PtCu alloyed dodecahedron nanoframes with excellent electrocatalytic performances for hydrogen evolution and oxygen reduction, J. Colloid Interface Sci. 539 (2019) 525–532, https://doi.org/10.1016/j.jcis.2018.12.066.

[45] D. Wu, W. Zhang, A. Lin, D. Cheng, Low Pt-content ternary PtNiCu nanoparticles with hollow interiors and accessible surfaces as enhanced multifunctional electrocatalysts, ACS Appl. Mater. Interfaces 12 (2020) 9600–9608, https://doi.org/10.1021/acsami.9b20076.

[46] P. Reiss, M. Protière, L. Liang, Core/shell semiconductor nanocrystals, Small 5 (2) (2009) 154–168, https://doi.org/10.1002/smll.200800841.

[47] J. Zou, M. Wu, S. Ning, H. Lin, X. Kang, S. Chen, Ru@Pt core–shell nanoparticles: impact of the atomic ordering of the Ru metal core on the electrocatalytic activity of the Pt shell, Research-article, ACS Sustain. Chem. Eng. 7 (2019) 9007–9016, https://doi.org/10.1021/acssuschemeng.9b01270.

[48] M. Sun, J. Dong, L. Yang, S. Zhao, C. Meng, Y. Song, G. Wang, Pt @ H-BN core–shell fuel cell Electrocatalysts with electrocatalysis confined under outer shells, Nano Res. 11 (6) (2018) 3490–3498, https://doi.org/10.1007/s12274-018-2029-5.

[49] X. Yuan, X. Jiang, M. Cao, L. Chen, K. Nie, Y. Zhang, Y. Xu, X. Sun, Y. Li, Intermetallic PtBi core/ultrathin Pt shell nanoplates for efficient and stable methanol and ethanol electro-oxidization, Nano Res. 12 (2) (2019) 429–436.

[50] S. Hu, Y. Liu, S. Wang, X. Zhang, Ultrathin Co 3 O 4 e Pt core-shell nanoparticles coupled with three-dimensional graphene for oxygen reduction reaction, Int. J. Hydrog. Energy 46 (17) (2021) 10303–10311, https://doi.org/10.1016/j.ijhydene.2020.12.137.

[51] X.X. Wang, M.T. Swihart, G. Wu, Achievements, challenges and perspectives on cathode catalysts in proton exchange membrane fuel cells for transportation, Nat. Catal. 2 (7) (2019) 578–589, https://doi.org/10.1038/s41929-019-0304-9.

[52] E.H. Fragal, V.H. Fragal, E.B. Tambourgi, A.F. Rubira, R. Silva, T. Asefa, Nanoporous carbons derived from metal-conjugated phosphoprotein/silica: efficient electrocatalysts for oxygen reduction and hydrazine oxidation reactions, J. Electroanal. Chem. 882 (2021) 114997, https://doi.org/10.1016/j.jelechem.2021.114997.

[53] E. Proietti, F. Jaouen, M. Lefèvre, N. Larouche, J. Tian, J. Herranz, J.-P. Dodelet, Iron-based cathode catalyst with enhanced power density in polymer electrolyte membrane fuel cells, Nat. Commun. 2 (1) (2011) 416, https://doi.org/10.1038/ncomms1427.

[54] R.K. Ahluwalia, X. Wang, L. Osmieri, J.-K. Peng, H.T. Chung, K.C. Neyerlin, Performance of polymer electrolyte fuel cell electrodes with atomically dispersed (AD) Fe-C-N ORR catalyst, J. Electrochem. Soc. 166 (14) (2019) F1096–F1104, https://doi.org/10.1149/2.0851914jes.

[55] G. Wu, K.L. More, C.M. Johnston, P. Zelenay, High-performance electrocatalysts for oxygen reduction derived from polyaniline, iron, and cobalt, Science 332 (6028) (2011) 443–447, https://doi.org/10.1126/science.1200832.

[56] M. Li, X. Bo, Y. Zhang, C. Han, A. Nsabimana, L. Guo, Cobalt and nitrogen co-embedded onion-like mesoporous carbon vesicles as efficient catalysts for oxygen reduction reaction, J. Mater. Chem. A 2 (30) (2014) 11672–11682, https://doi.org/10.1039/C4TA01078G.

[57] J.-Y. Choi, L. Yang, T. Kishimoto, F. Xiaogang, S. Ye, Z. Chen, D. Banham, Is the rapid initial performance loss of Fe/N/C non precious metal catalysts due to micropore flooding? Energy Environ. Sci. 10 (1) (2017) 296–305, https://doi.org/10.1039/C6EE03005J.

[58] C. Guo, Y. Wu, Z. Li, W. Liao, L. Sun, C. Wang, B. Wen, Y. Li, C. Chen, The oxygen reduction electrocatalytic activity of cobalt and nitrogen Co-doped carbon nanocatalyst synthesized by a flat template, Nanoscale Res. Lett. 12 (1) (2017) 144, https://doi.org/10.1186/s11671-016-1804-z.

[59] X.X. Wang, D.A. Cullen, Y.-T. Pan, S. Hwang, M. Wang, Z. Feng, J. Wang, et al., Nitrogen-coordinated single cobalt atom catalysts for oxygen reduction in proton exchange membrane fuel cells, Adv. Mater. 30 (11) (2018) 1706758, https://doi.org/10.1002/adma.201706758.

[60] P. Yin, T. Yao, Y. Wu, L. Zheng, Y. Lin, W. Liu, J. Huanxin, et al., Single cobalt atoms with precise N-coordination as superior oxygen reduction reaction catalysts, Angew. Chem. Int. Ed. 55 (36) (2016) 10800–10805, https://doi.org/10.1002/anie.201604802.

[61] L. Yang, J. Yu, Z. Wei, G. Li, L. Cao, W. Zhou, S. Chen, Co-N-doped MoO2 nanowires as efficient electrocatalysts for the oxygen reduction reaction and hydrogen evolution reaction, Nano Energy 41 (2017) 772–779, https://doi.org/10.1016/j.nanoen.2017.03.032.

[62] S. Kumar, D. Kumar, B. Kishore, S. Ranganatha, N. Munichandraiah, N.S. Venkataramanan, Electrochemical investigations of Co3Fe-RGO as a bifunctional catalyst for oxygen reduction and evolution reactions in alkaline media, Appl. Surf. Sci. 418 (2017) 79–86, https://doi.org/10.1016/j.apsusc.2016.11.130.

[63] R. Sirirak, B. Jarulertwathana, V. Laokawee, W. Susingrat, T. Sarakonsri, FeNi alloy supported on nitrogen-doped graphene catalysts by polyol process for oxygen reduction reaction (ORR) in proton exchange membrane fuel cell (PEMFC) cathode, Res. Chem. Intermed. 43 (5) (2017) 2905–2919, https://doi.org/10.1007/s11164-016-2802-6.

[64] M.A. Khalily, B. Patil, E. Yilmaz, T. Uyar, Atomic layer deposition of Pd nanoparticles on N-doped electrospun carbon nanofibers: optimization of ORR activity of Pd-based nanocatalysts by tuning their nanoparticle size and loading, ChemNanoMat 5 (12) (2019) 1540–1546, https://doi.org/10.1002/cnma.201900483.

[65] F.D. Sanij, P. Balakrishnan, P. Leung, A. Shah, S. Huaneng, X. Qian, Advanced Pd-based nanomaterials for electro-catalytic oxygen reduction in fuel cells: a review, Int. J. Hydrog. Energy 46 (27) (2021) 14596–14627, https://doi.org/10.1016/j.ijhydene.2021.01.185.

[66] M.G. Hosseini, R. Mahmoodi, V. Daneshvari-Esfahlan, Ni@Pd Core-Shell nanostructure supported on multi-walled carbon nanotubes as efficient anode nanocatalysts for direct methanol fuel cells with membrane electrode assembly prepared by catalyst coated membrane method, Energy 161 (2018) 1074–1084, https://doi.org/10.1016/j.energy.2018.07.148.

[67] N.R. Elezovic, P. Zabinski, P. Ercius, M. Wytrwal, V.R. Radmilovic, U.Č. Lačnjevac, N.V. Krstajic, High surface area Pd nanocatalyst on core-shell tungsten based support as a beneficial catalyst for low temperature fuel cells application, Electrochim. Acta 247 (2017) 674–684, https://doi.org/10.1016/j.electacta.2017.07.066.

[68] D. Bhalothia, D.-L. Tsai, S.-P. Wang, C. Yan, T.-S. Chan, K.-W. Wang, T.-Y. Chen, P.-C. Chen, Ir-oxide mediated surface restructure and corresponding impacts on durability of bimetallic NiOx@Pd nanocatalysts in oxygen reduction reaction, J. Alloys Compd. 844 (2020) 156160, https://doi.org/10.1016/j.jallcom.2020.156160.

[69] W. Zhou, M. Li, L. Zhang, S.H. Chan, Supported PtAu catalysts with different nano-structures for ethanol electrooxidation, Electrochim. Acta 123 (2014) 233–239, https://doi.org/10.1016/j.electacta.2013.12.153.

[70] L. Sun, F. Wen, S. Li, Z. Zhang, High efficient RGO-modified Ni foam supported Pd nanoparticles(PRNF) composite synthesized using spontaneous reduction for hydrogen peroxide electroreduction and electrooxidation, J. Power Sources 481 (2021) 228878, https://doi.org/10.1016/j.jpowsour.2020.228878.

[71] M. Escudero-Escribano, A. Verdaguer-Casadevall, P. Malacrida, U. Grønbjerg, B.P. Knudsen, A.K. Jepsen, J. Rossmeisl, I.E.L. Stephens, I. Chorkendorff, Pt5Gd as a highly active and stable catalyst for oxygen electroreduction, J. Am. Chem. Soc. 134 (40) (2012) 16476–16479, https://doi.org/10.1021/ja306348d.

[72] J. Wang, B. Li, D. Yang, H. Lv, C. Zhang, Preparation optimization and single cell application of PtNi/C octahedral catalyst with enhanced ORR performance, Electrochim. Acta 288 (2018) 126–133, https://doi.org/10.1016/j.electacta.2018.09.005.

[73] N. Ramaswamy, S. Kumaraguru, G. Wenbin, R.S. Kukreja, K. Yu, D. Groom, P. Ferreira, High-current density durability of Pt/C and PtCo/C catalysts at similar particle sizes in PEMFCs, J. Electrochem. Soc. 168 (2) (2021) 24519, https://doi.org/10.1149/1945-7111/abe5ea.

[74] V.T.T. Ho, H.Q. Pham, T.H.T. Anh, T. Van Nguyen, K.A.N. Quoc, H.T.H. Vo, T.T. Nguyen, Highly stable Pt/ITO catalyst as a promising electrocatalyst for direct methanol fuel cells, CR Chim. 22 (11) (2019) 838–843, https://doi.org/10.1016/j.crci.2019.08.003.

[75] G.R. Mirshekari, C.A. Rice, Effects of support particle size and Pt content on catalytic activity and durability of Pt/TiO2 catalyst for oxygen reduction reaction in proton exchange membrane fuel cells environment, J. Power Sources 396 (2018) 606–614, https://doi.org/10.1016/j.jpowsour.2018.06.061.

[76] X. Deng, S. Yin, X. Wu, M. Sun, Z. Xie, Q. Huang, Synthesis of PtAu/TiO2 nanowires with carbon skin as highly active and highly stable electrocatalyst for oxygen reduction reaction, Electrochim. Acta 283 (2018) 987–996, https://doi.org/10.1016/j.electacta.2018.06.139.

[77] P. Yu, M. Pemberton, P. Plasse, PtCo/C cathode catalyst for improved durability in PEMFCs, J. Power Sources 144 (1) (2005) 11–20, https://doi.org/10.1016/j.jpowsour.2004.11.067.

[78] Y. Liu, L. Chen, T. Cheng, H. Guo, B. Sun, Y. Wang, Preparation and application in assembling high-performance fuel cell catalysts of colloidal PtCu alloy nanoclusters, J. Power Sources 395 (2018) 66–76, https://doi.org/10.1016/j.jpowsour.2018.05.055.

[79] M. Gong, D. Xiao, Z. Deng, R. Zhang, W. Xia, T. Zhao, X. Liu, et al., Structure evolution of PtCu nanoframes from disordered to ordered for the oxygen reduction reaction, Appl. Catal. B Environ. 282 (2021) 119617, https://doi.org/10.1016/j.apcatb.2020.119617.

[80] Y. Zhou, D. Zhang, Nano PtCu binary and PtCuAg ternary alloy catalysts for oxygen reduction reaction in proton exchange membrane fuel cells, J. Power Sources 278 (2015) 396–403, https://doi.org/10.1016/j.jpowsour.2014.12.088.

[81] T. Sakamoto, K. Asazawa, K. Yamada, H. Tanaka, Study of Pt-free anode catalysts for anion exchange membrane fuel cells, Catal. Today 164 (1) (2011) 181–185, https://doi.org/10.1016/j.cattod.2010.11.012.

[82] R. Nandan, G.K. Goswami, K.K. Nanda, Direct synthesis of Pt-free catalyst on gas diffusion layer of fuel cell and usage of high boiling point fuels for efficient utilization of waste heat, Appl. Energy 205 (2017) 1050–1058, https://doi.org/10.1016/j.apenergy.2017.08.118.

[83] J. Choi, J.-H. Jang, C.-W. Roh, S. Yang, J. Kim, J. Lim, S.J. Yoo, H. Lee, Gram-scale synthesis of highly active and durable octahedral PtNi nanoparticle catalysts for proton exchange membrane fuel cell, Appl. Catal. B Environ. 225 (2018) 530–537, https://doi.org/10.1016/j.apcatb.2017.12.016.

[84] S.M. Brkovic, M.P. Marceta Kaninski, P.Z. Lausevic, A.B. Saponjic, A.M. Radulovic, A.A. Rakic, I.A. Pasti, V.M. Nikolic, Non-stoichiometric tungsten-carbide-oxide-supported Pt—Ru anode catalysts for PEM fuel cells—from basic electrochemistry to fuel cell performance, Int. J. Hydrog. Energy 45 (27) (2020) 13929–13938, https://doi.org/10.1016/j.ijhydene.2020.03.086.

[85] Q. Gong, L. Wang, N. Yang, T. Lin, G. Xie, B. Li, Ternary PtSmCo NPs electrocatalysts with enhanced oxygen reduction reaction, J. Rare Earths 38 (12) (2020) 1305–1311, https://doi.org/10.1016/j.jre.2019.11.016.

[86] N. Lindahl, E. Zamburlini, L. Feng, H. Grönbeck, M. Escudero-Escribano, I.E.L. Stephens, I. Chorkendorff, C. Langhammer, B. Wickman, High specific and mass activity for the oxygen reduction reaction for thin film catalysts of sputtered Pt3Y, Adv. Mater. Interfaces 4 (13) (2017) 1700311, https://doi.org/10.1002/admi.201700311.

[87] N. Qin, A. Raissi, P. Brooker, Analysis of Fuel Cell Vehicle Developments, University of Central Florida, 2014, pp. 2–10.

[88] T. Yoshida, K. Kojima, Toyota MIRAI fuel cell vehicle and progress toward a future hydrogen society, Interface Mag. 24 (2) (2015) 45–49, https://doi.org/10.1149/2.f03152if.

SECTION F

Nanomaterials for automotive application

CHAPTER 29

Magnetic nanomaterials for electromagnetic interference shielding application

Seyyed Mojtaba Mousavi[a], Sonia Bahrani[b], and Gity Behbudi[c]
[a]Department of Chemical Engineering, National Taiwan University of Science and Technology, Taiwan
[b]Department of Medical Nanotechnology, School of Advanced Medical Sciences and Technology, Shiraz University of Medical Science, Shiraz, Iran
[c]Department of Chemical Engineering, University of Mohaghegh ardabili, Ardabil, Iran

Chapter outline

1. Introduction 607
2. Microwave absorption of magnetic carbon-based nanocomposites 609
3. Microwave absorption performance 611
4. Conclusions and outlook 615
References 616

1. Introduction

Recently, electromagnetic (EM) wave light in gigahertz (GHz) extend has been created as an upsetting peril to defense safety techniques, high-quality information, biological systems, and commercial appliances. When these EM waves interference with the signal of electronic instruments, they produce loud sound known as EMI contamination. In general, EMI contamination could be considered as an unfavorable result of modern engineering which is worrying for human health and causes numerous illnesses such as, trepidation, sleeping disorders, and headaches. In commercial instruments (e.g., microwave circuits designing and microwaves ovens), correspondence frameworks (e.g., laptop, Bluetooth, computer, and cell phone), and automotive industries (e.g., integrated electrical circuits), EMI contamination degenerates the suitable function and permanence of electronic equipment. This type of contamination has become a fundamental worldwide concern that must be alleviated by using EMI shielding materials [1–6]. Metals that own great electrical conduction are the most suitable materials for application in EMI shielding. However, they have disadvantages such as: not being economically viable, corrosion capacity, and high reflectivity. Polymer-based composite materials because of owning mechanical and/or magnetic, dielectric, thermal, and electrical properties are a good option in response to effective electromagnetic shielding. Also, conductive fillers such as graphene, carbon nanotubes (CNTs), and intrinsically conducting polymers (ICPs), can be used effectively as EMI shielding materials [7]. Different forms of carbon such as graphite, carbon foams, carbon nanocoils, carbon nanofibers (CNF), and CNT are used as absorbers of electromagnetic waves [8]. Controlling the interference of the EM radiations from strategic systems and electronic devices needs the expansion and manufacture of low-cost, lightweight, effective, and modern EMI shielding materials [9, 10]. The EMI shielding materials have various applications in different areas such as communication, military, electronics, electrical, aerospace–aeronautic, and home instruments [11–14]. Among the different applications, EMI shielding at microwave frequencies is very beneficial [15–19]. The most usual kind of EMI is the radio frequency limited area (104–1012 Hz) and produces from more sources including overhead power lines, electric motors, radio transmitters, and computer circuits. Generation of these waves is necessary to be prevented, since they interfere with digital devices, due to the sensitivity of electronics and increasing abundance, especially radio frequency devices. EMI shielding with polymer compounds has many advantages including cost-effectiveness, corrosion resistance, greater design freedom, and lighter weight [6, 20–22]. Attractive structures of nanocomposites, were the favorite of researchers, due to the simplicity of preparing, improved properties, including increased biodegradability, gas permeability, decreased flammability, high moduli, and increased heat resistance and strength [23–26]. Their applications include packaging applications, biomedical applications, biotechnology, electronics, automotive, and aerospace industries, CNT-containing nanomaterials, microelectromechanical systems, microelectronic, data storage technology, high-performance catalysts, sensors, engineering parts, high-density magnetic recording media, optical and phosphors amplifiers, solid-state laser media, optoelectronic devices, structural materials, highperformance ferroelectric devices, chemical sensors light energy conversion, structural materials for high-temperature applications, tooling, high-speed machinery, photo-electrochemical applications, thermally graded and wear-resistant coatings, direct methanol fuel cells, automotive step assists, and optical fibers [27–33] which benefits of the most noteworthy components for competition [34]. Microwave-Absorbing (MA) advancements are of critical significance for tending to the issue of electromagnetic obstruction and radiation to achieve interesting military and mechanical objectives. MA materials will not just give solid impedance coordinating as worked for shell or cover covering to lessen the surface impression of occurrence electromagnetic signs, yet also, designed dielectric or/and attractive imperfections to contain the communicated electromagnetic wave vitality to the extent possible [35–39]. Accordingly, extensive endeavors have been made in the recent decades fundamentally on two focuses: the plan of electric/attractive components [40–44] and the advancement of a few interfaces or permeable structures [45–48]. The substance will have certain particulars for devouring microwaves such as extremely fine,

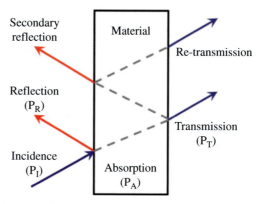

Fig. 1 Schematic of an incident electromagnetic (EM) wave through solid material [50].

lightweight, low thickness, and basic union absorbance over a wide recurrence range. Electromagnetic wave safeguards can have both conductive and attractive properties. Metallic is extremely substantial among conventional materials and may experience the ill effects of erosion, though polymers that have a low thickness, Although polymers are thin, they may separate and become less stable [49]. At a point when EM wave emerges on a substrate surface, it experiences three procedures: reflection, assimilation, and engendering, obeying optics rules, (Fig. 1). EM wave exists outside of a non-attractive dielectric object, the electrical power creates two separate electrical flows in the material, such as the conductive and uprooting flows, separately [50]. The plentiful hetero-interface produced by adaptable RGO pieces wrapping Ti3C2TX circles, and imperfections, such as fringes, heaping deficiencies, and usable surface classes, improve the effect of polarization bringing and effectiveness of EM ingestion. On account of its particular staggered development, enormous 3D cross-connecting, nuanced disappointment organize following, MWCNT/graphene froth shows adaptable, powerful permittivity and conductivity while acquiring unprecedented loss of polarization and conductivity [51]. The qualities of the lightweight, high vitality thickness and high assimilation limit of the delivered new nanomaterials are foreseen to fulfill the requests for the upsizing and softness of the devices [52–54].

2. Microwave absorption of magnetic carbon-based nanocomposites

CNTs through MAMs make impressive incentives due to mechanical opposition, viewpoint proportion, stable conductivity, low corruption, and lightweight weight. Writing shows that MWCNT can be bonded to SiO_2 compounds, and experiments have been performed with developing temperatures on their dielectric properties [55]. MWCNTs' efficiency and microwave-absorbing output have also been altered via sedimentation nanomaterials such as CDs on the surface of MWCNT [56]. At 473 K, CdS nanoparticles on MWCNTs (6 vol%) show 47 dB of microwave assimilation. Graphene-based materials, for example, rGO, were likewise chosen for solid microwave ingestion because of their lightweight element, rich oxygen practical gatherings, more prominent surface territory, and unique 2D structure [57–62]. Ferromagnetic items, for example, attractive metals and composites, have been appeared to own solid microwave absorption characteristics at room temperature; in any case, the more metals that are interesting, seem to have intelligent disappointment properties at high temperatures because of corruption of attractive properties. A few fundamental upgrades, including subbed ferrites [63] and 3D nickel connect systems [64] are needed. The retention properties at higher temperatures have been improved. Striking investigations were La-doped BiFeO3 (BFO) reasonable for La/Nd-doped BFO for 300–673 K, and X-band temperature ranges 373–673 K, which indicated intelligent misfortunes of 54 and 20 dB, respectively. In expansion, 3D polyvinylpyrrolidone (PVP) nickel chains showed solid microwave-engrossing properties for X groups at a temperature scope of 323–573 K, where 50 wt% heaps of 3D nickel chains displayed 64 dB at 573 K and arrived at the midpoint of <10 dB for the entire X band.

There are several metal oxides that are exceptionally employed as a high amount of temperature MAMs, which include Fe3O4, ZnO, and MnO2. The vast majority of such items are joined as carbon-based nanocomposites or earthenware production, such as a-MnO2 nanorods, Fe3O4/MWCNTs, ZnO-MWCNTs/SiO2, and others at higher temperatures have solid microwave ingestion. ZnO-MWCNTs/SiO2 has a reflectance missing of 20.7 dB at 323 K and arrived at the midpoint of <10 dB at 373–673 K, recommending roughly 90% of the radiation was devoured in the mishap. The permeable composite comprised of ZnO materials, ZrSiO4 layers, besides that Al dopes ZnO, brought about RL at 573 K and a size of 70 dB was in this way a remarkable decision for this capacity. An unconventional twofold ingestion design was displayed via Fe3O4-MWCNTs with a grape-like structure created for the X band zone and temperature scope of 323–473 K in whither the retention highlight can be balanced by differing temperatures somewhere in the range of 10 and 15 dB and 16 and 25 dB. Besides, different composites, for example, Fe3O4-ZnO2 yolk-shell structure, have empowering temperature attributes of up to 773 K [65]. Raman spectroscopy was utilized to group CNT microstructures through D- and G-band wave numbers and D- to G-band (ID: IG) quality ratios [66]. The D-band reflects sp3 hybridization, and the G-band reflects sp2 hybridization. The ID: IG esteem was 1.09, and the pinnacles of G- and D-band were based on 1581 and 1345 cm^{-1}, separately. What is, more is that for dissecting the CNT surface holding situations X-beam photoelectron spectroscopy (XPS) was utilized [67]. The CNPFE and CNPF electrical properties appear in Fig. 2.

Carbon-based materials and magnetic materials can improve dielectric loss or overall magnetic loss, respectively. Nevertheless, both substances cannot even get successful fitting impendence and strong microwave absorption. Notably, several articles have merged products focused on magnetic and carbon to achieve robust impedance matching by physical or chemical processes [69, 70]. Other content dependent on magnetic carbon

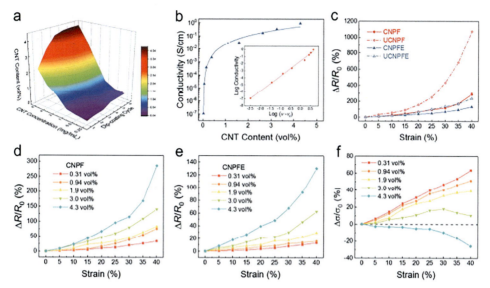

Fig. 2 (A) All through MAMs, CNTs make an impressive incentives for their lightweight weight, low 3D CNT material chart in CNPF as a component of CNT focus and Dip-covering period size. (B) Electric conductivity vs CNPF material. Inset: log–log conductivity plot as an element of (Őc). (C) Analysis of the rate increment in obstruction during CNPF, CNPFE, UCNPFE, and UCNPFE extending; The UCNPFE and UCNPF each have the equivalent CNT mass division as the separate CNPFE and CNPF. (D, E) A decrease in rate opposition as a strain includes for CNPFs (D) and CNPFEs (E) for explicit CNT substance. (F) Increase in rate conductance for CNT and CNPFEs quantities [68].

has been successfully synthesized, including materials such as 3D-Fe3O4/CNTs, Co3O4/MWCNTs/graphene, and Co/CNTs, as seen in Table 1.

3. Microwave absorption performance

As per EMI shielding fabrics protecting textures, perfect MA items show lower reflection and higher microwave ingestion contributions [71, 107–110]. The effectiveness of microwave ingestion was determined by RL utilizing the 2.118 GHz recurrence run curve technique (the estimation points of interest are given in S2). The test was watched on an aluminum plate, the impression of microwaves from the MA content was like the assimilation of microwaves in a genuine position [111]. Another estimation model of the MA proficiency was the prepared data transmission, for example, the recurrence range of RL values underneath dB [112–115]. Right now, endeavors and advances to acquire deformable MA matters are centered essentially around bendable and compressible materials, including Zhang bunch condensable sponges or nanocapsules [109], Bendable Polyimide/MXene Aerogels by Zhang group [114], Chou bunch bendable graphene/CNT/Fe3O4 films [111], and condensable graphene froths by our gathering. Nevertheless, all materials cannot accomplish a tolerable coordinating impendence and quick retention of

Table 1 Magnetic carbon-based nanocomposites for microwave absorption of EM waves.

Carbonaceous structure	Method	Thickness (mm)	RL (dB)	Frequency (GHz)	References
(CNT/Fe3O4/RGO)	[Ultrasonic method, solvothermal]	2.5	−50	8.7	[71]
(RGO/Fe3O4)	[Hummer, solvothermal]	2	−27	5.4	[72]
(Bowl like Fe3O4/RGO)	[Solvothermal]	2	−24	12.9	[73]
(Fe3O4/RGO composites)	(one-pot co-precipitation)	3.9	−44.6	6.6	[74]
(NiFe2O4 nanorod/graphene)	[One-step hydrothermal]	2	−29.2	16.1	[75]
(Popcorn like α-Fe2O3/3D G)	[Template assisted, co-precipitation]	1.4	−55.7	173	[76]
(α-Fe2O3 nanorod/graphene)	[chemical reduction]	2	−45	12.8	[77]
(Flaky CI/RGO)	[Modified Hummer, reflux]	3.87	−64.6	5.2	[78]
(α-Fe2O3 hydrogel/RGO)	[Two-step hydrothermal]	3	−33.5	0.712	[79]
(RGO/spherical CI)	[Wet chemical method]	3	−52.46	7.79,11.98	[80]
(RGO/NiFe2O4)	[One-step hydrothermal]	3	−39.7	9.2	[81]
[RGO/CoFe2O4ZnS]	(Co-precipitation, hydrothermal)	1.8	−43.2	−10.2,15.7	[82]
[RGO/CoFe2O4SnO2]	(Two steps hydrothermal)	1.6	−54.4	16.5	[83]
[Fe3O4//graphene]	(Two steps)	1.5	−29	8.12	[84]
[α-Fe2O3/γ-Fe2O3/RGO]	(Thermochemical reactions)	4	−13.6	3.76	[85]
[ZnFe2O4@graphene@TiO2)	[Hydrothermal]	2.5	−55.6	3.8	[86]
(TiO2/RGO/Fe2O3)	[Hydrothermal]	2	−44	14.8	[87]
(RGO/SiO2/Fe3O4)	[Two steps reaction]	4.5	−56.4	8.1	[88]
(RGO/ZnFe2O4)	[Solvothermal]	2.5	−41.1	9.4	[89]
(Co.5Ni.5Fe3O4/RGO)	[Chemical reduction]	2.5	−13.1	14.8	[90]
(Ni.8Zn.2Ce.06Fe1.94O4/GNS)	[Sol-gel deoxidation technique]	–	−37.4	12.3	[91]
(RGO/BaFe12O19/Fe3O4)	[Two step hydrothermal]	1.8	−46.04	15.6	[92]
(RGO@Fe3O4)	[Two-step process]	2	−56.25	12.62	[93]
(Pours Fe3O4/C)	[Hydrothermal process]	3	−31.75	12.88,−7.76	[94]
(Fe—C nanocapsules)	[Arc-discharge method]	3.1	−43.1	9.6	[95]

(CI/C)	[Hydrothermal]	1.3	−46.69	11.5	[96]
(Fe3O4 graphite)	[Molten salt route, temperature reduction]	4.8	−51	4.3	[97]
(EG/Fe3O4)	[Sintering route, solvothermal]	2.6	−24.8	6.8	[98]
(EG/Fe/Fe3O4)	[Chemical vapor deposition]	1.9	−42.4	9.36	[99]
(Epoxy graphitized nanosheet@Fe3O4-MnO2)	[Hydrothermal]	4.5	−31.7	5.875	[100]
(Flower like Fe3O4/MWCNTs)	[Acid treatment, hydrothermal]	0.9	−64	18	[101]
(MWCNTs/Fe2O4)	[In-situ growth method]	1.5	−30	5.7	[102]
(MWCNT/NiO-Fe2O4)	[Electroless plating]	2	−55	18	[103]
(Fe2O3/Fe3O4/MWCNTs)	[Hydrothermal]	2.5	44.1	10.4	[104]
(Fe-MWCNT)	[Chemical method]	4.27	−39	2.7	[105]
(Fe3O4/CF)	[Electrophoretic deposition]	1.7	−11	10.37–11.4	[106]

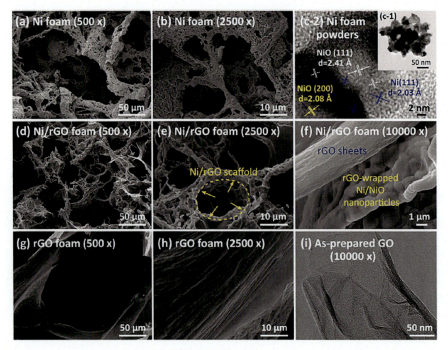

Fig. 3 Ni foam microstructure (A:C), (D–F) Ni-rGO foam (D:F), rGO foam (G:H) and as-prepare GO sheets (I) [117].

microwaves. Recently a few papers have blended attractive and carbon-based items to accomplish solid impedance coordinating through physical or substance forms. Highly attractive carbon subordinate substance includes jewelry, for example, 3D-Fe3O4/CNTs, Co3O4/MWCNTs/graphene, and nickel/graphene, which has likewise been effectively blended by Co/CNTs. Fig. 3A, B shows that the Ni froth key spine takes after a ceaseless wipe or reef arrange [116, 117].

The broader the spectrum concerning the eligible wavelength, the increased the potential in imitation of face up to the EM pulse. Fig. 4A, B shows the frequency dependency on EM bud absorption yield of Ni foam, yet rGO foam with specific frequency length thicknesses about 2–18 GHz. It is possible to show that the Ni foam amount can be reached on −35.28 dB at 8.32 GHz with maximum RL virtue, yet covers 5.46 GHz at 3 mm thick of EAB concerning. It also has to also keep remembered up to the expectation that no RL height is considered because the thicknesses are below 2 mm. This reaches a pinnacle RL ranking because of the EAB of 2.3 GHz at 5.5 mm thick and the rGO foam of −21.22 dB at 6.4 GHz. For comparison, the nominal RL values of rGO foam lime and Ni foam powder are −6 dB and −9.77 dB, respectively, beneath the ultima worth of about −10 dB. Hence that finding indeed illustrates the extraordinary effect over porous structure concerning electromagnetic suspense attenuation. As shown in Fig. 4C, D, the rGO/Ni foam has the most RL of −53.11 dB at 6.08 GHz together with an EAB of 4.91 GHz at

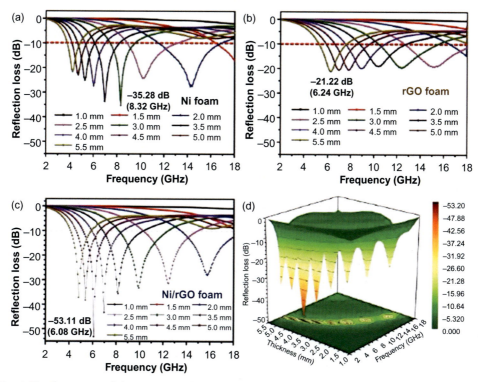

Fig. 4 The frequency of the EM wave absorption output: (A) Ni, (B) rGO, (C) rGO/Ni, and (D) rGO/Ni failure 3D reflection plots [117].

4.5 mm thickness, demonstrating excessive EM worry absorption capacity, altitude of 4.5 mm, and high EM suspense attention potent among the three foams [117].

To summarize, carbon-centered nanomaterials can cause severe dielectric loss, due to their excellent dielectric properties [118]. Also carbon-based nanomaterials will boost the theoretical portion of effect, because of owning a high degree of graphitization and good electrical conductivity. This progresses their microwave EM property and helps the electrical conductivity of carbon-dependent nanomaterials. Furthermore, the design of carbon-based nanomaterials with heterogeneous frameworks, elastic, large specific surface area, will surrogate intense interfacial polarization [119, 120]. In carbon-based nanomaterials, fairly good wave absorption characteristics of EM can find out eventually. It is difficult to achieve successful impedance matching because of the low magnetic failure of carbon-based nanomaterials. Hence it is always challenging to obtain excellent EM wave absorption efficiency for corresponding absorbers.

4. Conclusions and outlook

The EM shielding theory of microwave absorption was discussed in this chapter. First, the current reports in conductive loss and/or lack of polarization of known nanomaterial's (including magnetic carbon-based nanomaterials, magnetic-based nanomaterials, and other unique nanomaterials such as CNT, GO) dielectric and magnetic loss was summarized. Through electromagnetic-based nanoparticles, improved magnetic loss will affect either carbon materials or magnetic materials and will result in various encounters between magnetic-based carbon nanomaterials and magnetic-based nanomaterials, while at the same time providing extreme magnetic loss and dielectric loss, which would facilitate suitable matching impedances. Additionally, the composite's abundant heterointerfaces will promote absorption polarization and device activation and improve EM wave scattering. Many unique nanomaterials such as CNT and GO, in addition to magnetic nanomaterials, carbon nanomaterials and their composites, may also have significant polarization effects and healing by structural design.

References

[1] X. Jian, B. Wu, Y. Wei, S.X. Dou, X. Wang, W. He, N. Mahmood, Facile synthesis of Fe3O4/GCs composites and their enhanced microwave absorption properties, ACS Appl. Mater. Interfaces 8 (9) (2016) 6101–6109.

[2] S.R. Dhakate, K.M. Subhedar, B.P. Singh, Polymer nanocomposite foam filled with carbon nanomaterials as an efficient electromagnetic interference shielding material, RSC Adv. 5 (54) (2015) 43036–43057.

[3] R. Che, C. Zhi, C. Liang, X. Zhou, Fabrication and microwave absorption of carbon nanotubes/CoFe 2 O 4 spinel nanocomposite, Appl. Phys. Lett. 88 (3) (2006), 033105.

[4] P.R. Agarwal, R. Kumar, S. Kumari, S.R. Dhakate, Three-dimensional and highly ordered porous carbon–MnO 2 composite foam for excellent electromagnetic interference shielding efficiency, RSC Adv. 6 (103) (2016) 100713–100722.

[5] A. Abdollahifar, S.A. Hashemi, S.M. Mousavi, M. Rahsepar, Electromagnetic interference shielding effectiveness of reinforced composite with graphene oxide-lead oxide hybrid nanosheets, Radiat. Eff. Defects Solids 174 (9–10) (2019) 885–898.

[6] S.A. Hashemi, S.M. Mousavi, R. Faghihi, M. Arjmand, S. Sina, A.M. Amani, Lead oxide-decorated graphene oxide/epoxy composite toward X-ray radiation shielding, Radiat. Phys. Chem. 146 (2018) 77–85.

[7] P. Saini, Historical review of advanced materials for electromagnetic interference (EMI) shielding: conjugated polymers, carbon nanotubes, graphene based composites, Indian J. Pure Appl. Phys. 57 (5) (2019) 338–351.

[8] W. Xie, X. Zhu, S. Yi, J. Kuang, H. Cheng, W. Tang, Y. Deng, Electromagnetic absorption properties of natural microcrystalline graphite, Mater. Des. 90 (2016) 38–46.

[9] N. Das, T. Chaki, D. Khastgir, A. Chakraborty, Electromagnetic interference shielding effectiveness of conductive carbon black and carbon fiber-filled composites based on rubber and rubber blends, Adv. Polym. Technol. 20 (3) (2001) 226–236.

[10] A.P. Singh, P. Garg, F. Alam, K. Singh, R. Mathur, R. Tandon, A. Chandra, S. Dhawan, Phenolic resin-based composite sheets filled with mixtures of reduced graphene oxide, γ-Fe2O3 and carbon fibers for excellent electromagnetic interference shielding in the X-band, Carbon 50 (10) (2012) 3868–3875.

[11] J.D. Kraus, D.A. Fleisch, Electromagnetics with Applications, fifth ed., McGraw-Hill, New York, 1999.

[12] Z. Chen, C. Xu, C. Ma, W. Ren, H.M. Cheng, Lightweight and flexible graphene foam composites for high-performance electromagnetic interference shielding, Adv. Mater. 25 (9) (2013) 1296–1300.

[13] M.A. Kats, D. Sharma, J. Lin, P. Genevet, R. Blanchard, Z. Yang, M.M. Qazilbash, D. Basov, S. Ramanathan, F. Capasso, Ultra-thin perfect absorber employing a tunable phase change material, Appl. Phys. Lett. 101 (22) (2012) 221101.

[14] S.A. Hashemi, S.M. Mousavi, R. Faghihi, M. Arjmand, M. Rahsepar, S. Bahrani, S. Ramakrishna, C.W. Lai, Superior X-ray radiation shielding effectiveness of biocompatible polyaniline reinforced with hybrid graphene oxide-Iron tungsten nitride flakes, Polymers 12 (6) (2020) 1407.

[15] Z. Ye, S. Chaudhary, P. Kuang, K.-M. Ho, Broadband light absorption enhancement in polymer photovoltaics using metal nanowall gratings as transparent electrodes, Opt. Express 20 (11) (2012) 12213–12221.

[16] C. Hagglund, S.P. Apell, B. Kasemo, Maximized optical absorption in ultrathin films and its application to plasmon-based two-dimensional photovoltaics, Nano Lett. 10 (8) (2010) 3135–3141.

[17] N.I. Landy, S. Sajuyigbe, J.J. Mock, D.R. Smith, W.J. Padilla, Perfect metamaterial absorber, Phys. Rev. Lett. 100 (20) (2008) 207402.

[18] H. Li, L.H. Yuan, B. Zhou, X.P. Shen, Q. Cheng, T.J. Cui, Ultrathin multiband gigahertz metamaterial absorbers, J. Appl. Phys. 110 (1) (2011), 014909.

[19] F. Ding, Y. Cui, X. Ge, Y. Jin, S. He, Ultra-broadband microwave metamaterial absorber, Appl. Phys. Lett. 100 (10) (2012) 103506.

[20] J.A. Rogers, T. Someya, Y. Huang, Materials and mechanics for stretchable electronics, Science 327 (5973) (2010) 1603–1607.

[21] S.A. Hashemi, S.M. Mousavi, M. Arjmand, N. Yan, U. Sundararaj, Electrified single-walled carbon nanotube/epoxy nanocomposite via vacuum shock technique: effect of alignment on electrical conductivity and electromagnetic interference shielding, Polym. Compos. 39 (S2) (2018) E1139–E1148.

[22] A.A. Al-Ghamdi, F. El-Tantawy, New electromagnetic wave shielding effectiveness at microwave frequency of polyvinyl chloride reinforced graphite/copper nanoparticles, Compos. A: Appl. Sci. Manuf. 41 (11) (2010) 1693–1701.

[23] J. Hochepied, M. Pileni, Magnetic properties of mixed cobalt–zinc ferrite nanoparticles, J. Appl. Phys. 87 (5) (2000) 2472–2478.

[24] P. Gairola, L. Purohit, P. Gairola, P. Bhardwaj, S. Kaushik, Enhanced electromagnetic absorption in ferrite and tantalum pentoxide based polypyrrole nanocomposite, Prog. Nat. Sci.: Mater. Int. 29 (2) (2019) 170–176.

[25] S.M. Mousavi, F.W. Low, S.A. Hashemi, C.W. Lai, G. Younes, A. Savardashtaki, A. Babapoor, N. Rumjit, G. Mei, N. Amin, S.K. Tiong, Development of Graphene based Nanocomposites Toward Medical and Biological Applications, 2020.

[26] P.H.C. Camargo, K.G. Satyanarayana, F. Wypych, Nanocomposites: synthesis, structure, properties and new application opportunities, Mater. Res. 12 (1) (2009) 1–39.

[27] S.A. Hashemi, S.M. Mousavi, S. Bahrani, S. Ramakrishna, A. Babapoor, W.-H. Chiang, Coupled graphene oxide with hybrid metallic nanoparticles as potential electrochemical biosensors for precise detection of ascorbic acid within blood, Anal. Chim. Acta (2020).

[28] S.N. Rafeeq, M.M. Ismail, J.M. Sulaiman, Magnetic and dielectric properties of $CoFe_2O_4$ and $Co_xZn_{1-x}Fe_2O_4$ nanoparticles synthesized using sol-gel method, J. Magn. 22 (3) (2017) 406–413.

[29] Y. Kannan, R. Saravanan, N. Srinivasan, K. Praveena, K. Sadhana, Synthesis and characterization of some ferrite nanoparticles prepared by co-precipitation method, J. Mater. Sci. Mater. Electron. 27 (11) (2016) 12000–12008.

[30] R. Lakshmi, P. Bera, R. Chakradhar, B. Choudhury, S.P. Pawar, S. Bose, R.U. Nair, H.C. Barshilia, Enhanced microwave absorption properties of PMMA modified $MnFe_2O_4$–polyaniline nanocomposites, Phys. Chem. Chem. Phys. 21 (9) (2019) 5068–5077.

[31] S. Laurent, C. Henoumont, D. Stanicki, S. Boutry, E. Lipani, S. Belaid, R.N. Muller, L. Vander Elst, MRI Contrast Agents: From Molecules to Particles, Springer, 2017.

[32] T. Scepka, Noninvasive Control of Magnetic State in Ferromagnetic Nanodots by Hall Probe Magnetometry, Doctoral Degree, Slovak University of Technology, 2016.

[33] S.A. Hashemi, S.M. Mousavi, S. Bahrani, S. Ramakrishna, Integrated polyaniline with graphene oxide-iron tungsten nitride nanoflakes as ultrasensitive electrochemical sensor for precise detection of 4-nitrophenol within aquatic media, J. Electroanal. Chem. 114406 (2020).

[34] N. Gandhi, K. Singh, A. Ohlan, D. Singh, S. Dhawan, Thermal, dielectric and microwave absorption properties of polyaniline–CoFe2O4 nanocomposites, Compos. Sci. Technol. 71 (15) (2011) 1754–1760.
[35] M. Cao, C. Han, X. Wang, M. Zhang, Y. Zhang, J. Shu, H. Yang, X. Fang, J. Yuan, Graphene nanohybrids: excellent electromagnetic properties for the absorbing and shielding of electromagnetic waves, J. Mater. Chem. C 6 (17) (2018) 4586–4602.
[36] X. Liu, Z. Zhang, Y. Wu, Absorption properties of carbon black/silicon carbide microwave absorbers, Compos. Part B 42 (2) (2011) 326–329.
[37] K. Wu, T. Ting, G. Wang, W. Ho, C. Shih, Effect of carbon black content on electrical and microwave absorbing properties of polyaniline/carbon black nanocomposites, Polym. Degrad. Stab. 93 (2) (2008) 483–488.
[38] B. Quan, W. Shi, S.J.H. Ong, X. Lu, P.L. Wang, G. Ji, Y. Guo, L. Zheng, Z.J. Xu, Defect engineering in two common types of dielectric materials for electromagnetic absorption applications, Adv. Funct. Mater. 29 (28) (2019) 1901236.
[39] S.M. Mousavi, A. Babapoor, S.A. Hashemi, B. Medi, Adsorption and removal characterization of nitrobenzene by graphene oxide coated by polythiophene nanoparticles, Phys. Chem. Res. 8 (2) (2020) 225–240.
[40] X.-X. Wang, T. Ma, J.-C. Shu, M.-S. Cao, Confinedly tailoring Fe3O4 clusters-NG to tune electromagnetic parameters and microwave absorption with broadened bandwidth, Chem. Eng. J. 332 (2018) 321–330.
[41] Y. Wei, H. Liu, S. Liu, M. Zhang, Y. Shi, J. Zhang, L. Zhang, C. Gong, Waste cotton-derived magnetic porous carbon for high-efficiency microwave absorption, Comput. Commun. 9 (2018) 70–75.
[42] H. Wu, M. Qin, L. Zhang, NiCo2O4 constructed by different dimensions of building blocks with superior electromagnetic wave absorption performance, Compos. Part B 182 (2020) 107620.
[43] X. Liang, Z. Man, B. Quan, J. Zheng, W. Gu, Z. Zhang, G. Ji, Environment-stable co x Ni y encapsulation in stacked porous carbon Nanosheets for enhanced microwave absorption, Nano-Micro Lett. 12 (2020) 1–12.
[44] X. Cui, X. Liang, W. Liu, W. Gu, G. Ji, Y. Du, Stable microwave absorber derived from 1D customized heterogeneous structures of Fe3N@ C, Chem. Eng. J. 381 (2020) 122589.
[45] W. Li, L. Xu, X. Zhang, Y. Gong, Y. Ying, J. Yu, J. Zheng, L. Qiao, S. Che, Investigating the effect of honeycomb structure composite on microwave absorption properties, Compos. Commun. (2020).
[46] H. Zhang, Z. Jia, A. Feng, Z. Zhou, C. Zhang, K. Wang, N. Liu, G. Wu, Enhanced microwave absorption performance of sulfur-doped hollow carbon microspheres with mesoporous shell as a broadband absorber, Compos. Commun. (2020).
[47] H. Guan, D. Chung, Radio-wave electrical conductivity and absorption-dominant interaction with radio wave of exfoliated-graphite-based flexible graphite, with relevance to electromagnetic shielding and antennas, Carbon 157 (2020) 549–562.
[48] Y. Zhang, Y. Huang, T. Zhang, H. Chang, P. Xiao, H. Chen, Z. Huang, Y. Chen, Broadband and tunable high-performance microwave absorption of an ultralight and highly compressible graphene foam, Adv. Mater. 27 (12) (2015) 2049–2053.
[49] J. Dalal, S. Lather, A. Gupta, R. Tripathi, A.S. Maan, K. Singh, A. Ohlan, Reduced graphene oxide functionalized strontium ferrite in poly (3, 4-ethylenedioxythiophene) conducting network: a high-performance EMI shielding material, Adv. Mater. Technol. 4 (7) (2019) 1900023.
[50] X. Yin, L. Kong, L. Zhang, L. Cheng, N. Travitzky, P. Greil, Electromagnetic properties of Si–C–N based ceramics and composites, Int. Mater. Rev. 59 (6) (2014) 326–355.
[51] H. Chen, Z. Huang, Y. Huang, Y. Zhang, Z. Ge, B. Qin, Z. Liu, Q. Shi, P. Xiao, Y. Yang, Synergistically assembled MWCNT/graphene foam with highly efficient microwave absorption in both C and X bands, Carbon 124 (2017) 506–514.
[52] J.C. Shu, M.S. Cao, M. Zhang, X.X. Wang, W.Q. Cao, X.Y. Fang, M.Q. Cao, Molecular patching engineering to drive energy conversion as efficient and environment-friendly cell toward wireless power transmission, Adv. Funct. Mater. 30 (10) (2020) 1908299.
[53] Z. Pan, L. Yao, G. Ge, B. Shen, J. Zhai, High-performance capacitors based on NaNbO 3 nanowires/poly (vinylidene fluoride) nanocomposites, J. Mater. Chem. A 6 (30) (2018) 14614–14622.

[54] J.-C. Shu, X.-Y. Yang, X.-R. Zhang, X.-Y. Huang, M.-S. Cao, L. Li, H.-J. Yang, W.-Q. Cao, Tailoring MOF-based materials to tune electromagnetic property for great microwave absorbers and devices, Carbon (2020).

[55] H.-J. Yang, W.-Q. Cao, D.-Q. Zhang, T.-J. Su, H.-L. Shi, W.-Z. Wang, J. Yuan, M.-S. Cao, NiO hierarchical nanorings on SiC: enhancing relaxation to tune microwave absorption at elevated temperature, ACS Appl. Mater. Interfaces 7 (13) (2015) 7073–7077.

[56] Z.-L. Hou, M. Zhang, L.-B. Kong, H.-M. Fang, Z.-J. Li, H.-F. Zhou, H.-B. Jin, M.-S. Cao, Microwave permittivity and permeability experiments in high-loss dielectrics: caution with implicit Fabry-Pérot resonance for negative imaginary permeability, Appl. Phys. Lett. 103 (16) (2013) 162905.

[57] S.M. Mousavi, S.A. Hashemi, A.M. Amani, H. Saed, S. Jahandideh, F. Mojoudi, Polyethylene terephthalate/acryl butadiene styrene copolymer incorporated with oak shell, potassium sorbate and egg shell nanoparticles for food packaging applications: control of bacteria growth, physical and mechanical properties, Polym. Renewable Resour. 8 (4) (2017) 177–196.

[58] S.M. Mousavi, S.A. Hashemi, M. Arjmand, A.M. Amani, F. Sharif, S. Jahandideh, Octadecyl amine functionalized graphene oxide toward hydrophobic chemical resistant epoxy nanocomposites, ChemistrySelect 3 (25) (2018) 7200–7207.

[59] S.M. Mousavi, F.W. Low, S.A. Hashemi, N.A. Samsudin, M. Shakeri, Y. Yusoff, M. Rahsepar, C. W. Lai, A. Babapoor, S. Soroshnia, Development of hydrophobic reduced graphene oxide as a new efficient approach for photochemotherapy, RSC Adv. 10 (22) (2020) 12851–12863.

[60] A. Abdollahifar, S.A. Hashemi, S.M. Mousavi, M. Rahsepar, A.M. Amani, Fabrication of graphene oxide-lead oxide epoxy based composite with enhanced chemical resistance, hydrophobicity and thermo-mechanical properties, Adv. Polym. Technol. 37 (8) (2018) 3792–3803.

[61] M. Lu, X. Wang, W. Cao, J. Yuan, M. Cao, Carbon nanotube-CdS core–shell nanowires with tunable and high-efficiency microwave absorption at elevated temperature, Nanotechnology 27 (6) (2015), 065702.

[62] N. Goudarzian, S. Samiei, F. Safari, S.M. Mousavi, S.A. Hashemi, S. Mazraedoost, Enhancing the physical, mechanical, oxygen permeability and Photodegradation properties of styrene-acrylonitrile (SAN), butadiene rubber (BR) composite by silica nanoparticles, J. Environ. Treat. Tech. 8 (2) (2020) 718–726.

[63] B. Wen, M. Cao, M. Lu, W. Cao, H. Shi, J. Liu, X. Wang, H. Jin, X. Fang, W. Wang, Reduced graphene oxides: light-weight and high-efficiency electromagnetic interference shielding at elevated temperatures, Adv. Mater. 26 (21) (2014) 3484–3489.

[64] Y. Li, M.-s. Cao, D.-w. Wang, J. Yuan, High-efficiency and dynamic stable electromagnetic wave attenuation for La doped bismuth ferrite at elevated temperature and gigahertz frequency, RSC Adv. 5 (94) (2015) 77184–77191.

[65] D. Han, H. Mei, S. Xiao, W. Xue, Q. Bai, L. Cheng, CNT/SiC composites produced by direct matrix infiltration of self-assembled CNT sponges, J. Mater. Sci. 52 (14) (2017) 8401–8411.

[66] M. Dresselhaus, A. Jorio, R. Saito, Characterizing graphene, graphite, and carbon nanotubes by Raman spectroscopy, Annu. Rev. Condens. Matter Phys. 1 (1) (2010) 89–108.

[67] R. Haerle, E. Riedo, A. Pasquarello, A. Baldereschi, sp 2/s p 3 hybridization ratio in amorphous carbon from C 1 s core-level shifts: X-ray photoelectron spectroscopy and first-principles calculation, Phys. Rev. B 65 (4) (2001) 045101.

[68] K. Huang, M. Chen, G. He, X. Hu, W. He, X. Zhou, Y. Huang, Z. Liu, Stretchable microwave absorbing and electromagnetic interference shielding foam with hierarchical buckling induced by solvent swelling, Carbon 157 (2020) 466–477.

[69] S.M. Mousavi, S.A. Hashemi, A.M. Amani, H. Esmaeili, Y. Ghasemi, A. Babapoor, F. Mojoudi, O. Arjomand, Pb (II) removal from synthetic wastewater using Kombucha Scoby and graphene oxide/Fe_3O_4, Phys. Chem. Res. 6 (4) (2018) 759–771.

[70] M. Mousavi, A. Hashemi, O. Arjmand, A.M. Amani, A. Babapoor, M.A. Fateh, H. Fateh, F. Mojoudi, H. Esmaeili, S. Jahandideh, Erythrosine adsorption from aqueous solution via decorated graphene oxide with magnetic iron oxide nano particles: kinetic and equilibrium studies, Acta Chim. Slov. 65 (4) (2018) 882–894.

[71] R.C. Che, L.M. Peng, X.F. Duan, Q. Chen, X.L. Liang, Microwave absorption enhancement and complex permittivity and permeability of Fe encapsulated within carbon nanotubes, Adv. Mater. 16 (5) (2004) 401–405.

[72] X. Sun, J. He, G. Li, J. Tang, T. Wang, Y. Guo, H. Xue, Laminated magnetic graphene with enhanced electromagnetic wave absorption properties, J. Mater. Chem. C 1 (4) (2013) 765–777.

[73] H.-L. Xu, H. Bi, R.-B. Yang, Enhanced microwave absorption property of bowl-like Fe_3O_4 hollow spheres/reduced graphene oxide composites, J. Appl. Phys. 111 (7) (2012) 07A522.

[74] M. Zong, Y. Huang, Y. Zhao, X. Sun, C. Qu, D. Luo, J. Zheng, Facile preparation, high microwave absorption and microwave absorbing mechanism of RGO–Fe_3O_4 composites, RSC Adv. 3 (45) (2013) 23638–23648.

[75] M. Fu, Q. Jiao, Y. Zhao, Preparation of $NiFe_2O_4$ nanorod–graphene composites via an ionic liquid assisted one-step hydrothermal approach and their microwave absorbing properties, J. Mater. Chem. A 1 (18) (2013) 5577–5586.

[76] S. Yang, D.-w. Xu, P. Chen, H.-f. Qiu, X. Guo, J. Mater. Sci. Mater. Electron. 29 (2018) 19443–19453.

[77] L. Wang, X. Bai, M. Wang, Facile preparation, characterization and highly effective microwave absorption performance of porous α-Fe_2O_3 nanorod–graphene composites, J. Mater. Sci. Mater. Electron. 29 (4) (2018) 3381–3390.

[78] L. He, Y. Zhao, L. Xing, P. Liu, Z. Wang, Y. Zhang, Y. Wang, Y. Du, Preparation of reduced graphene oxide coated flaky carbonyl iron composites and their excellent microwave absorption properties, RSC Adv. 8 (6) (2018) 2971–2977.

[79] H. Zhang, A. Xie, C. Wang, H. Wang, Y. Shen, X. Tian, Novel rGO/α-Fe_2O_3 composite hydrogel: synthesis, characterization and high performance of electromagnetic wave absorption, J. Mater. Chem. A 1 (30) (2013) 8547–8552.

[80] Z. Zhu, X. Sun, H. Xue, H. Guo, X. Fan, X. Pan, J. He, Graphene–carbonyl iron cross-linked composites with excellent electromagnetic wave absorption properties, J. Mater. Chem. C 2 (32) (2014) 6582–6591.

[81] M. Zong, Y. Huang, X. Ding, N. Zhang, C. Qu, Y. Wang, One-step hydrothermal synthesis and microwave electromagnetic properties of RGO/$NiFe_2O_4$ composite, Ceram. Int. 40 (5) (2014) 6821–6828.

[82] N. Zhang, Y. Huang, M. Zong, X. Ding, S. Li, M. Wang, Synthesis of ZnS quantum dots and $CoFe_2O_4$ nanoparticles co-loaded with graphene nanosheets as an efficient broad band EM wave absorber, Chem. Eng. J. 308 (2017) 214–221.

[83] N. Zhang, Y. Huang, M. Zong, X. Ding, S. Li, M. Wang, Coupling $CoFe_2O_4$ and SnS_2 nanoparticles with reduced graphene oxide as a high-performance electromagnetic wave absorber, Ceram. Int. 42 (14) (2016) 15701–15708.

[84] W.-L. Song, X.-T. Guan, L.-Z. Fan, W.-Q. Cao, Q.-L. Zhao, C.-Y. Wang, M.-S. Cao, Tuning broadband microwave absorption via highly conductive Fe_3O_4/graphene heterostructural nanofillers, Mater. Res. Bull. 72 (2015) 316–323.

[85] L. Zhang, X. Yu, H. Hu, Y. Li, M. Wu, Z. Wang, G. Li, Z. Sun, C. Chen, Facile synthesis of iron oxides/reduced graphene oxide composites: application for electromagnetic wave absorption at high temperature, Sci. Rep. 5 (2015) 9298.

[86] Y. Wang, H. Zhu, Y. Chen, X. Wu, W. Zhang, C. Luo, J. Li, Design of hollow $ZnFe_2O_4$ microspheres@ graphene decorated with TiO_2 nanosheets as a high-performance low frequency absorber, Mater. Chem. Phys. 202 (2017) 184–189.

[87] B. Quan, G. Xu, D. Li, W. Liu, G. Ji, Y. Du, Incorporation of dielectric constituents to construct ternary heterojunction structures for high-efficiency electromagnetic response, J. Colloid Interface Sci. 498 (2017) 161–169.

[88] Y. Xu, Q. Wang, Y. Cao, X. Wei, B. Huang, Preparation of a reduced graphene oxide/SiO_2/Fe_3O_4 UV-curing material and its excellent microwave absorption properties, RSC Adv. 7 (29) (2017) 18172–18177.

[89] R. Shu, G. Zhang, J. Zhang, X. Wang, M. Wang, Y. Gan, J. Shi, J. He, Synthesis and high-performance microwave absorption of reduced graphene oxide/zinc ferrite hybrid nanocomposite, Mater. Lett. 215 (2018) 229–232.

[90] Y. Li, D. Li, J. Yang, H. Luo, F. Chen, X. Wang, R. Gong, Enhanced microwave absorption and surface wave attenuation properties of Co0. 5Ni0. 5Fe2O4 fibers/reduced graphene oxide composites, Materials 11 (4) (2018) 508.

[91] Y. Wang, Y. Huang, Q. Wang, M. Zong, Preparation and electromagnetic properties of graphene-supported Ni0. 8Zn0. 2Ce0. 06Fe1. 9O4 nanocomposite, Powder Technol. 249 (2013) 304–308.

[92] S. Jiao, M. Wu, X. Yu, H. Hu, Z. Bai, P. Dai, T. Jiang, H. Bi, G. Li, RGO/BaFe12O19/Fe3O4 nanocomposite as microwave absorbent with lamellar structures and improved polarization interfaces, Mater. Res. Bull. 108 (2018) 89–95.

[93] Q. Zhang, Z. Du, X. Huang, Z. Zhao, T. Guo, G. Zeng, Y. Yu, Tunable microwave absorptivity in reduced graphene oxide functionalized with Fe3O4 nanorods, Appl. Surf. Sci. 473 (2019) 706–714.

[94] C. Li, Y. Ge, X. Jiang, G.I. Waterhouse, Z. Zhang, L. Yu, Porous Fe3O4/C microspheres for efficient broadband electromagnetic wave absorption, Ceram. Int. 44 (16) (2018) 19171–19183.

[95] X. Zhang, X. Dong, H. Huang, B. Lv, J. Lei, C. Choi, Microstructure and microwave absorption properties of carbon-coated iron nanocapsules, J. Phys. D. Appl. Phys. 40 (17) (2007) 5383.

[96] H. Pourabdollahi, A.R. Zarei, Hydrothermal synthesis of carbonyl iron-carbon nanocomposite: characterization and electromagnetic performance, Results Phys. 7 (2017) 1978–1986.

[97] F. Peng, F. Meng, Y. Guo, H. Wang, F. Huang, Z. Zhou, Intercalating hybrids of sandwich-like Fe3O4–graphite: synthesis and their synergistic enhancement of microwave absorption, ACS Sustain. Chem. Eng. 6 (12) (2018) 16744–16753.

[98] Y. Zhao, L. Liu, K. Jiang, M. Fan, C. Jin, J. Han, W. Wu, G. Tong, Distinctly enhanced permeability and excellent microwave absorption of expanded graphite/Fe 3 O 4 nanoring composites, RSC Adv. 7 (19) (2017) 11561–11567.

[99] L. Liu, Z. He, Y. Zhao, J. Sun, G. Tong, Modulation of the composition and surface morphology of expanded graphite/Fe/Fe3O4 composites for plasmon resonance-enhanced microwave absorption, J. Alloys Compd. 765 (2018) 1218–1227.

[100] X. Su, J. Wang, B. Zhang, W. Chen, Q. Wu, W. Dai, Y. Zou, Enhanced microwave absorption properties of epoxy composites containing graphite nanosheets@ Fe3O4 decorated comb-like MnO2 nanoparticles, Mater. Res. Express 5 (5) (2018), 056305.

[101] S. Biswas, S.S. Panja, S. Bose, Unique multilayered assembly consisting of "flower-like" ferrite nanoclusters conjugated with MWCNT as millimeter wave absorbers, J. Phys. Chem. C 121 (26) (2017) 13998–14009.

[102] Y.-c. Sun, W.-y. Cui, J.-l. Li, J.-z. Wu, In-situ growth strategy to fabrication of MWCNTs/Fe3O4 with controllable interface polarization intensity and wide band electromagnetic absorption performance, J. Alloys Compd. 770 (2019) 67–75.

[103] L. Yu, X. Lan, C. Wei, X. Li, X. Qi, T. Xu, C. Li, C. Li, Z. Wang, MWCNT/NiO-Fe3O4 hybrid nanotubes for efficient electromagnetic wave absorption, J. Alloys Compd. 748 (2018) 111–116.

[104] L. Huang, X. Liu, R. Yu, Enhanced microwave absorption properties of rod-shaped Fe2O3/Fe3O4/MWCNTs composites, Prog. Nat. Sci.: Mater. Int. 28 (3) (2018) 288–295.

[105] F. Wen, F. Zhang, Z. Liu, Investigation on microwave absorption properties for multiwalled carbon nanotubes/Fe/co/Ni nanopowders as lightweight absorbers, J. Phys. Chem. C 115 (29) (2011) 14025–14030.

[106] K. Osouli-Bostanabad, H. Aghajani, E. Hosseinzade, H. Maleki-Ghaleh, M. Shakeri, High microwave absorption of nano-Fe3O4 deposited electrophoretically on carbon fiber, Mater. Manuf. Process. 31 (10) (2016) 1351–1356.

[107] H. Zhang, J. Zhang, H. Zhang, Numerical predictions for radar absorbing silicon carbide foams using a finite integration technique with a perfect boundary approximation, Smart Mater. Struct. 15 (3) (2006) 759.

[108] Z. Liu, G. Bai, Y. Huang, F. Li, Y. Ma, T. Guo, X. He, X. Lin, H. Gao, Y. Chen, Microwave absorption of single-walled carbon nanotubes/soluble cross-linked polyurethane composites, J. Phys. Chem. C 111 (37) (2007) 13696–13700.

[109] Y. Zhang, Y. Huang, H. Chen, Z. Huang, Y. Yang, P. Xiao, Y. Zhou, Y. Chen, Composition and structure control of ultralight graphene foam for high-performance microwave absorption, Carbon 105 (2016) 438–447.

[110] S.M. Mousavi, S.A. Hashemi, Y. Ghasemi, Decorated graphene oxide with OCTADECYL amine/SASOBIT/epoxy nanocomposite via vacuum shock technique: morphology and beams shielding, J. Chem. Technol. Metall. 54 (6) (2019).

[111] Y. Huang, H. Zhang, G. Zeng, Z. Li, D. Zhang, H. Zhu, R. Xie, L. Zheng, J. Zhu, The microwave absorption properties of carbon-encapsulated nickel nanoparticles/silicone resin flexible absorbing material, J. Alloys Compd. 682 (2016) 138–143.

[112] T. Zou, N. Zhao, C. Shi, J. Li, Microwave absorbing properties of activated carbon fibre polymer composites, Bull. Mater. Sci. 34 (1) (2011) 75–79.

[113] G. Wang, Z. Gao, S. Tang, C. Chen, F. Duan, S. Zhao, S. Lin, Y. Feng, L. Zhou, Y. Qin, Microwave absorption properties of carbon nanocoils coated with highly controlled magnetic materials by atomic layer deposition, ACS Nano 6 (12) (2012) 11009–11017.

[114] F. Meng, H. Wang, F. Huang, Y. Guo, Z. Wang, D. Hui, Z. Zhou, Graphene-based microwave absorbing composites: a review and prospective, Compos. Part B 137 (2018) 260–277.

[115] Y. Li, Z. Mao, R. Liu, X. Zhao, Y. Zhang, G. Qin, X. Zhang, Ultralight Fe@ C nanocapsules/sponge composite with reversibly tunable microwave absorption performances, Nanotechnology 28 (32) (2017) 325702.

[116] X. Zeng, X. Cheng, R. Yu, G.D. Stucky, Electromagnetic microwave absorption theory and recent achievements in microwave absorbers, Carbon (2020).

[117] Q. Liu, X. He, C. Yi, D. Sun, J. Chen, D. Wang, K. Liu, M. Li, Fabrication of ultra-light nickel/graphene composite foam with 3D interpenetrating network for high-performance electromagnetic interference shielding, Compos. Part B 182 (2020) 107614.

[118] S.M. Mousavi, S. Soroshnia, S.A. Hashemi, A. Babapoor, Y. Ghasemi, A. Savardashtaki, A.M. Amani, Graphene nano-ribbon based high potential and efficiency for DNA, cancer therapy and drug delivery applications, Drug Metab. Rev. 51 (1) (2019) 91–104.

[119] S.M. Mousavi, S.A. Hashemi, Y. Ghasemi, A.M. Amani, A. Babapoor, O. Arjmand, Applications of graphene oxide in case of nanomedicines and nanocarriers for biomolecules: review study, Drug Metab. Rev. 51 (1) (2019) 12–41.

[120] R. Azhdari, S.M. Mousavi, S.A. Hashemi, S. Bahrani, S. Ramakrishna, Decorated graphene with aluminum fumarate metal organic framework as a superior non-toxic agent for efficient removal of Congo red dye from wastewater, J. Environ. Chem. Eng. 7 (6) (2019) 103437.

CHAPTER 30

Graphene in automotive parts

Kuray Dericiler[a,b], Nargiz Aliyeva[a,b], Hadi Mohammadjafari Sadeghi[a,b], Hatice S. Sas[a,b], Yusuf Ziya Menceloglu[a,b], and Burcu Saner Okan[a,b]

[a]Sabanci University Integrated Manufacturing Technologies Research and Application Center & Composite Technologies Center of Excellence, Istanbul, Turkey
[b]Faculty of Engineering and Natural Sciences, Materials Science and Nano Engineering, Sabanci University, Tuzla, Istanbul, Turkey

Chapter outline

1. Introduction — 623
2. Transformation into lightweighting innovations — 629
3. Graphene in body and structural parts — 630
4. Coating applications of graphene — 632
5. Graphene in tire manufacturing — 634
6. Graphene in electronic parts of vehicles — 635
7. Graphene as a lubricating agent in fluids — 636
8. Graphene potential in electric vehicles — 642
9. Conclusions and outlook — 643
References — 645

1. Introduction

Mainly in the automotive industry, carbon fiber-reinforced plastic (CFRP) and glass fiber-reinforced plastic (GFRP) composites are utilized due to their low weight, flexible design, and high mechanical properties [1, 2]. As the reinforcing material mainly governs the strength and stiffness of the composite, the aspect ratio of the reinforcement is critical [3]. General applications of composites in the automotive parts are given in Fig. 1 [4]. Short fiber-reinforced thermoplastic components are used in several applications in the automotive industry, such as intake manifolds, clutch pedals, and accelerator pedals; however, their utilization in the structural (load-bearing) applications are limited due to low strength and stiffness [5, 6]. On the other hand, long fiber-reinforced composites offer both directional control (orientation) and a high aspect ratio, allowing the load transfer along with the fiber rather than intermittent transfer between fiber ends and matrix [7]. Long glass fiber-reinforced composites found several applications with polypropylene (PP), polyethylene terephthalate (PET), and polyamide (PA) matrices as an alternative to conventional metallic automotive components [8, 9].

Similarly, unidirectional laminates with preimpregnated (prepreg) tapes or sheets allow the production of high-performance thermoset composites for structural automotive parts. However, as no reinforcing fibers run transversely within or between the

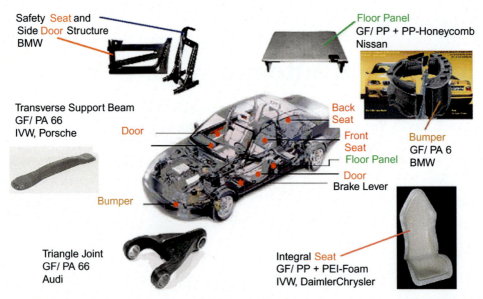

Fig. 1 Constituents of composite materials [4]. *(Reproduced with the permission of Springer Nature.)*

layers, the composite becomes vulnerable to delamination. Furthermore, these composites lack the necessary impact resistance due to the absence of a thermoplastic matrix that dissipates the impact by plastic deformation [4, 7]. Further requirements towards multifunctionality, fuel efficiency, and lightweight with environmentally friendly practices necessities new material solutions [10]. Integration of nanotechnology into composite materials has been extensively applied to meet the need of the automotive industry. The application of nanomaterials such as graphene can modify the composite materials' properties for improved performance.

Graphene, starting with its discovery in 2004 by Nobel laureates Andre Geim and Konstantin Novoselov, has attracted an increasing amount of interest by researchers to this day [11]. The outstanding properties of graphene, such as extremely high tensile strength and Young's modulus of 130.5 GPa and 1 TPa, respectively, combined with high theoretical thermal and electrical conductivity around 4840 W/mK, and 2000 S/cm, respectively, make it the material solution for a wide range of applications [12–14]. Furthermore, graphene's unique 2D isotropic structure with sp^2 hybridized carbon atoms offer both excellent physical properties as well as a high surface area around 1500 m^2/g and optical properties [12, 15]. However, these properties are very dependent on the structural integrity of the graphene and highly affected by the employed production techniques [16–18].

The main attraction of graphene for the automotive industry is graphene's mechanical performance with a high surface area, which opens an incredible potential for lightweight automotive parts. Currently, in light vehicles, plastics make up 50% of the volume while contributing only 10% of their weight, leading to lighter, more fuel-efficient vehicles with reduced greenhouse gas emissions [19] with a forecast of four times increase in the use of polymer and polymer composites by 2025 [20]. European Union (EU) has set targets concerning the CO_2 emissions of passenger cars and light commercial vehicles that aim to reduce the CO_2 emissions according to the EU's climate strategy [21]. The regulations require CO_2 emissions of as much as 95 g/km for passenger cars and 147 g/km for light commercial vehicles by 2021. Detailed plots regarding the CO_2 emissions factors for different vehicle types are given in Fig. 2 where "hybrid" referred to hybrid-gasoline vehicles. As seen in figure, L-category (mopeds and motorbikes) vehicles show lower CO_2 emission in comparison to light and heavy-duty vehicles. Besides the vehicle weight, used fuel type also affects the CO_2 Emissions. Therefore, producing lighter vehicles regardless of the category can result in drastic reduction in the CO_2 emissions.

Incorporation of graphene into automotive parts for lightweighting purposes can be achieved by various routes. Although the vehicles comprise a wide variety of materials from steel to glass to fluids and lubricants, it is much more suitable for graphene to be incorporated in plastics and polymer matrix composites (PMCs) due to the latter two offering favorable incorporation routes. PMCs present a combination of properties that no single material can provide by a broad spectrum of reinforcements and matrix materials. Moreover, the addition of graphene to the polymer matrix acts as a nucleating agent [23], reinforcement [24], and lubricating agent [25]. However, significant challenges remain for the efficient and homogenous distribution of graphene in the polymer matrix due to the tendency of graphene sheets to agglomerate, which forms weak points in the structure.

As the effect of the production method on the properties of graphene end product is significant, it is essential to investigate the main approaches for graphene production. Chemical vapor deposition (CVD) is based on exposing hydrocarbons on hot metal surfaces under vacuum to produce layers of carbon on the surface proven to be an extremely reliable method to produce single-layer high-quality graphene [17]. Moreover, the properties of the graphene can be tuned by changing process parameters and the substrate metal [26, 27]. However, the CVD technique is much proper for niche applications but not for mass production due to high manufacturing and operation cost. Another popular approach is liquid-phase exfoliation of graphite with the aid of surfactants or intercalating substances using microwave [28], ultrasonication [29], or shear mixing [30] in aqueous [31] or nonaqueous mediums [32]. This route generally yields few-layer graphene (FLG) with little to no oxidation, albeit at a slightly lower quality than CVD graphene [33, 34]. Over the years, oxidation and subsequent reduction of graphite to obtain graphene have also gained significant attention [35–37]. In the following years, Marcano et al. provided an improvement to the well-known Hummers' method [38] by excluding

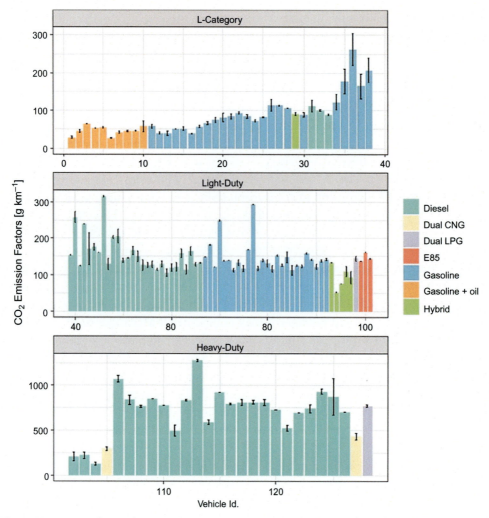

Fig. 2 CO_2 emission factors from L-category, light-duty, and heavy-duty vehicles [22]. *(Reproduced with the permission of Springer Nature.)*

sodium nitrate and including phosphoric acid, which prevented the generation of toxic gases and gave easy control over the reaction temperature [39, 40]. Subsequent controllable reduction with hydrazine or other reducing agents leads to varying oxidation degrees. Although the oxidation state leads to a decrease in the quality of the graphene, it contributes to incorporating graphene into the polymer matrix. Nonetheless, scalable graphene productions are required in order to compensate for the need of the automotive industry.

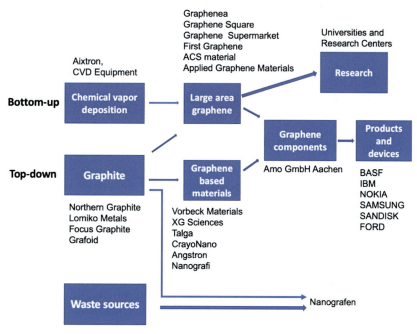

Fig. 3 Some selected graphene-related companies based on their production approaches and applications.

In recent years, the graphene market is extensively growing and there are various attempts for the commercialization of graphene and graphene-related technologies starting from supplying raw materials to intermediate products to systems and equipment to end products. Some selected graphene-related companies based on their production approaches (bottom-up and top-down) and applications are represented in Fig. 3. Graphenea, ACS Material, Graphene Supermarket procure large-area graphene by bottom-up methods, companies including Talga, Nanografi, Vorbeck Materials produce graphene-based materials. For instance, FORD produced graphene-reinforced parts in the front part of their F-max trucks by using thermoset polyurethane foams with the collaboration of XGSciences [41]. Tier 1 and Tier 2 companies or vehicle manufacturers need raw materials, semifinished products, and compounds. Specifically, Nanografen from Turkey produces graphene from recycled carbon produced from waste tires with the scale of 500 kg/month and adopts this technology for the development of lightweight automotive part production by environmentally friendly production processes with the automotive part producers and OEMs. Utilizing waste tires that would normally be discarded as landfill provides a great contribution to a circular economy as well as offers a meaningful end to the life cycle of tires.

Over the years, significant work has been carried out to achieve successful incorporation of graphene and graphene-related materials (GRMs) such as graphene oxide (GO), reduced graphene oxide (rGO), or functionalized graphene (f-Gr) to the polymer matrices. For instance, Li et al. produced foam-like structures from GO by hydrothermal method then incorporated 2 wt% GO during the in situ polymerization of polymeric precursors to obtain 3D Graphene/Polyamide 6 (PA6) nanocomposite, resulting in an increase of 400% in the thermal conductivity as well as antidripping properties for heat-retardance applications for the nanocomposite [42]. Furthermore, Duan et al. reduced GO powders during the in situ polymerization of polyamide 66 (PA66) to obtain 0.75 wt% rGO/PA66 nanocomposite with increased yield strength and tensile modulus [43]. Thus, in situ polymerization offers a good solution to the agglomeration and dispersion problem of graphene layers in the polymer matrix; the method requires a significant breakthrough to be economically feasible and scalable for the automotive industry.

The main concern arises from the weak interfacial attractions between the polymer matrix and the graphene layers. A perfect graphene structure without any surface functional groups such as hydroxyl, carboxyl, or epoxide groups is not capable of creating a load-carrying bridge between polymer chains in the matrix and graphene backbone. Therefore, it is significant to improve this interface with the functionalization or modification of the polymer or graphene structure. For instance, Scaffaro et al. produced GO using modified Tour's method [39], then modified GO with nanosilica (GOS) and fabricated GOS reinforced PA6 nanocomposites by melt compounding method, which yielded 180% and 210% increase in Young's modulus and tensile strength at 0.5 wt% GOS loading, respectively [44]. Furthermore, Wang et al. Utilized liquid-phase exfoliation and tannic acid modification methods to obtain tannic acid-modified graphene sheets (TAGS)/PA66 nanocomposite by solution mixing and drying method. As a result, fabricated nanocomposite demonstrated 15.9% and 118% improvement in Young's modulus and tensile strength at 0.5 wt% TAGS loading, respectively [45]. Recently, Dericiler et al. revealed that even with the addition of 0.3% GNP to the PA66 matrix by melt compounding, tensile and flexural strength were increased 30% and 20%, respectively, as well the tensile and flexural module displayed an improvement of more than 40% [23]. However, graphene or GRM addition to the polymer matrix cannot single-handedly maintain the required mechanical properties while providing a lightweighting effect for the automotive parts.

Due to regulations of governing bodies and authorities, the incorporation of graphene into automotive parts for lightweighting and strengthening purposes as well as providing better thermal and electrochemical performance is a trending topic. This chapter aims to bring a contribution to research efforts by incorporation of graphene in both potential and various commercial automotive applications such as body parts, coatings, tires, electronics, and fluids with and without other co-reinforcers. Additionally, the lightweighting concept with regards to graphene incorporation in cars and the feasible application of graphene for electric vehicles is presented in this chapter.

2. Transformation into lightweighting innovations

Lightweighting and cost-driven automotive industry questing novel material solutions have grown attraction to graphene. Commercially, automotive thermoplastics such as PP, PET, PA6, and PA66 are reinforced with GFs, CFs, and natural fibers (NFs), as previously discussed, and the parts are produced using injection molding, overmolding, and compression molding methods [46–48]. In order to reduce the weight, these plastics can be mixed with graphene and GRMs using extruders and other compounding methods [49]. Scientists and engineers have been working on adapting nanoscale materials such as graphene due to the advantages of their enhanced properties. Numerous companies and compounders are seeking new technologies to meet the demand from automotive all over the world. However, the challenges still remain for producing the right material for the proper application at an affordable cost and volume. In order to solve the issues, several steps have been taken. For instance, European Commission (EC) funded €1 billion to develop graphene-related technologies from both laboratory and industrial scale in 23 countries [50].

Automotive engineers swiftly adopt lightweight materials as polymers for the body, chassis, interior, power train, and under the hood applications. Common lightweighting applications in the automotive sector are given in Fig. 4 [51]. For instance, glass fiber (GF)-reinforced composites are used to fabricate interior headliners, airbag housings, and engine covers while the more the parts that require more mechanical strength such as chassis, bumpers, and roofs are made by carbon fiber (CF)-reinforced composites.

Fig. 4 Lightweighting applications in automotive parts [51]. *(Reproduced with the permission of Elsevier.)*

Additionally, applications such as door panels and seat backs can be reinforced with natural fibers. Like CF and GF, graphene is also being accepted in the automotive sector [52]. In one of the studies, Okan et al. demonstrated that the addition of GNP from recycling and upcycling waste tire to the PA66 and melt compounding them with GF demonstrated 23% improvement in the flexural modulus at 1% GNP loading and showed the proof of concept in the selected automotive part [53].

Graphene and its derivatives can be adapted in various car parts as bodies, tires, electronics, and fuels. However, some challenges are still present, such as composite production processes not being suitable to observe the exfoliation of graphene-based material and graphene-based composite characterization and monitoring under stresses such as crash conditions. The following sections describe the potential and current applications of graphene in the automotive industry with advantages and disadvantages.

3. Graphene in body and structural parts

Strength is the main requirement for making the body of cars safe for passengers. Indeed, it is required to manufacture nanostructured material that can provide high strength and give a high intensity of impact during an accident. Nanostructured materials can also reduce the weight of the car, which leads to fuel efficiency. Graphene-based polymer composites can be very effective as structural materials, and integration into the automotive sector is required. Fig. 5 provides an overview of the typical potential applications of graphene in the automotive sector such as smart adhesives, sensors for safety and pollutants detections, nanofluids, and functional textile [54].

Besides the impact strength, other improvements on mechanical properties as tensile and flexural strength can be enhanced by adding a small amount of graphene into a

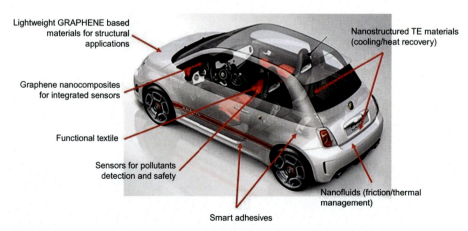

Fig. 5 Potential of graphene in automotive [54]. *(Reproduced with the permission of Elsevier.)*

polymer or fiber-reinforced composite system. For instance, Karataset al. fabricated CF/PA66 composites by melt compounding commercial graphene nanoplatelets (GNP) together, which yielded a 57% reduction in adhesive wear and 11% improved tensile strength with the addition of 0.5 wt% GNP [55]. Furthermore, Pan et al. explored the effect of graphene nanosheets (GNS) produced by Hummers' method and hydrazine reduction procedure on the fire retardance and mechanical properties of GF/PA6 with aluminum hypophosphite additive [56]. Obtained co-reinforced composite provided flame-retardant capabilities and improved bending strength by 44%; however, the tensile strength is decreased 38% by the addition of 1 wt% GNS. Moreover, Cho et al. explored another approach by functionalizing commercial GO with acyl chloride (AGO) and utilizing them together with carbon nanotube (CNT) during in situ polymerizations of PA66 to fabricate a coating over carbon fibers (CF), resulting in the improvement of interfacial shear strength and tensile strength by 160% and 136%, respectively, at 1 mg AGO and 0.5 mg CNT [57].

There are numerous studies, prototypes, and products where graphene is used to improve the mechanical performance of composite parts while the weight is decreasing. The bumper is one of the important car parts where strength is being crucial. At present, CF and GF-reinforced polymer composites are commonly utilized for automotive bumpers because of their high strengths [58]. However, high production cost, weight, and environmental concerns push researchers and companies to find alternative materials like graphene. Scientists at Sunderland University in the UK have been adding graphene to carbon-reinforced composite (bumper prototype). Results revealed that the prototype has capable of absorbing 40% more energy [59]. This work could lead to the creation of significantly lighter and safer cars. In addition, Elmarakbi et al. demonstrated the addition of 30 μm GNPs into the carbon-reinforced epoxy-based composite increased the absorbed energy for laminate with 16 layers, while small size (5 μm) GNPs affected the impact behavior adversely due to the agglomeration problem [60]. A synergistic effect of graphene with hybrid composites with GF and CF yields significant improvements in mechanical and tribological properties. In other work, Okan et al. prepared a hybrid compound containing graphene from recycled waste tire, glass fiber, and PA66 and produced automotive parts with weight reduction (10%). These graphene incorporated composites are recommended for different automotive applications as engine components, structural parts, and underhood components [53].

Polyurethane foams (PUF) are widely used in the manufacturing of transportation vehicles bringing benefits in terms of comfort, flexibility, protection, lightweightness, and free design option [61]. Typical applications of PUFs for the structural component including cross beams, pillars, roof joints, rocker rails, door frames, and bumper systems are given in Fig. 6 [62]. Seats, roofs, carpets, steering wheels, dashboards, seats, airbag covers, and headliners are other applications of PU using the car. A typical medium-sized car of 1000 kg total weight contains 100 kg of plastics, of which about 15 kg are PU [63].

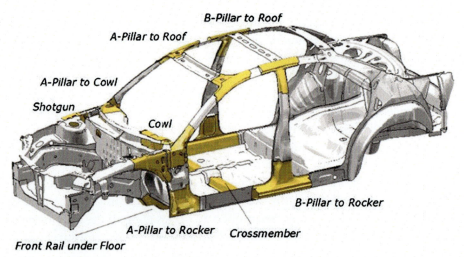

Fig. 6 Typical polyurethane foam applications for the structural automotive components [62]. *(Reproduced with the permission of Taylor and Francis Online.)*

However, PUs are suffering from low flame resistance and durability. By improving PU properties, the safety and cost-effectiveness of vehicles can be improved. PU has become a crucial and fundamental part of automotive design. Various halogenated free fillers, minerals, or chemicals have been added to PUF for better performance. Graphene and its derivatives are one of the solutions for flame retardancy, while mechanical properties are increased [64, 65]. For instance, Ye et al. encapsulated the fine expandable graphite in PMMA (pEG-PMMA) by emulsion polymerization to create compatibility. pEG-PMMA was added into rigid PUF to overcome for enhanced fire resistance and mechanical properties [66]. The usage of graphene for PUF was integrated into the automotive industry by Ford Motor Co. in 2018. Graphene would be used in foam materials to reduce noise and enhance the performance of the parts such as the fuel rail, pumps, belt-driven pulleys, and chain-driven gears on the front of engines. The tests eventuated with 17% quieter, 20% stronger, and 30% more heat-resistant [67].

4. Coating applications of graphene

Conventionally metallic or nonmetallic paints were used to paint automotive parts. Today, car paints with self-healing and dirt-repellent properties are desired for the automotive industry. Additionally, the amount of metal parts used for production has been dramatically reduced due to the increasing importance of vehicle lightening efforts. Despite this, the use of metal in cars, especially in engine parts, continues. Thus, the corrosion problem is an inevitable result. As in other sectors, car manufacturers continue to allocate large budgets to solve corrosion problems. Although there are various solutions, coating is one of the most common and simple methods used.

Graphene is an attractive 2D coating material because its remarkable physicochemical properties enable numerous applications. Graphene nanosheets can be independently or concertedly applied as the barrier against metal corrosion in the form of thin films, layered structures, or composites [68]. For instance, Al-Saadi et al. used the chemical vapor deposition (CVD) method to coat the nickel-copper alloy (Monel 400) with graphene against corrosion [69]. Results revealed that superior resistance to corrosion was obtained by replacing multilayer graphene coating with a protective oxide layer. The favorable effect of graphene on self-healing coating has been investigated in numerous studies, especially in recent years [70]. Self-healing materials have the ability to restore their structure after damage to increase the lifetime of materials and reduce the total cost of systems in long-term usage [71]. Ye et al. took advantage of a porous framework of graphene nanosheets containing porous polyhedral oligomeric silsesquioxane (POSS) to encapsulate the corrosion inhibitor as benzotriazole (BTA) [72]. The epoxy composite coatings revealed excellent self-healing capability due to the spontaneous release from graphene-based material and acted as a physical barrier of graphene that could increase the corrosion resistance. The addition of graphene can effectively improve its mechanical properties and physical shielding ability and increase its energy conversion efficiency. This ability makes graphene an ideal material for improving the corrosion resistance of self-healing coatings [73]. As mentioned in the introduction part, graphene derivatives obtained from chemical exfoliation of graphite have hydrophilic oxygen-containing functional groups on the sheet's surface. These functional groups can allow the dispersion of graphene in the matrix for anticorrosive coatings. However, the chance for covalent bonding for graphene derivatives is lower than the graphene. Additionally, the agglomeration problem, which is the main struggle to incorporate graphene in different applications, can affect the coating performance.

Graphene technology has also become a crucial component in constructing modern window films that are utilized in car windows too. These films control the amount of light that passes through the vehicle to protect both passengers and interior parts from extreme heat and UV exposure. The addition of graphene allows heat to quickly dissipate to the film's exterior, resulting in the efficient absorption of harmful infrared rays. Among the strength and conductivity properties, single-layer graphene has good optical transparency properties, allowing for improved glass and window films in the automotive industry [74]. STEK Automotive being one of many developers in the film protection market has been producing a premium nanoceramic window film (NEX). This film nanoceramic window film contains graphene alongside tungsten and antimony tin oxide to provide drivers with crystal clarity as well as UV and glare reduction. In addition, the infrared is absorbed; thus, vehicle interiors are protected from the heat [75].

Besides the anticorrosion and window film applications, graphene is integrated into the automotive paint coatings, where ceramic-based coatings are widely preferred. It is a viable addition to paints or coating materials because of its thin and flexible structure. Since ceramic-based coatings containing SiO_2 particles leave water spots after drying, graphene-based coatings can prevent this problem due to the hydrophobicity of graphene providing a high-water contact angle. Additionally, graphene having thermal and electrical conductivity offers antistatic properties. There are various products containing graphene for both the automotive industry and car owners. ArtDeShine, Glassparency, Ethos, SPS Graphene Coating, and IGL Ecoshine Renew F4 are examples of graphene-based coating applications [76].

5. Graphene in tire manufacturing

The automobile tire industry is one of the in which use of nanomaterials for many years. Nanofillers such as carbon black (CB) and nanooxides (silica, alumina) fillers, nanoclay, and carbon nanofibers (CNF) are used to improve the features [77]. Rubbers, which are viscoelastic polymers, have enabled many applications across a broad range of industrial fields by adding fillers, particularly in the tire industry. It is mainly reinforced with carbon black or nanosilica particles. While carbon black is still the most widely used reinforcing filler due to its high tensile strength and abrasion resistance, it provides higher stiffness, tear strength, and abrasion resistance to tire compounds [78, 79]. However, CB has been limited by the lack of simultaneous improvements in rolling resistance and wet traction in the tire industry.

In recent years, researchers have focused on enhancing tire performance to reduce rolling resistance, abrasion resistance, weight, and friction and increase wearability, wet traction, and safety air retention [80]. Moreover, the addition of different kinds of nanoscale materials boosts crosslinking between rubber molecules giving better properties [81]. The incorporation of graphene in tire treads, walls, and the inner linings has been working by companies to make tires lighter, provide better grip and reduce rolling resistance, as shown in Fig. 7. Graphene incorporated tires are already available for bicycles offered by the Italian company Vittoria. However, commercial automotive applications for graphene-reinforced tires are not yet available.

Graphene-enhanced rubber offers key performance improvements such as enhanced thermal conductivity, gas barrier properties, mechanical strength, and wear resistance. While rubber itself is an insulating material, it does not dissipate heat well. However, graphene's 2D structure enables it to form an overlapping network within the rubber to increase the tortuous path for gas diffusion, further reducing the chances of air leaks in the tires.

There are various production methods to prepare the graphene incorporated composites with well dispersed, improved mechanical and thermal properties: melt mixing,

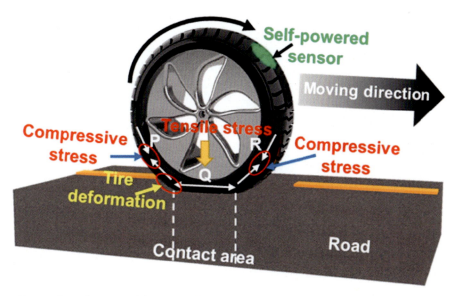

Fig. 7 Usage of graphene in rubber tires [82]. *(Reproduced with the permission of Springer Nature.)*

latex mixing, in situ polymerization, solution mixing, and mechanical mixing [83]. Although in situ polymerization, solution mixing, and latex mixing produce good dispersion and exfoliation of graphene, it is limited by industrial facilities to largely produce graphene/polymer composites. Direct mechanical mixing is widely used in industry due to its industrialized production and saving cost. For instance, Kang et al. produced natural rubber containing graphene, which was obtained from pristine graphite by direct mechanical mixing in order to enhance the dynamic properties. Increasing the content of graphene, decreasing the rolling resistance and the internal heat rise, which are desired for tires [84]. In another example, Wang et al. solved the agglomeration problem of CB and graphene (less than 1 wt%) resulting during twin-roll mixing by combining latex mixing with wet compounding. Incorporating rGO in the CB and natural rubber can also improve the hardness, thermal conductivity, and antioxidative thermal degradation [85]. In the end, it is clearly seen that the tire industry is open to integrate graphene into the tire manufacturing process. With the development of graphene production methods, raw materials can be obtained in a serial, inexpensive, and safe manner. This will lead to the acceleration of the mass production of graphene incorporated prototype parts.

6. Graphene in electronic parts of vehicles

Modern vehicles are being increasingly driven by electronics, sensors, radars, touch screens, and the desire to be lightweight as well as maintaining a certain strength. Graphene is an attractive option for electronic parts because of its enormous electron

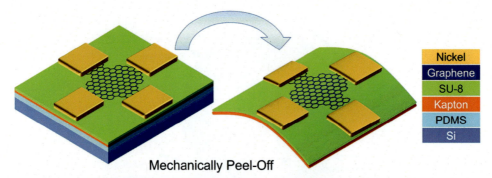

Fig. 8 Schematic view of Hall sensor fixed on and peeling from the rigid substrate [72]. *(Reproduced with the permission of Royal Society of Chemistry.)*

mobility, thermal conductivity, flexibility, and ultimately thin nature [86]. Cheng et al. developed a scalable method to fabricate self-aligned graphene transistors exhibiting a cut-off frequency of 427 GHz for a 67 nm channel length on glass. The results indicated that graphene field-effect transistors could potentially be cost-efficient for novel applications such as automotive radar in high-frequency analog electronics [66].

One of the other significant applications for vehicles is that sensors such as magnetic field sensors are widely utilized to detect position. Precision is required for this kind of application. In one of the studies, Kojimo et al. showed the graphene for Hall sensor detecting the presence and magnitude of a magnetic field using the Hall Effect because of high precision and good usability [87]. In another study, Wang et al. also produced flexible Hall sensors containing graphene by CVD method on copper foil. This sensor is recommended usage in electronic where precise position detection or current monitoring is required, such as road detection, line following, and pedestrian detection applications in vehicles [72]. A schematic representation of the Hall sensors and peeling from the surface is shown in Fig. 8. The results showed that graphene addition increased the sensitivity significantly as well as giving flexibility. Electronic devices such as touch screens, flexible displays, printable electronics, solid-state lighting, and thin-film photovoltaics have led to a rapidly growing market for flexible transparent conductors.

7. Graphene as a lubricating agent in fluids

Improving energy consumption rates and efficiency of the thermal system in the automotive industry is one of the important fields of research due to its impact on eliminating energy waste, lowering greenhouse gas emissions, and its economic aspects. In this regard, increasing the heat transfer rate by passive devices which do not require any additional energy, and active methods that require external power are promising techniques to improve the thermal performance of the systems. Some conventional techniques to intensify the cooling performance, such as utilizing thermal fins or microchannels [88–90], and some show restrictions in the airflow, however, alternative strategies should be proposed to improve the heat transfer rate.

Taking the advantages of nanofluids as a passive way to increase the efficiency of thermal systems has engaged a wide range of scientists and researchers recently after the word "Nanofluid" was coined by Choi and Eastman [91] in 1995. Thermophysical properties of conventional fluids such as water and oil used to dissipate the thermal loads in the thermal absorption systems of the automobile can be improved by the aim of nanofluids. In recent years, some of the research [89, 91–95] has been dedicated to investigating the effects of nanoparticles' size, weight, and distribution in the coolant to increase heat transfer rates.

Some of the drawbacks of using nanoparticles in the mixtures are the rheological problems, pressure drop in the channels, erosion of the pipelines, demand for more pumping, and stability issues of the mixture, especially in nanofluids heavy particles, which tends to settle quickly. The heat transfer characteristics of nanofluids critically depend on the uniform dissipation of the particles in the base fluid and the way of preparation of the mixture. Thermal conductivity, the heat capacity of the fluid, and the size of the heat transfer area can be improved by employing the nanofluid in the systems such as automobile radiators instead of conventional fluids. Graphene, CNTs, GNPs, and GO are some of the carbon-based nanomaterials demonstrating extreme thermal conductivity, and it was the subject of many kinds of research [94–97]. These materials are encounter in a broad spectrum of industrial and medical sciences requests ranging from thermal-related applications in automobiles, heat pipes, heat sinks, refrigerators, lubrications, and sensors to health-related issues and cosmetic industries [93].

The main reason for having high thermal conductivity for carbon-based nanofluids is their small interfacial thermal resistance and high aspect ratio. For example, the thermal conductivity of the single-layer graphene reaches 5000 W/mK, and its surface area is about 2630 m^2/g theoretically [98]. However, only these values are difficult to experimentally validate and realistically are measured in the ranges of 3000–5300 W/mK and 270–1550 m^2/g, respectively [99, 100]. Graphene-based nanofluids have the advantages of longer suspension time and high aspect ratio, lower corrosion, small pressure drop, lower clogging, lower pumping energy consumption, and higher energy saving [94]. Although carbon-based nanoparticles can introduce high thermal conductivity to the working fluid, the high production cost of the graphene materials is the downside should compromise. Intensifying the turbulence in the working GNP-water nanofluids due to the interaction of the particles in the fluid leads to enhancement of the heat transfer rate. An experimental study by Keklikcioglu et al. [101] shows that GNP-water nanofluids are effective than TiO_2-water nanofluids in terms of Nusselt number and thermal conductivity. Especially at high Reynolds numbers, the difference in GNP-water nanofluid's heat transfer compared to water is significant. The wall shear stress, density, and viscosity

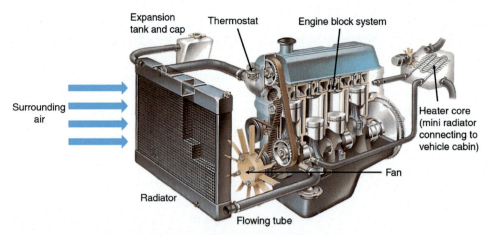

Fig. 9 Schematic setup of the cooling process in the automobile radiator using nanofluid as a coolant [102]. *(Reproduced with the permission of Springer Nature.)*

of the working fluid will increase as the nanoparticles are added to the liquid, which will increase the pressure drop. They could obtain the same value of friction factor for GNP-water nanofluids and TiO_2-water nanofluids for different nanoparticle weight fractions and Reynolds numbers. If the financial aspect is not considered, the use of GNP is preferred. An experimental study was conducted on the effect of using multiwalled carbon nanotube (MWCNT) as nanoparticles dispersed in the water/ethylene glycol (EG) employed in the vehicle engine cooling system. They used "Proton Kelisa 1000cc" as the real testing engine and examined the experiment for 0.1%, 0.25%, 0.50% of MWCNT in the base fluid. The result of the investigation reveals that the heat transfer rate in the automobile radiator is proportionally related to the nanoparticles' volume concentration and the Reynolds number of the passing flow. The authors could reach 196.3% enhancement in the average heat transfer coefficient of the system by using a 0.5%volume concentration of nanoparticles in the water/EG [90]. Fig. 9 demonstrates a setup of a conventional car radiator taking the advantages of nanofluid to enhance the thermal efficiency of the car engine.

The effect of GNPs in enhancing the convection heat transfer coefficient in the automobile radiator is studied by Selvam et al. [103]. The research results show an increase of 104% in the heat transfer coefficient by adding 0.5 vol% of GNP to the base fluid. The study proposes GNP/water-EG nanofluid as an innovative working fluid in automotive industries. In another similar study by Amiri et al., the use of GNP/water-EG nanofluid showed a significant enhancement in the convection heat transfer coefficient without any considerable pressure drop with respect to GNP's concentrations [104]. The production of nanofluids with the combination of EG and GNP provides advantages such as lowering the corrosion rates in the working area due to its usage as functional groups and increase the dispersion rate in the base fluid. The thermal performance of the vehicle radiator was experimentally examined by the Kılınç et al. using the three distinctive working fluids [105]. They employed GNRs, GO and compared the results when the automobile only uses pure water as a coolant. The variable parameters in their experiment were inlet temperatures and inlet flow rate. The study results show a significant enhancement in the overall heat transfer coefficients by utilizing only 0.01% and 0.02% volume concentrations of nanoparticles. In addition, comparing the nanofluid's effects with each other represent that the use of GNRs gives better results than GO. Ball milling is one of the powder grinding techniques used in GNP production to enhance the dispersion and thermal conductivity of nanofluids [106]. The stability of the graphene in the nanofluid increased by separating GO powder in the distilled water using ultrasonication and hydrazine hydrate. The result of the study by Wang et al. shows the optimal design operation condition for the heat pump system uses graphene in the nanofluid [107].

In order to increase the thermal efficiency in the tube, some numerical studies were conducted as well. According to a numerical investigation of heat transfer in a system that uses GO/water-EG nanofluids as the working flow, Sajjad et al. found that the average heat transfer coefficient enhanced in the horizontal tube by 13% when the nanofluid passes the tube in the laminar regime and the weight concentration of the nanoparticles is 0.1% [108]. The engine performance can be improved by the effective methods of removing waste heat from it. Bharadwaj et al. could increase the heat capacity of the coolant fluid by employing hybrid nanofluids in the car's radiator [109]. For this means, the thermal performance of nanofluids consist of carboxyl graphene and GO was studied numerically using ANSYS FLUENT. It was shown that these carbon-based materials could be employed as an appropriate nanoparticle to use in the car industry, which will lead to a decrease in the size of the radiator. Zhang et al. used the nanographite particles in the heavy-duty diesel engine to improve the cooling capability of the vehicle [110]. The results demonstrate a 15% increase in the heat transfer capacity using 3% weight nanographite in the working fluid. The nanofluids, including 0.025, 0.05, 0.075, and 0.1 wt% of GNP in the turbulent flow conditions, were used to analyze the heat transfer rate and pressure drop within a horizontal tube [111]. The maximum obtained heat transfer coefficient was 160% higher than that of the distilled water, proportional to the flow rate and the heat flux. On the other hand, the maximum amount of pressure drop was reported as 14.6%. These findings indicate proficiency in using GNP in thermal systems, which can be alternative options for conventional coolant fluid.

The preparation of the nanofluids can be done using various techniques. The two-step method knows as one of the economical ways of producing nanofluids. In the first stage of this method, the mechanical/chemical processes are applied to prepare dry powder, which is then mixed with based fluid, usually water or EG. This mixing can be accelerated by different techniques such as ultrasonic or magnetic vibrations leads to reduction of agglomeration in nanofluids. Moreover, as the agglomeration degree decreases in the

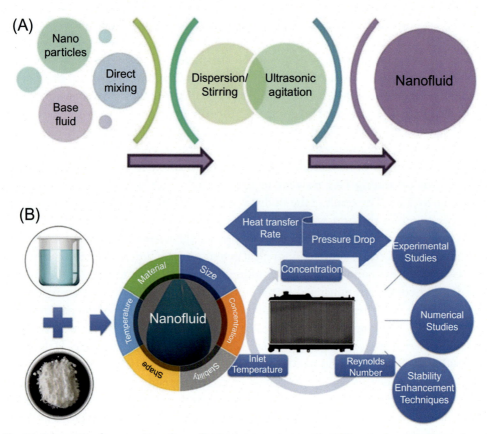

Fig. 10 Schematic of preparation of nanofluid using two-step method (A) and schematic illustration of using nanofluids in the automotive radiator and effective parameters in improving the performance of the cooling system (B) [112, 113]. *(Reproduced with the permission of Elsevier.)*

nanofluid, its thermal conductivity increases. In the next level, the use of surfactants can improve the stability of the nanofluid. On the other hand, the preparation of GO with a robust character to be dissolved and generate stable colloid solutions has been explored by using different methods. One of these methods is the Hummers' method, where sulfuric acid (H_2SO_4) and potassium permanganate ($KMnO_4$) are employed as acid media and oxidant, respectively [94, 112]. A simple schematic of the two-step method and the application of nanofluid in the car cooling system, as well as effective thermophysical properties of nanofluid in the cooling process, is represented in Fig. 10.

In one of the studies, Kimiagar et al. examined the effect of temperature and concentration of GO nanosheets (GONs) on thermal conductivity of reduced GONs (rGONs) using Hummers' method and obtained reach a 17.8% enhancement in the thermal conductivity of rGO nanofluid at the 0.05 wt% loading and 328.15 K test temperature [114].

Another study by Rashidi et al. utilized carbon quantum dots (CQDs) due to their small size and high stability to increase the thermal conductivity and convective heat transfer coefficient by 5.7% and 16.2%, respectively, in vehicular radiator nanofluidic systems [92]. Furthermore, Selvam et al. used a noncovalent method to produce graphene-based suspension nanofluids and utilized the nanofluids in the car radiator to investigate heat transfer rate and pressure drop within the radiator tubes to increase the Reynolds number and convective heat transfer coefficient by the increasing concentration of GNP/water-EG suspension [115]. Also, enhancement in nanofluid inlet temperature and the amount of flow rate could improve the efficiency of the thermal system. In another study, Selvam et al. explored the effects of using GNP as suspensions in water-EG mixture with respect to viscosity and flow regimes and concluded that the thermal conductivity increased linearly while the viscosity remained persistent [116]. In another study by Naveen et al., the thermal performance of an automotive car radiator with water-EG based nanocoolant was experimentally investigated at 0.1, 0.2, and 0.3 vol% of graphene nanoparticles [117]. It was found out that the convective heat transfer coefficient and Nusselt number improved by 66.22% and 53.4%, respectively, when a high concentration of nanofluid with an elevated flow rate was used in the radiator.

Another promising GRM for nanofluidic systems is nitrogen-doped crumpled graphene which offers high surface area, electrical, and thermal conductivity. In one of the recent studies, Amiri et al. synthesized nitrogen-doped crumpled graphene for a coolant in the car engine that operates in wide temperature ranges [118]. The findings indicated that the heat transfer coefficient of the fluid improved up to 83% in the radiator by using these nanosheets in the water-EG nanofluid systems.

The effect of toxicity nanomaterials, especially carbon-based nanomaterials, and their potential danger on the living organism and environment was studied by Esquivel-Gaon et al. [119]. Lubrication oil, which is used for mechanical transmitting, can be improved in terms of friction properties and antiwear capacity with the help of GNS. The friction reduction leads to a decrease in the vehicles' energy usage. Therefore, GNS can reduce vehicles' carbon emissions by improving their lubrication fluid life by reducing friction factors. The creation of a sustainable automotive industry is possible using safe nanomaterials such as nanoadditives based on rGO, which could diminish the transportation effects on the environment.

This section elaborates on some theoretical and empirical investigations on using graphene-based nanofluid as a coolant in industrial cases, especially in the automotive industry section. The nanofluid properties such as thermal conductivity, its density, viscosity, heat transfer coefficient, and their corresponding effect on heat transfer improvement were examined. These kinds of coolant properties are susceptible to their characterizations such as shape, dimension, volumetric concentrations, etc. Therefore, some contradictory results were reported in some cases in the literature due to the great sensitivity of nanofluid to their design and preparation properties. Thus, further experimental and numerical analysis are needed to cover this lack of knowledge in the literature. This section aims to highlight some advantages of using graphene-based nanofluids as coolants employed in heat exchangers in automobiles. The models that create the comprehensive relationship between the nanoscale effect of these materials and their macroscopic rheology and properties would expedite the transition from small-scale applications to large-scale manufacturing.

8. Graphene potential in electric vehicles

As the global trends move towards decreasing CO_2 emissions and fossil fuel usage, electric vehicles (EVs) and hybrid vehicles gain popularity. Currently, there are over 30 EV models on the market, and more models are on the horizon [120]. Lithium-ion batteries (LIBs) are currently the most suitable energy storage device for powering electric vehicles (EVs) owing to their attractive properties, including high energy efficiency, lack of memory effect, long cycle life, high energy density, and high power density [121]. Graphene has also been used in lithium-ion batteries (for both anode and cathode material) to improve the performance rate and stability of LIBs due to its high surface area ratio, stable chemical properties, and satisfactory electrical and thermal conductivity [122]. For instance, Son et al. demonstrated that the coating of silicon oxide (SiO_2) nanoparticles with graphene as anode material improved the cycle life and fast charging capability to provide efficient conductive pathways [123]. Gao et al. wrapped $LiFePO_4$ (LFP) with 2% graphene and graphene nanoribbons (GNRs) into LFP by spray deposition and vacuum filtration methods allowing an effective conductive network [124]. Readers are directed to see Refs. [122, 125] to get detailed knowledge of graphene's effect on LIB electrodes. Another aspect regarding the LIBs is the proper termination of the batteries at the end of their life cycle. As the development of EVs and hybrid vehicles grows, more than 100,000 tons of LFP will be used based on the predictions [126, 127]. The continuous usage of LFPs, combined with the lack of environmental consciousness, can lead to adverse effects on the environment [128, 129]. To overcome these problems, Song et al. fabricated LFP/graphene composites from spent LIBs for EV applications [130]. Life cycle assessment (LCA) analysis is an excellent method to gain insight into the environmental impact and carbon footprint of production techniques and the end-of-life process of the products. LCA analyses were done for EVs, mainly for the batteries such as LiS [131], MoS_2-LIB [132], nanomaterial-enhanced LIBs [133]; however, there are no studies regarding the LCA analysis of graphene in EVs.

One of the other ways to improve battery life and EV range is to decrease the weight of the EVs. As explained in previous sections, there are numerous ways to obtain automotive parts using graphene-based nanocomposites. Recently, researchers have found a way to decrease cable weight which EVs would benefit the most. In their study, Park et al. fabricated graphene-assisted CNT-Cu composite wire with a significantly higher current density of 4.42 A.cm/g compared to other CNT-Cu wires [134]. Moreover, obtained wire provided ~5 higher specific strengths than commercial Cu wire. This increase in the current density and specific strength is obviously more pronounced in EVs due to their immense wiring and can offer a significant range increase with the reduced vehicle weight. Improvement of the energy storage system as LIBs by enhancing performance (i.e., incorporation of graphene) and reducing fabrication times of graphene-enhanced parts should be adapted in the automotive industry, especially for electric vehicles. Although considerable research works on increasing the performance of lithium-ion batteries and graphene-based nanocomposites, fabrication of these parts on an industrial scale remains a challenge.

9. Conclusions and outlook

The discovery of graphene has created large echoes in the scientific community back in 2004. Now during the time in-between, researchers and engineers worldwide have grasped the intricacies of the properties, production, and integration of graphene and GRMs into the materials that we use every day. However, the demands, especially in the automotive industry, are high; therefore, reliable, consistent, and scalable graphene production is required to further augment the implementation of graphene in vehicles. At this point in technology, graphene not only needs to be better but much better and feasible than the current materials in the industry to replace them. As the automotive industry giants such as Ford pave the way to integrate graphene into their production lines, further commercialization and utilization by suppliers will commence. Thus, the future remains bright for graphene in the automotive industry, especially when the lawmakers of the countries around the globe stay determined to reduce CO_2 emissions and help reduce global warming. One of the key routes to reduce greenhouse gas emissions is to drop the use of fossil fuels and go electric or significantly lower the use of fossil fuels. In both cases, with the current technology in mind, it is required to lighten the vehicles while maintaining safety ratings, thus providing strong and light materials. In the recent past, the global automotive lightweighting market offered huge opportunities in market share and is expected to grow even more by 2024. This presents a great chance for automotive and parts manufacturers to jump in, considering the incredible effect of graphene part lightweighting. Therefore, the integration of graphene and GRMs into the automotive industry to keep up with lawmakers' agendas will play a critical role. The market is still somewhat inexperienced in graphene and integration of graphene into the products; however, with appetizing grants from the European Union's Graphene Flagship and Framework Programme's aimed at development and application areas, surely graphene can find a position in the automotive industry in the near future.

List of symbols and abbreviations

2D	two dimensional
AGO	functionalized graphene oxide with acyl chloride
BTA	benzotriazole
CB	carbon black
CF	carbon fiber
CFRP	carbon fiber-reinforced plastic
CNF	carbon nanofiber
CNT	carbon nanotube
CO_2	carbon dioxide
CQDs	carbon quantum dots
CVD	chemical vapor deposition
EC	European commission
EG	ethylene glycol
EU	European union
EV	electric vehicle
f-Gr	functionalized graphene
FLG	few-layer graphene
GF	glass fiber
GFRP	glass fiber-reinforced plastic
GHz	gigahertz
GNPs	graphene nanoplatelets
GNR	graphene nanoribbon
GNS	graphene nanosheets
GO	graphene oxide
GONs	graphene oxide nanosheets
GOS	modified go with nanosilica
GPa	gigapascal
GRM	graphene-related material
H_2SO_4	sulfuric acid
$KMnO_4$	potassium permanganate
LCA	life cycle assessment
LFP	lithium iron phosphate
LIBs	lithium-ion batteries
Monel 400	nickel-copper alloy
MoS_2	molybdenum disulfide
MWCNT	multiwalled carbon nanotube
NF	natural fiber
PA	polyamide
PA6	polyamide 6
PA66	polyamide 66
PET	polyethylene terephthalate
PMC	polymer matrix composite
PMMA	polymethyl methacrylate
POSS	porous polyhedral oligomeric silsesquioxane
PP	polypropylene
PU	polyurethane
PUF	polyurethane foam

PVA	polyvinyl alcohol
rGO	reduced graphene oxide
rGON	reduced graphene oxide nanosheet
SiO$_2$	silicon oxide
TAGS	tannic acid-modified graphene sheets
TPa	terapascal
UV	ultraviolet

References

[1] A.R. Abu Talib, A. Ali, M.A. Badie, N. Azida Che Lah, A.F. Golestaneh, Developing a hybrid, carbon/glass fiber-reinforced, epoxy composite automotive drive shaft, Mater. Des. 31 (1) (2010) 514–521, https://doi.org/10.1016/j.matdes.2009.06.015.

[2] Y.J. Jeon, J.M. Yun, D.Y. Kim, S.I. Na, S.S. Kim, Moderately reduced graphene oxide as hole transport layer in polymer solar cells via thermal assisted spray process, Appl. Surf. Sci. 296 (2014) 140–146, https://doi.org/10.1016/j.apsusc.2014.01.061.

[3] D. Hull, T.W. Clyne, An Introduction to Composite Materials, Cambridge University Press, 1996, https://doi.org/10.1017/CBO9781139170130.

[4] K. Friedrich, A.A. Almajid, Manufacturing aspects of advanced polymer composites for automotive applications, Appl. Compos. Mater. 20 (2) (2013) 107–128, https://doi.org/10.1007/s10443-012-9258-7.

[5] J. Osborne, Automotive composites-in touch with lighter and more flexible solutions, Met. Finish. 111 (2) (2013) 26–30, https://doi.org/10.1016/S0026-0576(13)70159-4.

[6] P. Anandakumar, M.V. Timmaraju, R. Velmurugan, Development of efficient short/continuous fiber thermoplastic composite automobile suspension upper control arm, Mater. Today: Proceed. 39 (2020) 1187–1191, https://doi.org/10.1016/j.matpr.2020.03.543.

[7] T.-W. Chou, Microstructural Design of Fiber Composites, Cambridge University Press, 2005.

[8] I.G. Lee, D.H. Kim, K.H. Jung, H.J. Kim, H.S. Kim, Effect of the cooling rate on the mechanical properties of glass fiber reinforced thermoplastic composites, Compos. Struct. 177 (2017) 28–37, https://doi.org/10.1016/j.compstruct.2017.06.007.

[9] N. Ramli, N. Mazlan, Y. Ando, Z. Leman, K. Abdan, A.A. Aziz, N.A. Sairy, Natural fiber for green technology in automotive industry: a brief review, in: Iop Conference Series: Materials Science and Engineering, vol. 368, IOP Publishing, 2018, p. 12012.

[10] Market Publishers, Automotive Composite Market Size, Share & Trends Analysis Report by Product (Polymer Matrix, Metal Matrix, Ceramic Matrix), By Application (Exterior, Structural & Powertrain), And Segment Forecasts, 2020.

[11] K.S. Novoselov, A.K. Geim, S.V. Morozov, D. Jiang, Y. Zhang, S.V. Dubonos, I.V. Grigorieva, A.A. Firsov, Electric field in atomically thin carbon films, Science 306 (5696) (2004) 666–669, https://doi.org/10.1126/science.1102896.

[12] R. Raccichini, A. Varzi, S. Passerini, B. Scrosati, The role of graphene for electrochemical energy storage, Nat. Mater. 14 (3) (2015) 271–279, https://doi.org/10.1038/nmat4170.

[13] A.K. Geim, K.S. Novoselov, The rise of graphene, Nat. Mater. 6 (3) (2007) 183–191, https://doi.org/10.1038/nmat1849.

[14] C. Lee, X. Wei, J.W. Kysar, J. Hone, Measurement of the elastic properties and intrinsic strength of monolayer graphene, Science 321 (5887) (2008) 385–388, https://doi.org/10.1126/science.1157996.

[15] S. Bae, H. Kim, Y. Lee, X. Xu, J.S. Park, Y. Zheng, J. Balakrishnan, T. Lei, H. Ri Kim, Y.I. Song, Y.J. Kim, K.S. Kim, B. Özyilmaz, J.H. Ahn, B.H. Hong, S. Iijima, Roll-to-roll production of 30-inch graphene films for transparent electrodes, Nat. Nanotechnol. 5 (8) (2010) 574–578, https://doi.org/10.1038/nnano.2010.132.

[16] D.R. Dreyer, S. Park, W. Bielawski, R.S. Ruoff, The Chemistry of Graphene Oxide, 2010, https://doi.org/10.1039/b917103g.

[17] Y. Zhang, L. Zhang, C. Zhou, Review of chemical vapor deposition of graphene and related applications, Acc. Chem. Res. 46 (10) (2013) 2329–2339, https://doi.org/10.1021/ar300203n.

[18] N. Liu, F. Luo, H. Wu, Y. Liu, C. Zhang, J. Chen, One-step ionic-liquid-assisted electrochemical synthesis of ionic-liquid-functionalized graphene sheets directly from graphite, Adv. Funct. Mater. 18 (10) (2008) 1518–1525, https://doi.org/10.1002/adfm.200700797.

[19] K. Swift, Plastics and Polymer Composites in Light Vehicles, 2020 (American Chemistry Council).

[20] J. Walkowiak, W. Papacz, A. Czulak, Manufacturing technologies of composite parts and subassemblies of automotive vehicles, IOP Conf. Ser.: Mater. Sci. Eng. 421 (3) (2018), https://doi.org/10.1088/1757-899X/421/3/032029.

[21] Joint Research Centre - - Institute for Environment and Sustainability, ILCD Handbook—General Guide on LCA - Provisons and Action Steps, 2010, https://doi.org/10.2788/94987.

[22] M. Clairotte, R. Suarez-Bertoa, A.A. Zardini, B. Giechaskiel, J. Pavlovic, V. Valverde, B. Ciuffo, C. Astorga, Exhaust emission factors of greenhouse gases (GHGs) from European road vehicles, Environ. Sci. Eur. 32 (1) (2020) 1–20.

[23] K. Dericiler, H.M. Sadeghi, Y.E. Yagci, H.S. Sas, B.S. Okan, Experimental and numerical investigation of flow and alignment behavior of waste tire-derived graphene nanoplatelets in PA66 matrix during melt-mixing and injection, Polymers 13 (6) (2021) 949, https://doi.org/10.3390/polym13060949.

[24] H. Fukushima, L.T. Drzal, Nylon - exfoliated graphite nanoplatelet (xGnP) nanocomposites with enhanced mechanical, electrical and thermal properties, in: 2006 NSTI Nanotechnology Conference And Trade Show—NSTI Nanotech 2006 Technical Proceedings, vol. 1, 2006, pp. 282–285. November 2014.

[25] N.A. Ismail, N.W.M. Zulkifli, Z.Z. Chowdhury, M.R. Johan, Functionalization of graphene-based materials: effective approach for enhancement of tribological performance as lubricant additives, Diam. Relat. Mater. 115 (January) (2021) 108357, https://doi.org/10.1016/j.diamond.2021.108357.

[26] Q.H. Wang, Z. Jin, K.K. Kim, A.J. Hilmer, G.L.C. Paulus, C.J. Shih, M.H. Ham, J.D. Sanchez-Yamagishi, K. Watanabe, T. Taniguchi, J. Kong, P. Jarillo-Herrero, M.S. Strano, Understanding and controlling the substrate effect on graphene electron-transfer chemistry via reactivity imprint lithography, Nat. Chem. 4 (9) (2012) 724–732, https://doi.org/10.1038/nchem.1421.

[27] X. Li, A. Reina, X. Jia, J. Ho, D. Nezich, H. Son, V. Bulovic, M.S. Dresselhaus, J. Kong, W. Cai, J. An, S. Kim, J. Nah, D. Yang, R. Piner, A. Velamakanni, I. Jung, E. Tutuc, S.K. Banerjee, L. Colombo, R.S. Ruoff, Large-area synthesis of high-quality and uniform graphene films on copper foils, Science 324 (5932) (2009) 1312–1314.

[28] S. Vadahanambi, J.H. Jung, I.K. Oh, Microwave syntheses of graphene and graphene decorated with metal nanoparticles, Carbon 49 (13) (2011) 4449–4457, https://doi.org/10.1016/j.carbon.2011.06.038.

[29] M. Lotya, P.J. King, U. Khan, S. De, J.N. Coleman, High-concentration, surfactant-stabilized graphene dispersions. pdf, ACS Nano 4 (6) (2010) 3155–3162.

[30] P. King, A. Crossley, C. Backes, S. Barwich, P. May, H. Pettersson, K.R. Paton, O.M. Istrate, G.S. Duesberg, A. O'Neill, E. Long, T.J. Pennycook, R.J. Smith, V. Nicolosi, N. McEvoy, T. Higgins, J. Coelho, B.M. Sanchez, U. Khan, E. Varrla, C. Boland, M. Moebius, I. Ahmed, C. Downing, M. Lotya, J.N. Coleman, P. Puczkarski, S.E. O'Brien, E.K. McGuire, Scalable production of large quantities of defect-free few-layer graphene by shear exfoliation in liquids, Nat. Mater. 13 (6) (2014) 624–630, https://doi.org/10.1038/nmat3944.

[31] H.R. Kim, S.H. Lee, K.H. Lee, Scalable production of large single-layered graphenes by microwave exfoliation 'in deionized water', Carbon 134 (2018) 431–438, https://doi.org/10.1016/j.carbon.2018.04.014.

[32] U. Khan, H. Porwal, A. O'Neill, K. Nawaz, P. May, J.N. Coleman, Solvent-exfoliated graphene at extremely high concentration, Langmuir 27 (15) (2011) 9077–9082, https://doi.org/10.1021/la201797h.

[33] M. Choucair, P. Thordarson, J.A. Stride, Gram-scale production of graphene based on solvothermal synthesis and sonication, Nat. Nanotechnol. 4 (1) (2009) 30–33, https://doi.org/10.1038/nnano.2008.365.

[34] Y. Arao, M. Kubouchi, High-rate production of few-layer graphene by high-power probe sonication, Carbon 95 (2015) 802–808, https://doi.org/10.1016/j.carbon.2015.08.108.

[35] B.C. Brodie, On the atomic weight o f graphite, Philos. Trans. A Math .Phys. Eng. Sci. (1859).

[36] L. Staudenmaier, Verfahren zur Darstellung der Graphitsäure, Ber. Dtsch. Chem. Ges. 31 (2) (1898) 1481–1487, https://doi.org/10.1002/cber.18980310237.

[37] H.L. Poh, F. Šaněk, A. Ambrosi, G. Zhao, Z. Sofer, M. Pumera, Graphenes prepared by Staudenmaier, Hofmann and Hummers methods with consequent thermal exfoliation exhibit very different electrochemical properties, Nanoscale 4 (11) (2012) 3515–3522, https://doi.org/10.1039/c2nr30490b.

[38] W.S. Hummers, R.E. Offeman, Preparation of graphitic oxide, J. Am. Chem. Soc. 80 (6) (1958) 1339, https://doi.org/10.1021/ja01539a017.

[39] D.C. Marcano, D.V. Kosynkin, J.M. Berlin, A. Sinitskii, Z. Sun, A. Slesarev, L.B. Alemany, W. Lu, J.-M. Tour, Improved synthesis of graphene oxide, ACS Nano 4 (8) (2010) 4806–4814, https://doi.org/10.1021/nn1006368.

[40] D.C. Marcano, D.V. Kosynkin, J.M. Berlin, A. Sinitskii, Z. Sun, A.S. Slesarev, L.B. Alemany, W. Lu, J.M. Tour, Correction to: improved synthesis of graphene oxide, ACS Nano (2018), https://doi.org/10.1021/acsnano.8b00128. (2010) 4: 8 (4806–4814) DOI:https://doi.org/10.1021/nn1006368), *ACS Nano*, Vol. 12, No. 2, 2078.

[41] R. Peleg, Ford to release Graphene-Enhanced Cars by the End of 2018, 2018, from https://www.graphene-info.com/ford-release-graphene-enhanced-cars-end-2018.

[42] X. Li, L. Shao, N. Song, L. Shi, P. Ding, Enhanced thermal-conductive and anti-dripping properties of polyamide composites by 3D graphene structures at low filler content, Compos. A: Appl. Sci. Manuf. 88 (2016) 305–314, https://doi.org/10.1016/j.compositesa.2016.06.007.

[43] X. Duan, B. Yu, T. Yang, Y. Wu, H. Yu, T. Huang, In situ polymerization of nylon 66/reduced graphene oxide nanocomposites, J. Nanomater. 2018 (2018), https://doi.org/10.1155/2018/1047985.

[44] R. Scaffaro, A. Maio, A green method to prepare nanosilica modified graphene oxide to inhibit nanoparticles re-aggregation during melt processing, Chem. Eng. J. 308 (2017) 1034–1047, https://doi.org/10.1016/j.cej.2016.09.131.

[45] Y.Z. Wang, X.F. Gao, H.H. Liu, J. Zhang, X.X. Zhang, Green fabrication of functionalized graphene via one-step method and its reinforcement for polyamide 66 fibers, Mater. Chem. Phys. 240 (October 2019) (2020) 122288, https://doi.org/10.1016/j.matchemphys.2019.122288.

[46] A. Zoller, P. Escalé, P. Gérard, Pultrusion of bendable continuous fibers reinforced composites with reactive acrylic thermoplastic ELIUM® resin, Front. Mater. 6 (December) (2019) 1–9, https://doi.org/10.3389/fmats.2019.00290.

[47] A. Ayadi, M. Deléglise-Lagardère, C.H. Park, P. Krawczak, Analysis of impregnation mechanism of weft-knitted commingled yarn composites by staged consolidation and laboratory X-ray computed tomography, Front. Mater. 6 (October) (2019) 1–18, https://doi.org/10.3389/fmats.2019.00255.

[48] P. Krawczak, A. Maffezzoli, Editorial: advanced thermoplastic composites and manufacturing processes, 7 (July) (2020) 1–2, https://doi.org/10.3389/fmats.2020.00166.

[49] A. Elmarakbi, W. Azoti, State of the art on graphene Lightweighting nanocomposites for automotive applications, in: Experimental Characterization, Predictive Mechanical and Thermal Modeling of Nanostructures and Their Polymer Composites, Elsevier Inc., 2018, https://doi.org/10.1016/B978-0-323-48061-1.00001-4.

[50] L. Nickels, Improving Composites with 'Wonder Material', 2016.

[51] S. Kumar, R.S. Bharj, Emerging composite material use in current electric vehicle: a review, Mater. Today: Proceed. 5 (14) (2018) 27946–27954, https://doi.org/10.1016/j.matpr.2018.10.034.

[52] B. Ravishankar, S.K. Nayak, M.A. Kader, Hybrid composites for automotive applications- -a review, J. Reinf. Plast. Compos. 38 (18) (2019) 835–845.

[53] B.S. Okan, Y. Menceloğlu, B.G. Ozunlu, Y.E. Yagci, Graphene from waste tire by recycling technique for cost-effective and light-weight automotive plastic part production, AIP Conf. Proc. 2205 (January) (2020), https://doi.org/10.1063/1.5142961.

[54] A. Elmarakbi, W. Azoti, Mechanical prediction of graphene-based polymer nanocomposites for energy-efficient and safe vehicles, in: Experimental Characterization, Predictive Mechanical and Thermal Modeling of Nanostructures and their Polymer Composites, Elsevier, 2018, pp. 159–177, https://doi.org/10.1016/B978-0-323-48061-1.00004-X.

[55] E. Karatas, O. Gul, N.G. Karsli, T. Yilmaz, Synergetic effect of graphene nanoplatelet, carbon fiber and coupling agent addition on the tribological, mechanical and thermal properties of polyamide 6,6 composites, Compos. B. Eng. 163 (July 2018) (2019) 730–739, https://doi.org/10.1016/j.compositesb.2019.01.014.

[56] Y. Pan, N. Hong, J. Zhan, B. Wang, L. Song, Y. Hu, Effect of graphene on the fire and mechanical performances of glass Fiber-reinforced polyamide 6 composites containing aluminum hypophosphite, Polym. Plast. Technol. Eng. 53 (14) (2014) 1467–1475, https://doi.org/10.1080/03602559.2014.909483.

[57] B.G. Cho, J.E. Lee, S.H. Hwang, J.H. Han, H.G. Chae, Y. Park, Bin, Enhancement in mechanical properties of polyamide 66-carbon fiber composites containing graphene oxide-carbon nanotube hybrid nanofillers synthesized through in situ interfacial polymerization, Compos. Part A Appl. Sci. Manuf. 135 (May) (2020) 105938, https://doi.org/10.1016/j.compositesa.2020.105938.

[58] A. John, S. Alex, A review on the composite materials used for automotive bumper in passenger vehicles, Int. j. eng. Manag. Res. (IJEMR) 4 (4) (2014) 98–101.

[59] J. Crosse, Graphene: The Breakthrough Material that Could Transform Cars, 2018, from https://www.autocarpro.in/news-international/graphene-breakthrough-material-transform-cars-29453.

[60] A. Elmarakbi, R. Ciardiello, A. Tridello, F. Innocente, B. Martorana, F. Bertocchi, F. Cristiano, M. Elmarakbi, G. Belingardi, Effect of graphene nanoplatelets on the impact response of a carbon fibre reinforced composite, Mater. Today Commun. 25 (2020) 101530.

[61] N.V. Gama, A. Ferreira, A. Barros-Timmons, Polyurethane foams: past, present, and future, Materials 11 (10) (2018) 1841.

[62] P. Zhou, E. Beeh, M. Kriescher, H.E. Friedrich, G. Kopp, Dynamic bending behaviour of magnesium alloy rectangular thin-wall beams filled with polyurethane foam, Int. J. Crashworthiness 21 (6) (2016) 597–613.

[63] Isopa. Polyurethane Transportation Aplications. Polyurethanes in Passenger Cars, The Benefits of Polyurethanes in Transportation..

[64] L. Shi, Z.-M. Li, M.-B. Yang, B. Yin, Q.-M. Zhou, C.-R. Tian, J.-H. Wang, Expandable graphite for halogen-free flame-retardant of high-density rigid polyurethane foams, Polym.-Plast. Technol. Eng. 44 (7) (2005) 1323–1337.

[65] X.-C. Bian, J.-H. Tang, Z.-M. Li, Z.-Y. Lu, A. Lu, Dependence of flame-retardant properties on density of expandable graphite filled rigid polyurethane foam, J. Appl. Polym. Sci. 104 (5) (2007) 3347–3355.

[66] R. Cheng, J. Bai, L. Liao, H. Zhou, Y. Chen, L. Liu, Y.-C. Lin, S. Jiang, Y. Huang, X. Duan, High-frequency self-aligned graphene transistors with transferred gate stacks, Proc. Natl. Acad. Sci. 109 (29) (2012) 11588–11592.

[67] Ford, Cell Phones, Sporting Goods, And Soon, Cars: Ford Innovates with "Miracle" Material, Powerful Graphene for Vehicle Parts, 2018, from https://media.ford.com/content/fordmedia/fna/us/en/news/2018/10/09/ford-innovates-with-miracle-material-powerful-graphene-for-vehicle-parts.html.

[68] R. Zhang, X. Yu, Q. Yang, G. Cui, Z. Li, The role of graphene in anti-corrosion coatings: a review, Constr. Build. Mater. 294 (2021) 123613.

[69] S. Al-Saadi, R.K.S. Raman, M.R. Anisur, S. Ahmed, J. Crosswell, M. Alnuwaiser, C. Panter, Graphene coating on a nickel-copper alloy (Monel 400) for microbial corrosion resistance: electrochemical and surface characterizations, Corros. Sci. 182 (2021) 109299.

[70] G. Cui, C. Zhang, A. Wang, X. Zhou, X. Xing, J. Liu, Z. Li, Q. Chen, Q. Lu, Research Progress on self-healing polymer/graphene anticorrosion coatings, Prog. Org. Coat. 155 (2021) 106231.

[71] Y. Du, D. Li, L. Liu, G. Gai, Recent achievements of self-healing graphene/polymer composites, Polymers 10 (2) (2018) 114.

[72] Z. Wang, M. Shaygan, M. Otto, D. Schall, D. Neumaier, Flexible hall sensors based on graphene, Nanoscale 8 (14) (2016) 7683–7687.

[73] S. Daradmare, S. Raj, A.R. Bhattacharyya, S. Parida, Factors affecting barrier performance of composite anti-corrosion coatings prepared by using electrochemically exfoliated few-layer graphene as filler, Compos. Part B 155 (2018) 1–10.

[74] B. Wang, B.V. Cunning, S.Y. Park, M. Huang, J.Y. Kim, R.S. Ruoff, Graphene coatings as barrier layers to prevent the water-induced corrosion of silicate glass, ACS Nano 10 (11) (2016) 9794–9800, https://doi.org/10.1021/acsnano.6b04363.

[75] T. Barkan, Stek Nex: Window Films with Game-Changing Graphene Cooling Technology, 2020, The Graphene Council, from https://www.thegraphenecouncil.org/blogpost/1501180/363316/STEK-NEX-window-films-with-game-changing-graphene-cooling-technology.

[76] 7 Best Graphene Coating for Cars 2020–2021, (Review), Graphene Uses (2021), from https://www.grapheneuses.org/best-graphene-coating-for-cars/..

[77] M.S. Jayalakshmy, R.K. Mishra, Applications of carbon-based nanofiller-incorporated rubber composites in the fields of tire engineering, flexible electronics and EMI shielding, in: Carbon-Based Nanofiller and their Rubber Nanocomposites, Elsevier, 2019, pp. 441–472.

[78] M.E. Spahr, R. Rothon, Carbon black as 14, in: Fillers for Polymer Applications, 2017, p. 261.

[79] R. Gómez-Hernández, Y. Panecatl-Bernal, M.Á. Méndez-Rojas, High yield and simple one-step production of carbon black nanoparticles from waste tires, Heliyon 5 (7) (2019), e02139.

[80] A.S. Malani, A.D. Chaudhari, R.U. Sambhe, A review on applications of nanotechnology in automotive industry, Int. J. Mech. Mechatron. Eng. 10 (1) (2015) 36–40.

[81] X. Zhang, L. Cai, A. He, H. Ma, Y. Li, Y. Hu, X. Zhang, L. Liu, Facile strategies for green tire tread with enhanced filler-matrix interfacial interactions and dynamic mechanical properties, Compos. Sci. Technol. 203 (2021) 108601.

[82] D. Maurya, S. Khaleghian, R. Sriramdas, P. Kumar, R.A. Kishore, M.G. Kang, V. Kumar, H.C. Song, S.Y. Lee, Y. Yan, J.M. Park, S. Taheri, S. Priya, 3D printed graphene-based self-powered strain sensors for smart tires in autonomous vehicles, Nat. Commun. 11 (1) (2020) 1–10, https://doi.org/10.1038/s41467-020-19088-y.

[83] X. Liu, L.-Y. Wang, L.-F. Zhao, H.-F. He, X.-Y. Shao, G.-B. Fang, Z.-G. Wan, R.-C. Zeng, Research progress of graphene-based rubber nanocomposites, Polym. Compos. 39 (4) (2018) 1006–1022.

[84] H. Kang, Y. Tang, L. Yao, F. Yang, Q. Fang, D. Hui, Fabrication of graphene/natural rubber nanocomposites with high dynamic properties through convenient mechanical mixing, Compos. Part B 112 (2017) 1–7.

[85] J. Wang, K. Zhang, Z. Cheng, M. Lavorgna, H. Xia, Graphene/carbon black/natural rubber composites prepared by a wet compounding and latex mixing process, Plast. Rubber. Compos. 47 (9) (2018) 398–412.

[86] V.B Mohan, K. Lau, D. Hui, D. Bhattacharyya, Graphene-based materials and their composites: a review on production, applications and product limitations, Compos. Part B Eng. 142 (2018) 200–220, https://doi.org/10.1016/j.compositesb.2018.01.013.

[87] E. Kojima, K. Kano, H. Wado, N. Iwamori, Magnetic Field Sensor of Graphene for Automotive Applications, 2017.

[88] D.P. Kulkarni, R.S. Vajjha, D.K. Das, D. Oliva, Application of aluminum oxide nanofluids in diesel electric generator as jacket water coolant, *applied thermal engineering*, Vol. 28, Nos. 14–15 (2008) 1774–1781.

[89] I. Mudawar, Two-phase microchannel heat sinks: theory, applications, and limitations, J. Electron. Packag. 133 (4) (2011).

[90] B. M'hamed, N.A.C. Sidik, M.F.A. Akhbar, R. Mamat, G. Najafi, Experimental study on thermal performance of MWCNT nanocoolant in Perodua Kelisa 1000cc radiator system, Int. Commun. Heat Mass Transf. 76 (2016) 156–161.

[91] S.U.S. Choi, J.A. Eastman, Enhancing Thermal Conductivity of Fluids with Nanoparticles, Argonne National lab, IL (United States), 1995.

[92] A. Rashidi, B. Ghobadian, G. Najafi, M.H. Khoshtaghaza, N.A.C. Sidik, A. Yadegari, H.W. Xian, Experimental investigation of conduction and convection heat transfer properties of a novel nanofluid based on carbon quantum dots, Int. Commun. Heat Mass Transf. 90 (2018) 85–92.

[93] A. Arshad, M. Jabbal, Y. Yan, D. Reay, A review on graphene based nanofluids: preparation, characterization and applications, J. Mol. Liq. 279 (2019) 444–484, https://doi.org/10.1016/j.molliq.2019.01.153.

[94] E. Sadeghinezhad, M. Mehrali, R. Saidur, M. Mehrali, S.T. Latibari, A.R. Akhiani, H.S.C. Metselaar, A comprehensive review on graphene nanofluids: recent research, development and applications, Energy Convers. Manag. 111 (2016) 466–487.

[95] W. Yu, H. Xie, X. Wang, X. Wang, Significant thermal conductivity enhancement for nanofluids containing graphene nanosheets, Phys. Lett. A 375 (10) (2011) 1323–1328.

[96] T.T. Baby, S. Ramaprabhu, Enhanced convective heat transfer using graphene dispersed nanofluids, Nanoscale Res. Lett. 6 (1) (2011) 1–9.

[97] Y. Gao, H. Wang, A.P. Sasmito, A.S. Mujumdar, Measurement and modeling of thermal conductivity of graphene nanoplatelet water and ethylene glycol base nanofluids, Int. J. Heat Mass Transf. 123 (2018) 97–109.

[98] A.A. Balandin, S. Ghosh, W. Bao, I. Calizo, D. Teweldebrhan, F. Miao, C.N. Lau, Superior thermal conductivity of single-layer graphene, Nano Lett. 8 (3) (2008) 902–907.

[99] S. Ghosh, D.L. Nika, E.P. Pokatilov, A.A. Balandin, Heat conduction in graphene: experimental study and theoretical interpretation, New J. Phys. 11 (9) (2009) 95012.

[100] U.K. Sur, Graphene: a rising star on the horizon of materials science, Int. J. Electrochem. Sci. 2012 (2012).

[101] O. Keklikcioglu, T. Dagdevir, V. Ozceyhan, Heat transfer and pressure drop investigation of graphene nanoplatelet-water and titanium dioxide-water nanofluids in a horizontal tube, Appl. Therm. Eng. 162 (July) (2019) 114256, https://doi.org/10.1016/j.applthermaleng.2019.114256.

[102] H.W. Xian, N.A.C. Sidik, G. Najafi, Recent state of nanofluid in automobile cooling systems, J. Therm. Anal. Calorim. 135 (2) (2019) 981–1008.

[103] C. Selvam, R.S. Raja, D.M. Lal, S. Harish, Overall heat transfer coefficient improvement of an automobile radiator with graphene based suspensions, Int. J. Heat Mass Transf. 115 (2017) 580–588.

[104] A. Amiri, R. Sadri, M. Shanbedi, G. Ahmadi, S.N. Kazi, B.T. Chew, M.N.M. Zubir, Synthesis of ethylene glycol-treated graphene nanoplatelets with one-pot, microwave-assisted functionalization for use as a high performance engine coolant, Energy Convers. Manag. 101 (2015) 767–777.

[105] F. Kılınç, E. Buyruk, K. Karabulut, Experimental investigation of cooling performance with graphene based nano-fluids in a vehicle radiator, Heat Mass Transf. 56 (2) (2020) 521–530.

[106] B. Munkhbayar, M. Bat-Erdene, B. Ochirkhuyag, D. Sarangerel, B. Battsengel, H. Chung, H. Jeong, An experimental study of the planetary ball milling effect on dispersibility and thermal conductivity of MWCNTs-based aqueous nanofluids, Mater. Res. Bull. 47 (12) (2012) 4187–4196.

[107] F. Wang, F. Wang, X. Fan, Z. Lian, Experimental study on an inverter heat pump with HFC125 operating near the refrigerant critical point, Appl. Therm. Eng. 39 (2012) 1–7.

[108] M. Sajjad, M.S. Kamran, R. Shaukat, M.I.M. Zeinelabdeen, Numerical investigation of laminar convective heat transfer of graphene oxide/ethylene glycol-water nanofluids in a horizontal tube, Eng. Sci. Technol., Int. J. 21 (4) (2018) 727–735.

[109] B.R. Bharadwaj, K. Sanketh Mogeraya, D.M. Manjunath, B.R. Ponangi, K.S. Rajendra Prasad, V. Krishna, CFD analysis of heat transfer performance of graphene based hybrid nanofluid in radiators, IOP Conf. Ser.: Mater. Sci. Eng. 346 (1) (2018), https://doi.org/10.1088/1757-899X/346/1/012084.

[110] K.J. Zhang, D. Wang, F.J. Hou, W.H. Jiang, F.R. Wang, J. Li, G. LIU, W. ZHANG, Characteristic and experiment study of HDD engine coolants, Chin. Intern. Combust. Engine Eng. 1 (2007) 17.

[111] E. Sadeghinezhad, M. Mehrali, S. Tahan Latibari, M. Mehrali, S.N. Kazi, C.S. Oon, H.S.C. Metselaar, Experimental investigation of convective heat transfer using graphene nanoplatelet based nanofluids under turbulent flow conditions, Ind. Eng. Chem. Res. 53 (31) (2014) 12455–12465.

[112] N. Arora, M. Gupta, An updated review on application of nanofluids in flat tubes radiators for improving cooling performance, Renew. Sust. Energ. Rev. 134 (2020) 110242.

[113] F. Abbas, H.M. Ali, T.R. Shah, H. Babar, M.M. Janjua, U. Sajjad, M. Amer, Nanofluid: potential evaluation in automotive radiator, J. Mol. Liq. 297 (2020) 112014.

[114] S. Kimiagar, N. Rashidi, Thermal conductivity of nanofluids containing microwave hydrothermal reactor reduced graphene oxide nanosheets, High Temp.-High Press. 46 (2017) 35–43.

[115] C. Selvam, D. Mohan Lal, S. Harish, Enhanced heat transfer performance of an automobile radiator with graphene based suspensions, Appl. Therm. Eng. 123 (2017) 50–60, https://doi.org/10.1016/j.applthermaleng.2017.05.076.

[116] C. Selvam, T. Balaji, D.M. Lal, S. Harish, Convective heat transfer coefficient and pressure drop of water-ethylene glycol mixture with graphene nanoplatelets, Exp. Thermal Fluid Sci. 80 (2017) 67–76.

[117] N.S. Naveen, P.S. Kishore, Experimental investigation on heat transfer parameters of an automotive car radiator using graphene/water-ethylene glycol coolant, J. Dispers. Sci. Technol. 0 (0) (2020) 1–13, https://doi.org/10.1080/01932691.2020.1840999.

[118] A. Amiri, M. Shanbedi, B.T. Chew, S.N. Kazi, K.H. Solangi, Toward improved engine performance with crumpled nitrogen-doped graphene based water–ethylene glycol coolant, Chem. Eng. J. 289 (2016) 583–595.

[119] M. Esquivel-Gaon, N.H.A. Nguyen, M.F. Sgroi, D. Pullini, F. Gili, D. Mangherini, A.I. Pruna, P. Rosicka, A. Sevcu, V. Castagnola, In vitro and environmental toxicity of reduced graphene oxide as an additive in automotive lubricants, Nanoscale 10 (14) (2018) 6539–6548, https://doi.org/10.1039/c7nr08597d.

[120] EV Models | The Driven. (n.d.)..

[121] Y. Ding, Z.P. Cano, A. Yu, J. Lu, Z. Chen, Automotive Li-ion batteries: current status and future perspectives, Electrochem. Energy Rev. 2 (1) (2019) 1–28.

[122] X. Chen, Y. Tian, Review of graphene in cathode materials for lithium-ion batteries, Energy & Fuels 35 (5) (2021) 3572–3580.

[123] I.H. Son, J.H. Park, S. Park, K. Park, S. Han, J. Shin, S.-G. Doo, Y. Hwang, H. Chang, J.W. Choi, Graphene balls for lithium rechargeable batteries with fast charging and high volumetric energy densities, Nat. Commun. 8 (1) (2017) 1–11.

[124] L. Gao, Y. Jin, X. Liu, M. Xu, X. Lai, J. Shui, A rationally assembled graphene nanoribbon/graphene framework for high volumetric energy and power density Li-ion batteries, Nanoscale 10 (16) (2018) 7676–7684.

[125] R.-P. Luo, W.-Q. Lyu, K.-C. Wen, W.-D. He, Overview of graphene as anode in lithium-ion batteries, J. Electron. Sci. Technol. 16 (1) (2018) 57–68.

[126] S. Kalluri, M. Yoon, M. Jo, H.K. Liu, S.X. Dou, J. Cho, Z. Guo, Feasibility of cathode surface coating Technology for High-Energy Lithium-ion and Beyond-Lithium-ion Batteries, Adv. Mater. 29 (48) (2017), https://doi.org/10.1002/adma.201605807.

[127] Y. Liu, Z. Tai, T. Zhou, V. Sencadas, J. Zhang, L. Zhang, K. Konstantinov, Z. Guo, H.K. Liu, An all-integrated anode via interlinked chemical bonding between double-shelled–yolk-structured silicon and binder for Lithium-ion batteries, Adv. Mater. 29 (44) (2017) 1–11, https://doi.org/10.1002/adma.201703028.

[128] J. Wang, X. Sun, Olivine LiFePO4: the remaining challenges for future energy storage, Energy Environ. Sci. 8 (4) (2015) 1110–1138, https://doi.org/10.1039/c4ee04016c.

[129] D. Di Lecce, R. Verrelli, J. Hassoun, Lithium-ion batteries for sustainable energy storage: recent advances towards new cell configurations, Green Chem. 19 (15) (2017) 3442–3467, https://doi.org/10.1039/c7gc01328k.

[130] W. Song, J. Liu, L. You, S. Wang, Q. Zhou, Y. Gao, R. Yin, W. Xu, Z. Guo, Re-synthesis of nanostructured LiFePO4/graphene composite derived from spent lithium-ion battery for booming electric vehicle application, J. Power Sources 419 (March) (2019) 192–202, https://doi.org/10.1016/j.jpowsour.2019.02.065.

[131] Y. Deng, J. Li, T. Li, X. Gao, C. Yuan, Life cycle assessment of lithium sulfur battery for electric vehicles, J. Power Sources 343 (2017) 284–295, https://doi.org/10.1016/j.jpowsour.2017.01.036.

[132] Y. Deng, J. Li, T. Li, J. Zhang, F. Yang, C. Yuan, Life cycle assessment of high capacity molybdenum disulfide lithium-ion battery for electric vehicles, Energy 123 (2017) 77–88, https://doi.org/10.1016/j.energy.2017.01.096.

[133] L. Oliveira, M. Messagie, S. Rangaraju, M. Hernandez, J. Sanfelix, J. Van Mierlo, Life cycle assessment of nanotechnology in batteries for electric vehicles, in: Emerging Nanotechnologies in Rechargeable Energy Storage Systems, Elsevier Inc., 2017, https://doi.org/10.1016/B978-0-323-42977-1.00007-8.

[134] M. Park, D.M. Lee, M. Park, S. Park, D.S. Lee, T.W. Kim, S.H. Lee, S.K. Lee, H.S. Jeong, B.H. Hong, S. Bae, Performance enhancement of graphene assisted CNT/cu composites for lightweight electrical cables, Carbon 179 (2021) 53–59, https://doi.org/10.1016/j.carbon.2021.03.055.

CHAPTER 31

Toxicity/risk assessment of nanomaterials when used in the automotive industry

S. Sathish[a], S. Rathish Kumar[b], K.C. Sekhar[a], and B. Chandar Shekar[c]

[a]Department of Physics, School of Basic and Applied Sciences, Central University of Tamil Nadu, Thiruvarur, Tamil Nadu, India
[b]Department of Biotechnology, Sri Ramakrishna College of Arts and Science, Coimbatore, Tamil Nadu, India
[c]Department of Physics, Kongunadu Arts and Science College, Coimbatore, Tamil Nadu, India

Chapter outline

1. Introduction — 653
2. Impact of nanomaterial in the automotive industry — 654
   2.1 Engine — 654
   2.2 Radiator coolant — 655
   2.3 Fuel cells — 657
   2.4 Windows and displays — 657
   2.5 Body parts — 657
   2.6 Coatings/paints — 658
3. Nanotoxicity — 658
4. Role of nanomaterials with their toxicity — 659
   4.1 Platinum (Pt) nanoparticles — 659
   4.2 Silicon dioxide (SiO_2) nanoparticles — 661
   4.3 Titanium dioxide (TiO_2) nanoparticles — 663
   4.4 Copper oxide (CuO) nanoparticles — 666
5. Conclusions — 669
   References — 670

1. Introduction

Nanotechnology is very fascinating field with potential role in the automotive industry. Introducing nanotechnology leads to the betterment of developments in the automobile industry compared to existing technology in the areas of engines, radiators, body coatings, tires, indoor designs, mirrors, and displays. It is playing a vital role toward automotive industry through enhancing performance to avail a flexible with lighter and stronger body parts for less fuel consumption and safety to humans over a long period. In order to make more efficient, less fuel/power consumption, faster, and safe, the automobile industry depends on flexible nanomaterials with the following properties are essential, i.e., stronger, bearable heat/pressure, long time sustainability, lightweight, flexibility with easy handling. These demands can be solved by using nanotechnology via flexible metals,

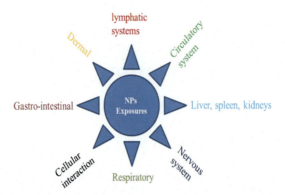

Fig. 1 Routes of NPs exposure into human body.

metal alloys, metal oxides, polymers, and hybrid composites, etc. Due to their unique properties with nanoscale size can provide excellent instruments and tools for various parts in the automotive industry. Even though the essential of NPs in the innovating improvements have manifested but till now the inevitability of NPs exposure with hazardous/toxicity to the environmental and living organisms is questionable. There are many ways of exposure of NPs into human as shown in Fig. 1.

The mitigation management is required to monitor as well as to do treatment to overcome/reduce the risk factors from hazardous/toxicity of NPs while using NPs in designing, manufacturing and during commercial utilization to prevent the environmental and living organisms. In this chapter, we deal with the toxicity/risk assessment of nanoparticles when used in automotive industry. First part explains about the impact of NPs in the automotive industry. Second part is discussing with the general toxicity of NPs. Third part explains about toxicity of NPs when used in automotive industry. Fourth part summarizes this chapter.

2. Impact of nanomaterials in the automotive industry

The role of nanoparticles usage with innovative techniques in the automotive industry is in its infancy. The nanostructured materials based on metals, semiconductors, and insulators/polymers with their unique properties make them potential applications in various parts of the automotive industry. Fig. 2 shows the applications of nanomaterials in the automotive industry.

2.1 Engine

In vehicle engine, nanomaterials are used in engine oil, transmission oil and radiator coolant to enhance the heat transfer removal [1]. Nanoaluminum is light weight, durable and powerful material that can produce efficient engines [2]. Copper and gold nanoparticles act as a liquid productive film to make sustainable automotive engines [3]. Boris Zhmud

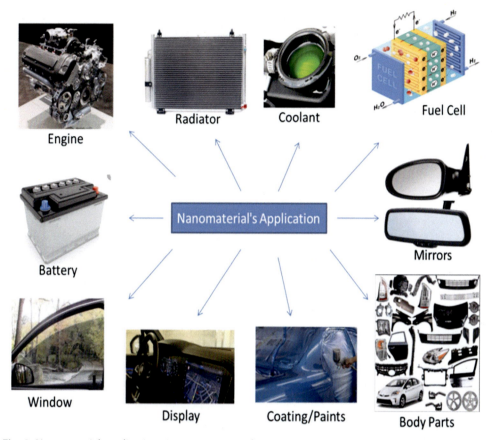

Fig. 2 Nanomaterial applications in automotive industry.

and BogdanPasalskiy, discussed the current advances for industrial perceptive based on nanomaterials (fullerenes, nanodiamonds, ultradispersed boric acid and polytetrafluoroethylene (PTFE)) in lubricants properties, engine oils, and greases [4]. MeenaLaad and Vijay Kumar S.Jatti, investigated the additives of TiO_2 nanoparticles in engine oil is significantly enhanced the lubricating properties of engine oil by reducing the friction and wear rate [5].

2.2 Radiator coolant

Nanofluids are nanoparticles with colloidal suspensions in a fluid. Nano fluids are playing a major role as a coolant for automotive radiators due to their thermo physical properties. It facilitates various cooling devices for fields such as transformers, computers, electronics, and vehicles. It also enhanced the thermo physical properties (heat transfer enhancement) of cooling systems by using nanoparticles with base fluids for

Table 1 Examples of nanoparticles and fluids for nanofluids.

Materials	Examples
Metals	Al, Cu, Ag, Au, Fe, Pt
Nonmetals	Graphite, carbon nanotubes, diamond, fullerene
Oxides ceramics	Al_2O_3, CuO, TiO_2, SiO_2
Carbide	SiC
Nitrides	AiN, SiN
Natural nanomaterial	Nanocellulose (CNC)
Phase change material (PCM)	Parafin
Magnetite Nanoparticles	Fe_3O_4, Fe_2O_3
Semiconductor	ZnO
Base fluids (Solvents)	Water, acetone, alcohol, ether, ethylene glycol (EG), propylene glycol (PEG), diethylene glycol (DEG), engine oils

automotive applications are reported in the literature [6]. The existing nanofluids depend on nanoparticles like metals, nonmetals, carbides, oxide ceramics, nitides, and base fluids like water (or other coolants), oil (and other lubricants), polymer solutions, bio-fluids, and other common fluids, such as paraffin. Table 1 shows the examples of different nanoparticles and fluids for nanofluids.

Ethylene glycol (EG), Diethylene glycol (DEG) and propylene glycol (PEG) are extensively utilized as antifreezes in automotive industries [7, 8]). Several metal or metal oxide nanoparticles such as aluminum oxide (Al_2O_3), silver (Ag), Copper oxide (CuO), Copper (Cu), Cobalt(II, III) oxide (Co_3O_4), Iron (II, III) oxide or Magnetite (Fe_3O_4), Iron(III) oxide or ferric oxide (Fe_2O_3), silicon carbide (SiC), silicon dioxide (SiO_2), titanium dioxide (TiO_2), zinc oxide (ZnO), nanodiamond, graphite and carbon-nanotubes (CNT) have been used and reported for the preparation of nanofluids. Nanofluids of Al_2O_3/water, SiO_2/water, CuO/water-EG and TiO_2/water-PEG based nanocoolants for automotive applications were reported by Gaurav Saxena and Poonam Soni [6]. Winifred NdukuMutuku., investigated on nanoparticles of (Al_2O_3), titanium dioxide (TiO_2), copper oxide (CuO) with ethylene glycol as the base fluid as a coolant for automotive radiator [9]. Sharma et al. reported the optimistic performance of Al_2O_3/water-EG as an engine coolant and heat exchanger to enhance heat transfer coefficient and reduces warm-up timings, which in turns less fuel consumption and reduction in emissions [10]. Ahmed et al. reported on the improvement of car engine radiator performance by using a coolant of TiO_2-water nanofluid and results were compared to the vehicle engine system FIAT DOBLO 1.3 MJTD ENG [11]. To get an enhanced efficiency of radiator, nanocellulose (CNC) to be added with the base fluids. F. Benedict et al., investigated the performance of hybrid form of metal oxide (Al_2O_3 and TiO_2) with or without plant base-extracted crystalline nanocellulose (CNC)/water-EG for better heat transfer nanofluid in radiator applications than presently existing radiator coolant like EG-distilled water [12].

2.3 Fuel cells

Fuel cells are an alternative to batteries in electric cars. In fuel cells, hydrogen ions are produced by introducing catalysts into fuels (hydrogen or methonal) for the automotive industry. So far, platinum (Pt) is the best material to use as a catalyst [13]. Metals such as Fe, Co, Ni, and Cu are introduced into Pt to form alloys to increase the catalytic performance of fuel cells [14]. Due to an expensive of Pt, replacements of other materials are needed. The cobalt-graphene (Co-Gr) catalyst is an alternative solution to replace platinum for the catalyst in fuel cells because of its excellent oxygen reduction reaction, durability and cost-effective. Yang et al., reported the performance of the Co-Gr nanomaterials toward hydrogenation of hydroxymethylfurfural (HMF) intodimethylfuran (DMF), which leads to excellent stability and catalytic activity [15].

2.4 Windows and displays

Nanomaterials proved the better production in glazing, windows, and displays in the automotive industry. Nano-coatings of Polycarbonate (PC) enhance the properties of standard glass for vehicle windows and headlight covers due to its much lighter, safer, and more flexible design to create a scratch resistance, abrasive resistance, and water-resistant surface or prevent water vapor condensing (hydrophobic) on glass during humid and other climate conditions [16]. Boentoro et al., reported a study on antimony doped tin oxide film (SnO_2:Sb) on glass and silicon oxide film (SiO_2) on polycarbonate for the analysis of scratch resistance [17]. To get more enhanced scratch resistance, polysiloxane or acrylate paints are coated with polycarbonate with Al_2O_3 nanoparticles placed into the substrate matrix during the hardening process for high abrasion resistance with excellent transparency. Also, the ultra-thin Al_2O_3 thin material with hydrophobic and oleophobic layers improves the dirt free, scratch resistance, and water repellent mirror surface to avoid discomfort during driving in day and night due to sunlight and glare [18].

2.5 Body parts

In vehicles, to ensure safety for passengers during a crash, the ultra-high strength steel body parts to be produced by plastic materials are bonded with aluminum to form a corrosion protective layer of boron steel for excellent sensitivity to temperature, strain rate, and formability [19,20]. To get the high tensile strength and ductility of boron steel, Al-Si —with ductile Fe-rich intermetallic compound coatings were added [21].

2.6 Coatings/paints

Nanoparticles are decorated on the paint on the exterior of the vehicle body via the following formulations to increase the surface hardness, scratch resistance, cracking resistance, and long-term appearance and durability [22,23]
(a) Hydroxyl + melamine = ether formulation,
(b) Carbamate + melamine = urethane formulation,
(c) Epoxy + carboxylic acid = ester formulation,
(d) Hydroxyl + isocyanate = urethane formulation and
(e) Silane + water = siloxane formulation

3. Nanotoxicity

Based on literature surveys and scientific inputs, it is evident that nanoparticles used in the automobile industry could cause oxidative stress which consequently causes an imbalance between intracellular pro-and antioxidants, decreased proliferation, loss of cell viability, protein carbonylation, apoptosis, ERS, the upregulation of MDA, and no and proinflammatory responses. Also, due to excess intracellular stress, it causes damage to the mitochondrial membrane potential, copy number of mitochondria, ATP synthesis and biogenesis associated with pulmonary tuberculosis, lung cancer and emphysema. Also results in induced oxidative stress and DNA damage. Fig. 3 shows the ways of NPs

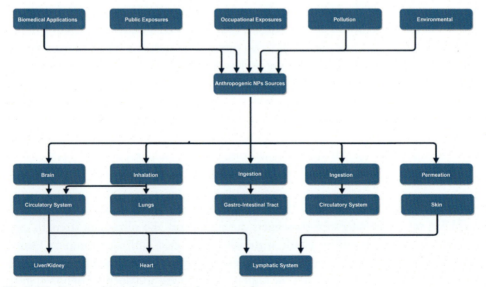

Fig. 3 NPs Entry and translocation into human body.

exposure into the human body. Exposure to nanoparticles regularly, even at smaller doses, might lead to the development of many diseases, tumors and has the potential to affect internal organs like brain, heart and intestinal mucosa.

4. Role of nanomaterials with their toxicity
4.1 Platinum (Pt) nanoparticles

Platinum nanoparticles are one of the scientific tools that are widely explored. Their physical and chemical properties make them an ideal metal for various research applications. They possess a large number of catalytic applications, such as their use in automotive catalytic converters, petrochemical cracking catalysts, etc., that are due to their property of having a very large surface area. These platinum nanoparticles are widely studied because of their antioxidant, antimicrobial, and anticancer properties [24]. It is mostly used in fuel cells in which oxidation reduction is frequently used as the cathode reaction and is used in a wide variety of reactions as a catalyst, including oxygen reduction [25]. In general, metallic nanoparticles possess potential cell-damaging activity because they have negative zeta potential. The size, shape, surface charges, and morphology of the nanoparticles are responsible for their antibacterial activity. Pt nanoparticles, being metallic nanoparticles, also have a negative zeta potential and thus have cell damaging properties. This cell damaging property of the Pt nanoparticles is responsible for their antibacterial activity against pathogens [26]. Significant antibacterial activity against *Pseudomonas aeruginosa* and *Staphylococcus aureus* [24] and antifungal activity against different pathogenic fungi including *C. fulvum, D. bryoniae, C. acutatum* and *P. capsica* by platinum nanoparticles have been absorbed in a study by [27]. Also, Pt nanoparticles have the capability to impair the downstream pathways that leads to inflammation and it can reduce intracellular reactive oxygen species (ROS) levels [28]. The platinum NPs exposure and their toxicity are shown in Fig. 4.

Toxicokinetics of the platinum nanoparticle is strongly dependent on the size of the particle [29]. For the toxicity study of platinum nanoparticles [29] exposed two different types of cell lines, which are lung epithelial cells and human endothelial cells [30]. The results of their study showed that the both exposed cells could possibly uptake platinum nanoparticles. The author concluded that the cytotoxicity of platinum nanoparticles is dependent on the particle shape because the entry of the particles and their cytotoxicity potential are determined by the shape of the particles [30]. In contrast, to determine the potential applications of Pt nanoparticles to consumer products [31], investigated the toxicity of platinum nanoparticles on THP-1 cells. The results from this study indicated that Pt nanoparticles could cause oxidative stress which consequently causes an imbalance between intracellular pro-and antioxidants, decreased proliferation, loss of cell viability, protein carbonylation, apoptosis, ERS, the upregulation of MDA, NO and proinflammatory responses. Also, due to excess intracellular stress, it caused damage to the

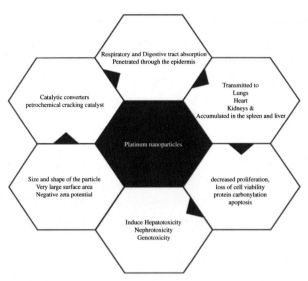

Fig. 4 Toxicity of platinum NPs exposure.

mitochondrial membrane potential, copy number of mitochondria, ATP synthesis and biogenesis [31]. The encapsulation of the Pt nanoparticles in liposomes may increase the cytotoxic effect of the particles. Coming to genotoxicity, Pt nanoparticles inhibit the functions of *Taq* DNA polymerase, thus causing DNA structure damage. The transition of Pt nanoparticles to Pt2+ along with the above-mentioned DNA damaging activity could cause mutagenicity. It was also found that genotoxic stress causes the activation of p53 with subsequent activation of p21 in platinum nanoparticle treated cells, which leads to apoptosis [32].

Platinum nanoparticles can enter into the body by respiratory and digestive tract absorption and can even penetrate through the epidermis, but there is no evidence available for entry through absorption by the skin. After the nanoplatinum exposure, they can be transmitted to other nearby organs such as the lungs, heart, or kidneys, though they mostly accumulate in the spleen and liver. It has the potential to induce hepatotoxicity and nephrotoxicity via the intravenous route of exposure [29]. Wistar rats were orally exposed to platinum nanoparticles. Those rats received platinum nanoparticles at a daily dose of 10, 50, or 100 mg/kg for consecutive 30 days. The results of this study showed that the Pt nanoparticles caused organ weight changes in rats. Also, exposed rats exhibited increased protein levels in tissues, altered serum urea creatine levels and a decrease in albumin. In the liver of the rats, it was found that the alkaline phosphatase and aspartate transaminase activities were reduced whereas the alanine amino transferase level was raised in both tissues and serum. This study showed that morphological lesions, including inflammation and cellular degeneration, were caused by the exposure of platinum nanoparticles [33]. An increased level of proinflammatory cytokines in bronchoalveolar lavage was observed when animals were exposed to platinum nanoparticles intratracheally [29]. Absorption of the platinum nanoparticles was not observed by dermal exposure, but they can penetrate through the skin. Mauro et al. applied the platinum nanoparticles to the outer surface of the skin for 24 h (which was obtained as a surgical waste) [34]. The results of their study proved that the platinum nanoparticle can penetrate through the skin barrier and metal penetration could be significantly increased even by a minor injury to the skin. Platinum nanoparticles can be used as a potential therapeutic tool. Safe therapy for cancer could be developed with the help of Pt nanoparticles by selective induction of cytotoxicity in the target cancer cells [31]. According to Nejdl et al., it is concluded that an alternative treatment for cancer may be provided by the effective antitumor activity of Pt nanoparticles [35]. Cell type and physiochemical parameters are found to be correlated with the biological responses of platinum. So, further detailed investigations are needed into their regulated utilization in various consumer applications, like automobiles, cosmetics, or as biomedical therapeutics [36].

4.2 Silicon dioxide (SiO_2) nanoparticles

SiO_2 ceramic wax containing SiO_2 nanoparticles suspended in a polymer resin is more of a sealant. Its nanostructure helps it to penetrate into and seal all the fine pores on the surface of a car's finish, resulting in a hard glass-like crystal clear protection layer. It can be applied to chrome, plastic, paint, metal, fiberglass, and glass parts of automobiles as it has the ability to resist UV light, wear and tear, airborne contaminants, and waterborne particles like road salt. The UV absorbance at the wavelength of 290–400 nm increased as the nano-SiO_2 content of the coatings increased [37]. Gläsel et al. [38] reported the use of siloxane encapsulated SiO_2 nanoparticles to produce scratch and abrasion resistant films. Additives of nano silica are added to tires of cars mainly containing synthetic or natural rubber. It helps in the reduction of roll resistance of the tire by 20% and also shortens the breaking distance by 10% (European Parliament 2008 (OECD 2014). Hence, the use of such additives plays an important role in reducing environmental burdens such as fuel usage as well as increasing driving safety. The potential reduction in environmental burdens is estimated at 5–10% over the life cycle of tires. It is assumed that from the use of nanosilica in paints and polymers, which are the main application areas, no releases take place during use and in the waste treatment processes [39]. SiO_2 nanoparticles have been in use for many years as fillers for tire rubber, which helps in enhancing tire performance and durability. The limitations of polycarbonate glazing, such as its poor scratch resistance and low UV shielding, can be lessened by the addition of a transparent coating containing silica nanoparticles. According to [40], it is predicted that by 2020, around 20% of all vehicle glazing will be made from nano-enhanced PC. The SiO_2 NPs exposure and their toxicity are shown in Fig. 5.

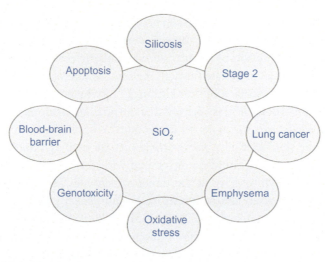

Fig. 5 Toxicity of SiO_2 NPs exposure.

Certain materials may become toxic when they are in nanometer size though nontoxic in bulk form. Toxicologist believes that due to the very small size but with high surface area, SiO_2 nanoparticles may cause different adverse effects when compared to micron-sized silica [41]. Effects of silica exposure, especially crystalline silica (0.5–10 μm), on human health have been widely studied. SiO_2 nanoparticles showed cytotoxic effects in a few in vitro studies. According to Leung et al. [42], crystalline silica exposure induces silicosis (a fibrotic lung disease) and found to be associated with pulmonary tuberculosis, lung cancer and emphysema. All cells in the pulmonary system, lung epithelial cells and pleural mesothelial cells exhibited susceptibility to SiO_2-induced oxidative stress and damages [43]. Napierska suggested that exposure to silica nanoparticles induced toxic effects in vitro and in vivo. The review Please provide appropriate numbered citation in the place of "by Napierska et al. [44] concluded that physiochemical properties of silica nanoparticles differ and may cause different effects based on the method of production. Silica nanoparticles induced genotoxicity in human tumor cell lines and the extent of the effect inversely correlated with the size of nano particle. At low concentrations (2.5–10 μg/mL) of SiO_2, DNA strand breaks were observed especially in cell lines that are derived from skins. Exposure to silica nanoparticles (70 nm; 10–90 μg/mL for 24 h) resulted in the elevation of oxidative DNA damage in HaCaT cells [43]. Cellular components could be damaged by oxidative stress induced by nanoparticles and to cell death through apoptosis [45]. Significant cytotoxic effects were observed only at or above the concentration of 25 μg/mL. Only in a size and dose-dependent manner, SNPs induces oxidative stress and mediated apoptosis [18]. Incubation with 5 or 25 μg/mL of 50-nm SNPs showed the aggregation of mouse blood platelets in 3 min [46] while in [47] such aggregation was observed in 15 min (10 μg/mL of 10 nm). A few reports revealed that SiO$_2$ penetration into skin barrier and followed by its localization at lymph nodes when dermal exposure for continuous 3 day was made [48]. Followed by dermal exposure of SiO$_2$ nanoparticles for more than 28 days resulted in systemic absorption into the brain and liver [43]. Translocation of SiO$_2$ NPs greater than 75 nm in size did not occur, whereas SiO$_2$ NPs less than 75 nm in size could be absorbed through the skin [49]. Hirai et al. [48] analyzed the well-dispersed amorphous silica NPs of 70 nm in diameter for skin penetration and cellular localizations in a mice. They concluded that topical exposures of SiO$_2$ NPs for a very long duration could lead to the penetration of SiO$_2$ NPs into the skin barrier and be transported to the lymph nodes. These results showed that to investigate the penetration activity and the harmful effects following exposure to silica nanoparticles through the skin would require more studies [50]. Ingestion is considered a major uptake route of silica NPs besides inhalation into the human body [51]. In human endothelial cell line dose-dependent cytotoxicity (by MTT and LDH assay) of monodisperse amorphous SNPs (16–335 nm) was reported by [52]. The particle size of the SiO$_2$nano particle was strongly related to the toxicity; higher toxicity was significantly showed by higher smaller particles and also the exposed cells were affected faster. Gerloff et al. [53] investigated with the human epithelial cell lines of colon Caco-2 for the cytotoxic and DNA damaging properties of amorphous fumed SiO$_2$ nanoparticles (14 nm). Cell morality, total glutathione depletion and notable DNA damage was observed when exposure to SNPs for up to 24 h was made. [54] addressed the toxicity of nano- and micron-sized silica particles which ranges from (14 nm and 1–5 μm, respectively) in vitro and in vivo. In vitro analysis was performed using RAW 264.7 cells, which were exposed to both particle sizes for 24 h, and the cell viability was decreased in dose-dependent manner; however, apoptosis was observed only after treatment with nanoparticles. In vivo, mice were treated with 5 mg/kg silica particles via oropharyngeal aspiration. Again, size-dependent toxicity of silica was reported. Thereby it implies that nano-sized particulates leads to the greater pulmonary injury and neutrophilic infiltration than micro-sized particulates [54].

The harmful effects of SNPs have mainly been studied in terms of exposure via inhalation, after acute or sub-acute exposure; other exposure routes like skin, blood, and gastrointestinal tract routes also studied in detail [41]. Data is insufficient to identify and classify the harmful effects that SiO$_2$ NPs exposure and demonstrating certain conditions for the use of such materials safely is currently not possible due to the lack of realistic exposure and epidemiological data.

4.3 Titanium dioxide (TiO$_2$) nanoparticles

TiO$_2$ is one of the safest nanoparticles and useful to a wide range of applications because of its nano size, it has distinctive physical and chemical properties. It is considered a potential nanoparticle because of its availability and low cost. It possesses the photocatalytic activity that makes it a multipurpose nanoparticle. Self-cleaning, air-purifying coatings were developed from the TiO_2 nanoparticle with photocatalytic and hydrophobic property. In Europe, 25% of the production of nano TiO_2 are used by the paint and coating industry making it as the second end user of nano TiO_2[55]. Photocatalytic activity of nano TiO_2 has been industrially utilized for the production of anti-fogging car mirrors [56]. The lubricating properties of the engine oil can be enhanced by mixing TiO_2 nanoparticles in the oil. It significantly reduces the wear rate and friction in the engine [5]. They also found that lubrication enhancing activity of TiO_2 nanoparticles is due to the great solubility and stability property it possess by dispersion analysis using UV spectrometer. Ahmed et al. [11] studied the usage of TiO_2 water nanofluid's in car engines as a cooler and it was found that depending on the quantity of TiO_2 nanoparticle added to the base fluid heat transfer rate of the engine is significantly enhanced. They concluded that it was because when compared to pure water TiO_2 has higher thermal resistance, thermal conductivity, lower specific gravity and aspect specific ratio. TiO_2 nanoparticles also shows antibacterial and antibiofilm activity against the *Citrobacterfreundii* and *Streptococcus mutans* [57]. The applications of TiO_2 NPs and their exposure/toxicity are shown in Fig. 6.

Exposure to TiO_2 nanoparticles regularly even at smaller dose might lead to develop many diseases, tumors and has the potential to affect the internal organs like brain, heart and intestinal mucosa[58]. Materials only with low-molecular weight and an appropriate water partition coefficient could be able to penetrate the tough outermost layer of skin (i.e.) stratum corneum of human beings. Therefore, under normal conditions, it is

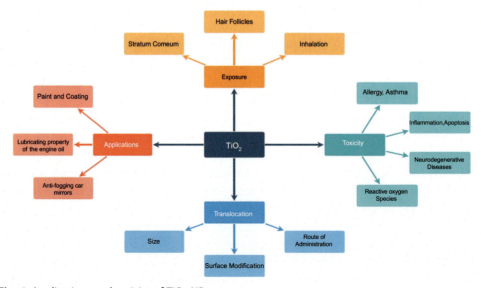

Fig. 6 Applications and toxicity of TiO_2 NPs.

impossible for inorganic materials to penetrate the layer of stratum corneum of human under normal conditions. It is also found that TiO$_2$ nanoparticles of size less than 100 nm did not penetrate the skin even though the stratum corneum layer was damaged. Some of the particles found penetrated into hair follicles but not into the dermis [59]. Cytoplasms of epithelial and endothelial cells of the rats were found to contain free particles of TiO$_2$ nanoparticles after subjecting the rats to TiO$_2$ exposure by inhalation [60]. When new born and weanling (2-week-old) rats were exposed to TiO$_2$ nanoparticles (12 mg/m^3; 5.6 h/day for 3 days) showed the increased expression of lung neurotrophins, which play an important role in the development of childhood asthma. This study reveals that TiO$_2$ exposure to children may lead to the risk of developing asthma [61]. Inhalation exposure to TiO$_2$ nanoparticles leads to the deposition of the particle in lung and can cause chronic inflammation followed by tumor development. The inhaled TiO$_2$ nanoparticles may also translocated to other tissues which may lead to allergy, asthma, and so on [62]. Administration routes, size, and surface modification of the TiO$_2$ nanoparticle are the parameters that regulates the translocation process of TiO$_2$ NPs into the brain. Once the TiO$_2$ nanoparticles find its way to the brain region, it would affect the major CNS cells including the glial cells and the neurons. TiO$_2$ nanoparticles would induce the reactive oxygen species (ROS), inflammation and apoptosis, which may further disturb the CNS functions or even induce neurodegenerative diseases and cause cell death too. The cell viability, cell cycle, cell morphology, antioxidant capability and other cellular components would be affected when glial cells and neurons were incubated with TiO$_2$ nanoparticles in some in vitro studies [63]. Titanium metal or its alloy made orthopedic and dental implants commonly forms a few nanometers thick layer amorphous TiO$_2$. Wear exposed implants such as knee joints can release relevant amount of nanometer TiO$_2$ debris under particular physiological conditions and mechanical stress. These released TiO$_2$ nanoparticle might get translocated into other internal organs and surrounding tissues. It is increasingly being reported that these debris particles are associated with systemic diseases and inflammation [64].

In vivo, TiO$_2$ nanoparticles were found to accumulate in different internal organs such as lungs, liver, spleen, heart, alimentary tract, and so on after being exposed to the TiO$_2$ nanoparticle mainly by injection, inhalation and oral administration [58]. The size of the TiO$_2$ nanoparticle has an important role in determining both toxicity and accumulation level of the nanoparticle in the organs of human body [65]. He concluded that smaller nanoparticles of size 25 nm were found in different organs like spleen and to a lesser extent, in the lungs and kidneys, whereas larges particles are mainly found only in the liver of the mice after one-time oral administration. In the photo-thermal therapy (PTT), which is a therapeutic method to induce hyperthermia in malignant tumor cells, heat efficacy was enhanced by performing energy conversion with the help of TiO$_2$ nanoparticles. The very important properties of TiO$_2$ nanoparticles used in this technique are high optical absorbance and low-cytotoxicity. Polyethylene glycol

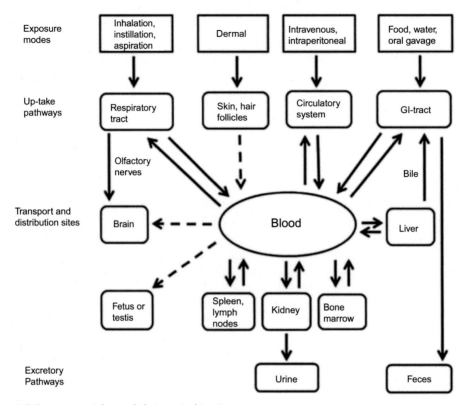

Fig. 7 TiO$_2$ nanoparticles and their toxicokinetics.

(PEG)-TiO$_2$ nanoparticles were synthesized for increasing water dispersibility and biocompatibility. The effect of such nanoparticles on reducing melanoma tumor size was assessed experimentally [66].

Fig. 7 shows the various possible accumulation sites of TiO$_2$ nanoparticle and its toxicokinetic where dotted lines represent uncertainties [67].

Since the toxicokinetics studies including absorption, distribution, metabolism, excretion and the specific organ toxicity of the TiO$_2$ nanoparticle is not known it is likely impossible to conclude regarding the safety and exposure limits of nano TiO$_2$ particles. In addition, the molecular mechanism of nano TiO$_2$ by which it may cause cancer and other diseases are unclear. Therefore, it is suggested that TiO$_2$ nanoparticles should be used precautiously with great care.

4.4 Copper oxide (CuO) nanoparticles

In general, nanoparticles have different properties from their bulk form. Particularly, nanoparticles like CuO have high surface area because of their spherical structure.

Because of their unique physiochemical properties CuO nanoparticles are used widely in different fields such as gas electrodes, lithium-ion battery electrodes, and so on. It is chemically stable, inexpensive, and abundant [68]. Cuo nanoparticles exhibit high thermal conductivity when mixed with appropriate mixture of base fluid. Godley et al. [69] used nanofluids, which is a combination of Ethylene Glycol + water (50:50) and nanoparticles as a coolant for car radiators. They observed increased overall heat transfer efficient along with heat transfer rate of the radiator with different concentration of the coolant mixed with the nano CuO particle. CuO nanoparticles are also used in the production of copper-oxide brake nanofluid (CBN). Copper-oxide brake nanofluid showed higher boiling temperature, higher conductivity and higher viscosity in a study done by [70]. It also possesses properties like anti-wear and having low coefficient of friction. Such properties make it a suitable additive material for lubricants (nano fluids) which helps in decreasing contact pressures and creation of tribo-films in machineries [71]. Highlighting property of this CuO nanoparticle is that its antimicrobial property. It has been reported that with suitable dose and incubation period, Cuo nanoparticles have the ability to kill 99.9% of Gram-negative and Gram-positive bacteria [72]. Antimicrobial packaging films with improved mechanical barrier, antimicrobial and UV- light blocking properties were produced while incorporating this CuO nano particle into agar polymer by [73]. In their study, it is also found that it has improved water vapor barrier and UV light-blocking properties when incorporated into various carbohydrate-based biopolymers. The copper oxide nanoparticles have become one of the ideal materials for producing gas-sensing devices which allows to detect several gases like nitrogen oxides, carbon oxides, hydrogen sulfide, and ammonia. These copper-oxide based nano sensors have its applications in wide range of industrial fields. In various fields like automotive industry, exhaled breath analyzers and environmental pollution detectors these copper oxide based nano sensors are already in use. In automotive industry it can be used for determining the quality of the air. Rydosz [74] has verified the gas-sensing property of the CuO nanoparticles by exposure to various gases such as hydrogen sulfide, volatile organic compounds, and so on.

The toxicological outcomes of the copper oxide nanoparticles are dependent on the route of exposure of the particles. After exposure these nanoparticles can pass via tissue interstitially, passing through the cell membranes and subsequently enter into the blood circulatory system of the body. These nanoparticles after entering into the body it can translocate into other organs and has the capacity to accumulate in it. In the translocation and accumulation of the CuO nanoparticles from the exposure site to other organs, both circulatory and lymphatic system plays an important role [75]. Inhalation, ingestion, dermal exposure and through blood stream are some of the main routes for the entry of CuO nanoparticles into the body [76]. Among them inhalation is the chief route of entrance for the CuO nanoparticles. They penetrate into the lungs during inhalation and can cause inflammation by interacting with the epithelium. Nano copper oxide particles can enter to the other organs of the body like CNS via the olfactory bulb which is considered as one of the hazardous routes [77]. They can easily translocate between cells and cross cell

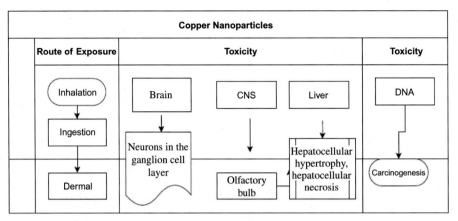

Fig. 8 Flowchart of CuO NPs exposure and their toxicity.

membrane thereby it leads to cell disruptions because of its small size. Fig. 8 shows the flowchart of CuO NPs exposure and their toxicity.

A study done by De Jong et al. [78], reported that copper nanoparticles at the dose of 512 mg/kg b.w induced histopathological alterations in the liver, stomach and bone marrow. The liver showed hepatocellular hypertrophy, hepatocellular necrosis, Kupffer cell hypertrophy and inflammation composed of mixed inflammatory cells. An increased incidence and severity of submucosal glandular inflammation with eosinophilic granulocytes in stomach was observed in their study. Several changes in bone marrow like decreased erythroid elements and increased myeloid elements were occurred. Genomic DNA in the cells of different organs could be affected by copper oxide nanoparticles significantly which suggests us that these nanoparticles can trigger carcinogenesis. Also, brains of the rats after exposing to copper oxide nanoparticles showed pathological changes of neurons in the ganglion cell layer of the and also in the basal nuclei. Particularly, it damages the front cortex, neuronal apoptosis and the basal ganglia [79]. Rani et al. [80] showed the histopathological examinations in rat lungs that are exposed to copper nanoparticles. Results suggested that the copper oxide nanoparticles caused the dose dependent pulmonary toxicity and reduced the antioxidant capacity in rats. Copper oxide nanoparticles could potentially increase the intracellular ROS level and cause the loss of mitochondrial membrane and leakiness of lysosomal membrane. Hence it was evident that it could effectively induce oxidative stress along with causing damage on mitochondria and lysosomes in human blood lymphocytes [81]. Based on the size of the copper oxide nanoparticles, they get adhere to the cell membrane or internalized into the human skin epidermal cells. CuO nanoparticles larger than the size of 500 nm remained outside the cell whereas nanoparticles of size 30–100 nm were internalized into the cytoplasm, vacuole and even the nucleus of the cell [82]. In their study, they demonstrated that CuO nanoparticles has genotoxic and cytotoxic effect on human skin epidermal cells. It is also concluded that the mode of the cell death was by the ROS-trigged mitochondrial pathway mediated apoptosis. Prabhu et al. [83] investigated the neurotoxicity of copper nanoparticles of different sizes and concentrations. After exposing the primary cultures of dorsal root ganglion of neonatal rat pups to 10–100 µm copper nanoparticles for 24 h, the neurons became detached from the substratum and showed disruptive neurite growth. Much work should be done to study the toxicokinetic and toxicodynamic of the CuO nanoparticle in order to define criteria for sustainable applications in different fields.

5. Conclusions

The implementation of nanotechnology has notably increased in the field of automobile industry because of physiochemical properties of nanoparticles. Automobile industries are anticipating new nanomaterials with innovative technology to develop a smart vehicle. Even though nanomaterials are very fascinating research field toward automobile industries however nanoparticle's exposure to human is a prime concern. Exposure of nanoparticles and its toxicity has occurred since ancient times. Due to the size, nanoparticles enter through respiratory and digestive tract and translocate from entry portals into circulatory and lymphatic systems and ultimately to tissues and organs.

Platinum nanoparticles possess a large number of catalytic applications such as automotive catalytic convertors, petrochemical cracking catalyst, and so on, and they enter into body through respiratory and digestive tract absorption and even through epidermis. After the exposure, they can get transmitted to lungs, heart or kidneys and they get accumulated in spleen and liver. They induce hepatotoxicity and nephrotoxicity. SiO_2 nanoparticles are used in chromo plastics, paints, and metal, fiberglass and glass parts of automobiles as it is having the ability to resist UV light, wear and tear, contaminants and waterborne particles. A physiochemical property of silica nanoparticles differ and causes different effects based on their preparation method. Exposure to silica nanoparticles results in oxidative DNA damage and genotoxicity. It is also reported that oxidative stress induced by silica nanoparticles are responsible in cell death via apoptosis. TiO_2 is considered a potential nanoparticle because of its availability and low cost. It possesses photocatalytic activity and it is very much employed in self-cleaning and air purifying. Automobile industry uses TiO_2 for coatings and anti-fogging car mirrors. The lubricating properties of the engine oil can be enhanced by mixing TiO_2 nanoparticles in the oil due to its solubility and stability property. Exposure to TiO_2 nanoparticles regularly even at smaller dose might lead to develop many diseases, tumors and has the potential to affect the internal organs like brain, the heart and the intestinal mucosa. CuO have high surface area because of their spherical structure. Copper- oxide brake nanofluid showed the higher boiling temperature, higher conductivity and higher viscosity. It is a suitable additive material for lubricants (nano fluids) which helps in decreasing contact pressures and creation of tribo-films. Inhalation, ingestion, dermal exposure and through blood stream are some of the main routes for the entry of CuO nanoparticles into the body. After exposure these nanoparticles can pass via tissue interstitially, passing through the cell membranes and subsequently enter into the blood circulatory system.

Advanced research on properties of nanoparticles especially on their size, crystalline nature, morphology and other aspects relating with the toxic effect to mankind should be of serious concern. The introduction of innovative ideas with combined team of various streams for the minimization of nanoparticles exposure to human beings in the automobile industry paves to healthy human life.

References

[1] N.A.C. Sidik, M.N.A.W.M. Yazid, R. Mamat, A review on the application of nanofluids in vehicle engine cooling system, Int. Commun. Heat Mass Transfer 68 (2015) 85–90. http://umpir.ump.edu.my/id/eprint/10468/1/A%20Review%20on%20the%20Application%20of%20Nanofluids%20in%20Vehicle%20Engine%20Cooling%20System.pdf.

[2] R. Surendran, N. Manibharathi, A. Kumaravel, Wear properties enhancement of Aluminium alloy with addition of Nano alumina, FME Trans. 45 (2017) 83–88. https://www.mas.bg.ac.rs/_media/istrazivanje/fme/vol45/1/13_rsurendran_et_al.pdf.

[3] M.K.A. Ali, H. Xianjun, Improving the tribological behavior of internal combustion engines via the addition of nanoparticles to engine oils, Nanotechnol. Rev. 4 (4) (2015) 347–358, https://doi.org/10.1515/ntrev-2015-0031. https://www.degruyter.com/document/doi/10.1515/ntrev-2015-0031/pdf.

[4] B. Zhmud, B. Pasalskiy, Nanomaterials in lubricants: An industrial perspective on current research, Lubricants 1 (2013) 95–101. https://www.mdpi.com/2075-4442/1/4/95.

[5] M. Laad, V.K.S. Jatti, Titanium oxide nanoparticles as additives in engine oil, J. King Saud Univ. Eng. Sci. 30 (2018) 116–122, https://doi.org/10.1016/j.jksues.2016.01.008.

[6] G. Saxena, P. Soni, Nano coolants for automotive applications: a review, Nano Trends: A J. Nanotechnol. App. 20 (2018) 9–22. http://techjournals.stmjournals.in/index.php/NTs/article/view/78.

[7] S. Rebsdat, D. Mayer, Ethylene glycol, in: Ullmann's Encyclopedia of Industrial Chemistry, 13, Wiley, Hoboken, NJ, USA, 2000, pp. 531–546. http://www.ugr.es/~tep028/pqi/descargas/Industria%20quimica%20organica/tema_5/etilenglicol_a10_101.pdf.

[8] J.P. Yadav, B.R. Singh, Study on performance evaluation of automotive radiator, SAMRIDDHI A J. Phys. Sci. Eng. Technol. 2 (2015) 47–56. https://smsjournals.com/index.php/SAMRIDDHI/article/view/1193.

[9] W.N. Mutuku, Ethylene glycol (EG)-based nanofluids as a coolant for automotive radiator, Mutuku, Asia Pac. J. Comput. Eng. 3 (1) (2016) 1–15. https://apjcen.springeropen.com/articles/10.1186/s40540-016-0017-3.

[10] V. Sharma, R. Nirmal Kumar, K. Thamilarasan, G. Vijay Bhaskar, B. Devra, Heat reduction FromIc engine by using Al_2O_3 Nanofluid in engine cooling system, Am. J. Eng. Res. 3 (2014) 173–177. http://www.ajer.org/papers/v3(4)/W034173177.pdf.

[11] S.A. Ahmed, M. Ozkaymak, A. Sözen, T. Menlik, A. Fahed, Improving car radiator performance by using TiO_2-water nanofluid, Int. J. Eng. Sci. Technol. 21 (2018) 996–1005. https://www.mendeley.com/catalogue/2e8dafc0-8f88-33bf-a26c-ac1757cb64dc/.

[12] F. Benedict, A. Kumar, K. Kadirgama, H.A. Mohammed, D. Ramasamy, M. Samykano, R. Saidur, Thermal performance of hybrid-inspired coolant for radiator application, Nano 10 (1100) (2020) 1–34. https://www.mdpi.com/2079-4991/10/6/1100/htm.

[13] O.T. Holton, J.W. Stevenson, The role of platinum in proton exchange membrane fuel cells, Platinum Metals Rev. 57 (4) (2013) 259–271. https://www.technology.matthey.com/article/57/4/259-271/.

[14] X. Ren, Q. Lv, L. Liu, B. Liu, Y. Wang, A. Liu, G. Wu, Current progress of Pt and Pt-based electrocatalysts used for fuel cells, Sustainable Energy Fuels 4 (2020) 15–30. https://pubs.rsc.org/en/content/articlelanding/2020/se/c9se00460b.

[15] F. Yang, J. Mao, S. Li, J. Yin, J. Zhou, W. Liu, Cobalt–graphene nanomaterial as an efficient catalyst for selective hydrogenation of 5-hydroxymethylfurfural into 2,5-dimethylfuran, Cat. Sci. Technol. 9 (2019) 1329–1333. https://pubs.rsc.org/en/content/articlelanding/2019/cy/c9cy00330d#!divAbstract.

[16] C. Seubert, K. Nietering, M. Nichols, R. Wykoff, S. Bollin, An overview of the scratch resistance of automotive coatings: exterior clearcoats and polycarbonate hardcoats, Coatings 2 (4) (2012) 221–234. https://www.mdpi.com/2079-6412/2/4/221.

[17] W. Boentoro, A. Pflug, B. Szyszka, Scratch resistance analysis of coatings on glass and polycarbonate, Thin Solid Films (2009) 3121–3125.

[18] M. Shafique, X. Luo, Nanotechnology in transportation vehicles: An overview of its applications, environmental, health and safety concerns, Materials (Basel) 12 (15) (2019) 2493. https://www.ncbi.nlm.nih.gov/pmc/articles/PMC6696398/.

[19] B. Francesca, et al., Investigation of the high strength steel Al-Si coating during hot stamping operations, Key Eng. Mater. 410–411 (2009) 289–296, https://doi.org/10.4028/www.scientific.net/kem.410-411.289. trans tech publications, ltd. https://www.scientific.net/KEM.410-411.289.

[20] J.H. Jang, J.H. Lee, B.D. Joo, Y.H. Moon, Flow characteristics of aluminum coated boron steel in hot press forming, Trans. Nonferrous Metals Soc. China 19 (2009) 913–916. http://www.ysxbcn.com/down/upfile/soft/2009812/27-p0913.pdf.

[21] Z.-X. Gui, W.K. Liang, Y.S. Zhang, Formability of aluminum-silicon coated boron steel in hot stamping process, Trans. Nonferrous Metals Soc. China 24 (2014) 1750–1757. https://www.sciencedirect.com/science/article/abs/pii/S1003632614632490.

[22] N.K. Akafuah, S. Poozesh, A. Salaimeh, G. Patrick, K. Lawler, K. Saito, Evolution of the automotive body coating process—a review, Coatings 6 (2) (2016) 1–24. https://www.mdpi.com/2079-6412/6/2/24.

[23] H. Lotfizadeh, S. Rezazadeh, M.R. Fathollahi, J. Jokar, A.A. Mehrizi, B. Soltannia, The effect of silver nanoparticles on the automotive-based paint drying process: an experimental study, Int. J. Adv. Multidis. Eng. Sci. 2 (1) (2018) 7–14. http://article.sapub.org/10.5923.j.james.20180201.02.html.

[24] M. Jeyaraj, S. Gurunathan, M. Qasim, M. Kang, J. Kim, A comprehensive review on the synthesis, characterization, and biomedical application of platinum nanoparticles, Nano 9 (12) (2019) 1719, https://doi.org/10.3390/nano9121719.

[25] L. Forbes, Controlling the Growth And Catalytic Activity of Platinum Nanoparticles Using Peptide and Polymer Ligands, 2013. UC san Diego. ProQuest ID: Forbes_ucsd_0033D_13500. Merritt ID: ark:/20775/bb3653383p. Retrieved from https://escholarship.org/uc/item/8r68x2f3.

[26] A. Chwalibog, E. Sawosz, A. Hotowy, J. Szeliga, S. Mitura, et al., Visualization of interaction between inorganic nanoparticles and bacteria or fungi, Int. J. Nanomedicine 1085 (2010), https://doi.org/10.2147/ijn.s13532.

[27] P. Velmurugan, J. Shim, K. Kim, B. Oh, Prunus × yedoensis tree gum mediated synthesis of platinum nanoparticles with antifungal activity against phytopathogens, Mater. Lett. 174 (2016) 61–65, https://doi.org/10.1016/j.matlet.2016.03.069.

[28] D. Pedone, M. Moglianetti, E. De Luca, G. Bardi, P. Pompa, Platinum nanoparticles in nanobiomedicine, Chem. Soc. Rev. 46 (16) (2017) 4951–4975, https://doi.org/10.1039/c7cs00152e.

[29] E. Czubacka, S. Czerczak, Are platinum nanoparticles safe to human health? Med. Pr. 70 (4) (2019) 487–495, https://doi.org/10.13075/mp.5893.00847.

[30] A. Elder, H. Yang, R. Gwiazda, X. Teng, S. Thurston, H. He, G. Oberdörster, Testing nanomaterials of unknown toxicity: An example based on platinum nanoparticles of different shapes, Adv. Mater. 19 (20) (2007) 3124–3129, https://doi.org/10.1002/adma.200701962.

[31] S. Gurunathan, M. Jeyaraj, H. La, H. Yoo, Y. Choi, J. Do, et al., Anisotropic platinum nanoparticle-induced cytotoxicity, apoptosis, inflammatory response, and transcriptomic and molecular pathways in human acute Monocytic leukemia cells, Int. J. Mol. Sci. 21 (2) (2020) 440, https://doi.org/10.3390/ijms21020440.

[32] P. Asharani, N. Xinyi, M. Hande, S. Valiyaveettil, DNA damage and p 53-mediated growth arrest in human cells treated with platinum nanoparticles, Nanomedicine 5 (1) (2010) 51–64, https://doi.org/10.2217/nnm.09.85.

[33] O. Adeyemi, F. Sulaiman, M. Akanji, H. Oloyede, A. Sulaiman, A. Olatunde, et al., Biochemical and morphological changes in rats exposed to platinum nanoparticles, Comp. Clin. Pathol. 25 (4) (2016) 855–864, https://doi.org/10.1007/s00580-016-2274-5.

[34] M. Mauro, M. Crosera, C. Bianco, G. Adami, T. Montini, P. Fornasiero, et al., Permeation of platinum and rhodium nanoparticles through intact and damaged human skin, J. Nanopart. Res. 17 (6) (2015), https://doi.org/10.1007/s11051-015-3052-z.

[35] L. Nejdl, J. Kudr, A. Moulick, D. Hegerova, B. Ruttkay-Nedecky, J. Gumulec, et al., Platinum nanoparticles induce damage to DNA and inhibit DNA replication, PLoS One 12 (7) (2017), https://doi.org/10.1371/journal.pone.0180798, e0180798.

[36] C. Labrador-Rached, R. Browning, L. Braydich-Stolle, K. Comfort, Toxicological implications of platinum nanoparticle exposure: stimulation of intracellular stress, inflammatory response, and Akt signaling in vitro, J. Toxicol. 2018 (2018) 1–8, https://doi.org/10.1155/2018/1367801.

[37] S. Zhou, L. Wu, J. Sun, W. Shen, The change of the properties of acrylic-based polyurethane via addition of nano-silica, Prog. Org. Coat. 45 (1) (2002) 33–42, https://doi.org/10.1016/s0300-9440(02)00085-1.

[38] H. Gläsel, F. Bauer, H. Ernst, M. Findeisen, E. Hartmann, H. Langguth, et al., Preparation of scratch and abrasion resistant polymeric nanocomposites by monomer grafting onto nanoparticles, 2 characterization of radiation-cured polymeric nanocomposites, Macromol. Chem. Phys. 201 (18) (2000) 2765–2770, https://doi.org/10.1002/1521-3935(20001201)201:18<2765::aid-macp2765>3.0.co;2-9.

[39] T. Zimmermann, D. Jepsen, A. Reihlen, Use of Nanomaterials in Tires—Environmental Relevance and Emissions, 2018, pp. 1–16. https://www.bmu.de/fileadmin/Daten_BMU/Download_PDF/Nanotechnologie/nanodialog_5_fd2_abschlussbericht_en_bf.pdf.

[40] S. Will, Nanotechnology in the automotive industry, AZoNano 2020 (2021). https://www.azonano.com/article.aspx?ArticleID=3031.

[41] D. Napierska, L. Thomassen, D. Lison, J. Martens, P. Hoet, The nanosilica hazard: another variable entity, Part. Fibre Toxicol. 7 (1) (2010) 39, https://doi.org/10.1186/1743-8977-7-39.

[42] J. Michael Berg, A. Romoser, D. Figueroa, C. Spencer West, C. Sayes, Comparative cytological responses of lung epithelial and pleural mesothelial cells following in vitro exposure to nanoscale SiO2, Toxicol. in Vitro 27 (1) (2013) 24–33.

[43] H. Nabeshi, T. Yoshikawa, K. Matsuyama, Y. Nakazato, K. Matsuo, A. Arimori, et al., Systemic distribution, nuclear entry and cytotoxicity of amorphous nanosilica following topical application, Biomaterials 32 (11) (2011) 2713–2724, https://doi.org/10.1016/j.biomaterials.2010.12.042.

[44] C. Leung, I. Yu, W. Chen, Silicosis, Lancet 379 (9830) (2012) 2008–2018, https://doi.org/10.1016/s0140-6736(12)60235-9.

[45] P. Fu, Q. Xia, H. Hwang, P. Ray, H. Yu, Mechanisms of nanotoxicity: generation of reactive oxygen species, J. Food Drug Anal. 22 (1) (2014) 64–75, https://doi.org/10.1016/j.jfda.2014.01.005.

[46] A. Nemmar, P. Yuvaraju, S. Beegam, J. Pathan, E. Kazzam, B. Ali, Oxidative stress, inflammation, and DNA damage in multiple organs of mice acutely exposed to amorphous silica nanoparticles, Int. J. Nanomedicine 919 (2016), https://doi.org/10.2147/ijn.s92278.

[47] J. Corbalan, C. Medina, Jacoby, Malinski, M. Radomski, C. Medina, Amorphous silica nanoparticles aggregate human platelets: potential implications for vascular homeostasis, Int. J. Nanomedicine 631 (2012), https://doi.org/10.2147/ijn.s28293.

[48] T. Hirai, T. Yoshikawa, T. Yoshida, M. Uji, H. Nabeshi, Y. Yoshioka, et al., The safety assessment of amorphous nanosilica following dermal exposure, Toxicol. Lett. 205 (2011) S283, https://doi.org/10.1016/j.toxlet.2011.05.960.

[49] F. Rancan, Q. Gao, C. Graf, S. Troppens, S. Hadam, S. Hackbarth, et al., Skin penetration and cellular uptake of amorphous silica nanoparticles with variable size, surface functionalization, and colloidal stability, ACS Nano 6 (8) (2012) 6829–6842, https://doi.org/10.1021/nn301622h.

[50] S. An, H. Ryu, N. Seong, B. So, H. Seo, J. Kim, et al., Evaluation of silica nanoparticle toxicity after topical exposure for 90 days, Int. J. Nanomedicine 127 (2014), https://doi.org/10.2147/ijn.s57929.

[51] G. Oberdörster, E. Oberdörster, J. Oberdörster, Nanotoxicology: An emerging discipline evolving from studies of ultrafine particles, Environ. Health Perspect. 113 (7) (2005) 823–839, https://doi.org/10.1289/ehp.7339.

[52] D. Napierska, L. Thomassen, V. Rabolli, D. Lison, L. Gonzalez, M. Kirsch-Volders, et al., Size-dependent cytotoxicity of monodisperse silica nanoparticles in human endothelial cells, Small 5 (7) (2009) 846–853, https://doi.org/10.1002/smll.200800461.

[53] K. Gerloff, C. Albrecht, A. Boots, I. Forster, R. Schins, Cytotoxicity and oxidative DNA damage by nanoparticles in human intestinal Caco-2 cells, Nanotoxicology (2009) 1–10, https://doi.org/10.1080/17435390903276933.

[54] H. Kim, E. Ahn, B. Jee, H. Yoon, K. Lee, Y. Lim, Nanoparticulate-induced toxicity and related mechanism in vitro and in vivo, J. Nanopart. Res. 11 (1) (2009) 55–65, https://doi.org/10.1007/s11051-008-9447-3.

[55] N. Mueller, B. Nowack, Exposure modeling of engineered nanoparticles in the environment, Environ. Sci. Technol. 42 (12) (2008) 4447–4453, https://doi.org/10.1021/es7029637.

[56] M. Montazer, S. Seifollahzadeh, Enhanced self-cleaning, antibacterial and UV protection properties of Nano TiO2 treated textile through enzymatic pretreatment, Photochem. Photobiol. 87 (4) (2011) 877–883, https://doi.org/10.1111/j.1751-1097.2011.00917.x.

[57] D. Achudhan, S. Vijayakumar, B. Malaikozhundan, M. Divya, M. Jothirajan, K. Subbian, et al., The antibacterial, antibiofilm, antifogging and mosquitocidal activities of titanium dioxide (TiO2) nanoparticles green-synthesized using multiple plants extracts, J. Environ. Chem. Eng. 8 (6) (2020) 104521, https://doi.org/10.1016/j.jece.2020.104521.

[58] E. Baranowska-Wójcik, D. Szwajgier, P. Oleszczuk, A. Winiarska-Mieczan, Effects of titanium dioxide nanoparticles exposure on human health—a review, Biol. Trace Elem. Res. 193 (1) (2019) 118–129, https://doi.org/10.1007/s12011-019-01706-6.

[59] M. Senzui, T. Tamura, K. Miura, Y. Ikarashi, Y. Watanabe, M. Fujii, Study on penetration of titanium dioxide (TiO2) nanoparticles into intact and damaged skin in vitro, J. Toxicol. Sci. 35 (1) (2010) 107–113, https://doi.org/10.2131/jts.35.107.

[60] M. Geiser, B. Rothen-Rutishauser, N. Kapp, S. Schürch, W. Kreyling, H. Schulz, et al., Ultrafine particles cross cellular membranes by nonphagocytic mechanisms in lungs and in cultured cells, Environ. Health Perspect. 113 (11) (2005) 1555–1560, https://doi.org/10.1289/ehp.8006.

[61] M. Scuri, B. Chen, V. Castranova, J. Reynolds, V. Johnson, L. Samsell, et al., Effects of titanium dioxide nanoparticle exposure on Neuroimmune responses in rat airways, J. Toxic. Environ. Health A 73 (20) (2010) 1353–1369, https://doi.org/10.1080/15287394.2010.497436.

[62] M. Skocaj, M. Filipic, J. Petkovic, S. Novak, Titanium dioxide in our everyday life; is it safe? Radiol. Oncol. 45 (4) (2011), https://doi.org/10.2478/v10019-011-0037-0.

[63] B. Song, J. Liu, X. Feng, L. Wei, L. Shao, A review on potential neurotoxicity of titanium dioxide nanoparticles, Nanoscale Res. Lett. 10 (1) (2015), https://doi.org/10.1186/s11671-015-1042-9.

[64] D. Cadosch, E. Chan, O. Gautschi, L. Filgueira, Metal is not inert: role of metal ions released by biocorrosion in aseptic loosening-current concepts, J. Biomed. Mater. Res. A 91A (4) (2009) 1252–1262, https://doi.org/10.1002/jbm.a.32625.

[65] J. Wang, G. Zhou, C. Chen, H. Yu, T. Wang, Y. Ma, et al., Acute toxicity and biodistribution of different sized titanium dioxide particles in mice after oral administration, Toxicol. Lett. 168 (2) (2007) 176–185, https://doi.org/10.1016/j.toxlet.2006.12.001.

[66] M. Behnam, F. Emami, Z. Sobhani, O. Koohi-Hosseinabadi, A. Dehghanian, S. Zebarjad, M. Moghim, A. Oryan, Novel combination of silver nanoparticles and carbon nanotubes for Plasmonic photo thermal therapy in melanoma Cancer model, Adv. Pharm. Bull. 8 (1) (2018) 49–55.

[67] H. Shi, R. Magaye, V. Castranova, J. Zhao, Titanium dioxide nanoparticles: a review of current toxicological data, Part. Fibre Toxicol. 10 (1) (2013) 15, https://doi.org/10.1186/1743-8977-10-15.

[68] F. Nishino, M. Jeem, L. Zhang, K. Okamoto, S. Okabe, S. Watanabe, Formation of CuO nano-flowered surfaces via submerged photo-synthesis of crystallites and their antimicrobial activity, Sci. Rep. 7 (1) (2017), https://doi.org/10.1038/s41598-017-01194-5.

[69] M. Godley, B. Tomar, A. Tripathi, Investigation of automobile radiator using Nanofluid-CuO/water mixture as coolant, Int. J. Adv. Sci. Eng. Inf. Technol. 2 (2350–0328) (2015) 1136–1145. Retrieved 9

March 2021, from http://www.ijarset.com/upload/2015/december/9_IJARSET_bhanupratapsingh.pdf.
[70] M. Kao, C. Lo, T. Tsung, Y. Wu, C. Jwo, H. Lin, Copper-oxide brake nanofluid manufactured using arc-submerged nanoparticle synthesis system, J. Alloys Compd. 434-435 (2007) 672–674, https://doi.org/10.1016/j.jallcom.2006.08.305.
[71] L. Pena-Paras, J. Gutiérrez, M. Irigoyen, M. Lozano, M. Velarde, D. Maldonado-Cortes, J. Taha-Tijerina, Study on the anti-wear properties of metal-forming lubricants with TiO2 and CuO nanoparticle additives, IOP Conf. Ser.: Mater. Sci. Eng. 400 (2018), https://doi.org/10.1088/1757-899x/400/6/062022, 062022.
[72] A. Lazary, I. Weinberg, J. Vatine, A. Jefidoff, R. Bardenstein, G. Borkow, N. Ohana, Reduction of healthcare-associated infections in a long-term care brain injury ward by replacing regular linens with biocidal copper oxide impregnated linens, Int. J. Infect. Dis. 24 (2014) 23–29, https://doi.org/10.1016/j.ijid.2014.01.022.
[73] L. Jaiswal, S. Shankar, J. Rhim, Methods in microbiology, in: 43–60, Academic Press, 2021.
[74] A. Rydosz, The use of copper oxide thin films in gas-sensing applications, Coatings 8 (12) (2018) 425, https://doi.org/10.3390/coatings8120425.
[75] S. Naz, A. Gul, M. Zia, Toxicity of copper oxide nanoparticles: a review study, IET Nanobiotechnol. 14 (1) (2019) 1–13, https://doi.org/10.1049/iet-nbt.2019.0176.
[76] D. Docter, D. Westmeier, M. Markiewicz, S. Stolte, S. Knauer, R. Stauber, The nanoparticle biomolecule corona: lessons learned – challenge accepted? Chem. Soc. Rev. 44 (17) (2015) 6094–6121, https://doi.org/10.1039/c5cs00217f.
[77] Y. Liu, Y. Gao, L. Zhang, T. Wang, J. Wang, F. Jiao, et al., Potential health impact on mice after nasal instillation of Nano-sized copper particles and their translocation in mice, J. Nanosci. Nanotechnol. 9 (11) (2009) 6335–6343, https://doi.org/10.1166/jnn.2009.1320.
[78] W. De Jong, E. De Rijk, A. Bonetto, W. Wohlleben, V. Stone, A. Brunelli, et al., Toxicity of copper oxide and basic copper carbonate nanoparticles after short-term oral exposure in rats, Nanotoxicology 13 (1) (2018) 50–72, https://doi.org/10.1080/17435390.2018.1530390.
[79] L.I. Privalova, B.A. Katsnelson, N.V. Loginova, V.B. Gurvich, V.Y. Shur, I.E. Valamina, O.H. Makeyev, et al., Subchronic toxicity of copper oxide nanoparticles and its attenuation with the help of a combination of bioprotectors, Int. J. Mol. Sci. 15 (7) (2014) 12379–12406, https://doi.org/10.3390/ijms150712379.
[80] V. Rani, A. Kumar, C. Kumar, A. Reddy, Pulmonary toxicity of copper oxide (CuO) nanoparticles in rats, J. Med. Sci. 13 (7) (2013) 571–577, https://doi.org/10.3923/jms.2013.571.577.
[81] E. Assadian, M. Zarei, A. Gilani, M. Farshin, H. Degampanah, J. Pourahmad, Toxicity of copper oxide (CuO) nanoparticles on human blood lymphocytes, Biol. Trace Elem. Res. 184 (2) (2017) 350–357, https://doi.org/10.1007/s12011-017-1170-4.
[82] S. Alarifi, D. Ali, A. Verma, S. Alakhtani, B. Ali, Cytotoxicity and genotoxicity of copper oxide nanoparticles in human skin keratinocytes cells, Int. J. Toxicol. 32 (4) (2013) 296–307, https://doi.org/10.1177/1091581813487563.
[83] B.M. Prabhu, S.F. Ali, R.C. Murdock, S.M. Hussain, M. Srivatsan, Copper nanoparticles exert size and concentration dependent toxicity on somatosensory neurons of rat, Nanotechnology 4 (2020) 150–160, https://doi.org/10.3109/17435390903337693.

CHAPTER 32

Nanolubricant additives

Mohamed Kamal Ahmed Ali[a], Mohamed A.A. Abdelkareem[a], Ahmed Elagouz[a], and Hou Xianjun[b]

[a]Automotive and Tractors Engineering Department, Faculty of Engineering, Minia University, El-Minia, Egypt
[b]Hubei Key Laboratory of Advanced Technology for Automotive Components, Wuhan University of Technology, Wuhan, China

Chapter outline

1. Introduction — 675
2. Preparation of nanolubricants — 679
   2.1 Dispersion stability — 679
   2.2 Surfactants — 681
   2.3 Characterization of dispersion stability — 681
3. Tribological and thermophysical performance of nanolubricant additives — 686
4. Mechanisms of nanolubricant additives — 692
   4.1 Rolling bearing effect — 692
   4.2 Tribofilm formation — 693
   4.3 Microstructure transformation — 694
   4.4 Surface repairing effect — 695
5. Role of nanolubricants in improving vehicle engines performance — 697
   5.1 Fuel economy in vehicle engines — 700
   5.2 Effect of nanolubricants on exhaust emissions — 702
6. Conclusions and recommendations — 704
   Acknowledgments — 706
   References — 706

1. Introduction

In the 21st century, one of the major worldwide challenges is the development of new energy technologies due to serious problems of energy generation and utilization [1, 2]. In the transportation sector, the global energy problem can be related to insufficient fossil fuel supplies and massive levels of emissions resulting from internal combustion engines due to increasing fuel consumption. We now use 13 TW of energy per year. It is interesting to consider that the global energy market is predicted to be nearly 30 and 46 TW by 2050 and 2100, respectively [3]. In recent years, Nanotechnologies have been considered one of the most recommended alternatives to solve energy problems [4]. The term "Nanotechnology" describes the process technologies and analytical techniques for materials at the nanoscale (millionth of a millimeter) range [5]. Nanotechnology applications in the automotive industry have provided principal roles in the global economy.

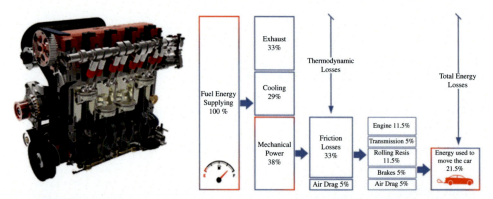

Fig. 1 Breakdown of Energy use in automotive engines. *(Reproduced from Ref. M.A. Abdelkareem, L. Xu, M.K.A. Ali, A. Elagouz, J. Mi, S. Guo, et al., Vibration energy harvesting in automotive suspension system: a detailed review, Appl. Energy 229 (2018) 672–99, with permission, © Elsevier, 2018.)*

Nanotechnology has numerous applications in automotive, such as pollution sensing, energy conversion, and friction/wear reduction for automobiles (nanotribology) [6]. From an energy-saving point of view, Skjoedt et al. [7] reported the reduction of engine friction by 10% for all US passenger cars (in 2007) would result in a fuel economy of 3.4 billion gallons. Furthermore, the losses in the United States caused by tribology ignorance in various applications were estimated to be around 6% of the total national product, or $420 billion per year [8]. Hence, frictional power losses are one of the critical challenges in mechanical systems, including automotive engines.

In terms of energy consumption and environmental protection, vehicles account for approximately 19% of global energy consumption and 23% of total greenhouse-gas emissions each year [9, 10]. The research on fuel economy and reducing total frictional power losses has become a major research aspect in automotive engine performance via improving tribological performance [11, 12]. Hence, the ability of nanotribology to enhance fuel economy is critical in automotive engines worldwide. Total power generated by the engine is reduced in the range of 12%–19% due to the frictional losses, as shown in Fig. 1 [13]. The lubrication is classified into three general regimes: boundary, mixed-elastohydrodynamic, and hydrodynamic, as shown in Fig. 2 [4]. The piston ring assembly contributes about 40%–50% of total frictional losses in automotive engines, so improving its tribological efficiency is critical to lowering total frictional losses [14]. To address this issue, most designers and researchers have focused on Nanotechnology in automotive engines as the primary strategy for reducing frictional power losses, excessive heat generation, and contact surface wear, eventually leading to improved engine efficiency.

Recent literature reports that 40%–55% of the engine friction losses have been associated with the total friction of the piston assembly, as shown in Fig. 3A. If such friction

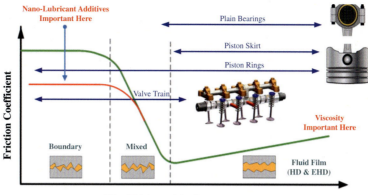

Fig. 2 Stribeck curve representing different lubrication regimes for automotive engines. *(Reprinted (adapted) from Ref. M.K.A. Ali, H. Xianjun, F. Essa, M.A. Abdelkareem, A. Elagouz, S. Sharshir, Friction and wear reduction mechanisms of the reciprocating contact interfaces using nanolubricant under different loads and speeds, J. Tribol. 140 (2018).)*

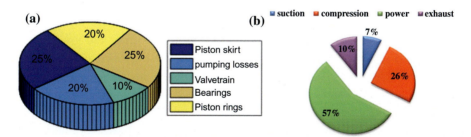

Fig. 3 The distribution of the frictional power losses in an engine and friction losses % in various strokes. *(Reprinted (adapted) from Ref. M.K.A. Ali, H. Xianjun, R. FiifiTurkson, M. Ezzat, An analytical study of tribological parameters between piston ring and cylinder liner in internal combustion engines, Proc. Inst. Mech. Eng. Pt. K J. Multi-body Dyn. 230 (2016) 329–49; P. Mishra, A review of piston compression ring tribology, Tribol. Ind. 36 (2014).)*

power loss could be comprehensively reduced, it will be positively reflected on an enhanced fuel-saving and could increase the life span for engine parts [15, 16]. Thus, existing and future automotive engines would require more efficient engine oils. This situation poses a new challenge for researchers and designers in terms of improving the tribological characteristics of automotive engines while also lowering fuel and lube oil consumption [17]. As shown in Fig. 3B, the maximum amount of friction loss is 57% during the power stroke and 26% during the compression stroke of an engine cycle. Different friction mechanisms may occur due to the sliding action between the piston ring and the cylinder liner during one engine's working cycle, mainly because of the variations

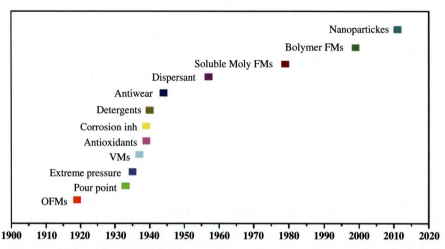

Fig. 4 Temporal evolution of the uses of lubrication enhancers. *(Adapted from Ref. H. Spikes, Friction modifier additives, Tribol. Lett. 60 (2015) 5.)*

in speed, load, and opposite surface effects [18]. The most extreme lubrication conditions (boundary lubrication) occur during the compression stroke at the top dead center (TDC), particularly for the top ring, due to the high temperature and low viscosity of the lubricant [19]. With increasing engine speed, the friction force also increases due to the increased lube-oil shear stress [20]. Friction coefficient values at TDC and BDC locations ranged from 0.10 to 0.15, and mid-stroke values ranged from 0.05 to 0.10. Therefore, these values depend on the surface texture and material, the lubricant, and other operating conditions [21]. A piston ring may undergo boundary, mixed and full film lubrication (hydrodynamic) in one piston stroke [22]. Mixed and boundary lubrication regimes show the highest friction and occur near TDC and BDC [23].

Manufacturing technology of lubricants uses additives to improve lube oil behavior. Additives were the main way used to upgrade lubricant properties. Generally, lubricant enhancers were first used in the first two decades of the 19th century. They began with organic friction modifiers (OFMs) as antiwear, followed by other additives, and finished with nanoparticles as in Fig. 4 [24]. Additive components are generally classified into chemically active types and chemically inert. Chemically active additives form protective films on metals to minimize wear and pollutants and avoid viscosity increases. Dispersants, detergents, antiwear, oxidation inhibitors, intense pressure (EP) agents, friction modifiers (FM), and rust and corrosion inhibitors are examples of chemically active additives. Chemically inert additives improve the lubricant physical properties (e.g., emulsifiers, foam inhibitors, demulsifiers, pour point depressants, viscosity modifiers, etc.) [25]. Formulating a lubricant does not involve blending all the best additives, leading to undesirable interactions between additives. Additives interact in a variety of ways, both in the bulk oil (forming complexes) and on surfaces (competition with other additives for surface sites), resulting in synergies or antagonisms which greatly complicate the task of oil formulation [26].

Current and future automotive engines would need more efficient engine oils, posing a new challenge for designers and researchers worldwide looking for ways to improve the tribological properties of these engines [27]. Accordingly, the studies on frictional power loss reduction obtained prominent consideration as a promising trend in improving the performance of automobile engines for fuel economy [4, 28]. Additionally, new efficiency and emission standards imposed on vehicles have been the principal driving power behind the advancement of eco-friendly and more fuel-efficient lube oil over the years [29]. The use of nanoparticles as nanolubricant additives is a new concept combining both liquid and solid lubrication benefits. Nanolubricants are an attractive oil modification mechanism commonly used because no major hardware modifications are needed.

2. Preparation of nanolubricants

The incorporation of nanomaterials into lube oils always results in problems uniformly disperse. Nanolubricants are prepared via two major techniques: (i) single-step direct evaporation, in which nanomaterials are directly fabricated in the base liquids, and (ii) the two-step method, in which the nanomaterials are synthesized and subsequently dispersed in the base liquids [30]. Nanomaterials can be dispersed by various physical techniques (magnetic stirrers, ultrasonic baths, ultrasonic probes, ball mills, and high-shear mixers) [31]. The recent literature also reported the chemical methods (modification agents of nanoparticles surface) [32]. Moreover, additional techniques for preparing nanolubricants in a single step, such as the green synthesis by pulsed laser technique and modified magnetron sputtering system [33]. The physical properties of nanomaterials (e.g., morphological shape, grain size, and concentration) have been proved to have significant influences on nanolubricantdispersion stability [34]. Nanomaterials with high surface activities and large surface areas are most likely to create intimate contact with each other, leading to nanomaterials aggregation and reducing the dispersion stability. In addition, a high-concentration nanomaterial increases the density of nanoparticles in the base liquid. It thus makes the nanoparticles closer to each other, which increases the attraction forces of Van Der Waals and leads to a high level of nanomaterial agglomeration [35].

2.1 Dispersion stability

Despite the ability of lubricants blended with nanomaterials to enhance the tribological mechanism, such nanolube oils are still in progress and have yet to reach utility-scale commercialized lube oil production lines. Nanolubricant science is concerned with the stability of nanoparticles since it remains a great challenge to achieve long-term

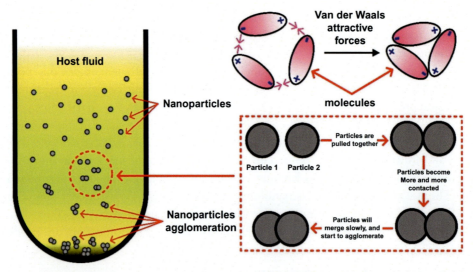

Fig. 5 Agglomeration of nanoparticles under Van der Waals attractive forces.

well-suspended nanolubricants. The limited dispersion stability of these nanolubricants, owing to chemical reactions between nanomaterials themselves and between nanoparticles and the host fluid and its additives, is one of these challenges [36]. The poor dispersion stability of nanolubricantsis mainly caused by the Van der Waals attractive forces [37]. After a while, due to the Van der Waals effect, the nanoparticles are agglomerated, which both increases the particle grain size and detaches the nanoparticles from the host fluid, and thereby, these nanoparticles will then settle down in the bottom of the container due to the gravitational force. The agglomeration of nanoparticles under the Van der Waals phenomena is illustrated in Fig. 5.

Derjaguin, Verwey, Landau, and Overbeek (DVLO) suggested a hypothesis to deal with nanolubricants/nanofluids dispersion stability [38]. This hypothesis presents that colloidal stability is characterized by the attractive van der Waals forces and electrical repulsion forces between the nanoadditives. We can see that the Van der Waals effect forces particles to be pulled toward each other by molecule attraction, resulting in direct surface contact. This also means that as more molecules are closely contacted, the binding force becomes greater and stronger. The nanoparticles would then slowly agglomerate together and sink to the container floor. On the other side, the electrical double layer repulsive force allows reseparation of the agglomerated nanoparticles thanks to the steric and electrostatic repulsion mechanisms [39]. Keeping the nanoparticles under continuous repulsion allows increasing their dispersion stability, and thereby this allows nanolube oils to hold their desired thermophysical and tribological potentials [40]. In this regard, researchers are still focusing on developing long-lasting, robust, durable, and stable nanofluids for lubrication purposes by enhancing nanomaterial dispersion stability in host fluids.

2.2 Surfactants

Researchers successfully suggested using surfactants, commonly known as dispersants, to improve the stability of nanolubricants. Dispersants prevent nanoparticles from agglomerate since they reduce the surface tension of the base oil and enhance nanoparticle immersion. Table 1 reports a comprehensive literature review to assess and summarize the results of the conducted studies on the dispersion stability of the nanoadditives using different surfactants. Surfactant with a functional group attached to oleic acid is proved to reduce attractive forces and increase the electrostatic repulsion of nanomaterials which effectively prevents agglomeration and increases dispersion stability [14, 57]. Gulzar et al. [43] evaluated the stability of the Al_2O_3/TiO_2-Therminol-55 nanofluid mixed with the oleic acid as a surfactant. The results concluded that oleic acid with a long chain of fatty acids improves the dispersion stability by creating a coating layer on the nanoparticles, resulting in reducing their attractive forces and preventing agglomeration. Several major mechanisms are responsible for that, including but are not limited to steric hindrance, electrostatic repulsion, the formation of an electrical adsorption layer, polarization change, polymeric chain interaction, and the incorporation of grafted polymers on the surface of nanoparticles.

2.3 Characterization of dispersion stability

The dispersion performance of the nanolubricants was investigated by sedimentation photograph capturing method, electron microscopy methods, UV–vis spectroscopy, dynamic light scattering (DLS), and zeta potential analyses [41]. The analysis of the dispersion stability of the Cu nanoadditives modified with bis(2-ethylhexyl) phosphate (HDEHP) as surfactant is reported, as shown in Fig. 6. From the sedimentation images, it can be observed that the Cu-PAO6 with the HDEHP presented no sedimentation. The results from the ultraviolet-visible (UV–vis) spectroscopy confirmed the stability of Cunanolubricant, and the analysis also suggested 60 days of stability could be obtained for a lower concentration of Cu nanomaterials [31]. The dispersion stability enhancer mechanism is the formation of an electrical adsorption layer on the Cu surface, which increases the electrostatic repulsion forces and reduces agglomeration.

Wang et al. [58] reported that Ag@graphene with laser irradiation (L-Ag@rGO) presented notable self-dispersed characteristics than Ag@GO. Their results also explained that the suspension of L-Ag@rGO showed a 60 days dispersion stability, while Ag@GO dispersion remained stable for only 30 days, as observed in Fig. 7A and B. Furthermore, the absorption coefficient of Ag@GO nanoadditives decreased quickly with the static time. At the same time, the L-Ag@rGO nanoadditives in base oil prepared a higher absorption coefficient during 60 days, as noted in Fig. 7C and D.

As shown in Fig. 8, Ali and Xianjun [41] proposed improving the dispersion stability of the hybrid nanolubricant of Al_2O_3/TiO_2-PAO6 by using HDEHP surfactant. With increasing settling time, the dynamic light scattering (DLS) results illustrated that the

Table 1 A comprehensive review of studies associated with nanolubricantdispersion stability using different surfactants.

			Fluid, surfactant, and nanomaterial details						
Year	Ref.	Nanomaterial	Grain size	Concentration	Host fluid	Surfactant (concentration/weight)	Ultrasonication time (device)	Stability time	Findings
2021	[41]	Al_2O_3/TiO_2	25–30 nm	0.1 wt%	PAO6[a] oil	HDEHP[b] (5 wt%)	6 h (ultrasonication)	70 days	Due to its high charge density and steric repulsion, HDEHP provided electrostatic and electrosteric stabilization by its strong chemisorption on the nanoparticle surfaces, which enhanced the stability of their dispersion by generating steric hindrance and electrostatic repulsion
		TiO_2	25–30 nm	0.1 wt%	PAO6 oil	HDEHP (5 wt%)		–	
		Al_2O_3	25–30 nm	0.1 wt%	PAO6 oil	HDEHP (5 wt%)		50 days	
2020	[42]	TiO_2	25 nm	0.1 wt%	PAO6 oil	2-Octyl dodecyl gallic acid ester [ODG]	12 h (magnetic stirrer)	5 days	Surface modification, in situ surface polymerization using ODG, increases the TiO_2 NPs dispersion stability in nonpolar oils due to the alkyls that help slow down the sedimentation kinetic
2020	[31]	Cu	10–20 nm	0.07–0.10 wt%	PAO6 oil	HDEHP (2 wt%)	4 h (ultrasonic probe)	60 days	The nanolubricant stabilization is improved by forming an electrical adsorption layer on the Cu surface, which increases the electrostatic repulsion forces and reduces agglomeration
2019	[43]	Al_2O_3/TiO_2	15–25 nm (TiO_2), <80 nm (Al_2O_3)	0.05–0.5 wt%	Therminol-55 hybrid	Oleic acid	4 h magnetic stirrer and followed by 2 h ultrasonic probe	7 days	Oleic acid with a long chain of fatty acids improves the dispersion stability by creating a coating layer on the nanoparticles, resulting in reducing their attractive forces and preventing agglomeration
2019	[44]	Al_2O_3	20 nm	1.0 vol%	Ethylene glycol (EG)	Nonionic polymer of Polyvinylpyrrolidone [PVP]	0.5 h magnetic stirring and followed by 2 h ultrasonic vibration	30 days	The Al_2O_3– EG nanofluids with the PPV surfactant achieved better stabilization and homogenization than the SDS due to the polymeric chain interaction
						Anionic material of sodium dodecyl sulfate [SDS]		10 days	
2019	[45]	MWCNT	2–5 nm (diameter)	0.1–0.5 vol%	Water	–	10–80 min (ultrasonic probe)	30 days	The increase of the ultrasonic time to 60 min leads to increased sample stability while extending the ultrasonic time furthermore deteriorates the stability
2018	[46]	Graphene oxide (GO)	~0.8 nm (thickness)	0.2–1.0 wt%	Hydrophobic ionic liquid [EMIM][TF_2N]	Ionic liquid unit with carboxylic acid group[c] [ILC_9-COOH]	2 h	5 days	GO sheets were coated with ILC_9-COOH, and thereby the modified GO contains positively charged butylimidazole moieties on its surfaces. The high affinity of the surfactant, which provides sufficient steric and electrostatic repulsions resulted in stable colloidal suspension

Year	Ref	Nanoparticle	Size	Concentration	Base fluid	Surfactant	Processing	Stability	Observation
2018	[47]	SiO_2	10–15 nm	0.1 wt%	PAO4 oil	[P8888][DEHP]^d (1.04 wt%)	60 min (ultracentrifugation)	10 days	If used as a nano additive in PAO, the SiO_2 NPs and [P8888][DEHP] have a stable dispersion in the presence of IL, even in high temperatures, for a longer duration
2017	[48]	TiO_2/SiO_2	50 nm	0.75 wt%	Palm TMP ester	—	30 min (ultrasonic probe)	3 days	Without a surfactant, the TiO_2/SiO_2 offered noticeable dispersion, which not only provided a robust suspension but also reduced friction and wears in comparison with blank Palm TMP Ester
2017	[49]	TiO_2	30 nm	0.5–2 wt%	Water	—	10 min (ultrasonic vibration)	3 days	After 3 days, TiO_2 nanolubricant showed has good stability and dispersibility
2016	[28]	Al_2O_3/TiO_2	12–20 nm	0.25 wt%	5 W-30	Oleic acid (1.75 wt%)	4 h (magnetic stirrer)	14 days	Oleic acid facilitated the mixing of nanoparticles to make a soluble nanofluid engine oil
2016	[50]	CuO Al_2O_3	30–50 nm 10 nm	0.05–0.15 wt% 0.05–0.15 wt%	Deionized water	SDBS (0.05–0.2 wt%) SDBS ^{e} (0.05–0.2 wt%)	— —	— —	The SDBS surfactant helps improve the dispersion stability and thermal conductivity and reduces the viscosity of both the CuO_2 and Al_2O_3 nanofluids
2016	[51]	Al_2O_3	45 nm	0.3–1.0 vol%	Therminol 66 (diathermic oil)	Oleic acid (90% purity)	30 min magnetic stirring and followed by a 10 min sonication	8 days	Oleic acid enhances stability. However, excessive surfactant determines sedimentation because of the formed double chain in nanoparticle clusters
2014	[52]	Al_2O_3/TiO_2	75 nm	0.05–0.1 wt%	Carrier oil	Silane coupling agent^f (1.5 wt%)	30 min (ultrasonication)	5 days	Due to the incorporation of grafted polymers on the surface of nanoparticles, the modified Al_2O_3/TiO_2 nanofluid greatly increases dispersion stability
2014	[53]	Al_2O_3	78 nm	0.1 wt%	Base oil	KH-560 (1.5 wt%)	30 min (magnetic stirring)	20 days	The agglomerate of Al_2O_3 nanolubricant is effectively prevented when the Al_2O_3 surface changes from hydrophilic to hydrophobic after modification with KH-560. The Al_2O_3-KH-560 nanofluid obtained homogeneous dispersion and thereby good stability.
2013	[54]	MWCNT	20–30 nm (diameter)	1.27 vol%	Deionized water	SDBS and TritonX-100^g	(high-shear mixer)	—	Surfactants are used for the convenient stability of nanofluids
2012	[55]	Al_2O_3/TiO_2	79–103 nm	0.05–0.5 wt%	Mineral lube oil	Acrylic acid (1 g)	30 min (magnetic stirring)	10 days	For several weeks, the TiO_2/Al_2O_3 nanolubricant was observed to be homogeneously dispersed with significantly improved dispersion stabilization due to the presence of grafted polymers

Continued

Table 1 A comprehensive review of studies associated with nanolubricantdispersion stability using different surfactants—cont'd

			Fluid, surfactant, and nanomaterial details					
Year	Ref.	Nanomaterial	Grain size	Concentration	Host fluid	Surfactant (concentration/weight)	Ultrasonication time (device)	Stability time	Findings
2007	[56]	Fe_3O_4	9–10 nm	1.75–1 vol%	Deionized water	Oleate sodium and PEG-4000	1 h (centrifugation at 6400g)	60 days	Surfactant improves the sedimentation and keeps nanoparticles from being aggregated for about 60 days

[a]Polyalphaolefin (a synthetic oil).
[b]Bis(2-ethylhexyl) phosphate (HDEHP).
[c]A mixture of 1-butylimidazole and 10-bromodecanoic acid (1:1) was heated in an oil bath for 9 h under 120°C to fabricate the carboxylic acid group surfactant, 1-butyl-3-(9-carboxydecyl)-1H-imidazol-3-ium bromide (ILC_9–COOH).
[d]Tetraoctylphosphoniumbis(2-ethylhexyl) phosphate ([P8888][DEHP]).
[e]SDBS: dispersant sodium dodecylbenzene sulfonate.
[f]The silane coupling agent is 3-glycidoxypropyltrimethoxysilane, which is also named KH-560.
[g]The surfactant was prepared from a 20:1 mass ratio of a binary mixture of a nonionic surfactant of TritonX-100 and an anionic surfactant of sodium dodecyl benzene sulfonate (SDBS).

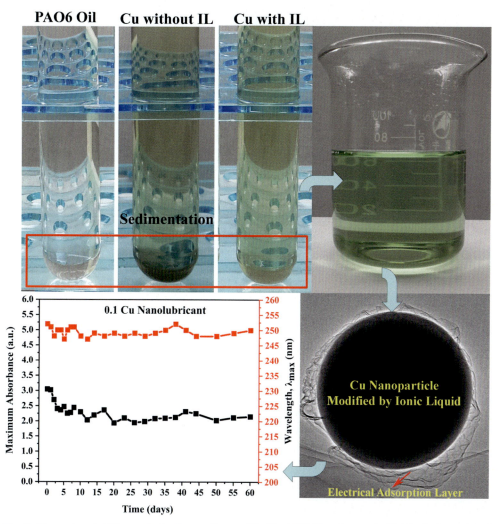

Fig. 6 Dispersion stability analysis of nano-Cu modified by HDEHP surfactant in PAO6 oil. *(Reproduced from Ref. M.K.A. Ali, H. Xianjun, Colloidal stability mechanism of copper nanomaterials modified by bis (2-ethylhexyl) phosphate dispersed in polyalphaolefin oil as green nanolubricants. J. Colloid Interface Sci. 578 (2020) 24–36, with permission,© Elsevier, 2020.)*

average diameter of the hybrid Al_2O_3/TiO_2 nanoadditives hardly changed for 55-days with a much lower hydrodynamic diameter than the individual nanoadditives, which obtained a noticeable increase in their particle diameters. After 70 days, as shown in Fig. 8, we can observe that the TiO_2 nanolubricant presented the lowest stability with noticeable agglomerated particles due to the higher density and lower specific surface area

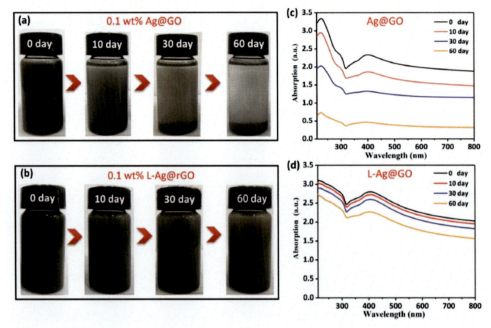

Fig. 7 Dispersion stability of Ag@GO and L-Ag@rGO nanoadditives versus different static times. *(Reproduced from Ref. L. Wang, P. Gong, W. Li, T. Luo, B. Cao, Mono-dispersed ag/graphene nanocomposite as lubricant additive to reduce friction and wear, Tribol. Int. 146 (2020) 106228 with permission,© Elsevier, 2020.)*

TiO$_2$ nanoparticles compared to the Al$_2$O$_3$ nanolubricant. The main dispersing mechanism responsible for stability enhancement is electrosteric stabilization as an advantage of using the HDEHP, which generates steric hindrance and electrostatic repulsion to reduce nanomaterial agglomeration. In Ref. [47], a 1 wt% of the tetraoctylphosphonium bis(2-ethylhexyl) phosphate ([P8888][DEHP]) as a surfactant was considered to be blended in nano-SiO$_2$ in PAO 4 oil, and the results of the SiO$_2$ nanolubricant presented a good stable dispersion.

3. Tribological and thermophysical performance of nanolubricant additives

In recent years, researchers have been focusing with increasing interest in nanomaterials and their use in many fields related to energy due to their unusual mechanical, electrical, and piezoelectric properties. Some nanomaterials are added to engine lube-oils (nanoadditives) to improve the tribological properties by forming a protective layer on the surfaces and creating a rolling effect between sliding surfaces [12, 27, 59]. Nanoparticles are

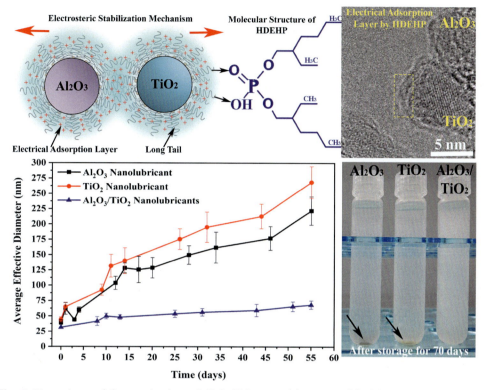

Fig. 8 Dispersion stability evaluation of Al_2O_3/TiO_2 nanoadditives modified by HDEHP surfactant. (Reproduced from Ref. M.K.A. Ali, H. Xianjun, Role of bis(2-ethylhexyl) phosphate and Al2O3/TiO2 hybrid nanomaterials in improving the dispersion stability of nanolubricants, Tribol. Int. 155 (2021) 106767 with permission,© Elsevier, 2021.)

cohesive and soft particles representing the third body between two surfaces to reduce friction and corrosion. There are many forms of using nanoparticles, as they can serve as lubricating powders in engine oils or as solid material additives in solid lubrication [60]. Nanomaterials are powders and materials optimized at the nanoscale of 1–100 nm [61]. Some of the commonly used nanoadditives in lube oils are Al_2O_3, TiO_2, Cu, MoS_2, ZnO, Ag, WS_2, graphene, carbon nanotube, etc.

Ali et al. [14] studied the tribological characteristics for piston ring-cylinder liner contact during various lubrication regimes using Al_2O_3 and TiO_2 nanomaterials (8–12 nm) into the 5 W-30 engine oil. The experiments were conducted with a reciprocating test-bench under different conditions (185–340 N) normal load and (0.25–0.66 m/s) sliding speed. The findings showed that as compared to engine oil, the boundary friction, and wear rate improved by 35% and 41%, respectively, as shown in Fig. 9. Other studies found that adding 80 nm Al_2O_3 nanoparticles at a 5 wt% concentration decreased the

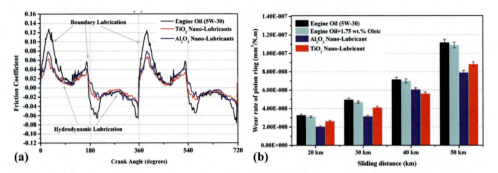

Fig. 9 Tribologicalperformance of piston ring assembly versus crank angle with and without nanoadditives under the boundary lubrication regime. *(Reproduced from Ref. M.K.A. Ali, H. Xianjun, L. Mai, C. Qingping, R.F. Turkson, C. Bicheng, Improving the tribological characteristics of piston ring assembly in automotive engines using Al2O3 and TiO2 nanomaterials as nano-lubricant additives, Tribol. Int. 103 (2016) 540–54 with permission, © Elsevier, 2016.)*

wear rate to 10–7 mm^3/Nm, compared to $0.7 * 10^{-3}$ mm^3/Nm for the Polytetrafluoroethylene (PTFE) [62]. Mohan et al. [61] investigated the tribological behavior of an engine using Al_2O_3 nanoparticles (20 nm &concentrations of 0.25, 0.50, and 0.75 wt %) into 20 W-40 engine oil. The findings indicate that a 0.5 wt% concentration of Al_2O_3 nanolubricant provided the best tribology behavior under starved lubrication. The friction coefficient decreased by 49.1% and 21.6% during submerged and starved lubrication conditions, respectively, compared to the reference lube oil, while wear depth decreased by 20.1% and 31.1%, respectively. Vasheghani et al. [63] reported that the viscosity and thermal conductivity for Al_2O_3 nanoparticles (20 nm and 3 wt%) into engine oil was increased by 31%–37% and 36%–38%, respectively, compared to engine oil. Additionally, the existence of TiO_2 in lube oil at a volume concentration of 0.01 showed a 40% rise in the load-bearing ability [64]. The major motivations for selecting Al_2O_3 and TiO_2 nanoparticles as nanoadditives in lube oils are eco-friendly, higher thermal stability, and inexpensive [65, 66]. Moreover, Al_2O_3 and TiO_2 nanoparticles offer excellent tribological properties as a solid lubricant at high temperatures [27]. TiO_2 and Al_2O_3 nanoadditives are acting as effective catalysts (oxygen donating/absorbing), which means they could play important roles in lowering engine exhaust emissions via oxygen absorption for NOx reduction or oxidation of CO and HC [4, 67].

Hybrid nanoadditives displayed properties not found in individual additives and also a synergistic effect between the nanoparticles [68]. The tribological behavior of an Al_2O_3/TiO_2 nanocomposite mixture with lube oil was demonstrated by Luo et al. [52]. The finding showed that Al_2O_3/TiO_2 nanolubricant could significantly decrease friction coefficient by 20.51% and enhance the antiwear property with an optimal concentration of 0.1 wt%.Moreover, the frictional power of piston ring conjunction was also decreased by 39%–53% for the Al_2O_3/TiO_2 hybrid nanoadditives [28]. Another research found

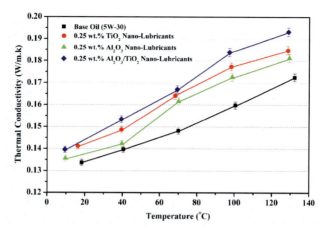

Fig. 10 Influence of Al_2O_3, TiO_2, and Al_2O_3/TiO_2 nanoadditives on thermal conductivity. *(Reprinted from Ref. M.K.A. Ali, H. Xianjun, R.F. Turkson, Z. Peng, X. Chen, Enhancing the thermophysical properties and tribologicalbehaviour of engine oils using nano-lubricant additives, RSC Adv. 6 (2016) 77913–24.)*

that using hybrid Al_2O_3/TiO_2 nanoparticles with an 80 nm diameter and a 0.1 wt%, the friction coefficient was decreased by 14.7% [55]. The thermophysical properties of nanolubricants are important factors for heat transfer performance and provide better resistance to lubricant thinning and film strength retention on the worn surfaces in the engine. The thermophysical tests by Ali et al. [11] showed that the viscosity index improved by 2% using Al_2O_3/TiO_2 nanolubricants. Besides, thermal conductivity was improved by 12%–16% in the temperature ranged from 10 to 130°C compared to the commercial oil, as shown in Fig. 10. Another study also showed that using Al_2O_3 nanolubricants improves the thermal conductivity by 9.5% [69].

Among various nanoadditives, ZnO offers potential advantages such as a large surface area, high surface energy, strong adsorption, high diffusion, easy sintering, good tribological characteristics, environment-friendly, and low melting point [70, 71]. Elagouz et al. [72] investigated the frictional performance of piston ring assembly using nano-ZnO (20 nm and 0.6 wt%) as an engine oil additive. The results showed that using ZnO nanolubricants under different loads and speeds reduces friction coefficient and wear rate by 23% and 88%, respectively. Battez et al. [73] reported that the nano-ZnO could produce effective tribofilms that give superior tribological performance. Silver (Ag) additives are used as a lubricant because of their low shear strength, rapid recovery, and recrystallization properties which allow it to have low friction in the relative motion of sliding contacts [74]. Twist et al. [75] investigatedthe impact of nano-Ag (0.5 wt%) blended in SAE 15 W40 motor oil at extremely high temperatures between 180°C and 300°C. The results showed a considerably reduced wear contrasted with the engine oil. Another study also showed that using Ag nanoadditives improves the antiwear

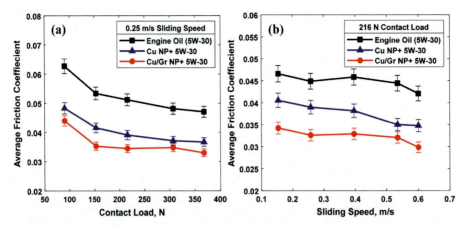

Fig. 11 Effect of Cu/graphene nanolubricants on the friction properties. *(Reprinted from Ref. M.K.A. Ali, X. Hou, M.A. Abdelkareem, Anti-wear properties evaluation of frictional sliding interfaces in automobile engines lubricated by copper/graphene nanolubricants, Friction 8 (2020) 905–16.)*

properties and thermal conductivity [76–78]. The tribological results by Ali et al. [57] also indicated that the friction coefficient of the nanolubricants, including nano-Cu and Cu/graphene, reduced versus the sliding speeds and contact loads, compared to reference oil (Fig. 11). Based on the experimental tests by Padgurskas et al. [79], the Cu nanoadditives are more effective in the mixed and boundary lubrication than full film lubrication. This is related to the potential interaction of the rubbing surfaces for producing tribofilm layer by nanoadditives. Pisal and Chavan [80] investigated the tribological properties of nano-CuO (25–55 nm and 0.2–1 wt%) into 20 W-40 oil using oleic acid as a surface modifier. Results indicate that the CuO nanolubricant reduced the friction coefficient by 66% and wear by 79% compared to the 20 W-40 oil.

One such new material demonstrating high potential with respect to tribological properties is graphene and carbon nanomaterials. Since Japanese scientists discovered carbon nanotubes in 1991, they have been widely used in different applications. The cylindrical fullerene is called a carbon nanotube. The carbon nanotubes (CNTs) area hexagonal group of carbon atoms formed thinly with a hollow-shaped cylinder. There are three structural forms of CNT single (SWCNT)—double (DWCNT)—multiwall carbon nanotubes (MWCNT), as shown in Fig. 12A [81]; CNT lengths range from a few hundred nanometers to several micrometers while diameters to several nanometers. There are several kinds of carbon fibers, each with its own morphology and properties. The final structure-property relationships of carbon fibers are determined by the processing and treatment conditions. This variation in the diameter of fibrous graphitic materials is summarized in Fig. 12B [82].

Carbon nanomaterials assuredly have a great interest as nanoadditives. Carbon NPs have special properties that make them desirable for tribological applications, such as high

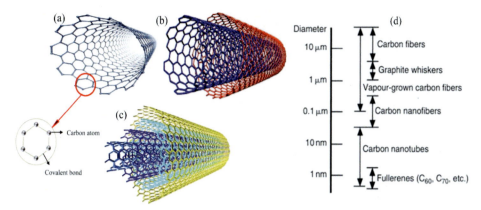

Fig. 12 Forms of carbon nanotubes; (A) SWCNT, (B) DWCNT, (C) MWCNT. *(Reproduced from Ref. I. Rafique, A. Kausar, Z. Anwar, B. Muhammad, Exploration of epoxy resins, hardening systems, and epoxy/carbon nanotube composite designed for high performance materials: a review, Polym.-Plast. Technol. Eng. 55 (2016) 312–33 with permission, © Taylor & Francis, 2016.)*

tensile and flexural powers, high elastic modulus, and a high aspect ratio [83]. MWCNT nanolubricants can significantly decrease the friction by up to 25%–40% and the wear by up to 56% at different sliding conditions compared to the engine oils [84, 85]. Tian et al. [86] studied the effect of Al2O3/MWCNT (1 vol%) hybrid nano additives into 10 W-40 oil on thermal conductivity. The results showed that the thermal conductivity improved by 13%–30% at different temperatures from 25°C to 65°C compared to the base oil. The investigations of new materials for tribological applications are still ongoing. One such novel material demonstrating high potential with respect to tribological properties is graphene. The fundamental physical properties of graphene, especially mechanical strength, tensile stress, thermal conductivity, and aspect ratio, are enormously high, which promises its potential for tribological applications [10, 87]. Microscopic morphology and characterization of graphene nanosheets are shown in Fig. 13. A sharp reflection and another weak reflection in the XRD pattern of graphene are located at $2\theta = 26.381°$ and at about $2\theta = 54.542°$ which are attributed to (002) and (0 0 4) crystal planes of hexagonal graphite [88].

In recent experimental studies regarding graphene nanolubricant, Ali et al. [88] investigated the frictional performance of graphene nanoadditives in lube oil using the piston ring/liner reciprocating tribometer. The graphene has a diameter of 5–10 μm and a thickness of 3–10 nm. The concentrations of graphene ranged from 0.03 to 0.6 wt%. The experiments were carried out at a variety of sliding velocities (0.154–0.6 m/s), with normal loads varying from 90 to 368 N. In the boundary lubrication regime, the graphene nanolubricant decreases the friction coefficient and wear rate by 29%–35% percent and 22%–29%, respectively. Multilayer graphene flakes may be used as solid lubrication for stainless steel [89]. Friction can be decreased by six times with a low graphene flake

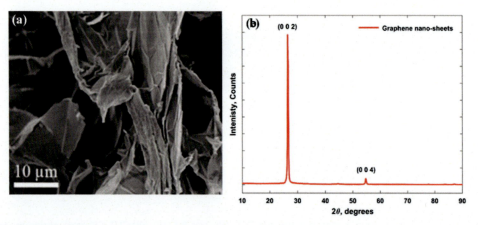

Fig. 13 Morphology and nanostructures of graphene nanosheets. *(Reproduced from Ref. M.K.A. Ali, H. Xianjun, M.A. Abdelkareem, M. Gulzar, A. Elsheikh, Novel approach of the graphene nanolubricant for energy saving via anti-friction/wear in automobile engines, Tribol. Int. 124 (2018) 209–29 with permission,© Elsevier, 2018.)*

concentration. Since graphene is a two-dimensional material, it easily shears when moving the sliding surfaces and thus provides low friction. Gu et al. [90] investigated the tribological behavior of graphene nanoadditives in PAO 6 oil. The results showed clearly that the friction coefficient and wear rate declined by 73.1% and 97.8%, respectively.

4. Mechanisms of nanolubricant additives

4.1 Rolling bearing effect

The effect of spherical nanoparticles allows a roll between sliding surfaces. If the grain size of the particles is greater than the asperities of the surfaces, they convert partial rolling friction into pure sliding friction [91]. In most cases, the surface roughness is greater than the medium diameter of nanoparticles. Nanoparticles with spherical morphologies will internally roll among friction surface asperities, turning friction from sliding to rolling friction (see Fig. 14). Nanoparticles can be deposited on frictional surfaces to form a physical tribofilm, which compensates for substance mass loss [92]. Nanoparticles reduce compressive stress concentrations associated with high contact pressure by bearing compressive force raise [93]. Many studies [14, 94, 95] reported that nanoadditives actas rolling bearings during friction between rubbing surfaces, reducing friction and increasing antiwear properties. Moreover, polymer nanoparticles can lead to the formation of an ultra-thin tribofilm into the contact pairs and have been associated with a ball bearing under load loads. However, under high loads, the rolling bearing effect loses its ability to reduce friction [96]. According to the results of Refs. [4, 12, 28, 73, 97, 98], oil film thickness significantly impacts lubrication mechanisms. When the oil film thickness

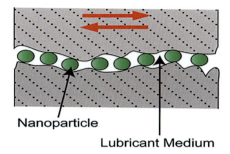

Fig. 14 Rolling bearing effect of nanoparticles in lube oils. *(Reprinted from Ref. I.E. Uflyand, V.A. Zhinzhilo, V.E. Burlakova, Metal-containing nanomaterials as lubricant additives: state-of-the-art and future development, Friction 7 (2019) 93–116.)*

approached the particle grain size, the rolling bearing mechanism was the dominant antifriction mechanism. However, nanoparticles usually form a transfer film when the film is thin.

4.2 Tribofilm formation

Generally, tribofilm is described as a coating layer generated due to the friction process, which is deposited on the rubbing surfaces but has various chemical compositions. The rate of tribofilm formation should be higher than the wear rate elimination to protect the frictional surfaces [99]. The self-replenishment is required to keep a tribofilm formation with enough adhesion to a substrate [60]. Due to adsorption or chemical reactions, nanoparticles can effectively produce a protective layer on the contact interfaces. This is due to their high specific surface area and surface energy. The tribofilm layer reduces metal-to-metal contact and plays a major role in lubrication. Nanoparticles that have low chemical activity (e.g., graphene, copper, and silver) always form a physical adsorption tribofilm [100]. Rotoi et al. [101] analyzed the components of the tribofilm created by WS_2 nanoadditives at 100°C using XPS/SIMS analysis. As Fig. 15 shows, the tribofilm can mainly be divided into four layers. The top layer consisted of exfoliated WS_2 nanoparticles, whereas the WS_2, WO_3, iron oxide, sulfide Fe and W were included in the second. In the third layer, the WO_3, Fe, and W were found, while the bottom layer consisted of Fe and W.

Nanomaterials become mechanically unstable during high loads, causing broken and exfoliation to form adhesion and third bodies on rubbing interfaces [102]. The interaction of the asperities increases the contact pressure, which will enhance the contact local contact temperature [4]. Hence, these sliding conditions guide the tribochemical reaction among the tribopair and nanomaterials, producing the tribofilm formation on the rubbing surfaces to separate the asperities [14]. Many research studies have been provided to present the mechanism of action of various nanolubricants in tribofilm formation by

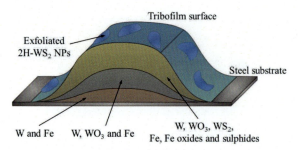

Fig. 15 Tribofilm formation mechanism generated by WS_2 nanolubricant at 100°C. *(Reprinted from Ref. Ratoi M, Niste VB, Walker J, Zekonyte J. Mechanism of action of WS2 lubricant nanoadditives in high-pressure contacts. Tribol. Lett. 2013; 52:81–91.Ratoi M, Niste VB, Walker J, Zekonyte J. Mechanism of action of WS2 lubricant nanoadditives in high-pressure contacts. Tribology letters. 2013; 52:81-91.)*

HRTEM or FESEM analysis [28, 79, 88, 103, 104]. Ali et al. [1] presented that the excellent tribological performance of automotive engines lubricated by Al_2O_3/TiO_2 nanolubricants is attributed to the tribofilm formation of Al_2O_3/TiO_2 nanoadditives on the rubbing piston ring surface as illustrated in Fig. 16. Recent investigations also revealed a good potential for utilizing ionic liquids as lubricant additives in the tribofilm formation on the rubbing surfaces in automotive engines [105].

4.3 Microstructure transformation

During a friction process, subjecting to high pressure and heat might destroy the original microstructure, and a new microstructure could be formed. As a consequence, variations in tribological conduct appear as the microstructure changes. During the friction, the microstructure of nanocarbon materials appears to be arranged, and hence the antiwear/friction properties are improved [106]. Results in [107] showed that, during the tribological tests, the transformation of fullerene into graphene resulted in a strong lubrication effect with the transition in the microstructure. In Ref. [96], the findings showed that when the applied load was low, long carbon nanotubes appeared to be shortened, allowing them to be entered into the contact surfaces and form a protective layer easily. Nanotubes are inhibited in the meantime, and the rolling effect is improved with smaller nanotube lengths. According to the research carried out in [108], the graphene nanoadditive with higher exfoliation tended to overlap and order under high applied load and shear action. As shown in Fig. 17A, it then developed a stable and lamellar adsorption film parallel to the sliding track that improved the tribological performance of the contacted surfaces significantly. However, the frictional tendency of graphene is quite the contrary (Fig. 17B). The friction interfaces were further scratched because, after long friction, the integrated and oriented graphene layers were damaged. Hence, the microstructure transformation shows a key role in tribological performance, particularly for carbon nanoadditives.

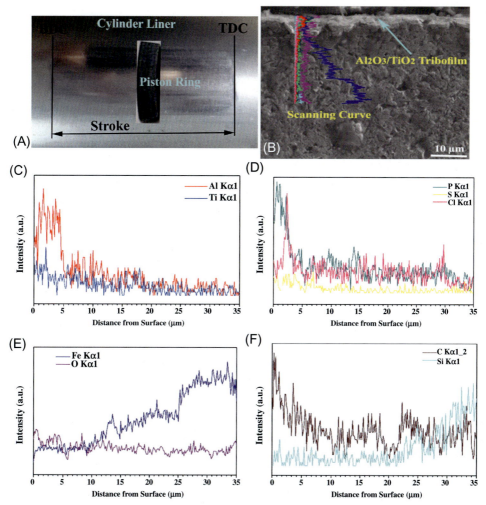

Fig. 16 Tribofilm formation formed on the worn ring piston surface lubricated by Al$_2$O$_3$/TiO$_2$ nanolubricants. *(Reproduced from Ref. M.K.A. Ali, P. Fuming, H.A. Younus, M.A.A. Abdelkareem, F.A. Essa, A. Elagouz, et al., Fuel economy in gasoline engines using Al2O3/TiO2 nanomaterials as nanolubricant additives, Appl. Energy 211 (2018) 461–78 with permission,© Elsevier, 2018.)*

4.4 Surface repairing effect

The rubbing surfaces are usually smooth, and their asperities make contact with each other during the boundary lubrication, leading to high wear and friction properties. In the boundary lubrication, the contact area between the rubbing surfaces is explained by the scale of surface asperities because of an intermittent ultra-thin oil film. These sliding events may benefit from the nanoscale dimension of the nanoadditives in the lube oils by filling of the valleys among asperities by nanoadditives (mending effect) as shown in Fig. 18, which

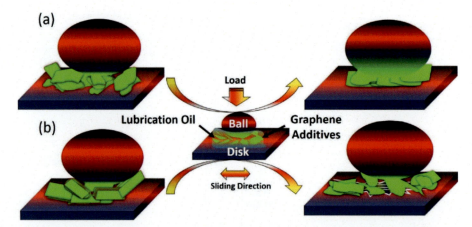

Fig. 17 Microstructure transformation of graphene nanoadditives during the friction process. (Reproduced from Ref. J. Zhao, J. Mao, Y. Li, Y. He, J. Luo, Friction-induced nano-structural evolution of graphene as a lubrication additive, Appl. Surf. Sci. 434 (2018) 21–7 with permission,© Elsevier, 2018.)

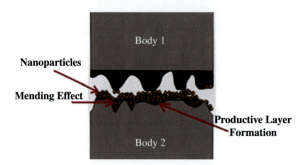

Fig. 18 Mending effect of nanoparticles into engine oils between the sliding surfaces. (Reproduced from Ref. S. Bhaumik, R. Maggirwar, S. Datta, S. Pathak, Analyses of anti-wear and extreme pressure properties of castor oil with zinc oxide nano friction modifiers, Appl. Surf. Sci. 449 (2018) 277–86 with permission, © Elsevier, 2018.)

supports the tribopair surface to be smoothened for decreasing the boundary friction coefficient and wear rate [4]. Furthermore, the deposition of nanoadditives resulted in the covering of the asperities, reducing contact between tribo-pairs and improving load-bearing capacity. Ali et al. [88] used SEM to examine the morphological characteristics of a rubbing liner surface and reported that graphene nanoadditives introduced the mending effect by filling the asperities completely (see Fig. 19). According to Ref. [48], TiO_2/SiO_2 nanocomposites have outstanding antifriction and antiwear properties. SEM, EDS, and AFM morphological analyzes verified that the friction was improved by the mending and polishing effect of the TiO_2 and SiO_2 nanocomposites. The polishing effect of nanodiamond particles increased the surface hardness, and the surfaces were very smooth [109]. Several studies have reported the synergistic effect of nanoadditives by the formation of tribofilm, rolling, mending, and the polishing effects on the friction interfaces.

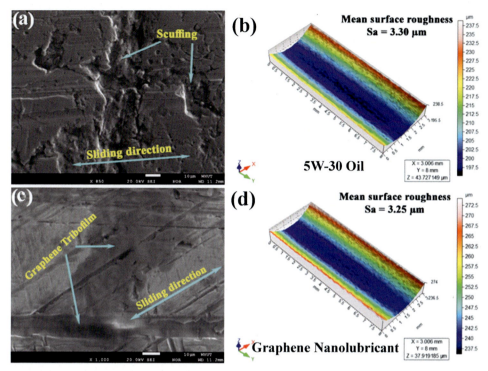

Fig. 19 Worn surface morphologies and 3D surface roughness of the cylinder liner lubricated graphene nanolubricant. *(Reproduced from Ref. M.K.A. Ali, H. Xianjun, M.A. Abdelkareem, M. Gulzar, A. Elsheikh, Novel approach of the graphene nanolubricant for energy saving via anti-friction/wear in automobile engines, Tribol. Int. 124 (2018) 209–29 with permission, © Elsevier, 2018.)*

5. Role of nanolubricants in improving vehicle engines performance

In automotive engines, frictional power losses, typically due to the relative motion of contact interfaces, are among the leading causes of excessive fuel consumption. Hence, lubrication has an essential impact on engine performance. Nanolubricants recently have been a research focus with highly promising tribological and thermophysical properties [11, 14, 72]. This will help to minimize frictional losses, thus improve fuel economy and reduce gas emissions with the use of nanomaterial additives into engine oils. The friction coefficient reduction by 10% in vehicle engines has the potential to achieve fuel economy by about 1% [110]. However, this is a critical task, and only a few studies have reported engine performance measurements under nanolubrication. The challenge is how nanolubricants are able not only to save fuel but also to control gas emissions. The lubrication system of combustion engines has both direct and indirect effects on their mechanical and thermal efficiencies as well as the exhaust gasses. Therefore, lubrication durability and nanolubricant dispersion stability, along with a fair compromise between their positives and negatives on engine overall performance, are the main challenging mission. Table 2

Table 2 A comprehensive summary of studies associated with measurements of automotive engine performance lubricated by nanolubricants.

Year	Ref.	Nanolubricant details				Engine details	Condition	
		Nanomaterial/base oil	Grain size	Concentration	Morphology			
2021	[111]	Graphite/SAE-30	<20 μm	0.3 wt%	Sheets	Single cylinder 4-stroke diesel engine with power rate of 4.8 HP @ 1500 rpm		500 W load 1500 W load 2500 W load 4000 W load
		Nanographite/SAE-30	~50 nm	0.3 wt%				500 W load 1500 W load 2500 W load 4000 W load
2018	[112, 113]	Al_2O_3/SAE15W40	40–46 nm	0.3 vol%	Spherical	Single cylinder 4-stroke diesel engine with power rating of 3.5 kW @ 3000 rpm		500 W load 1500 W load 2500 W load
				0.9 vol%				500 W load 1500 W load 2500 W load
		SiO_2/SAE15W40	40–46 nm	0.3 vol%	Spherical			500 W load 1500 W load 2500 W load
				0.9 vol%				500 W load 1500 W load 2500 W load
2018	[1]	Al_2O_3-TiO_2/5 W-30	8–12 nm (Al_2O_3) and 10 nm (TiO_2)	0.1 wt%	Spherical	4-Cylinder 4-stroke gasoline engine with 1600 cm^3 swept volume, max. Power of 85 kW @ 5600 rpm, and max. Torque of 160 Nm @ 4000–4500 rpm.	50% load 100% load	1000 rpm 2500 rpm 4000 rpm 1000 rpm 2500 rpm 4000 rpm
2018	[88]	Graphene (Gr)/5 W-30	5–10 μm (diameter)/3–10 nm (thickness)	0.03–0.6 wt%	Nanosheets	4-Cylinder 4-stroke gasoline engine with 1600 cm^3 swept volume, max. Power of 85 kW @ 5600 rpm, and max. Torque of 160 Nm @ 4000–4500 rpm.	50% load 75% load	1000 rpm 2500 rpm 3500 rpm 1000 rpm 2500 rpm 3500 rpm
2017	[114]	MoS_2/5 W30	52 nm	0.5 wt%	Spherical	Diesel engine with 1900 cm^3 and a max. Torque of 600 N.m @ 10,000 rpm.	40°C (cooling water temp.) 95°C (cooling water temp.)	1500 rpm 2500 rpm 3500 rpm 1500 rpm 2500 rpm 3500 rpm

Abbreviations: *BP*: brake power; *BSFC*: brake specific fuel consumption; *BTE*: brake thermal efficiency; *ET*: engine torque; *HC*: hydrocarbons; *NEDC*: new european driving cycle; *TEPL*: Total friction power losses.
[a]The percentage of change in fuel consumption in this study was calculated from the fuel consumption data in kg/h.
[b]The percentage of change here was calculated for the NO emissions.

Engine performance									
BSFC	BTE	BP	ET	TFPL	CO_2	CO	HC	NO_x	Findings
3.84%	28.77%	–	–	–	−3.47%	−44.26%	5.08%	8.74%	The nanographite oil demonstrated a 15.2% decrease in overall fuel consumption, an 18% increase in brake specific fuel consumption, a 22% increase in brake thermal efficiency, a 28.2% increase in heat transfer, and a 5% reduction in exhaust gas temperature when compared to the base oil
8.38%	7.02%	–	–	–	−4.52%	−52.00%	5.90%	4.92%	
10.51%	5.68%	–	–	–	−2.71%	−39.88%	4.43%	1.19%	
5.86%	6.56%	–	–	–	−0.56%	−6.53%	3.56%	0.66%	
13.87%	73.06%	–	–	–	−8.25%	46.26%	9.07%	16.60%	
22.92%	24.68%	–	–	–	−6.60%	67.27%	10.18%	8.66%	
24.26%	14.84%	–	–	–	−5.43%	26.60%	8.23%	2.09%	
14.59%	17.66%	–	–	–	−2.96%	8.33%	5.43%	1.44%	
18.43%	−18.30%	–	–	–	–	–	–	–	Engine tests revealed that Al_2O_3 outperformed SiO_2, with Al_2O_3 minimizing fuel consumption at various engine loads, thus providing the highest brake thermal efficiency Nanoparticles are more effective in engine performance at lower volume fractions
12.59%	−14.32%	–	–	–	–	–	–	–	
6.94%	−7.56%	–	–	–	–	–	–	–	
−2.10%	2.00%	–	–	–	–	–	–	–	
−9.04%	8.29%	–	–	–	–	–	–	–	
−7.32%	8.21%	–	–	–	–	–	–	–	
−19.41%	19.90%	–	–	–	–	–	–	–	
−14.09%	12.38%	–	–	–	–	–	–	–	
−7.69%	8.09%	–	–	–	–	–	–	–	
−24.08%	23.63%	–	–	–	–	–	–	–	
−19.75%	16.70%	–	–	–	–	–	–	–	
−10.21%	22.44%	–	–	–	–	–	–	–	
11.90%	–	11.48%	9.99%	–	–	–	–	–	The Al_2O_3/TiO_2 nanolubricant demonstrated a 1.7%–2.5% enhancement in engine mechanical efficiency and a 16%–20% reduction in fuel consumption (4 L/100 km fuel saving). The enhanced engine performance has been associated with a ~6.5% reduction in total engine friction power losses
0.85%	–	3.01%	2.71%	–	–	–	–	–	
0.68%	–	5.36%	5.47%	–	–	–	–	–	
2.11%	–	6.51%	3.63%	6.47%	–	–	–	–	
1.02%	–	1.16%	0.08%	4.64%	–	–	–	–	
1.08%	–	4.60%	3.99%	3.46%	–	–	–	–	
4.56%	–	2.81%	4.22%	16.98%	4.11%	−5.41%	−6.85%	1.74%	The Gr nanolubricant boosted the engine power and torque by 7%–10%, which was associated with a 6% drop in total engine friction power losses. Fuel consumption recorded a 17% reduction under the NEDC. Exhaust gases (CO_2, HC, and NO_x) demonstrated a ~2.8%–5.4% reduction which was linked to the enhancement in the cylinder liner heat transfer
3.91%	–	3.87%	3.82%	5.70%	1.38%	36.84%	31.25%	−29.81%	
5.42%	–	4.08%	4.05%	4.16%	33.79%	–	−45.00%	81.19%	
–	–	–	–	–	2.80%	−88.89%	28.46%	0.83%	
–	–	–	–	–	3.38%	40.74%	36.36%	6.27%	
–	–	–	–	–	4.67%	−39.13%	14.75%	16.23%	
12.87%[a]	–	–	–	–	–	−7.69%	−1.08%	−1.71%[b]	Measurement of fuel consumption using the MoS_2–SAE5W30 nanolubricant has shown fuel savings of 5%–10% under some operational conditions and a 1% fuel saving was achieved under road tests. Negative effects were seen on the exhaust emissions for most of the tested conditions
7.53%	–	–	–	–	–	−146.9%	−5.34%	5.48%	
8.05%	–	–	–	–	–	−59.87%	−28.94%	−2.56%	
4.54%	–	–	–	–	–	−40.29%	−37.61%	13.51%	
−6.68%	–	–	–	–	–	−10.13%	−31.56%	−4.85%	
0.16%	–	–	–	–	–	−32.91%	−14.91%	−7.91%	

contains a comprehensive summary of studies associated with measurements of the performance of automotive internal combustion engines under the use of nanolubrication. The effects of nanolubricants on mechanical, thermal, and exhaust efficiencies under several working conditions are evaluated in the following sections.

5.1 Fuel economy in vehicle engines

From the data in Table 2, we can clearly see that most of the nanomaterials were found to have a spherical morphology to achieve the rolling effect mechanism and hence reduce the friction and wear between the frictional engine components. In a study by Singh et al. [111], nanographite was blended into the SAE-30 engine oil and then used as the primary lubricant in a single-cylinder 4-stroke diesel engine. As Fig. 20 shows, the engine test results using the graphite nanolubricant demonstrated a 15.2% decrease in overall fuel consumption, an 18% increase in brake specific fuel consumption, a 22% increase in brake thermal efficiency, a 28.2% increase in heat transfer, and a 5% reduction in exhaust gas temperature when compared to the plain SAE-30. These positive results are due to the formation of an ultra-thin tribofilm layer on the sliding engine interfaces, which not only reduces friction and wear but also enhances the engine's brake thermal efficiency because of the increased heat equivalent to the brake engine power.

The hybrid Al_2O_3/TiO_2 nanolubricant was proposed by Ali et al. [1], and the performance measurements of a 4-cylinder 4-stroke gasoline engine were conducted. As shown in Fig. 21, the Al_2O_3/TiO_2 nanolubricants demonstrated a 1.7%–2.5% enhancement in engine mechanical efficiency and a 16%–20% reduction in fuel consumption (2.4 L/100 km fuel saving in economic speed). The enhanced engine performance has been associated with a ~6.5% reduction in total engine friction power losses. Engine results also reported that the Al_2O_3/TiO_2 nanolubricants accelerated the warm-up phase by 24%. The rolling effect and tribofilm formation of the Al_2O_3 and TiO_2 nanomaterials on the sliding contacts (piston ring and cylinder liner) is the dominant mechanisms responsible for improving the engine characteristics. In Fig. 22, we can see a complete thermal analysis of a gasoline engine lubricated by the Al_2O_3/TiO_2 nanolubricants compared to 5 W-30 oil [115]. Using the Al_2O_3/TiO_2 nanolubricants, the results demonstrated a 9%–14% increase in the heat transfer and a 3.9%–8.6% increase in the brake thermal efficiency compared to the plain 5 W-30 oil. This achieved thermal stability enables the Al_2O_3/TiO_2 nanolubricants to be used in broad temperature ranges, which optimizes the drain ranges of the lube oil and reduces the maintenance costs. A 17% reduction in the cumulative fuel mass consumption could be achieved using the graphene nanolubricant as reported in [88]. This was due to the reduction in the friction coefficient and wear ratefor the cylinder liner and ring surfaces due to the formed tribofilm coating resulted from the graphene nanoadditives.

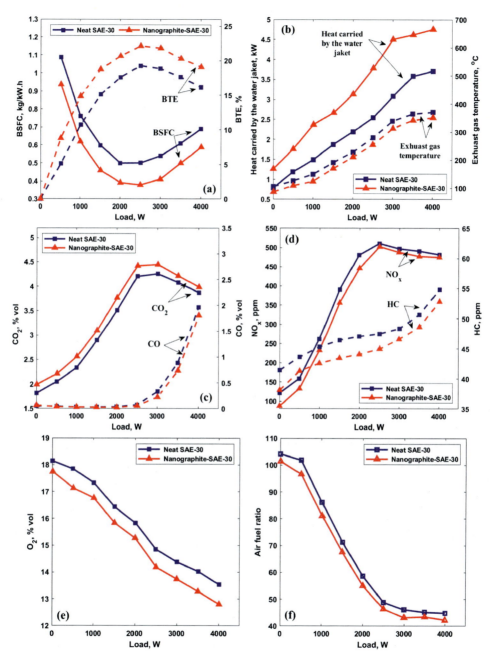

Fig. 20 Comparison of engine characteristics for a neat engine oil of 5 W-30 and graphite nanolubricant. *(Reproduced from Ref. J.P. Singh, S. Singh, T. Nandi, S.K. Ghosh, Development of graphitic lubricant nanoparticles based nanolubricant for automotive applications: Thermophysical and tribological properties followed by IC engine performance, Powder Technol. 387 (2021) 31–47 with permission, © Elsevier, 2021.)*

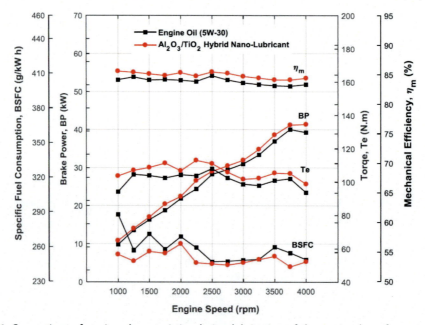

Fig. 21 Comparison of engine characteristics during lubrication of the engine by reference oil and Al_2O_3/TiO_2 nanolubricants. *(Reproduced from Ref. M.K.A. Ali, P. Fuming, H.A. Younus, M.A.A. Abdelkareem, F.A. Essa, A. Elagouz, et al., Fuel economy in gasoline engines using Al2O3/TiO2 nanomaterials as nanolubricant additives, Appl. Energy 211 (2018) 461–78 with permission,© Elsevier, 2018.)*

5.2 Effect of nanolubricants on exhaust emissions

Nanolubricants have previously been proved to influence the heat transfer properties of the engine and thus to have an effect on exhaust gases. In Fig. 20 and Table 2, complete comparisons of the engine exhaust gases (CO_2, CO, NOx, and HC) at different engine loads are given for the base engine oil of 5 W-30 and a modified graphite nanolubricant. Comparing to the reference lube oil, the emissions have all been reduced in the case of the nanolubricant except the CO_2 emissions [111]. In Fig. 23, the effect of graphene nanolubricant on exhaust emissions of a 4-cylinder petrol engine is presented. Data showed that the CO_2 emissions have decreased by 3.4%–4.66%, and the NOx emissions from graphene nanolubricant have decreased by 3%–5% compared with the reference oil [92]. This is due to the enhanced heat transfer properties resulting from the formation of tribofilm layers on the sliding surfaces.

Sgroi et al. [114] conducted measurements of the engine emissions under the use of MoS_2 nanolubricant (see Fig. 24). Measurement of fuel consumption using the MoS_2-SAE5W30 nanolubricant has shown a maximum fuel saving of 2%–5% at steady-state conditions. As seen in Fig. 24, the MoS_2 nanoadditives explained slight variations in

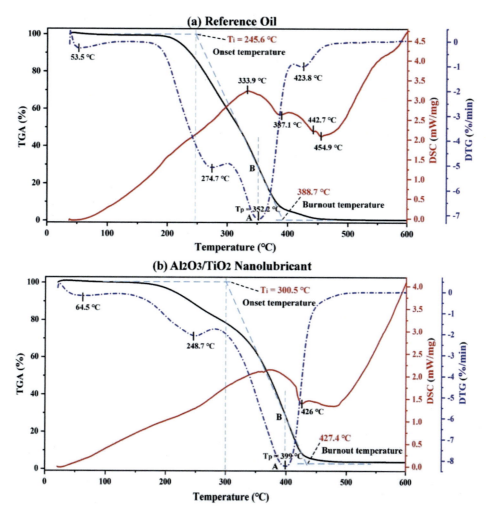

Fig. 22 Thermal analysis results (TGA/DTG/DSC) for a base engine oil of 5 W-30 and Al_2O_3/TiO_2 nanolubricants at a heating rate of 10°C/min in an N_2 environment. *(Reproduced from Ref. M.K.A. Ali, H. Xianjun, Improving the heat transfer capability and thermal stability of vehicle engine oils using Al2O3/TiO2 nanomaterials, Powder Technol. 363 (2020) 48–58 with permission,© Elsevier, 2020.)*

the emissions values for most tested conditions and cooling temperatures compared to the reference oil. The HC and CO values exhibited a not negligible but not shown increase at the DPF outlets of the engine, whereas, in most circumstances, NOx was reduced, particularly at low cooling temperatures and low engine points (1500 × 1, 2000 × 2, and 2500 × 3). More research is required to fill the gap in understanding how and what the main mechanisms are responsible for reducing engine emissions when lubricated by nanolubricants. Researchers can also focus on investigating novel hybrid nanolubricants to enhance engine performance and emissions further.

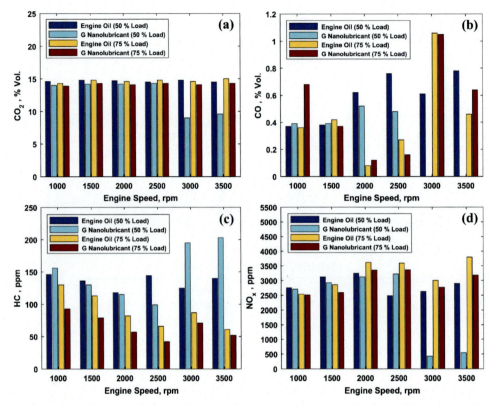

Fig. 23 Engine exhaust emission results versus engine speed for the engine base oil (5 W-30) and the graphene nanolubricant. *(Reproduced from Ref. M.K.A. Ali, H. Xianjun, M.A. Abdelkareem, M. Gulzar, A. Elsheikh, Novel approach of the graphene nanolubricant for energy saving via anti-friction/wear in automobile engines, Tribol. Int. 124 (2018) 209–29 with permission, © Elsevier, 2018.)*

6. Conclusions and recommendations

Nanolubricants have been proposed in many studies to improve the tribological and thermophysical properties of engine oils. The following inference can be taken from the current comprehensive literature review:

- The dispersion stability of nanomaterials in lube oils is critical for long-term improvement of the thermal conductivity and rheology performance. Functionalization of nanomaterials appears to be a more powerful current process for achieving stable suspensions than using surfactants. The use of zeta-potential and pH may also aid in the development of stable suspensions. As a result, more emphasis should be placed on investigating all factors influencing the dispersion stability of nanomaterials in lube oils.
- Nanolubricantshave excellent tribological performance, which means they can improve antifriction/wear properties and increase the life of engine components.

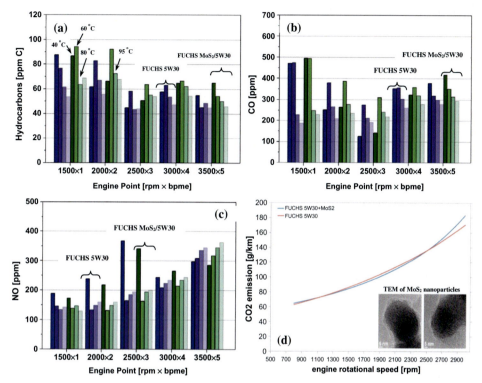

Fig. 24 Exhaust emission results for the MoS$_2$ nanolubricant. *(Reproduced from M.F. Sgroi, M. Asti, F. Gili, F.A. Deorsola, S. Bensaid, D. Fino, et al., Engine bench and road testing of an engine oil containing MoS2 particles as nano-additive for friction reduction, Tribol. Int. 105 (2017) 317–25 with permission,© Elsevier, 2017.)*

The use of nanolubricants can improve heat transfer by enhancing the thermal conductivity of the lubrication oil. However, rheology behavior must be considered because an increase in viscosity can increase frictional power loss (viscous friction).

- The physical interaction among the nanoadditives and the frictional interfaces was main responsible for improving antifriction properties (rolling bearing effect, microstructure transformation, and surface repairing effect). This is owing to the spherical morphology of nanoadditives, which can convert sliding friction to rolling friction. Antiwear properties were more closely linked to the chemical interaction between nanoadditives, lube oils, and substrate interfaces, which played a key role in the protective tribofilmlayer on the rubbing interfaces.
- The engine torque and engine brake power improved when the engine was lubricated with nanolubricants. The key reason is that the total frictional power losses decreased because of the improving tribological performance. The engine's mechanical and thermal efficiency also increased. As a result, the cumulative fuel

consumption of the engine was reduced by 4–21% under different running conditions. New cost-effective methods for mass-producing nanolubricants should be presented, paving the way for commercialization.
— Only a few studies had reported measurements of exhaust emissions when the engine was lubricated with nanolubricants. These studies revealed that when the engine was lubricated with nanolubricants instead of conventional oil, the exhaust emissions were reduced. Fundamentally, more research is required to explain how nanolubricants affect engine exhaust emissions during cold starts and low engine load/speed.

Acknowledgments

The authors would like to express their deep appreciation for the support by the Minia University and the Wuhan University of Technology.

References

[1] M.K.A. Ali, P. Fuming, H.A. Younus, M.A.A. Abdelkareem, F.A. Essa, A. Elagouz, et al., Fuel economy in gasoline engines using Al2O3/TiO2 nanomaterials as nanolubricant additives, Appl. Energy 211 (2018) 461–478.
[2] K. Holmberg, P. Andersson, N.-O. Nylund, K. Mäkelä, A. Erdemir, Global energy consumption due to friction in trucks and buses, Tribol. Int. 78 (2014) 94–114.
[3] U. Sahaym, M.G. Norton, Advances in the application of nanotechnology in enabling a 'hydrogen economy', J. Mater. Sci. 43 (2008) 5395–5429.
[4] M.K.A. Ali, H. Xianjun, F. Essa, M.A. Abdelkareem, A. Elagouz, S. Sharshir, Friction and wear reduction mechanisms of the reciprocating contact interfaces using nanolubricant under different loads and speeds, J. Tribol. 140 (2018).
[5] H. Presting, U. König, Future nanotechnology developments for automotive applications, Mater. Sci. Eng. C 23 (2003) 737–741.
[6] I. Srivastava, A. Kotia, S.K. Ghosh, M.K.A. Ali, Recent advances of molecular dynamics simulations in Nanotribology, J. Mol. Liq. 116154 (2021).
[7] M. Skjoedt, R. Butts, D.N. Assanis, S.V. Bohac, Effects of oil properties on spark-ignition gasoline engine friction, Tribol. Int. 41 (2008) 556–563.
[8] B.N. Persson, Sliding Friction: Physical Principles and Applications: Springer Science & Business Media, 2013.
[9] M.K.A. Ali, H. Xianjun, R. FiifiTurkson, M. Ezzat, An analytical study of tribological parameters between piston ring and cylinder liner in internal combustion engines, Proc. Inst. Mech. Eng. Pt. K J. Multi-body Dyn. 230 (2016) 329–349.
[10] A. Erdemir, G. Ramirez, O.L. Eryilmaz, B. Narayanan, Y. Liao, G. Kamath, et al., Carbon-based tribofilms from lubricating oils, Nature 536 (2016) 67–71.
[11] M.K.A. Ali, H. Xianjun, R.F. Turkson, Z. Peng, X. Chen, Enhancing the thermophysical properties and tribologicalbehaviour of engine oils using nano-lubricant additives, RSC Adv. 6 (2016) 77913–77924.
[12] M.K.A. Ali, H. Xianjun, A. Elagouz, F. Essa, M.A. Abdelkareem, Minimizing of the boundary friction coefficient in automotive engines using Al2O3 and TiO2 nanoparticles, J. Nanopart. Res. 18 (2016) 377.
[13] M.A. Abdelkareem, L. Xu, M.K.A. Ali, A. Elagouz, J. Mi, S. Guo, et al., Vibration energy harvesting in automotive suspension system: a detailed review, Appl. Energy 229 (2018) 672–699.

[14] M.K.A. Ali, H. Xianjun, L. Mai, C. Qingping, R.F. Turkson, C. Bicheng, Improving the tribological characteristics of piston ring assembly in automotive engines using Al2O3 and TiO2 nanomaterials as nano-lubricant additives, Tribol. Int. 103 (2016) 540–554.

[15] H. Rahnejat, Tribology and Dynamics of Engine and Powertrain: Fundamentals, Applications and Future Trends, Elsevier, 2010.

[16] M.K.A. Ali, H. Xianjun, M.A. Abdelkareem, A.H. Elsheikh, Role of nanolubricants formulated in improving vehicle engines performance, in: IOP Conference Series: Materials Science and Engineering, IOP Publishing, 2019, p. 022015.

[17] A. Elagouz, M.K.A. Ali, H. Xianjun, M.A. Abdelkareem, Techniques used to improve the tribological performance of the piston ring-cylinder liner contact, in: IOP Conference Series: Materials Science and Engineering, IOP Publishing, 2019, p. 022024.

[18] P. Andersson, J. Tamminen, C.-E. Sandström, Piston ring tribology, in: A literature survey VTT Tiedotteita-Research Notes, 2002, p. 2178.

[19] A. Wolff, Simulation based study of the system piston–ring–cylinder of a marine two-stroke engine, Tribol. Trans. 57 (2014) 653–667.

[20] S. Kunkel, M. Werner, G. Wachtmeister, Setting up a measuring device to determine the friction of the piston assembly, SAE Int. J. Mater. Manuf. 4 (2011) 340–351.

[21] V. Durga, N. Rao, B. Boyer, H. Cikanek, D. Kabat, Influence of surface characteristics and oil viscosity on friction behaviour of rubbing surfaces in reciprocating engines, in: Fall Technical Conference ASME-ICE, 1998, pp. 23–35.

[22] M. Priest, C. Taylor, Automobile engine tribology—approaching the surface, Wear 241 (2000) 193–203.

[23] N.W. Bolander, B.D. Steenwyk, A. Kumar, F. Sadeghi, Film thickness and friction measurement of piston ring cylinder liner contact with corresponding modeling including mixed lubrication, in: ASME 2004 Internal Combustion Engine Division Fall Technical Conference, American Society of Mechanical Engineers, 2004, pp. 811–821.

[24] H. Spikes, Friction modifier additives, Tribol. Lett. 60 (2015) 5.

[25] L.R. Rudnick, Lubricant Additives: Chemistry and Applications, CRC Press, 2009.

[26] P. Willermet, Some engine oil additives and their effects on antiwear film formation, Tribol. Lett. 5 (1998) 41–47.

[27] M.K.A. Ali, H. Xianjun, Improving the tribological behavior of internal combustion engines via the addition of nanoparticles to engine oils, Nanotechnol. Rev. 4 (2015) 347–358.

[28] M.K.A. Ali, H. Xianjun, L. Mai, C. Bicheng, R.F. Turkson, C. Qingping, Reducing frictional power losses and improving the scuffing resistance in automotive engines using hybrid nanomaterials as nano-lubricant additives, Wear 364 (2016) 270–281.

[29] K. Holmberg, A. Erdemir, The impact of tribology on energy use and CO2 emission globally and in combustion engine and electric cars, Tribol. Int. 135 (2019) 389–396.

[30] A. Kotia, K. Chowdary, I. Srivastava, S.K. Ghosh, M.K.A. Ali, Carbon nanomaterials as friction modifiers in automotive engines: recent progress and perspectives, J. Mol. Liq. 113200 (2020).

[31] M.K.A. Ali, H. Xianjun, Colloidal stability mechanism of copper nanomaterials modified by bis(2-ethylhexyl) phosphate dispersed in polyalphaolefin oil as green nanolubricants, J. Colloid Interface Sci. 578 (2020) 24–36.

[32] M. Rafiq, Y. Lv, C. Li, A review on properties, opportunities, and challenges of transformer oil-based nanofluids, J. Nanomater. 2016 (2016).

[33] A. Asadi, F. Pourfattah, I. MiklósSzilágyi, M. Afrand, G. Żyła, H. SeonAhn, et al., Effect of sonication characteristics on stability, thermophysical properties, and heat transfer of nanofluids: a comprehensive review, Ultrason. Sonochem. 58 (2019) 104701.

[34] X. Hou, H. Liu, X. Li, H. Jiang, Z. Tian, M.K.A. Ali, An experimental study and mechanism analysis on improving dispersion stability performance of Al2O3 nanoparticles in base synthetic oil under various mixing conditions, J. Nanopart. Res. 23 (2020) 1–16.

[35] X. Hou, H. Jiang, M.K.A. Ali, H. Liu, D. Su, Z. Tian, Dispersion behavior assessment of the molybdenum disulfide nanomaterials dispersed into poly alpha olefin, J. Mol. Liq. 311 (2020) 113303.

[36] D. Dey, P. Kumar, S. Samantaray, A review of nanofluid preparation, stability, and thermo-physical properties, Heat Transf.-Asian Res. 46 (2017) 1413–1442.

[37] N. Ali, J.A. Teixeira, A. Addali, A review on Nanofluids: fabrication, stability, and Thermophysical properties, J. Nanomater. 2018 (2018) 6978130.

[38] T. Missana, A. Adell, On the applicability of DLVO theory to the prediction of clay colloids stability, J. Colloid Interface Sci. 230 (2000) 150–156.

[39] J. Hong, D. Kim, Effects of aggregation on the thermal conductivity of alumina/water nanofluids, Thermochim. Acta 542 (2012) 28–32.

[40] O. Arthur, M.A. Karim, An investigation into the thermophysical and rheological properties of nanofluids for solar thermal applications, Renew. Sust. Energ. Rev. 55 (2016) 739–755.

[41] M.K.A. Ali, H. Xianjun, Role of bis(2-ethylhexyl) phosphate and Al2O3/TiO2 hybrid nanomaterials in improving the dispersion stability of nanolubricants, Tribol. Int. 155 (2021) 106767.

[42] F.T. Hong, A. Schneider, S.M. Sarathy, Enhanced lubrication by core-shell TiO2 nanoparticles modified with gallic acid ester, Tribol. Int. 146 (2020) 106263.

[43] O. Gulzar, A. Qayoum, R. Gupta, Experimental study on stability and rheological behaviour of hybrid Al2O3-TiO2 Therminol-55 nanofluids for concentrating solar collectors, Powder Technol. 352 (2019) 436–444.

[44] Y. Zhai, L. Li, J. Wang, Z. Li, Evaluation of surfactant on stability and thermal performance of Al2O3-ethylene glycol (EG) nanofluids, Powder Technol. 343 (2019) 215–224.

[45] A. Asadi, I.M. Alarifi, V. Ali, H.M. Nguyen, An experimental investigation on the effects of ultrasonication time on stability and thermal conductivity of MWCNT-water nanofluid: finding the optimum ultrasonication time, Ultrason. Sonochem. 58 (2019) 104639.

[46] T. He, Q. Dai, W. Huang, X. Wang, Colloidal suspension of graphene oxide in ionic liquid as lubricant, App. Phy. A 124 (2018) 777.

[47] B.T. Seymour, W. Fu, R.A.E. Wright, H. Luo, J. Qu, S. Dai, et al., Improved lubricating performance by combining oil-soluble hairy silica nanoparticles and an ionic liquid as an additive for a Synthetic Base oil, ACS Appl. Mater. Interfaces 10 (2018) 15129–15139.

[48] M. Gulzar, H.H. Masjuki, M.A. Kalam, M. Varman, N.W.M. Zulkifli, R.A. Mufti, et al., Dispersion stability and Tribological characteristics of TiO2/SiO2 nanocomposite-enriched biobased lubricant, Tribol. Trans. 60 (2017) 670–680.

[49] W. Xia, J. Zhao, H. Wu, S. Jiao, Z. Jiang, Effects of oil-in-water based nanolubricant containing TiO2 nanoparticles on the tribologicalbehaviour of oxidised high-speed steel, Tribol. Int. 110 (2017) 77–85.

[50] M.A. Khairul, K. Shah, E. Doroodchi, R. Azizian, B. Moghtaderi, Effects of surfactant on stability and thermo-physical properties of metal oxide nanofluids, Int. J. Heat Mass Transf. 98 (2016) 778–787.

[51] G. Colangelo, E. Favale, P. Miglietta, M. Milanese, A. de Risi, Thermal conductivity, viscosity and stability of Al2O3-diathermic oil nanofluids for solar energy systems, Energy 95 (2016) 124–136.

[52] T. Luo, X. Wei, H. Zhao, G. Cai, X. Zheng, Tribology properties of Al2O3/TiO2 nanocomposites as lubricant additives, Ceram. Int. 40 (2014) 10103–10109.

[53] T. Luo, X. Wei, X. Huang, L. Huang, F. Yang, Tribological properties of Al2O3 nanoparticles as lubricating oil additives, Ceram. Int. 40 (2014) 7143–7149.

[54] J. Wang, J. Zhu, X. Zhang, Y. Chen, Heat transfer and pressure drop of nanofluids containing carbon nanotubes in laminar flows, Exp. Thermal Fluid Sci. 44 (2013) 716–721.

[55] W. Li, S. Zheng, Q. Chen, B. Cao, A new method for surface modification of TiO2/Al2O3 nanocomposites with enhanced anti-friction properties, Mater. Chem. Phys. 134 (2012) 38–42.

[56] R.Y. Hong, Z.Q. Ren, Y.P. Han, H.Z. Li, Y. Zheng, J. Ding, Rheological properties of water-based Fe3O4 ferrofluids, Chem. Eng. Sci. 62 (2007) 5912–5924.

[57] M.K.A. Ali, X. Hou, M.A. Abdelkareem, Anti-wear properties evaluation of frictional sliding interfaces in automobile engines lubricated by copper/graphene nanolubricants, Friction 8 (2020) 905–916.

[58] L. Wang, P. Gong, W. Li, T. Luo, B. Cao, Mono-dispersed ag/graphene nanocomposite as lubricant additive to reduce friction and wear, Tribol. Int. 146 (2020) 106228.

[59] L. Mai, Y. Dong, L. Xu, C. Han, Single nanowire electrochemical devices, Nano Lett. 10 (2010) 4273–4278.

[60] M.K.A. Ali, H. Xianjun, Tribological characterization of M50 matrix composites reinforced by TiO2/graphene nanomaterials in dry conditions under different speeds and loads, Mater. Res. Express 6 (2019), 1165d6.

[61] N. Mohan, M. Sharma, R. Singh, N. Kumar, Tribological Properties of Automotive Lubricant SAE 20W-40 Containing Nano-Al2O3 Particles, SAE Technical Paper, 2014.

[62] S.E. McElwain, T.A. Blanchet, L.S. Schadler, W.G. Sawyer, Effect of particle size on the wear resistance of alumina-filled PTFE micro-and nanocomposites, Tribol. Trans. 51 (2008) 247–253.

[63] M. Vasheghani, E. Marzbanrad, C. Zamani, M. Aminy, B. Raissi, T. Ebadzadeh, et al., Effect of Al2O3 phases on the enhancement of thermal conductivity and viscosity of nanofluids in engine oil, Heat Mass Transf. 47 (2011) 1401–1405.

[64] K.G. Binu, B.S. Shenoy, D.S. Rao, R. Pai, A variable viscosity approach for the evaluation of load carrying capacity of oil lubricated journal bearing with TiO2 nanoparticles as lubricant additives, Procedia Mater. Sci. 6 (2014) 1051–1067.

[65] A.H. Elsheikh, S.W. Sharshir, M.E. Mostafa, F.A. Essa, M.K. Ahmed Ali, Applications of nanofluids in solar energy: a review of recent advances, Renew. Sust. Energ. Rev. 82 (2018) 3483–3502.

[66] K. Fangsuwannarak, K. Triratanasirichai, Improvements of palm biodiesel properties by using nano-TiO2 additive, exhaust emission and engine performance, Romanian Rev. Precis. Mech. Opt. Mechatron. 43 (2013) 111–118.

[67] V.W. Khond, V. Kriplani, Effect of nanofluid additives on performances and emissions of emulsified diesel and biodiesel fueled stationary CI engine: a comprehensive review, Renew. Sust. Energ. Rev. 59 (2016) 1338–1348.

[68] J. Sarkar, P. Ghosh, A. Adil, A review on hybrid nanofluids: recent research, development and applications, Renew. Sust. Energ. Rev. 43 (2015) 164–177.

[69] K. Bagavathi, Investigation of Alumina Additive in Lubricant Oil for Enhanced Engine Performance, Universiti Malaysia Pahang, 2012.

[70] Q. Jianhua, Z. Yu, W. Lingling, X. Jinjuan, Study on lubrication properties of modified Nano ZnO in base oil, China Pet. Process. Petrochemical Technol. 13 (2011) 14–18.

[71] L. Gara, Q. Zou, Friction and wear characteristics of oil-based ZnOnanofluids, Tribol. Trans. 56 (2013) 236–244.

[72] A. Elagouz, M.K.A. Ali, H. Xianjun, M.A. Abdelkareem, M.A. Hassan, Frictional performance evaluation of sliding surfaces lubricated by zinc-oxide nanoadditives, Surf. Eng. 36 (2020) 144–157.

[73] A.H. Battez, J.F. Rico, A.N. Arias, J.V. Rodriguez, R.C. Rodriguez, J.D. Fernandez, The tribologicalbehaviour of ZnO nanoparticles as an additive to PAO6, Wear 261 (2006) 256–263.

[74] H. Ghaednia, M.S. Hossain, R.L. Jackson, Tribological performance of silver nanoparticle–enhanced polyethylene glycol lubricants, Tribol. Trans. 59 (2016) 585–592.

[75] C.P. Twist, I. Bassanetti, M. Snow, M. Delferro, H. Bazzi, Y.-W. Chung, et al., Silver-organic oil additive for high-temperature applications, Tribol. Lett. 52 (2013) 261–269.

[76] M.S. Khan, M.S. Sisodia, S. Gupta, M. Feroskhan, S. Kannan, K. Krishnasamy, Measurement of tribological properties of cu and ag blended coconut oil nanofluids for metal cutting, Int. J. Eng. Sci. Technol. 22 (2019) 1187–1192.

[77] F. Jamil, H.M. Ali, Chapter 6—Applications of hybrid nanofluids in different fields, in: H.M. Ali (Ed.), Hybrid Nanofluids for Convection Heat Transfer, Academic Press, 2020, pp. 215–254.

[78] D. Li, B. Hong, W. Fang, Y. Guo, R. Lin, Preparation of well-dispersed silver nanoparticles for oil-based nanofluids, Ind. Eng. Chem. Res. 49 (2010) 1697–1702.

[79] J. Padgurskas, R. Rukuiza, I. Prosyčevas, R. Kreivaitis, Tribological properties of lubricant additives of Fe, Cu and Co nanoparticles, Tribol. Int. 60 (2013) 224–232.

[80] A.S. Pisal, D. Chavan, Experimental Investigation of Tribological Properties of Engine oil with CuO Nanoparticles, 49, Research in Mechanical Engineering (ICTARME), 2014, pp. 49–53.

[81] I. Rafique, A. Kausar, Z. Anwar, B. Muhammad, Exploration of epoxy resins, hardening systems, and epoxy/carbon nanotube composite designed for high performance materials: a review, Polym.-Plast. Technol. Eng. 55 (2016) 312–333.

[82] K. Friedrich, A.K. Schlarb, Tribology of Polymeric Nanocomposites: Friction and Wear of Bulk Materials and Coatings, Elsevier, 2011.

[83] P.L. Dickrell, S.B. Sinnott, D.W. Hahn, N.R. Raravikar, L.S. Schadler, P.M. Ajayan, et al., Frictional anisotropy of oriented carbon nanotube surfaces, Tribol. Lett. 18 (2005) 59–62.

[84] N. Salah, A. Alshahrie, N.D. Alharbi, M.S. Abdel-wahab, Z.H. Khan, Nano and micro structures produced from carbon rich fly ash as effective lubricant additives for 150SN base oil, J. Mater. Res. Technol. 8 (2019) 250–258.

[85] K. Lijesh, S. Muzakkir, H. Hirani, Experimental tribological performance evaluation of nano lubricant using multi-walled carbon nano-tubes (MWCNT), Int. J. Appl. Eng. Res. 10 (2015) 14543–14550.

[86] X.-X. Tian, R. Kalbasi, C. Qi, A. Karimipour, H.-L. Huang, Efficacy of hybrid nano-powder presence on the thermal conductivity of the engine oil: an experimental study, Powder Technol. 369 (2020) 261–269.

[87] H.-J. Choi, S.-M. Jung, J.-M. Seo, D.W. Chang, L. Dai, J.-B. Baek, Graphene for energy conversion and storage in fuel cells and supercapacitors, Nano Energy 1 (2012) 534–551.

[88] M.K.A. Ali, H. Xianjun, M.A. Abdelkareem, M. Gulzar, A. Elsheikh, Novel approach of the graphene nanolubricant for energy saving via anti-friction/wear in automobile engines, Tribol. Int. 124 (2018) 209–229.

[89] D. Berman, A. Erdemir, A.V. Sumant, Reduced wear and friction enabled by graphene layers on sliding steel surfaces in dry nitrogen, Carbon 59 (2013) 167–175.

[90] W. Gu, K. Chu, Z. Lu, G. Zhang, S. Qi, Synergistic effects of 3D porous graphene and T161 as hybrid lubricant additives on 316 ASS surface, Tribol. Int. 107072 (2021).

[91] S. Tarasov, A. Kolubaev, S. Belyaev, M. Lerner, F. Tepper, Study of friction reduction by nanocopper additives to motor oil, Wear 252 (2002) 63–69.

[92] G. Liu, X. Li, B. Qin, D. Xing, Y. Guo, R. Fan, Investigation of the mending effect and mechanism of copper nano-particles on a tribologically stressed surface, Tribol. Lett. 17 (2004) 961–966.

[93] L. Rapoport, V. Leshchinsky, I. Lapsker, Y. Volovik, O. Nepomnyashchy, M. Lvovsky, et al., Tribological properties of WS2 nanoparticles under mixed lubrication, Wear 255 (2003) 785–793.

[94] S. Sia, E.Z. Bassyony, A.A. Sarhan, Development of SiO2 nanolubrication system to be used in sliding bearings, Int. J. Adv. Manuf. Technol. 71 (2014) 1277–1284.

[95] H.-S. Kim, J.-W. Park, S.-M. Park, J.-S. Lee, Y.-Z. Lee, Tribological characteristics of paraffin liquid with nanodiamond based on the scuffing life and wear amount, Wear 301 (2013) 763–767.

[96] L. Liu, Z. Fang, A. Gu, Z. Guo, Lubrication effect of the paraffin oil filled with functionalized multi-walled carbon nanotubes for bismaleimide resin, Tribol. Lett. 42 (2011) 59–65.

[97] Y. Bao, J. Sun, L. Kong, Tribological properties and lubricating mechanism of SiO2 nanoparticles in water-based fluid, in: IOP Conference Series: Materials Science and Engineering, IOP Publishing, 2017, p. 012025.

[98] W. Dai, B. Kheireddin, H. Gao, H. Liang, Roles of nanoparticles in oil lubrication, Tribol. Int. 102 (2016) 88–98.

[99] M.K.A. Ali, H. Xianjun, M50 matrix sintered with nanoscale solid lubricants shows enhanced self-lubricating properties under dry sliding at different temperatures, Tribol. Lett. 67 (2019) 71.

[100] J. Zhao, Y. Huang, Y. He, Y. Shi, Nanolubricant additives: a review, Friction (2020) 1–27.

[101] M. Ratoi, V.B. Niste, J. Walker, J. Zekonyte, Mechanism of action of WS2 lubricant nanoadditives in high-pressure contacts, Tribol. Lett. 52 (2013) 81–91.

[102] H. Kato, K. Komai, Tribofilm formation and mild wear by tribo-sintering of nanometer-sized oxide particles on rubbing steel surfaces, Wear 262 (2007) 36–41.

[103] H. Wu, J. Zhao, W. Xia, X. Cheng, A. He, J.H. Yun, et al., A study of the tribological behaviour of TiO2 nano-additive water-based lubricants, Tribol. Int. 109 (2017) 398–408.

[104] L. Rapoport, M. Lvovsky, I. Lapsker, V. Leshchinsky, Y. Volovik, Y. Feldman, et al., Slow release of fullerene-like WS2 nanoparticles from Fe– Ni graphite matrix: a self-lubricating nanocomposite, Nano Lett. 1 (2001) 137–140.

[105] Y. Zhou, J. Qu, Ionic liquids as lubricant additives: a review, ACS Appl. Mater. Interfaces 9 (2017) 3209–3222.

[106] R. Li, Y. Wang, J. Zhang, J. Zhang, Origin of higher graphitization under higher humidity on the frictional surface of self-mated hydrogenated carbon films, Appl. Surf. Sci. 494 (2019) 452–457.

[107] L. Joly-Pottuz, B. Vacher, N. Ohmae, J. Martin, T. Epicier, Anti-wear and friction reducing mechanisms of carbon nano-onions as lubricant additives, Tribol. Lett. 30 (2008) 69–80.
[108] J. Zhao, J. Mao, Y. Li, Y. He, J. Luo, Friction-induced nano-structural evolution of graphene as a lubrication additive, Appl. Surf. Sci. 434 (2018) 21–27.
[109] H.Y. Chu, W.C. Hsu, J.F. Lin, Scuffing mechanism during oil-lubricated block-on-ring test with diamond nanoparticles as oil additive, Wear 268 (2010) 1423–1433.
[110] Y. Morita, S. Jinno, M. Murakami, N. Hatakeyama, A. Miyamoto, A computational chemistry approach for friction reduction of automotive engines, Int. J. Engine Res. 15 (2014) 399–405.
[111] J.P. Singh, S. Singh, T. Nandi, S.K. Ghosh, Development of graphitic lubricant nanoparticles based nanolubricant for automotive applications: Thermophysical and tribological properties followed by IC engine performance, Powder Technol. 387 (2021) 31–47.
[112] A. Kotia, S. Borkakoti, S.K. Ghosh, Wear and performance analysis of a 4-stroke diesel engine employing nanolubricants, Particuology 37 (2018) 54–63.
[113] A. Kotia, R. Kumar, A. Haldar, P. Deval, S.K. Ghosh, Characterization of Al2O3-SAE 15W40 engine oil nanolubricant and performance evaluation in 4-stroke diesel engine, J. Braz. Soc. Mech. Sci. Eng. 40 (2018) 38.
[114] M.F. Sgroi, M. Asti, F. Gili, F.A. Deorsola, S. Bensaid, D. Fino, et al., Engine bench and road testing of an engine oil containing MoS2 particles as nano-additive for friction reduction, Tribol. Int. 105 (2017) 317–325.
[115] M.K.A. Ali, H. Xianjun, Improving the heat transfer capability and thermal stability of vehicle engine oils using Al2O3/TiO2 nanomaterials, Powder Technol. 363 (2020) 48–58.

CHAPTER 33

Nanofluids as coolants

Zafar Said[a,b,c], Maham Sohail[a], and Arun Kumar Tiwari[d]
[a]Sustainable and Renewable Energy Engineering Department, University of Sharjah, Sharjah, United Arab Emirates
[b]Research Institute for Sciences and Engineering, University of Sharjah, Sharjah, United Arab Emirates
[c]U.S.-Pakistan Center for Advanced Studies in Energy (USPCAS-E), National University of Sciences and Technology (NUST), Islamabad, Pakistan
[d]Mechanical Engineering Department, Institute of Engineering & Technology, Dr. A.P.J. Abdul Kalam Technical University, Lucknow, Uttar Pradesh, India

Chapter outline

1. Introduction   713
    1.1 Nanofluid-based coolants and efficiency   715
    1.2 Engine cooling system   717
    1.3 Nanofluid for automotive applications   719
2. Numerical and experimental studies   727
3. Challenges and future outlook   728
4. Conclusion   732
References   732

1. Introduction

The advancement in automotive technology has resulted in enhanced thermal demands, and better cooling behavior is expected. Several methods have been advised to improve automotive radiators' heat transfer performance, such as active and passive methods [1]. The active approach requires a power source, whereas the passive approach is autonomous to external sources. Passive techniques are considered to have minimal operating expenses, and superior stability, particularly with compact devices [2], such as adding fins, microchannel, and turbulators are the passive techniques utilized to improve radiators' cooling rate, that is broadened to their restrictions [3]. Conventional coolants have shown superior heat transfer characteristics as a result of lower thermal conductivities, corrosion, clogging, and high pressure in heat exchangers and pipes. Novel and innovative coolants are needed for heat removal, and later, progress in colloid and interface science resulted in colloidal suspensions such as nanofluids [4–6]. Nanofluids have become a theme of significance, in recent years by nanosized (nanomaterials) in base fluids such as water, ethylene glycol, and oil [7–10]. Compared to the conventional base fluids they show better thermal conductivity [11–14]. Nanofluids were introduced by Choi [15], and comprehensive publications on the thermophysical properties of nanofluids have been published by many researchers [4, 9, 16, 17]. Several applications of thermo fluid systems, including automotive cooling systems, cooling of heat exchanging devices

[18–20], cooling of electronics, lubrication [10, 21], thermal storage, solar water heating [19, 22, 23], and cooling in machining [24]. Nanofluids possess exceptional heat transfer coefficients compared to the standard fluids owing to greater surface area [25–28]. Scholars have reported that Brownian motion is one of the main factors that aid in heat transfer increment [29, 30]. Xuan and Roetzel [31] reported that the slight perturbations cause Brownian motion in temperature and velocity formulation.

The exponential increment witnesses the increasing concern of the automotive industry on nanofluid coolants in published patents and market reports. The significant feature of nanofluids is the remarkable thermal conductivity of nanoparticles than conventional heat transfer fluids [32]. Due to the better thermal diffusivity, nanofluids are considered suitable coolants, which can be used in any system. Radiator performs a significant function in hindering the engine from high temperature due to combustion. A radiator has been designed with louvered fins so that the extra heat transfer at the surface area can be produced and disrupt the expansion of the boundary layer developed along the surface, as shown in Fig. 1 [33]. Nanofluids as coolants have significant effects in internal combustion engines such as energy-efficient, reducing automotive emissions because of reduced fuel consumption, and lessening global warming [34]. Nanofluids have a great capability to augment locomotive and large-scale engine cooling rates, decreasing the mass and sophistication of thermal control systems [3, 33]. Nanofluids have been employed innumerous automobile applications such as coolants, fuel additives, oils,

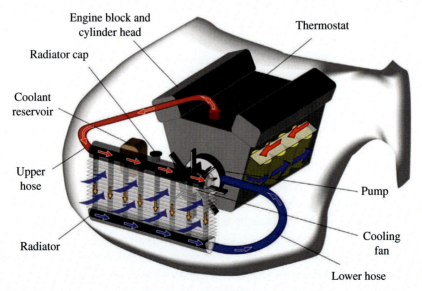

Fig. 1 Automotive engine and radiator. *(Copyright Elsevier 2017 (Lic#5080690296348), N.A. Che Sidik, M.N.a. Witri Mohd Yazid, R. Mamat, recent advancement of nanofluids in engine cooling system, Renew. Sust. Energ. Rev. 75 (2017) 137–144.)*

and refrigerants [17]. The high thermal conductivity and high volume concentrations of nanoparticles result in increased viscosity and pumping power, restricting nanofluids' use in heat transfer applications [35]. Automotive coolants using nanofluids have been reported in the literature. Said et al. [36] suggested that oxide-based nanofluids are promising for automotive applications because of their anticorrosive properties. Leong et al. [37] reported superior thermophysical characteristics and heat transfer performance in the car radiator using nanofluid-based coolants. Wen et al. [38] reported that when multi-walled carbon nanotubes (MWCNT)-based nanofluids increase the heat rate through copper. To improve engine efficiency, nanoparticles have been added into standard coolants such as water, ethylene glycol, and glycerol and their effectiveness has been accredited by several scholars.

Several peculiar properties are necessary for coolants in automotive applications. A coolant ought to have remarkable thermal conductivity, specific heat, low viscosity, and freezing point regarding the thermophysical properties. It should be nonhazardous, inert, and should not corrode. Accessible automobiles coolant are generally described by a 50:50 mass solution of EG and water as they possess low thermal conductivity. It must be advisable to utilize coolants with excellent thermal conduction to enhance the heat performance of the cooling system. Regardless of the enhancement in thermophysical properties may hinder the application of nanofluid coolants. Some significant investigations on the properties and their impact on the efficiency of nanofluid coolants, engine cooling circuit, and nanofluids coolants for automotive applications are described in the following sections.

1.1 Nanofluid-based coolants and efficiency

The thermal conduction increment by nanofluid-based coolants is normally associated with a surge in pressure decay [39]. Trade-off among higher heat transfer and significant pressure loss have to be taken into account by presenting performance index. Behabadi et al. [40] measured heat transfer and pressure drop properties of nanofluid coolant (CuO-oil) flow in a tube and defined an efficiency coefficient below to measure the overall advantages of nanofluid-based coolants:

$$\eta_I = \frac{\frac{h^*}{h_{RT,bf}}}{\frac{\Delta P^*}{\Delta P_{RT,bf}}} \tag{1}$$

h^* and ΔP^* represents the heat transfer coefficient and pressure drop, respectively, whereas $h_{RT,bf}$ and $\Delta P_{RT,bf}$ represents the traditional oil coolant flowing within a circular tube. Performance index larger than 1 suggests that heat carrier augmentation by either nanofluids or straightened tubes surmount the increased pressure decay. It can be observed from the Fig. 2 that the performance index is greater for nanofluids for 1 wt% and 2 wt%

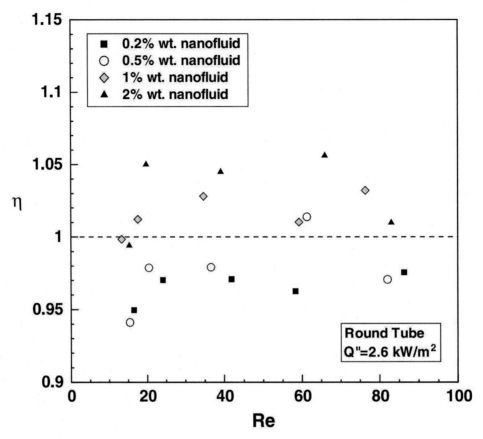

Fig. 2 Reynolds number for various nanoparticle loadings of nanofluid with varying performance index. (Copyright Elsevier 2011 (Lic#5080690501829), P. Razi, M.A. Akhavan-Behabadi, M. Saeedinia, Pressure drop and thermal characteristics of CuO–base oil nanofluid laminar flow in flattened tubes under constant heat flux, Int. Commun. Heat Mass Transf. 38(7) (2011) 964–971.)

particle loadings, and the record performance index of 1.056 is achieved at Reynolds number of 65.9 for 2 wt%. Ferrouillat et al. [41] experimentally investigated the effect of nanoparticle shape on the overall energy performances of SiO_2-water and ZnO-water nanofluids for cooling applications. Minor increment in Nusselt number was reported with nanofluid-based coolants. Fig. 3 shows that the experimental data corresponds well with the predictions of the correlations within ±10%. They also developed a second efficiency coefficient (η_{II}) to study the coolants performance. It is expressed as:

$$\eta_{II} = \frac{\frac{Q_{nf}}{W_{nf}}}{\frac{Q_{bf}}{W_{bf}}} \qquad (2)$$

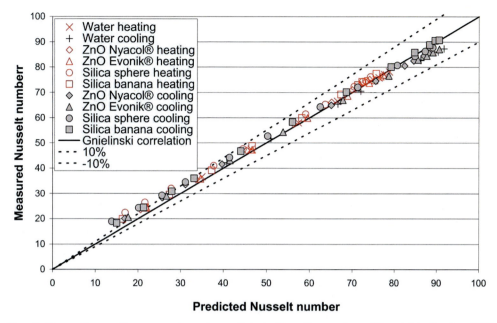

Fig. 3 Comparison of measured and predicted Nusselt number. *(Copyright Elsevier 2013 (Lic#5080710176915), S. Ferrouillat, et al., Influence of nanoparticle shape factor on convective heat transfer and energetic performance of water-based SiO2 and ZnO nanofluids, Appl. Therm. Eng. 51(1) (2013) 839–851.)*

$$Q = \epsilon C_{min}\left(T_{nf,in} - T_{a,in}\right) \quad (3)$$

$$W = V_{nf}\Delta P_{nf} \quad (4)$$

Results showed that $\eta_{II} > 1$, that is, coolants with energy effectiveness greater than base fluid, just for ZnO nanoparticles with shape factor above 3.

1.2 Engine cooling system

A conventional cooling system including radiator, fan, water pump, coolant tank, regulator. Once the engine is started during the warming-up process, the coolant is pumped from the bottom part of the radiator reservoir to the engine, where the coolant gets heat up [42]. Normally, the coolant temperature in operational surroundings is about 90°C. The fan improves thermal behavior in the radiator when a high temperature occurs. It is helpful to decrease warming through period, as hydrocarbon (HC) and carbon monoxide (CO) ejections are higher than in a heated state [43]. Nanofluids helps in reducing the warming-up time. Consider Al_2O_3-water + EG nanofluids, whose effective specific heat is lesser than the base fluid and increased density. Therefore, nanofluid's smaller volumetric heat capacity regarding base liquid implies a rapid warm-up phase leads to exceptional efficiency and decrement in emissions.

The ϵ-NTU approach determines coolants' heat carrier and exit temperature that flows through heat exchangers. Suppose the radiators model is demonstrated for both wind and coolant sections to identify the impacts of nanofluid coolants.

The air mass flow rate across the radiator is expressed as:

$$\dot{m}_a = \rho_a A_{c,a} u_a \tag{5}$$

where ρ_a represents air density, $A_{c,a}$ represents the flow area in the airside of the radiator and u_a refers to air velocity.

On airside, the convective heat transfer coefficient (h_a) can be expressed in terms of the Colburn factor (j_a), specific heat capacity ($C_{p,a}$), and the Prandtl number (Pr) as:

$$h_a = \frac{j_a \rho_a u_a C_{p,a}}{Pr^{2/3}} \tag{6}$$

The fins efficiency in the radiator can be computed as:

$$\eta_{fin} = \frac{\tanh(mL_{fin})}{mL_{fin}} \tag{7}$$

$$m = \sqrt{\frac{2h_a}{k_{fin} t}} \tag{8}$$

In this equation, L_{fin} is the fin length, k_{fin} is the thermal conductivity, and t represents a thickness of fins. In order to take into account the overall heat transfer efficiency on the air side, the overall surface effectiveness is computed by [44]:

$$\eta_0 = 1 - \left(1 - \eta_{fin}\right) A_{ft} \tag{9}$$

where A_{ft} represents the ratio between fin area and heat transfer surface.

The convective heat transmission coefficient of the coolant side is expressed as:

$$h_{nf} = \frac{Nu_{nf} k_{nf}}{D_{h,nf}} \tag{10}$$

where Nu_{nf} represents coolants' Nusselt number, which relies on the system characteristic and hydraulic tube diameter [42, 45].

Then, ϵ-NTU approach is implemented, with the number of heat transfer units (NTU) can be expressed as:

$$NTU = \frac{U_a \alpha_a V_r}{C_{min}} \tag{11}$$

With $C_{min} = \min(C_a, C_{nf})$ refers to the smallest heat capacity ability.

Nanofluid or air heat capacity is determined as:

$$C = \dot{m} C_p \qquad (12)$$

The effectiveness of heat exchanger (ϵ) for the real unmixed cross-flow pattern with interminable passages is observed as [44]:

$$\epsilon = 1 - \exp\left(\frac{1}{C^*}\right)(NTU)^{0.22}\left[\exp\left[-C^*(NTU)^{0.78}\right] - 1\right] \qquad (13)$$

The heat flux switched in the radiator is calculated as:

$$\dot{Q} = \epsilon C_{min}\left(T_{nf,in} - T_{a,in}\right) \qquad (14)$$

where $T_{nf,\,in}$ and $T_{a,\,in}$ refers to coolant and air entrance temperature, respectively.

1.3 Nanofluid for automotive applications

Numerous advancements have been carried out by enhancing the performance and lowering the difficulty of thermal management systems by implementing nanofluids in automotive applications. The increase in the heat transfer rate employing nanofluids compared with conventional fluid decreases the system's size and improves productivity [46]. The quantity of energy consumption reduces and conserves the energy for other uses by increasing system performance efficiency. Advancement of nanofluids can be the novel cooling thermal fluid technology [47].

The relations mentioned in the previous section are likely to develop nanofluids with remarkable heat transmitted performance while maintaining lower pumping power. Several investigators worked on the nanofluids application in automotive radiators. Bai et al. [48] numerically analyzed the heat transfer rate of Al_2O_3, TiO_2, Cu, and Al water-based nanofluids in a cooling system reported that Cu-water-based nanofluid exhibited a remarkable heat transfer coefficient of 46% more than pure water with the penalty of increased pumping power. Kumar et al. [49] investigated the performance of a car radiator using nanofluids along with louvered fins attached on the sides of tube (Fig. 4). Nanoparticles like Al_2O_3, ZnO, and CuO are dispersed in the base coolant comprise of 60:40 of EG:water. It is observed that nanofluids possess outstanding thermal conductivity and the maximum reported of 0.4355 W/mK for CuO-based nanofluids (Fig. 5). It can be seen from the Fig. 5, the viscosity enhances with the higher loading of nanoparticles. The viscosity increases by 14% for an increment in concentration from 0.05% to 5%. The Nusselt (Nu) number dependency with volume concentration and Reynolds (Re) number for CuO-based nanofluids is shown in Fig. 6. It is reported that the Nu boosts linearly with Re at all volume fractions. Leong et al. [37] examined the behavior of a car radiator using Cu-EG-based nanofluids. It is observed from the Figs. 7 and 8 that the heat transfer performance is enhanced with higher Reynolds number and volume

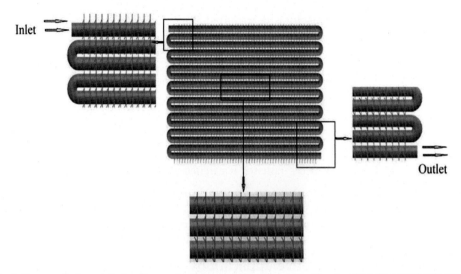

Fig. 4 Geometrical model of radiator with louvered fins. *(Copyright Elsevier 2020 (Lic#5080710360570), A. Kumar, M.A. Hassan, P. Chand, Heat transport in nanofluid coolant car radiator with louvered fins, Powder Technol. 376 (2020) 631–642.)*

concentration. Heat transfer improvement was reached up to 3.8% at 2 vol% of copper nanoparticles at 6000 and 5000 Reynolds number for air and coolant.

Hussein et al. [50] studied the heat transfer performance of a car radiator using SiO_2-water nanofluid. The experimental setup is displayed in Fig. 9. Friction factor was observed to reduce with increasing flow rate and increases with increasing volume concentration. It was also observed that the Nusselt number improves with increased flow rate, nanoparticle loading, and entrance temperature. Fig. 10 (A and B) shows heat transfer enhancement experimentally and numerically with changing volume fraction and inlet temperature to the radiator. The maximum heat transfer increment is observed around 46% numerically and around 42% experimentally for 2.5 vol%. The heat transfer improvement is improved from 39% to 56% as the temperature increases.

Peyghambarzadeh et al. [51] examined the heat transfer performance of Al_2O_3-water/EG-based nanofluids in a car radiator with louvered fins and flat tubes. The experimental design and schematic are displayed in Fig. 11. It is reported that the Nusselt number enhanced around 40% with rising volume fraction and Reynolds number for nanofluids. By adding 1 vol% of Al_2O_3 nanoparticles into water or EG, an increment of about 40% is reported than that of pure water and EG data (Fig. 12).

Heris et al. [52] experimented with a car radiator's heat transfer performance with CuO-water + EG nanofluids as a coolant. The experimental setup is displayed in Fig. 13. A car radiator with 40 vertical tubes was used as a heat exchanger with circular-shaped cross-section. It is observed from Fig. 14, Nusselt number enhances with increased

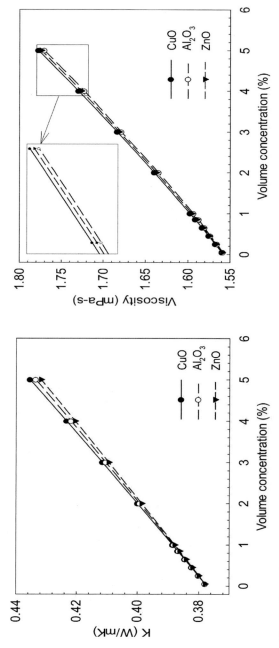

Fig. 5 Thermal conductivity and viscosity dependence on nanoparticle concentration. *(Copyright Elsevier 2020 (Lic#5080710360570), A. Kumar, M.A. Hassan, P. Chand, Heat transport in nanofluid coolant car radiator with louvered fins, Powder Technol. 376 (2020) 631–642.)*

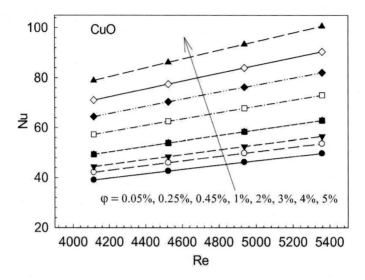

Fig. 6 Nusselt number variation with concentration and Reynolds number for CuO. *(Copyright Elsevier 2020 (Lic#5080710360570), A. Kumar, M.A. Hassan, P. Chand, Heat transport in nanofluid coolant car radiator with louvered fins, Powder Technol. 376 (2020) 631–642.)*

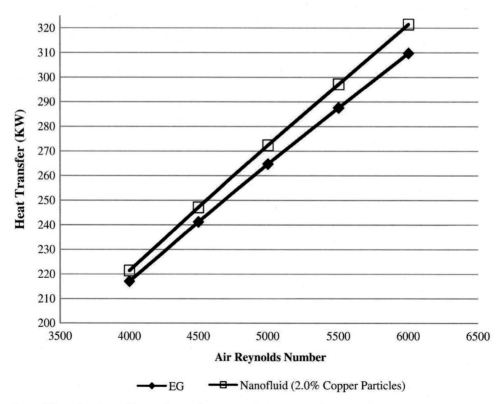

Fig. 7 Effect of air Reynolds number and nanoparticle loading on heat transfer. *(Copyright Elsevier 2010 (Lic#5080710588126), K.Y. Leong, et al., Performance investigation of an automotive car radiator operated with nanofluid-based coolants (nanofluid as a coolant in a radiator), Appl. Therm. Eng. 30(17) (2010) 2685–2692.)*

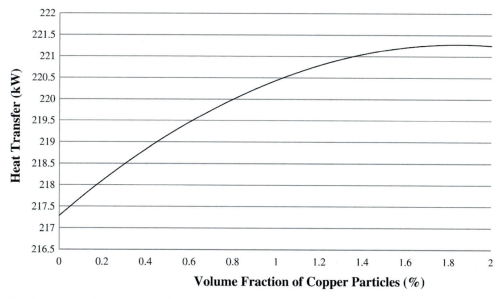

Fig. 8 Heat transfer at constant Reynolds number with respect to copper volume. *(Copyright Elsevier 2010 (Lic#5080710588126), K.Y. Leong, et al., Performance investigation of an automotive car radiator operated with nanofluid-based coolants (nanofluid as a coolant in a radiator), Appl. Therm. Eng. 30(17) (2010) 2685–2692.)*

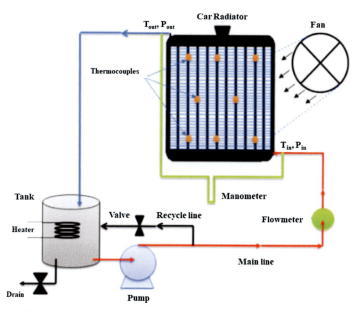

Fig. 9 Experimental setup. *(Copyright Elsevier 2014 (Lic#5080710747642), A.M. Hussein, R.A. Bakar, K. Kadirgama, Study of forced convection nanofluid heat transfer in the automotive cooling system. Case Stud. Therm. Eng. 2 (2014) 50–61.)*

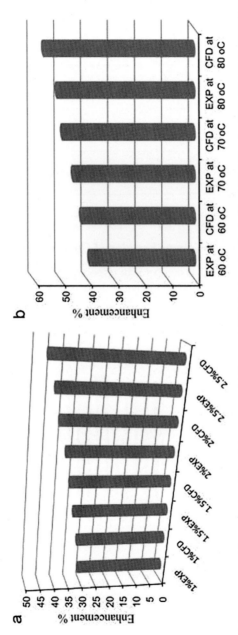

Fig. 10 Enhancement with respect to (A) volume concentration and (B) inlet temperature. (Copyright Elsevier 2014 (Lic#5080710747642), A.M. Hussein, R.A. Bakar, K. Kadirgama, Study of forced convection nanofluid heat transfer in the automotive cooling system. Case Stud. Therm. Eng. 2 (2014) 50–61.)

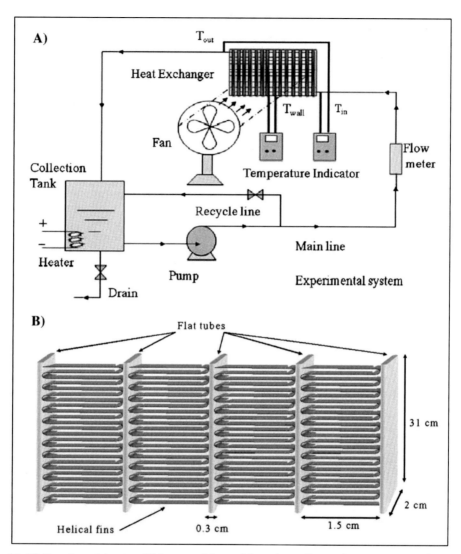

Fig. 11 (A) Experimental setup. (B) Louvered fin and flat tube radiator schematic. *(Copyright Elsevier 2011 (Lic#5080710887099), S.M. Peyghambarzadeh, et al., Experimental study of heat transfer enhancement using water/ethylene glycol based nanofluids as a new coolant for car radiators, Int. Commun. Heat Mass Transf. 38(9) (2011) 1283–1290.)*

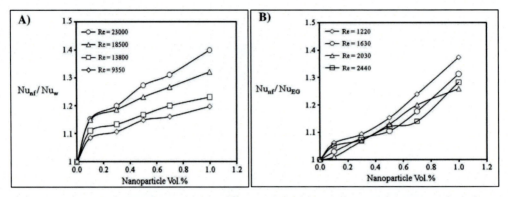

Fig. 12 Variations of Nusselt number at different Reynolds numbers and nanoparticle volume concentration of (A) water-based nanofluids (B) EG-based nanofluids. *(Copyright Elsevier 2011 (Lic#5080710887099), S.M. Peyghambarzadeh, et al., Experimental study of heat transfer enhancement using water/ethylene glycol based nanofluids as a new coolant for car radiators, Int. Commun. Heat Mass Transf. 38(9) (2011) 1283–1290.)*

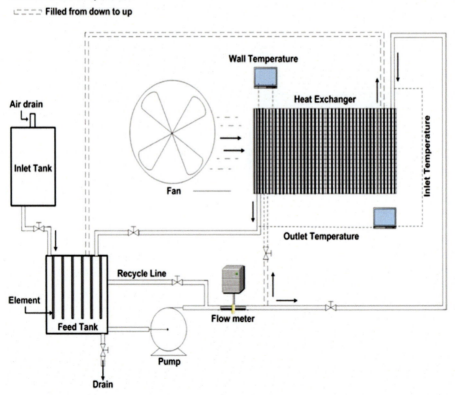

Fig. 13 Experimental setup. *(Copyright Elsevier 2014 (Lic#5080711075105), S.Z. Heris, et al., Experimental study of heat transfer of a Car radiator with CuO/ethylene glycol-water as a coolant, J. Dispers. Sci. Technol. 35(5) (2014) 677–684.)*

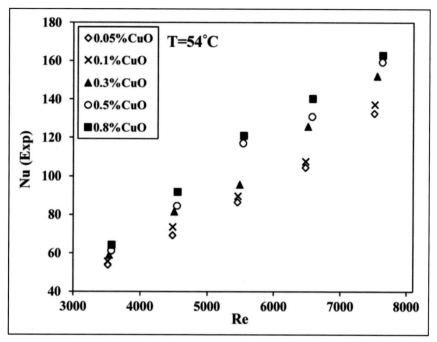

Fig. 14 Variations of Nusselt number of nanofluids depending on Reynolds number. *(Copyright Elsevier 2014 (Lic#5080711075105), S.Z. Heris, et al., Experimental study of heat transfer of a Car radiator with CuO/ethylene glycol-water as a coolant, J. Dispers. Sci. Technol. 35(5) (2014) 677–684.)*

volume fraction and Reynolds number. They reported that the dispersion effect and chaotic movement of the nanoparticles in high flow rates increased the mixing fluctuations and enhanced the heat transfer coefficient. The maximum Nusselt number was reported up to 55% for 0.8 vol% of CuO-based nanofluids.

2. Numerical and experimental studies

Recent advancements in computational schemes have attracted scientists in computational physics. In contrast to experimental studies, computational or numerical projections permit significant studies to be studied at a minimal price. Numerical investigations on nanofluids have been developed as a substitute method to validate exceptional behavior to conventional heat transfer liquids. Lv et al. [53] projected the increment in thermal performance for inner combustion mechanism using Cu-nanoparticles and observed an increment in heat transfer coefficient of 46% as compared to the fluid. Vajjha et al. [54] examined the performance of oxide nanoparticles using water and ethylene glycol mixture of an automotive radiator computationally and reported the increment of heat transfer coefficient for 3 vol% of 36.6% for Al_2O_3 nanofluid and 49.7% for CuO nanofluids compared to the base fluid. Further numerical studies on the nanofluids as coolants are presented in Table 1.

Experimental study using nanofluid for the performance evaluation in the automotive cooling system was developed by Tzeng et al. [64]. It was studied that the behavior of CuO, Al_2O_3 nanoparticles, and antifoam when combined with transmission oil. They reported that CuO-based nanofluid shows a remarkable heat transfer effect. The recent experimental research on a vehicle cooling system was carried out by researcher Ali [65]. They investigated convection heat transfer in automotive radiators employed with Al_2O_3-H_2O nanofluid. It was reported that the heat transfer coefficient achieved a maximum at 1 vol% concentration. Further increment in volume concentration would have a negative impact on the radiator cooling system performance. Summary of experimental studies has been presented in Table 2.

Numerical findings have shown that heat transfer performance with volume fraction enhances. In many studies, CuO-based nanofluids have shown remarkable performance as coolants and are considered as potential coolants. Experimental findings have shown different types of radiators used with nanofluids of various volume concentrations. Al_2O_3-water-based nanofluids have exhibited remarkable heat transfer performance. Radiators with different geometries have made improvements in Nusselt number and overall heat transfer coefficients. The maximum efficiency enhancement reported of 45% for radiator with 34 vertical tubes.

3. Challenges and future outlook

Researchers have studied the extensive use of nanofluids in automotive applications. But the advancement is restricted by some challenges which must be given considerable attention shortly. Serious concerns like thermal conductivity, nanoparticles' Brownian motion, and changing thermophysical property with temperature, should be considered with convective heat transfer in nanofluids. Experimental investigations in the convective heat transfer of nanofluids are needed. Metallic nanoparticles with various geometries and particle loadings for heat transfer increment in laminar, transition, and turbulent regions should be considered for future convective investigations.

Nanofluids are produced either by one step that produces and disperse the nanoparticles into base fluids or a two-step technique that involves creating nanoparticles and dispersing them into a base fluid. The base fluids consist of ions and reaction effects that are complicated to detach from the fluid. Another significant challenge in producing nanofluids is agglomerating into bigger clusters of particles, which negatively affects the benefits of high surface area nanoparticles. To overcome this concern, surfactants are added to the nanofluids. More studies should focus on stability as it is the most significant step for the advancement of nanofluids. Despite having the investigations on

Table 1 Numerical studies on the nanofluids as coolants.

Researchers	Nanofluids	Numerical technique	Volume concentration (vol%)	Findings
Tijani et al. [55]	Al_2O_3-water + EG CuO-water + EG	ANSYS	0.05–0.3	Maximum heat transfer performance was reported for CuO-based nanofluids
Hatami et al. [56]	CuO-water/EG TiO_2-water/EG Al_2O_3-water/EG Fe_3O_4-water/EG	ANSYS	0–6	Maximum enhancement in cooling is reported for TiO_2-EG with the highest volume fraction
Dawood et al. [57]	TiO_2-water	FLUENT 6.3.26	1–4	Nusselt number and Friction factor is dependent on volume fraction, inlet temperature, and Reynolds number
Shedid et al. [58]	Al_2O_3-water CuO-water	FLUENT 14.0	1–7	Heat transfer coefficient enhancement of 45% and 38% for Al_2O_3 and CuO nanofluids was reported
Das et al. [59]	Al_2O_3-water + EG CuO-water + EG	FLUENT	0.01–0.1	Al_2O_3-based nanofluids showed increment in heat transfer coefficient and decrement in pumping power
Aravindkumar et al. [60]	CuO-water SnO-water SiC-water BeO-water	ε-NTU	–	CuO-water nanofluid can be used as a promising coolant due to better heat transfer characteristics
Huminic and Huminic [61]	CuO-EG	ANSYS CFX—12.0	0–4	Heat transfer coefficient increases with increasing volume concentration, Brownian motion, and Reynolds number
Delavari et al. [62]	Al_2O_3-water Al_2O_3-EG	CFD	0–1	Nusselt number for two-phase method was 10%–45% greater than single-phase approach
Huminc and Huminic [63]	Cu-EG	ANSYS CFX—12.0	0–2	Maximum improvement of heat transfer coefficient of 82% was obtained

Table 2 Experimental studies on the nanofluids as coolants.

Researchers	Nanofluids	Radiator type	Volume concentration (vol%)	Findings
Filho et al. [66]	Graphene-water + EG Silver-water + EG	Automotive radiator	0.01, 0.05, 0.1	Heat transfer rate was improved up to 4.7% for Ag nanofluids, and thermohydraulic performance was reduced with graphene nanofluids
Naraki et al. [67]	CuO-water	Crossflow automotive radiator	0.04–0.4	Overall heat transfer coefficient improves with increasing volume fraction, and maximum heat transfer coefficient was reported up to 8% than base fluid
Reddy et al. [68]	TiO_2-water/EG	Automobile radiator	0.1, 0.3, and 0.5	Heat transfer rate was improved up to 37% at lower volume fractions and thermal conductivity was improved by about 3%
Goudarzi et al. [69]	Al_2O_3-EG	Automobile radiator with wire coil inserts	0.08, 0.5, 1	Nusselt number was enhanced by 13% for 1 vol%
Hafiz et al. [70]	ZnO-water	Aluminum car radiator	0.01–0.3	Improvement in heat transfer performance was reported up to 46% as compared to water at 0.2 vol%
Peyghambarzadeh et al. [71]	CuO water Fe_2O_3-water	Finned tube automobile radiator	0.15–0.65	Overall heat transfer coefficient improves, and fluid inlet temperature reduces.

Table 2 Experimental studies on the nanofluids as coolants—cont'd

Researchers	Nanofluids	Radiator type	Volume concentration (vol%)	Findings
Elsaid et al. [72]	Al_2O_3-water + EG Co_3O_4-water + EG	Radiator FIAT-128	0.02–0.2	Nusselt number is improved by 31.8%.
Peyghambarzadeh et al. [73]	Al_2O_3-water	Radiator with 34 vertical tube	0.1–1	Heat transfer efficiency can be enhanced up to 45%
Yadav et al. [74]	Al_2O_3-water	Radiator with metallic duct and exhaust fan	0–0.2	Heat transfer improvements up to 44.29%
Hussein et al. [75]	TiO_2-water SiO_2-water	Automobile radiator	1–2	Highest Nusselt number improvements of 11% and 22.5% were attained for TiO_2 and SiO_2
Tzeng et al. [64]	CuO-engine oil Al_2O_3-engine oil	—	—	CuO showed an effective heat transfer effect for all rotating speeds
Kole and Dey [76]	Al_2O_3-EG	Car engine coolant	0.001–0.035	Maximum increment of thermal conductivity was 11.25% at 3.5 vol%
Ebrahimi et al. [77]	SiO_2-water	Automobile radiator	0.1, 0.2, 0.4	Maximum heat transfer enhancement was reported 9.3%
Nieh et al. [78]	Al_2O_3-water + EG TiO_2-water + EG	Air-cooled radiator	0.2, 0.5, 1	Maximum efficiency factor was 27.2% with 2 vol%of TiO_2-water + EGnanofluids

nanofluids property, a few features are under development phase or have still to be investigated numerically and experimentally. The improved nanofluids' stability and longer durability will bring a revolution in heat transfer applications.

4. Conclusion

This book chapter outlines the applications of nanofluids as automotive coolants. Several studies represented that nanofluids have a potential application in the advancement of coolant radiators. The heat transfer performance of nanofluids is a function of the type of base fluids, the ratio of EG:water mixture, nanoparticles' type, temperature, volume fraction, size and shape, and flow characteristics. Current investigations on nanofluids as automotive coolant is still at its developing stage and needs further advancement. The problems for the nanofluids in the radiator cooling are considered along with the unique application area and future extent of the nanofluids.

References

[1] D. Wen, et al., Review of nanofluids for heat transfer applications, Particuology 7 (2) (2009) 141–150.
[2] S. Liu, M. Sakr, A comprehensive review on passive heat transfer enhancements in pipe exchangers, Renew. Sust. Energ. Rev. 19 (2013) 64–81.
[3] T. Alam, M.-H. Kim, A comprehensive review on single phase heat transfer enhancement techniques in heat exchanger applications, Renew. Sust. Energ. Rev. 81 (2018) 813–839.
[4] Z. Said, et al., Recent advances on nanofluids for low to medium temperature solar collectors: energy, exergy, economic analysis and environmental impact, Prog. Energy Combust. Sci. 84 (2021) 100898.
[5] M. Sheikholeslami, et al., Recent progress on flat plate solar collectors and photovoltaic systems in the presence of nanofluid: A review, J. Clean. Prod. (2021) 126119.
[6] A.K. Tiwar, et al., A review on the application of hybrid nanofluids for parabolic trough collector: Recent progress and outlook, J. Clean. Prod. (2021) 126031.
[7] Z. Said, Thermophysical and optical properties of SWCNTs nanofluids, Int. Commun. Heat Mass Transf. 78 (2016) 207–213.
[8] Z. Said, et al., Acid-functionalized carbon nanofibers for high stability, thermoelectrical and electrochemical properties of nanofluids, J. Colloid Interface Sci. 520 (2018) 50–57.
[9] Z. Said, S. Arora, E. Bellos, A review on performance and environmental effects of conventional and nanofluid-based thermal photovoltaics, Renew. Sust. Energ. Rev. 94 (2018) 302–316.
[10] Z. Said, et al., A comprehensive review on minimum quantity lubrication (MQL) in machining processes using nano-cutting fluids, Int. J. Adv. Manuf. Technol. 105 (5) (2019) 2057–2086.
[11] Z. Said, et al., Heat transfer, entropy generation, economic and environmental analyses of linear fresnel reflector using novel rGO-Co3O4 hybrid nanofluids, Renew. Energy 165 (2021) 420–437.
[12] Z. Said, et al., Performance enhancement of a flat plate solar collector using titanium dioxide nanofluid and polyethylene glycol dispersant, J. Clean. Prod. 92 (2015) 343–353.
[13] Z. Said, et al., Energy and exergy efficiency of a flat plate solar collector using pH treated Al2O3 nanofluid, J. Clean. Prod. 112 (2016) 3915–3926.
[14] Z. Said, et al., Thermophysical properties of Single Wall carbon nanotubes and its effect on exergy efficiency of a flat plate solar collector, Sol. Energy 115 (2015) 757–769.
[15] S.U. Choi, J.A. Eastman, Enhancing Thermal Conductivity of Fluids with Nanoparticles, Argonne National Lab, IL (United States), 1995.

[16] M. Sajid Hossain, et al., Spotlight on available optical properties and models of nanofluids: a review, Renew. Sust. Energ. Rev. 43 (2015) 750–762.

[17] M. Gupta, et al., A review on thermophysical properties of nanofluids and heat transfer applications, Renew. Sust. Energ. Rev. 74 (2017) 638–670.

[18] L. Sundar, et al., Heat transfer of rGO/Co3O4 hybrid nanomaterial based nanofluids and twisted tape configurations in a tube, J. Therm. Sci. Eng. Appl. (2020) 1–41.

[19] M. Gupta, V. Singh, Z. Said, Heat transfer analysis using zinc ferrite/water (hybrid) nanofluids in a circular tube: an experimental investigation and development of new correlations for thermophysical and heat transfer properties, Sustain. Energy Technol. Assess. 39 (2020) 100720.

[20] Z. Said, et al., Heat transfer enhancement and life cycle analysis of a Shell-and-tube heat exchanger using stable CuO/water nanofluid, Sustain. Energy Technol. Assess 31 (2019) 306–317.

[21] X. Wang, et al., Vegetable oil-based nanofluid minimum quantity lubrication turning: academic review and perspectives, J. Manuf. Process. 59 (2020) 76–97.

[22] A.A. Hachicha, et al., On the thermal and thermodynamic analysis of parabolic trough collector technology using industrial-grade MWCNT based nanofluid, Renew. Energy 161 (2020) 1303–1317.

[23] A.A. Hachicha, et al., A review study on the modeling of high-temperature solar thermal collector systems, Renew. Sust. Energ. Rev. 112 (2019) 280–298.

[24] R. Saidur, K.Y. Leong, H.A. Mohammed, A review on applications and challenges of nanofluids, Renew. Sust. Energ. Rev. 15 (3) (2011) 1646–1668.

[25] Z. Said, et al., Evaluating the optical properties of TiO2 Nanofluid for a direct absorption solar collector, Numer. Heat Transf. A: Appl. 67 (9) (2015) 1010–1027.

[26] A.K. Tiwari, et al., 4S consideration (synthesis, sonication, surfactant, stability) for the thermal conductivity of CeO2 with MWCNT and water based hybrid nanofluid: an experimental assessment, Colloids Surf. A Physicochem. Eng. Asp. 610 (2021) 125918.

[27] A.K. Tiwari, et al., Experimental and numerical investigation on the thermal performance of triple tube heat exchanger equipped with different inserts with WO3/water nanofluid under turbulent condition, Int. J. Therm. Sci. (2021) 106861.

[28] L. Syam Sundar, et al., Heat transfer of rGO/CO3O4 hybrid nanomaterial-based Nanofluids and twisted tape configurations in a tube, J. Therm. Sci. Eng. Appl. 13 (3) (2021), 031004.

[29] L.S. Sundar, et al., Heat transfer, energy, and exergy efficiency enhancement of nanodiamond/water nanofluids circulate in a flat plate solar collector, J. Enhanc. Heat Transf. (2021) 28(2).

[30] L.S. Sundar, et al., Experimental investigation of thermo-physical properties, heat transfer, pumping power, entropy generation, and exergy efficiency of nanodiamond + Fe3O4/60: 40% water-ethylene glycol hybrid nanofluid flow in a tube, Therm. Sci. Eng. Prog. 21 (2021) 100799.

[31] Y. Xuan, W. Roetzel, Conceptions for heat transfer correlation of nanofluids, Int. J. Heat Mass Transf. 43 (19) (2000) 3701–3707.

[32] P. Keblinski, et al., Mechanisms of heat flow in suspensions of nano-sized particles (nanofluids), Int. J. Heat Mass Transf. 45 (4) (2002) 855–863.

[33] N.A. Che Sidik, M.N.A.W.M. Yazid, R. Mamat, Recent advancement of nanofluids in engine cooling system, Renew. Sust. Energ. Rev. 75 (2017) 137–144.

[34] Y. Mukkamala, Contemporary trends in thermo-hydraulic testing and modeling of automotive radiators deploying nano-coolants and aerodynamically efficient air-side fins, Renew. Sust. Energ. Rev. 76 (2017) 1208–1229.

[35] T. Ambreen, M.-H. Kim, Heat transfer and pressure drop correlations of nanofluids: a state of art review, Renew. Sust. Energ. Rev. 91 (2018) 564–583.

[36] Z. Said, et al., Enhancing the performance of automotive radiators using nanofluids, Renew. Sust. Energ. Rev. 112 (2019) 183–194.

[37] K.Y. Leong, et al., Performance investigation of an automotive car radiator operated with nanofluid-based coolants (nanofluid as a coolant in a radiator), Appl. Therm. Eng. 30 (17) (2010) 2685–2692.

[38] D. Wen, Y. Ding, Experimental investigation into convective heat transfer of nanofluids at the entrance region under laminar flow conditions, Int. J. Heat Mass Transf. 47 (24) (2004) 5181–5188.

[39] Y. Li, et al., Experimental investigation on heat transfer and pressure drop of ZnO/ethylene glycol-water nanofluids in transition flow, Appl. Therm. Eng. 93 (2016) 537–548.

[40] P. Razi, M.A. Akhavan-Behabadi, M. Saeedinia, Pressure drop and thermal characteristics of CuO–base oil nanofluid laminar flow in flattened tubes under constant heat flux, Int. Commun. Heat Mass Transf. 38 (7) (2011) 964–971.
[41] S. Ferrouillat, et al., Influence of nanoparticle shape factor on convective heat transfer and energetic performance of water-based SiO2 and ZnO nanofluids, Appl. Therm. Eng. 51 (1) (2013) 839–851.
[42] D.R. Ray, D.K. Das, Superior performance of nanofluids in an automotive radiator, J. Therm. Sci. Eng. Appl. 6 (4) (2014).
[43] M.V. Millan, S. Samuel, Nanofluids and thermal management strategy for automotive application, SAE Tech. Pap. (2015).
[44] F.C. McQuiston, J.D. Parker, J.D. Spitler, Heating, Ventilating, and Air Conditioning: Analysis and Design, John Wiley & Sons, 2004.
[45] A. Bianco, et al., Crosstalk minimization in microring-based wavelength routing matrices, in: IEEE Global Telecommunications Conference—GLOBECOM 2011, 2011.
[46] H.A. Mohammed, et al., Convective heat transfer and fluid flow study over a step using nanofluids: a review, Renew. Sust. Energ. Rev. 15 (6) (2011) 2921–2939.
[47] S.K. Das, et al., Nanofluids: Science and Technology, John Wiley & Sons, 2007.
[48] M. Bai, Z. Xu, J. Lv, Application of nanofluids in engine cooling system, SAE Tech. Pap. (2008).
[49] A. Kumar, M.A. Hassan, P. Chand, Heat transport in nanofluid coolant car radiator with louvered fins, Powder Technol. 376 (2020) 631–642.
[50] A.M. Hussein, R.A. Bakar, K. Kadirgama, Study of forced convection nanofluid heat transfer in the automotive cooling system, Case Stud. Therm. Eng. 2 (2014) 50–61.
[51] S.M. Peyghambarzadeh, et al., Experimental study of heat transfer enhancement using water/ethylene glycol based nanofluids as a new coolant for car radiators, Int. Commun. Heat Mass Transf. 38 (9) (2011) 1283–1290.
[52] S.Z. Heris, et al., Experimental study of heat transfer of a Car radiator with CuO/ethylene glycol-water as a coolant, J. Dispers. Sci. Technol. 35 (5) (2014) 677–684.
[53] Z. Said, et al., Recent advances on the fundamental physical phenomena behind stability, dynamic motion, thermophysical properties, heat transport, applications, and challenges of nanofluids, Phys. Rep. (2021), https://doi.org/10.1016/j.physrep.2021.07.002. In press.
[54] R.S. Vajjha, D.K. Das, D.R. Ray, Development of new correlations for the Nusselt number and the friction factor under turbulent flow of nanofluids in flat tubes, Int. J. Heat Mass Transf. 80 (2015) 353–367.
[55] A.S. Tijani, A.S.B. Sudirman, Thermos-physical properties and heat transfer characteristics of water/anti-freezing and Al2O3/CuO based nanofluid as a coolant for car radiator, Int. J. Heat Mass Transf. 118 (2018) 48–57.
[56] M. Hatami, et al., Investigation of engines radiator heat recovery using different shapes of nanoparticles in H2O/(CH2OH)2 based nanofluids, Int. J. Hydrog. Energy 42 (16) (2017) 10891–10900.
[57] A.M. Hussein, et al., Numerical study on turbulent forced convective heat transfer using nanofluids TiO2 in an automotive cooling system, Case Stud. Therm. Eng. 9 (2017) 72–78.
[58] M. Elsebay, et al., Numerical resizing study of Al2O3 and CuO nanofluids in the flat tubes of a radiator, Appl. Math. Model. 40 (13) (2016) 6437–6450.
[59] R.S. Vajjha, D.K. Das, P.K. Namburu, Numerical study of fluid dynamic and heat transfer performance of Al2O3 and CuO nanofluids in the flat tubes of a radiator, Int. J. Heat Fluid Flow 31 (4) (2010) 613–621.
[60] N. Aravindkumar, et al., Computational and numerical analysis of radiator with different tube structures and nano fluid as coolant, Mater. Today Proceed. (2020).
[61] G. Huminic, A. Huminic, Numerical analysis of laminar flow heat transfer of nanofluids in a flattened tube, Int. Commun. Heat Mass Transf. 44 (2013) 52–57.
[62] V. Delavari, H. Hashemabadi, CFD simulation of heat transfer enhancement of Al2O3/water and Al2O3/ethylene glycol nanofluids in a car radiator, Appl. Therm. Eng. 73 (2014) 378–388.
[63] G. Huminic, A. Huminic, The cooling performances evaluation of nanofluids in a compact heat exchanger, SAE Tech. Pap. (2012).

[64] S.C. Tzeng, C.W. Lin, K.D. Huang, Heat transfer enhancement of nanofluids in rotary blade coupling of four-wheel-drive vehicles, Acta Mech. 179 (1) (2005) 11–23.
[65] M. Ali, A.M. El-Leathy, Z. Al-Sofyany, The effect of Nanofluid concentration on the cooling system of vehicles radiator, Adv. Mech. Eng. 6 (2014) 962510.
[66] E.M. Cárdenas Contreras, G.A. Oliveira, E.P. Bandarra Filho, Experimental analysis of the thermo-hydraulic performance of graphene and silver nanofluids in automotive cooling systems, Int. J. Heat Mass Transf. 132 (2019) 375–387.
[67] M. Naraki, et al., Parametric study of overall heat transfer coefficient of CuO/water nanofluids in a car radiator, Int. J. Therm. Sci. 66 (2013) 82–90.
[68] S. Devireddy, C.S.R. Mekala, V.R. Veeredhi, Improving the cooling performance of automobile radiator with ethylene glycol water based TiO2 nanofluids, Int. Commun. Heat Mass Transf. 78 (2016) 121–126.
[69] K. Goudarzi, H. Jamali, Heat transfer enhancement of Al2O3-EG nanofluid in a car radiator with wire coil inserts, Appl. Therm. Eng. 118 (2017) 510–517.
[70] H.M. Ali, et al., Experimental investigation of convective heat transfer augmentation for car radiator using ZnO–water nanofluids, Energy 84 (2015) 317–324.
[71] S.M. Peyghambarzadeh, et al., Experimental study of overall heat transfer coefficient in the application of dilute nanofluids in the car radiator, Appl. Therm. Eng. 52 (1) (2013) 8–16.
[72] A.M. Elsaid, Experimental study on the heat transfer performance and friction factor characteristics of Co3O4 and Al2O3 based H2O/(CH2OH)2 nanofluids in a vehicle engine radiator, Int. Commun. Heat Mass Transf. 108 (2019) 104263.
[73] S.M. Peyghambarzadeh, et al., Improving the cooling performance of automobile radiator with Al2O3/water nanofluid, Appl. Therm. Eng. 31 (10) (2011) 1833–1838.
[74] P. Chaurasia, et al., Heat transfer augmentation in automobile radiator using Al2O3–water based nanofluid, SN Appl. Sci. 1 (3) (2019) 257.
[75] A.M. Hussein, et al., Heat transfer enhancement using nanofluids in an automotive cooling system, Int. Commun. Heat Mass Transf. 53 (2014) 195–202.
[76] M. Kole, T.K. Dey, Thermal conductivity and viscosity of Al2O3nanofluid based on car engine coolant, J. Phys. D. Appl. Phys. 43 (31) (2010) 315501.
[77] M. Ebrahimi, et al., Experimental investigation of force convection heat transfer in a car radiator filled with SiO2-water nanofluid, Int. J. Eng. Trans. B: Appl. 27 (2014) 333–340.
[78] H.-M. Nieh, T.-P. Teng, C.-C. Yu, Enhanced heat dissipation of a radiator using oxide nano-coolant, Int. J. Therm. Sci. 77 (2014) 252–261.

CHAPTER 34

Nanomaterials in automotive fuels

Arun Kumar Tiwari[a], Amit Kumar[a], and Zafar Said[b]
[a]Mechanical Engineering Department, Institute of Engineering & Technology, Dr. A.P.J. Abdul Kalam Technical University, Lucknow, Uttar Pradesh, India
[b]Sustainable and Renewable Energy Engineering Department, University of Sharjah, Sharjah, United Arab Emirates

Chapter outline

1. Introduction — 737
2. Nanomaterials impact on fuel properties — 738
3. Metal oxide nanomaterials application in automotive fuels — 740
   3.1 CeO_2 nanoparticles — 740
   3.2 Al_2O_3 — 742
   3.3 TiO_2 — 743
   3.4 Other metal oxide nanoparticles — 745
4. Conclusion — 746
References — 746

1. Introduction

Hydrocarbon fuels used in automobile vehicles are one of the significant sources of greenhouse gas (GHG) emissions. Diesel engines are using heavily in automobile vehicle transportation release harmful emissions. To reduce the harmful GHG emission, alternate fuels can be used in automobiles. Different researchers are focusing on using metal additives to reduce engine emissions. These additives not only reduce harmful emissions but also increase the combustion performance parameters. Applying nanoscale additives of energetic metal particles to liquid fuels is an exciting concept that has not yet been fully exploited. These formulated nanofuels offer shorter ignition times, reduced combustion times, and rapid oxidation leading to complete combustion. The total calorific value of the liquid fuel is increased by the high energy density of the metal nanoparticles, which enhances the efficiency of the engine by increasing the output power. So far, different nanomaterials as additives were used by the researchers, such as aluminum oxide (Al_2O_3), silver oxide (Ag_2O), magnesium (Mg), cerium oxide (CeO_2), Multi-wall carbon nanotubes (MWCNTs), graphene oxide (GO), iron oxide (Fe_3O_4), silicon (Si), copper oxide (CuO), zinc oxide (ZnO), etc. Different types of nanoparticles at different particles concentration affects the performance parameters (indicated power, Brake power, engine torque, brake specific fuel consumption (BSFC), emission of toxic gasses and brake thermal efficiency (BTE), etc.) of internal combustion (IC) engine of the automobile. Further researches have been conducted on mixing nanoparticles with water emulsion technology to enhance energy efficiency and reduce emissions [1]..

2. Nanomaterials impact on fuel properties

Its properties simply determine the quality of fuel. Nanoparticles addition in the base fuel is proved to be an appreciated technique for the improvement of fuel properties. The researchers measured significant enhancement in different fuel properties like cetane number, kinematic viscosity, flash point, calorific value, and density while adding nanoparticles [2–8].

Vedagiri et al. [8] inspected the performance, combustion, and exhaust exhalation parameters while suspension of cerium oxide (CeO_2) and zinc oxide (ZnO) into the biodiesel and got a significant amount of enhancement in brake thermal efficiency (BTE) of biodiesel fuel. The enhancement in BTE they obtained was 30.2% for CeO_2 and 30.51% for ZnO. Praveena et al. [7] analyzed the engine performance characteristics while nanosuspension of cerium oxide (CeO_2) and zinc oxide (ZnO) into the grapeseed oil methyl ester and got a significant amount of enhancement in BTE of biodiesel fuel. The enhancement in BTE they obtained was approximately 29%. Karthikeyan et al. [6] examined the effect of cerium oxide (CeO_2) emulsion into the grapeseed oil methyl ester (GSOME) and compared the fuel properties at different mixture concentrations with diesel and CeO_2. Their experiment concluded increment in thermal performance parameters with a significant amount of reduction in harmful emission gases.

Soner et al. [4] investigate the nanodiesel fuel by dispersing the aluminum oxide (Al_2O_3) and copper oxide (CuO) nanoparticles. The blend was prepared by ultrasonicator and homogenizer. Their experimental study concluded that Al_2O_3 blend diesel produced reduces HC, CO, and NO_x emission by 13%, 11%, and 6%, respectively. Ang et al. [5] disperse the aluminum oxide (Al_2O_3), silicon oxide (SiO_2), and carbon nanotubes (CNTs) nanoparticles diesel and analyzed the performance parameters, combustion characteristics, and exhaust exhalation of a 4-stroke YANMAR TF120M CI engine. Their experimental study concluded that the BTE was increased by 19.85% as a result of the high calorific value of the CNTs blend. Al_2O_3 and SiO_2 blend exhibit the reduction of 1.76 times in hydrocarbon (HC) emission because of shorter ignition delay, negatively affecting NO_x emission. Silva et al. [9] experimented to test the engine's performance and exhaust exhalation for TiO_2 blend diesel fuel. Their study obtained a 22% decrement in BSFC at maximum load and an 18% and 25% reduction in HC and CO emission. Rashmi et al. [3] investigate the CuO blend diesel nanofuel applications in a CI engine. They tested the engine performance and different loads and compare the engine's performance parameters and exhaust exhalation at altered CuO and diesel blend ratios (shown in Fig. 1). Devarajan et al. [10] disperse the silver oxide (AgO) nanoparticles into palm stearin biodiesel (PSBD) at various mass friction. They concluded that the AgO

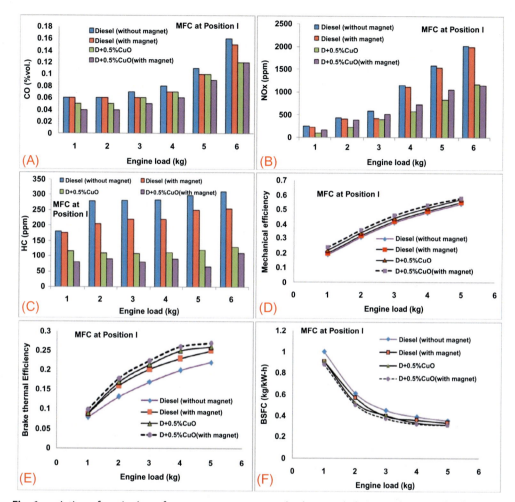

Fig. 1 variation of engine's performance parameters and exhaust exhalation with engine load (A) CO emission (B) NO_x emission (C) HC emission (D) Mechanical efficiency (E) BTE (F) BSFC. *(Copyright Elsevier 2017 (Lic#5083571298374), R.R. Sahoo, A. Jain, Experimental analysis of nanofuel additives with magnetic fuel conditioning for diesel engine performance and emissions, Fuel 236 (2019) 365–72. https://doi.org/10.1016/j.fuel.2018.09.027).*

nanoparticles suspension in the PSBD enhance the BTE and reduced the BSFC. They also found that 20 nm AgO nanoparticles at 10 ppm concentration reduced the exhaust emission.

Perumal et al. [11] examined the Pongamia biodiesel and copper oxide nanoparticles based fuel and concluded the enhancement of BTE by 4.01% while a 1% reduction in BSFC. Rao et al. [12] experimented on nanoparticle blend biodiesel fuel. The Aluminum oxide hydroxide (AlO(OH)) nanoparticles (25, 50 nm size and 100 ppm) are dispersed in biodiesel emulsion fuels BD5W (Blend of 93% biodiesel +1% Tween80 1% + Span80 + 5% water), and BD10W. The prepared fuel was used in a single-cylinder, 4-strokes direct injection CI engine. They obtained a 6% improvement in BTE with a 42% reduction in NO_x emission for BD10W100 compared to biodiesel. Nanthagopal et al. [13] analyzed the effect of ZnO and TiO_2 nanoparticles dispersed in *Calophyllum inophyllum* biodiesel in a water-cooled, 2-cylinder, 4-strokeCI engine. They observed a 7%–29% reduction in NO_x emission. Sivakumar et al. [14] analyzed the Al_2O_3 nanoparticles blend Pongamia methyl ester (PME)fuels in a CI engine. The experiment was conducted on three different nanoparticles and fuel mixture i.e., B25 (75% diesel and 25% PME), B25A50, and B25A100. From this experimental study, they obtain better performance parameters with reduced exhaust emission. Hosseini et al. [15] examined the performance parameters of diesel fuel mixed with biodiesel and carbon nanotubes (CNTs) in a CI engine. It was noticed that dispersion of CNTs into the blended fuel has a significant amount of enhancement in brake thermal power (3.67%) and BTE (8.12%) with a reduction in CO, HC, and shoot exhaust emission.

3. Metal oxide nanomaterials application in automotive fuels

The main objective of nanoparticle addition to the automotive fuel is to provide an increased surface-to-volume ratio with the enhanced reactive surface. These nanoparticles act as the chemical catalyst to prepare a better air-fuel mixture to enhance combustion performance.

3.1 CeO₂ nanoparticles

Gharehghani et al. [16] examined the performance and exhaust exhalation characteristics of water and cerium oxide (CeO_2) blend diesel–biodiesel using a single-cylinder diesel engine. The experimental shows B5W7m enhanced the BTE by 13.5% with a 14% reduction in NO_x emission. Annamalai et al. [17] investigated the cerium oxide (CeO_2) blend Lemongrass Oil (LGO) emulsion fuel (5% of water, 93% of LGO, and 2% of span8 surfactant) in CI engine with constant speed. The experimental analysis obtained a 24.80% BTE with a reduction in unburnt hydrocarbon (HC), CO, and NO_x emission by 15.69%, 26%, and 20.3%, respectively, concerning neat diesel. Sathiyamoorthi et al. [18] studied the mutual influence of exhaust gas recirculation (EGR) and nanoemulsion on the combustion performance and exhaust emission characteristics in a 4-stroke, CI engine fueled with a neat LGO-diesel-DEE (diethyl ether) blend. The experimental study determined an increment of 2.4% and 10.8% in BTE and BSFC, respectively. They also found a reduction of 30.72% and 11.2% in NO_x emission and smoke emission, respectively, with emulsified LGO25 (75 vol% diesel +25 vol% LGO) and cerium oxide blended emulsified LGO25 fuels (Fig. 2).

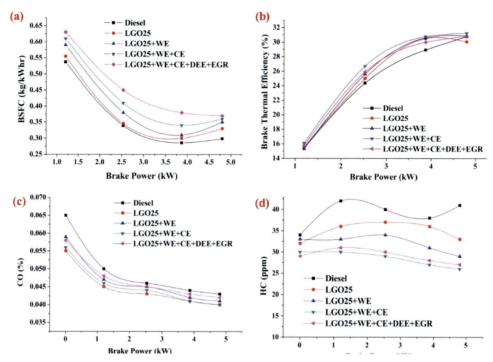

Fig. 2 Variation of different parameters to Brake Power (A) BSFC (B) BTE (C) Carbon mono-oxide (CO) emission (D) Hydrocarbon emission. *(Copyright Elsevier 2017 (Lic#5083580153342), R. Sathiyamoorthi, G. Sankaranarayanan, K. Pitchandi, Combined effect of nanoemulsion and EGR on combustion and emission characteristics of neat lemongrass oil (LGO)-DEE-diesel blend fuelled diesel engine, Appl. Therm. Eng. 112 (2017) 1421–32. https://doi.org/10.1016/j.applthermaleng.2016.10.179).*

Khalife et al. [19] investigate the performance of cerium oxide (CeO_2) dispersed diesel/biodiesel blend fuel (B5) in a 4-stroke engine. Ananda et al. [20] investigated ethanol-based biodiesel fuel performance parameters and emission characteristics blend with cerium oxide (CeO_2). The experimental study obtained the maximum BTE of 30.98% at a 10 kW load. They also obtained a reduction in HC, CO, and NO_x emission. Janakiramana et al. [21] investigate the performance parameters of B20 (20% Garcinia gummi-gutta biodiesel +80% diesel) fuel blend with three different nanoparticles, i.e., CeO_2, TiO_2, and ZrO_2 in as Kirloskar TAF-1 engine. Experimentally, they concluded that all nanoparticles blend diesel reduced the overall CO emission. They also obtained an enhancement of 6.05% in BTE for 25 ppm of TiO_2 blend with B20. Ananda et al. [22] examined performance parameters and exhaust exhalation characteristics of Karanja oil methyl ester (KOME) biodiesel and CeO_2 mixed KOME biodiesel in military heavy-duty 38.8 l, 12 cylinder, 4-strokes, 585 kW supercharged, CIDI engine. Experimentally they obtained a 5% enhancement in engine performance with 14% to 25%.

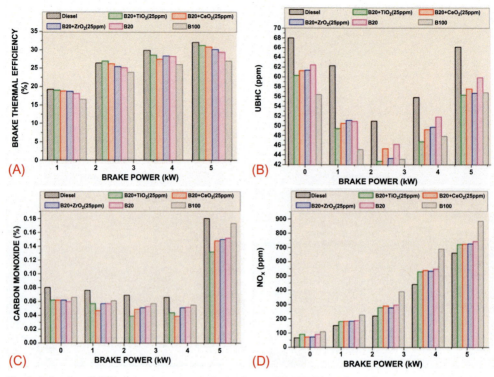

Fig. 3 Behavior of different parameters to Brake Power (A) BTE (B) Hydrocarbon emission (C) Carbon mono-oxide (CO) emission (D) NO_x emission. *(Copyright Elsevier 2017 (Lic#5083580262905), S. Janakiraman, T. Lakshmanan, V. Chandran, L. Subramani, Comparative behavior of various nano additives in a DIESEL engine powered by novel Garcinia gummi-gutta biodiesel, J. Clean. Prod. 245 (2020) 118940. https://doi.org/10.1016/j.jclepro.2019.118940).*

Sadiya et al. [23] conducted a study on waste cooking oil/biodiesel blend with CeO_2 nanoparticles and $Ce_{0.5}CO_{0.5}$ nano-composite oxide and observed in the reduction of CO, NO_x and unburnt hydrocarbon (Fig. 3).

3.2 Al_2O_3

Tosun et al. [24] examined the performance and exhaust exhalation characteristics of Al_2O_3 nanoparticles and diesel-soybean biodiesel-hydrogen blend fuel in a 4-stroke, 4-cylinder DICI engine. Soudagar et al. [25] examined the effect of aluminum oxide nanoparticles (Al_2O_3) with honge oil methyl ester and diesel fuel blend. The aluminum oxide nanoparticles were dispersed into the nanofuel blends at different concentrations. The honge oil methyl ester (B20) (20% biodiesel +80% diesel) nanofuel blend according to ASTM D6751–15 standards. The experiment was conducted at a constant speed and different brake power, and they obtained a 10.57% increment in BTE with an 11.65% reduction in BSFC. They furthermore found a decrement of 48.83%, 26.72%, and 22.84% in CO, HC, and smoke emission, respectively. Aalam et al. [26] investigated the performance of diesel and 25% of zizipus jujube methyl ester blended fuel (ZJME25) along with Al_2O_3 nanoparticles additive in a CRDI engine. Their experimental study achieved an improvement in BTE with a reduction in harmful exhaust emissions. Rakhi et al. [27] disperse the aluminum (Al), iron (Fe), and boron (B) nanoparticles in diesel to investigate the performance parameters and exhaust emission and obtained the reduction in NO_x emission. Ramesh et al. [28] examined the biodiesel blend poultry litter oil B20 with aluminum oxide additives in a diesel engine and obtained reduced exhaust emission. Raju et al. [29] used tamarind seed methyl ester blend diesel fuel with Al_2O_3 and MWCNTs as fuel additives to study nanoparticles' impact in a diesel engine to analyze its performance. Experimental results found the BTE of 35.74% for TSME20 ANP60 (tamarind seed methyl ester +60 ppm of alumina oxide) biodiesel blend. They also found a reduction of 87.4% and 56.6% in smoke and CO emission, respectively.

3.3 TiO_2

Kumar et al. [30] examined the performance and exhaust exhalation parameters of orange peel oil biodiesel with nano-emulsion of TiO_2 particles. The experiment was performed on the biodiesel sample OOME (Orange oil methyl ester), OOME-T50 (Orange oil methyl ester +50 ppm TiO_2 Nanoemulsion), and OOME-T100 (Orange oil methyl ester +100 ppm TiO_2 Nanoemulsion) found an enhancement of 1.4% and 3.0% in BTE for OOME-T50 and OOME-T100 respectively as compared to OOME. The reduction in NO_x, HC, CO, and smoke they obtained was 9.7%, 16.0%, 18.4%, and 24.2%, respectively, for the OOME-T100 biodiesel sample. Senthil et al. [31] examined the exhaust exhalation characteristics of neat diesel with TiO_2 nanoparticles as additive of 50 ppm and 100 ppm concentrations in a 4-stroke, single-cylinder diesel engine. Experimental results found a reduction of 2.3% in unburnt hydrocarbons, 4.1% in NO_x emission, and 2.2% in CO emission at 100 ppm nanoparticles concentration. Yuvarajan et al. [32] performed a comparative study to analyze the emission parameters using diesel, mustard oil methyl ester, and nanoparticle mixed mustard oil methyl ester in a 4-stroke CI engine. The experimental result shows a significant amount of reduction in CO, HC, and NO_x emission. Karthikeyan et al. [33] conducted a comparative study of the emission and performance parameters using diesel, biodiesel and TiO_2 nanoparticles mixed fuel of different nanoparticles concentrations (25, 50, 75, and 100 ppm) in a two-cylinder diesel engine at different load conditions. They experimentally concluded the better performance with less diesel engine emission using TiO_2 nanoparticles additives with blend fuels. Praveen et al. [34] used *Calophyllum inophyllum* biodiesel blends with TiO_2 nanoadditives and exhaust gas recirculation (EGR) in a single-cylinder compression ignition

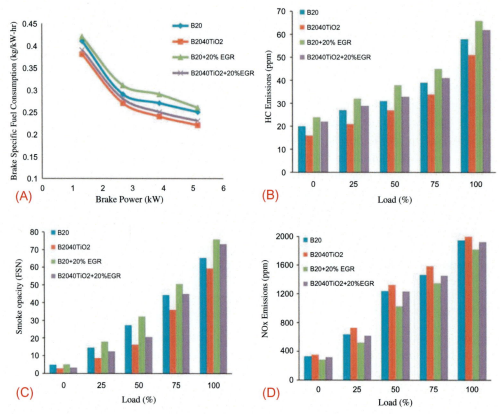

Fig. 4 Variation of different parameters to load(%) (A) BSFC (B) Hydrocarbon emission (C) smoke opacity (D) NO_x emission. *(Copyright Elsevier 2017 (Lic#501659302), A. Praveen, G. Lakshmi Narayana Rao, B. Balakrishna, Performance and emission characteristics of a diesel engine using Calophyllum Inophyllum biodiesel blends with TiO2 nanoadditives and EGR, Egypt. J. Pet. 27 (2018) 731–8. https://doi.org/10.1016/j.ejpe.2017.10.008).*

engine. The experimentation is carried out on B20 (20% *C. inophyllum* biodiesel +80% diesel), B2040TiO$_2$, B20 + 20%EGR and B2040TiO$_2$ + 20% EGR blend biodiesel fuel. The experimental study claimed substantial enhancements in BTE and reduction in exhaust emission (shown in Fig. 4).

Karisathan et al. [35] analyzed the emission characteristics of TiO$_2$ nanoparticles dispersed water-diesel emulsion fuel in a CRDI engine and demonstrated a significant performance enhancement using TiO$_2$ mixed emulsified diesel fuel. Sankoc et al. [36] prepared a fuel blend of TiO$_2$ nanoparticles-diesel-biodiesel and n-butanol. The blended fuel's performance parameters and exhaust exhalation characteristics were tested in a CI engine at 1400 rpm and 2800 rpm. The experimental result showed an increment of 10.20% and 9.47% in engine torque and power.

3.4 Other metal oxide nanoparticles

Mehregan et al. [37] studied the influence of nanoparticles (Co_3O_4 and Mn_2O_3) additives over the NO_x emission in a CI engine using blended diesel/biodiesel fuel, and experimentation showed the better performance for Co_3O_4 nanoparticles as compared to Mn_2O_3. Vedagiri et al. [38] analyzed the performance and exhaust emission characteristics of grapeseed oil methyl ester blend fuel with zinc oxide (ZnO) nanoparticles additive in a CI engine at 5.2 kW load and 1500 rpm. The experimental result showed an enhancement of BTE from 28.17% to 29.3%, reducing brake-specific fuel consumption from 0.3258 kg/k-Wh to 0.3128 kg/k-Wh. Kalaimurugan et al. [39] examined the engine performance and exhaust exhalation characteristics of different biodiesel blend with RuO_2 nanoparticles additive in the air-cooled, single-cylinder, 4-strokes engine. The result showed the enhancement of BTE and reduction in CO pollutants for RuO_2 nanoparticles additions in different biodiesel blends. Srinidhi et al. [40] examined the performance of a blend of *Azadirachta indica* biodiesel (25%) and conventional diesel (75%) with different concentrations of nickel oxide (25, 50, 75, and 100 ppm) nanoadditives in a diesel engine. The maximum increment in BTE obtained was 7.86% at a 100 ppm concentration of NiO (shown in Fig. 5).

Tamilvanan et al. [41] examined the engine performance parameters and exhaust exhalation characteristics of biodiesel mixed with copper nanoparticles compared to diesel. The experimental result showed a significant improvement in BTE and a considerable amount of reduction in NO_x emission. But the result obtained for biodiesel is slightly

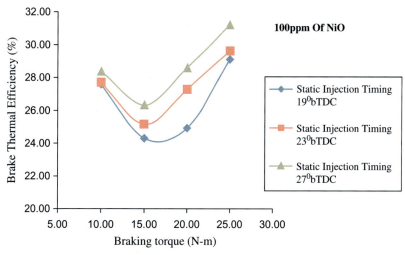

Fig. 5 Variation of BTHE to braking torque at 100 ppm of NiO. *(Copyright Elsevier 2017 (Lic#5083580727724), C. Srinidhi, A. Madhusudhan, S.V. Channapattana, Effect of NiO nanoparticles on performance and emission characteristics at various injection timings using biodiesel-diesel blends, Fuel 235 (2019) 185–93. https://doi.org/10.1016/j.fuel.2018.07.067).*

lesser than the diesel. Venu et al. [42] analyzed biodiesel's engine performance, and emission characteristics blend with zirconium oxide (Zr_2O_3) nanoadditive at 25 ppm concentration in a 4.4 kW DICI engine. The Experimental result measured a significant reduction in NO_x, CO, and unburnt hydrocarbon with the addition of zirconium oxide nanoparticles.

4. Conclusion

The demand for automotive vehicles is increasing rapidly. The potential use of nanoparticles additive in the fuel blend reduces the harmful greenhouse gasses (GHG) emission and contributes to enhancing engine performance. Earlier studies have shown that the use of additive nanoparticles has improved fuel properties, engine combustion parameters, engine performance parameters, and exhaust emission characteristics. Further, the use of nanoparticles with blended biodiesel/diesel could improve the fuel properties like calorific value, cetane number, kinematic viscosity, and density. The high calorific value of the fuel stimulates a higher BTE by lowering the BSFC.

References

[1] V.B. Pedrozo, I. May, W. Guan, H. Zhao, High efficiency ethanol-diesel dual-fuel combustion: a comparison against conventional diesel combustion from low to full engine load, Fuel 230 (2018) 440–451, https://doi.org/10.1016/j.fuel.2018.05.034.

[2] C.S. Aalam, C.G. Saravanan, Effects of nano metal oxide blended Mahua biodiesel on CRDI diesel engine, Ain. Shams. Eng. J. 8 (2017) 689–696, https://doi.org/10.1016/j.asej.2015.09.013.

[3] R.R. Sahoo, A. Jain, Experimental analysis of nanofuel additives with magnetic fuel conditioning for diesel engine performance and emissions, Fuel 236 (2019) 365–372, https://doi.org/10.1016/j.fuel.2018.09.027.

[4] S. Gumus, H. Ozcan, M. Ozbey, B. Topaloglu, Aluminum oxide and copper oxide nanodiesel fuel properties and usage in a compression ignition engine, Fuel 163 (2016) 80–87, https://doi.org/10.1016/j.fuel.2015.09.048.

[5] A.F. Chen, M. Akmal Adzmi, A. Adam, M.F. Othman, M.K. Kamaruzzaman, A.G. Mrwan, Combustion characteristics, engine performances and emissions of a diesel engine using nanoparticle-diesel fuel blends with aluminium oxide, carbon nanotubes and silicon oxide, Energy Convers. Manag. 171 (2018) 461–477, https://doi.org/10.1016/j.enconman.2018.06.004.

[6] S. Karthikeyan, A. Elango, P. Marimuthu, A. Prathima, Performance, combustion and emission characteristic of a marine engine running on grape seed oil biodiesel blends with nano additive, 43 (2014).

[7] V. Praveena, M.L.J. Martin, V.E. Geo, Experimental characterization of CI engine performance, combustion and emission parameters using various metal oxide nanoemulsion of grapeseed oil methyl ester, J. Therm. Anal. Calorim. 139 (2020) 3441–3456, https://doi.org/10.1007/s10973-019-08722-7.

[8] P. Vedagiri, L.J. Martin, E.G. Varuvel, T. Subramanian, Experimental study on NOx reduction in a grapeseed oil biodiesel-fueled CI engine using nanoemulsions and SCR retrofitment, Environ. Sci. Pollut. Res. 27 (2020) 29703–29716, https://doi.org/10.1007/s11356-019-06097-8.

[9] R. D'Silva, K.G. Binu, T. Bhat, Performance and emission characteristics of a C.I. engine Fuelled with diesel and TiO2 nanoparticles as fuel additive, Mater. Today Proc. 2 (2015) 3728–3735, https://doi.org/10.1016/j.matpr.2015.07.162.

[10] Y. Devarajan, D.B. Munuswamy, A. Mahalingam, Investigation on behavior of diesel engine performance, emission, and combustion characteristics using nano-additive in neat biodiesel, Heat Mass Transf. Und Stoffuebertragung 55 (2019) 1641–1650, https://doi.org/10.1007/s00231-018-02537-2.

[11] V. Perumal, M. Ilangkumaran, The influence of copper oxide nano particle added pongamia methyl ester biodiesel on the performance, combustion and emission of a diesel engine, Fuel 232 (2018) 791–802, https://doi.org/10.1016/j.fuel.2018.04.129.

[12] M. Srinivasa Rao, R.B. Anand, Performance and emission characteristics improvement studies on a biodiesel fuelled DICI engine using water and AlO(OH) nanoparticles, Appl. Therm. Eng. 98 (2016) 636–645, https://doi.org/10.1016/j.applthermaleng.2015.12.090.

[13] K. Nanthagopal, B. Ashok, A. Tamilarasu, A. Johny, A. Mohan, Influence on the effect of zinc oxide and titanium dioxide nanoparticles as an additive with Calophyllum inophyllum methyl ester in a CI engine, Energy Convers. Manag. 146 (2017) 8–19, https://doi.org/10.1016/j.enconman.2017.05.021.

[14] M. Sivakumar, N. Shanmuga Sundaram, R. Ramesh Kumar, M.H. Syed Thasthagir, Effect of aluminium oxide nanoparticles blended pongamia methyl ester on performance, combustion and emission characteristics of diesel engine, Renew. Energy 116 (2018) 518–526, https://doi.org/10.1016/j.renene.2017.10.002.

[15] S.H. Hosseini, A. Taghizadeh-Alisaraei, B. Ghobadian, A. Abbaszadeh-Mayvan, Performance and emission characteristics of a CI engine fuelled with carbon nanotubes and diesel-biodiesel blends, Renew. Energy 111 (2017) 201–213, https://doi.org/10.1016/j.renene.2017.04.013.

[16] A. Gharehghani, S. Asiaei, E. Khalife, B. Najafi, M. Tabatabaei, Simultaneous reduction of CO and NOx emissions as well as fuel consumption by using water and nano particles in diesel–biodiesel blend, J. Clean. Prod. 210 (2019) 1164–1170, https://doi.org/10.1016/j.jclepro.2018.10.338.

[17] M. Annamalai, B. Dhinesh, K. Nanthagopal, P. SivaramaKrishnan, J. Isaac JoshuaRamesh Lalvani, M. Parthasarathy, et al., An assessment on performance, combustion and emission behavior of a diesel engine powered by ceria nanoparticle blended emulsified biofuel, Energy Convers. Manag. 123 (2016) 372–380, https://doi.org/10.1016/j.enconman.2016.06.062.

[18] R. Sathiyamoorthi, G. Sankaranarayanan, K. Pitchandi, Combined effect of nanoemulsion and EGR on combustion and emission characteristics of neat lemongrass oil (LGO)-DEE-diesel blend fuelled diesel engine, Appl. Therm. Eng. 112 (2017) 1421–1432, https://doi.org/10.1016/j.applthermaleng.2016.10.179.

[19] E. Khalife, M. Tabatabaei, B. Najafi, S.M. Mirsalim, A. Gharehghani, P. Mohammadi, et al., A novel emulsion fuel containing aqueous nano cerium oxide additive in diesel–biodiesel blends to improve diesel engines performance and reduce exhaust emissions: part I – experimental analysis, Fuel 207 (2017) 741–750, https://doi.org/10.1016/j.fuel.2017.06.033.

[20] C. Ananda Srinivasan, C.G. Saravanan, M. Gopalakrishnan, Emission reduction on ethanol–gasoline blend using cerium oxide nanoparticles as fuel additive, Part. Sci. Technol. 36 (2018) 628–635, https://doi.org/10.1080/02726351.2017.1287791.

[21] S. Janakiraman, T. Lakshmanan, V. Chandran, L. Subramani, Comparative behavior of various nano additives in a DIESEL engine powered by novel Garcinia gummi-gutta biodiesel, J. Clean. Prod. 245 (2020) 118940, https://doi.org/10.1016/j.jclepro.2019.118940.

[22] A. Pandey, M. Nandgaonkar, U. Pandey, S. Suresh, Experimental investigation of the effect of karanja oil biodiesel with cerium oxide nano particle fuel additive on lubricating oil tribology and engine wear in a heavy duty 38.8L,780 HP military CIDI diesel engine, SAE Technical Paper 2018-01-1753, 2018, https://doi.org/10.4271/2018-01-1753.

[23] S. Akram, M.W. Mumtaz, M. Danish, H. Mukhtar, A. Irfan, S.A. Raza, et al., Impact of cerium oxide and cerium composite oxide as nano additives on the gaseous exhaust emission profile of waste cooking oil based biodiesel at full engine load conditions, Renew. Energy 143 (2019) 898–905, https://doi.org/10.1016/j.renene.2019.05.025.

[24] E. Tosun, M. Özcanlı, Hydrogen enrichment effects on performance and emission characteristics of a diesel engine operated with diesel-soybean biodiesel blends with nanoparticle addition, Eng. Sci. Technol. Int. J. 24 (2021) 648–654, https://doi.org/10.1016/j.jestch.2020.12.022.

[25] M.E.M. Soudagar, N.N. Nik-Ghazali, M.A. Kalam, I.A. Badruddin, N.R. Banapurmath, M.A. Bin Ali, et al., An investigation on the influence of aluminium oxide nano-additive and honge oil methyl

ester on engine performance, combustion and emission characteristics, Renew. Energy 146 (2020) 2291–2307, https://doi.org/10.1016/j.renene.2019.08.025.
[26] C.S. Aalam, C.G. Saravanan, M. Kannan, Experimental investigations on a CRDI system assisted diesel engine fuelled with aluminium oxide nanoparticles blended biodiesel, Alexandria Eng. J. 54 (2015) 351–358, https://doi.org/10.1016/j.aej.2015.04.009.
[27] R.N. Mehta, M. Chakraborty, P.A. Parikh, Nanofuels: combustion, engine performance and emissions, Fuel 120 (2014) 91–97, https://doi.org/10.1016/j.fuel.2013.12.008.
[28] D.K. Ramesh, J.L. Dhananjaya Kumar, S.G. Hemanth Kumar, V. Namith, P. Basappa Jambagi, S. Sharath, Study on effects of alumina nanoparticles as additive with poultry litter biodiesel on performance, combustion and emission characteristic of diesel engine, Mater. Today Proc. 5 (2018) 1114–1120, https://doi.org/10.1016/j.matpr.2017.11.190.
[29] V. Dhana Raju, P.S. Kishore, K. Nanthagopal, B.Ashok. An experimental study on the effect of nanoparticles with novel tamarind seed methyl ester for diesel engine applications, Energy Convers. Manag. 164 (2018) 655–666, https://doi.org/10.1016/j.enconman.2018.03.032.
[30] A.M. Kumar, M. Kannan, G. Nataraj, A study on performance, emission and combustion characteristics of diesel engine powered by nano-emulsion of waste orange peel oil biodiesel, Renew. Energy 146 (2020) 1781–1795, https://doi.org/10.1016/j.renene.2019.06.168.
[31] J. Senthil Kumar, B.R. Ramesh Bapu, R. Gugan, Emission examination on nanoparticle blended diesel in constant speed diesel engine, Pet. Sci. Technol. 38 (2020) 98–105, https://doi.org/10.1080/10916466.2019.1683579.
[32] D. Yuvarajan, M. Dinesh Babu, N. BeemKumar, K.P. Amith, Experimental investigation on the influence of titanium dioxide nanofluid on emission pattern of biodiesel in a diesel engine, Atmos. Pollut. Res. 9 (2018) 47–52, https://doi.org/10.1016/j.apr.2017.06.003.
[33] P. Karthikeyan, G. Viswanath, Effect of titanium oxide nanoparticles in tamanu biodiesel operated in a two cylinder diesel engine, Mater. Today Proc. 22 (2020) 776–780, https://doi.org/10.1016/j.matpr.2019.10.138.
[34] A. Praveen, G. Lakshmi Narayana Rao, B. Balakrishna, Performance and emission characteristics of a diesel engine using Calophyllum Inophyllum biodiesel blends with TiO2 nanoadditives and EGR, Egypt. J. Pet. 27 (2018) 731–738, https://doi.org/10.1016/j.ejpe.2017.10.008.
[35] N. Karisathan Sundararajan, A.R.B. Ammal, Improvement studies on emission and combustion characteristics of DICI engine fuelled with colloidal emulsion of diesel distillate of plastic oil, TiO2 nanoparticles and water, Environ. Sci. Pollut. Res. 25 (2018) 11595–11613, https://doi.org/10.1007/s11356-018-1380-0.
[36] I. Örs, S. Sarıkoç, A.E. Atabani, S. Ünalan, S.O. Akansu, The effects on performance, combustion and emission characteristics of DICI engine fuelled with TiO2 nanoparticles addition in diesel/biodiesel/n-butanol blends, Fuel 234 (2018) 177–188, https://doi.org/10.1016/j.fuel.2018.07.024.
[37] M. Mehregan, M. Moghiman, Experimental investigation of the distinct effects of nanoparticles addition and urea-SCR after-treatment system on NOx emissions in a blended-biodiesel fueled internal combustion engine, Fuel 262 (2020) 116609, https://doi.org/10.1016/j.fuel.2019.116609.
[38] P. Vedagiri, L.J. Martin, E.G. Varuvel, Characterization study on performance, combustion and emission of nano additive blends of grapeseed oil methyl ester fuelled CI engine with various piston bowl geometries, Heat Mass Transf. Und Stoffuebertragung (2019), https://doi.org/10.1007/s00231-019-02740-9.
[39] K. Kalaimurugan, S. Karthikeyan, M. Periyasamy, G. Mahendran, T. Dharmaprabhakaran, Performance, emission and combustion characteristics of RuO2 nanoparticles addition with neochloris oleoabundans algae biodiesel on CI engine, Energy Sources, Part A Recover Util. Environ. Eff. 00 (2019) 1–15, https://doi.org/10.1080/15567036.2019.1694102.
[40] C. Srinidhi, A. Madhusudhan, S.V. Channapattana, Effect of NiO nanoparticles on performance and emission characteristics at various injection timings using biodiesel-diesel blends, Fuel 235 (2019) 185–193, https://doi.org/10.1016/j.fuel.2018.07.067.
[41] A. Tamilvanan, Effects of nano-copper additive on performance, combustion and emission characteristics of *Calophyllum inophyllum* biodiesel in CI engine, J. Therm. Anal. Calorim. 136 (2019) 317–330, https://doi.org/10.1007/s10973-018-7743-4.
[42] H. Venu, P. Appavu, Experimental studies on the influence of zirconium nanoparticle on biodiesel–diesel fuel blend in CI engine, Int. J. Ambient Energy 0750 (2019), https://doi.org/10.1080/01430750.2019.1611653.

CHAPTER 35

Nanomaterials for electromagnetic interference shielding application

Arun Kumar Tiwari[a], Amit Kumar[a], and Zafar Said[b]
[a]Mechanical Engineering Department, Institute of Engineering & Technology, Dr. A.P.J. Abdul Kalam Technical University, Lucknow, Uttar Pradesh, India
[b]Sustainable and Renewable Energy Engineering Department, University of Sharjah, Sharjah, United Arab Emirates

Chapter outline

1. Introduction — 749
2. EMI and their potential receptor in automotives — 750
3. Electromagnetic interference shielding — 750
4. Nanomaterials for EMI shielding in automotive applications — 751
   4.1 Metal-based nanomaterials — 751
   4.2 Metal oxide-based nanomaterials — 757
   4.3 Carbon-based nanomaterials — 759
5. Conclusion — 760
References — 765

1. Introduction

With recent technological advances, the electrical and electronic system has become an integral part of the automobile industry, and their demands are increasing dramatically. These systems include communication devices, safety systems, cruise control systems, mobile media, DC motors and controllers, and infotainment systems, including wireless headsets. Because of the limited space in automobiles, the physical size and weight of these electric and electronic systems have decreased. Nowadays, electronic systems turn out to be wireless to eliminate the complexity and improve the system's reliability. Introducing so many electrical and electronic systems into a confined space generates electromagnetic interference problems (EMI). The EMI radiation of these systems affects the performance of each other. Because of improper EMI control, it may cause system malfunction and even failure.

Electromagnetic interference (EMI) is a phenomenon in which harmful electromagnetic radiation energy is transmitted from one electronic device to other devices via a conducted or radiated path, or both. In automobiles, EMI can adversely affect the performance of the electronic system nearby. To prevent such unwanted EMI radiation, a layer of EMI reflective material (EMI shield) is provided at the external surface of an electronic device [1]. EMI shielding is a process of reflection or absorption of harmful electromagnetic waves to prevent their penetration into electronic devices [2]. Charge carriers carry out the reflecting mechanism in electrically conductive shielding materials, while the absorbing mechanism is performed by magnetic and electric dipoles of shielding materials [3]. Metal sheets had been used for EMI shielding applications for the past few years, but due to their high cost, high weight, and corrosion risk, they are replacing them with nanobased materials [4, 5]. Due to continuous increment in EM radiations from different sources, the demand for EMI shielding materials has been triggered in the military, electronic industry, communication, automotive, and aerospace industries.

2. EMI and their potential receptor in automotives

The installation of additional new automobiles such as in-vehicle communication, onboard GPS, wireless charging, and touchscreen infotainment is very common in modern-day automobiles to attract customers. But unfortunately, these electronic devices are responsible for undesirable EMI emissions. This EMI emission further increases by using mobile phones, laptops, tablets, and other wireless devices or equipment near automobiles. These EMI emissions threaten the performance of devices having integrated circuits and other nearby electronic devices. The malfunction of these necessary electronic devices can be very dangerous, like losing control of the vehicle, which can endanger human life. The internal system, as shown in Fig. 1, can be malfunctioned due to the EMI emission. These EMI emissions can be generated from external sources such as a garage door opener, Bluetooth devices, high power transmitters such as radio and TV towers, and third-party navigation, etc.

3. Electromagnetic interference shielding

Electromagnetic interference radiation produced by electronic devices can be mitigated by using EMI shielding materials [6]. EMI shielding in an automobile is the reflection or/ and absorption of electromagnetic radiation to protect the electronic devices from malfunctioning. The materials used as a barrier against the penetration of these electromagnetic radiations are called shielding materials. These materials stop the propagation of electromagnetic radiation by reflection and/or absorption of EM radiation or by removing electromagnetic signals such that the EM waves do not interfere in the functioning and durability of electronic devices. Generally, highly conductive materials such as metals are used to prevent EM radiation due to having their high reflective properties. But due to the lack of flexibility, high cost, and heaviness, the metals have very limited use in shielding materials. Also, ferromagnetic materials have an intrinsic cut-off frequency in the low GHz range, so that they cannot be used as EMI shielding material on a wide GHz frequency [7–12].

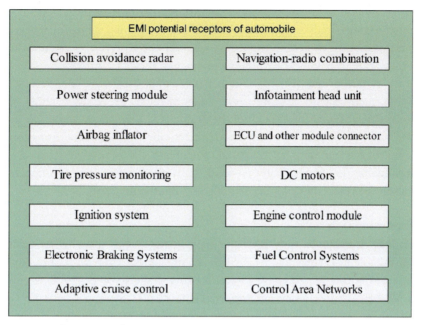

Fig. 1 EMI potential receptor of an automobile.

4. Nanomaterials for EMI shielding in automotive applications

The EMI shielding capability of the materials depends upon their properties, such as complex permittivity, electrical conductivity, EMI attenuation loss, and complex permeability [13]. Highly reflective materials due to the presence of mobile charge carriers are generally used in EMI shielding. Metals are the best suitable option for EMI shielding due to having large free electrons. But metals have limited use in EMI shielding application because of the low intrinsic cut-off frequency. So nanomaterials can be the alternative options. Researchers recommend different nanomaterials such as carbon-based nanomaterials (Single-wall carbon nanotubes, Multiwall carbon nanotubes, graphene, and other conductive polymers), metal-based nanomaterials, metal oxide-based nanomaterials, etc. due to having extraordinary properties such as lightweight, high surface area, and high corrosion resistance.

Nanomaterials are the most suitable alternatives of metal for EMI shielding applications in automobiles. Different types of following nanomaterials can be used for proper EMI shielding in automotive.

4.1 Metal-based nanomaterials

Metal-based nanomaterials are heavily used in EMI shielding applications for their high conductive property. The metals with high thermal conductivity are the first choice for nanomaterials synthesis to prepare shielding materials to protect the electronic devices from electromagnetic radiations. These metals can be formed in nanoparticles as well as nanowires according to requirements. Silver is the most preferred metal for nanomaterial synthesis due to having the highest thermal conductivity in metals as well as good electrical and optical properties [14]. These metals can be formed in nanoparticles as well as nanowires according to requirements. Nanomaterials are used in EMI shielding applications while distributing them into the matrix [15]. It has been shown from various studies that nanowires, due to having a high aspect ratio, can generate more conductive pathways within nanocomposite compared to nanoparticles even at very low concentrations. This enables nanocomposites to reach the critical value at a significantly lower concentration, to make them highly flexible and lightweight [16].

4.1.1 Silver nanomaterials
Kim et al. [17] developed a new EMI shielding material with silver nanoparticles/poly (styrene-b-butadiene-b-styrene) (NP/SBS) having high thermal conductivity and stretchability. Their experimental study found an average of 69 dB EMI shielding efficiency (EMI-SE) at a frequency ranging from 8 to 12 GHz having a 66.5 wt% concentration of silver nanoparticles. They also exhibited that the higher flexibility and stretchability of Ag NP/SBS composite can be used to develop EMI shielding materials for advanced electronic devices in the future, such as hand watches and printed electronics. Arjmand et al. [18] synthesized the silver nanowires by AC electrodeposition of Ag into the porous Al_2O_3 templates. By using the solution processing technique, these silver nanowires were embedded to generate a nanocomposite. Shamy et al. [19] developed a new nanocomposite using silver quantum dots and polyvinyl alcohol and got a 56 dB EMI-SE at 8 GHz frequency and 1 mm thick film (Shown in Fig. 2C). The developed nanocomposite has good EM reflection and absorption to shield 99.99% EM radiation in the X-band range.

Lim et al. [20] fabricated a flexible nanoporous silver (Ag) membrane with high EMI-SE. They found that a 1.2 μm thick film of Ag nanoporous membrane exhibits a high EMI-SE of approximately 53 dB with 99.99% EMI radiation protection at a frequency range of 0.5–18 GHz. The Ag nanoporous membrane also exhibits high electromechanical stability and durability with more remarkable heat dissipation ability. Lee et al. [15] fabricated silver nanowires coated cellulose papers and investigated EMI-SE. They found that the Ag nanowire coated cellulose paper exhibit a high EMI-SE of 48.6 dB at GHz. Weng et al. [21] developed a lightweight, high-performance AgNW/Mxene hybrid sponge that exhibits a good EMI-SE of 52.6 dB at 49.5 g/cm^3. Ji et al. [22] developed a hybrid membrane made up of AgNP/crosslinking polyacrylonitrile (CPAN) nanofiber, and a 53 μm thick AgNP/CPAN membrane 90 dB EMI-SE was achieved. Aepuru et al. [23] developed a low-cost and environment-friendly Ag/3D graphene aerogel for EMI shielding applications. This film was able to restrict 99.99%

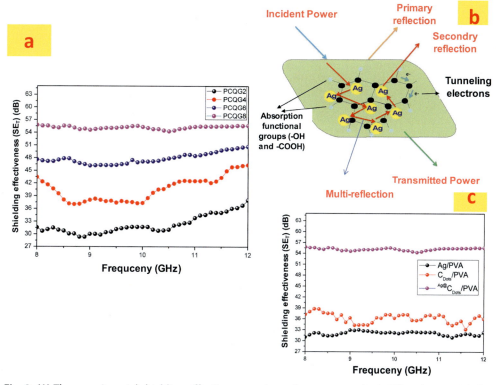

Fig. 2 (A) The experimental shielding effectiveness values of nanocomposite in X-band range (B) The transmission mechanism, internal reflection, and absorption of electromagnetic wave with the nanocomposite films. (C) The experimental shielding effectiveness of Ag quantum dots/PVA nanocomposites for comparison. (Copyright Elsevier (Lic # 5083610658687), A.G. El-Shamy, Polyvinyl alcohol and silver decorated carbon quantum-dots for new nano-composites with application electromagnetic interface (EMI) shielding, Prog. Org. Coat. 146 (2020) 105747. https://doi.org/10.1016/j.porgcoat.2020.105747.)

of EMI radiations. They found a maximum EMI-SE of 32 dB for Ag/GA-20 having an average film thickness of 0.8 mm (shown in Fig. 3).

4.1.2 Copper nanomaterial

Copper has very good thermal and electrical conductivity. It is 100 times cheaper and 1000 times more abundant than silver metal [24]. Copper is extensively used as an alternative to silver nanomaterials in various applications like OLEDs, sensors, EMI shielding, solar cells. Pengfei et al. [25] fabricated the bacterial cellulose/copper (BC/Cu) nanocomposite having thermal, mechanical, and conductive properties. Different characterization techniques (FTIR, XRD) have been used to analyze the morphological characteristics of the nanocomposite. They found that BC/Cu nanocomposites have good thermal conductivity

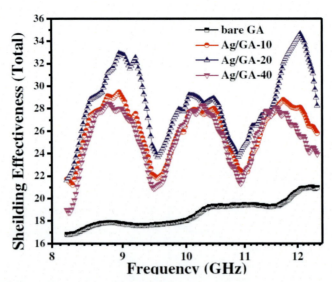

Fig. 3 EMI shielding effectiveness (dB) of bare GA and Ag/Gas. *(Copyright Elsevier (Lic # 5083610792716), R. Epuru, M. Ramalinga Viswanathan, B.V.B. Rao, H.S. Panda, S. Sahu, P.K. Sahoo, Tailoring the performance of mechanically robust highly conducting silver/3D graphene aerogels with superior electromagnetic shielding effectiveness, Diam. Relat. Mater. 109 (2020) 108043. https://doi.org/10.1016/j.diamond.2020.108043.)*

(0.026 S/m), high mechanical properties (41.4 MPa) and exhibit good EMI-SE of 55 dB. Arranz et al. [26] used hybrid poly(vinylidene fluoride) and Cu nanoparticles at varying concentration of Cu nanoparticles and performed their structural and morphological characterization (shown in Fig. 4) to analyzed their mechanical behavior. They obtained a 99% of shielding effect with a 40 dB EMI-SE at 3.06 GHz frequency.

Al-Saleh et al. [27] developed copper nanowire (CuNW)/polystyrene (PS) nanocomposite powder. They developed a 210 μm thick layer of (CuNW/PS) nanocomposite for experimental study. They obtained an effective EMI-SE of 27 dB at 1.3 vol% CuNW and 2.1 vol% PS in the X-band frequency range with 54% absorption contribution to overall shielding. Hou et al. [28] developed an ultrathin flexible copper/graphene (Cu/Gr) nanolayer composite to protect the electronic devices from the Terahertz frequency range Electromagnetic radiations. A very thin Cu/Gr nanolayer composite material possesses good flexibility, high stability, and high EMI-SE. Cu/Gr nanolayer composite of 160 nm thickness can able to prevent approximately 100% EMI radiations. Their experimental study obtained a 60.95 dB of EMI-SE at 0.1 to 1.0 THz frequency range.

4.1.3 Aluminum and other nanoparticles

Arranz et al. [29] synthesized poly(vinylidene fluoride) and Al nanoparticles based nanocomposite. The synthesized nanocomposite was able to prevent up to 90% of

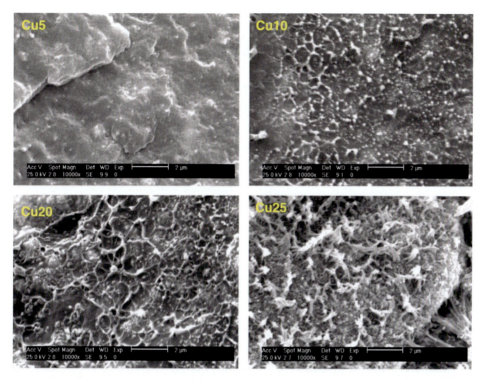

Fig. 4 SEM pictures of PVDF/Cu hybrids with different contents of Cu nanoparticles. *(Copyright Elsevier (Lic # 5083610923313), J. Arranz-Andrés, E. Pérez, M.L. Cerrada, Hybrids based on poly(vinylidene fluoride) and cu nanoparticles: characterization and EMI shielding. Eur. Polym. J. 2012; 48:1160–8. https://doi.org/10.1016/j.eurpolymj.2012.04.006.)*

electromagnetic radiation. The XRD and SEM techniques were used to characterize the nanocomposite to analyze their morphological properties (shown in Fig. 5). Based on their experimental study, they obtain a 20 dB EMI-SE at a frequency range of 2 GHz. Therefore, they suggested using this nanocomposite in place of pure metal EMI shield.

Gleb et al. [30] studied iron and cobalt nanoparticles' magnetic and structural properties having a nanoparticle diameter of 8 nm for electromagnetic shielding applications. TEM and XRD techniques were used to characterize the metallic properties in the ethylene matrix. Jalali et al. [31] fabricated a carbon fiber-reinforced polymer matrix with the addition of different metallic nanoparticles like cobalt, iron, iron oxide, and nickel as low volume concentration filler materials of different particle sizes. The material exhibits excellent electromagnetic absorption properties. Based on their experimental study, they concluded that iron nanoparticles of 50 nm particle size could improve the EMI-SE of carbon fiber reinforced polymer matrix up to 15 dB at a frequency range of 8.2–12.4 GHz. Li et al. [32] developed a porous polymer nanocomposite (density <0.26 g/cm^3) embedded with ionic gold nanoparticles (NPs) known as poly(dimethyl

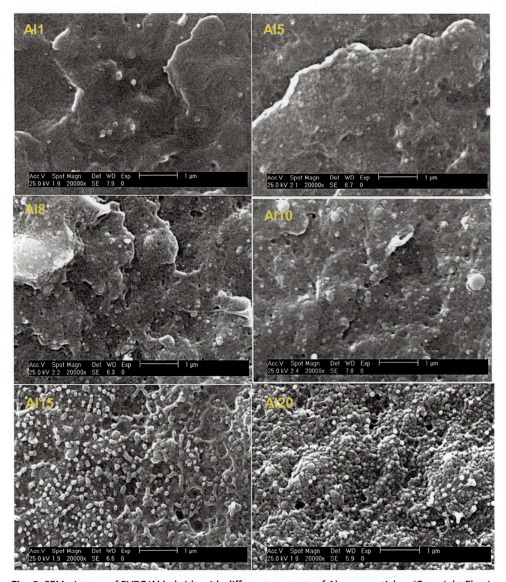

Fig. 5 SEM pictures of PVDF/Al hybrids with different contents of Al nanoparticles. *(Copyright Elsevier (Lic # 5083611074168), J. Arranz-Andrés, N. Pulido-González, C. Fonseca, E. Pérez, M.L. Cerrada, Lightweight nanocomposites based on poly(vinylidene fluoride) and Al nanoparticles: structural, thermal and mechanical characterization and EMI shielding capability. Mater. Chem. Phys. 142 (2013) 469–78. https://doi.org/10.1016/j.matchemphys.2013.06.038.)*

diallyl ammonium chloride) (PIPD-g-PDDA) composite nanofibers. A 20 μm thick layer of the nanocomposite obtained an EMI-SE of 64.9 dB at the frequency range of 250–1.5 GHz. The addition of gold nanoparticles also increases the mechanical

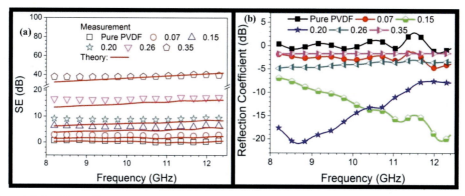

Fig. 6 (A) Shielding effectiveness. (B) Reflection coefficient. *(Copyright Elsevier (Lic #5083611247714), H. Gargama, A.K. Thakur, S.K. Chaturvedi, Polyvinylidene fluoride/nanocrystalline iron composite materials for EMI shielding and absorption applications, J. Alloys Compd. 654 (2016) 209–15. https://doi.org/10.1016/j.jallcom.2015.09.059.)*

properties and electrical conductivity. The EMI shielding properties of the material are enhanced by increasing the absorption of electromagnetic energy with multiple internal reflections. Gargama et al. [33] studied the EMI shielding property of polymer-metal composite (PMC) based nanocrystalline iron (n-Fe) embedded polyvinylidene fluoride (PVDF). They calculate the permeability and EMI-SE of the synthesized material in the X-band frequency range (8.2–12.4 GHz). The obtained result confirm the enhancement in the EMI-SE (shown in Fig. 6).

Gouda et al. [34] developed MNP-embedded carboxymethyl cellulose (CMC) electrospun nanofibers with metallic fillers of different materials like iron, copper, cadmium, zinc, and cobalt nanoparticles. They analyzed the EMI shielding properties of the metals nanoparticles embedded carboxymethyl cellulose (CMC) sample and obtained a 20 dB EMI-SE at a frequency range of 1–18 GHz. Kumar et al. [34] developed a carbon foam embedded with Nickel (Ni) nanoparticles. Different characterization techniques such as SEM, XRD, vector network analyzer, Raman spectroscopy, and vibration sample magnetometer were used to analyze the different mechanical and electrical properties. They found that Nickel embedded carbon foam exhibit excellent electromagnetic properties at a frequency range of 8.2–12.4 GHz.

4.2 Metal oxide-based nanomaterials

Metallic oxide nanoparticle such as Fe_3O_4 has very good magnetic properties and good chemical stability, which can be widely used in EMI shielding applications [35]. Gholampoor et al. [36] fabricated 25 nm thick Fe_3O_4 embedded carbon fiber-based material for EMI shielding applications. Based on their experimental study, they found the maximum reflection loss (RL) of -10.21 dB at a frequency range of 10.12 GHz and effective absorption bandwidth of 2 GHz for a 2 nm thick sample. It also exhibits a − 23 dB of EMI-SE at a frequency range of 8.2–12.4 GHz.

Nasouri et al. [37] fabricated Fe_3O_4 nanoparticles embedded polyvinylpyrrolidone (PVP) composite nanofibers (FCNFs) (Fig. 7) for the potential EMI shielding application of 8.2–12.4 GHz frequency range and obtained the EMI-SE of 22 dB for 150 to 500 nm size nanoparticles. Li et al. [38] fabricated polyvinylidene Fluoride (PVDF)/polystyrene (PS)/high-density polyethylene (HDPE) ternary blends displayed a core-shell structure nanocomposite containing multiwall carbon nanotubes (MWCNTs) and ferrous oxide (Fe_3O_4). They found the EMI-SE of above 20 dB over the whole X-band frequency, but the maximum EMI-SE is obtained at 0.1 vol% Fe_3O_4 and 0.1 vol% MWCNT, which is 25 dB at a frequency range of 9.5 GHz. Ding et al. [39] investigated iron oxide impregnated kenaf fiber-based nanocomposites to collect the data of EMI-SE at the frequency of 9–11 GHz. They found that on increasing the iron content in the nanocomposite, EMI-SE increases. Saini et al. [40] fabricated a highly conductive and dielectric fabric consist of $BaTiO_3$ and Fe_3O_4 nanoparticles of 15–25 nm size coated with poly (aniline) (PANI) matrix. At the frequency range of Ku-band (12.4–18 GHz), they obtained the total shielding effectiveness of −15.3 dB which was further increased to −16.8 dB and −19.4 dB for $BaTiO_3$ and Fe_3O_4 nanoparticles respectively. Singh et al. [41] fabricated the Co-doped ZnO-poly(vinyl alcohol) (PVA) nanocomposite. They found that nanocomposite filled with 2 wt% concentration of $Co_{0.5}Zn_{0.95}O$ in PAV exhibits the reflection loss of −47.31 dB at the frequency of 10.4 GHz. Rajavel et al. [42] developed a

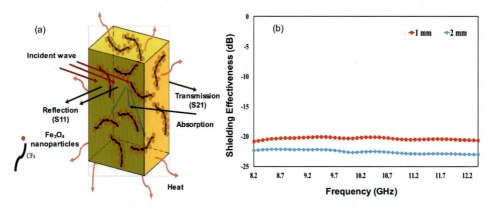

Fig. 7 (A) Schematic illustration of an EMI shielding composite (B) SE as a function of frequency for epoxy-based nano-Fe3O4/CFs composites with the thicknesses of 1 and 2 mm. (Copyright Elsevier (Lic # 5083611499415), M. Gholampoor, F. Movassagh-Alanagh, H. Salimkhani, Fabrication of nano-Fe3O4 3D structure on carbon fibers as a microwave absorber and EMI shielding composite by modified EPD method, Solid State Sci. 64 (2017) 51–61. https://doi.org/10.1016/j.solidstatesciences.2016.12.005.)

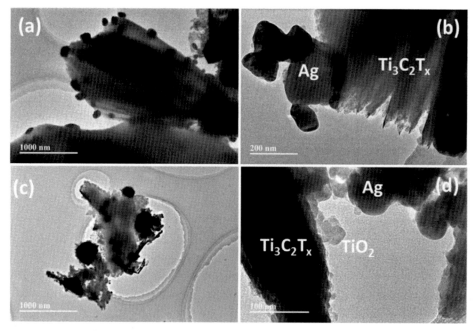

Fig. 8 TEM image of Ag-MXene hybrid nanostructures, [42].

smart multilayered two-dimensional Mxene/metal-oxides/Ag ternary hybrid nanostructure. They found that Nb_2CT_x/Nb_2O_5-Ag hybrid nanostructure (shown in Fig. 8) achieved the EMI-SE of 72.04 dB and 68.76 dB in Ku-band and X-band frequency region respectively at a 1 mm thick layer. Nakhaei et al. [43] fabricated TiO_2/SiO_2 core-shell nanofibers. Different characterization techniques such as FTIR, XRD, SEM, and EDS were used to analyze nanofiber morphology and structural properties. Their study obtained EMI-SE of 132 dB at 7 GHz and 131 dB at the 8.7 GHz frequency range.

4.3 Carbon-based nanomaterials

Lu et al. [44] fabricated a lightweight, highly efficient hybrid nanocomposite material of MWCNT decorated ZnO nanocrystal. The nanocomposite exhibits a minimum reflection loss RL = −20.7 dB for a 2.5 mm thick sample at the X-band frequency range. Li et al. [45] fabricated a single-wall carbon nanotube (SWCNTs)-polymer composite and tested it in the frequency of 10 MHz to 1.5 GHz. They obtained a maximum EMI-SE of 49 dB at 10 MHz with 15 wt% SWCNT. The nanocomposite also exhibits 15–20 dB EMI-SE at 500 MHz to 1.5 GHz. Fletcher et al. [46] fabricated MWCNT embedded lightweight fluorocarbon nanocomposite. The MWCNT was embedded into the

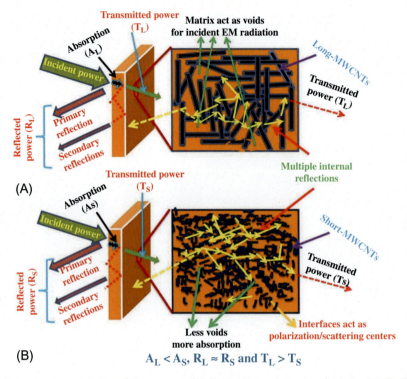

Fig. 9 Schematic representation of the proposed EMI shielding mechanism for l- & s-MCNTs filled PPC nanocomposites. *(Copyright Elsevier (Lic # 5083620160273), P. Verma, P. Saini, V. Choudhary, Designing of carbon nanotube/polymer composites using melt recirculation approach: effect of aspect ratio on mechanical, electrical and EMI shielding response, Mater. Des. 88 (2015) 269–77. https://doi.org/10.1016/j.matdes.2015.08.156.)*

nanocomposite at a very low concentration to achieve a certain level of conductivity and energy shielding to surpass the EMI shielding. Verma et al. [47] describe the mechanical and electrical properties of MWCNTs and the electromagnetic properties of polypropylene random copolymer (PPC) to analyze the electromagnetic shielding behavior. Fig. 9 shows the EMI shielding behavior of the nanocomposite. Their study found that l-MWCNT and s-MWCNT-based PPC nanocomposite provide −27 dB and −37 dB of EMI-SE, respectively. Table 1 further summarizes the EMI shielding applications of different nanomaterials.

5. Conclusion

EMI radiation is very harmful to both the human body and electronic devices. The EMI emissions threaten the performance of devices having integrated circuits and other nearby electronic devices. The malfunction of these necessary electronic devices can be very

Table 1 Summary of different nanomaterials used in EMI shielding applications.

Materials	Frequency	Thickness	Total SE (dB)	References
3D carbon fiber mats/nano-Fe_3O_4	8.2–12.4 GHz	–	62.6	[48]
3D flower-like Co_3O_4-rGO hybrid architectures	11.5 GHz	4.4 mm	RL = −61	[49]
Ag-decorated highly porous poly (vinyl alcohol)/Fe_2O_3 nanofibrous	0.5–18 GHz	20 μm	45	[50]
AgNP deposited on functionalized graphene (Ag@FRGO)	8.2–12.4 GHz	2 mm	35	[51]
AgNP-decorated graphene	5.8–8.2GHz 8.2–12.4GHz 12.4–18GHz	81μm	24 27–24 21	[52]
AgNW/CNT	1 GHz	161.1 μm	23.8	[53]
AgNW/PANI	8.2–12.4 GHz	13.3 μm	50	[54]
Au nanoparticle-MWCNT/PANI	8–12 GHz	2 mm	−16	[55]
AuNP-MWCNT powder	12 GHz	0.5 mm	26.7	[56]
Barium ferrite, PANI, and MWCNT nanocomposite	8.2 GHz 12.4 GHz	4.5 mm	SE_R = 5.14 SE_A = 36.4	[57]
Barium ferrite-decorated reduced graphene oxide ($BaFe_{12}O_{19}$@RGO)	18 GHz	3 mm	32	[58]
Carbon aerogels-supported goethite (α-FeOOH) nanoneedles and nanoflowers	12.5 GHz	2 mm	34	[59]
Carbon fiber/RGO/nickel composite	12 GHz	3–3.3 mm	61	[60]
Carbon nanofiber/magnetite (Fe_3O_4)	12.14 GHz	2.5 mm	SER = −32	[61]
Carbon spheres@MnO_2 core-shell nanocomposites	8–18 GHz	2 mm	16–23	[62]
Carbon/MnO_2 nanoparticles	10.7 GHz	2.5 mm	45	[63]
Carbon-encapsulated cobalt nanoparticles	7.54 GHz	3 mm	RL = −52	[64]
Carbon-encapsulated nanoscale iron/iron-carbide/graphite particles (FeC) (PVDF/FeC composite)	18 GHz	5 mm	23.9	[65]
Chemically reduced vs photo-reduced clay-Ag-PPy ternary nanocomposites	4.7–7.7 MHz	–	16–31	[66]
Cobalt (II)-porphine-anchored MWCNTs	18 GHz	2 mm	18.5	[67]
$CoFe_2O_4$ nanoparticles/CNT	0.5–1 GHz	78 μm	30	[68]

Continued

Table 1 Summary of different nanomaterials used in EMI shielding applications—cont'd

Materials	Frequency	Thickness	Total SE (dB)	References
$CoFe_2O_4$ nanoparticles/PANI	8.1 GHz	4 mm	RL = −28.4	[69]
CoNi alloy nanoparticles coated on biocarbon nanofibers	12.4 GHz	2.5 mm	30.7	[70]
Cu nanoparticles-decorated Pd-doped MWCNTs	12.4–18GHz	200μm	35	[71]
Cubic Ni Fe_2O_4 particles on graphene-PANI	12.5 GHz	2.5 mm	RL = −50.5	[72]
Expanded graphite/γ—Fe_2O_3/SiO_2	8 GHz	2 mm	55	[73]
Expanded graphite-nanoferrite (γ—Fe_2O_3)-fly ash composites	15 GHz	2.5 mm	105	[74]
Fe@C nanocapsules	7 GHz	2 mm	RL = −62	[75]
Fe@FeO core-shell nanoparticles	3.4 GHz	3 mm	RL > -20	[76]
Fe@SiO core-shell nanoparticles	11.3 GHz	1.8 mm	RL < -21.2	[76]
Fe_3O_4 clusters-nitrogen-doped graphene	6 GHz	1.8 mm	RL = −53.6	[77]
Fe_3O_4 nanoparticles-incorporated mesocarbon microbeads and MWCNTs (MCMBs/MWCNTs)	8.2–12.4 GHz	0.5 mm	80	[78]
Fe_3O_4@MWCNTs	8–12 GHz	2.5 mm	25	[79]
Fe_3O_4@rGO	8.2–12.4 GHz	1 mm	26.4	[80]
Fe_3O_4-decorated PANI/SWCNH	14.5 GHz	2 mm	−29.7	[81]
Ferrite-grafted PANI nanofibers	30 MHz	2 mm	20	[82]
Ferrous-ferric oxide-decorated reduced graphene oxide/single-walled carbon nanohorn	14–20 GHz	10 mm	−35.83	[83]
Gaphene/Fe_3O_4-incorporated PANI	15 GHz	2.5 mm	SE_A = 26	[84]
Gaphene@Fe_3O_4@SiO_2@PANI	12.5GHz	2.5 mm	RL = −40.7	[85]
Graphene decorated with graphene quantum dots and AgNP	8 GHz	–	43	[86]

Table 1 Summary of different nanomaterials used in EMI shielding applications—cont'd

Materials	Frequency	Thickness	Total SE (dB)	References
Graphene nanoplate/ Fe$_3$O$_4$@BaTiO$_3$	1–20 GHz	2.6 mm	26.7–33.3	[87]
Graphene nanosheets and NiFe$_2$O$_4$-rGO nanohybrids	9.2 GHz	2.7 mm	50	[88]
Graphene/Fe$_3$O$_4$ nanoparticles	8–12 GHz	1.8 mm	13	[89]
Graphene/magnetite nanoparticles	8.2–12.4 GHz	2.9 mm	63.2	[90]
Graphene/nickel nanoparticles	8–12 GHz	0.7 mm	40	[91]
Graphene/TiO$_2$ nanoparticles	12–18 GHz	2 mm	11	[92]
Graphene@ Fe$_3$O$_4$	9.6 GHz	2.5 mm	17.8	[93]
Graphene@ Fe$_3$O$_4$@WO$_3$@PANI	9.4 GHz	4 mm	RL = −46.7	[94]
Graphene-Mn Fe$_2$O$_4$ NP-MWCNT	18 GHz	5.5 mm	38	[95]
Graphite/copper nanoparticles	20 GHz	2 mm	70	[96]
Hollow spherical Co Fe$_2$O$_4$ and CNTs	11.7 GHz	2 mm	RL = −32.8	[97]
Iron nanoparticles on expanded graphite	1.5 GHz	–	105	[98]
Magnetic nanoparticles-decorated graphene oxide sheets	18 GHz	6 mm	−27	[99]
Magnetite-based nanomaterials	8.2–12.4 GHz	0.14 mm	1.2–1.4	[100]
MnO$_2$ nanorod-doped PANI	8.2–12.4 GHz 12.4–18 GHz	169 μm	35 39	[101]
MoS$_2$-graphene modified with Fe$_3$O$_4$ nanoparticles	5.9 GHz	2.5 mm	RL = −45.8	[102]
MoS$_2$-reduced graphene oxide /Fe$_3$O$_4$	8.0–12.0 GHz	–	8.27	[103]
MWCNT/manganese zinc ferrite nanoparticle	10 GHz	2 mm	44	[104]
MWCNTs and Ni—Fe nanoparticles	18 GHz	2.5 mm	−35	[105]
MWCNTs, nanosized Fe$_3$O$_4$, and Fe	35 GHz	9 mm	SE$_A$ = 100	[106]
MWCNTs/silica nanoparticle/polystyrene microsphere/PANI	8.2–12.4 GHz	1 mm	24.5	[107]

Continued

Table 1 Summary of different nanomaterials used in EMI shielding applications—cont'd

Materials	Frequency	Thickness	Total SE (dB)	References
MWNT- Fe_3O_4 hybrid nanoparticles	18 GHz	0.9 mm	−60	[108]
MWNT-grafted Fe_3O_4 nanoparticles	18 GHz	–	−32.5	[109]
Nano-NiFe alloy/expanded graphite	300 kHz–10 MHz	–	54–70	[110]
Ni nanoparticle/graphene	18 GHz	–	29.4	[111]
Nickel-plated MWCNTs	1.2 GHz	3 mm	16	[112]
Ni-P-W-coated nanocenosphere	1 GHz	–	20.22	[113]
PANI nanorod/graphene@Fe_3O_4@C	11.4 GHz	3 mm	RL = −44.2	[114]
PANI/nanomagnetic particles	8–12 GHz	<1 mm	90	[115]
PANI/$Ni_{0.5}Zn_{0.5}Fe_2O_4$ nanoparticles	11 GHz	2.5 mm	RL = −39.5	[5]
PANI@nano- Fe_3O_4@CFs	8.2–18 GHz	3 mm	29	[116]
PANI-coated graphite oxide/γ—Fe_3O_4/$BaTiO_3$	800–3000 MHz	–	31–37	[117]
PANI-coated MWCNTs/maghemite (γ—Fe_2O_3) nanoparticle	0.8–2.5 GHz	2 mm	28.2–34.1	[118]
PPy/Fe_3O_4-ZnO nanoparticles	9.96 GHz	2 mm	32.53	[119]
PPy/barium strontium titanate/rGO/Fe_3O_4 nanoparticles	8.2–12.4 GHz	2.5 mm	48	[120]
PPy/ferrite nanoparticles	12.4–18 GHz	1.25 mm	31.9	[121]
rGO, γ—Fe_2O_3, and carbon fibers	8.2–12.4 GHz	0.4 mm	45.26	[122]
rGO/Fe_3O_4 nanodisc	2–10 GHz	4.6 µm	11.2	[123]
Single-layer graphene/Fe_3O_4 nanofillers	8.2–12.4 GHz	0.3 mm	15	[124]
SWCNTs and Co Fe_2O_4 nanocrystals	12.9 GHz	2 mm	RL = −30.7	[125]
Thermally reduced graphene oxide/magnetic carbonyl iron	11.7 GHz	4 mm	40	[126]
Ti_3SiC_2/nano Cu powders	8.2–12.4 GHz	2 mm	27	[127]
Tin oxide nanoparticles@rGO	10.5 GHz	4 mm	62	[128]
Zinc oxide particles-coated MWCNTs	8–14 GHz	–	5	[129]
γ—Fe_2O_3/PPy nanoparticles	17–18 GHz	2 mm	10.10	[130]

dangerous, like losing control of the vehicle, which can endanger human life. In this chapter, the authors discussed different nanomaterials used to block the EMI radiation released from internal and external sources to vehicles. So far, different nanomaterials such as carbon-based nanomaterials (Single-wall carbon nanotubes, Multiwall carbon nanotubes, graphene, and other conductive polymers), metal-based nanomaterials, metal oxide-based nanomaterials, etc. have been used in EMI shielding materials due to having extraordinary properties such as lightweight, high surface area, and high corrosion resistance.

References

[1] J. Ma, M. Zhan, K. Wang, Ultralightweight silver nanowires hybrid polyimide composite foams for high-performance electromagnetic interference shielding, ACS Appl. Mater. Interfaces 7 (2015) 563–576, https://doi.org/10.1021/am5067095.

[2] X. Ouyang, W. Huang, E. Cabrera, J. Castro, L.J. Lee, Graphene-graphene oxide-graphene hybrid nanopapers with superior mechanical, gas barrier and electrical properties, AIP Adv. 5 (2015) 0–8, https://doi.org/10.1063/1.4906795.

[3] P.J. Bora, K.J. Vinoy, P.C. Ramamurthy, G. Madras, Electromagnetic interference shielding efficiency of MnO2 nanorod doped polyaniline film, Mater. Res. Express 4 (2017), https://doi.org/10.1088/2053-1591/aa59e3.

[4] M. Chen, L. Zhang, S. Duan, S. Jing, H. Jiang, M. Luo, et al., Highly conductive and flexible polymer composites with improved mechanical and electromagnetic interference shielding performances, Nanoscale 6 (2014) 3796–3803, https://doi.org/10.1039/c3nr06092f.

[5] N.N. Ali, Y. Atassi, A. Salloum, A. Charba, A. Malki, M. Jafarian, Comparative study of microwave absorption characteristics of (polyaniline/NiZn ferrite) nanocomposites with different ferrite percentages, Mater. Chem. Phys. 211 (2018) 79–87, https://doi.org/10.1016/j.matchemphys.2018.02.017.

[6] X. Jian, B. Wu, Y. Wei, S.X. Dou, X. Wang, W. He, et al., Facile synthesis of Fe3O4/GCs composites and their enhanced microwave absorption properties, ACS Appl. Mater. Interfaces 8 (2016) 6101–6109, https://doi.org/10.1021/acsami.6b00388.

[7] S.B. Kondawar, P.R. Modak, Theory of EMI shielding, in: Materials for Potential EMI Shielding Applications, 2020, pp. 9–25, https://doi.org/10.1016/b978-0-12-817590-3.00002-6. Elsevier.

[8] R. Wilson, G. George, K. Joseph, An introduction to materials for potential EMI shielding applications: status and future, Mater. Potential EMI Shield. Appl. (2020) 1–8, https://doi.org/10.1016/b978-0-12-817590-3.00001-4. Elsevier.

[9] M. Nasreen Taj, B. DarukaPrasad, N. Rama Rao, H. Nagabhushana, K.S. Anantharaju, M.V. Murugendrappa, Polymer—silicate nanocomposites: package material for nanodevices as an EMI shielding, Mater Today Proc. (2021), https://doi.org/10.1016/j.matpr.2020.12.1159.

[10] R.K. Bheema, K.K. Vuba, N. Etakula, K.C. Etika, Enhanced thermo-mechanical, thermal and EMI shielding properties of MWNT/MAgPP/PP nanocomposites prepared by extrusion, Compos. Part C Open Access 4 (2021) 100086, https://doi.org/10.1016/j.jcomc.2020.100086.

[11] B.K. Choi, H.J. Lee, W.K. Choi, M.K. Lee, J.H. Park, J.Y. Hwang, et al., Effect of carbon fiber content on thermal and electrical conductivity, EMI shielding efficiency, and radiation energy of CMC/PVA composite papers with carbon fibers, Synth. Met. 273 (2021) 116708, https://doi.org/10.1016/j.synthmet.2021.116708.

[12] P. Gahlout, V. Choudhary, EMI shielding response of polypyrrole-MWCNT/polyurethane composites, Synth. Met. 266 (2020) 116414, https://doi.org/10.1016/j.synthmet.2020.116414.

[13] J.W. Gooch, J.K. Daher, Electromagnetic Shielding and Corrosion Protection for Aerospace Vehicles, 2007, https://doi.org/10.1007/978-0-387-46096-3.

[14] Y. Sun, Silver nanowires—unique templates for functional nanostructures, Nanoscale 2 (2010) 1626–1642, https://doi.org/10.1039/c0nr00258e.

[15] T.W. Lee, S.E. Lee, Y.G. Jeong, Highly effective electromagnetic interference shielding materials based on silver nanowire/cellulose papers, ACS Appl. Mater. Interfaces 8 (2016) 13123–13132, https://doi.org/10.1021/acsami.6b02218.

[16] Y.H. Yu, C.C.M. Ma, C.C. Teng, Y.L. Huang, S.H. Lee, I. Wang, et al., Electrical, morphological, and electromagnetic interference shielding properties of silver nanowires and nanoparticles conductive composites, Mater. Chem. Phys. 136 (2012) 334–340, https://doi.org/10.1016/j.matchemphys.2012.05.024.

[17] E. Kim, D.Y. Lim, Y. Kang, E. Yoo, Fabrication of a stretchable electromagnetic interference shielding silver nanoparticle/elastomeric polymer composite, RSC Adv. 6 (2016) 52250–52254, https://doi.org/10.1039/c6ra04765c.

[18] M. Arjmand, A.A. Moud, Y. Li, U. Sundararaj, Outstanding electromagnetic interference shielding of silver nanowires: comparison with carbon nanotubes, RSC Adv. 5 (2015) 56590–56598, https://doi.org/10.1039/c5ra08118a.

[19] A.G. El-Shamy, Polyvinyl alcohol and silver decorated carbon quantum-dots for new nanocomposites with application electromagnetic interface (EMI) shielding, Prog. Org. Coat. 146 (2020) 105747, https://doi.org/10.1016/j.porgcoat.2020.105747.

[20] G.H. Lim, N. Kwon, E. Han, S. Bok, S.E. Lee, B. Lim, Flexible Nanoporous silver membranes with unprecedented high effectiveness for electromagnetic interference shielding, J. Ind. Eng. Chem. 93 (2021) 245–252, https://doi.org/10.1016/j.jiec.2020.09.030.

[21] C. Weng, G. Wang, Z. Dai, Y. Pei, L. Liu, Z. Zhang, Buckled AgNW/MXene hybrid hierarchical sponges for high-performance electromagnetic interference shielding, Nanoscale 11 (2019) 22804–22812, https://doi.org/10.1039/c9nr07988b.

[22] H. Ji, R. Zhao, N. Zhang, C. Jin, X. Lu, C. Wang, Lightweight and flexible electrospun polymer nanofiber/metal nanoparticle hybrid membrane for high-performance electromagnetic interference shielding, NPG Asia Mater 10 (2018) 749–760, https://doi.org/10.1038/s41427-018-0070-1.

[23] R. Aepuru, M. Ramalinga Viswanathan, B.V.B. Rao, H.S. Panda, S. Sahu, P.K. Sahoo, Tailoring the performance of mechanically robust highly conducting silver/3D graphene aerogels with superior electromagnetic shielding effectiveness, Diam. Relat. Mater. 109 (2020) 108043, https://doi.org/10.1016/j.diamond.2020.108043.

[24] S. Ye, A.R. Rathmell, Z. Chen, I.E. Stewart, B.J. Wiley, Metal nanowire networks: the next generation of transparent conductors, Adv. Mater. 26 (2014) 6670–6687, https://doi.org/10.1002/adma.201402710.

[25] P. Lv, A. Wei, Y. Wang, D. Li, J. Zhang, L.A. Lucia, et al., Copper nanoparticles-sputtered bacterial cellulose nanocomposites displaying enhanced electromagnetic shielding, thermal, conduction, and mechanical properties, Cellulose 23 (2016) 3117–3127, https://doi.org/10.1007/s10570-016-1030-y.

[26] J. Arranz-Andrés, E. Pérez, M.L. Cerrada, Hybrids based on poly(vinylidene fluoride) and cu nanoparticles: characterization and EMI shielding, Eur. Polym. J. 48 (2012) 1160–1168, https://doi.org/10.1016/j.eurpolymj.2012.04.006.

[27] M.H. Al-Saleh, G.A. Gelves, U. Sundararaj, Copper nanowire/polystyrene nanocomposites: lower percolation threshold and higher EMI shielding, Compos. Part A Appl. Sci. Manuf. 42 (2011) 92–97, https://doi.org/10.1016/j.compositesa.2010.10.003.

[28] S. Hou, W. Ma, G. Li, Y. Zhang, Y. Ji, F. Fan, et al., Excellent terahertz shielding performance of ultrathin flexible cu/graphene nanolayered composites with high stability, J. Mater. Sci. Technol. 52 (2020) 136–144, https://doi.org/10.1016/j.jmst.2020.04.007.

[29] J. Arranz-Andrés, N. Pulido-González, C. Fonseca, E. Pérez, M.L. Cerrada, Lightweight nanocomposites based on poly(vinylidene fluoride) and Al nanoparticles: structural, thermal and mechanical characterization and EMI shielding capability, Mater. Chem. Phys. 142 (2013) 469–478, https://doi.org/10.1016/j.matchemphys.2013.06.038.

[30] G.Y. Yurkov, A.S. Fionov, A.V. Kozinkin, Y.A. Koksharov, Y.A. Ovtchenkov, D.A. Pankratov, et al., Synthesis and physicochemical properties of composites for electromagnetic shielding applications: a polymeric matrix impregnated with iron- or cobalt-containing nanoparticles, J. Nanophotonics 6 (2012), https://doi.org/10.1117/1.jnp.6.061717, 061717.

[31] M. Jalali, S. Dauterstedt, A. Michaud, R. Wuthrich, Electromagnetic shielding of polymer-matrix composites with metallic nanoparticles, Compos. Part B Eng. 42 (2011) 1420–1426, https://doi.org/10.1016/j.compositesb.2011.05.018.

[32] J. Li, H. Liu, J. Guo, Z. Hu, Z. Wang, B. Wang, et al., Flexible, conductive, porous, fibrillar polymer-gold nanocomposites with enhanced electromagnetic interference shielding and mechanical properties, J. Mater. Chem. C 5 (2017) 1095–1105, https://doi.org/10.1039/c6tc04780g.

[33] H. Gargama, A.K. Thakur, S.K. Chaturvedi, Polyvinylidene fluoride/nanocrystalline iron composite materials for EMI shielding and absorption applications, J. Alloys Compd. 654 (2016) 209–215, https://doi.org/10.1016/j.jallcom.2015.09.059.

[34] M. Gouda, A.A. Hebeish, A.I. Aljaafari, New route for development of electromagnetic shielding based on cellulosic nanofibers, J. Ind. Text. 46 (2017) 1598–1615, https://doi.org/10.1177/1528083715627166.

[35] K. Zhu, Y. Ju, J. Xu, Z. Yang, S. Gao, Y. Hou, Magnetic nanomaterials: chemical design, synthesis, and potential applications, Acc. Chem. Res. 51 (2018) 404–413, https://doi.org/10.1021/acs.accounts.7b00407.

[36] M. Gholampoor, F. Movassagh-Alanagh, H. Salimkhani, Fabrication of nano-Fe3O4 3D structure on carbon fibers as a microwave absorber and EMI shielding composite by modified EPD method, Solid State Sci. 64 (2017) 51–61, https://doi.org/10.1016/j.solidstatesciences.2016.12.005.

[37] K. Nasouri, A.M. Shoushtari, Fabrication of magnetite nanoparticles/polyvinylpyrrolidone composite nanofibers and their application as electromagnetic interference shielding material, J. Thermoplast. Compos. Mater. 31 (2018) 431–446, https://doi.org/10.1177/0892705717704488.

[38] L.y. Li, S.l. Li, Y. Shao, R. Dou, B. Yin, M.b. Yang, PVDF/PS/HDPE/MWCNTs/Fe3O4 nanocomposites: effective and lightweight electromagnetic interference shielding material through the synergetic effect of MWCNTs and Fe3O4 nanoparticles, Curr. Appl. Phys. 18 (2018) 388–396, https://doi.org/10.1016/j.cap.2018.01.014.

[39] Z. Ding, S.Q. Shi, H. Zhang, L. Cai, Electromagnetic shielding properties of iron oxide impregnated kenaf bast fiberboard, Compos. Part B Eng. 78 (2015) 266–271, https://doi.org/10.1016/j.compositesb.2015.03.044.

[40] P. Saini, V. Choudhary, N. Vijayan, R.K. Kotnala, Improved electromagnetic interference shielding response of poly(aniline)-coated fabrics containing dielectric and magnetic nanoparticles, J. Phys. Chem. C 116 (2012) 13403–13412, https://doi.org/10.1021/jp302131w.

[41] H. Singh, D. Kumar, K.K. Sawant, N. Devunuri, S. Banerjee, Co-doped ZnO–PVA nanocomposite for EMI shielding, Polym.—Plast. Technol. Eng. 55 (2016) 149–157, https://doi.org/10.1080/03602559.2015.1070869.

[42] K. Rajavel, Y. Hu, P. Zhu, R. Sun, C. Wong, MXene/metal oxides-ag ternary nanostructures for electromagnetic interference shielding, Chem. Eng. J. 399 (2020) 125791, https://doi.org/10.1016/j.cej.2020.125791.

[43] O. Nakhaei, N. Shahtahmassebi, M. Rezaee Roknabadi, M. Behdani, Co-electrospinning fabrication and study of structural and electromagnetic interference-shielding effectiveness of TiO2/SiO2 core–shell nanofibers, Appl. Phys. A Mater. Sci. Process. 122 (2016) 1–10, https://doi.org/10.1007/s00339-016-0072-1.

[44] M.M. Lu, W.Q. Cao, H.L. Shi, X.Y. Fang, J. Yang, Z.L. Hou, et al., Multi-wall carbon nanotubes decorated with ZnO nanocrystals: mild solution-process synthesis and highly efficient microwave absorption properties at elevated temperature, J. Mater. Chem. A 2 (2014) 10540–10547, https://doi.org/10.1039/c4ta01715c.

[45] N. Li, Y. Huang, F. Du, X. He, X. Lin, H. Gao, et al., Electromagnetic interference (EMI) shielding of single-walled carbon nanotube epoxy composites, Nano Lett. 6 (2006) 1–5, https://doi.org/10.1021/nl0602589.

[46] A. Fletcher, M.C. Gupta, K.L. Dudley, E. Vedeler, Elastomer foam nanocomposites for electromagnetic dissipation and shielding applications, Compos. Sci. Technol. 70 (2010) 953–958, https://doi.org/10.1016/j.compscitech.2010.02.011.

[47] P. Verma, P. Saini, V. Choudhary, Designing of carbon nanotube/polymer composites using melt recirculation approach: effect of aspect ratio on mechanical, electrical and EMI shielding response, Mater. Des. 88 (2015) 269–277, https://doi.org/10.1016/j.matdes.2015.08.156.

[48] Y. Zhan, Z. Long, X. Wan, J. Zhang, S. He, Y. He, 3D carbon fiber mats/nano-Fe$_3$O$_4$ hybrid material with high electromagnetic shielding performance, Appl. Surf. Sci. 444 (2018) 710–720, https://doi.org/10.1016/j.apsusc.2018.03.006.

[49] J. Ma, X. Wang, W. Cao, C. Han, H. Yang, J. Yuan, et al., A facile fabrication and highly tunable microwave absorption of 3D flower-like Co$_3$O$_4$-rGO hybrid-architectures, Chem. Eng. J. 339 (2018) 487–498, https://doi.org/10.1016/j.cej.2018.01.152.

[50] H.R. Kim, B.S. Kim, I.S. Kim, Fabrication and EMI shielding effectiveness of ag-decorated highly porous poly(vinyl alcohol)/Fe$_2$O$_3$ nanofibrous composites, Mater. Chem. Phys. 135 (2012) 1024–1029, https://doi.org/10.1016/j.matchemphys.2012.06.008.

[51] S.C. Lin, C.C.M. Ma, S.T. Hsiao, Y.S. Wang, C.Y. Yang, W.H. Liao, et al., Electromagnetic interference shielding performance of waterborne polyurethane composites filled with silver nanoparticles deposited on functionalized graphene, Appl. Surf. Sci. 385 (2016) 436–444, https://doi.org/10.1016/j.apsusc.2016.05.063.

[52] Y. Chen, Y. Li, M. Yip, N. Tai, Electromagnetic interference shielding efficiency of polyaniline composites filled with graphene decorated with metallic nanoparticles, Compos. Sci. Technol. 80 (2013) 80–86, https://doi.org/10.1016/j.compscitech.2013.02.024.

[53] H.Y. Choi, T.W. Lee, S.E. Lee, J.D. Lim, Y.G. Jeong, Silver nanowire/carbon nanotube/cellulose hybrid papers for electrically conductive and electromagnetic interference shielding elements, Compos. Sci. Technol. 150 (2017) 45–53, https://doi.org/10.1016/j.compscitech.2017.07.008.

[54] F. Fang, Y.Q. Li, H.M. Xiao, N. Hu, S.Y. Fu, Layer-structured silver nanowire/polyaniline composite film as a high performance X-band EMI shielding material, J. Mater. Chem. C 4 (2016) 4193–4203, https://doi.org/10.1039/c5tc04406e.

[55] E.J. Jelmy, S. Ramakrishnan, N.K. Kothurkar, EMI shielding and microwave absorption behavior of au-MWCNT/polyaniline nanocomposites, Polym. Adv. Technol. 27 (2016) 1246–1257, https://doi.org/10.1002/pat.3790.

[56] R. Kumaran, S.D. Kumar, N. Balasubramanian, M. Alagar, V. Subramanian, K. Dinakaran, Enhanced electromagnetic interference shielding in a au-MWCNT composite nanostructure dispersed PVDF thin films, J. Phys. Chem. C 120 (2016) 13771–13778, https://doi.org/10.1021/acs.jpcc.6b01333.

[57] M.H. Zahari, B.H. Guan, C.E. Meng, M.F.C. Mansor, L.K. Chuan, EMI shielding effectiveness of composites based on barium ferrite, PANI, and MWCNT, Prog. Electromagn. Res. M 52 (2016) 79–87, https://doi.org/10.2528/PIERM16080701.

[58] M. Verma, A.P. Singh, P. Sambyal, B.P. Singh, S.K. Dhawan, V. Choudhary, Barium ferrite decorated reduced graphene oxide nanocomposite for effective electromagnetic interference shielding, Phys. Chem. Chem. Phys. 17 (2015) 1610–1618, https://doi.org/10.1039/c4cp04284k.

[59] C. Wan, Y. Jiao, T. Qiang, J. Li, Cellulose-derived carbon aerogels supported goethite (α-FeOOH) nanoneedles and nanoflowers for electromagnetic interference shielding, Carbohydr. Polym. 156 (2017) 427–434, https://doi.org/10.1016/j.carbpol.2016.09.028.

[60] X. Bian, L. Liu, H. Li, C. Wang, Q. Xie, et al., Construction of three-dimensional graphene interfaces into carbon fiber textiles for increasing deposition of nickel nanoparticles: flexible hierarchical magnetic textile, Iopscience.Iop.Org (2016), https://doi.org/10.1088/1361-6528/28/4/045710.

[61] Z. Durmus, A. Durmus, M.Y. Bektay, H. Kavas, I.S. Unver, B. Aktas, Quantifying structural and electromagnetic interference (EMI) shielding properties of thermoplastic polyurethane–carbon nanofiber/magnetite nanocomposites, J. Mater. Sci. 51 (2016) 8005–8017, https://doi.org/10.1007/s10853-016-0069-3.

[62] H. Wang, Z. Zhang, C. Dong, G. Chen, reports YW-S, 2017 undefined, Carbon spheres@ MnO$_2$ core-shell nanocomposites with enhanced dielectric properties for electromagnetic shielding, Sci. Rep. 7 (2017) 15841, https://doi.org/10.1038/s41598-017-16059-0.

[63] P.R. Agarwal, R. Kumar, S. Kumari, S.R. Dhakate, Three-dimensional and highly ordered porous carbon-MnO2 composite foam for excellent electromagnetic interference shielding efficiency, RSC Adv. 6 (2016) 100713–100722, https://doi.org/10.1039/c6ra23127f.

[64] D. Zhang, F. Xu, J. Lin, Z. Yang, M. Zhang, Electromagnetic characteristics and microwave absorption properties of carbon-encapsulated cobalt nanoparticles in 2-18-GHz frequency range, Carbon N Y 80 (2014) 103–111, https://doi.org/10.1016/j.carbon.2014.08.044.

[65] R. Kumar, H.K. Choudhary, S.P. Pawar, S. Bose, B. Sahoo, Carbon encapsulated nanoscale iron/iron-carbide/graphite particles for EMI shielding and microwave absorption, Phys. Chem. Chem. Phys. 19 (2017) 23268–23279, https://doi.org/10.1039/c7cp03175k.

[66] I. Ebrahimi, M.P. Gashti, Chemically reduced vs photo-reduced clay-Ag-polypyrrole ternary nanocomposites: comparing thermal, optical, electrical and electromagnetic shielding properties, Mater. Res. Bull. 83 (2016) 96–107, https://doi.org/10.1016/j.materresbull.2016.05.024.

[67] R. Rohini, K. Lasitha, S. Bose, Epoxy composites containing cobalt(II)-porphine anchored multi-walled carbon nanotubes as thin electromagnetic interference shields, adhesives and coatings, J. Mater. Chem. C 4 (2016) 352–361, https://doi.org/10.1039/c5tc03098f.

[68] G.H. Lim, S. Woo, H. Lee, K.S. Moon, H. Sohn, S.E. Lee, et al., Mechanically robust magnetic carbon nanotube papers prepared with CoFe2O4 nanoparticles for electromagnetic interference shielding and magnetomechanical actuation, ACS Appl. Mater. Interfaces 9 (2017) 40628–40637, https://doi.org/10.1021/acsami.7b12147.

[69] M.M. Ismail, S.N. Rafeeq, J.M.A. Sulaiman, A. Mandal, Electromagnetic interference shielding and microwave absorption properties of cobalt ferrite CoFe2O4/polyaniline composite, Appl. Phys. A Mater. Sci. Process. 124 (2018) 1–12, https://doi.org/10.1007/s00339-018-1808-x.

[70] X. Huang, B. Dai, Y. Ren, J. Xu, C. Zhao, Controllable synthesis and electromagnetic interference shielding properties of magnetic CoNi alloy nanoparticles coated on biocarbon nanofibers, J. Mater. Sci. Mater. Electron. 26 (2015) 2584–2588, https://doi.org/10.1007/s10854-015-2727-7.

[71] A. Kumar, A.P. Singh, S. Kumari, A.K. Srivastava, S. Bathula, S.K. Dhawan, et al., EM shielding effectiveness of Pd-CNT-cu nanocomposite buckypaper, J. Mater. Chem. A 3 (2015) 13986–13993, https://doi.org/10.1039/c4ta05749j.

[72] P. Liu, Y. Huang, X. Zhang, Cubic NiFe2O4 particles on graphene-polyaniline and their enhanced microwave absorption properties, Compos. Sci. Technol. 107 (2015) 54–60, https://doi.org/10.1016/j.compscitech.2014.11.021.

[73] M. Mishra, A.P. Singh, P. Sambyal, S. Teotiaa, S.K. Dhawan, Facile synthesis of phenolic resin sheets consisting expanded graphite/γ-Fe2O3/SiO2 composite and its enhanced electromagnetic interference shielding properties, Indian J. Pure Appl. Phys. 52 (2014) 478–485.

[74] M. Mishra, A.P. Singh, S.K. Dhawan, Expanded graphite-nanoferrite-fly ash composites for shielding of electromagnetic pollution, J. Alloys Compd. 557 (2013) 244–251, https://doi.org/10.1016/j.jallcom.2013.01.004.

[75] Y. Li, R. Liu, X. Pang, X. Zhao, Y. Zhang, G. Qin, et al., Fe@C nanocapsules with substitutional sulfur heteroatoms in graphitic shells for improving microwave absorption at gigahertz frequencies, Carbon N Y 126 (2018) 372–381, https://doi.org/10.1016/j.carbon.2017.10.040.

[76] J. Zhu, S. Wei, N. Haldolaarachchige, D.P. Young, Z. Guo, Electromagnetic field shielding polyurethane nanocomposites reinforced with core-shell Fe-silica nanoparticles, J. Phys. Chem. C 115 (2011) 15304–15310, https://doi.org/10.1021/jp2052536.

[77] X. Wang, S. Yu, Y. Wu, H. Pang, S. Yu, Z. Chen, et al., The synergistic elimination of uranium (VI) species from aqueous solution using bi-functional nanocomposite of carbon sphere and layered double hydroxide, Chem. Eng. J. 342 (2018) 321–330, https://doi.org/10.1016/j.cej.2018.02.102.

[78] A. Chaudhary, R. Kumar, S. Teotia, S.K. Dhawan, S.R. Dhakate, S. Kumari, Integration of MCMBs/MWCNTs with Fe3O4 in a flexible and light weight composite paper for promising EMI shielding applications, J. Mater. Chem. C 5 (2017) 322–332, https://doi.org/10.1039/c6tc03241a.

[79] H. Zhang, G. Zhang, J. Li, X. Fan, Z. Jing, J. Li, et al., Lightweight, multifunctional microcellular PMMA/Fe3O4@MWCNTs nanocomposite foams with efficient electromagnetic interference shielding, Compos. Part A Appl. Sci. Manuf. 100 (2017) 128–138, https://doi.org/10.1016/j.compositesa.2017.05.009.

[80] Y. Zhan, J. Wang, K. Zhang, Y. Li, Y. Meng, N. Yan, et al., Fabrication of a flexible electromagnetic interference shielding Fe3O4@reduced graphene oxide/natural rubber composite with segregated network, Chem. Eng. J. 344 (2018) 184–193, https://doi.org/10.1016/j.cej.2018.03.085.

[81] R. Bera, A.K. Das, A. Maitra, S. Paria, S.K. Karan, B.B. Khatua, Salt leached viable porous Fe3O4 decorated polyaniline – SWCNH/PVDF composite spectacles as an admirable electromagnetic

shielding efficiency in extended Ku-band region, Compos. Part B Eng. 129 (2017) 210–220, https://doi.org/10.1016/j.compositesb.2017.07.073.

[82] W. Wang, S.P. Gumfekar, Q. Jiao, B. Zhao, Ferrite-grafted polyaniline nanofibers as electromagnetic shielding materials, J. Mater. Chem. C 1 (2013) 2851–2859, https://doi.org/10.1039/c3tc00757j.

[83] R. Bera, A. Maitra, S. Paria, S.K. Karan, A.K. Das, A. Bera, et al., An approach to widen the electromagnetic shielding efficiency in PDMS/ferrous ferric oxide decorated RGO–SWCNH composite through pressure induced tunability, Chem. Eng. J. 335 (2018) 501–509, https://doi.org/10.1016/j.cej.2017.10.178.

[84] K. Singh, A. Ohlan, V.H. Pham, R.B. Balasubramaniyan, S. Varshney, J. Jang, et al., Nanostructured graphene/Fe3O4 incorporated polyaniline as a high performance shield against electromagnetic pollution, Nanoscale 5 (2013) 2411–2420, https://doi.org/10.1039/c3nr33962a.

[85] L. Wang, J. Zhu, H. Yang, F. Wang, Y. Qin, T. Zhao, et al., Fabrication of hierarchical graphene@Fe3O4@SiO2@polyaniline quaternary composite and its improved electrochemical performance, J. Alloys Compd. 634 (2015) 232–238, https://doi.org/10.1016/j.jallcom.2015.02.062.

[86] N.V. Lakshmi, P. Tambe, EMI shielding effectiveness of graphene decorated with graphene quantum dots and silver nanoparticles reinforced PVDF nanocomposites, Compos. Interfaces 24 (2017) 861–882, https://doi.org/10.1080/09276440.2017.1302202.

[87] L. Jin, X. Zhao, J. Xu, Y. Luo, D. Chen, G. Chen, The synergistic effect of a graphene nanoplate/Fe3O4@BaTiO3 hybrid and MWCNTs on enhancing broadband electromagnetic interference shielding performance, RSC Adv. 8 (2018) 2065–2071, https://doi.org/10.1039/c7ra12909b.

[88] F. Ren, Y. Shi, P. Ren, X. Si, H. Wang, Cyanate ester resin filled with graphene nanosheets and NiFe2O4-reduced graphene oxide nanohybrids for efficient electromagnetic interference shielding, Nano 12 (2017), https://doi.org/10.1142/S1793292017500667.

[89] K. Yao, J. Gong, N. Tian, Y. Lin, X. Wen, Z. Jiang, et al., Flammability properties and electromagnetic interference shielding of PVC/graphene composites containing Fe3O4 nanoparticles, RSC Adv. 5 (2015) 31910–31919, https://doi.org/10.1039/c5ra01046b.

[90] F. Sharif, M. Arjmand, A.A. Moud, U. Sundararaj, E.P.L. Roberts, Segregated hybrid poly(methyl methacrylate)/graphene/magnetite nanocomposites for electromagnetic interference shielding, ACS Appl. Mater. Interfaces 9 (2017) 14171–14179, https://doi.org/10.1021/acsami.6b13986.

[91] H.J. Im, G.H. Jun, D.J. Lee, H.J. Ryu, S.H. Hong, Enhanced electromagnetic interference shielding behavior of graphene Nanoplatelet/Ni/wax nanocomposites, J. Mater. Chem. C 5 (2017) 6471–6479, https://doi.org/10.1039/c7tc01405h.

[92] A. Kumar, R. Anant, K. Kumar, S.S. Chauhan, S. Kumar, R. Kumar, Anticorrosive and electromagnetic shielding response of a graphene/TiO2-epoxy nanocomposite with enhanced mechanical properties, RSC Adv. 6 (2016) 113405–113414, https://doi.org/10.1039/c6ra15273b.

[93] B. Shen, W. Zhai, M. Tao, J. Ling, W. Zheng, Lightweight, multifunctional polyetherimide/graphene@Fe3O4 composite foams for shielding of electromagnetic pollution, ACS Appl. Mater. Interfaces 5 (2013) 11383–11391, https://doi.org/10.1021/am4036527.

[94] Y. Wang, X. Wu, W. Zhang, C. Luo, J. Li, Q. Wang, 3D heterostructure of graphene@Fe3O4@WO3@PANI: preparation and excellent microwave absorption performance, Synth. Met. 231 (2017) 7–14, https://doi.org/10.1016/j.synthmet.2017.06.013.

[95] R.K. Srivastava, P. Xavier, S.N. Gupta, G.P. Kar, S. Bose, A.K. Sood, Excellent electromagnetic interference shielding by graphene- MnFe2O4-multiwalled carbon nanotube hybrids at very low weight percentage in polymer matrix, Chem. Select 1 (2016) 5995–6003, https://doi.org/10.1002/slct.201601302.

[96] A.A. Al-Ghamdi, F. El-Tantawy, New electromagnetic wave shielding effectiveness at microwave frequency of polyvinyl chloride reinforced graphite/copper nanoparticles, Compos. Part A Appl. Sci. Manuf. 41 (2010) 1693–1701, https://doi.org/10.1016/j.compositesa.2010.08.006.

[97] S. Zhang, Z. Qi, Y. Zhao, Q. Jiao, X. Ni, Y. Wang, et al., Core/shell structured composites of hollow spherical CoFe2O4 and CNTs as absorbing materials, J. Alloys Compd. 694 (2017) 309–312, https://doi.org/10.1016/j.jallcom.2016.09.324.

[98] Z. Xu, Y. Huang, Y. Yang, J. Shen, T. Tang, R. Huang, Dispersion of iron nano-particles on expanded graphite for the shielding of electromagnetic radiation, J. Magn. Magn. Mater. 322 (2010) 3084–3087, https://doi.org/10.1016/j.jmmm.2010.05.034.

[99] P.K.S. Mural, S.P. Pawar, S. Jayanthi, G. Madras, A.K. Sood, S. Bose, Engineering nanostructures by decorating magnetic nanoparticles onto graphene oxide sheets to shield electromagnetic radiations, ACS Appl. Mater. Interfaces 7 (2015) 16266–16278, https://doi.org/10.1021/acsami.5b02703.

[100] J.A. Marins, B.G. Soares, H.S. Barud, S.J.L. Ribeiro, Flexible magnetic membranes based on bacterial cellulose and its evaluation as electromagnetic interference shielding material, Mater. Sci. Eng. C 33 (2013) 3994–4001, https://doi.org/10.1016/j.msec.2013.05.035.

[101] Bora P, Vinoy K, ... PR–MR 2017 undefined , Electromagnetic interference shielding efficiency of MnO_2 nanorod doped polyaniline film. IopscienceIopOrg n.d.

[102] Y. Wang, Y. Chen, X. Wu, W. Zhang, C. Luo, J. Li, Fabrication of MoS_2-graphene modified with Fe_3O_4 particles and its enhanced microwave absorption performance, Adv. Powder Technol. 29 (2018) 744–750, https://doi.org/10.1016/j.apt.2017.12.016.

[103] J. Prasad, A.K. Singh, J. Shah, R.K. Kotnala, K. Singh, Synthesis of MoS_2-reduced graphene oxide/Fe_3O_4 nanocomposite for enhanced electromagnetic interference shielding effectiveness, Mater. Res. Express 5 (2018), https://doi.org/10.1088/2053-1591/aac0c2, 055028.

[104] C.H. Phan, M. Mariatti, Y.H. Koh, Electromagnetic interference shielding performance of epoxy composites filled with multiwalled carbon nanotubes/manganese zinc ferrite hybrid fillers, J. Magn. Magn. Mater. 401 (2016) 472–478, https://doi.org/10.1016/j.jmmm.2015.10.067.

[105] V. Bhingardive, S. Suwas, S. Bose, New physical insights into the electromagnetic shielding efficiency in PVDF nanocomposites containing multiwall carbon nanotubes and magnetic nanoparticles, RSC Adv. 5 (2015) 79463–79472, https://doi.org/10.1039/c5ra13901e.

[106] Y. Liu, D. Song, C. Wu, J. Leng, EMI shielding performance of nanocomposites with MWCNTs, nanosized Fe_3O_4 and Fe, Compos. Part B Eng. 63 (2014) 34–40, https://doi.org/10.1016/j.compositesb.2014.03.014.

[107] S. Meer, A. Kausar, T. Iqbal, Synthesis of multi-walled carbon nanotube/silica nanoparticle/polystyrene microsphere/polyaniline based hybrids for EMI shielding application, Fullerenes, Nanotubes, Carbon Nanostruct. 24 (2016) 507–519, https://doi.org/10.1080/1536383X.2016.1195816.

[108] S. Biswas, S.S. Panja, S. Bose, A novel fluorophore-spacer-receptor to conjugate MWNTs and ferrite nanoparticles to design an ultra-thin shield to screen electromagnetic radiation, Mater. Chem. Front. 1 (2017) 132–145, https://doi.org/10.1039/c6qm00074f.

[109] S.P. Pawar, D.A. Marathe, K. Pattabhi, S. Bose, Electromagnetic interference shielding through MWNT grafted Fe_3O_4 nanoparticles in PC/SAN blends, J. Mater. Chem. A 3 (2015) 656–669, https://doi.org/10.1039/c4ta04559a.

[110] W. Liu, Y.A. Huang, L. Wei, Y. Zhai, R.L. Zhang, T. Tang, et al., A Mössbauer investigation of nano-NiFe alloy/expanded graphite for electromagnetic shielding, Nucl. Sci. Tech. 27 (2016), https://doi.org/10.1007/s41365-016-0127-1.

[111] S.P. Pawar, S. Stephen, S. Bose, V. Mittal, Tailored electrical conductivity, electromagnetic shielding and thermal transport in polymeric blends with graphene sheets decorated with nickel nanoparticles, Phys. Chem. Chem. Phys. 17 (2015) 14922–14930, https://doi.org/10.1039/c5cp00899a.

[112] Y.J. Yim, K.Y. Rhee, S.J. Park, Electromagnetic interference shielding effectiveness of nickel-plated MWCNTs/high-density polyethylene composites, Compos. Part B Eng. 98 (2016) 120–125, https://doi.org/10.1016/j.compositesb.2016.04.061.

[113] S. Vynatheya, J.R.N. Kumar, R. Madhusudhana, L.C. Sagar, V.B. Raju, Preparation and characterization of electroless Ni-P-W coated nanocenosphere/ABS composite for EMI shielding application, Mater. Today Proc. 4 (2017) 12130–12137, https://doi.org/10.1016/j.matpr.2017.09.141. Elsevier Ltd.

[114] L. Wang, Q.K. Liu, J.P. Ma, D.Y. Bin, M(II)-coordination polymers (M = Zn and cd) constructed from 1,2-bis[4-(pyridin-3-yl)phenoxy]ethane and 1,4-benzenedicarboxylic acid, J. Mol. Struct. 1084 (2015) 1–8, https://doi.org/10.1016/j.molstruc.2014.12.014.

[115] N. El Kamchi, B. Belaabed, J.-L. Wojkiewicz, S. Lamouri, T. Lasri, et al., Hybrid polyaniline/nano-magnetic particles composites: high performance materials for EMI shielding, Wiley Online Libr. 127 (2013) 4426–4432, https://doi.org/10.1002/app.38036.

[116] F. Movassagh-Alanagh, A. Bordbar-Khiabani, A. Ahangari-Asl, Three-phase PANI@nano-Fe3O4@CFs heterostructure: fabrication, characterization and investigation of microwave absorption and EMI shielding of PANI@nano-Fe3O4@CFs/epoxy hybrid composite, Compos. Sci. Technol. 150 (2017) 65–78, https://doi.org/10.1016/j.compscitech.2017.07.010.

[117] Y.E. Moon, J. Yun, H.I. Kim, Synergetic improvement in electromagnetic interference shielding characteristics of polyaniline-coated graphite oxide/γ-Fe2O3/BaTiO3 nanocomposites, J. Ind. Eng. Chem. 19 (2013) 493–497, https://doi.org/10.1016/j.jiec.2012.09.002.

[118] J. Yun, K.H. Il, Erratum: electromagnetic interference shielding effects of polyaniline-coated multi-wall carbon nanotubes/maghemite nanocomposites (Polymer Bulletin), Polym. Bull. 68 (2012) 561–573, https://doi.org/10.1007/s00289-011-0651-4. 2012; 69:261. doi:https://doi.org/10.1007/s00289-012-0768-0.

[119] A. Olad, S. Shakoori, Electromagnetic interference attenuation and shielding effect of quaternary epoxy-PPy/Fe3O4-ZnO nanocomposite as a broad band microwave-absorber, J. Magn. Magn. Mater. 458 (2018) 335–345, https://doi.org/10.1016/j.jmmm.2018.03.050.

[120] P. Sambyal, S.K. Dhawan, P. Gairola, S.S. Chauhan, S.P. Gairola, Synergistic effect of polypyrrole/BST/RGO/Fe3O4 composite for enhanced microwave absorption and EMI shielding in X-band, Curr. Appl. Phys. 18 (2018) 611–618, https://doi.org/10.1016/j.cap.2018.03.001.

[121] S. Varshney, S.K. Dhawan, Improved electromagnetic shielding performance of lightweight compression molded Polypyrrole/ferrite composite sheets, J. Electron. Mater. 46 (2017) 1811–1820, https://doi.org/10.1007/s11664-016-5233-7.

[122] A.P. Singh, P. Garg, F. Alam, K. Singh, R.B. Mathur, R.P. Tandon, et al., Phenolic resin-based composite sheets filled with mixtures of reduced graphene oxide, γ-Fe2O3 and carbon fibers for excellent electromagnetic interference shielding in the X-band, Carbon N Y 50 (2012) 3868–3875, https://doi.org/10.1016/j.carbon.2012.04.030.

[123] Y. Yang, M. Li, Y. Wu, T. Wang, E.S.G. Choo, J. Ding, et al., Nanoscaled self-alignment of Fe3O4 nanodiscs in ultrathin rGO films with engineered conductivity for electromagnetic interference shielding, Nanoscale 8 (2016) 15989–15998, https://doi.org/10.1039/c6nr04539a.

[124] B.V. Bhaskara Rao, P. Yadav, R. Aepuru, H.S. Panda, S. Ogale, S.N. Kale, Single-layer graphene-assembled 3D porous carbon composite with PVA and Fe3O4 nano-fillers: an interface-mediated superior dielectric and EMI shielding performance, Phys. Chem. Chem. Phys. 17 (2015) 18353–18362, https://doi.org/10.1039/C5CP02476E.

[125] G. Li, L. Sheng, L. Yu, K. An, W. Ren, X. Zhao, Electromagnetic and microwave absorption properties of single-walled carbon nanotubes and CoFe2O4 nanocomposites, Mater. Sci. Eng. B Solid-State Mater. Adv. Technol. 193 (2015) 153–159, https://doi.org/10.1016/j.mseb.2014.12.008.

[126] Y. Chen, Z.H. Bin, Y. Huang, Y. Jiang, W.G. Zheng, Z.Z. Yu, Magnetic and electrically conductive epoxy/graphene/carbonyl iron nanocomposites for efficient electromagnetic interference shielding, Compos. Sci. Technol. 118 (2015) 178–185, https://doi.org/10.1016/j.compscitech.2015.08.023.

[127] Y. Liu, X. Jian, X. Su, F. Luo, J. Xu, J. Wang, et al., Electromagnetic interference shielding and absorption properties of Ti3SiC2/nano cu/epoxy resin coating, J. Alloys Compd. 740 (2018) 68–76, https://doi.org/10.1016/j.jallcom.2018.01.017.

[128] M. Mishra, A.P. Singh, B.P. Singh, S.K. Dhawan, Performance of a nanoarchitectured tin oxide@reduced graphene oxide composite as a shield against electromagnetic polluting radiation, RSC Adv. 4 (2014) 25904–25911, https://doi.org/10.1039/c4ra01860e.

[129] W.L. Song, M.S. Cao, B. Wen, Z.L. Hou, J. Cheng, J. Yuan, Synthesis of zinc oxide particles coated multiwalled carbon nanotubes: dielectric properties, electromagnetic interference shielding and microwave absorption, Mater. Res. Bull. 47 (2012) 1747–1754, https://doi.org/10.1016/j.materresbull.2012.03.045.

[130] J. Azadmanjiri, P. Hojati-Talemi, G.P. Simon, K. Suzuki, C. Selomulya, Synthesis and electromagnetic interference shielding properties of iron oxide/polypyrrole nanocomposites, Polym. Eng. Sci. 51 (2011) 247–253, https://doi.org/10.1002/pen.21813.

CHAPTER 36

Automotive coolants

Zafar Said[a,b,c], Maham Sohail[a], and Arun Kumar Tiwari[d]

[a]Sustainable and Renewable Energy Engineering Department, University of Sharjah, Sharjah, United Arab Emirates
[b]Research Institute for Sciences and Engineering, University of Sharjah, Sharjah, United Arab Emirates
[c]U.S.-Pakistan Center for Advanced Studies in Energy (USPCAS-E), National University of Sciences and Technology (NUST), Islamabad, Pakistan
[d]Mechanical Engineering Department, Institute of Engineering & Technology, Dr. A.P.J. Abdul Kalam Technical University, Lucknow, Uttar Pradesh, India

Chapter outline

1. Introduction — 773
2. Features of advanced cooling system — 776
   - 2.1 Temperature setpoint — 776
   - 2.2 High temperature set point — 777
   - 2.3 Low temperature set point — 777
   - 2.4 Split cooling system — 777
   - 2.5 Precision cooling system — 778
   - 2.6 Controlled engine cooling and elements — 778
3. Numerical studies and correlations — 779
   - 3.1 Coolant heat transfer coefficient — 779
   - 3.2 Tube-side heat transfer coefficient — 779
   - 3.3 Air heat transfer coefficient — 780
   - 3.4 Heat transfer through tube wall — 780
   - 3.5 Overall heat transfer coefficient — 780
   - 3.6 Automotive engine and radiator thermodynamics — 780
   - 3.7 Assumptions — 781
4. Experimental studies — 781
5. Advancements in automotive cooling using nanotechnology — 786
6. Challenges and outlook — 788
7. Conclusion — 790
References — 790

1. Introduction

Numerous scientific progress has been made pioneered to sustain the conditions set out on minimal fuel consumption and CO_2 emanation in automobiles. Automobile coolant generates about 30% of the total energy to the engine [1]. Cooling is one of the most important technical challenges facing numerous industries such as automobiles, electronics, and manufacturing. Engine cooling is essential and promising for researchers and engineers from the beginning of this technical field. Recent advancements to upsurge automobile mass and enhance engine performance with supercharging or turbocharging place higher need on the engine cooling system. Several studies are developed to handle increased heat load and to reduce the cost, size, and weight of the engine coolants, which aids in improving fuel economy and decline in automotive emissions [2]. Conventional cooling system varies depending on the coolant pump, and radiator fan pushed off the engine's crankshaft. The dependency of the fan and pump on the engine velocity often permitted the thermal control system to overcool the fluid, and, degrading total efficiency [3]. Innovative automotive cooling system restores the standard wax-based thermostat flexible regulator that improves the mechanical pump and radiator fan with pc-controlled servo-motor devices [4]. The progress in the automotive industry is persistently engaged in an effective and economical approach to finding excellent automotive design in various aspects such as performance, fuel consumption, cost, and safety.

Automotive cooling systems comprise the radiator, water pump, cooling fan, pressure lid, and thermostat. A radiator is a vital device in the vehicle's cooling system. Radiators are heat exchangers that transfer heat or thermal energy from one form to another for cooling and heating. The main objective is to radiate the excessive heat from the engine to the environment to ensure consistent engine operation. Radiator cooling of automobiles is evolving with time to design the engine increasingly efficient and eco-friendly. Radiators are employed for cooling internal engines, such as automobiles, piston–engine aircraft, railway engines, and many more. Literature has reported that radiator's performances, comprising thermal dissipation rate, pressure drop in coolant and air, mostly rely on the factors such as inlet temperature, coolant volume flow rate, and air velocity [5]. Coolant flow between the radiator and engine is maintained through a thermostat, adjusting the coolant flow across the radiator to maintain engine temperature in close proximity to a prearranged regular operation point. Expanding the size of the cooling system and increasing the engine operating temperature may have a negative impact on the heat transfer and reduction in the lifetime of the cooling system. There are basically two techniques to boost the thermal behavior of radiators consisting of active and passive techniques. Active techniques are those in which the external power is needed, whereas passive techniques consist of unique surface geometries and fluid additives. Passive techniques are more considered to be promising due to their reliability, cost, and safety. The passive technique includes coated surfaces, rough and enlarged surfaces, inserts, swirl flow, and coiled tubes. These surfaces will result in the elevated efficiency in heat exchangers, which is promising for the automotive industry [6]. Automotive radiators utilize finned surfaces of different designs outside the passages through which the coolant flows. These fins increase the surface area accessible for heat to flow to the air. The fin geometry and quantity play an important part in the radiators' heat transfer. The plain fins have been used in automotive, an effort of utilizing structured airfoil fins were tested in the preliminary research phase. The fin structured is varied with distinct cross-sections like round, box, elliptical, and triangular shapes.

At the primitive phase of automobiles, air-based coolants were far more widespread than water-based coolants [7]. The brief history of the vehicle growth and truck radiator cooling states that it was found that many ages have been expanded to enhance the cooling ability in vehicle engines. Right at the beginning of the initial stage, the copper and brass material was utilized as radiators. Both were replaced with aluminum in the second stage during the late 20th century to make the automobile engine cooling system light weight to enhance efficiency, it can cope with heat intensely in spite of its many shortcomings. In the third stage, copper and brass made their way back to the radiator because aluminum radiators cause corrosion and are expensive to refurbish. In the years to come, the radiator materials will continue to strengthen the quality. Researchers have started investigations on synthetic coolants in the radiator to advance the water, like anti-freezing ability.

Conventional car radiators are typically air-cooled cross-flow heat exchangers. The hot coolant is enforced to flow downwards across one side of the radiator to the other side through the parallel tubes at the hotside. On the cold side, ambient air is enforced to pass through the finned tubes to eradicate the coolant heat. The heat generated by either an engine or a thermal form is transferred to the radiator by coolant. The coolant and cooled air is transported orthogonally to one another without mixing in the radiator [8]. The vehicle's air-cooled heat exchangers such as radiator, AC condenser and evaporator, and charge air cooler have a significant part in the design and mass of its front-end module, which strongly affects automobile aerodynamic performance. It requests innovative design devices that can imply the improved solution and the fundamental reason for a performance augmentation. The heat transfer and fluid-dynamic performance of an automotive radiator depend on thermal fluids mass flow. The heat transfer performance reduces with increasing air inlet temperature, reducing the cooling temperature difference. It is noteworthy to state that the slight effect of air inlet temperature on the overall heat transfer coefficient. The choice of coolant fluid is reliant on the ecological requirements, while toxicity limitations are restricted in certain applications. The overall heat transfer coefficient is a function of the flow arrangement and the coolant flow rate. The actual heat transfer rate depends on the overall heat transfer coefficient, the temperature difference between the coolant and air flowing into radiator, the mass flow rates, specific heat of air and coolant, the radiator's surface area, and geometry. The key components of the industrial coolants comprise 30–70 vol% EG, and the additional inhibitors contain molybdate, phosphate, nitrate, benzoate, and silicate [9].

Superior thermophysical characteristics of coolant in automotive thermal management are suitable for performance and protection. Water has extensively used as a coolant because of the remarkable heat transfer rate. Propylene Glycol (PG) based coolants are becoming popular for heavy-duty engines due to reduced propylene glycol toxicity compared to Ethylene Glycol (EG). Propylene Glycol based coolants have been used by automobile and heavy-duty truck manufacturers in America [10]. Numerous investigations have described the findings of vehicle testing with Propylene Glycol against Ethylene Glycol coolant solutions. Using both PG and EG with water downgrades the heat transfer in addition to pressure drop performance. PG is similar in its physical characteristics to EG and generated substantial interest as a base material for coolant antifreeze in automotive and light-duty applications. PG has a different chemical composition which makes it less toxic than EG. PG offers distinct features over EG as a coolant base fluid, especially in toxicological and environmental properties [11]. JuGer and Crook [10] experimentally studied the heat transfer performance of heavy-duty radiators with Propylene Glycol/water and Ethylene Glycol/water-based coolants. It was observed that the performance of PG/water is degraded at higher flow rates to a greater degree than ethylene glycol-based coolants. Gollin and Bjork [12] studied the heat transfer and hydraulic performance of pure water, PG/water, and EG/water-based coolants with different ratios. The most effective coolants with the lowest pressure drop were reported for water. It was observed that the pressure drop associated with 50:50 PG/water is slightly higher than 50:50 EG/water, and the thermal performance is observed to be similar for 70:30 PG/water and 70:30 EG/water coolants. Coughenour et al. [13] reported the performance of propylene glycol coolants at high ambient temperatures and under high load conditions on the engine dynamometer and compared it with ethylene glycol-based coolants. It was reported that propylene glycol-based coolants showed better performance. Nanofluids have emerges as potential prospect coolants in heat transfer applications.

2. Features of advanced cooling system

A sophisticated cooling system has the purpose of advancing fuel consumption and reducing emanation. The following three parameters should be considered for bringing the engine cooling system to fuel efficiency and emissions:
1. Engine's frictional losses.
2. Auxiliary energy needs to run the cooling system.
3. Combustion boundary situation, for instance, combustion-chamber temperature, charge density, and temperature.

Sophisticated engine-cooling systems have merits over traditional systems by incorporating characteristics in the systems layout, module technologies, and operational approach. These features enable a sophisticated cooling system to affect the engine-coolant elements to the engine efficiency with higher productivity [14].

2.1 Temperature setpoint

Regulating coolant temperature to a specified setpoint is the widespread methodology for the engine cooling system. It is normally supposed that the coolant temperature is characteristic of the metal temperature at a steady-state, and for a certain pace and load at a specific place in the engine. As the working demands revolve around the highest temperature at which the exhaust-regulator bridge area can work, it is appropriate to regulate the engine-cooling system according to the metal temperature rather than the coolant temperature for engine safety.

2.2 High temperature set point

Improving the set point of operating temperature is a widely used path between the studies of engine-cooling systems. Increasing the engine's operating temperature has several advantages, as it affects the components that affect the casualties in engine and the cooling system's efficacy, with emissions. The increment in operating temperature will enhance the engine temperature, reducing frictional losses and enhancing fuel efficiency.

The effect of rising operating temperature is reported in several investigations. Finlay et al. [15] increased the temperature up till 195°C by adjusting the outlet temperature to 150°C to minimize the frictional losses, followed by a 4%–6% increment to fuel consumption. Improving the operating temperature also substantially affects the efficiency of cooling system. Enhancing the operating temperature improves the efficacy of the heat transfer in the radiator, permitting a low coolant flow rate for a particular power, and, reducing the pumping demands. The important aspect of increasing operating temperature is its effect on the producing emissions by-product such as carbon monoxide (CO) and unburned hydrocarbon (uHC). Chanfreau et al. [16] observed a reduction of 15% in CO output and a 17% decline in hydrocarbons in a drive-cycle test, with a greater operating temperature set point. The same author stated decrement in NO_x, pointing out that the decreased frictional losses and efficiency might contribute parameters in controlling NO_x levels while maintaining a similar maximal in-cylinder pressure and temperature levels.

2.3 Low temperature set point

Reducing the temperature set point is contrary to the advantage that is accomplished by improving operating temperature. There are advantages in lowering the temperatures, including superior engine volumetric effectiveness and lesser charge temperature, two parameters that affect the incineration technique and its output in fuel efficiency and emissions. Through dropping the coolant temperature at 50°C, Finlay et al. [15] stated that knock protection enhances up to $2°$ of spark angle, suggesting minimal charge temperature and the ability to recalibrate spark planning for better fuel efficiency and reduced emissions.

2.4 Split cooling system

The split cooling system is another kind of engine-cooling system by combining system development and operation technique is being used to increase the possibility. The split cooling system offers a distinct advantage as it enables each section of engine to function at its optimal temperature points, boosting the overall engine efficiency. The preferred thermal condition for the engine is to have the running head cooler. Volumetric efficiency improves by running the head cooler, improving the air trapped mass, though at low temperature. This increased air trapped mass at lower temperature enables swift combustion, lowering CO, HC, and NO_x emissions, while enhancing the output power. Higher block temperature would decrease frictional losses, benefit fuel efficiency advances, and ultimately lower the pressure and temperature, which strongly affects NO_x formation.

2.5 Precision cooling system

The precision cooling system is incorporated into the coolant gallery design and the functional layout. In a precision cooling system, thermally vital areas, such as the exhaust-valve bridge, high coolant flow speeds to encourage heat transfer with reasonable temperature differences or large heat flux, effectively decreasing temperatures around these sections. This can be obtained by decreasing the coolant passage's cross-sectional area in these areas to achieve high flow speed. The important design features of a precision cooling system include the gallery sizing and pairing of the pump to guarantee that the heat-elimination process can meet their straining obstruction on the operating temperature in sensitive areas. Clough [17] reported a decline in pumping power consumption of 54% and the difference between the metal temperature of 100°C with a precise cooling system.

2.6 Controlled engine cooling and elements

The traditional engine cooling system still being applied on innovative engines is the passive approach explicitly intended for minimalism and cost-effectiveness. Oversizing the current engine-cooling system to deal with the potential of overheating the engine in certain remote situations leading to further inadequacies in the cooling system, improving the cooling system's power demand. Controllable elements, like the electric pump in the engine cooling system permitted Chanfreau et al. [18] to function with a coolant temperature of 110°C, as compared to 90°C in a traditional engine, with fuel reduction of 2%–5%, CO reduction of 20%, and 10% decline in hydrocarbons. The electrical modules, like water pump and flow valve, perform as a convenient engine-cooling system in mass-produced automobiles have not been employed in practical applications due to elevated operating temperature. Engine-cooling system needs to increase the engine temperature to attain remarkable heat-transfer effectiveness to compensate low efficiency of powering electrical modules for automobile [14].

3. Numerical studies and correlations

Several numerical analyses have been proposed by researchers and scholars for an automotive thermal control system. Researcher Vaughan [19] developed a nonlinear analysis for hydraulic solenoid regulators. Henry et al. [20] proposed an automotive power train cooling system analysis for light-truck vehicles. Frick et al. [21] presented an array of numerical models describing hydraulic-driven heat exchangers for vehicle coolants. Trivedi et al. [22] studied numerically the heat transfer and fluid flow of an automotive radiator using ANSYS 12.1 software. It was observed that by varying the pitch of the tube, the heat transfer rate of the radiator could be improved. Torregrosa et al. [1] investigated and developed a thermo-hydrodynamic model to analyze the performance of a diesel engine during the warm-up phase. It was observed that the model projections fits well with the experimental data. It was reported that substantial advances can be attained by applying cooling systems adequately flexible to monitor the total coolant volume.

3.1 Coolant heat transfer coefficient

The coolant flowing in the radiator has been supposed to internal pipe flow and the heat transfer coefficient linked with this flow. Hausen [23] proposed an expression for computing Nusselt numbers for laminar coolant flow as:

$$Nu_c = 3.66 + \frac{0.0668 \left(\frac{D_{h,c}}{Y_l}\right) Re_c Pr_c}{1 + 0.04 \left[\left(\frac{D_{h,c}}{Y_l}\right) Re_c Pr_c\right]^{2/3}} \quad (1)$$

The Nusselt number depends on the coolant flow conditions. The flow can be either laminar, transitional, or turbulent, portrayed by Reynolds number.

The Dittus and Boelter's [24] expression can be computed for Nusselt number for fully developed turbulent coolant flow as:

$$Nu_c = 0.023 Re_c^{0.8} Pr_c^{0.4} \quad (2)$$

3.2 Tube-side heat transfer coefficient

The tube-side heat transfer coefficient can be computed from Gnielinski expression [25] as:

$$h_i = \left(\frac{k_w}{d_i}\right) \frac{(Re_{Di} - 1000) Pr(f_i/2)}{1 + 12.7\sqrt{f_i/2}\left(Pr^{2/3} - 1\right)} \quad (3)$$

$$f_i = [1.58 \ln(Re_{Di}) - 3.28]^{-2} \quad (4)$$

Re_{Di} represents the tube-side Reynolds number depending on the tube hydraulic diameter.

3.3 Air heat transfer coefficient

An outer tube surface characterizes the air-side heat transfer with attached plate fins. Heat transfer coefficients for plate-fin surfaces are obtained using the Colburn factor (j_h), expressed as:

$$j_h = \frac{h_a}{\rho V_a c_p}\left(\frac{c_p \mu}{k}\right)^{2/3} \tag{5}$$

The efficiency of the plate fins is expressed as:

$$\eta_f = \frac{\tan h(0.5mb)}{0.5mb} \tag{6}$$

Where $m = \left(\dfrac{2h_a}{k_f \delta_f}\right)^{1/2}$.

3.4 Heat transfer through tube wall

The resistance to conduction of the rectangular cross-section is computed as:

$$R_{t,cond} = \frac{1}{k_t}\int_{s_1}^{s_2}\frac{ds}{A(s)} \tag{7}$$

where s represents the spatial variable, and $A(s)$ refers to the area.

3.5 Overall heat transfer coefficient

It is computed by adding resistances in series for the coolant, tube wall, and air heat transfer. It is expressed as:

$$\frac{1}{UA} = \frac{1}{h_c A_{t,i}} + R_{t,cond} + \frac{1}{h_a\left(A_{t,o} + \eta_f A_{fin}\right)} \tag{8}$$

3.6 Automotive engine and radiator thermodynamics

The dynamic performance may be presented with reduced order two-node lumped factor heat analysis. The engine and radiator temperature dynamic behaviors can be expressed by Salah and his co-workers [26] as:

$$C_e \dot{T}_e = Q_{in} - C_{pc}\dot{m}_c(T_e - T_r) \tag{9}$$

$$C_r \dot{T}_r = -Q_O + C_{pc}\dot{m}_c(T_e - T_r) - \varepsilon C_{pa}\dot{m}_a(T_e - T_\infty) \qquad (10)$$

The variables Q_{in} and Q_O refers to thermal load created during combustion and the radiator thermal loss, respectively.

3.7 Assumptions

Heat transfer evaluation of a radiator as an automotive coolant should consider some assumptions such as:
1. It works under steady-state conditions with a constant flow rate and coolant temperatures at the inlet and time-independent inside the radiator.
2. No thermal energy sources and sinks in radiator walls or coolant.
3. The fluid flow rate is delivered evenly through the radiator on each side.
4. Kinetic energy and potential energy changes are insignificant.

4. Experimental Studies

Literature has shown the experimental study of the thermal and fluid-dynamic performance of automotive radiators. Sahoo et al. [27] investigated the energetic and exergetic performance of rectangular fin and flat tube automotive radiator with 25% of PG and EG-based brine solutions. By increasing volume fraction, the heat transfer rate and effectiveness reduce for EG-based coolant, and for PG, it reduces at first, and increases and then again declines, as shown in Fig. 1. It is also seen from Fig. 2 that the brine concentration of 25 vol% PG shows maximum performance index with the least pumping power. Cao and Kengskool [28] presented the application of heat pipes for heavy-duty engines. The cooling load of the radiator can be improved, and the electricity consumption of the fan can be decreased for better energy efficiency. Oliva et al. [29] presented parametric analyses on automotive radiators through comprehensive rating and design heat exchanger models. Lin et al. [8] investigated theoretically the impact of ambient and inlet temperature of the coolant radiator based on the specific dissipation (SD) expression. It was reported that SD could be employed to assess the cooling system behavior as the influence of ambient and coolant radiator entrance temperature changes on SD are minimal. John [30] investigated an engine cooling system employing a passive heat load accumulator. Heat load accumulator is considered a phase change material that keeps the heat produced at uttermost times and disintegrates stored heat during lowered heat load conditions. The schematic is shown in Fig. 3. It was observed that it helps in the reduction of load on the cooling system and can manage extreme loads, and allows rapid warm-up during engine start. Chen et al. [31] experimentally studied the heat transfer properties of a tube and fin vehicle radiator. It was observed that the radiator's thermal dissipation rate increases as there is an increment in inlet coolant temperature, coolant volume flow rate, and cooling air velocity. Salah et al. [32] developed a nonlinear robust

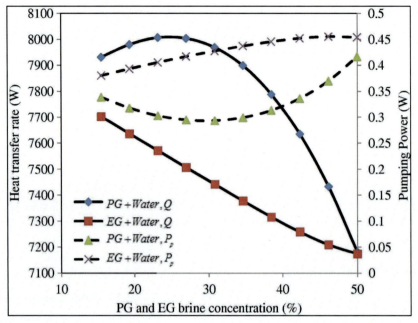

Fig. 1 Heat transfer rate and pumping power at different brine concentrations. *(Copyright Elsevier 2017 (Lic#5080670198059), R.R. Sahoo, P. Ghosh, J. Sarkar, Energy and exergy comparisons of water based optimum brines as coolants for rectangular fin automotive radiator, Int. J. Heat Mass Transf. 105 (2017): 690–696.)*

method by regulating a hydraulic pump and fan to adjust the engine coolant temperature. Fig. 4 displays the dynamic model describing the transient behavior of the hydraulic-based advanced vehicle thermal control system. Mounika et al. [33] investigated the heat transfer rate of an automobile radiator with EG as coolant and reported an increment in Nusselt number as the Reynolds number increments. An increment in the overall heat transfer [34] coefficient of 91% is reported when the Reynolds number increases. Witry et al. reported that the heat rate of radiators performs a significant part in the general performance of automotive cooling. They developed an aluminum roll-bonding technique for the production of heat exchangers. The overall tube side flow arrangement is shown in Fig. 5. The CFD results showed that the increment on each side is reported for a patterned plate heat exchanger. It is observed from Fig. 6, water is reported as a better heat transfer agent inside the plate as the heat transfer coefficient of internal water is more than the external air. Vithayasai et al. [35] investigated the performance of a louvered fin and flat tube automobile radiator installed in a wind tunnel and heat was exchanged between a hot water and a cold air. The schematic is shown in Fig. 7. The louvered fin and tube design of heat exchanger is displayed in Fig. 8. It was observed from the figure that the thermal performance of radiator increases with air stream velocity for all voltage supplies.

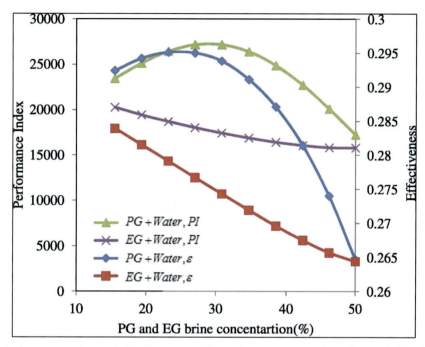

Fig. 2 Performance index and effectiveness at different brine concentrations. *(Copyright Elsevier 2017 (Lic#5080670198059), R.R. Sahoo, P. Ghosh, J. Sarkar, Energy and exergy comparisons of water based optimum brines as coolants for rectangular fin automotive radiator, Int. J. Heat Mass Transf. 105 (2017): 690–696.)*

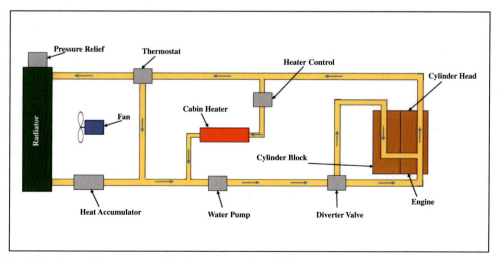

Fig. 3 Engine cooling system with a heat accumulator [30].

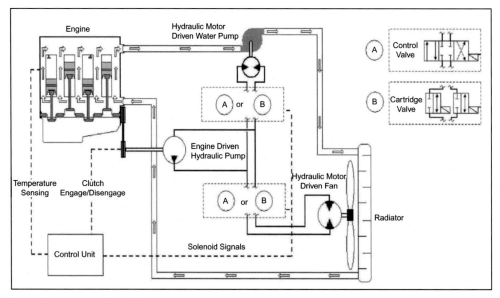

Fig. 4 The advanced automotive cooling system includes a hydraulic-driven coolant pump and fan, valves, and sensors. *(Copyright Elsevier 2009 (Lic#5080670323872), M.H. Salah, et al., Hydraulic actuated automotive cooling systems—nonlinear control and test. Control, Eng. Pract. 17(5) (2009) 609–621.)*

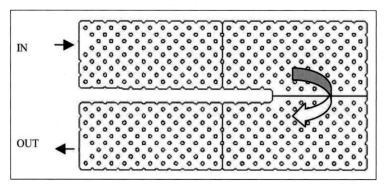

Fig. 5 Overall tube side flow arrangement. *(Copyright Elsevier 2005 (Lic#5080670323872), A. Witry, M.H. Al-Hajeri, A.A. Bondok, Thermal performance of automotive aluminium plate radiator, Appl. Therm. Eng. 25(8) (2005) 1207–1218.)*

Fig. 9 displays a pressure drop in the air stream. It is observed that electric field influences the pressure drop marginally for all velocities. Gehres [36] investigated the heat transfer behavior of various engine coolants, including 50% EG-water. It was reported that the nucleate boiling occur on approximately 60% of the cylinder surface and that 80% of the heat is transmitted from the engine to the coolant. Beard and Smith [37] investigated analytically and wind tunnel test outcomes for a 1.51 engine radiator, with the Reynolds

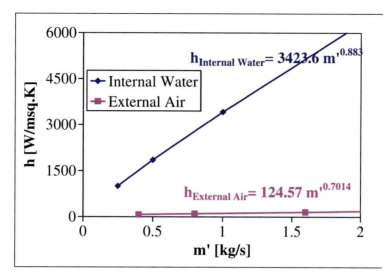

Fig. 6 Average heat transfer coefficient per heat exchanger flow side. *(Copyright Elsevier 2005 (Lic#5080670323872), A. Witry, M.H. Al-Hajeri, A.A. Bondok, Thermal performance of automotive aluminium plate radiator, Appl. Therm. Eng. 25(8) (2005) 1207–1218.)*

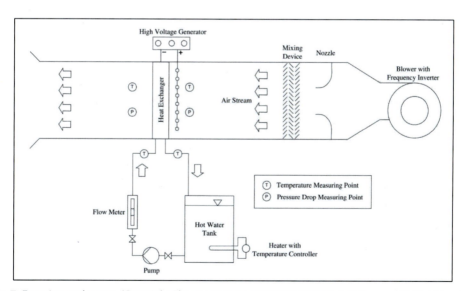

Fig. 7 Experimental setup. *(Copyright Elsevier 2006 (Lic#5080670617526), S. Vithayasai, T. Kiatsiriroat, A. Nuntaphan, Effect of electric field on heat transfer performance of automobile radiator at low frontal air velocity, Appl. Therm. Eng. 26(17) (2006) 2073–2078.)*

Fig. 8 Louvered fin and flat tube automotive radiator. *(Copyright Elsevier 2006 (Lic#5080670617526), S. Vithayasai, T. Kiatsiriroat, A. Nuntaphan, Effect of electric field on heat transfer performance of automobile radiator at low frontal air velocity, Appl. Therm. Eng. 26(17) (2006) 2073–2078.)*

number below 5000. They assessed experimental and systematic data of thermal dissipation. The water-based coolant side Reynolds number ranged from 3900 to 9100 for 3 in. and 1 in. core intensity, respectively. Eitel et al. [38] compared the performance of aluminum vs copper-brass radiator cores for commercial vehicles. The aluminum radiator with 9.4 kg mass has been demonstrated to be 10% lower than copper-brass core radiator. Cozzone [39] introduced relative findings for General Motors 1994 3.8L V6 engine using PG-water and EG-water based coolants. It was reported that the PG-coolant has increased heat transfer coefficient owing to nuclear boiling.

5. Advancements in automotive cooling using nanotechnology

Global demands for fuel consumption and minimal emanations for production and transport developed cost-effective, excellent performance, and lightweight materials to replace metals. Nanocomposites are considered a final solid matrix containing nanoobjects, for instance, a pioneered group of polymeric-based materials manifesting advanced mechanical, thermal, and processing characteristics to substitute metals in automotive applications. The nanoparticles' productivity is so that the material added usually is about 0.5%–5% by mass. Developing this technology only to minor structural components such as front and back, cowl vent grills, regulator/timing jacket, and truck beds can save billions of kilograms of mass annually [40]. One of the significant parameters is the morphology of nanoparticles which improves the thermal performance in the cooling devices. Other factors such as size, shape, and concentration of nanoparticles and temperature of nanofluids have a substantial impact on the thermal performance. Several investigations have reported the utilization of working fluids and nanofluids as coolants in automotive applications. Recently synthesized coolants based on thermally conductive nanoparticles show potential in automobile radiator thermal performance enhancement. The application of nanofluids as coolants in thermal performance depends on the thermophysical

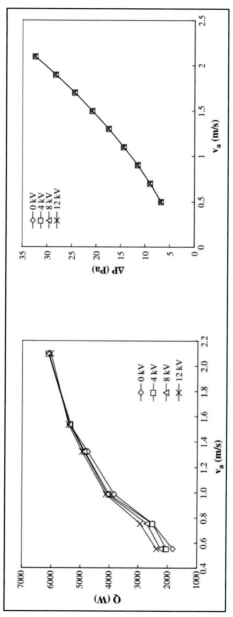

Fig. 9 Effect of electric field on the automobile radiator's heat transfer rate and pressure drop. (Copyright Elsevier 2006 (Lic#5080670617526), S. Vithayasai, T. Kiatsiriroat, A. Nuntaphan, Effect of electric field on heat transfer performance of automobile radiator at low frontal air velocity, Appl. Therm. Eng. 26(17) (2006) 2073–2078.)

properties of nanofluids, and its properties depend on the nanoparticle morphology. Such improvements reduce fuel-specific consumption, improve engine durability, and reduce maintenance expenses [41]. The design and production of automobiles may be influenced by nanotechnology and the technologies associated with up to 60% in a decade. Table 1 summarizes the advances in the studies of nanocoolants for automotive applications.

The advancement of heterogeneous catalysis was essential to decrease emanations from vehicle engine combustion from the last decades. The catalysis has been enhanced by employing catalytically mobile nanoparticles overextremely permeable foundation material with an excellent surface area for automotive emission cleansing. Application of permeable nanocomposites is associated with pollution filtrate, which mechanically and/or by catalytic reaction suppress emission of soot particles or toxic gases [53].

6. Challenges and outlook

Until today, there have been many investigations on heat transfer, and advancements are developing, notably in cooling systems. There is a restriction in the conventional coolants' growth is the weak thermal conductivity of the conventional heat transfer fluids. The thermal conductivity of the solids is considerably greater than that of fluids. It is believed that the thermal conductivity of fluids that contain suspended solid nanoparticles could be considerably more significant than that of conventional liquids such as water, ethylene glycol, and oil, etc. Nanofluids are supposed to have remarkable thermal conductivity than conventional fluids and are considered as promising as automobile coolants in automotive and transportation applications. Nowadays, nanofluids have become a critical topic in the field of automotive radiators due to its prospect of enhanced heat transfer performance.

In the automobile industry, coolant corrosion is a major concern. Substantial investigations have been reported on coolant corrosion and inhibition. Several coolants for aluminum and cast-iron engine blocks have been introduced and are accessible in the marketplace. Up to this point, all coolants and inhibitors are built to preserve the conventional materials. In a conventional coolant, engine materials often rust more quickly at elevated temperatures. These inorganic inhibitors react with metals to create salt films (phase films) accumulating on the surface. Elevated temperature is beneficial in the production of these phase films due to the salt's solubility. In a magnesium engine, specific components are expected to be produced from aluminum alloys. As aluminum alloy is conventional material, the most promising commercial coolants must not be eroding to aluminum. Song et al. [9] tested the effectiveness of magnesium alloy AZ91D in commercial coolants. It was reported that the experimented industrial coolants are corrosive to the magnesium alloys for general and galvanic rust. Among investigated commercial coolants, Toyota's durable coolant showed remarkable performance. Some factors that

Table 1 Summary of the studies of nanocoolants for automotive applications.

Researchers	Nanocoolants	Applications	Findings
Nagib et al. [42]	Al_2O_3-water	Car radiator for 4-stroke gasoline engine	Increasing nanofluid concentration and high fluid circulation rates can improve heat transfer and reduce CO and NO_x emissions
Leong et al. [43]	Cu-EG	Automotive cooling system	Heat transfer rate is increased by 45.2% with an increasing volume concentration of nanoparticles
Raja et al. [44]	Al_2O_3-water	Diesel engine coolant	Overall heat transfer coefficient was increased with increasing volume concentration and resulted in NO_x emissions by 12.5%
Hussein et al. [45]	TiO_2-water SiO_2-water	Automobile radiator	Heat transfer for SiO_2-water and TiO_2-water was increased up to 32% and 20% with increasing nanofluid volume concentration
Esfe et al. [46]	ZnO + DWCNT-EG	Internal combustion engines	Maximum thermal conductivity was enhanced to 24.9%. It was reported from the economic analysis that hybrid nanofluids is cost-effective
Hussein et al. [47]	SiO_2-water ZnO-water	Automobile radiator	The thermal performance of ZnO-water nanofluids was superior to SiO_2-water nanofluids. The effectiveness of 13.9% and 16% were observed for SiO_2-water and ZnO-water nanofluids, respectively
Kumar et al. [48]	Al_2O_3-water	Wavy fin radiator coolant	The spherical nanoparticles exhibited better performance than brick and platelet nanofluids as radiator coolants
Sahoo et al. [49]	CuO + CNT + graphene-water ternary hybrid nanofluid	Radiator coolant	The ternary hybrid nanofluid increases the irreversibility of the system with coolant flow rate and air velocity. Heat transfer rate was improved up to 19.35%
Delavari et al. [50]	Al_2O_3-water Al_2O_3-EG	Automotive radiator	Two-phase approach has a better prediction for nanofluids Nusselt number as compared to single-phase approach
Abbasi et al. [51]	CuO-water Al_2O_3-water	Cooling car engines	Maximum heat transfer coefficient enhancement was reported for

Continued

Table 1 Summary of the studies of nanocoolants for automotive applications—cont'd

Researchers	Nanocoolants	Applications	Findings
Liu et al. [52]	TiO_2-water Au-water Nanodiamond-engine oil	Friction tester and diesel engine	Al_2O_3 and then followed by TiO_2, Au, and CuO nanofluids Nanofluid improves the engine performance by maximizing the engine power and torque up to 1.15% and 1.18%, respectively, and decrement in the fuel consumption up to 1.27% as compared to the engine oil

restricts the use of nanofluids in such systems such as increased pumping power due to enhanced viscosity and stability, should be considered for future advancement.

7. Conclusion

Automobile technology has been increasing rapidly and requires high-efficiency engines. Novel operating attributes from cooling systems are described and presented for their potential to enhance fuel-efficiency, and reduce emissions. Several investigations have described the utilization of working fluids and nanofluids as coolants in radiators and engines. The utilization of nanofluids in automotive radiators as a coolant is an excellent option for heat transfer advancement. Investigations are still under development and will maintain to be explored. The described characteristics can improve fuel economy or emissions, but the engine protection has restricted the advantage of employing an individual design or operating feature.

References

[1] A.J. Torregrosa, et al., Assessment of the influence of different cooling system configurations on engine warm-up, emissions and fuel consumption, Int. J. Automot. Technol. 9 (4) (2008) 447–458.
[2] Z. Said, Enhancing the performance of automotive radiators using nanofluids, Renew. Sust. Energ. Rev. 112 (2019) 183–194, https://doi.org/10.1016/j.rser.2019.05.052.
[3] M.W. Wambsganss, Thermal management concepts for higher-efficiency heavy vehicles, SAE Trans. (1999) 41–47.
[4] J. Wagner, et al., Enhanced automotive engine cooling systems-a mechatronics approach, Int. J. Veh. Des. 28 (1–3) (2002) 214–240.
[5] M. Sortor, Vehicle cost reduction through cooling system optimization, SAE Trans. (1993) 249–257.
[6] W.M. Kays, A.L. London, Compact Heat Exchangers, 1984.
[7] J. Mackerle, Air-Cooled Automotive Engines, 1972.
[8] C. Lin, J. Saunders, S. Watkins, The effect of changes in ambient and coolant radiator inlet temperatures and coolant flowrate on specific dissipation, SAE Trans. (2000) 760–771.
[9] G. Song, D.H. StJohn, Corrosion of magnesium alloys in commercial engine coolants, Mater. Corros. 56 (1) (2005) 15–23.

[10] J.J. JuGer, R.F. Crook, Heat transfer performance of propylene glycol versus ethylene glycol coolant solutions in laboratory testing, SAE Trans. 108 (1999) 71–81.

[11] R.D. Hercamp, R.D. Hudgens, Aqueous propylene glycol coolant for heavy duty engines, SAE Trans. (1990) 105–135.

[12] M. Gollin, D. Bjork, Comparative performance of ethylene glycol/water and propylene glycol/water coolants in automobile radiators, SAE Tech. Pap. (1996).

[13] G.E. Coughenour, et al., High temperature, high load performance of propylene glycol engine coolants in modern gasoline engines, SAE Trans. (1995) 790–802.

[14] H. Pang, C.J. Brace, Review of engine cooling technologies for modern engines, Proc. Inst. Mech. Eng. D: J. Auto. Eng. 218 (11) (2004) 1209–1215.

[15] I. Finlay, et al., The Influence of Coolant Temperature on the Performance of a Four Cylinder 1100cc Engine Employing a Dual Circuit Cooling, Heat and Mass transfer in Gasoline Engines, 1989.

[16] M. Chanfreau, et al., Advanced engine cooling thermal management system on a dual voltage 42v-14v minivan, SAE Trans. (2002) 107–114.

[17] M. Clough, Precision cooling of a four valve per cylinder engine, SAE Trans. (1993) 1555–1566.

[18] M. Chanfreau, et al., The Need for an Electrical Water Valve in a rmal Management Intelligent System (THEMIS™), SAE Trans. (2003) 243–252.

[19] N. Vaughan, J. Gamble, The Modeling and Simulation of a Proportional Solenoid Valve, 1996.

[20] R.R. Henry, J. Koo, C. Richter, Model development, simulation and validation, of power train cooling system for a truck application, SAE Tech. Pap. (2001).

[21] P. Frick, et al., A hydraulic fan driven heat exchanger for automotive cooling systems, in: ASME 2006 International Mechanical Engineering Congress and Exposition, 2006.

[22] P. Trivedi, N. Vasava, Effect of variation in pitch of tube on heat transfer rate in automobile radiator by CFD analysis, Int. J. Eng. Adv. Technol. 1 (6) (2012) 180–183.

[23] H. Hausen, Heat Transfer in Counterflow, Parallel Flow and Cross Flow, McGraw-Hill Book Company, 1983. xx+ 515, 23 x 16 cm, illustrated, pounds sterling 45. 25.

[24] F.W. Dittus, Heat transfer in automobile radiators of the tubler type, Univ. Calif. Pubs. Eng. 2 (1930) 443.

[25] V. Gnielinski, New equations for heat and mass transfer in turbulent pipe and channel flow, Int. Chem. Eng. 16 (2) (1976) 359–368.

[26] M.H. Salah, et al., Nonlinear-control strategy for advanced vehicle thermal-management systems, IEEE Trans. Veh. Technol. 57 (1) (2008) 127–137.

[27] R.R. Sahoo, P. Ghosh, J. Sarkar, Energy and exergy comparisons of water based optimum brines as coolants for rectangular fin automotive radiator, Int. J. Heat Mass Transf. 105 (2017) 690–696.

[28] Y. Cao, K. Kengskool, An automotive radiator employing wickless heat pipes, in: MAESC Conference, Memphis, TN, 2009.

[29] C. Oliet, et al., Parametric studies on automotive radiators, Appl. Therm. Eng. 27 (11–12) (2007) 2033–2043.

[30] J. Vetrovec, Engine cooling system with a heat load averaging capability, SAE Tech. Pap. (2008).

[31] J. Chen, D. Wang, L. Zheng, Experimental study of operating performance of a tube-and-fin radiator for vehicles, Proc. Inst. Mech. Eng. Pt. D J. Automobile Eng. 215 (8) (2001) 911–918.

[32] M.H. Salah, et al., Hydraulic actuated automotive cooling systems—nonlinear control and test, Control. Eng. Pract. 17 (5) (2009) 609–621.

[33] P. Mounika, R. Sharma, P. Srinivas Kishore, Performance Analysis of Automobile Radiator, 3, 2016, pp. 35–38.

[34] A. Witry, M.H. Al-Hajeri, A.A. Bondok, Thermal performance of automotive aluminium plate radiator, Appl. Therm. Eng. 25 (8) (2005) 1207–1218.

[35] S. Vithayasai, T. Kiatsiriroat, A. Nuntaphan, Effect of electric field on heat transfer performance of automobile radiator at low frontal air velocity, Appl. Therm. Eng. 26 (17) (2006) 2073–2078.

[36] E. Gehres, An analysis of engine cooling in modern passenger cars, SAE Tech. Pap. (1963).

[37] R. Beard, G. Smith, A method of calculating the heat dissipation from radiators to cool vehicle engines, SAE Tech. Pap. (1971).

[38] J. Eitel, et al., The aluminum radiator for heavy duty trucks, SAE Tech. Pap. (1999).

[39] G.E. Cozzone, Effect of coolant type on engine operating temperatures, SAE Tech. Pap. (1999).
[40] J. Garces, D. Moll, J. Bicerano, R. Fibiger, D.G. McLeod, Adv. Mat. 12 (2000) 1835.
[41] N.A.C. Sidik, M.N.A.W.M. Yazid, R. Mamat, Recent advancement of nanofluids in engine cooling system, Renew. Sust. Energ. Rev. 75 (2017) 137–144.
[42] M. Nagib, et al., Experimental study for enhancement the cooling system and exhaust gasses for gasoline automotive engine using Nano-fluid, Eng. Res. J. 161 (2019) 96–111.
[43] K. Leong, et al., Performance investigation of an automotive car radiator operated with nanofluid-based coolants (nanofluid as a coolant in a radiator), Appl. Therm. Eng. 30 (17–18) (2010) 2685–2692.
[44] M. Raja, et al., Effect of Heat Transfer Enhancement and NOx Emission Using Al_2O_3/Water Nanofluid as Coolant in CI Engine, 2013.
[45] A.M. Hussein, et al., Heat transfer augmentation of a car radiator using nanofluids, Heat Mass Transf. 50 (11) (2014) 1553–1561.
[46] M.H. Esfe, et al., Experimental evaluation, new correlation proposing and ANN modeling of thermal properties of EG based hybrid nanofluid containing ZnO-DWCNT nanoparticles for internal combustion engines applications, Appl. Therm. Eng. 133 (2018) 452–463.
[47] H. Maghrabie, H. Mousa, Thermal performance intensification of car radiator using SiO_2/water and ZnO/water nanofluids, J. Therm. Sci. Eng. Appl. (2021) 1–25.
[48] V. Kumar, R.R. Sahoo, Exergy and energy analysis of a wavy fin radiator with variously shaped nanofluids as coolants, Heat Transfer Asian Res. 48 (6) (2019) 2174–2192.
[49] R. Rekha Sahoo, Effect of various shape and nanoparticle concentration based ternary hybrid nanofluid coolant on the thermal performance for automotive radiator, Heat Mass Transf. 57 (5) (2021) 873–887.
[50] V. Delavari, S.H. Hashemabadi, CFD simulation of heat transfer enhancement of Al_2O_3/water and Al_2O_3/ethylene glycol nanofluids in a car radiator, Appl. Therm. Eng. 73 (1) (2014) 380–390.
[51] M. Abbasi, Z. Baniamerian, Analytical simulation of flow and heat transfer of two-phase nanofluid (stratified flow regime), Int. J. Chem. Eng. 2014 (2014).
[52] H. Liu, M. Bai, Y. Qu, The Impact of Oil-Based Diamond Nanofluids on Diesel Engine Performance, Springer Berlin Heidelberg, Berlin, Heidelberg, 2013.
[53] Z. Said, et al., Recent advances on the fundamental physical phenomena behind stability, dynamic motion, thermophysical properties, heat transport, applications, and challenges of nanofluids, Phys. Rep. (2021), https://doi.org/10.1016/j.physrep.2021.07.002. In press.

Index

Note: Page numbers followed by *f* indicate figures and *t* indicate tables.

A

Abrasion resistance properties, 364–365
Accumulative roll bonding (ARB), 170
Acrylate coatings, 362–364
Advanced high-strength steels (AHSS)
    effect, 295–296
    sustainability, 293–295
AHP. *See* Aluminum hypophosphite (AHP)
Aircraft materials processes, 280
Air heat transfer coefficient, 780
Airplane brake system, 144*f*
Air pollutant emissions, 530
Alloy, 277
    composition, 540
Aluminum hypophosphite (AHP), 109
Aluminum nanoparticles, 754–757
Aluminum nitride (AlN), 156–157
Aluminum oxide (Al_2O_3), 156–157, 182–183, 349, 656–657, 742–743
Aluminum trihydroxide (ATH), 110*f*, 111–112
Amorphous silicon (a-Si) solar cell, 472–473
Anisotropic light, 34*f*
Anti-corrosion coatings, 361–362
Anticorrosive polymer coating, 324–332
Antifouling component, 333
Antimicrobial fabrics, 279–280
Anti-reflection coatings, 364–365
Anti-scratch coatings, 361–362
ATH. *See* Aluminum trihydroxide (ATH)
Austenite-reverted-transformation (ART), 307*f*
Automobile industry
    coatings, 124–126
    fiber-reinforced nanocomposite, 223
    nanomaterials, 38
    polymer nanocomposites, 38–40
    renewable materials, 106–107
Automotive cooling systems, 774
    challenges, 788–790
    experimental studies, 781–786
    high temperature set point, 777
    low temperature set point, 777
    nanotechnology, 786–788
    numerical studies and correlations, 779–781
    temperature setpoint, 776–777
Automotive engine, 714*f*, 780–781
Automotive fuels
    metal oxide nanomaterials, 740–746
    nanomaterials, 738–740
Automotive hydraulic brake tube, 277
Automotive industry, 9
Automotive materials, 16–17
Automotive radiators, 713–714

B

Ball milling, 562
Bandgaps, 488–489
Batteries, 454–455
    application, 26–27
    grain boundary, 25–26
    lithium ion battery, 22–23
    metal oxides, 24–25
    model composites, 22–23
Battery-supercapacitor hybrid (BSH), 453
Bearings, 278
Benzotriazole (BTA), 633
Bimetallic nanocatalyst, 594
Binder systems, 347
Biodegradable natural fibers, 87
Bio-derived polyurethane foams
    structural characterizations, 109–111
    synthesis and characterizations, 107–108
Biodiesel blend, 741–742
Body parts, 657
Boron carbide (B_4C), 156–157
Bottom-up methods, 562–565
Brake specific fuel consumption (BSFC), 737–738
Brake system, 142–147
Brake thermal efficiency (BTE), 738

C

Cadmium telluride (CdTe) solar cell, 473
Caliper, 420
Capsule-based healing process, 350*f*
Carbonaceous materials, 612–613*t*
Carbon-based nanomaterials, 759–760
Carbon-based nanoparticles, 567–568

Carbon black (CB), 51
Carbon–carbon nanocomposites (CCNCs)
    application, 137–138
    brake systems, 142–147
    components, 134f
    exhaust nozzles, 147–150
    fabrication, 133
    production methods, 135–136
    properties, 136–137
    technologies, 138–142
Carbon dots (CDs), 387–388
Carbon emission control, 279
Carbon fiber (CF), 131–132, 218, 629–630
Carbon-fiber reinforced plastic (CFRP), 413
Carbon monoxide (CO) oxidation, 519–520
Carbon nanofiber (CNF), 47–50
Carbon nanostructures, 546–547
Carbon nanotube (CNT), 57–58
    dispersion, 231–235
    electric vehicles, 229–231
    Joule heating, 239–242
    long-term stability, 242–244
    nanocomposites, 235–239
    temperature coefficient of resistance, 235–239
Catalyst model, 17
    model composites, 28–31
    titania (TiO_2) photocatalysts, 28–31
    types of, 32f
Catalytic converters, 422
    diesel and learn-burn gasoline engines, 551–555
    internal combustion engine, 535–537
CCNCs. See Carbon–carbon nanocomposites (CCNCs)
Cellulose nanofibers, 94
Ceramic, 418–419
Ceramic lite, 419
Ceramic matrix, 5
Ceramic Plus, 419
Ceria-based catalysts, 515
Cerium oxide (CeO_2), 740–742
Cerium salts, 377
CF. See Cobalt ferrite (CF)
Chemical vapor deposition (CVD), 625–626
Coatings/paints, 658
Cobalt ferrite (CF), 333
Coil coated steel, 347–348
Combustion performance, 569–573
Commercialization, of polymer nanocomposite, 58–60

Compo casting, 173
Composite materials, 17–18
Compression molding, 46–47
Concentrated solar cells, 477–478
Concentrating photovoltaic (CPV), 477–478
Condensation, 563
Conductive coatings, 431–432
Conductive nanopaints
    advantages, 441–443
    characteristics and use, 432–436
    commercial use, 439–441
    development, 437–439
Coolant heat transfer coefficient, 779
Copper, 185–187
Copper indium gallium di-selenide (CIGS), 474–475
Copper nanomaterial, 753–754
Copper nanopaints, 443
Copper nanoparticles (Cu-NP)
    applications, 280–281
    limitations, 281
    plant extracts, 275–276
    properties, 270–272
    synthesis, 272–276
Copper oxide (CuO_x), 272–273, 666–669
Co-precipitation method, 24–25, 564
Corrosion inhibitors
    inorganic, 376–384
    magnetic nanogel hybrid, 319–321
    organic, 384–395
    polymer-magnetic nanoparticles (MNPs) hybrids, 322–323
Covalent binding, 8–9

D

Density functional theory (DFT), 521
Deposition processes, 171–172
Derivative thermogravimetric analysis (DTGA), 76–77
Derjaguin, Verwey, Landau, and Overbeek (DVLO), 680
Disc-disc brake systems, 142
Disintegrated melt deposition (DMD), 173–175, 202–203
Dispersion, 231–244
Dispersion stability, 679–680
    characterization, 681–686
Displays, 657

DTGA. *See* Derivative thermogravimetric analysis (DTGA)
Dual phase (DP) steels, 299
Dye-sensitized solar cell (DSSC), 468–469, 476–477

E
Electrical driving controls, 278–279
Electrically conductive coating, 431–432
Electric conductivity, 444
Electrode materials, 456
Electrodes, 460–461, 500–501
Electromagnetic interference (EMI), 337, 607–609, 749–750
   effectiveness, 754f
   nanomaterials, 751–760
   potential receptor, 750
   shielding, 607–609, 750
   solid material, 609f
Eley–Rideal (E-R) mechanism, 519–520
Encapsulated healing, 412–413
Energy density, 462–463
Engine, 654–655
   coatings, 279
   cooling system, 717–719, 778
   oils, 687–688
   performance, 569–573
   tribology, 687–688
Epoxy-zeolite coatings, 356
Equal channel angular pressing (ECAP), 169
Ethylene glycol (EG), 775–776
European emission standards, 512f
Exhaust emissions, 531–533
   characteristics, 740
Exhaust gases, 511
Exhaust nozzle, 147–150
Exhaust system, 533–534
Expendable graphite, 81–82

F
Fiber-reinforced nanocomposite (FNC)
   airplane industries, 223
   applications, 224
   automobile industries, 223
   chemical industry, 224
   design and manufacturing, 222
   failure analysis, 220
   graphene oxide, 216f
   mechanical characterization, 218–219
   military devices, 224
   morphological characterization, 220–222
   physical properties, 218
   rheological characterization, 220
Flame retardancy mechanism, 79–81
Flame-retardant behavior, 119–124
Flame-retardant rigid polyurethane foams (FR-RPUF), 107
Flame-retardants (FRs)
   polyols, 69–70
   polyurethane foams, 69–71
   relevance of, 66–68
Flammability behavior, 79–81
Four-way catalyst, 536
Friction stir processing (FSP), 159, 170–171, 203–204, 205f
Fuel cells, 12–13, 581–583, 657
   advantages and challenges, 596–599
   development, 591–596
   nanocatalyst, 583–585
Fuel consumption, 776
Fuel economy, 700–701

G
Gas separation and storage, 279
Gear box, 278
Glass fiber (GF), 629–630
Global warming potential (GWP), 531t
Grain boundary, 25–26
Grain refinement, 289–293
Graphene, 56–57, 609–610
   body and structural parts, 630–632
   coating applications, 632–634
   electric vehicles, 642–643
   electronic parts of vehicles, 635–636
   lubricating agent, 636–642
Graphene oxide (GO), 333, 359–360
Greenhouse gas (GHG), 737–738
Green magnetic nanoparticles (MNPs), 323

H
Halloysite nanocontainers, 410
Heal lite, 419–420
Heal Plus, 420
Heat exchanger, 276–277
Heat transfer coefficient, 780
Heat transfer systems, 276–277, 714–715
Heat transfer through tube wall, 780

Heterogeneous nanomaterials, 588
Hierarchical electrode, 456
High-density polyethylene (HDPE), 39–40
High-energy ball milling, 171
High entropy materials, 187–190
High pressure torsion (HPT), 169–170
Hole transporting materials (HTMs), 500–501
Hydrolysis, 563
Hydro marine coating, 421
Hydrothermal method, 563–564

I
Injection molding, 47
Inorganic corrosion inhibitors, 376–384
In situ impregnation process, 8
In situ synthesis, 42, 175
Intrinsic healing, 413–414

J
Joule heating
    carbon nanotube (CNT), 239–242
    long-term stability, 242–244
    polymer nanocomposites, 239–242

L
Lamellar double hydroxides (LDH), 390
Langmuir–Hinshelwood (L-H) mechanism, 519–520
Layered material-reinforced polymer nanocomposites, 52–58
Lean NO_x trap (LNT), 553–555
Lightweighting, 629–630
Lightweight vehicles, 10–11
Liquid-state processing, 172–175
Lithium-ion battery (Li-ion) battery, 22–23
Lotus coatings, 421
Low-density materials, 73
Low-density polyethylene (LDPE), 39–40
Lubrication
    stribeck curve, 677f
    temporal evolution, 678f

M
Magnesium, 183–185
Magnesium alloys, 359
Magnetic carbon-based nanocomposites, 609–611
Magnetic fluids, 332–333
Magnetic nanogel hybrid, 319–321
Magnetic nanoparticles (MNPs)
    anticorrosion coating, 324–332
    direct corrosion inhibitors, 319–323
    electromagnetic absorbing coatings, 337–338
    network formation, 326–332
    smart coatings, 334–337
    textiles coatings, 338
Mechanical milling, 159
Medium-Mn steels (MMnS), 304–308
Melt deposition method, 204f
Melt extrusion process, 41f
Melt intercalation, 41
Mesoporous silica nanocontainers (MSNs), 411
Metal-based nanomaterials, 751–757
Metal-based nanoparticles, 562–567
Metal complexes, 16
Metal hybrid thin films, 324
Metallic nano-coating, 350–352
Metallic nanocomposite (MNC), 156–160
    automotive application, 192–194
    characteristics, 182–190
    coatings, 190–192
    corrosion behavior, 180–181
    ductility, 178
    hardness, 178–179
    strength, 176–177
    thermal properties, 181
    wear behavior, 179–180
Metal matrix composites, 5–6, 200
Metal oxide-based catalysts, 544–546
Metal oxide-based nanomaterials, 757–759
Metal oxide nanomaterials, 740–746
    automotive fuels, 740–746
Metal oxides, 24–25
Metal oxide-supported copper nanoparticles (Cu-NP), 273–275
Microstructure transformation, 694
Microvascular healing systems, 413
Microwave absorption
    magnetic carbon-based nanocomposites, 609–611
    performance, 611–615
Microwave-assisted method, 565
MMnS. *See* Medium-Mn steels (MMnS)
Model composites, 22–23
Monocrystalline silicon solar cell, 471
Morphological tuning, 538
Multicrystalline silicon solar cell, 472
Multi-walled carbon nanotubes (MWCNTs), 57–58, 568

MXenes, 493–495
   perovskite layer, 498–500
   recombination reactions, 495–497
   transition metal, 498

N

Nanoalloys, 257
   automobile industry, 263–265
   automobile parts, 259–260*t*
   manufacturing process, 261–262
   mechanical properties, 262
Nanobelt solar cell, 481
Nanocatalysts, 583–585
   carbon nanostructures, 546–547
   metal oxide-based catalysts, 544–546
   nonplatinum group metals, 542–544
   perovskite catalyst, 547–548
   platinum group metals, 537–542
   types, characteristics and synthesis, 585–590
Nano clay, 53–56
Nanocoatings, 405–409
   advantages, 407
Nanocomposite materials
   auto components, 209–211
   characterization, 206–208
   fabrication, 201–206
Nanocomposites
   ceramic matrix, 5
   challenges, 9–10
   vs. conventional composites, 164–165
   costs and benefits, 13
   fuel, 12–13
   ingredients, 5*f*
   lightweight vehicles, 10–11
   metal matrix, 5–6
   nanofillers, 6–7
   polymer matrix, 6
   processing, 167–175
   synthesis techniques, 7–9
   in tribology, 12
   in tyres, 11–12
Nanocones solar cell, 480
Nanocontainers, 409–411
Nanocrystal based solar cells, 475
Nanofiber-reinforced polymer composites (NPRPC), 44–47
Nanofibers, 217–218
Nano fibrillated cellulose (NFC), 49–50
Nanofillers, 6–7, 324–332
Nanofluid-based coolants, 715–717
   automotive applications, 719–727
   challenges, 728–732
   numerical and experimental studies, 727–728
Nanofuels
   application, 569–573
   nanoparticles, 562–568
   properties, 568–569
Nano-grain steel, 291
Nanographene platelets (NGPs), 11–12
Nano-infused fluids, 279
Nanolayers, 4–5
Nanolubricants
   exhaust emissions, 702–703
   mechanisms, 692–696
   preparation, 679–686
   tribological and thermophysical performance, 686–692
   vehicle engines performance, 697–703
Nanomaterials, 411–414
   in automobile industry, 38
   automotive fuels, 738–740
   automotive industry, 654–658
   toxicity, 659–669
Nanoparticle-reinforced polymer composites, 50–52
Nanoparticles, 4–5
Nanopillar solar cell, 480–481
Nanorods solar cell, 481
Nanosteel, strengthening mechanisms of, 287–289
Nanostructured platinum (Pt) based catalysts, 592–594
Nanotechnology-based solar cell, 479–481
Nanotoxicity, 658–659
Nanotubes, 4–5
Nanotubes solar cell, 479–480
Nanowires solar cell, 479
Natural fiber-reinforced nanocomposites (NFRNs), 92–95
   applications, 95–98
Natural fibers, 88–92
Natural polymer, 86
Negative thermal expansion (NTE), 18
Nonplatinum group metals, 542–544

O

OFMs. *See* Organic friction modifiers (OFMs)
Open mold process, 44–45

Organic coatings, 347–348
  mechanical wear, 353–355
Organic corrosion inhibitors, 384–395
Organic friction modifiers (OFMs), 678–679
Oxygen storage components (OSCs), 537, 548–551

P

PA6. *See* Polyamide 6 (PA6)
Pencil hardness test, 347
Percolation theory, 231–235
Perovskite-based catalysts, 547–548
Perovskite solar cells (PSC), 488–491
  encapsulation, 501–502
  high cost, 492–493
  recombination reactions, 491
  stability, 492
Photoluminescence (PL), 271
Photovoltaic (PV), 467, 487
Plasma electrolytic oxidation (PEO), 392–393
Platinum (Pt)
  free nanocatalyst, 594–596
  nanoparticles, 659–661
Platinum group metals (PGMs), 537–542
  wash-coat composition, 539
Polyamide 6 (PA6), 138
Polyhedral oligomeric silsesquioxane (POSS), 51–52
Polymer-based solar cells, 475–476
Polymer composites, 40f
Polymeric coating, 345–346
  applications, 350–365
  sol–gel method, 347–348
Polymeric nanocomposites, 239–242
  automobile industry, 38–40
  car parts, 39f
  classification, 43–58
  commercialization, 58–60
  high heat distortion temperature, 42
  importance of, 37–38
  modulus and dimension stability, 42
  parts of vehicles, 40t
  processing, 40–42
  scratch and mar resistance, 43
  toughness and rheology, 43
Polymer intercalation, 41–42
Polymerization process, 9
Polymer-magnetic nanoparticles (MNPs) hybrids, 322–323
Polymer matrix, 6

Polymer-supported copper nanoparticles (Cu-NP), 275
Poly/multicrystalline cells, 472
Polyolefins, 39–40
Polyols
  characterizations, 70–71
  structural characterizations, 71–72
  synthesis, 69–70
Polytetrafluoroethylene (PTFE), 687–688
Polyurethanes (PUs)
  foams
    characterizations, 70–71
    closed-cell content, 73–74
    compressive strength, 74–76
    density, 73
    flame-retardant behavior, 119–124
    low density, 111
    mechanical properties, 74–76, 115–119
    morphology and cellular characteristics, 112–113
    morphology and cellular structure, 74
    physical properties, 73–74
    significance, 107
    synthesis, 69–70
    thermal behavior, 76–78
  importance, 68–69
Potential nanoreinforcements, 166–167
Powder metallurgy, 168
Powder processing, 168
Power density, 462–463
Precision cooling system, 778
Propylene glycol (PG), 775–776
Proton exchange membrane fuel cell (PEMFC), 582–583, 584f
PSC. *See* Perovskite solar cells (PSC)
PTFE. *See* Polytetrafluoroethylene (PTFE)
Pultrusion, 47
PUs. *See* Polyurethanes (PUs)

Q

Quenching and partitioning (Q&P) steels, 301–302
  elements, 302–304
  heat treatment, 303f

R

Radiator, 714f
Radiator coolant, 655–657
Radiator thermodynamics, 780–781
Reinforcement, 5
Resin transfer molding (RTM), 45–46

Rietveld method, 17–18
Rolling bearing effect, 692–693

S

Scratch resistance
   acrylate coatings, 362–364
   bio-based materials, 356–358
   epoxy-zeolite coatings, 356
   magnesium alloys, 359
   nano, micro, and macro level, 347
   properties, 364–365
   silica nanoparticles, 355–356
   viscoelastic effects, 346
   waterborne polymers, 352–353
Selective catalytic reduction (SCR), 516–519, 551–553
Self-healing nanocoatings, 407–409
   advantages, 422
   commercial use, 418–421
   disadvantages, 422
   environmental impacts, 417–418
   health impacts, 418
   nanocontainers, 409–411
Self-healing process, 372–373, 412–414
   investigation, 414–416
Separators, 461–462
Severe plastic deformation (SPD), 168, 290–291
Silica (SiO_2), 349
Silica-based nanoparticles, 51–52, 355–356
Silicon carbide (SiC), 159–160
Silicon dioxide (SiO_2), 661–663
Silver nanomaterials, 752–753
Single crystalline silicon solar cell, 471
Single-walled carbon nanotubes (SWCNTs), 57–58, 567–568
Smart coatings, 387
Solar cells
   classification, 471–478
   efficiency, 470
   nanomaterials, 479–481
   nanotechnology, 479–481
   working, 468–471
Sol-gel method, 8, 348–350, 563
   polymeric coating, 347–348
Solid-phase method, 24
Solid polymer electrolyte (SPE), 462
Solid-state processing, 168–171
Solution casting, 7–8
Sonication-assisted co-precipitation method, 565
Sound absorption, 114–115
SPD. *See* Severe plastic deformation (SPD)
SPE. *See* Solid polymer electrolyte (SPE)
Split cooling system, 777–778
Spray casting, 173–175
Squeeze infiltration process, 175
Steels
   first-generation advanced high-strength steels, 297–299
   second-generation advanced high-strength steels, 299–300
   third-generation advanced high-strength steels, 300–311
Stir casting method, 172, 203f
Structural supercapacitor, 462–463
Supercapacitor, 454–455
   component, 460–462
   hybridization, 455–457
   supercapacitor, 455–457
Surface repairing effect, 695–696
Surfactants, 681

T

Tantalum carbide (TaC), 156–157
Temperature coefficient of resistance (TCR), 235–239
Textiles coatings, 338
Thermal conductivity, 439
Thermal degradation stability, 242–243
Thermal expansion, 18–22
Thermogravimetric analysis (TGA), 70–71, 76–77
Thermomechanical processes, 291–292
Thermoplastic polyolefin (TPO), 39–40
Three-way catalyst (TWC), 513–516, 535
Ticlopidine (TiC), 156–157
Titanium (Ti), 187
   nanoalloy, 257
      automobile industry, 263–265
      automobile parts, 259–260t
      manufacturing process, 261–262
      mechanical properties, 262
Titanium diboride (TiB2), 156–157
Titanium dioxide (TiO_2), 28–31, 349, 663–666, 743–744
Topcoat, 420–421
Top-down methods, 562
Toxicity, 659–669
Transformation-induced plasticity (TRIP), 308–309
Tribofilm formation, 693–694
Tribology, 12

TRIP. *See* Transformation-induced plasticity (TRIP)
TRIP-aided bainitic ferrite (TBF) steels, 309–311
Tube-side heat transfer coefficient, 779–780
Twinning-induced-plasticity (TWIP), 308–309
2-Mercaptoethanol (2-ME), 69–70
Two-way catalyst, 535
Tyres, 11–12

U
Ultrafine-grained microstructure, 302
Ultrafine-grained steels
   heat treatment, 293
   micro-alloying, 293
Ultrasonic cavitation, 173, 205*f*

V
Vehicles engine weight, 421–422

W
Wash-coat compositions, 548–551
Wet-on-wet-on-wet (3W) process, 372, 373*f*
Wheel, 420
Windows, 657
Windshields, 421
Wood–plastic composites (WPCs), 93

X
X-band frequency, 754
X-ray diffraction (XRD)
   and automotive materials, 16–17
   Rietveld method, 17–18
   thermal expansion, 18–22

Z
Zinc oxide (ZnO), 656–657
Zirconia (ZrO_2), 349, 361–362, 515–516

Printed in the United States
by Baker & Taylor Publisher Services